T0180035

Advanced Textbooks in Control and Signal Processing

Series editors
Michael J. Grimble, Glasgow, UK
Michael A. Johnson, Oxford, UK
Linda Bushnell, Seattle, WA, USA

More information about this series at http://www.springer.com/series/4045

Victor Manuel Hernández-Guzmán
Ramón Silva-Ortigoza

Automatic Control
with Experiments

 Springer

Victor Manuel Hernández-Guzmán
Universidad Autonoma de Queretaro,
Facultad de Ingenieria
Querétaro, Querétaro, Mexico

Ramón Silva-Ortigoza
Instituto Politécnico Nacional, CIDETEC
Mexico City, Mexico

ISSN 1439-2232 ISSN 2510-3814 (electronic)
Advanced Textbooks in Control and Signal Processing
ISBN 978-3-030-09330-3 ISBN 978-3-319-75804-6 (eBook)
https://doi.org/10.1007/978-3-319-75804-6

© Springer International Publishing AG, part of Springer Nature 2019
Softcover re-print of the Hardcover 1st edition 2019
This work is subject to copyright. All rights are reserved by the Publisher, whether the whole or part of
the material is concerned, specifically the rights of translation, reprinting, reuse of illustrations, recitation,
broadcasting, reproduction on microfilms or in any other physical way, and transmission or information
storage and retrieval, electronic adaptation, computer software, or by similar or dissimilar methodology
now known or hereafter developed.
The use of general descriptive names, registered names, trademarks, service marks, etc. in this publication
does not imply, even in the absence of a specific statement, that such names are exempt from the relevant
protective laws and regulations and therefore free for general use.
The publisher, the authors and the editors are safe to assume that the advice and information in this book
are believed to be true and accurate at the date of publication. Neither the publisher nor the authors or
the editors give a warranty, express or implied, with respect to the material contained herein or for any
errors or omissions that may have been made. The publisher remains neutral with regard to jurisdictional
claims in published maps and institutional affiliations.

This Springer imprint is published by the registered company Springer Nature Switzerland AG
The registered company address is: Gewerbestrasse 11, 6330 Cham, Switzerland

To Judith, my parents, and my brothers.
To the memory of my grand-parents.

Victor.

To my wonderful children – Ale, Rhomy,
Robert, and Rhena – and to my mother.

Ramón.

Foreword

Control systems are described by differential equations; hence, mathematics is an important tool for the analysis and design of control systems. This is the reason why most books on control systems traditionally have a strong mathematics content. However, control systems are also part of engineering and it is practical engineering problems that have motivated the development of control systems as a science. As the influence of advanced mathematics on the subject has grown over time, the modern control design techniques become less comprehensible each time to practitioners. Moreover, this problem is so important that it is also present on basic control courses, i.e., courses on classical control. Because of this situation, several control system scientists have pointed out the necessity of reducing the gap between theory and practice.

Automatic control with experiments is a book intended to reduce the gap between theory and practice in control systems education. The book focuses on classical control techniques and modern linear control techniques. The first chapters of the book are devoted to theoretical aspects of these control techniques, whereas the last chapters are devoted to practical applications of this theory. Moreover, several theoretical examples in the first chapters of the book are intended to be employed in the experiments reported in the latter chapters of the book.

Practical applications presented in the book include feedback electronic circuits (amplifiers with dead-zone, sinusoidal oscillators, and regenerative radio-frequency receivers), brushed DC motors, a magnetic levitation system, the ball and beam system, a mechanism including flexibility, and some systems including pendulums. All theoretical and practical aspects that are necessary to design and to experimentally test the complete control system are described for each prototype: modeling, plant construction and instrumentation, experimental identification, controller design, practical controller implementation, and experimental tests of the complete control system.

Another important feature of the book is that the reader is instructed how to build her/his own experimental prototypes using cheap components. The main objective of this is that the reader has his or her own experimental platforms. In this respect it is the authors' experience that the use of scholars' facilities is restricted in both time and space. Thus, this proposal of the book is attractive.

The electronic components employed are basic and the authors know that today's technology offers more powerful alternatives. However, the subject of the book is automatic control, not electronics or programming languages. Hence, it is the intention of the authors not to divert the reader's attention from automatic control to electronics or computer programming. Thus, the electronics and programming are kept as simple as possible. It is the authors' belief that once the reader understands automatic control she/he will be capable of translating the simple designs in the book to sophisticated platforms based on modern advanced electronics and programming technologies.

Instituto Tecnologico de la Laguna   Victor Santibanez
Division de Estudios de Posgrado e Investigacion
Torreon, Coah., Mexico

Preface

Automatic control is one of the disciplines that support the technologically advanced lifestyle that we know today. Its applications are present in almost all the activities performed by humans in the twenty-first century. From the Hubble spatial telescope and spacecrafts, to the fridge at home used for food preservation. From residential water tanks to large industries producing all the products demanded by people: automobiles, aircrafts, food, drinks, and medicines, to name but some.

Although it is known that applications of automatic control have existed for more than 2000 years, the Industrial Revolution motivated its development as scientific and technological knowledge oriented toward the solution of technological problems. Since then, automatic control has been instrumental in rendering human activities more efficient, increasing the quality and repeatability of products.

It is for this reason that courses on automatic control have become common in academic programs on electrical engineering, electronics, mechanics, chemistry and, more recently, mechatronics and robotics. However, the fact that conventional automatic control techniques are based on mathematics has traditionally posed difficulties for education in this subject: to learn to design automatic control systems the student is required to understand how to solve ordinary, linear, differential equations with constant coefficients using Laplace transforms. This is an important obstacle because this subject is commonly difficult for most undergraduate students. The problem becomes worse because in automatic control the most important part of solving a differential equation is the physical interpretation of a solution, which is difficult for undergraduate students because most do not even understand how to find the solution.

Another difficulty in automatic control education is how to teach students to relate abstract mathematical results to the practical issues in a control system. How do they implement a controller given in terms of the Laplace transform, i.e., as a transfer function in practice? How do they implement a controller using digital or analog electronics? How do they take into account sensors and power amplifier gains? How do they determine the gain of a pulse width modulation-based power amplifier? What are the effects of these gains in a control system?

The problems related to the practice described in the previous paragraph have been traditionally solved using commercial teaching prototypes. However, this has two drawbacks: (1) this equipment is excessively expensive and (2) many practical issues in control systems remain "invisible" for students. This is because this equipment is designed under the premise that it is not necessary for an automatic control student to know how to solve practical issues related to electronics and programming, for instance, that are present in several components of a control system. This is a case of how do we build a power amplifier? How do we design and implement a controller using operational amplifiers or a microcontroller? How do we build our own sensors?

The present textbook offers undergraduate students and professors teaching material that is intended to solve some of the above-mentioned difficulties. To render the learning of theoretical aspects easier, a chapter devoted to solving ordinary, linear, with differential equations with constant coefficients using Laplace transforms is included. Although this chapter may be seen as a course on differential equations, the main difference with respect to a mathematics course is that our book is intended to help students to interpret the solution of differential equations. Furthermore, effects that the differential equation parameters have on the solution waveform are highlighted. According to the experience of the authors, automatic control textbooks in the literature merely present a compendium of solutions of differential equation and they do not succeed in making students reason what they are doing. To overcome this problem, in the present textbook, we resort to explaining differential equations through examples that every undergraduate student has observed in real life, i.e., we resort to students' everyday experience to understand the meaning of mathematical results.

Difficulties related to practical aspects in control systems are overcome through applications in several experimental control systems. Each one of these examples is studied using the same procedure. First, tasks performed using the control systems under study is described. Then, a complete explanation is given to the reader on how to build each one of components of that control system. After that, it is shown how to obtain the corresponding mathematical model and it is explained how to experimentally estimate the numerical values of the system mathematical model parameters. Automatic control techniques studied in the first chapters of the book are then used to mathematically design the corresponding controller. It is also explained in detail how to practically implement the designed controller using either digital or analog electronics, and, finally, results obtained when testing the designed control system experimentally are presented.

The present textbook is organized as follows. Chapter 1 presents a general view of automatic control systems. The aim is to explain to the reader the main ideas behind designing automatic control systems. This is achieved using several practical examples whose main tasks are well understood by most people: a position control system, a steering control system, a video camera recording control system, etc. A brief history of automatic control is also presented and related to the content of this book. The idea is to render the reader capable of identifying reasons why each automatic control tool and concept has been developed. Chapter 2 is devoted to physical system modeling. This chapter is oriented toward physical systems that are common in electrical, electronics, mechanics, and mechatronics engineering. One important reason for including this subject is that the reader realizes that control systems are described by ordinary, linear, differential equations with constant coefficients. This motivates the solution of differential equations in Chap. 3, as this is instrumental to understanding how a control system responds and what the designer has to modify in a control system to achieve the desired response.

Mathematical tools employed to design classical and modern control systems are presented in Chaps. 4 to 7: stability criteria and the steady-state error (Chap. 4), the root locus method (Chap. 5), the frequency response approach (Chap. 6), and the state variables approach (Chap. 7). Exposition of these subjects is oriented toward their application to practical examples presented in subsequent chapters of the book. Hence, several examples in the first chapters of the book deal with the design of controllers that are practically implemented and experimentally tested in the later chapters.

Chapter 8 is included to study the theory required to understand some interesting phenomena appearing during experiments when controlling some of the mechanisms in the last chapters of the book. This is the case of: (a) large overshoots observed even when all closed-loop poles are real, and (b) limit cycles in the Furuta pendulum. Furthermore, a methodology useful for selecting controller gains such that limit cycles are avoided is proposed using ideas in this chapter.

The structure of Chaps. 9 to 16 is the same, as they have the same objective: the application of control techniques in Chaps. 4 to 8 to analyze and design practical control systems. The designed controllers and the complete control systems are practically built, employing low-cost components. Finally, experimental results obtained when testing the complete control systems are presented.

In Chap. 9, several feedback electronic circuits are studied and designed. Among them are some sine-wave oscillator circuits based on operational amplifiers (audio-frequency) and bipolar transistors (radio-frequency), in addition to some power amplifiers and a regenerative radio-frequency receiver. In Chaps. 10 and 11 velocity and position are controlled respectively in a permanent magnet brushed DC motor. Position of a mechanical system with flexibility is controlled in Chap. 12. A magnetic levitation system is controlled in Chap. 13 whereas a ball and beam system, a well-known mechanical system in the automatic control literature, is controlled in Chap. 14. Finally, in Chaps. 15 and 16, two mechanisms including a pendulum are controlled: the Furuta pendulum and the inertia wheel pendulum.

The authors hope the readers this material find useful.

Querétaro, Mexico   Victor Manuel Hernández-Guzmán
México City, Mexico   Ramón Silva-Ortigoza

Acknowledgments

The first author acknowledges the work of his coauthor; his collaboration has been inestimable, not only in the elaboration of this book, but also in the diverse research activities that the authors have performed since they were PhD students. Special thanks to Dr Hebertt Sira-Ramírez from CINVESTAV-IPN, México City, my PhD advisor, and to Dr Víctor Santibáñez from Instituto Tecnológico de la Laguna, Torreón, México, for his collaboration. The first author also thanks his students: Dr Jorge Orrante Sakanassi, Dr Valentín Carrillo Serrano, Dr Fortino Mendoza Mondragón, Moises Martínez Hernández, Dr Mayra Antonio Cruz, Dr Celso Márquez Sánchez, and Dr José Rafael García Sánchez. In particular, Dr Carrillo Serrano has performed some of experiments in the book and he has helped with the construction of some experimental prototypes. My deepest thanks to him. Very special thanks to Dr Luis Furtado, Prof Michael Johnson, and Kiruthika Kumar, from Springer, whose help throughout the whole publishing process has been very important for the authors.

Ideas that have motivated this book have arisen during the undergraduate and the graduate automatic control courses that the first author has taught at Facultad de Ingeniería of Universidad Autónoma de Querétaro, where he has been based since 1995. He acknowledges this University for supporting him through the years, and the Mexican Researcher's National System (SNI-CONACYT) for financial support since 2005. A special acknowledgment is deserved by my wife Judith, my parents, Raul and Estela, and my brothers, Raul and Gustavo.

The second author acknowledges and thanks the first author for his invitation to participate in the creation of this book and for other ambitious academic and research projects. Special thanks to Drs Gilberto Silva-Ortigoza and Hebertt Sira-Ramírez, researchers at Benemérita Universidad Autónoma de Puebla and CINVESTAV-IPN, the former has been his mentor throughout his entire professional formation and the latter was his mentor during his graduate years. He also acknowledges the important academic collaboration of Drs Hind Taud (CIDETEC-IPN), Griselda Saldaña González (Universidad Tecnológica de Puebla) and Mariana Marcelino Aranda (UPIICSA-IPN). The second author is grateful to CIDETEC of Instituto Politécnico Nacional, the Research Center where he has been based since

2006, SIP and programs EDI and COFAA from Instituto Politécnico Nacional, for financial support, and to CONACYT's Mexican Researcher's National System (SNI). A special mention is deserved by my children, Ale, Rhomy, Robert, and Rhena. They are the inspiration I need to improve myself every day.

Querétaro, Mexico   Victor Manuel Hernández-Guzmán
México City, Mexico   Ramón Silva-Ortigoza

Contents

Chapter 1
Introduction

1.1 The Human Being as a Controller

Everybody has been a part of a control system at some time. Some examples of this are when driving a car, balancing a broomstick on a hand, walking or standing up without falling, taking a glass to drink water, and so on. These control systems, however, are not automatic control systems, as a person is required to perform a role in it. To explain this idea, in this section some more technical examples of control systems are described in which a person performs a role.

1.1.1 Steering a Boat

A boat sailing is depicted in Fig. 1.1. There, the boat is heading in a direction that is different from the desired course indicated by a compass. A sailor, or a human pilot, *compares* these directions to obtain a *deviation*. Based on this information, she/he *decides* what *order* must be sent to her/his *arms* to apply a suitable *torque* on the *rudder wheel*. Then, through *a mechanical transmission*, the *rudder* angle is modified, rendering it possible, by action of the water flow hitting the rudder, to apply a *torque* to the boat. Finally, by the action of this torque, the boat rotates toward the desired course. This secession of actions continuously repeats until the boat is heading on the desired course. A *block diagram* of this process is shown in Fig. 1.2.

In this block diagram, the fundamental concept of *feedback* in control systems is observed. Feedback means *to feed again* and refers to the fact that the resulting action of the control system, i.e., the boat actual course, is measured to be compared with the desired course and, on the basis of such a comparison, a corrective action (torque on the rudder wheel) is commanded again, i.e., fed again, trying to render

© Springer International Publishing AG, part of Springer Nature 2019
V. M. Hernández-Guzmán, R. Silva-Ortigoza, *Automatic Control with Experiments*,
Advanced Textbooks in Control and Signal Processing,
https://doi.org/10.1007/978-3-319-75804-6_1

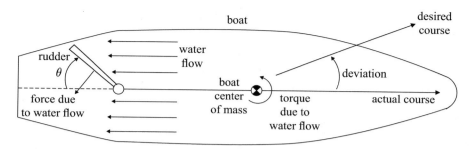

Fig. 1.1 Steering a boat

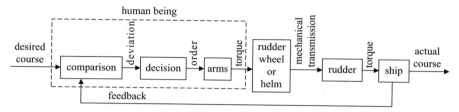

Fig. 1.2 Actions performed when steering a ship

zero deviation between the desired and actual courses. It is said that the human being performs as a controller: she/he evaluates the actual deviation and, based on this information, commands a corrective order until the actual course reaches the desired course. Although arms act as actuators, notice that the rudder wheel and the mechanical transmission suitably amplify torque generated by the arms to actuate the rudder.

This control system is not an automatic control system because a human being is required to perform the task. Suppose that a large ship is engaged in a long trip, i.e., traveling between two harbors on two different continents. In such a case, it is preferable to replace the human brain by a computer. Moreover, as the ship is very heavy, a powerful motor must be used to actuate the rudder. Thus, a machine (the computer) must be used to control the ship by controlling another machine (the rudder motor). In such a case, this control system becomes an automatic control system.

1.1.2 Video Recording While Running

In some instances, video recording must be performed while a video camera is moving, for example, when it is placed on a car driving on an irregular terrain, on a boat on the sea surface, or attached to a cameraman who is running to record a scene that is moving. The latter situation is depicted in Fig. 1.3. The cameraman must direct

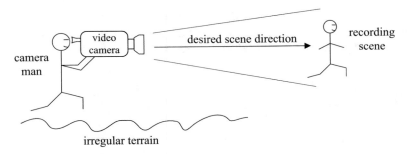

Fig. 1.3 Video recording while running

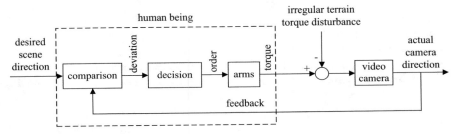

Fig. 1.4 Actions performed when video recording while running

the video camera toward the scene. However, because of the irregular terrain and the natural movement of the cameraman who is running, the arms transmit an undesired vibration to the video camera. A consequence of this vibration is deterioration of the image recording. Although the cameraman may try to minimize this effect by applying a corrective torque to the video camera, as shown in Fig. 1.4, the natural limitations of the human being render it difficult to record well-defined images.

Thus, it is necessary to replace the human being in this control system by more precise mechanisms, i.e., design of an automatic control system is required This requires the use of a computer to perform the *comparison* and *decision* tasks, and the combined use of tracks and a motor to actuate the camera. Because of the high capability of these machines to perform fast and precise actions, high-quality images can be recorded because vibration induced on the video camera can be suitably compensated for [1, 2].

1.2 Feedback Is Omnipresent

The main feature of control systems introduced in the previous section and the automatic control systems with which this book is concerned is *feedback*. This is the capability of a control system to command corrective actions until the desired response is accomplished. On the other hand, feedback is not an invention of human beings: it is rather a concept that human being has learned from nature, where feedback is omnipresent. Examples in this section are intended to explain this.

1.2.1   A Predator–Prey System

This is a fundamental feedback system in all ecosystems. Predators need to eat prey to survive. Then, if there are many predators, the number of prey diminishes because many predators require lots of food. Then, a reduced number of prey ensures that the number of predators also diminishes because of the lack of food. As the number of predators diminishes, the number of prey increases, because there are fewer predators to eat prey. Hence, there will be a point in time where the number of prey is so large and the number of predators is so small that the number of predators begins to increase, because of the food abundance. Thus, at some point in the future, the number of predators will be large and the number of prey will be small, and the process repeats over and over again.

Feedback exists in this process because the number of predators depends on the number of prey and vice versa. Notice that, because of this process, the number of predators and prey are kept within a range that renders possible sustainability of the ecosystem. Too many predators may result in prey extinction which, eventually, will also result in predator extinction. On the other hand, too many prey may result in extinction of other species the prey eat and, hence, prey and predator extinction results again.

The reader may wonder whether it is possible for the number of predators and prey to reach constant values instead of oscillating. Although this is not common in nature, the question is Why? This class of question can be answered using Control Theory, i.e., the study of (feedback) automatic control systems.

1.2.2   Homeostasis

Homeostasis is the ability of an organism to maintain its internal equilibrium. This means that variables such as arterial blood pressure, oxygen, CO_2 and glucose concentration in blood, in addition to the relations among carbohydrates, proteins, and fats, for instance, are kept constant at levels that are good for health.

1.2.2.1   Glucose Homeostasis [3, 4]

Glucose concentration in the blood must be constrained to a narrow band of values. Glucose in the blood is controlled by the pancreas by modifying the concentrations of insulin and glucagon (see Fig. 1.5). When the glucose concentration increases, the

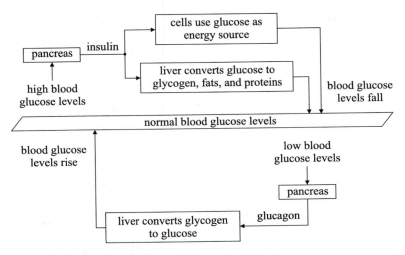

Fig. 1.5 Glucose homeostasis: regulation of glucose in the blood

pancreas delivers more insulin and less glucagon, which has the following effects: (*i*) it favors transportation of glucose from blood to cells, (*ii*) it increases demand for glucose in cells, (*iii*) it stimulates the liver for glucose consumption to produce glycogen, fats, and proteins. The effect of this set of actions is a reduction of glucose concentration in the blood to safe, healthy levels.

On the other hand, if the glucose concentration in the blood diminishes, the pancreas delivers more glucagon and less insulin, which has the following effects: (*i*) it stimulates the liver cells to produce glucose, which is delivered into the blood, (*ii*) it stimulates the degradation of fats into fatty acids and glycerol, which are delivered into the blood, *iii*) it stimulates the liver to produce glucose from glycogen, which is delivered into the blood. The effect of this set of actions is the incrementation of the glucose concentration in the blood to safe, healthy levels.

The glucose regulation mechanisms described above are important because the blood glucose level changes several times within a day: it increases after meals and it decreases between meals because cells use or store glucose during these periods of time. Thus, it is not difficult to imagine that glucose homeostasis performs as a perturbed control system equipped with an efficient regulator: the pancreas.

1.2.2.2 Psychological Homeostasis [3]

According to this concept, homeostasis regulates internal changes for both physiological and psychological reasons, which are called *necessities*. Thus, the life of an organism can be defined as the constant search for and equilibrium between necessities and their satisfaction. Every action searching for such an equilibrium is a behavior.

Fig. 1.6 Regulation of body
temperature

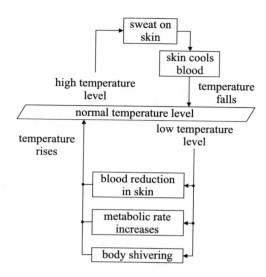

1.2.2.3   Body Temperature [3, 5]

Human beings measure their body temperatures, using temperature sensors in their
brains and bodies.

A body temperature decrement causes a reduction of blood supply to the skin
to avoid heat radiation from the body to the environment, and the metabolic rate is
increased by the body shivering to avoid hypothermia.

When the body temperature increases, the sweat glands in the skin are stimulated
to secrete sweat onto the skin which, when it evaporates, cools the skin and blood
(Fig. 1.6).

On the other hand, fever in a human being is considered when the body's
temperature is above 38°C. Fever, however, is a body's natural defense mechanism
against infectious diseases, as high temperatures help the human body to overcome
microorganisms that produce diseases. This, however, results in body weakness,
because of the energy employed in this process. Thus, when this is not enough,
medical assistance is required.

1.3   Real-Life Applications of Automatic Control

In this section, some examples of real-life applications are presented to intuitively
understand the class of technological problems with which automatic control is
concerned, how they are approached, and to stress the need for automatic control.

1.3.1 A Position Control System

The task to be performed by a position control system is to force the position or orientation of a body, known as the load, to track a desired position or orientation. This is the case, for instance, of large parabolic antennae, which have to track satellites while communicating with them. A simplified description of this problem is presented in the following where the effect of gravity is assumed not to exist because movement is performed in a horizontal plane (Fig. 1.7).

A permanent magnet brushed DC motor is used as the actuator. The motor shaft is coupled to the load shaft by means of a gear box. The assembly works as follows. If a positive voltage is applied at the motor terminals, then a counter-clockwise torque is applied to the load. Hence, the load starts moving counter-clockwise. If a negative voltage is applied at the motor terminals, then a clockwise torque is applied to the load and the load starts moving clockwise. If a zero voltage is applied at the motor terminals, then a zero torque is applied to the load and the load has a tendency to stop.

Let θ represent the actual angular position of the load. The desired load position is assumed to be known and it is designated as θ_d. The control system objective is that θ approaches θ_d as fast as possible. According to the working principle described above, one manner to accomplish this is by applying at the motor terminals an electric voltage v that is computed according to the following law:

$$v = k_p\, e, \tag{1.1}$$

where $e = \theta_d - \theta$ is the system error and k_p is some positive constant. The mathematical expression in (1.1) is to be computed using some low-power electronic equipment (a computer or a microcontroller, for instance) and a power amplifier must also be included to satisfy the power requirements of the electric motor. It is assumed in this case that the power amplifier has a unit voltage gain, but that the electric current gain is much larger (see Chap. 10, Sect. 10.2). According to Fig. 1.8, one of the following situations may appear:

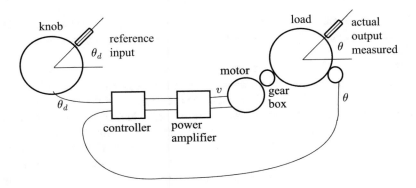

Fig. 1.7 A position control system

Fig. 1.8 Load must always
move such that $\theta \to \theta_d$. (**a**)
$\theta_d > \theta$, $v > 0$, load moves
counter-clockwise. (**b**)
$\theta_d < \theta$, $v < 0$, load moves
clockwise

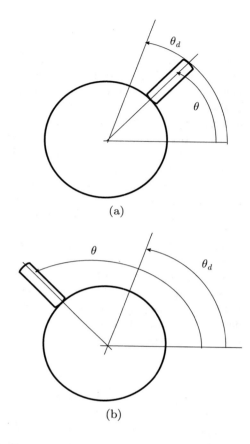

(a)

(b)

- If $\theta < \theta_d$, then $v > 0$ and the load moves counter-clockwise such that θ approaches θ_d.
- If $\theta > \theta_d$, then $v < 0$ and the load moves clockwise such that θ approaches θ_d again.
- If $\theta = \theta_d$, then $v = 0$ and the load does not move and $\theta = \theta_d$ stands forever.

According to this reasoning it is concluded that the law presented in (1.1) to compute the voltage to be applied at the motor terminals has the potential to work well in practice.

The expression in (1.1) is known as a *control law* or, simply, as a *controller*. A block diagram showing the component interconnections in the position control system is presented in Fig. 1.9. Notice that the construction of the controller in (1.1) requires knowledge of the actual load position θ (also known as the *system output* or *plant output*), which has to be used to compute the voltage v to be applied at the motor terminals (also known as the *plant input*). This fact defines the basic control concepts of *feedback* and a *closed-loop system*. This means that the control system compares the actual load position (θ, the *system output*) with the desired load

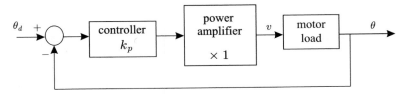

Fig. 1.9 Block diagram of a position control system

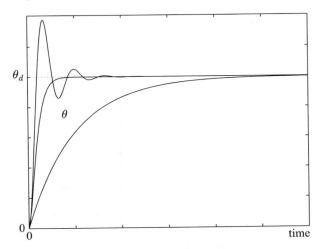

Fig. 1.10 Three possible transient responses in a one-degree-of-freedom position control system

position (θ_d, *system desired output* or *system reference*) and applies, at the motor terminals, a voltage v, which depends on the difference between these variables (see (1.1)).

The *system error* is defined as $e = \theta_d - \theta$. Hence, the *steady state error*[1] is zero because, as explained above, $\theta_d - \theta = 0$ can stand forever. However, the term *steady state* means that this will be achieved once the time is long enough such that the load stops moving (i.e., when the system *steady-state response* is reached). Hence, a zero steady-state error does not describe how the load position θ evolves as it approaches θ_d. This evolution is known as the *system transient response*.[2] Some examples of the possible shapes of the transient response are shown in Fig. 1.10. If k_p in (1.1) is larger, then voltage v applied at the motor terminals is larger and, thus, the torque applied to the load is also larger, forcing the load to move faster. This means that less time is required for θ to reach θ_d. However, because of a faster load movement combined with load inertia, θ may reach θ_d with a nonzero load velocity $\dot{\theta} \neq 0$. As

---

[1] See Chap. 4.
[2] See Chaps. 3, 5 and 6.

Fig. 1.11 Simple pendulum

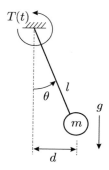

a consequence, the load position continues growing and the sign of $\theta_d - \theta$ changes. Thus, the load position θ may perform several oscillations around θ_d before the load stops moving. It is concluded that k_p has an important effect on the transient response and it must be computed such that the transient response behaves as desired (a fast response without oscillations). Moreover, sometimes this requirement cannot be satisfied just by adjusting k_p and the control law in (1.1) must be modified, i.e., another controller must be used (see Chaps. 5, 6, 7 and 11). Furthermore, the steady-state error may be different to zero ($\theta \neq \theta_d$ when the motor stops) because torque disturbances (static friction at either the motor or the load shafts may make the load deviate from its desired position). This means that even the search for a zero or, at least, a small enough steady-state error may be a reason to select a new controller.

Stability is a very important property that all control systems must possess. Consider the case of a simple pendulum (see Fig. 1.11). If the pendulum desired position is $\theta_d = 0$, it suffices to let the pendulum move (under a zero external torque, $T(t) = 0$) just to find that the pendulum oscillates until the friction effect forces it to stop at $\theta = 0$. It is said that the pendulum is stable at $\theta_d = 0$ because it reaches this desired position in a steady state when starting from any initial position that is close enough. On the other hand, if the desired position is $\theta_d = \pi$, it is clear that, because of the effect of gravity, the pendulum always moves far away from that position, no matter how close to $\theta_d = \pi$ the initial position selected. It is said that the pendulum is unstable at $\theta_d = \pi$. Notice that, according to this intuitive description, the position control system described above is unstable if control law in (1.1) employs a constant negative value for k_p: in such a case, the load position θ would move far away from θ_d. Thus, k_p also determines the stability of the closed-loop control system and it must be selected such that closed-loop stability is ensured.[3] It must be remarked that in the case of an unstable closed-loop system (when using a negative k_p) a zero steady-state error will never be accomplished, despite the control law in (1.1), indicating that the motor stops when $\theta = \theta_d$. The reason for this is that, even if $\theta = \theta_d$ since the beginning, position measurements θ always have a significant noise content, which would render $\theta \neq \theta_d$ in (1.1) and this would be enough to move θ far away from θ_d.

---

[3] See Chaps. 3, 4, 5, 6, 7.

Another important factor to be taken into account in a position control system is the desired load trajectory. It is clear that tracking a constant value for θ_d is easier than the case where θ_d changes very fast, i.e., when the time derivative of θ_d is large or in the case when the second time derivative of θ_d is large. Hence, the control law in (1.1) must be designed such that the closed-loop control system behaves satisfactorily under any of the situations described above. When θ_d is different to any of these situations, it is assumed that the control system will behave correctly if it behaves well in any of the situations described above. This is the main idea behind the design of the system *steady-state error* studied in Chap. 4.

Thus, the three basic specifications for a closed-loop control system are the transient response, the steady-state response, or steady=state error, and the stability. One controller must be designed such that, by satisfying these three fundamental specifications, a fast and well-damped system response is obtained, the load position reaches the desired position in the steady state and the closed-loop system is stable. To achieve these goals, the automatic control techniques studied in this book require knowledge and study of the mathematical model of the whole closed-loop control system. According to Chap. 2, this mathematical model is given as ordinary differential equations, which are assumed to be linear and with constant coefficients. This is the reason why Chap. 3 is concerned with the study of this class of differential equations. The main idea is to identify properties of differential equations that determine their stability in addition to their transient and steady-state responses. This will allow the design of a controller as a component that suitably modifies properties of a differential equation such that the closed-loop differential equation behaves as desired. This is the rationale behind the automatic control system design tools presented in this book.

The control techniques studied in this book can be grouped as classical or modern. Classical control techniques are presented in Chaps. 3, 4, 5, 6 and there are two different approaches: *time response techniques* (Chap. 5) and *frequency response techniques* (Chap. 6). *Classical control techniques* rely on the use of the Laplace transform to solve and analyze ordinary linear differential equations. Classical time response techniques study the solution of differential equations on the basis of transfer function poles and zero locations (Chap. 3) and the main control design tool is Root Locus (Chap. 5). Classical frequency response techniques exploit the fundamental idea behind the Fourier transform: (linear) control systems behave as filters; a system response is basically obtained by filtering the command signal, which is applied at the control system input. This is why the fundamental analysis and design tools in this approach are Bode and polar plots (Chap. 6), which are widely employed to analyze and design linear filters (low-pass, high-pass, band-pass, etc.). Some experimental applications of the classical control techniques are presented in Chaps. 9, 10, 11, 12, 13, and 14.

On the other hand, the *modern control techniques* studied in this book are known as the *state variables* approach (Chap. 7) which, contrary to classical control tools, allow the study of the *internal* behavior of a control system. This means that the state variables approach provides more information about the system to be controlled, which can be exploited to improve performance. Some examples of the experimental application of this approach are presented in Chaps. 15 and 16.

1.3.2 Robotic Arm

A robotic arm is shown in Fig. 1.12. A robotic arm can perform several tasks, for instance:

- Take a piece of some material from one place to another to assemble it together with other components to complete complex devices such as car components.
- Track a pre-established trajectory in space to solder two pieces of metal or to paint surfaces. Assume that the pre-established trajectory is given as six dimensional coordinates parameterized by time, i.e., three for the robot tip position $[x_d(t), y_d(t), z_d(t)]$ and three for the robot tip orientation $[\alpha_{1d}(t), \alpha_{2d}(t), \alpha_{3d}(t)]$. Thus, the control objective is that the actual robot tip position $[x(t), y(t), z(t)]$ and orientation $[\alpha_1(t), \alpha_2(t), \alpha_3(t)]$ reach their desired values as time grows, i.e., that:

$$\lim_{t \to \infty} [x(t), y(t), z(t)] = [x_d(t), y_d(t), z_d(t)],$$

$$\lim_{t \to \infty} [\alpha_1(t), \alpha_2(t), \alpha_3(t)] = [\alpha_{1d}(t), \alpha_{2d}(t), \alpha_{3d}(t)].$$

Fig. 1.12 A commercial robotic arm (with permission of Crustcrawler Robotics)

Fig. 1.13 A
two-degrees-of-freedom
robotic arm

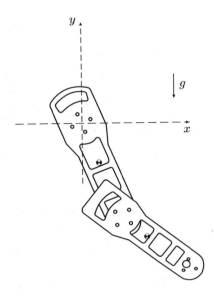

To be capable of efficiently accomplishing these tasks in three-dimensional space, a robotic arm has to be composed of at least seven bodies joined by at least six joints, three for position and three for orientation. Then, it is said that the robotic arm has at least six degrees of freedom. However, to simplify the exposition of ideas, the two-degrees-of-freedom robotic arm depicted in Fig. 1.13 is considered in the following.

Two bodies, called the arm and the forearm, move in a coordinated fashion to force the robotic arm tip to track a desired trajectory in space to perform the tasks described above. To achieve this goal, two permanent magnet brushed DC motors are employed.[4] The first motor is placed at the shoulder, i.e., at the point in Fig. 1.13 where the x and y axes intersect. The stator of this motor is fixed at some point that never moves (the robot base), whereas the motor shaft is fixed to the arm. The second motor is placed at the elbow, i.e., at the point joining the arm and forearm. The stator of this motor is fixed to the arm, whereas the shaft is fixed to the forearm. This allows the arm to move freely with respect to the robot base[5] and the forearm to move freely with respect to the arm. Hence, any point can be reached by the robot tip, as long as it belongs to a plane where the robot moves and is placed within the robot's reach.

---

[4]Permanent magnet brushed DC motors have been replaced by more efficient permanent magnet synchronous motors in recent industrial robot models. However, control of permanent magnet synchronous motors is complex and, hence, they are not suitable for an introductory example of automatic control systems.

[5]The robot base is selected to be the origin of the coordinate frame defining the robot tip position.

One way to define the trajectory to be tracked by the robot tip is as follows. The robot tip is manually taken through all points defining the desired trajectory. While this is performed, the corresponding angular positions at the shoulder and elbow joints are measured and recorded. These data are used as the desired positions for motors at each joint, i.e., θ_{1d}, θ_{2d}. Then, the robot is made to track the desired trajectory by using a control scheme similar to that described in Sect. 1.3.1 for each motor. Two main differences exist between the control law in (1.1) and the control law used for the motor at the shoulder, with angular position θ_1:

$$v_1 = k_{p1}(\theta_{1d} - \theta_1) - k_{d1}\dot{\theta}_1 + k_{i1} \int_0^t (\theta_{1d}(r) - \theta_1(r))dr, \qquad (1.2)$$

and for the motor at the elbow, with angular position θ_2:

$$v_2 = k_{p2}(\theta_{2d} - \theta_2) - k_{d2}\dot{\theta}_2 + k_{i2} \int_0^t (\theta_{2d}(r) - \theta_2(r))dr. \qquad (1.3)$$

These differences are the velocity feedback term, with the general form $-k_d\dot{\theta}$, and the integral term, with the general form $k_i \int_0^t (\theta_d(r) - \theta(r))dr$. Notice that the term $-k_d\dot{\theta}$ has the same form as viscous friction, i.e., $-b\dot{\theta}$ with b the viscous friction coefficient.[6] Hence, adding velocity feedback to control laws in (1.2) and (1.3) is useful to increase the system damping, i.e., in order to reduce the oscillations pointed out in section (1.3.1), allowing fast robot movements without oscillations. On the other hand, the integral term ensures that position θ reaches the desired position θ_d, despite the presence of gravity effects. This can be seen as follows. Suppose that:

$$v_1 = k_{p1}(\theta_{1d} - \theta_1) - k_{d1}\dot{\theta}_1, \qquad (1.4)$$

$$v_2 = k_{p2}(\theta_{2d} - \theta_2) - k_{d2}\dot{\theta}_2. \qquad (1.5)$$

are used instead of (1.2) and (1.3). If $\theta_{1d} = \theta_1$ and $\dot{\theta}_1 = 0$, then $v_1 = 0$. Hence, if gravity exerts a torque at the shoulder joint this motor cannot compensate for such a torque (because $v_1 = 0$) and, thus, $\theta_{1d} \neq \theta_1$ results. On the other hand, if $\theta_{1d} \neq \theta_1$, the term $k_{i1} \int_0^t (\theta_{1d}(r) - \theta_1(r))dr$ adjusts voltage v_1 until $\theta_{1d} = \theta_1$ again and this can remain forever because the integral is not zero, despite its integrand being zero. A similar analysis yields the same results for the joint at the elbow, i.e., the use of (1.2) and (1.3) is well justified.

---

[6]See Sect. 2.1.2.

The main problem for the use of (1.2) and (1.3) is the selection of the controller gains $k_{p1}, k_{p2}, k_{d1}, k_{d2}, k_{i1}, k_{i2}$ and the automatic control theory has been developed to solve this kind of problem. Several ways of selecting these controller gains are presented in this book and are known as proportional integral derivative (PID) control tuning methods.

1.3.3 Automatic Steering of a Ship

Recall Sect. 1.1.1 where the course of a boat was controlled by a human being. This situation was depicted in Figs. 1.1 and 1.2. Consider now the automatic control problem formulated at the end of Sect. 1.1.1: where the course of a large ship is to be controlled using a computer and a permanent magnet brushed DC motor to actuate a rudder. The block diagram of this control system is depicted in Fig. 1.14.

Mimicking Sect. 1.3.1, controller 1 is designed to perform the following computation:

$$\theta_d = k_{p1}(\text{desired course} - \text{actual ship course}). \tag{1.6}$$

The desired course angle shown in Fig. 1.1 is defined as positive. Thus, if k_{p1} is positive θ_d is also positive. Suppose that the rudder angle θ, defined as positive when described as in Fig. 1.1, reaches $\theta_d > 0$. Then, water hitting the rudder will produce a torque T_2 on the ship, which is defined as positive when applied as in Fig. 1.1. This torque is given as a function $\delta(\theta, s)$[7] depending on the rudder angle θ and the water speed s. Torque T_2 produces a ship rotation such that the *actual course* approaches the *desired course*.

The reader can verify, following the above sequence of ideas, that in the case where the desired course is negative, it is reached again by the actual course if $k_{p1} > 0$. This means that a positive k_{p1} is required to ensure that the control system is stable. Moreover, for similar reasons to the position control system, as $k_{p1} > 0$ is larger, the ship rotates faster and several oscillations may appear before

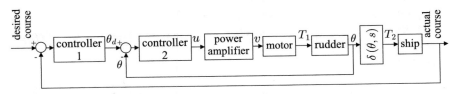

Fig. 1.14 Block diagram for the automatic steering of a ship

---

[7]This function is determined by ship construction, cannot be modified by a controller design and it is not computed during controller evaluation.

settling at the desired course. On the contrary, as $k_{p1} > 0$ is smaller the ship rotates
more slowly and the desired course is reached after a long period of time. Thus,
the transient behavior of the actual course is similar to that shown in Fig. 1.10 as
$k_{p1} > 0$ is changed. Finally, it is not difficult to verify that the control system is
unstable if k_{p1} is negative.

In this system controller 2 is designed to perform the following computation:

$$u = k_{p2}(\theta_d - \theta), \quad k_{p2} > 0, \tag{1.7}$$

whereas the power amplifier just amplifies u by a positive factor to obtain voltage v
to be applied at the motor actuating on the rudder. Following ideas in Sect. 1.3.1, it
is not difficult to realize that θ reaches θ_d as time grows because $k_{p2} > 0$ whereas
instability is produced if $k_{p2} < 0$. As explained before, the ship may oscillate if
$k_{p1} > 0$ is large. Moreover, for similar reasons, the rudder may also oscillate several
times before θ reaches θ_d if $k_{p2} > 0$ is large. A combination of these oscillatory
behaviors may result in closed-loop system instability despite $k_{p1} > 0$ and $k_{p2} > 0$.
Thus, additional terms must be included in expressions (1.6) and (1.7) for controller
1 and controller 2. The question is, what terms? Finding an answer to this class of
questions is the reason why control theory has been developed and why this book
has been written. See Chap. 14 for a practical control problem which is analogous
to the control problem in this section.

1.3.4 A Gyro-Stabilized Video Camera

Recall the problem described at the end of Sect. 1.1.2, i.e., when proposing to use
a computer to perform the *comparison* and *decision* tasks, and a combination of
some tracks and a motor to actuate on the camera, to compensate for video camera
vibration when recording a scene that is moving away. To fully solve this problem,
the rotative camera movement must be controlled on its three main orientation axes.
However, as the control for each axis is similar to the other axes and for the sake
of simplicity, the control on only one axis is described next. The block diagram of
such an automatic control system is presented in Fig. 1.15.

A gimbal is used to provide pivoted support to the camera allowing it to rotate in
three orthogonal axes. See Fig. 1.16. A measurement device known as gyroscope, or

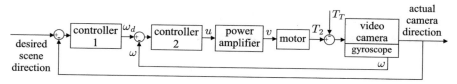

Fig. 1.15 A gyro-stabilized video camera control system

Fig. 1.16 A gimbal

camera is placed here

gyro for short, measures the *inertial* angular velocity ω of the camera. This velocity is not a velocity measured with respect to the cameraman's arms (or the vehicle to which the video camera is attached), but it is camera velocity measured with respect to a coordinate frame that is fixed in space. Thus, this velocity is independent of the movement of the cameraman (or the movement of the vehicle to which the video camera is attached). As a simple example, consider the camera orientation control problem in a single axis. Controller 2 may be designed to perform the following mathematical operation:

$$u = k_{p2}(\omega_d - \omega),\tag{1.8}$$

where k_{p2} is a positive number. Suppose that $\omega_d = 0$, then $u = -k_{p2}\omega$. This means that in the case where, because of the vibration induced by a negative torque $T_T < 0$, i.e. $-T_T > 0$, the camera is instantaneously moving with a large positive velocity $\omega > 0$, then a negative voltage $u = -k_{p2}\omega < 0$ will be commanded to the motor, which will result in a negative generated torque $T_2 < 0$ intended to compensate for $-T_T > 0$, i.e., resulting in a zero camera angular velocity again. The reader can verify that a similar result is obtained if a negative camera velocity $\omega < 0$ is produced by a $T_T > 0$. Notice that, according to (1.8), u adjusts until the camera velocity $\omega = \omega_d = 0$. Hence, the vibration produced by the disturbance torque T_T is eliminated. Although solving this problem is the main goal of this control system, the camera must also be able to record the scene when located in another direction if required. Recall that the scene is moving. This is the main reason for Controller 1 which, for simplicity, is assumed to perform the following mathematical operation:

$$\omega_d = k_{p1}(\text{desired scene direction} - \text{actual camera direction}),$$

where k_{p1} is a positive constant. Hence, Controller 1 determines the angular velocity that the camera must reach to keep track of the desired scene. Observe that ω_d adjusts until the actual camera direction equals the desired scene direction. However, according to the above definition, ω_d would be zero when the actual camera direction equals the desired scene direction. Notice that this is not possible when these variables are equal but not constant, i.e., when the camera must remain moving at a nonzero velocity. For this reason, the following expression is preferable for Controller 1:

$$\omega_d = k_{p1}(\text{desired scene direction} - \text{actual camera direction})$$

$$+k_{i1} \int_0^t (\text{desired scene direction} - \text{actual camera direction})dt.$$

This allows that $\omega_d \neq 0$, despite the actual camera direction equals the desired scene direction because an integral is constant, but not necessarily zero, when its integrand is zero. Moreover, computing u as in (1.8) forces the camera velocity ω to reach ω_d even if the latter is not zero in a similar manner to the actual position reaching the desired position in Sect. 1.3.1. However, computing u as in (1.8) would result in a zero voltage to be applied at motor terminals when $\omega_d = \omega$ and, hence, the motor would tend to stop, i.e., $\omega_d = \omega$ cannot be satisfied when $\omega_d \neq 0$. For this reason, the following expression is preferable for Controller 2:

$$u = k_{p2}(\omega_d - \omega) + k_{i2} \int_0^t (\omega_d - \omega)dt.$$

This allows the voltage commanded to motor u to be different from zero despite $\omega_d = \omega$. Thus, this expression for Controller 2 computes a voltage to be commanded to the motor to force the camera velocity to reach the desired velocity ω_d, which is computed by Controller 1 such that the actual camera direction tracks the direction of the desired scene. Again, a selection of controller gains $k_{p1}, k_{i1}, k_{p2}, k_{i2}$, is one reason why automatic control theory has been developed.

1.4 Nomenclature in Automatic Control

The reader is referred to Fig. 1.17 for a graphical explanation of the concepts defined below.

- **Plant**. The mechanism, device or process to be controlled.
- **Output**. Variable or property of the plant that must be controlled.
- **Input**. Variable or signal that, when adjusted, produces important changes in the plant output.

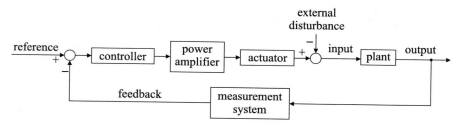

Fig. 1.17 Nomenclature employed in a closed-loop control system

- **Desired output or reference**. Signal that represents the behavior that is desired at the plant output.
- **Feedback**. Use of plant output measurements to verify whether it behaves as the desired output or reference.
- **Controller**. Device that computes the value of the signal to be used as the input to the plant to force the plant output to behave as the desired output or reference.
- **System**. Assembly of devices with connections among them. The plant is also designated as a system.
- **Control system**. Assembly of devices that includes a plant and a controller.
- **Closed-loop system**. A system provided with feedback. In a closed-loop system, the output of every system component is the input of some other system component. Thus, it is important to specify what input or what output is referred to.
- **Open-loop system**. A system that is not provided with feedback. Although the plant may be naturally provided with some kind of feedback, it is common to designate it as an open-loop system when not controlled using feedback.
- **Measurement system**. A system that is devoted to measure the required signals to implement a closed-loop system. Its main component is a sensor.
- **Actuator**. A device that applies the input signal to the plant. It is a device that requires high levels of power to work.
- **Power amplifier**. A device that receives a weak signal from a controller and delivers a powerful signal to the actuator.
- **External disturbance**. A signal that is external to the control system having deleterious effects on the performance of the closed-loop system. This signal may appear at the plant input or output or even at any component of the control system.

1.5 History of Automatic Control

The previous sections of this chapter described the objectives pursued when designing a control system along with some useful ideas on how to achieve them. In this section, a brief history of automatic control is presented. The reader is expected to realize that the development of fundamental tools and concepts in automatic control have been motivated by practical engineering problems arising when introducing new technologies [10]. In this exposition, the reader is referred to specific chapters in this book where the corresponding tools and ideas are further studied. The content of the present section is based on information reported in [6]. The reader is referred to that work for more details on the history of automatic control.

Automatic control has existed and has been applied for more than 2000 years. It is known that some water clocks were built by Ktesibios by 270 BC, in addition to some ingenious mechanisms built at Alexandria and described by Heron. However, from the engineering point of view, the first important advance in automatic control was attributed to James Watt in 1789 when introducing a velocity regulator for its improved steam-engine.[8] However, despite its importance Watt's velocity regulator had several problems. Sometimes, the velocity oscillated instead of remaining constant at a desired value or, even worse, sometimes the velocity increased without a limit.[9] Working toward a solution of these problems, between 1826 and 1851, J.V. Poncelet and G.B. Airy showed that it was possible to use differential equations to represent the steam-engine and the velocity regulator working together (see Chap. 2 for *physical system modeling*).

By those years, mathematicians knew that the *stability* of a differential equation was determined by the location of the roots of the corresponding characteristic polynomial equation, and they also knew that instability appeared if some root had a positive real part (see Chap. 3, Sect. 3.4). However, it was not simple to compute the roots of a polynomial equation and sometimes this was even not possible. In 1868, J.C. Maxwell showed how to establish the stability of steam-engines equipped with Watt's velocity regulator just by analyzing the system's differential equation coefficients. Nevertheless, this result was only useful for second-, third-, and fourth-order differential equations. Later, between 1877 and 1895, and independently, E.J. Routh and A. Hurwitz conceived a method of determining the stability of arbitrary order systems, solving the problem that Maxwell had left open. This method is known now as the *Routh criterion* or the *Routh–Hurwitz criterion* (see Chap. 4, Sect. 4.3).

Many applications related to automatic control were reported throughout the nineteenth century. Among the most important were temperature control, pressure control, level control, and the velocity control of rotative machines. On the other hand, several applications were reported where steam was used to move large guns and as actuators in steering systems for large ships [9, 11]. It was during this period when the terms servo-motor and servo-mechanism were introduced in France to describe a movement generated by a servo or a slave device. However, despite this success, most controllers were simple on-off. People such as E. Sperry and M.E. Leeds realized that performance could be improved by smoothly adjusting the power supplied to the plant as the controlled variable approached its desired value. In 1922, N. Minorsky presented a clear analysis of position control systems and introduced what we know now as the *PID controller* (see Chap. 5, Sect. 5.2.5). This controller was conceived after observing how a ship's human pilot controls heading [12].

On the other hand, distortion by amplifiers had posed many problems to telephony companies since 1920.[10] It was then that H.S. Black found that distortion

---

[8]Since flooding in coal and tin mines was a major problem, the first practical steam-engine was introduced in England by T. Newcomen in 1712 to improve ways of pumping water out of such mines.

[9]This behavior is known now as instability, see Chap. 3, Sect. 3.4.5.

[10]As the wired telephony lines expanded because of the growing number of users, signal attenuation was stronger because of the wires' electrical resistance. The solution was to use several signal amplifiers at selected points in the telephony network. These amplifiers, however, produced significant signal distortion and the sound quality deteriorated.

is reduced if a small quantity of signal at the amplifier output is fed back to its input. During this work, Black was helped by H. Nyquist who, in 1932, published these experiences in a report entitled "Regeneration Theory" where he established the basis of what we know now as *Nyquist Analysis* (see Chap. 6, Sect. 6.4).

During the period 1935–1940, telephony companies wanted to increase the bandwidth of their communication systems to increase the number of users. To accomplish this, it was necessary for the telephonic lines to have a good frequency response characteristic (see Chap. 6). Motivated by this problem, H. Bode studied the relationship between a given attenuation characteristic and the minimum associated phase shift. As a result, he introduced the concepts of *gain margin* and *phase margin* (see Chap. 6, Sect. 6.5) and he began to consider the point $(-1, 0)$ in the complex plane as a critical point instead of point $(+1, 0)$ introduced by Nyquist. A detailed description of Bode's work appeared in 1945 in his book "Network Analysis and Feedback Amplifier Design."

During World War II, work on control systems focused on some important problems [8]. The search for solutions to these problems motivated the development of new ideas on mechanism control. G.S. Brown from the Massachusetts Institute of Technology showed that electrical and mechanical systems can be represented and manipulated using *block diagrams* (see Chap. 4, Sect. 4.1) and A.C. Hall showed in 1943 that defining blocks as *transfer functions* (see Chap. 3), it was possible to find the equivalent transfer function of the complete system. Then, the Nyquist Stability criterion could be used to determine gain and phase margins.

Researchers at the Massachusetts Institute of Technology employed phase lead circuits (see Chap. 11, Sect. 11.2.2) at the direct path to improve the performance of the closed-loop system, whereas several internal loops were employed in the UK to modify the response of the closed-loop system.

By the end of World War II, the frequency response techniques, based on the Nyquist methodology and Bode diagrams, were well established, describing the control system performance in terms of bandwidth, resonant frequency, phase margin, and gain margin (see Chap. 6). The alternative approach to these techniques relied on the solution of the corresponding differential equations using Laplace transform, describing control system performance in terms of rise time, overshoot, steady-state error, and damping (see Chap. 3, Sect. 3.3). Many engineers preferred the latter approach because the results were expressed in "real" terms. But this approach had the drawback that there was no simple technique for designers to use to relate changes in parameters to changes in the system response. It was precisely the Root Locus method [7] (see Chap. 5, Sects. 5.1 and 5.2), introduced between 1948 and 1950 by W. Evans, that allowed designers to avoid these obstacles. Hence, what we know now as the *Classical Control Techniques* were well established by that time, and they were orientated to systems represented by ordinary, linear differential equations with constant coefficients and single-input single-output.

Then, the era of supersonic and space flights arrived. It was necessary to employ detailed physical models represented by differential equations that could be linear or nonlinear. Engineers working in the aerospace industries found, following the ideas of Poincaré, that it was possible to formulate general differential equations in terms of a set of first-order differential equations: the *state variable approach* was conceived (see Chap. 7). The main promoter of this approach was R. Kalman who introduced the concepts of controllability and observability around 1960 (see Sect. 7.7). Kalman also introduced what is today known as the Kalman filter, which was successfully employed for guidance of the Apollo space capsule, and the linear quadratic regulator, or LQR control, which is an optimal controller minimizing the system time response and the input effort required for it.

After 1960, the state space approach was the dominant subject for about two decades, leading to I. Horowitz, who continued to work on the frequency response methods, to write in 1984: "modern PhD.s seem to have a poor understanding of even such a fundamental concept of bandwidth and not the remotest idea of its central importance in feedback theory. It is amazing how many are unaware that the primary reason for feedback in control is uncertainty."

In fact, control systems in practice are subject to parameter uncertainties, external disturbances, and measurement noise. An important advantage of classical control methods was that they were better suited than the state variable approach to coping with these problems. This is because classical design control methods are naturally based on concepts such as bandwidth, and gain and phase margins (see Chap. 6, Sects. 6.5 and 6.6.1) which represent a measure of the robustness[11] of the closed-loop system. Furthermore, there was a rapid realization that the powerful results stemming from the state variable approach were difficult to apply in general industrial problems because the exact models of processes are difficult to obtain and sometimes, it is not possible to obtain them. In this respect, K. Astrom and P. Eykoff wrote in 1971 that an important feature of the classical frequency response methods is that they constitute a powerful technique for system identification allowing transfer functions to be obtained that are accurate enough to be used in the design tasks. In modern control, the models employed are parametric models in terms of state equations and this has motivated interest in parameter estimation and related techniques. Moreover, the state variable approach has been demonstrated to be a very powerful tool for the analysis and design of nonlinear systems.

Finally, new problems have arisen since then in the study of control systems theory, which have motivated the introduction of diverse new control techniques, some of them still under development. For instance, nonlinearities found in servomechanisms have motivated the study of nonlinear control systems. Control of supersonic aircrafts, which operate under wide variations in temperature, pressure, velocity, etc., has motivated the development of adaptive control techniques. The use of computers in modern navigation systems has resulted in the introduction of discrete time control systems, etc.

---

[11]Robustness is a property of a control system that indicates that the closed-loop system performance is not affected or it is just a little affected by external disturbances, parameter uncertainties, and measurement noise.

1.6   Experimental Prototypes

It has been stated in the previous sections that automatic control has developed to solve important engineering problems. However, teaching automatic control techniques requires students to experimentally apply their new knowledge of this subject. It is not possible to accomplish this using either industrial facilities or high-technology laboratories. This is the reason why teaching automatic control relies on construction of some *experimental prototypes*. An experimental prototype is a device that has two main features: (i) it is simple enough to be built and put into operation using low-cost components, and (ii) its model is complex enough to allow some interesting properties to appear such that the application of the control techniques under study can be demonstrated. A list of experimental prototypes used in this book is presented in the following and the specific control techniques tested on them are indicated:

- Electronic oscillators based on operational amplifiers and bipolar junction transistors (Chap. 9). Frequency response, Nyquist stability criterion, and Routh stability criterion.
- Permanent magnet brushed DC motors (Chaps. 10 and 11). Several basic controller designs using time response: proportional, proportional–derivative, proportional–integral, proportional–integral–derivative, phase lead controllers, two-degrees-of-freedom controllers.
- Mechanism with flexibility (Chap. 12). Frequency response for experimental identification and root locus for controller design.
- Magnetic levitation system (Chap. 13). PID controller design using linear approximation of a nonlinear system and the root locus method.
- Ball and beam system (Chap. 14). Design of a multi-loop control system using frequency response (Nyquist criterion and Bode diagrams) and the root locus method.
- Furuta pendulum (Chap. 15). Design of a linear state feedback controller using the state variable approach. A linear approximation of a nonlinear system is employed.
- Inertia wheel pendulum (Chap. 16). Design of two state feedback controllers. One of these controllers is designed on the basis of the complete nonlinear model of the inertial wheel pendulum and it is employed to introduce the reader to the control of nonlinear systems.

1.7   Summary

In the present chapter, the main ideas behind closed-loop control have been explained and the objectives of designing a control system were also described. A brief history of automatic control has been presented to show the reader that all concepts and tools in control systems have been motivated by solving important technological problems. This historical review has been related to content of this book.

1.8   Review Questions

1. What objectives can the reader give for automatic control?
2. Can the reader make a list of equipment at home that employs feedback?
3. Investigate how a pendulum-based clock works. How do you think that feedback appears in the working principle of these clocks?
4. Why is Watt's velocity regulator for a steam engine historically important?
5. What does instability of a control system mean?
6. What do you understand by the term "fast response"?
7. Why is it stated that an inverted pendulum is unstable?
8. Why did the frequency response approach develop before the time response approach?

References

1. M. K. Masten, Inertially stabilized platforms for optical imaging systems, *IEEE Control Systems Magazine*, pp. 47–64, February 2008.
2. T. L. Vincent, Stabilization for film broadcast cameras, *IEEE Control Systems Magazine*, pp. 20–25, February 2008.
3. Homeostasis (In Spanish), *Wikipedia, the free encyclopedia*, available at https://es.wikipedia.org/wiki/Homeostasis, July 2017.
4. A. Adserá-Bertran, The human body. What is homeostasis? (In Spanish), *enciclopediasalud.com*, available at http://www.enciclopediasalud.com/categorias/cuerpo-humano/articulos/que-es-la-homeostasis-ejemplos-de-homeostasis, November 2009.
5. Fratermindad Muprespa, The health corner. Fever (In Spanish), available at http://www.rincondelasalud.com/es/articulos/salud-general_la-fiebre_20.html#prim, 2017.
6. S. Bennett, A brief history of automatic control, *IEEE Control Systems Magazine*, pp. 17–25, June 1996.
7. G. W. Evans, The story of Walter R. Evans and his textbook Control-Systems Dynamics, *IEEE Control Systems Magazine*, pp. 74–81, December 2004.
8. D. A. Mindell, Anti-aircraft fire control and the development of integrated systems at Sperry, 1925–1940, *IEEE Control Systems Magazine*, pp. 108–113, April 1995.
9. S. W. Herwald, Recollection of the early development of servomechanism and control systems, *IEEE Control Systems Magazine*, pp. 29–32, November 1984.
10. D. S. Bernstein, Feedback control: an invisible thread in the history of technology, *IEEE Control Systems Magazine*, pp. 53–68, April 2002.
11. W. Oppelt, On the early growth of conceptual thinking in the control system theory – the German role up to 1945, *IEEE Control Systems Magazine*, pp. 16–22, November 1984.
12. S. Bennett, Nicolas Minorsky and the automatic steering of ships, *IEEE Control Systems Magazine*, pp. 10–15, November 1984.

Chapter 2
Physical System Modeling

Automatic control is interested in mathematical models describing the evolution in time of system variables as a response to excitation. Such mathematical models are also known as dynamical models and systems represented by these models are called dynamical systems. The dynamical models that are useful for classical control techniques consist of differential equations, in particular, ordinary differential equations where the independent variable is time. Partial differential equations arise when several independent variables exist, for instance, a flexible body where vibration has to be described with respect to both position in the body and time. However, this class of problem is outside the scope of classical control; hence, partial differential equations will not be used to model physical systems in this book.

On the other hand, ordinary differential equations can be linear or nonlinear, time-invariant or time-variant. Because of the mathematical complexity required to study nonlinear and time-variant differential equations, classical control constrains its study to linear invariant (also called constant coefficients) ordinary differential equations. As real physical systems are intrinsically nonlinear and time-variant, it is common to consider several simplifications when modeling physical systems in classical control. For instance, model nonlinearities are approximated by linear expressions and parameters that change with time (such as resistance in an electric motor) are assumed to be constant in time, but the design of the controller must be *robust*. This means that the controller must ensure good performance of the designed closed-loop control system, despite the changes in such a parameter.

It is important to remark that model simplifications are only valid if the simplified model is still accurate enough to represent the true behavior of the physical system. This idea is captured by stating that a good model must be simple enough to render possible its mathematical study, but it must also be complex enough to describe the important properties of the system. This is the approach employed in the present chapter to model physical systems, following many ideas in [1].

© Springer International Publishing AG, part of Springer Nature 2019
V. M. Hernández-Guzmán, R. Silva-Ortigoza, *Automatic Control with Experiments*,
Advanced Textbooks in Control and Signal Processing,
https://doi.org/10.1007/978-3-319-75804-6_2

2.1 Mechanical Systems

In mechanical systems there are three basic phenomena: bodies, flexibility, and friction. The main idea is to model each one of these phenomena and then connect them to obtain the mathematical model of the complete mechanical system. Furthermore, there are two different classes of mechanical systems: translational and rotative. In the following, we study both of them.

2.1.1 Translational Mechanical Systems

In Fig. 2.1 a rigid body is shown (without flexibility) with mass m (a positive quantity), which moves with velocity v under the effect of a force F. It is assumed that no friction exists between the body and the environment. The directions shown for force and velocity are defined as positive for this system component. Under these conditions, the constitutive function is given by Newton's Second Law [2], pp. 89:

$$F = ma, \tag{2.1}$$

where $a = \frac{dv}{dt}$ is the body acceleration.

In Fig. 2.2, a spring is shown, which deforms (either compresses or stretches) with velocity v under the effect of a force F. Notice that, although force F is applied at point 1 of the spring, the same force but in the opposite direction must also be considered at point 2 to render spring deformation possible. Moreover, $v = v_1 - v_2$ where v_1 and v_2 are velocities of points 1 and 2 of spring respectively. The directions shown for forces and velocities are defined as positive for this system component. Notice that velocity v indicates that point 1 approaches point 2. It is assumed that the spring has no mass, whereas deformation is not permanent and does not produce heat.

Fig. 2.1 A rigid body with mass m in a translational mechanical system

Fig. 2.2 A spring in a translational mechanical system

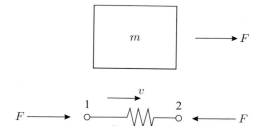

The net spring deformation is given as:

$$x = \int_0^t v\, dt + x(0),\qquad(2.2)$$

where $x = x_1 - x_2$ with x_1 and x_2 the positions of points 1 and 2 on the spring respectively, whereas $x(0) = 0$ represents the spring deformation when it is neither compressed nor stretched. According to the direction of velocity v defined in Fig. 2.2, it is concluded that the net deformation x is positive when the spring is compressed. In the case of a linear spring, the constitutive function is given by Hooke's Law [2], pp. 640:

$$F = k\, x,\qquad(2.3)$$

where k is a positive constant known as the spring stiffness constant.

Friction always appears when two bodies are in contact while a relative movement exists between them. In such a case, a force must be applied to maintain the relative movement. Kinetic energy converts into heat because of friction. The constitutive function of friction is given as:

$$F = \varphi(v),$$

where v is the relative velocity between the bodies and F is the applied force. A very important specific case is viscous friction, which is represented by the linear constitutive function :

$$F = bv,\quad \text{viscous friction}\qquad(2.4)$$

where b is a positive constant known as the viscous friction coefficient. This kind of friction appears, for instance, when a plate with several holes moves inside a closed compartment with air, as shown in Fig. 2.3. Because air must flow through the holes as relative movement exists with velocity v, it is necessary to apply a force F to maintain the velocity. It is defined $v = v_1 - v_2$ where v_1 and v_2 are velocities of bodies 1 and 2 respectively. The directions of forces and velocities shown in Fig. 2.3 are defined as positive. Although the device shown in Fig. 2.3 may not be physically placed between the two bodies, the effect of viscous friction is commonly present and must be taken into account.

There are other kinds of friction that often appear in practice. Two of them are static friction and Coulomb friction. The respective constitutive functions are given as:

$$F = \pm F_s|_{v=0},\quad \text{static friction},\qquad(2.5)$$

$$F = F_c\, \text{sign}(v),\quad \text{sign}(v) = \begin{cases} +1, & \text{if } v > 0 \\ -1, & \text{if } v < 0 \end{cases}, \text{Coulomb friction.}\qquad(2.6)$$

Fig. 2.3 Viscous friction appears as a result of opposition to air flow through the holes

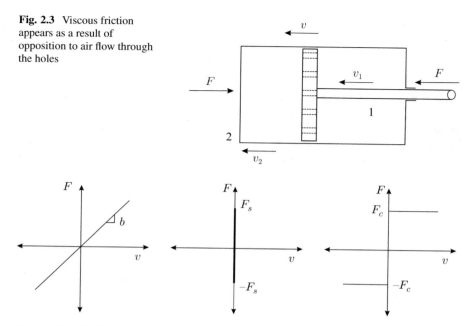

Fig. 2.4 Constitutive functions defined in (2.4), (2.5) and (2.6)

Static friction is represented by a force that prevents movement only just before this starts. This is the reason why constant F_s is evaluated at $v = 0$ in (2.5). Coulomb friction, in contrast, only appears when movement has started ($v \neq 0$) and it is constant for velocities having the same sign. The constitutive functions in (2.4), (2.5), and (2.6) are depicted in Fig. 2.4.

As the static friction force is not zero as velocity tends to zero, it is responsible for some of the problems in position control systems: the difference between the desired position and position actually reached by the mechanism is different from zero, which is known as "a nonzero steady state error." On the other hand, static and Coulomb friction have nonlinear and discontinuous constitutive functions; hence, they can be handled using control techniques developed for neither linear systems nor "smooth" nonlinear systems. This is the reason why these kinds of friction are not often considered when modeling control systems. However, it must be remarked that the effects of these frictions are always appreciable in experimental situations. The reader is referred to [4] to know how to measure the parameters of some friction models.

Once the constitutive functions of each system component have been established, they must be connected to conform the mathematical model of more complex systems. The key to connecting components of a mechanical system is to assume that all of them are connected by means of rigid joints. This means that velocities on both sides of a joint are equal and that, invoking Newton's Third Law, each action produces a reaction that is equal in magnitude but opposite in direction. This is clarified in the following by means of several examples.

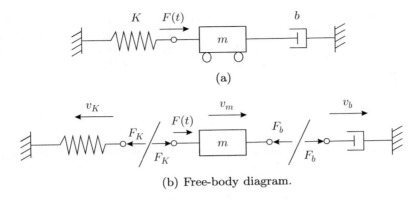

(a)

(b) Free-body diagram.

Fig. 2.5 A mass-spring-damper system

Example 2.1 Consider the mass-spring-damper system shown in Fig. 2.5a where an external force $F(t)$ is applied on mass m. This system is, perhaps, the simplest mechanical system from the modeling point of view, but it is very important from the control point of view because it represents the fundamental behavior of position control systems.

A free-body diagram is shown in Fig. 2.5b. Notice that two forces, equal in magnitude but with opposite directions, exist at each joint between two components. This is to satisfy Newton's Third Law [2], pp. 88, which establishes that for each action (from A on B) there is one corresponding reaction (from B on A). Wheels under the mass indicate that no friction exists between the mass and the floor. This also means that gravity has no effect. From the friction modeling point of view, it is equivalent to eliminating a damper at the right and replace it by friction between the mass and the floor. The friction model between the mass and the floor is identical to the friction model due to a damper as considered in the following.

The nomenclature employed is defined as:

- $v_m = \frac{dx_m}{dt}$, where x_m is the mass position.
- $v_b = \frac{dx_b}{dt}$, where x_b is the position of the damper's mobile side.
- $v_K = \frac{dx_K}{dt}$, where x_K is the position of the spring's mobile side.

The following expressions are found from Fig. 2.5b, and (2.1), (2.3), (2.4):

$$\text{Mass:} \qquad m\frac{dv_m}{dt} = F(t) + F_K - F_b. \qquad (2.7)$$

$$\text{Spring:} \qquad Kx_K = F_K. \qquad (2.8)$$

$$\text{Damper:} \qquad bv_b = F_b. \qquad (2.9)$$

Notice that, according to the sign definitions shown in Fig. 2.1, forces actuating on mass m having the same direction as v_m in Fig. 2.5b appear affected by a positive sign in (2.7); therefore, F_b appears affected by a negative sign. Signs on both sides of expressions in (2.8) and (2.9) are given according to the sign definitions shown in Figs. 2.2 and 2.3 respectively. Also notice that one side of both spring and damper has no movement and the mobile sides have a velocity and a force applied in the same direction as in Figs. 2.2 and 2.3. Finally:

$$v_K = \frac{dx_K}{dt},\qquad (2.10)$$

has been defined. Assuming that all of the system components are connected by rigid joints and according to Fig. 2.5b, the following can be written:

$$v_K = -v_m = -v_b. \qquad (2.11)$$

Hence, it is concluded that:

$$x_K = -x_m + c_1 = -x_b + c_2,$$

where c_1 and c_2 are two constants. If $x_K = 0$, $x_m = 0$ and $x_b = 0$ are defined as the positions of those points at the spring's mobile side, the center of mass m and the damper's mobile side respectively, when the system is at rest with $F(t) = 0$, then $c_1 = c_2 = 0$ and hence:

$$x_K = -x_m = -x_b.$$

Using these facts, (2.7), (2.8), (2.9), can be written as:

$$m\frac{d^2x_m}{dt^2} + b\frac{dx_m}{dt} + Kx_m = F(t). \qquad (2.12)$$

This expression is an ordinary, linear, second-order, constant coefficients differential equation representing the mass-spring-damper system mathematical model. Thus, assuming that the parameters m, b, and K are known, the initial conditions $x_m(0)$, $\frac{dx_m}{dt}(0)$ are given, and the external force $F(t)$ is a known function of time, this differential equation can be solved to find the mass position $x_m(t)$ and the velocity $\frac{dx_m}{dt}(t)$ as functions of time. In other words, the mass position and velocity can be known at any present or future time, i.e., for all $t \geq 0$. Finally, when analyzing a differential equation, it is usual to express it with an unitary coefficient for the largest order time derivative, i.e., (2.12) is written as:

$$\ddot{x}_m + \frac{b}{m}\dot{x}_m + \frac{K}{m}x_m = \frac{1}{m}F(t), \qquad (2.13)$$

where $\ddot{x}_m = \frac{d^2x_m}{dt^2}$ and $\dot{x}_m = \frac{dx_m}{dt}$.

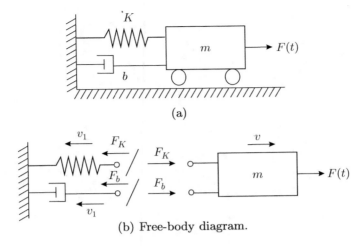

(a)

(b) Free-body diagram.

Fig. 2.6 A mass-spring-damper system

Example 2.2 A mass-spring-damper system is depicted in Fig. 2.6a. Both spring and damper are fixed to the same side of the mass m. The purpose of this example is to show that the mathematical model in this case is identical to that obtained in the previous example. The nomenclature is defined as:

- $v = \frac{dx}{dt}$, where x is the mass position.
- $v_1 = \frac{dx_K}{dt} = \frac{dx_b}{dt}$, where $x_K = x_b$ stands for the spring's and damper's mobile sides.

According to Fig. 2.6b and (2.1), (2.3), (2.4), the following can be written:

$$\text{Mass:} \quad m\frac{dv}{dt} = F(t) + F_K + F_b. \tag{2.14}$$

$$\text{Spring:} \quad Kx_K = F_K.$$

$$\text{Damper:} \quad bv_1 = F_b.$$

Notice that, according to the sign definition shown in Fig. 2.1, all forces in (2.14) appear with a positive sign because all of them have the same direction as v in Fig. 2.6b. On the other hand, as the joint is rigid and according to the sign definition in Fig. 2.6b:

$$v = -v_1. \tag{2.15}$$

Proceeding as in the previous example, if $x = 0$, $x_K = 0$ and $x_b = 0$ are defined respectively, as the center of mass m and the spring's and damper's mobile sides when the whole system is at rest with $F(t) = 0$, then:

$$x_K = x_b = -x. \tag{2.16}$$

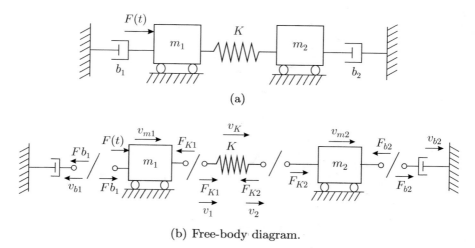

(a)

(b) Free-body diagram.

Fig. 2.7 Two bodies connected by a spring

Notice that combining expressions in (2.14) and using (2.16), (2.15) yield:

$$\ddot{x} + \frac{b}{m}\dot{x} + \frac{K}{m}x = \frac{1}{m}F(t),$$

where $\ddot{x} = \frac{d^2x}{dt^2}$ and $\dot{x} = \frac{dx}{dt}$. This mathematical model is identical to that shown in (2.13) if $x = x_m$ is assumed.

Example 2.3 Two bodies connected by a spring, with an external force $F(t)$ applied on the body at the left, are shown in Fig. 2.7a. This situation arises in practice when body 1 transmits movement to body 2 and they are connected through a piece of some flexible material. It is remarked that flexibility is not intentionally included, but it is an undesired problem that appears because of the finite stiffness of the material in which the piece that connects bodies 1 and 2 is made. It is often very important to study the effect of this flexibility on the achievable performance of the control system; hence, it must be considered at the modeling stage. The dampers are included to take into account the effects of friction between each body and the floor.

The free-body diagram is depicted in Fig. 2.7b. The nomenclature employed is defined as:

- $v_{m1} = \frac{dx_{m1}}{dt}$, where x_{m1} is the position of body 1.
- $v_{m2} = \frac{dx_{m2}}{dt}$, where x_{m2} is the position of body 2.
- $v_{b1} = \frac{dx_{b1}}{dt}$, where x_{b1} is the position of damper 1's mobile side.
- $v_{b2} = \frac{dx_{b2}}{dt}$, where x_{b2} is the position of damper 2's mobile side.
- $v_1 = \frac{dx_1}{dt}$, where x_1 is the position of the spring side connected to body 1.
- $v_2 = \frac{dx_2}{dt}$, where x_2 is the position of the spring side connected to body 2.

Using Fig. 2.7b, as well as (2.1), (2.3), (2.4), the following is obtained:

$$\text{Mass 1:} \qquad m_1 \frac{dv_{m1}}{dt} = F(t) + F_{b1} - F_{K1}. \qquad (2.17)$$

$$\text{Mass 2:} \qquad m_2 \frac{dv_{m2}}{dt} = F_{K2} - F_{b2}. \qquad (2.18)$$

$$\text{Spring:} \qquad K x_K = F_{K1} = F_{K2}.$$

$$\text{Damper 1:} \qquad b_1 v_{b1} = F_{b1}.$$

$$\text{Damper 2:} \qquad b_2 v_{b2} = F_{b2}.$$

Forces having the same direction as v_{m1} and v_{m2}, actuating on each one of the bodies, appear with a positive sign, and a negative sign if this is not the case. Notice that it is also defined:

$$v_K = \frac{dx_K}{dt} = v_1 - v_2.$$

Assuming that all of the system components are connected by rigid joints then, according to Fig. 2.7b:

$$v_{m1} = -v_{b1} = v_1, \qquad (2.19)$$

$$v_{m2} = v_{b2} = v_2.$$

Hence, if $x_{m1} = 0$, $x_{m2} = 0$, $x_{b1} = 0$, $x_{b2} = 0$, $x_1 = 0$, $x_2 = 0$ are defined as those positions of the center of body 1, the center of body 2, damper 1's mobile side, damper 2's mobile side, the spring side fixed to body 1, and the spring side fixed to body 2 respectively, when the complete system is at rest with $F(t) = 0$, then:

$$x_{m1} = -x_{b1} = x_1, \qquad (2.20)$$

$$x_{m2} = x_{b2} = x_2.$$

Hence, the expressions in (2.17) and (2.18) can be written as:

$$\text{Mass 1:} \qquad m_1 \frac{dv_{m1}}{dt} = F(t) + b_1 v_{b1} - K x_K.$$

$$\text{Mass 2:} \qquad m_2 \frac{dv_{m2}}{dt} = K x_K - b_2 v_{b2}.$$

Use of (2.19), (2.20), yields:

$$\text{Mass 1:} \qquad m_1 \frac{d^2 x_{m1}}{dt^2} = F(t) - b_1 \frac{dx_{m1}}{dt} - K x_K.$$

$$\text{Mass 2:} \qquad m_2 \frac{d^2 x_{m2}}{dt^2} = K x_K - b_2 \frac{d x_{m2}}{dt}.$$

On the other hand, $x_K = x_1 - x_2$ according to paragraph after (2.2) and, thus:

$$\text{Mass 1:} \qquad m_1 \frac{d^2 x_{m1}}{dt^2} = F(t) - b_1 \frac{d x_{m1}}{dt} - K(x_1 - x_2).$$

$$\text{Mass 2:} \qquad m_2 \frac{d^2 x_{m2}}{dt^2} = K(x_1 - x_2) - b_2 \frac{d x_{m2}}{dt}.$$

Finally, using again (2.20), it is found that the mathematical model is given by the following pair of differential equations, which have to be solved simultaneously:

$$\text{Mass 1:} \qquad \frac{d^2 x_{m1}}{dt^2} + \frac{b_1}{m_1} \frac{d x_{m1}}{dt} + \frac{K}{m_1} (x_{m1} - x_{m2}) = \frac{1}{m_1} F(t).$$

$$\text{Mass 2:} \qquad \frac{d^2 x_{m2}}{dt^2} + \frac{b_2}{m_2} \frac{d x_{m2}}{dt} - \frac{K}{m_2} (x_{m1} - x_{m2}) = 0.$$

These differential equations cannot be combined into a single differential equation because variables x_{m1} and x_{m2} are linearly independent, i.e., there is no algebraic equation relating them.

Example 2.4 Two bodies connected to three springs and a damper are shown in Fig. 2.8a. The damper represents friction existing between the bodies and, because of that, it cannot be replaced by the possible friction appearing between each body and the floor (which would be the case if wheels were not placed under each body). The corresponding free-body diagram is depicted in Fig. 2.8b. The nomenclature employed is defined as:

- $\dot{x}_1 = \frac{dx_1}{dt}$, where x_1 is the position of body 1.
- $\dot{x}_2 = \frac{dx_2}{dt}$, where x_2 is the position of body 2.
- $v_{b1} = \frac{dx_{b1}}{dt}$, where x_{b1} is the position of the damper side connected to body 1.
- $v_{b2} = \frac{dx_{b2}}{dt}$, where x_{b2} is the position of the damper side fixed to body 2.
- $v_1 = \frac{dx}{dt}$, where x is the position of the spring side connected to body 1.
- $v_2 = \frac{dy}{dt}$, where y is the position of the spring side fixed to body 2.
- $v_{K1} = \frac{dx_{K1}}{dt}$, where x_{K1} is the position of the mobile side of the spring connected only to body 1.
- $v_{K3} = \frac{dx_{K3}}{dt}$, where x_{K3} is the position of the mobile side of the spring fixed only to body 2.

Using Fig. 2.8b, and (2.1), (2.3), (2.4), the following is found:

$$\text{Mass 1:} \qquad m_1 \ddot{x}_1 = F(t) + F_{K1} - F_{K2} - F_b. \tag{2.21}$$

$$\text{Mass 2:} \qquad m_2 \ddot{x}_2 = F_{K2} - F_{K3} + F_b. \tag{2.22}$$

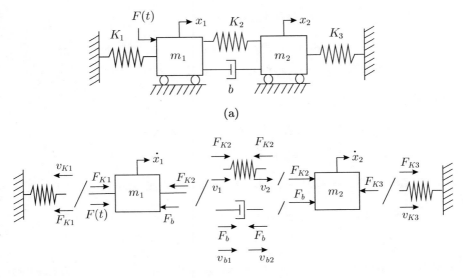

(a)

(b) Free-body diagram.

Fig. 2.8 A mechanical system with two bodies and three springs

Spring between bodies: $K_2 x_{K2} = F_{K2}.$

Spring at the left: $K_1 x_{K1} = F_{K1}.$

Spring at the right: $K_3 x_{K3} = F_{K3}.$

Damper: $b(v_{b1} - v_{b2}) = F_b.$

Assuming that all of the system components are connected by rigid joints then, according to Fig. 2.8b:

$$\dot{x}_1 = -v_{K1} = v_1 = v_{b1}, \tag{2.23}$$

$$\dot{x}_2 = v_{b2} = v_2 = v_{K3}.$$

Hence, if $x_1 = 0$, $x_2 = 0$, $x_{b1} = 0$, $x_{b2} = 0$, $x = 0$, $y = 0$, $x_{K1} = 0$, $x_{K3} = 0$ are defined as the positions of the center of body 1, the center of body 2, the side of the damper fixed to body 1, the side of the damper fixed to body 2, the side of the central spring that is fixed to body 1, the side of the central spring fixed to body 2, the mobile side of the spring at the left and the mobile side of the spring at the right respectively, when the complete system is at rest with $F(t) = 0$, then:

$$x_1 = -x_{K1} = x = x_{b1}, \tag{2.24}$$

$$x_2 = x_{b2} = y = x_{K3}.$$

Thus, the expressions in (2.21) and (2.22) can be written as:

Mass 1: $m_1\ddot{x}_1 = F(t) + K_1 x_{K1} - K_2 x_{K2} - b(v_{b1} - v_{b2})$.

Mass 2: $m_2\ddot{x}_2 = K_2 x_{K2} - K_3 x_{K3} + b(v_{b1} - v_{b2})$.

Use of (2.23), (2.24), yields:

Mass 1: $m_1\ddot{x}_1 = F(t) - K_1 x_1 - K_2 x_{K2} - b(\dot{x}_1 - \dot{x}_2)$.

Mass 2: $m_2\ddot{x}_2 = K_2 x_{K2} - K_3 x_2 + b(\dot{x}_1 - \dot{x}_2)$.

On the other hand, $x_{K2} = x_1 - x_2$ according to paragraph after (2.2), and hence:

Mass 1: $m_1\ddot{x}_1 = F(t) - K_1 x_1 - K_2(x_1 - x_2) - b(\dot{x}_1 - \dot{x}_2)$.

Mass 2: $m_2\ddot{x}_2 = K_2(x_1 - x_2) - K_3 x_2 + b(\dot{x}_1 - \dot{x}_2)$.

Finally, the mathematical model is given by the following pair of differential equations, which must be solved simultaneously:

Mass 1: $\ddot{x}_1 + \dfrac{b}{m_1}(\dot{x}_1 - \dot{x}_2) + \dfrac{K_1}{m_1}x_1 + \dfrac{K_2}{m_1}(x_1 - x_2) = \dfrac{1}{m_1}F(t)$.

Mass 2: $\ddot{x}_2 - \dfrac{b}{m_2}(\dot{x}_1 - \dot{x}_2) + \dfrac{K_3}{m_2}x_2 - \dfrac{K_2}{m_2}(x_1 - x_2) = 0$.

As in the previous example, these differential equations cannot be combined in a single differential equation because variables x_1 and x_2 are linearly independent, i.e., there is no algebraic equation relating these variables. Also notice that terms corresponding to the central spring and damper appear with the opposite sign in each one of these differential equations. This is because forces exerted by the central spring or the damper on body 1 are applied in the opposite direction to the force exerted on body 2.

2.1.2   Rotative Mechanical Systems

A rigid body (without flexibility) that rotates with angular velocity ω under the effect of a torque T is depicted in Fig. 2.9. It is assumed that no friction exists between the body and the environment. The directions shown for torque and angular velocity are defined as positive for this system component. According to Newton's Second Law [2], pp. 122:

$$T = I\alpha, \tag{2.25}$$

Fig. 2.9   A rotative rigid
body with inertia I

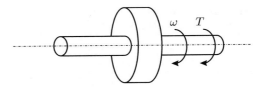

Fig. 2.10   A rotative spring

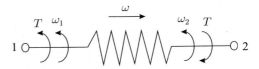

where $\alpha = \frac{d\omega}{dt}$ represents the body angular acceleration and I, a positive constant, stands for the body moment of inertia.

A rotative spring that deforms at an angular velocity ω, under the effect of a torque T, is shown in Fig. 2.10. Although torque T is applied at the spring's side 1, it must be considered that the same torque, but in the opposite direction, must be applied at the spring's side 2, to render possible deformation. The angular velocities on sides 1 and 2 respectively are defined $\omega = \omega_1 - \omega_2$ with ω_1 and ω_2

The directions shown for torque and velocities are defined as positive for this system component. It is assumed that the spring has no moment of inertia whereas deformation is not permanent and does not produce heat. The net spring angular deformation is given as:

$$\theta = \int_0^t \omega \, dt + \theta(0), \tag{2.26}$$

where $\theta = \theta_1 - \theta_2$ with θ_1, and θ_2 are the angular positions of the spring's sides 1 and 2 respectively, whereas $\theta(0) = 0$ represents deformation when the spring is neither compressed nor stretched. According to the direction defined for ω in Fig. 2.10, it is concluded that the net angular deformation θ is positive when $\theta_1 > \theta_2$. In the case of a linear rotative spring, the constitutive function is given by Hooke's Law:

$$T = k \, \theta, \tag{2.27}$$

where k is a positive constant known as the spring stiffness constant.

Friction in rotative mechanical systems (see Fig. 2.11) is modeled as in translational mechanical systems. The only difference is that, in rotative mechanical systems, friction is expressed in terms of angular velocity $\omega = \omega_1 - \omega_2$ and torque T:

$$T = b\omega, \quad \text{viscous friction}, \tag{2.28}$$

$$T = \pm T_s|_{\omega=0}, \quad \text{static friction},$$

Fig. 2.11 A damper in
rotative mechanical systems

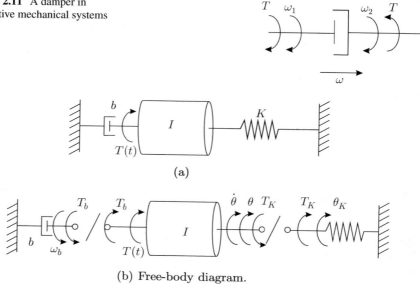

Fig. 2.11 A damper in
rotative mechanical systems

(a)

(b) Free-body diagram.

Fig. 2.12 A rotative mass-spring-damper system

$$T = T_c \, \text{sign}(\omega), \quad \text{sign}(\omega) = \begin{cases} +1, \text{ if } \omega > 0 \\ -1, \text{ if } \omega < 0 \end{cases}, \text{Coulomb friction.}$$

Example 2.5 A rotative body connected to two walls through a spring and a damper
is shown in Fig. 2.12a. An external torque $T(t)$ is applied to this body. The
damper represents friction between the body and the bearings supporting it, whereas
the spring represents the shaft flexibility. The free-body diagram is depicted in
Fig. 2.12b. The employed nomenclature is defined as:

- $\dot{\theta} = \frac{d\theta}{dt}$, where θ is the body's angular position.
- $\omega_b = \frac{d\theta_b}{dt}$, where θ_b is the angular position of the damper's mobile side.
- $\theta_K = \frac{d\theta_K}{dt}$, where θ_K is the angular position of the spring's mobile side.

Use of Fig. 2.12b, and (2.25), (2.27), (2.28), yields:

$$\begin{array}{lll} \text{Inertia:} & I\ddot{\theta} = T(t) + T_b - T_K. & (2.29) \\ \text{Spring:} & K\theta_K = T_K. & \\ \text{Damper:} & b\omega_b = T_b. & \end{array}$$

Notice that torques applied to the body with inertia I that have the same direction
as angular velocity $\dot{\theta}$, appear with a positive sign in (2.29). If this is not the case,
then a negative sign appears (see T_K, for instance). Assuming that all the system
components are connected by rigid joints, then, according to Fig. 2.12b:

$$\dot{\theta} = -\omega_b = \dot{\theta}_K. \tag{2.30}$$

If $\theta = 0$, $\theta_b = 0$, $\theta_K = 0$, are defined as the angular positions of the body, the damper's mobile side and the spring's mobile side respectively, when the complete system is at rest with $T(t) = 0$, then:

$$\theta = -\theta_b = \theta_K. \tag{2.31}$$

Hence, the expressions in (2.29) can be written as:

$$I\ddot{\theta} = T(t) + b\omega_b - K\theta_K,$$
$$= T(t) - b\dot{\theta} - K\theta.$$

Finally, the mathematical model is given by the following differential equation:

$$\ddot{\theta} + \frac{b}{I}\dot{\theta} + \frac{K}{I}\theta = \frac{1}{I}T(t). \tag{2.32}$$

Notice that this model is identical to that presented in (2.13) for a translational mass-spring-damper system if mass is replaced by inertia, angular position θ replaces position x_m, and external torque $T(t)$ replaces the external force $F(t)$.

Example 2.6 A simple pendulum is depicted in Fig. 2.13a, where θ stands for the pendulum angular position and $\dot{\theta} = \frac{d\theta}{dt}$. It is assumed that the pendulum mass m is concentrated in a single point and pendulum rod has a length l with no mass (or it is negligible compared with mass m). The corresponding free-body diagram is depicted in Fig. 2.13b. The pendulum can be assumed to be a rotative body with inertia:

$$I = ml^2,$$

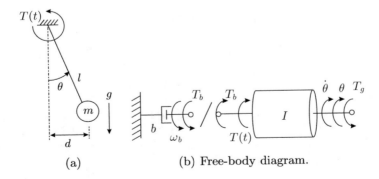

(a)

(b) Free-body diagram.

Fig. 2.13 Simple pendulum

which represents inertia of a particle with mass m describing a circular trajectory with a constant radius l [7], Ch. 9. It is assumed that three torques are applied to the pendulum: (1) the external torque $T(t)$, (2) the friction torque, T_b, which appears at the pendulum pivot, and (3) the torque due to gravity, T_g. According to Figs. 2.13a and b, torque T_g is applied in the opposite direction to that defined by θ and its magnitude is given as:

$$T_g = mgd, \quad d = l\sin(\theta),$$
$$= mgl\sin(\theta). \tag{2.33}$$

Use of Fig. 2.13b, and (2.25), (2.28), yields:

Inertia: $\quad I\ddot{\theta} = T(t) + T_b - T_g.$ \hfill (2.34)

Damper: $\quad b\omega_b = T_b.$

Notice that torques applied to I that have the same direction as angular velocity $\dot{\theta}$ appear affected by a positive sign in (2.34), or a negative sign if this is not the case (see for instance T_g). Assuming that all the system components are connected by rigid joints, then, according to Fig. 2.13b, it is found that:

$$\dot{\theta} = -\omega_b. \tag{2.35}$$

Hence, if $\theta = 0$, $\theta_b = 0$ (where $\omega_b = \dot{\theta}_b$) are defined as the angular positions of the pendulum and the damper's mobile side respectively, when the whole system is at rest with $T(t) = 0$, then:

$$\theta = -\theta_b. \tag{2.36}$$

This means that (2.34) and (2.33) can be written as:

$$I\ddot{\theta} = T(t) + b\omega_b - mgl\sin(\theta),$$
$$= T(t) - b\dot{\theta} - mgl\sin(\theta).$$

Finally, the mathematical model is given by the following differential equation:

$$ml^2\ddot{\theta} + b\dot{\theta} + mgl\sin(\theta) = T(t).$$

Notice that this is a nonlinear differential equation, because of the function $\sin(\theta)$. In Sect. 7.3, Chap. 7, it is explained how this kind of differential equation can be studied. Moreover, in Chaps. 13, 15 and 16 it is shown how to design controllers for systems represented by nonlinear differential equations.

Example 2.7 Two rotative bodies connected by a spring are shown in Fig. 2.14a. An external torque $T(t)$ is applied to the body on the left. The mechanical system

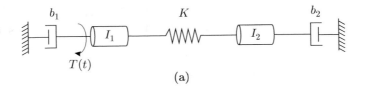

(a)

(b) Free-body diagram.

Fig. 2.14 Two rotative bodies connected by a spring

in Fig. 2.14a can be interpreted as a motor (body on the left), which moves a load (body on the right) through the motor shaft, exhibiting flexibility. This flexibility is not intentionally included, but is undesired shaft behavior arising from the finite stiffness of the material in which the motor shaft is made. In such a situation, it is very important to take into account this flexibility to study its effect on the closed-loop system performance. According to sign definitions shown in Figs. 2.9, 2.10, and 2.11, the corresponding free-body diagram is depicted in Fig. 2.14b. The employed nomenclature is defined as:

- $\omega_2 = \frac{d\theta_2}{dt}$, where θ_2 is the body 1 angular position.
- $\omega_5 = \frac{d\theta_5}{dt}$, where θ_5 is the body 2 angular position.
- $\omega_1 = \frac{d\theta_1}{dt}$, where θ_1 is the angular position of the mobile side of the damper connected to body 1.
- $\omega_6 = \frac{d\theta_6}{dt}$, where θ_6 is the angular position of the mobile side of the damper connected to body 2.
- $\omega_3 = \frac{d\theta_3}{dt}$ and $\omega_4 = \frac{d\theta_4}{dt}$, where θ_3 and θ_4 are the angular positions of the spring sides.

Use of Fig. 2.14b, and (2.25), (2.27), (2.28), yields:

$$\text{Inertia 1:} \qquad I_1\dot{\omega}_2 = T(t) + T_1 - T_3. \qquad (2.37)$$

$$\text{Inertia 2:} \qquad I_2\dot{\omega}_5 = T_4 - T_6. \qquad (2.38)$$

$$\text{Left damper:} \qquad b_1\omega_1 = T_1.$$

$$\text{Right damper:} \qquad b_2\omega_6 = T_6.$$

$$\text{Spring:} \qquad Kx_K = T_3 = T_4, \quad x_K = \theta_3 - \theta_4,$$

where the last expression is obtained according to the paragraph after (2.26). Notice that in equations (2.37) and (2.38), torques that have the same direction as ω_2 and

ω_5 appear to be affected by a positive sign and a negative sign if this is not the case. Assuming that all of the system components are connected by rigid joints then, according to Fig. 2.14b:

$$\omega_2 = -\omega_1 = \omega_3, \quad \omega_5 = \omega_6 = \omega_4. \tag{2.39}$$

If $\omega_2 = 0$, $\omega_5 = 0$, $\omega_1 = 0$, $\omega_6 = 0$, $\omega_3 = 0$, and $\omega_4 = 0$ are defined as the angular positions of body 1, body 2, the left damper mobile side, the right damper mobile side, and the spring sides respectively, when the whole system is at rest with $T(t) = 0$, then:

$$\theta_2 = -\theta_1 = \theta_3, \quad \theta_5 = \theta_6 = \theta_4. \tag{2.40}$$

Hence, (2.37) and (2.38) can be written as:

Inertia 1: $\quad I_1 \dot{\omega}_2 = T(t) + b_1 \omega_1 - K(\theta_3 - \theta_4).$

Inertia 2: $\quad I_2 \dot{\omega}_5 = K(\theta_3 - \theta_4) - b_2 \omega_6.$

Finally, using (2.39) and (2.40), it is found that the mathematical model is given by the following differential equations, which have to be solved simultaneously:

Inertia 1: $\quad I_1 \ddot{\theta}_2 + b_1 \dot{\theta}_2 + K(\theta_2 - \theta_5) = T(t).$

Inertia 2: $\quad I_2 \ddot{\theta}_5 + b_2 \dot{\theta}_5 - K(\theta_2 - \theta_5) = 0.$

These differential equations cannot be combined in a single differential equation because θ_2 and θ_5 are linearly independent, i.e., there is no algebraic equation relating these variables.

2.2 Electrical Systems

There are three basic phenomena in electrical systems: inductance, capacitance, and resistance. An electric current i flowing through an inductor under the effect of a voltage v that is applied at its terminals is depicted in Fig. 2.15. It is assumed that there are no parasitic effects, i.e., the inductor has neither internal electrical resistance nor capacitance. The directions shown for electric current and voltage are defined as positive for this system component.

Fig. 2.15 An inductor in electrical systems

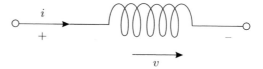

The magnetic flux linkage of the inductor is given as the product of the electric current i and a positive constant L, known as *inductance*, depending on the geometric shape of the conductor:

$$\lambda = Li. \tag{2.41}$$

According to Faraday's Law, [2], pp. 606, [3], pp. 325, flux linkage determines the voltage at the inductor terminals:

$$v = \frac{d\lambda}{dt} = L\frac{di}{dt}. \tag{2.42}$$

One capacitor is formed wherever two electrical conductors, with different electrical potentials and with an insulator material between them, are placed close enough to generate an electric field between them. These electrical conductors are called *plates*. An electric current i flowing through a capacitor under the effect of a voltage v applied at its terminals is shown in Fig. 2.16. It is assumed that no parasitic effects exist, i.e., there is no leakage current between the capacitor plates and there are no inductive effects due to capacitor plates. Because of the difference in potential between the plates, an electric charge is stored. It is common to use letter q to designate such an electric charge, which is the same, but has the opposite sign, at each plate. *Capacitance C* is a positive constant defined as a means of quantifying how much electric charge can be stored in a capacitor. The constitutive function of a capacitor is given as [3], pp. 121:

$$C = \frac{q}{v}. \tag{2.43}$$

Capacitance C depends on the geometric form of plates, the distance between them and the dielectric properties of the insulator placed between them. Recall that the electric current is defined as:

$$i = \frac{dq}{dt}. \tag{2.44}$$

Fig. 2.16 A capacitor in electrical systems

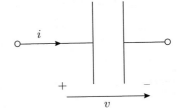

Fig. 2.17 An electric resistance in electrical systems

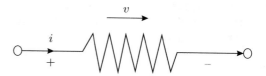

An electric *resistance* is a device requiring application of a voltage v at its terminals to maintain an electric current i flowing through it. The constitutive function of a linear resistance is defined by Ohm's Law [2], pp. 530:

$$v = iR, \tag{2.45}$$

where R is a positive constant known as the electric resistance. An electric current i flowing through an electric resistance when a voltage v is applied at its terminals is shown in Fig. 2.17. The directions shown for the variables are defined as positive.

Once the constitutive functions of each electrical system component have been defined, it is necessary to connect several of them to model more complex electrical systems. Kirchhoff's Laws are the key to establishing connections between the different components of the electrical systems.

Property 2.1 (Kirchhoff's Voltage Law) The algebraic sum of voltages around a closed path (or mesh) must be zero.

Property 2.2 (Kirchhoff's Current Law) The algebraic sum of all currents entering and exiting a node must equal zero.

Example 2.8 Consider the series resistor–capacitor (RC) circuit shown in Fig. 2.18a. Using Kirchhoff's voltage law (KVL):

$$v_i = Ri + v_0, \quad v_0 = \frac{q}{C},$$

As $i = \dot{q}$, then $Ri = RC\frac{dv_0}{dt}$; hence:

$$v_i = RC\dot{v}_0 + v_0.$$

Finally, the mathematical model is given as the following differential equation:

$$\dot{v}_0 + \frac{1}{RC}v_0 = \frac{1}{RC}v_i.$$

Example 2.9 Consider the series RC circuit shown in Fig. 2.18b. Using KVL:

$$v_i = \frac{q}{C} + v_0, \quad v_0 = Ri.$$

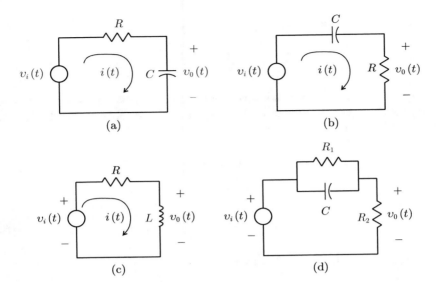

Fig. 2.18 Some first-order circuits

Differentiating once with respect to time:

$$\dot{v}_i = \frac{1}{C}i + \dot{v}_0,$$

$$= \frac{1}{RC}v_0 + \dot{v}_0,$$

where $i = \dot{q}$ and $v_0 = Ri$ have been used. Hence, the dynamic model is given as:

$$\dot{v}_0 + \frac{1}{RC}v_0 = \dot{v}_i.$$

Example 2.10 Consider the series RL circuit shown in Fig. 2.18c. Using KVL:

$$v_i = v_0 + Ri, \quad v_0 = L\frac{di}{dt}.$$

Differentiating once with respect to time:

$$\dot{v}_i = \dot{v}_0 + R\frac{di}{dt},$$

$$= \dot{v}_0 + \frac{R}{L}v_0,$$

where $v_0 = L\frac{di}{dt}$ has been used. Hence, the dynamic model is given as:

$$\dot{v}_0 + \frac{R}{L}v_0 = \dot{v}_i.$$

Example 2.11 Consider the series-parallel RC circuit shown in Fig. 2.18d. Define i_1, i_2, and i as the electric currents through R_1, C, and R_2 respectively. Hence, Kirchhoff's current law (KCL) can be used to obtain:

$$i = i_1 + i_2,$$
$$= \frac{v_c}{R_1} + \frac{dq}{dt},$$
$$= \frac{q}{R_1 C} + \frac{dq}{dt}, \tag{2.46}$$

On the other hand, KVL allows us to write:

$$\frac{q}{C} + R_2 i = v_i.$$

Replacing (2.46) and arranging terms yield:

$$\frac{dq}{dt} + \frac{1}{C}\left(\frac{R_1 + R_2}{R_1 R_2}\right)q = \frac{1}{R_2}v_i.$$

Applying the Laplace transform (see (3.2) in Chap. 3):

$$Q(s) = \frac{\frac{1}{R_2}}{s+a}V_i(s) + \frac{q(0)}{s+a}, \quad a = \frac{1}{C}\frac{R_1 + R_2}{R_1 R_2}, \tag{2.47}$$

where $q(0)$ stands for the initial charge in the capacitor. On the other hand:

$$v_0 = R_2 i,$$
$$= R_2\left(\frac{q}{R_1 C} + \frac{dq}{dt}\right),$$

where (2.46) has been used. Arranging and using the Laplace transform:

$$Q(s) = \frac{\frac{1}{R_2}}{s + \frac{1}{R_1 C}}V_0(s) + \frac{q(0)}{s + \frac{1}{R_1 C}}. \tag{2.48}$$

Equating (2.47) and (2.48) and after an algebraic procedure:

Fig. 2.19 A series RLC circuit

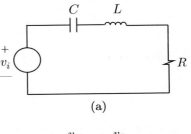

(a)

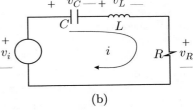

(b)

$$V_0(s) = \frac{s+b}{s+a} V_i(s) - \frac{v_c(0)}{s+a}, \quad b = \frac{1}{R_1 C}, \tag{2.49}$$

where $v_c(0) = \frac{q(0)}{C}$ represents the initial voltage at the capacitor. Notice that:

$$a > b > 0.$$

The expression in (2.49) represents the mathematical model of the circuit in Fig. 2.18d.

Example 2.12 A series RLC circuit is shown in Fig. 2.19a with a voltage v_i applied at its terminals. Connection of the elements of this circuit is established by KVL [8], pp. 67. Using sign definitions in Figs. 2.15, 2.16, 2.17 the electric diagram shown in Fig. 2.19b is obtained. According to this and (2.41), (2.42), (2.43), (2.44), (2.45), in addition to KVL, it is found that:

$$v_i = v_L + v_R + v_C,$$

$$v_C = \frac{q}{C}, \quad q = \int_0^t i(r)dr, \text{ if } q(0) = 0, \tag{2.50}$$

$$v_L = \frac{d\lambda}{dt}, \quad \lambda = Li, \tag{2.51}$$

$$v_R = iR, \quad i = \frac{dq}{dt}. \tag{2.52}$$

Notice that the electric current flowing through all of the circuit components is the same. Hence:

$$v_i = L\frac{d^2q}{dt^2} + R\frac{dq}{dt} + \frac{1}{C}q, \tag{2.53}$$

or, differentiating once with respect to time:

$$\frac{dv_i}{dt} = L\frac{d^2i}{dt^2} + R\frac{di}{dt} + \frac{1}{C}i. \tag{2.54}$$

The corresponding mathematical model is represented by either (2.53) or (2.54).

It is observed from (2.52) that there exists a (linear) algebraic relation between voltage and current in a resistance, i.e., Ohm's Law $v_R = iR$. Also notice that the electric resistance R is the constant relating electric current and voltage $\frac{v_R}{i} = R$. Because of this fact, it is natural to wonder whether similar relations might exist for electric current and voltage in an inductor and in a capacitor. However, from (2.50) and (2.51) it is clear that some mathematical artifice has to be employed, because such relations involve differentiation and integration. The mathematical artifice is the Laplace transform as it possesses the following properties:

$$\mathcal{L}\left\{\frac{dx}{dt}\right\} = sX(s), \quad \mathcal{L}\left\{\int_0^t x(r)dr\right\} = \frac{1}{s}X(s), \quad \text{if x(0)=0}, \tag{2.55}$$

where $X(s)$ stands for the Laplace transform of $x(t)$, i.e., $\mathcal{L}\{x(t)\} = X(s)$. Thus, using (2.50), (2.51), and the Laplace transform, and assuming that all of the initial conditions are zero, yield:

$$V_C(s) = \frac{1}{sC}I(s),$$

$$V_L(s) = sLI(s),$$

where $\mathcal{L}\{v_C(t)\} = V_C(s)$, $\mathcal{L}\{v_L(t)\} = V_L(s)$ and $\mathcal{L}\{i(t)\} = I(s)$. The above expressions represent the algebraic relations between electric current and voltage in an inductor and a capacitor. The factors relating these variables are called *impedance* which is represented as:

$$Z(s) = \frac{V(s)}{I(s)}. \tag{2.56}$$

Hence, the impedance in an inductor is:

$$Z_L(s) = \frac{V_L(s)}{I(s)} = sL, \tag{2.57}$$

and the impedance in a capacitor is:

$$Z_C(s) = \frac{V_C(s)}{I(s)} = \frac{1}{sC}. \tag{2.58}$$

Then, using the Laplace transform and the respective impedances, it is possible to write the mathematical model in (2.54) as:

$$V_i(s) = sLI(s) + RI(s) + \frac{1}{sC}I(s),$$

$$V_i(s) = Z_L(s)I(s) + RI(s) + Z_C(s)I(s),$$

$$V_i(s) = (Z_L(s) + R + Z_C(s))I(s),$$

which motivates the definition of the series circuit equivalent impedance as:

$$Z_t(s) = Z_L(s) + R + Z_C(s),$$

$$Z_t(s) = sL + R + \frac{1}{sC}, \quad Z_t(s) = \frac{V_i(s)}{I(s)}. \tag{2.59}$$

This result proves an important property of series electric circuits:

Property 2.3 The equivalent impedance of a series connected circuit is given as the addition of impedances of all of the series connected circuit components.

The concept of impedance is very important in electric circuits and it is a tool commonly used to model and to analyze circuits in electrical engineering. This means that expressions in (2.59) also represent the mathematical model of the circuit in Fig. 2.19a.

Example 2.13 A parallel connected electric circuit with an electric current i_i flowing through it is depicted in Fig. 2.20a. Models of the circuit components are connected, in this case using KCL [8], pp. 68. Using sign definitions presented in Figs. 2.15, 2.16, 2.17 the electric diagram shown in Fig. 2.20b is obtained. From this, using (2.41), (2.42), (2.43), (2.44), (2.45), and KCL it is found that:

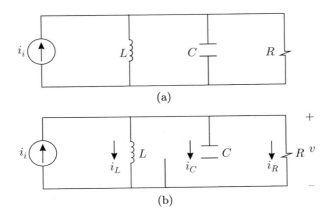

Fig. 2.20 A parallel connected electric circuit

$$i_i = i_L + i_R + i_C,$$

$$q = vC, \quad q = \int_0^t i_C(r)dr, \text{ if } q(0) = 0 \Rightarrow i_C = C\frac{dv}{dt}, \quad (2.60)$$

$$i_L = \frac{1}{L}\int_0^t v(r)dr, \quad v = L\frac{di_L}{dt}, \quad (2.61)$$

$$i_R = \frac{v}{R}. \quad (2.62)$$

Notice that the voltage is the same in all the circuit components. Hence, the mathematical model is given as:

$$i_i = \frac{1}{L}\int_0^t v(r)dr + \frac{v}{R} + C\frac{dv}{dt} \quad (2.63)$$

Using the Laplace transform in (2.60), (2.61), (see (2.55)) and assuming that all of the initial conditions are zero, yields:

$$I_C(s) = sCV(s),$$

$$I_L(s) = \frac{1}{sL}V(s),$$

where $\mathcal{L}\{i_C(t)\} = I_C(s)$, $\mathcal{L}\{i_L(t)\} = I_L(s)$ and $\mathcal{L}\{v(t)\} = V(s)$. The factor relating the Laplace transforms of electric current and voltage in a circuit element is called *admittance* and is represented as:

$$Y(s) = \frac{I(s)}{V(s)}. \quad (2.64)$$

This means that the admittance of an inductor is:

$$Y_L(s) = \frac{I_L(s)}{V(s)} = \frac{1}{sL}, \quad (2.65)$$

and the admittance of a capacitor is:

$$Y_C(s) = \frac{I_C(s)}{V(s)} = sC. \quad (2.66)$$

Then, using the Laplace transform and the definition of admittance, the mathematical model in (2.63) can be written as:

$$I_i(s) = \frac{1}{sL}V(s) + \frac{1}{R}V(s) + sCV(s),$$

$$I_i(s) = Y_L(s)V(s) + \frac{1}{R}V(s) + Y_C(s)V(s),$$

$$I_i(s) = (Y_L(s) + \frac{1}{R} + Y_C(s))V(s),$$

which motivates the definition of the parallel circuit equivalent admittance as:

$$Y_T(s) = Y_L(s) + \frac{1}{R} + Y_C(s),$$

$$Y_T(s) = \frac{1}{sL} + \frac{1}{R} + sC, \quad Y_T(s) = \frac{I_i(s)}{V(s)}. \tag{2.67}$$

This proves the following important property of parallel connected electric circuits:

Property 2.4 The parallel circuit equivalent admittance is the addition of admittances of all of the parallel connected circuit components.

The admittance concept is very important as a tool to model and to analyze electric circuits in electrical engineering. This means that expressions in (2.67) also represent the mathematical model of the circuit in Fig. 2.20a. From definitions of impedance and admittance in (2.56) and (2.64), it is clear that admittance is the inverse of impedance, i.e.:

$$Y(s) = \frac{1}{Z(s)} \tag{2.68}$$

This is also corroborated by comparing the definitions of impedance and admittance for an inductor and a capacitor shown in (2.57), (2.58), (2.65) and (2.66). Hence, according to (2.67) it is possible to write:

$$\frac{1}{Z_T(s)} = \frac{1}{Z_L(s)} + \frac{1}{R} + \frac{1}{Z_C(s)}, \quad Y_T(s) = \frac{1}{Z_T(s)},$$

where $Z_T(s)$ stands for the equivalent impedance of the parallel connected circuit shown in Fig. 2.20a. It is stressed that no relation exists between $Z_T(s)$ and $Z_t(s)$ defined in (2.59) as the equivalent impedance of the series connected circuit shown in Fig. 2.19a. In the case when n circuit elements are connected in parallel, it is possible to write:

$$\frac{1}{Z_T(s)} = \frac{1}{Z_1(s)} + \frac{1}{Z_2(s)} + \cdots + \frac{1}{Z_n(s)}. \tag{2.69}$$

When only two circuit elements are connected in parallel, it is easy to verify that the previous expression reduces to:

$$Z_T(s) = \frac{Z_1(s)Z_2(s)}{Z_1(s) + Z_2(s)}. \tag{2.70}$$

Fig. 2.21 Series-parallel RC circuit

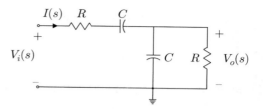

Example 2.14 Consider the electric circuit shown in Fig. 2.21. If $V_o(s)$ is the voltage at the terminals of the parallel connected circuit elements and $Z_p(s)$ is the total impedance of these parallel connected circuit elements, it can be written as follows:

$$Z_p(s) = \frac{V_o(s)}{I(s)}, \tag{2.71}$$

$$Z_p(s) = \frac{R\frac{1}{sC}}{R + \frac{1}{sC}} = \frac{R}{RCs + 1},$$

where (2.70) has been used to compute equivalent impedance of the parallel connected circuit elements. Notice that $I(s)$ is the total electric current flowing through the parallel connected circuit elements. On the other hand, if $Z_{sp}(s)$ is the equivalent impedance of the series-parallel circuit, i.e., impedance measured at terminals where the voltage $V_i(s)$ is applied, then it can be written as follows:

$$Z_{sp}(s) = \frac{V_i(s)}{I(s)}, \tag{2.72}$$

$$Z_{sp}(s) = R + \frac{1}{sC} + \frac{R}{RCs + 1},$$

The fact that the equivalent impedance of series connected circuit elements is computed by adding the impedances of all of the series connected elements whereas impedance of the two parallel connected elements is given in (2.71) has been taken into account. Use of (2.71) and (2.72) yields:

$$\frac{V_o(s)}{V_i(s)} = G_T(s) = \frac{Z_p(s)}{Z_{sp}(s)} = \frac{\frac{R}{RCs+1}}{R + \frac{1}{sC} + \frac{R}{RCs+1}},$$

and, thus:

$$\frac{V_o(s)}{V_i(s)} = G_T(s) = \frac{Ts}{T^2s^2 + 3Ts + 1}, \quad T = RC. \tag{2.73}$$

Notice that $G_T(s)$ is not an impedance as it is given as the ratio of two impedances or the ratio of two voltages.

Fig. 2.22 A *RC* phase-shift circuit

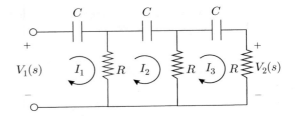

Example 2.15 Consider the electric circuit depicted in Fig. 2.22. Ratio $F(s) = \frac{V_2(s)}{V_1(s)}$ is computed in the following. Define three closed paths (or meshes) as shown in Fig. 2.22. It is assumed that the electric currents $I_1(s)$, $I_2(s)$ and $I_3(s)$ flow around each mesh. The KVL can be used using the following criterion:

Criterion 2.1 *Consider the i−th mesh. The algebraic sum of voltages at the circuit elements in the i−th mesh equals zero. Voltage at each circuit element is computed as the product of its impedance and the algebraic sum of the electric currents $I_1(s)$, $I_2(s)$ or $I_3(s)$, flowing through that circuit element. Let $I_i(s)$ be the electric current defining the i−th mesh. If $I_1(s)$, $I_2(s)$ or $I_3(s)$ has the same direction as $I_i(s)$ in that circuit element then $I_1(s)$, $I_2(s)$ or $I_3(s)$, is affected by a "+" whereas a "−" affects those electric currents with the opposite direction. A coefficient equal to zero affects the electric currents not flowing through that circuit element. The value of a voltage source is affected by a "+" if $I_i(s)$ flows through that voltage source from terminal "+" to terminal "−". A "−" is used if this is not the case.*

This criterion constitutes the basis of the so-called *mesh analysis method* employed for circuit analysis in electrical engineering. The reader is referred to [9] and [8] for a deeper explanation of this method. The use of this criterion in each one of the meshes shown in Fig. 2.22 yields:

$$I_1(s)\frac{1}{sC} + (I_1(s) - I_2(s))R = V_1(s), \tag{2.74}$$

$$(I_2(s) - I_1(s))R + I_2(s)\frac{1}{sC} + (I_2(s) - I_3(s))R = 0,$$

$$(I_3(s) - I_2(s))R + I_3(s)\frac{1}{sC} + V_2(s) = 0, \quad V_2(s) = RI_3(s).$$

From the third expression in (2.74) the following is obtained:

$$I_2(s)R = \frac{1}{R}\left(R + \frac{1}{sC}\right)V_2(s) + V_2(s). \tag{2.75}$$

Replacing this and $I_3(s) = \frac{V_2(s)}{R}$ in the second expression in (2.74), a new expression is obtained which allows $I_1(s)$ to be computed as a function of $V_2(s)$. Replacing this value of $I_1(s)$ and $I_2(s)$ given in (2.75) in the first expression in (2.74), $V_1(s)$ is obtained as a function of $V_2(s)$. Finally, suitably arranging terms, the following expression is obtained:

$$\frac{V_2(s)}{V_1(s)} = \frac{R^3 C^3 s^3}{R^3 C^3 s^3 + 6R^2 C^2 s^2 + 5RCs + 1} = F(s). \qquad (2.76)$$

2.3   Transformers

In this book, the term transformers is used to designate those components that manipulate the system variables in such a way that energy is conserved, i.e., they neither store nor dissipate energy; energy simply flows through them.

2.3.1   Electric Transformer

An electric transformer is composed of two inductors wound on the same ferromagnetic core (see Fig. 2.23). This means that inductors are magnetically coupled but electrically isolated.

Each inductor defines one port. v_1 stands for the voltage at the terminals of port 1 and i_1 represents the electric current flowing through port 1. v_2 stands for the voltage at the terminals of port 2 and i_2 represents the electric current flowing through port 2. The indicated directions are defined as positive. The dot placed at the upper side of each inductor is used to indicate that inductors are wound in the same direction. If this were not the case, dots would be placed on opposite sides of the inductors. The inductor at port 1 has n_1 turns and the inductor at port 2 has n_2 turns. As the inductors are magnetically coupled, flux linkage in inductor 1, λ_1, and flux linkage in inductor 2, λ_2, depend on the electric currents flowing through both inductors:

Fig. 2.23 Electric transformer

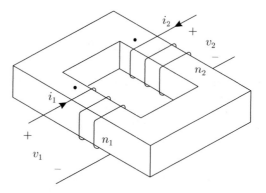

$$\lambda_1 = L_1 i_1 + M_{12} i_2, \tag{2.77}$$

$$\lambda_2 = M_{21} i_1 + L_2 i_2, \tag{2.78}$$

where L_1, L_2, are the inductances of inductors 1 and 2, whereas M_{12} and M_{21} are the mutual inductances between the two circuits. It is always true that $M_{12} = M_{21} = M$. The positive signs in terms with factors M_{12} and M_{21} in the previous expressions are due to the fact that the direction of electric currents defined as positives enter wounds through its sides labeled with a dot (see Sect. D.2 in Appendix D). Solving for i_2 in (2.77) and replacing it in (2.78) yield:

$$\lambda_2 = \frac{L_2}{M} \lambda_1 + \left(M - \frac{L_1 L_2}{M} \right) i_1. \tag{2.79}$$

The coupling coefficient is defined as:

$$k = \frac{M}{\sqrt{L_1 L_2}}.$$

If both circuits are not coupled at all, then $k = 0$, but $k = 1$ if both circuits are completely coupled such that flux linkage in inductor 1 equals flux linkage in inductor 2. This case is well approximated in practice if the core of both inductors is the same and it is made in a ferromagnetic material with a large magnetic permeability μ. Assuming that this is the case, i.e., $M = \sqrt{L_1 L_2}$, then (2.79) becomes:

$$\lambda_2 = \sqrt{\frac{L_2}{L_1}} \lambda_1. \tag{2.80}$$

On the other hand, from (2.78) and (2.77), using $M_{12} = M_{21} = M$, and (2.80), it is found that:

$$\frac{\lambda_2}{\lambda_1} = \frac{M i_1 + L_2 i_2}{L_1 i_1 + M i_2} = \sqrt{\frac{L_2}{L_1}}.$$

Hence, solving for the indicated ratio:

$$\frac{i_2}{i_1} = \frac{\sqrt{L_1 L_2} - M}{L_2 - \sqrt{\frac{L_2}{L_1}} M},$$

$$= \frac{\frac{\sqrt{L_1}}{\sqrt{L_2}} (\sqrt{L_1 L_2} - M)}{\sqrt{L_1 L_2} - M} = \sqrt{\frac{L_1}{L_2}}. \tag{2.81}$$

The inductance of each circuit is given as [3]:

$$L_1 = \mu \frac{n_1^2 A}{l},$$

$$L_2 = \mu \frac{n_2^2 A}{l},$$

where A, l, and μ are the core area, the core mean length, and the core magnetic permeability respectively. Hence:

$$\sqrt{\frac{L_1}{L_2}} = \frac{n_1}{n_2}. \tag{2.82}$$

Using this and (2.42) in (2.80) yields:

$$v_2 = \frac{n_2}{n_1} v_1,$$

and, from (2.81):

$$i_2 = \frac{n_1}{n_2} i_1.$$

This means that the power at both ports is the same, because:

$$w_2 = v_2 i_2 = \frac{n_2}{n_1} v_1 \frac{n_1}{n_2} i_1 = v_1 i_1 = w_1.$$

2.3.2   Gear Reducer

Two gear wheels with radii r_1 and r_2 are depicted in Fig. 2.24. Wheel 1 has radius r_1, n_1 teeth, and it rotates at an angular velocity ω_1 under the effect of an applied

Fig. 2.24 A gear reducer

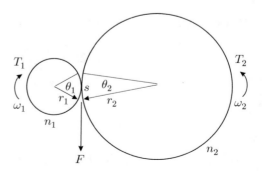

torque T_1. Wheel 2 has radius r_2, n_2 teeth, and it rotates at an angular velocity ω_2 under the effect of an applied torque T_2. The directions defined as positive for these variables are also shown. It is important to note that teeth allow wheels to rotate without slipping. Hence, the arch length s generated by an angle θ_1 at wheel 1 is equal to the arch length generated by the angle θ_2 at wheel 2. It is well known from basic geometry that:

$$s = \theta_1 \, r_1, \qquad s = \theta_2 \, r_2, \tag{2.83}$$

where θ_1 and θ_2 must be given in radians. Thus, it is concluded that:

$$\theta_1 \, r_1 = \theta_2 \, r_2. \tag{2.84}$$

It is also known that the tooth width α is the same in both wheels, which is necessary for teeth to engage. This implies that:

$$p_1 = \alpha n_1, \qquad p_2 = \alpha n_2,$$
$$p_1 = 2\pi r_1, \qquad p_2 = 2\pi r_2.$$

where p_1 and p_2 stand for wheel circumferences or perimeters. Combining these expressions, it is found that:

$$\frac{r_1}{r_2} = \frac{n_1}{n_2}. \tag{2.85}$$

From (2.84) and (2.85):

$$\theta_1 = \frac{n_2}{n_1} \theta_2.$$

Differentiating with respect to time:

$$\omega_1 = \frac{n_2}{n_1} \omega_2, \tag{2.86}$$

On the other hand, force F applied at the point where the wheels are in contact is the same for both wheels. This force satisfies:

$$T_1 = Fr_1, \qquad T_2 = Fr_2,$$

Solving each equation for F and equating the resulting expressions:

$$T_1 = \frac{r_1}{r_2} T_2 = \frac{n_1}{n_2} T_2. \tag{2.87}$$

Using (2.87) and (2.86), it is found that the mechanical power, w, is the same on both ports, i.e.:

$$w_1 = T_1\omega_1 = \frac{n_1}{n_2}T_2\frac{n_2}{n_1}\,\omega_2 = T_2\omega_2 = w_2.$$

2.3.3 Rack and Pinion

Consider the rack and pinion system depicted in Fig. 2.25. Port 1 is defined in terms of rotative variables, angular velocity ω_1 and torque T_1, whereas port 2 is defined in terms of translational variables, velocity v_2 and force F_2. By definition, torque T_1 and force F_2 are related as:

$$T_1 = rF_2, \tag{2.88}$$

where r is the pinion radius, whereas angular velocity ω_1 and velocity v_2 are related as (see (2.83)):

$$v_2 = r\omega_1. \tag{2.89}$$

Thus, the power at both ports is the same:

$$w_1 = T_1\omega_1 = rF_2\frac{1}{r}v_2 = F_2v_2 = w_2.$$

The directions of variables shown in Fig. 2.25 are defined as positive.

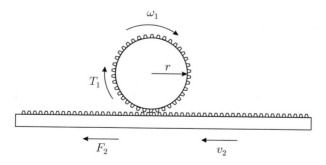

Fig. 2.25 A rack and pinion system

2.4 Converters

We label as converters those system components that constitute a connection between two systems having a different nature.

2.4.1 Armature of a Permanent Magnet Brushed Direct Current Motor

A square loop with area $A = l^2$, which is placed within a magnetic field B, is depicted in Fig. 2.26. An electric current i flows through the loop, with the indicated direction, under the effect of an external voltage source E. Resistance R and inductance L, shown in Fig. 2.26a, stand for the loop resistance and inductance. The force $\bar{F}$ exerted by a magnetic field $\bar{B}$ on an electric current of length l is given as [2], pp. 541:

$$\bar{F} = il\bar{u} \times \bar{B}, \tag{2.90}$$

Fig. 2.26 A permanent magnet brushed direct current (DC) motor. (**a**) Superior view. (**b**) Front view

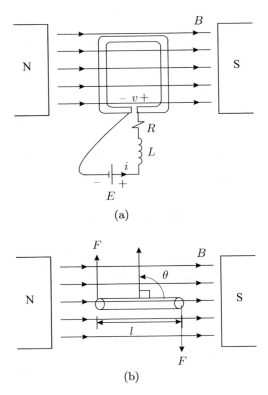

where $\bar{u}$ is a unitary vector defined in the same direction as the electric current. A bar is used on the top of variables to indicate that they represent vectors. Symbol "$\times$" stands for the standard vector product. Hence, under the conditions depicted in Fig. 2.26b, i.e., when $\theta = 90°$, the force exerted by the magnetic field on the left and right sides of the loop is given as:

$$F = ilB.$$

This means that the following torque is produced on the loop axis:

$$T = 2\frac{l}{2}F = iBl^2 = iBA.$$

Notice that $lF/2$ is torque because of the force exerted on each loop side and there are two applied forces on the loop with the same magnitude: one is applied on the left side and another on the right side. The reader can verify that, according to (2.90), forces on the rear and the front sides of the loop do not produce any torque. This means that the loop rotates at an angular velocity $\omega = -\dot{\theta}$, i.e., in the opposite direction defined for θ in Fig. 2.26b where θ is the angle between the direction of the magnetic field and a line orthogonal to the plane defined by the loop. According to Faraday's Law [3], pp. 325, this induces a voltage v at the loop terminals that is given according to (2.42), i.e.:

$$v = \frac{d\lambda}{dt},$$

where λ is the flux linkage in the loop and is given as:

$$\lambda = BA\cos(\theta).$$

Hence, using these expressions, the following is found:

$$v = -BA\dot{\theta}\sin(\theta) = BA\omega\sin(\theta).$$

Notice that, according to Lenz's Law [3], pp. 327, the induced voltage v has a polarity (see Fig. 2.26a) that is opposite to the change in electric current i. Now, assume that several loops are employed, each one of them having a different tilt angle, such that a mechanical commutator connects only that loop for which $\theta = 90°$. Then, the voltage at the commutator terminals is given as:

$$v = BA\omega.$$

This is a description of a permanent magnet brushed direct current (DC) motor. Port 1 is represented by electrical variables, i.e., voltage v and electric current i, whereas port 2 is represented by mechanical variables, i.e., angular velocity ω and torque T. The relationship among these port variables is given as:

$$v = BA\omega, \tag{2.91}$$

$$i = \frac{1}{BA}T.$$

If it is assumed that a torque T is applied at port 2 at an angular velocity ω, producing a voltage v and an electric current i at port 1, then a DC generator is obtained and both expressions in (2.91) are still valid. Notice that in both cases power at both ports is the same:

$$w_1 = vi = \frac{1}{BA}TBA\omega = T\omega = w_2.$$

The expressions in (2.91) can be written in a more general manner as:

$$v = k_e\omega, \tag{2.92}$$

$$T = k_m i, \tag{2.93}$$

where $k_e > 0$ and $k_m > 0$ are known as the counter-electromotive force constant and the torque constant respectively. These constants are introduced to take into account the fact that several loops are usually series connected and some other variants in the construction of practical DC motors and generators. The fact that power at both ports is the same:

$$w_1 = vi = k_e\omega\frac{1}{k_m}T = T\omega = w_2,$$

can be used to show that $k_e = k_m$ in any DC motor or generator. However, attention must be payed to the units employed for the involved variables: $k_e = k_m$ is true if these constants are expressed in the International System of Units.

2.4.2   Electromagnet

A bar made in iron, placed at a distance x from an electromagnet is shown in Fig. 2.27. The electromagnet is composed of a conductor wire wound, with N turns, on a core made in iron. An electric current i flows through the conductor under the effect of a voltage v applied at the conductor terminals. According to Ampere's Law [5], the magnetomotive force M is given as:

$$M = Ni. \tag{2.94}$$

The *reluctance* $\mathcal{R}$ of the magnetic circuit is defined through the constitutive relation:

$$M = \mathcal{R}\lambda, \tag{2.95}$$

Fig. 2.27 Force due to a
magnetic field

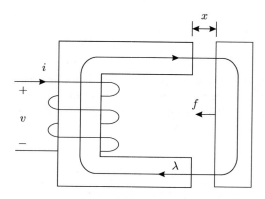

where λ represents the magnetic flux in the magnetic circuit and the reluctance is
given as:

$$\mathcal{R} = \frac{l}{\mu_i A} + \frac{2x}{\mu_a A} \approx \frac{2x}{\mu_a A}, \quad \mu_i \gg \mu_a, \tag{2.96}$$

where μ_i and μ_a stand for the magnetic permeability of iron and air respectively,
whereas l and x represent the magnetic circuit length and the air gap respectively.

The flux linkage is defined as $\psi = N\lambda$. According to Faraday's Law $v = \frac{d(N\lambda)}{dt}$.
Using this and (2.94), the power w supplied by the electric circuit is given as:

$$w = iv = \frac{M}{N} N\dot{\lambda} = M\dot{\lambda} = \dot{E},$$

where E is the energy supplied by the electric circuit, which is stored in the magnetic
circuit as:

$$E = \int_0^t M \frac{d\lambda}{dt} dt = \int_{\lambda(0)}^{\lambda(t)} M d\lambda.$$

Using (2.95), this can be written as:

$$E = \int_{\lambda(0)}^{\lambda(t)} \mathcal{R}\lambda d\lambda = \frac{1}{2}\mathcal{R}\lambda^2(t), \tag{2.97}$$

if $\lambda(0) = 0$. Notice that this represents the energy stored in the reluctance, i.e.,
reluctance stores energy. This must be stressed, as the common belief is that
reluctance is the magnetic analog of electric resistance, which is not true because
electric resistance *dissipates* energy and reluctance *stores* energy.

The stored energy E also represents the mechanical work performed by an
external force F to create a gap of width x. This means that F is applied in the
opposite direction to the magnetic force f, i.e., $F = -f$ and:

$$E = \int_0^x F ds = -\int_0^x f ds, \quad f = -\frac{\partial E}{\partial x}.$$

Using (2.97):

$$f = -\frac{1}{2}\lambda^2 \frac{\partial \mathcal{R}}{\partial x}.$$

This and (2.96) imply that the magnetic force is produced by the reluctance and it is applied in the sense where the reluctance, energy, and gap width decrease.

On the other hand, the inductance $L(x)$ depends on the bar position x, as explained in the following. According to (2.41), the flux linkage in the inductor is given as the product of inductance and the electric current, i.e., $\psi = L(x)i = N\lambda$. Moreover, according to (2.96) and (2.95), if x decreases, flux λ increases if the electric current is kept constant. According to $N\lambda = L(x)i$, an increase in flux λ only can be produced by an increase in inductance $L(x)$, i.e., inductance increases if x decreases.

From (2.97) and (2.95), it follows that:

$$E = \frac{1}{2}M\lambda = \frac{1}{2}Ni\frac{\psi}{N} = \frac{1}{2}i\psi = \frac{1}{2}L(x)i^2.$$

Finally, according to $f = -\frac{\partial E}{\partial x}$:

$$f = -\frac{1}{2}\frac{\partial L(x)}{\partial x}i^2. \tag{2.98}$$

It is stressed that f in (2.98) is applied in the sense where the air gap x decreases. The above-mentioned ideas are based on [10].

Example 2.16 Two rotative bodies connected by a rack are shown in Fig. 2.28a. An external torque $T(t)$ is applied to the body at the left. Hence, it may be assumed that the body at the left is a motor that transmits movement to the body at the right. Following the sign definitions shown in Figs. 2.1, 2.9, 2.3, and 2.25, the free-body diagrams shown in Figs. 2.28b and c are obtained. Figure 2.28b shows the connection of both rotative bodies by means of two rack and pinion systems, whereas Fig. 2.28c shows the free-body diagram of each rotative body.[1] The employed nomenclature is defined as follows:

- $v_m = \frac{dx_m}{dt}$, where x_m is the position of the rack with the mass m.
- $\omega_1 = \frac{d\theta_1}{dt}$, where θ_1 is the angular position of the rotative body 1.

---

[1] In a rack and pinion system, Newton's Third Law stating that to every action (force or torque), there is a corresponding reaction (force or torque) must be applied at only one port, as force or torque in the other port is simply a consequence of what happens in the first port. Also, see Example 2.17.

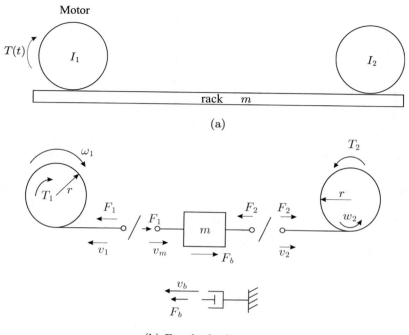

(a)

(b) Free-body diagram.

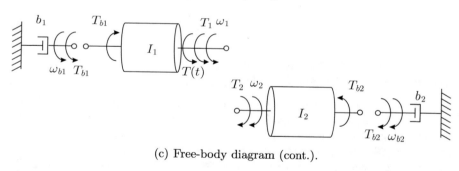

(c) Free-body diagram (cont.).

Fig. 2.28 Two rotative bodies connected by a rack

- $\omega_2 = \frac{d\theta_2}{dt}$, where θ_2 is the angular position of the rotative body 2.
- v_1 is the rack translational velocity produced as a consequence that body 1 rotates with angular velocity ω_1.
- v_2 is the rack translational velocity produced as a consequence that body 2 rotates at angular velocity ω_2.
- $\omega_{b1} = \frac{d\theta_{b1}}{dt}$, where θ_{b1} is the angular position of the mobile side of the damper connected to body 1 (friction between body 1 and its supports).

- $\omega_{b2} = \frac{d\theta_{b2}}{dt}$, where θ_{b2} is the angular position of the mobile side of the damper connected to body 2 (friction between body 2 and its supports).
- $v_b = \frac{dx_b}{dt}$, where x_b is the position of the mobile side of the damper connected to rack m. This damper represents the friction between rack and floor.

Using Figs. 2.28b, c, and (2.1), (2.25), (2.4), (2.28), (2.88), (2.89), the following is obtained:

$$\text{Inertia 1:} \qquad I_1 \frac{d\omega_1}{dt} = T(t) + T_1 - T_{b1}, \tag{2.99}$$

$$\text{Inertia 2:} \qquad I_2 \frac{d\omega_2}{dt} = T_2 - T_{b2}, \tag{2.100}$$

$$\text{Mass:} \qquad m \frac{dv_m}{dt} = F_1 - F_2 + F_b, \tag{2.101}$$

Left damper: $\qquad b_1 \omega_{b1} = T_{b1},$

Right damper: $\qquad b_2 \omega_{b2} = T_{b2},$

Central damper: $\qquad b v_b = F_b,$

Rack and pinion at the left: $\qquad T_1 = r F_1, \quad v_1 = r \omega_1, \tag{2.102}$

Rack and pinion at the right: $\qquad T_2 = r F_2, \quad v_2 = r \omega_2. \tag{2.103}$

Notice that forces with the same direction as ω_1, ω_2 and v_m are affected by a positive sign in (2.99), (2.100), (2.101), or a negative sign if this is not the case. Assuming that all of the system components are connected by rigid joints then, according to Figs. 2.28b and c, it is found that:

$$v_m = -v_1 = v_2 = -v_b, \tag{2.104}$$

$$\omega_1 = \omega_{b1}, \quad \omega_2 = \omega_{b2}. \tag{2.105}$$

According to this, if $x_m = 0, \theta_1 = 0, \theta_2 = 0, x_b = 0, \theta_{b1} = 0, \theta_{b2} = 0$ are defined as positions of the rack, body 1, body 2, and the mobile sides of the dampers connected to the rack and bodies 1 and 2 respectively, when the whole system is at rest with $T(t) = 0$, then it is concluded that:

$$x_m = -x_b, \tag{2.106}$$

$$\theta_1 = \theta_{b1}, \quad \theta_2 = \theta_{b2}.$$

Hence, expressions in (2.99), (2.100) and (2.101) can be rewritten as:

$$\text{Inertia 1:} \qquad I_1 \frac{d\omega_1}{dt} = T(t) + r F_1 - b_1 \omega_{b1}, \tag{2.107}$$

$$\text{Inertia 2:} \qquad I_2 \frac{d\omega_2}{dt} = rF_2 - b_2\omega_{b2}, \qquad (2.108)$$

$$\text{Mass:} \qquad m\frac{dv_m}{dt} = F_1 - F_2 + bv_b. \qquad (2.109)$$

Recall that $v_1 = -v_2$. Then, (2.102) and (2.103) can be used to find that $\omega_1 = -\omega_2$. Using this result, (2.105), and subtracting (2.108) to (2.107):

$$(I_1 + I_2)\frac{d\omega_1}{dt} = T(t) + r(F_1 - F_2) - (b_1 + b_2)\omega_1. \qquad (2.110)$$

On the other hand, employing $-v_m = v_1 = r\omega_1 = v_b$, (2.109) can be written as:

$$-mr\frac{d\omega_1}{dt} = F_1 - F_2 - bv_m.$$

Solving for $F_1 - F_2$ in this expression and replacing it in (2.110), the following is finally found:

$$(I_1 + I_2 + mr^2)\frac{d\omega_1}{dt} = T(t) - (b_1 + b_2 + r^2b)\omega_1.$$

This differential equation represents the mathematical model of the system in Fig. 2.28a. Contrary to several previous examples, in this case it is possible to combine three differential equations in a single differential equation. This is because the three variables ω_1, ω_2 are v_m are linearly dependent as they are related through the algebraic equation $-v_m = r\omega_1 = -r\omega_2$. Also notice how the radii of the pinions modify (i) the rack mass to express its effect on the equivalent inertia on the axis of body 1, and (ii) friction between the rack and floor to express the equivalent friction on the axis of body 1.

Example 2.17 Two rotative bodies connected through a gear reducer are shown in Fig. 2.29a. The external torque $T(t)$ is applied t the body at the left. This mechanical system can be seen as an electric motor (body at the left) transmitting movement to a mechanical load (body at the right) through a gear reducer. According to the sign definitions shown in Figs. 2.9, 2.11, and 2.24, the free-body diagram shown in Fig. 2.29b is obtained.[2] The employed nomenclature is defined as follows:

- $\omega = \frac{d\theta}{dt}$, where θ is the angular position of body 1.
- $\omega_4 = \frac{d\theta_4}{dt}$, where θ_4 is the angular position of body 2.
- $\omega_{b1} = \frac{d\theta_{b1}}{dt}$, where θ_{b1} is the angular position of the mobile side of the damper connected to body 1.

---

[2]In a gear reducer, Newton's Third Law, stating that to every action (torque) there is a corresponding reaction (torque), must be applied only at one port, as the torque at one port is a consequence of the torque at the other port. Also see Example 2.16.

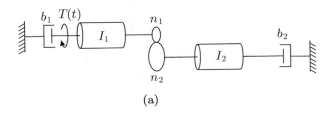

(a)

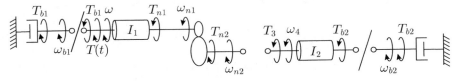

(b) Free-body diagram.

Fig. 2.29 Two rotative bodies connected by a gear reducer

- $\omega_{b2} = \frac{d\theta_{b2}}{dt}$, where θ_{b2} is the angular position of the mobile side of the damper connected to body 2.
- ω_{n1} and ω_{n2} are the angular velocities of the gear wheels that result from the movement of bodies 1 and 2.
- T_{n1} and T_{n2} are the torques appearing on gear wheels as consequences of the movement of bodies 1 and 2.

Using Fig. 2.29b, and (2.25), (2.28), (2.86), (2.87), the following expressions are obtained:

Body 1: $\qquad I_1 \dot{\omega} = T(t) - T_{b1} - T_{n1},$ $\qquad\qquad$ (2.111)

Body 2: $\qquad I_2 \dot{\omega}_4 = T_3 + T_{b2},$ $\qquad\qquad$ (2.112)

Damper at the left: $\qquad b_1 \omega_{b1} = T_{b1},$

Damper at the right: $\qquad b_2 \omega_{b2} = T_{b2},$

Gear reducer: $\qquad T_{n1} = \dfrac{n_1}{n_2} T_{n2}, \quad \omega_{n1} = \dfrac{n_2}{n_1} \omega_{n2}, \quad T_{n2} = T_3,$ (2.113)

Last expression is due to the fact that, according to Newton's Third Law, an action of A on B produces a reaction from B on A. Notice that in (2.111) and (2.112) torques that have the same direction as ω and ω_4 appear with a positive sign and a negative sign if this is not the case. Assuming that all of the system elements are connected by rigid joints, then from Fig. 2.29b, it is found that:

$$\omega = \omega_{b1} = -\omega_{n1}, \qquad \omega_4 = -\omega_{b2} = -\omega_{n2}. \qquad (2.114)$$

Hence, if $\theta = 0$, $\theta_4 = 0$, $\theta_{b1} = 0$, $\theta_{b2} = 0$, $\theta_{n1} = 0$, $\theta_{n2} = 0$ are defined as the angular positions of body 1, body 2, the mobile side of the damper at the left, the mobile side of the damper at the right, gear wheel 1, and gear wheel 2 when the whole system is at rest with $T(t) = 0$, then it is concluded that:

$$\theta = \theta_{b1} = -\theta_{n1}, \quad \theta_4 = -\theta_{b2} = -\theta_{n2}. \tag{2.115}$$

Thus, using (2.114), (2.115), (2.111), and (2.112), the following can be written:

Body 1: $$I_1 \dot{\omega} = T(t) - b_1 \omega - \frac{n_1}{n_2} T_3, \tag{2.116}$$

Body 2: $$I_2 \dot{\omega}_4 = T_3 - b_2 \dot{\theta}_4. \tag{2.117}$$

On the other hand, from (2.113) and (2.114) it is found that $\omega = \frac{n_2}{n_1} \omega_4$. Using this, solving for T_3 in (2.117), replacing it in (2.116), and rearranging terms:

$$\left(I_2 + \left(\frac{n_2}{n_1} \right)^2 I_1 \right) \ddot{\theta}_4 + \left(b_2 + \left(\frac{n_2}{n_1} \right)^2 b_1 \right) \dot{\theta}_4 = \frac{n_2}{n_1} T(t). \tag{2.118}$$

Notice that both differential equations in (2.116) and (2.117) have been combined in a single differential equation. This is possible because variables ω and ω_4 are linearly dependent, i.e., they are related by the algebraic equation $\omega = \frac{n_2}{n_1} \omega_4$. The above-mentioned differential equation represents the mathematical model of the system: if $T(t)$ is known, then θ_4, i.e., the position of body 2, can be computed solving this differential equation.

Example 2.18 In Fig. 2.30a, the mechanism in Fig. 2.28a (in Example 2.16) is shown again, but now it is assumed that the rack is flexible. This flexibility is not intentional, but it is an undesired problem that appears because of the finite stiffness of the steel of which the rack is made. The mathematical model in this case can be obtained taking advantage of the procedure in Example 2.16, i.e., to continue from (2.107), (2.108), (2.109), which are rewritten here for ease of reference:

Inertia 1: $$I_1 \frac{d\omega_1}{dt} = T(t) + r F_1 - b_1 \omega_{b1},$$

Inertia 2: $$I_2 \frac{d\omega_2}{dt} = r F_2 - b_2 \omega_{b2},$$

Mass: $$m \frac{dv_m}{dt} = F_1 - F_2 + b v_b.$$

However, it must be considered now that, according to Fig. 2.30b and the sign definitions shown in Fig. 2.2 and (2.3):

$$F_1 = K_1(x_A - x_B), \quad F_2 = K_2(x_C - x_D),$$

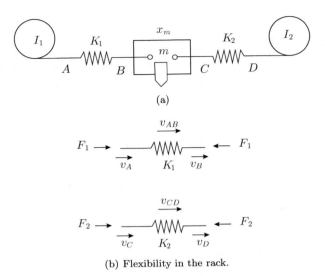

(a)

(b) Flexibility in the rack.

Fig. 2.30 A transmission system with a flexible rack

with:

- $v_A = \frac{dx_A}{dt}$ and $v_B = \frac{dx_B}{dt}$, where x_A and x_B are the positions of the sides of the spring on the left.
- $v_C = \frac{dx_C}{dt}$ and $v_D = \frac{dx_D}{dt}$, where x_C and x_D are the positions of the sides of the spring on the right.

Using this and (2.104), (2.105), the following is found:

$$\text{Inertia 1:} \qquad I_1 \frac{d\omega_1}{dt} = T(t) + r K_1(x_A - x_B) - b_1\omega_1, \qquad (2.119)$$

$$\text{Inertia 2:} \qquad I_2 \frac{d\omega_2}{dt} = r K_2(x_C - x_D) - b_2\omega_2. \qquad (2.120)$$

$$\text{Mass:} \qquad m\ddot{x}_m = K_1(x_A - x_B) - K_2(x_C - x_D) - b\dot{x}_m. \qquad (2.121)$$

On the other hand, according to (2.102), (2.103), (2.104), (2.106) and assuming that all the system components are connected by rigid joints:

$$x_B = x_C = x_m, \qquad x_A = -r\theta_1, \qquad x_D = r\theta_2, \qquad (2.122)$$

where θ_1 and θ_2 are angular positions of gear wheels 1 and 2 defined according to Fig. 2.28b. Replacing (2.122) in (2.119), (2.120), and (2.121):

$$\text{Inertia 1:} \qquad I_1\ddot{\theta}_1 + b_1\dot{\theta}_1 - r K_1(-r\theta_1 - x_m) = T(t),$$

$$\text{Inertia 2:} \qquad I_2\ddot{\theta}_2 + b_2\dot{\theta}_2 - r K_2(x_m - r\theta_2) = 0,$$

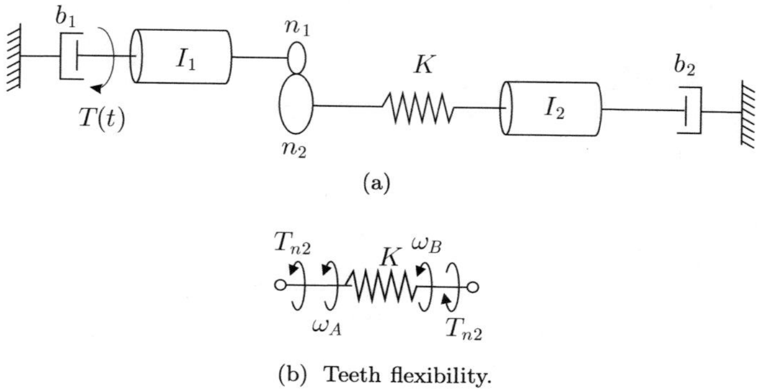

(a)

(b) Teeth flexibility.

Fig. 2.31 Transmission system with a flexible gear reducer

Mass: $m\ddot{x}_m + b\dot{x}_m - K_1(-r\theta_1 - x_m) + K_2(x_m - r\theta_2) = 0.$

The mathematical model is given by these three differential equations, which must be solved simultaneously. In this case, it is not possible to combine these three differential equations in a single differential equation because variables θ_1, θ_2, and x_m are not linearly dependent. This means that there is no algebraic expression relating these three variables. Notice that, from the modeling point of view, this is the effect introduced by the springs considered in this example. Compare this mathematical model with that obtained in Example 2.16.

Example 2.19 Two rotative bodies connected through a gear reducer and a spring are depicted in Fig. 2.31a. The external torque $T(t)$ is applied to the body on the left. This example can be seen as a motor (body on the left) that transmits movement to a mechanical load (body on the right) through a gear reducer that exhibits teeth flexibility. This flexibility is an undesired behavior that, however, appears in practice because of the finite stiffness of steel, the material in which the gear reducer is made. This modeling problem can be solved taking advantage of the procedure in Example 2.17 which has been used to model the mechanical system in Fig. 2.29a. We can go back to expressions in (2.116) and (2.117), which are rewritten here for ease of reference:

$$\text{Body 1:} \qquad I_1\dot{\omega} = T(t) - b_1\omega - \frac{n_1}{n_2}T_{n2}, \qquad (2.123)$$

$$\text{Body 2:} \qquad I_2\dot{\omega}_4 = T_{n2} - b_2\dot{\theta}_4, \qquad (2.124)$$

but now, according to Figs. 2.31b, 2.10 and (2.27):

$$T_{n2} = K(\theta_A - \theta_B),$$

where $\omega_A = \dot{\theta}_A$ and $\omega_B = \dot{\theta}_B$ with θ_A and θ_B, the angular position of both sides of the spring. On the other hand, according to (2.113), it is found that $\theta = \frac{n_2}{n_1}\theta_A$, because $\omega_A = -\omega_{n2}$ and $\omega_{n1} = -\theta$, where θ and θ_A are the angular positions of body 1 and the side of the spring connected to gear wheel 2 (see Fig. 2.31b). Also notice that $\theta_B = \theta_4$. Replacing the above-mentioned expressions in (2.123) and (2.124), it is found that:

Body 1:
$$I_1\ddot{\theta} + b_1\dot{\theta} + \frac{n_1}{n_2}K\left(\frac{n_1}{n_2}\theta - \theta_4\right) = T(t),$$

Body 2:
$$I_2\ddot{\theta}_4 + b_2\dot{\theta}_4 - K\left(\frac{n_1}{n_2}\theta - \theta_4\right) = 0.$$

These differential equations must be solved simultaneously and they represent the corresponding mathematical model. Notice that these differential equations cannot be combined into a single differential equation because variables θ and θ_4 are linearly independent, i.e., there is no algebraic expression relating both of these variables. This is the main difference with respect to the problem solved in Example 2.17 and it is due to the introduction of the spring in the present example.

Example 2.20 (Taken from [6], pp. 122) An electro-mechanical system is shown in Fig. 2.32. There is a permanent magnet brushed DC motor that is used to actuate on a body with the mass M through a gear reducer and a rack and pinion system. The body M moves on a horizontal plane; hence, it is not necessary to consider the effect of gravity. According to the sign definitions shown in Figs. 2.9, 2.1, 2.11, 2.3, 2.24, 2.25, 2.26, the free-body diagrams in Fig. 2.33 are obtained. The electrical DC motor

Fig. 2.32 Servomechanism

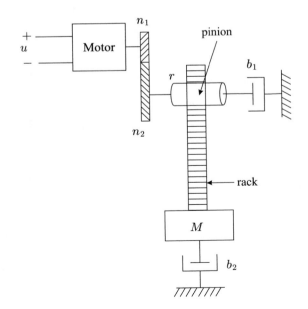

Fig. 2.33 Free-body
diagrams corresponding to
the servomechanism in
Fig. 2.32

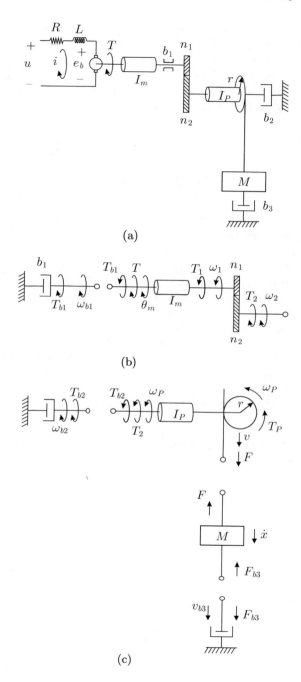

components are analyzed in Fig. 2.33a. The mechanical DC motor components and the gear reducer are analyzed in Fig. 2.33b. The rack and pinion system components are analyzed in Fig. 2.33c.[3] Notice that friction at motor bearings, friction at pinion bearings, and friction between body M and its supports (or the floor) are considered. The nomenclature employed is defined as follows:

- $\dot{\theta}_m = \frac{d\theta_m}{dt}$, where θ_m is the motor angular position.
- $\omega_P = \frac{d\theta_P}{dt}$, where θ_P is the pinion angular position.
- $\dot{x} = \frac{dx}{dt}$, where x is the body M position.
- $\omega_{b1} = \frac{d\theta_{b1}}{dt}$, where θ_{b1} is the angular position of the mobile side of the damper connected to the motor.
- $\omega_{b2} = \frac{d\theta_{b2}}{dt}$, where θ_{b2} is the angular position of the mobile side of the damper connected to the pinion.
- $v_{b3} = \frac{dx_{b3}}{dt}$, where x_{b3} is the position of the mobile side of the damper connected to body M.
- ω_1 and ω_2 are the angular velocities of the gear wheels that result from the movement of the motor and the pinion.
- u is the voltage applied at the motor terminals and i is the electric current flowing through the motor.
- F, T_P, and v are the force, torque, and translational velocity produced at the rack and pinion system as a result of movement of the pinion and mass M.

Using Fig. 2.33, and (2.1), (2.25), (2.4), (2.28), (2.86), (2.87), (2.88), (2.89), the following expressions are obtained:

$$\text{Motor inertia:} \qquad I_m \ddot{\theta}_m = T - T_1 - T_{b1}, \qquad (2.125)$$

$$\text{Pinion inertia:} \qquad I_P \ddot{\theta}_P = T_2 + T_{b2} + T_P, \qquad (2.126)$$

$$\text{Mass } M: \qquad M\ddot{x} = -F - F_{b3}, \qquad (2.127)$$

$$\text{Damper at motor:} \qquad b_1 \omega_{b1} = T_{b1},$$

$$\text{Damper at pinion:} \qquad b_2 \omega_{b2} = T_{b2},$$

$$\text{Damper at mass } M: \qquad b_3 v_{b3} = F_{b3},$$

$$\text{Gear reducer:} \qquad T_1 = \frac{n_1}{n_2} T_2, \quad \omega_1 = \frac{n_2}{n_1} \omega_2, \qquad (2.128)$$

$$\text{Rack and pinion:} \qquad v = r\omega_P, \quad T_P = rF. \qquad (2.129)$$

Notice that, in (2.125), (2.126), and (2.127), forces that have the same direction as $\dot{\theta}_m$, ω_P and $\dot{x}$ appear affected by a positive sign and a negative sign if this is not the case. Assuming that all the system components are connected by rigid joints, then, according to Fig. 2.33, it is found that:

---

[3] See footnotes in Examples 2.17 and 2.16.

$$v = \dot{x} = v_{b3}, \tag{2.130}$$

$$\dot{\theta}_m = \omega_{b1} = -\omega_1, \quad \omega_2 = \omega_{b2} = -\omega_P, \tag{2.131}$$

Hence, if $x = 0$, $x_{b3} = 0$, $\theta_m = 0$, $\theta_{b1} = 0$, and $\theta_{b2} = 0$ are defined as positions of the body M (the mobile side of the damper connected to mass M, the motor rotor, the mobile side of the damper connected to the motor, and the mobile side of the damper connected to the pinion respectively), when the whole system is at rest with $T = 0$, it is concluded that:

$$x = x_{b3}, \tag{2.132}$$

$$\theta_m = \theta_{b1}. \tag{2.133}$$

Hence, (2.125), (2.126), and (2.127), can be written as:

Motor inertia: $\qquad I_m \ddot{\theta}_m = T - \dfrac{n_1}{n_2} T_2 - b_1 \omega_{b1},$

Pinion inertia: $\qquad I_P \ddot{\theta}_P = T_2 + b_2 \omega_{b2} + rF,$

Mass M: $\qquad M\ddot{x} = -F - b_3 v_{b3}.$

Using (2.130) and (2.131):

Motor inertia: $\qquad I_m \ddot{\theta}_m = T - \dfrac{n_1}{n_2} T_2 - b_1 \dot{\theta}_m, \tag{2.134}$

Pinion inertia: $\qquad I_P \ddot{\theta}_P = T_2 - b_2 \omega_P + rF, \tag{2.135}$

Mass M: $\qquad M\ddot{x} = -F - b_3 \dot{x}. \tag{2.136}$

On the other hand, according to (2.128), (2.129), (2.130), and (2.131), it is found that $\dot{\theta}_m = \dfrac{n_2}{n_1 r}\dot{x}$. Using this, solving for T_2 from (2.135) and replacing it in (2.134):

$$I_m \frac{n_2}{n_1 r}\ddot{x} + b_1 \frac{n_2}{n_1 r}\dot{x} = T - \frac{n_1}{n_2}[I_P \dot{\omega}_P + b_2 \omega_P - rF]. \tag{2.137}$$

From (2.129) and (2.130), it is obtained that $\dot{x} = r\omega_P$. Using this in (2.137) and replacing F from (2.136):

$$I_m \frac{n_2}{n_1 r}\ddot{x} + b_1 \frac{n_2}{n_1 r}\dot{x} = T - \left(\frac{n_1}{n_2 r}I_P + \frac{n_1}{n_2}rM\right)\ddot{x} - \left(\frac{n_1}{n_2 r}b_2 + \frac{n_1}{n_2}rb_3\right)\dot{x}.$$

Rearranging, the following is obtained:

$$\left(I_m \left(\frac{n_2}{n_1 r}\right)^2 + I_P \frac{1}{r^2} + M\right)\ddot{x} + \left(b_1 \left(\frac{n_2}{n_1 r}\right)^2 + b_2 \frac{1}{r^2} + b_3\right)\dot{x} = \frac{n_2}{n_1 r}T.$$

$$\tag{2.138}$$

On the other hand, applying KVL (see Example 2.12) to the electric circuit shown in Fig. 2.33a yields:

$$L\frac{di}{dt} + Ri + e_b = u. \tag{2.139}$$

From (2.92), (2.93) and $\dot{\theta}_m = \frac{n_2}{n_1 r}\dot{x}$, it is found that the induced voltage and the generated torque are given as:

$$e_b = k_e\dot{\theta}_m = k_e\frac{n_2}{n_1 r}\dot{x}, \quad T = k_m i.$$

Using this result, (2.139) and (2.138), it is finally found that the mathematical model is given by the following two differential equations, which have to be solved simultaneously:

$$L\frac{di}{dt} + Ri + k_e\frac{n_2}{n_1 r}\dot{x} = u, \tag{2.140}$$

$$\left(I_m\left(\frac{n_2}{n_1 r}\right)^2 + I_P\frac{1}{r^2} + M\right)\ddot{x} + \left(b_1\left(\frac{n_2}{n_1 r}\right)^2 + b_2\frac{1}{r^2} + b_3\right)\dot{x} = \frac{n_2}{n_1 r}k_m i. \tag{2.141}$$

Another way of representing this mathematical model using a single differential equation is by combining (2.140) and (2.141). To this aim, differentiate (2.141) once with respect to time. This produces $\frac{di}{dt}$ to appear on the right-hand side of the resulting equation. Then, replace $\frac{di}{dt}$ from (2.140). Thus, a differential equation is obtained in terms of $\frac{d^3x}{dt^3}$, $\ddot{x}$, $\dot{x}$ and i. Solve (2.141) for i and replace it to obtain a third-order differential equation in the unknown variable x with voltage u as the excitation function. Thus, the mathematical model can be expressed by a single differential equation, which can be solved for the mass position x once the voltage applied to motor u is known. The reader is encouraged to solve this problem, proceeding as indicated above.

2.5 Case Study: A DC-to-DC High-Frequency Series Resonant Power Converter

Much electronic equipment at present employs a variety of electronic circuits that require several different levels of DC voltage. Several values of DC voltage can be obtained using a large electric transformer with several tabs in the secondary winding. Then, a rectifier filter circuit can be placed at each one of these tabs to obtain a different DC voltage level at the output of each circuit. However, this method is no longer utilized because it requires voluminous inductors and capacitors, aside from the fact that it produces large energy losses.

This problem is currently solved by using several power electronic circuits, which, using electronic commutation devices, are capable of delivering different DC voltage levels from a unique DC power supply. These circuits are known as *DC-to-DC power electronic converters* and they solve the above-described problem, reducing both equipment volume and energy losses. In fact, the impressive equipment miniaturization that it is known today is possible thanks to employment of power electronic converters.

A particular class of power electronic converters is that known as *resonant converters* which are divided into *series* and *parallel* according to the way in which their components are connected (see Fig. 2.34). The DC-to-DC series resonant power electronic converter depicted in Fig. 2.34a works as follows. The commutation network is composed of an inverter delivering a square alternating current (AC) power voltage (taking values $\pm E$) at the input of the series resonant circuit. See switches on the left of Fig. 2.35. In the most simple operating mode, the frequency of the AC voltage delivered by the inverter is equal to the resonance frequency of the series inductor capacitor (LC) circuit. Hence, the transistors comprising the commutation network are turned on and off (Q_1, Q_3 and Q_2, Q_4) when the electric current through them is zero. This is very useful because it avoids transistor stress. On the other hand, because the series LC circuit works at resonance, the electric current i through the circuit has a sinusoidal waveform,

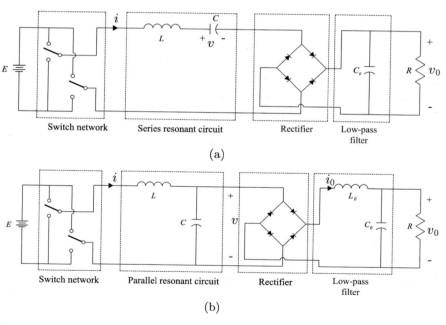

Fig. 2.34 DC-to-DC resonant power electronic converters. (**a**) Series. (**b**) Parallel

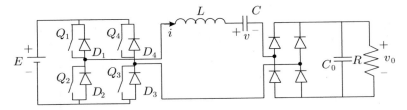

Fig. 2.35 A DC-to-DC series resonant power electronic converter showing devices comprising the inverter

Fig. 2.36 A simplified circuit of a DC-to-DC series resonant power electronic converter

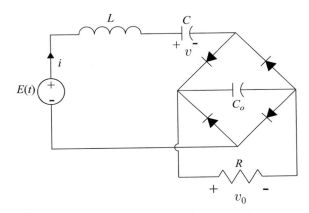

which is rectified and filtered by the circuit on the right in Fig. 2.34a. Hence, the circuit load, represented by the resistance R, receives the DC voltage v_0.

In Fig. 2.36, the electric diagram of a DC-to-DC series resonant power electronic converter is shown, representing the inverter as a square AC voltage source $E(t)$ taking the values $\pm E$. Notice that, because of the presence of the rectifying diodes, the circuit operation can be divided in two cases: 1) when $i > 0$, Fig. 2.37a, and 2) when $i < 0$, Fig. 2.37b. The equations for $i > 0$ are obtained from Fig. 2.37a. Applying KVL to the closed path on the left, the following is obtained:

$$E(t) = L\frac{di}{dt} + v + v_0, \quad i > 0.$$

It is important to stress that the direction of i and the polarity of v_0 have been taken into account. On the other hand, according to (2.50), the voltage and electric current at the resonant capacitor are related by:

$$C\frac{dv}{dt} = i, \quad i > 0.$$

Finally, applying KCL to the node that is common to both capacitors, it is found that:

Fig. 2.37 Equivalent circuits
for (**a**) $i > 0$ and (**b**) $i < 0$

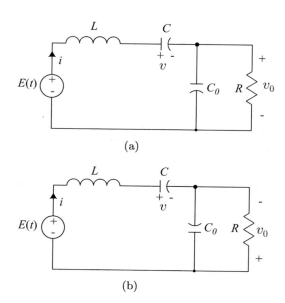

$$i = C_0 \frac{dv_0}{dt} + \frac{v_0}{R}, \quad i > 0.$$

On the other hand, the equations for $i < 0$ are obtained from Fig. 2.37b. Applying
KVL to the closed path on the left, the following is obtained:

$$E(t) = L\frac{di}{dt} + v - v_0, \quad i < 0.$$

Notice that v_0 is affected by a "−" because its polarity is opposite to the direction
of electric current i. However, voltage and electric current at the resonant capacitor
satisfy again:

$$C\frac{dv}{dt} = i, \quad i < 0.$$

Finally, applying KCL to the node that is common to both capacitors yields:

$$i = -\frac{v_0}{R} - C_0\frac{dv_0}{dt}, \quad i < 0.$$

Notice that the correct direction of the electric current through the capacitor C_0 and
resistance is from "+" to "−", i.e., it is opposite to the direction indicated for i.

Analyzing these two sets of equations, it is possible to use a unique set of
equations representing both cases $i > 0$ and $i < 0$:

$$E(t) = L\frac{di}{dt} + v + v_0 sign(i), \tag{2.142}$$

$$C\frac{dv}{dt} = i, \tag{2.143}$$

$$abs(i) = C_0\frac{dv_0}{dt} + \frac{v_0}{R}, \tag{2.144}$$

where:

$$sign(i) = \begin{cases} +1, & i > 0 \\ -1, & i < 0 \end{cases}, \tag{2.145}$$

whereas $abs(i)$ stands for the absolute value of i, i.e., $abs(i) = i$ if $i > 0$ and $abs(i) = -i$ if $i < 0$. The equations in (2.142)–(2.144) constitute the DC-to-DC high-frequency series resonant power electronic converter mathematical model. The study of this class of power electronic converters is continued in Sect. 3.9, Ch. 3. Analyzing (2.142)–(2.144) will explain how this circuit works and why it receives the name *high-frequency*. For more information on modeling this class of power electronic converter, the reader is referred to [11].

2.6 Summary

The mathematical model of a physical system is a differential equation or a set of differential equations describing the evolution of the important variables of the physical system when this is subject to particular conditions. Such a description is obtained when solving the corresponding differential equations, as in Chap. 3.

In the remainder of this book, the mathematical model of the system to be controlled is employed to design a controller (another differential equation, in general). The objective is that, when connecting the controller in feedback with the system to be controlled, a new differential equation is obtained (the closed-loop system differential equation) whose mathematical properties (studied in Chap. 3) ensure that the variable to be controlled evolves over time as desired.

2.7 Review Questions

1. What are the fundamental laws of physics that are useful for connecting components in mechanical and electrical systems?
2. Why is the mathematical model in Example 2.16 given by a single first-order differential equation, whereas the mathematical model in Example 2.7 is given by two second-order differential equations? Notice that there are three bodies in Example 2.16 and two bodies are involved in Example 2.7.

3. Why is the term corresponding to θ_4, i.e., the position term, absent in the mathematical model in (2.118)?

4. Why are first-order differential equations the mathematical models of RL and RC circuits, whereas the mathematical model of a RLC circuit is a second-order differential equation?

5. It is often said that an inductor (and in general any electric circuit that is long enough) has properties that are analogous to the properties of momentum in mechanical systems. Can you explain what this statement refers to? Have you observed that a spark appears when a highly inductive circuit is disconnected? Do you remember what Newton's First Law states?

6. It has been shown that $k_e = k_m$ in a permanent magnet brushed DC motor. On the other hand, it is clear that the electrical subsystem must deliver energy to the mechanical subsystem to render motion possible. Can you give an example of a situation in a DC motor where the mechanical subsystem delivers energy to the electrical subsystem? What is the role of the fact that $k_e = k_m$ in this case?

7. It has been shown that a gear box is the mechanical analog of an electric transformer. Can you list the relationship between electric currents and voltages at both windings of an electric transformer? Is it possible to establish some similar relationship between velocities and torques on both sides of a gear box? Can you list the advantages, disadvantages, and applications of these features in gear boxes and electric transformers?

8. According to the previous question, why does an automobile run slower when climbing a hill, but it runs faster on flat surfaces?

9. If an electric transformer can increase voltage just by choosing a suitable rate n_1/n_2, why are transistor-based electronic circuits employed to amplify signals in communication systems instead of electric transformers?

10. If the power is the same at both ports in electric transformers and gear boxes, where are the power losses?

2.8   Exercises

1. Consider the mass-spring-damper system shown in Fig. 2.6a which was studied in Example 2.2. Suppose that this mechanism is rotated clockwise 90°. This means that, now, the effect of gravity appears to be on the direction in which both the spring and the damper move, i.e., it can be assumed that the mass hangs from the roof through a spring and a damper. Show that the corresponding mathematical model is given as:

$$\ddot{y} + \frac{b}{m}\dot{y} + \frac{K}{m}y = \frac{1}{m}F(t),$$

where $y = x - x_0$ with $x_0 = mg/K$ and g the gravity constant. What does this mean?

Fig. 2.38 Electric circuit of
Exercise 5

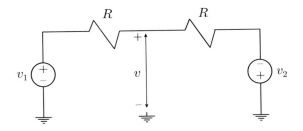

2. Consider the mechanical system in Example 2.17. Assume that an external disturbance torque $T_p(t)$ is applied to the body on the right. Also assume that the body at the left is the rotor of a permanent magnet brushed DC motor, i.e., that the applied torque $T(t)$ is torque generated by this electric motor, as in Example 2.20. Find the mathematical model from the voltage applied at the motor terminals to the position of the body on the right. Compare with the mathematical model in (10.9) and (10.10), chapter 10.
3. Perform on (2.140), (2.141), the algebraic operations suggested in the last paragraph of Example 2.20.
4. Perform the algebraic operations referred to in the previous exercise on the mathematical model obtained in Exercise 2 in this section.
5. Consider the electric circuit shown in Fig. 2.38. Show that voltage v is given as:

$$v = \frac{1}{2}(v_1 - v_2).$$

6. Consider the electric circuits shown in Fig. 2.39. Show that the mathematical models are given as:
   For the circuit in Fig. 2.39a:

$$\dot{v}_0 + \frac{R}{L}v_0 = \frac{R}{L}v_i.$$

   For the circuit in Fig. 2.39b:

$$V_0(s) = \frac{R_2}{R_1 + R_2} \frac{s + \frac{1}{R_2 C}}{s + \frac{1}{(R_1+R_2)C}} V_i(s) + \frac{R_1}{R_1 + R_2} \frac{1}{s + \frac{1}{(R_1+R_2)C}} v_c(0),$$

$$(2.146)$$

   where $v_c(0)$ is the initial voltage at the capacitor.
   For the circuit in Fig. 2.39c:

$$\ddot{v}_0 + \frac{R}{L}\dot{v}_0 + \frac{1}{LC}v_0 = \frac{1}{LC}v_i.$$

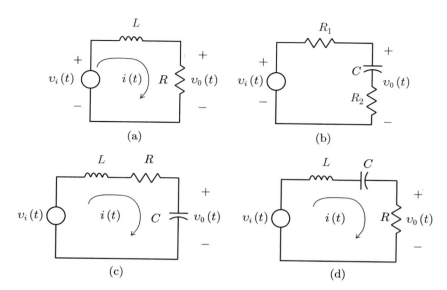

Fig. 2.39 Some electric circuits

Fig. 2.40 Electric diagram of a DC-to-DC buck power converter. (**a**) Converter construction using a transistor and a diode. (**b**) Ideal circuit of the converter

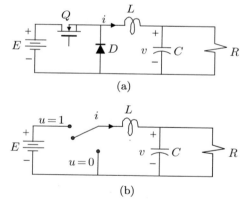

For the circuit in Fig. 2.39d:

$$\ddot{v}_0 + \frac{R}{L}\dot{v}_0 + \frac{1}{LC}v_0 = \frac{R}{L}\dot{v}_i.$$

7. An analog DC voltmeter is basically a permanent magnet brushed DC motor with a spring fixed between the rotor and a point at the stator. This means that such a DC motor is constrained to rotate no more than 180°. Based on this description, find the mathematical model of an analog DC voltmeter.

8. Consider the system shown in Fig. 2.40, which represents a *DC-to-DC buck power converter*. This converter is also known as *step-down converter*, because

the DC output voltage v is less than or equal to the supplied voltage E, i.e., $v \leq E$. Show that the mathematical model is given as:

$$L\frac{di}{dt} = -v + uE,$$

$$C\frac{dv}{dt} = i - \frac{v}{R},$$ (2.147)

where i, v stand for electric current through the inductor L and voltage at the terminals of capacitor C respectively. Constant E is the voltage of the power supply and R is the load resistance. Finally, u stands for the switch position, which is the control variable, taking only two discrete values: 0 or 1.

9. A *buck power converter DC motor system*, is shown in Fig. 2.41. It represents a suitable way of supplying power to a permanent magnet brushed DC motor. Given the following mathematical model for a permanent magnet brushed DC motor:

$$L_a\frac{di_a}{dt} = v - R_a i_a - k_e \omega,$$

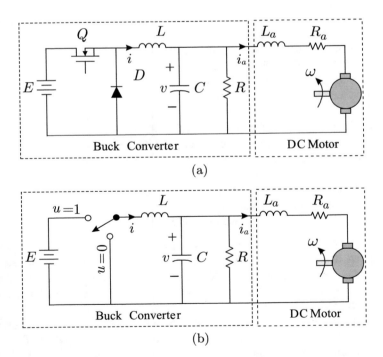

Fig. 2.41 Electric diagram of the buck power converter-DC Motor system. (**a**) Construction of the buck power converter DC motor system using a transistor and a diode. (**b**) Ideal representation of the buck power converter DC motor system

$$J\frac{d\omega}{dt} = k_m i_a - B\omega,$$

show that the complete mathematical model of the system *Buck power converter-DC Motor*, ideally represented in Fig. 2.41b, is given as:

$$L\frac{di}{dt} = -\upsilon + Eu,$$

$$C\frac{d\upsilon}{dt} = i - i_a - \frac{\upsilon}{R},$$

$$L_a\frac{di_a}{dt} = \upsilon - R_a i_a - k_e\omega,$$

$$J\frac{d\omega}{dt} = k_m i_a - B\omega,$$

where:

- i is the electric current through the Buck power converter inductor.
- υ is the converter output voltage, which also is the voltage applied at the motor terminals.
- u stands for the switch position, i.e., the control variable, taking only two discrete values 0 or 1.
- L represents the converter inductance.
- C represents the converter capacitance.
- E is the DC power supply voltage.
- R is the converter load resistance.
- i_a stands for the electric current through the DC motor.
- ω stands for the angular velocity of the DC motor.
- L_a represents the armature inductance of the DC motor.
- R_a represents the armature resistance of the DC motor.
- J is the inertia of the DC motor.
- k_e is the counter-electromotive force constant.
- k_m is the torque constant.
- B stands for the viscous friction coefficient.

10. Consider the series RLC circuit in Example 2.12, whose mathematical model is given in (2.53). The energy stored in the circuit is given as the magnetic energy stored in the inductor plus the electric energy stored in the capacitor, i.e., $E = \frac{1}{2}Li^2 + \frac{1}{2C}q^2$. Show that $\dot{E} = i\upsilon_i - Ri^2$. What does each term of this expression mean?

11. Consider the inertia-spring-damper system in Example 2.5 whose mathematical model is given in (2.32). The energy stored in the system is given as the kinetic energy plus the potential energy in the spring, i.e., $E = \frac{1}{2}I\dot{\theta}^2 + \frac{1}{2}K\theta^2$. Show that $\dot{E} = T\dot{\theta} - b\dot{\theta}^2$. What does each one of the terms in this expression mean?

12. Consider the mathematical model in (2.140), (2.141), for the electro-mechanical system in Example 2.20. The energy stored in this system is given as the magnetic energy stored in the armature inductor plus the kinetic energy of the mechanism. Is it possible to proceed as in the two previous examples to find a similar expression for $\frac{dE}{dt}$? What does each one of terms appearing in $\frac{dE}{dt}$ mean? Can the reader identify the terms representing the energy exchange between the electrical and the mechanical subsystems?

References

1. P. E. Wellstead, *Physical system modelling*, Academic Press, London, 1979.
2. M. Alonso and E. J. Finn, *Physics*, Addison-Wesley, Reading, Mass., 1992.
3. R. K. Wangsness, *Electromagnetic fields*, Wiley and Sons, New Jersey, 1986.
4. R. Kelly, J. Llamas, and R. Campa, A measurement procedure for viscous and coulomb friction, *IEEE Transactions on Instrumentation and Measurement*, vol. 49, no. 4, pp. 857–861, 2000.
5. C. T. A. Johnk, *Engineering electromagnetic fields and waves*, Wiley and Sons, New Jersey, 1987.
6. N. S. Nise, *Control systems engineering*, Wiley and Sons, New Jersey, 2008.
7. F. W. Sears, M. W. Zemansky, and H. D. Young, *University physics*, Addison-Wesley, Reading, Mass., 1976.
8. M. E. Van Valkenburg, *Network analysis*, Prentice-Hall, Upper-Saddle River, 1974.
9. W. H. Hayt and J. E. Kemmerly, *Engineering circuit analysis*, McGraw-Hill, 1971.
10. M. D. Bryant, *Electromechanics, sensors and actuators*, Course ME 344 Dynamic Systems and Control, The University of Texas at Austin, 2007.
11. R. Silva-Ortigoza, Control of resonant power converters based on differential flatness: design and construction, *MSc. Thesis*, CINVESTAV-IPN (Zacatenco), México City, February, 2002.

Chapter 3
Ordinary Linear Differential Equations

The automatic control techniques employed in classical control require knowledge of the mathematical model of the physical system to be controlled. As has been shown in Chap. 2, these mathematical models are differential equations. The controller is designed as another differential equation that must be connected in closed-loop to the system to be controlled. This results in another differential equation representing the closed-loop control system. This differential equation is forced by the controller to possess the mathematical properties that ensure that its solution evolves as desired. This means that the controlled variable evolves as desired. Hence, it is important to know the properties of a differential equation determining how its solution behaves in time. Although several different approaches exist to solve differential equations, the use of the Laplace transform is the preferred method in classical control. This is the reason why the Laplace transform method is employed in this chapter, to study linear ordinary differential equations with constant coefficients.

Let $f(t)$ be a function of time, the Laplace transform of $f(t)$ is represented by $F(s)$ and is computed as [2], pp. 185, [3], pp. 285:

$$F(s) = \mathcal{L}\{f(t)\} = \int_0^\infty f(t)e^{-st}dt. \tag{3.1}$$

The fundamental property of the Laplace transform employed to solve linear ordinary differential equations with constant coefficients is the following [2], pp. 192, [3], pp. 293:

$$\mathcal{L}\left\{\frac{d^n f(t)}{dt^n}\right\} =$$
$$s^n F(s) - s^{n-1} f(0) - s^{n-2} f^{(1)}(0) - \cdots - s f^{(n-2)}(0) - f^{(n-1)}(0), \tag{3.2}$$

© Springer International Publishing AG, part of Springer Nature 2019
V. M. Hernández-Guzmán, R. Silva-Ortigoza, *Automatic Control with Experiments*,
Advanced Textbooks in Control and Signal Processing,
https://doi.org/10.1007/978-3-319-75804-6_3

where the exponent in parenthesis indicates the order of the derivative with respect to time. The following property is also useful [2], pp. 193,[3], pp. 293:

$$\mathcal{L}\left\{\int_0^t f(\tau)d\tau\right\} = \frac{F(s)}{s}. \tag{3.3}$$

Other important properties of the Laplace transform are the final value theorem [1], pp. 25, [3], pp. 304:

$$\lim_{t\to\infty} f(t) = \lim_{s\to 0} sF(s), \tag{3.4}$$

which is only valid if the indicated limits exist, and the initial value theorem [3], pp. 304:

$$f(0^+) = \lim_{t\to 0^+} f(t) = \lim_{s\to\infty} sF(s). \tag{3.5}$$

3.1   First-Order Differential Equation

Consider the following linear, ordinary, first-order, differential equation with constant coefficients:

$$\dot{y} + a\,y = k\,u, \quad y(0) = y_0, \tag{3.6}$$

where k and a are real constants, y_0 is a real constant known as the initial condition or initial value of $y(t)$ and $\dot{y} = \frac{dy}{dt}$. The solution of this differential equation is a function of time $y(t)$, which satisfies (3.6) when replaced. With this aim, it is required that constants a, k, y_0, and the function of time u be known. Use of (3.2) in (3.6) yields:

$$sY(s) - y_0 + a\,Y(s) = k\,U(s). \tag{3.7}$$

Solving for $Y(s)$:

$$Y(s) = \frac{k}{s+a}U(s) + \frac{1}{s+a}y_0. \tag{3.8}$$

As stated above, to continue, it is necessary to know u as a function of time. Assume that:

$$u = A, \quad U(s) = \frac{A}{s}, \tag{3.9}$$

where A is a constant. Hence, (3.8) can be written as:

$$Y(s) = \frac{kA}{s(s+a)} + \frac{1}{s+a} y_0. \tag{3.10}$$

According to the partial fraction expansion method [2], Ch. 4, [3], pp. 319, if $a \neq 0$, then it is possible to write:

$$\frac{kA}{s(s+a)} = \frac{B}{s+a} + \frac{C}{s}, \tag{3.11}$$

where B and C are constants to be computed. The value of B can be computed, multiplying both sides of (3.11) by the factor $(s+a)$ and then evaluating the resulting expressions at $s = -a$:

$$\left. \frac{kA}{s(s+a)}(s+a) \right|_{s=-a} = \left. \frac{B}{s+a}(s+a) \right|_{s=-a} + \left. \frac{C}{s}(s+a) \right|_{s=-a},$$

i.e.,

$$B = \left. \frac{kA}{s} \right|_{s=-a} = -\frac{kA}{a}. \tag{3.12}$$

The value of C is computed multiplying both sides of (3.11) by the factor s and then evaluating the resulting expressions at $s = 0$:

$$\left. \frac{kA}{s(s+a)}s \right|_{s=0} = \left. \frac{B}{s+a}s \right|_{s=0} + \left. \frac{C}{s}s \right|_{s=0},$$

i.e.,

$$C = \left. \frac{kA}{s+a} \right|_{s=0} = \frac{kA}{a}. \tag{3.13}$$

Replacing (3.11), (3.12), (3.13) in (3.10) yields:

$$Y(s) = -\frac{kA}{a}\frac{1}{s+a} + \frac{kA}{a}\frac{1}{s} + \frac{y_0}{s+a}. \tag{3.14}$$

From tabulated Laplace transform formulas [4], Ch. 32, it is known that:

$$\mathcal{L}\{e^{\beta t}\} = \frac{1}{s-\beta}, \tag{3.15}$$

where β is any real constant. Using (3.15) it is found that the inverse Laplace transform of (3.14) is given as:

$$y(t) = -\frac{kA}{a}e^{-at} + \frac{kA}{a} + y_0 e^{-at}, \tag{3.16}$$

which is the solution of (3.6).

Example 3.1 To verify that (3.16) is the solution of (3.6) proceed as follows. First, compute the first time derivative of $y(t) = -\frac{kA}{a}e^{-at} + \frac{kA}{a} + y_0 e^{-at}$, i.e.,

$$\dot{y}(t) = kAe^{-at} - ay_0 e^{-at},$$

and replace $\dot{y}(t)$ and $y(t)$ in (3.6) to obtain:

$$\dot{y}(t) + ay(t) = kAe^{-at} - ay_0 e^{-at} + a\left(-\frac{kA}{a}e^{-at} + \frac{kA}{a} + y_0 e^{-at}\right) = kA = ku,$$

because $u = A$ has been defined in (3.9). This means that the expression for $y(t)$ given in (3.16) satisfies (3.6); thus, it has been verified that (3.16) is the solution of (3.6). This procedure must be followed to verify the solution of any differential equation.

The solution given in (3.16) can be decomposed into two parts:

$$y(t) = y_n(t) + y_f(t), \tag{3.17}$$

$$y_n(t) = \left(-\frac{kA}{a} + y_0\right)e^{-at}, \tag{3.18}$$

$$y_f(t) = \frac{kA}{a}, \tag{3.19}$$

where $y_n(t)$ and $y_f(t)$ are called *the natural response* and *the forced response* respectively. The forced response $y_f(t)$ receives this name because it is caused by the presence of the function of time u. On the other hand, the natural response $y_n(t)$ receives this name because it is produced by the structure (i.e., the nature) of the differential equation, i.e., by terms $\dot{y} + ay$ on the left hand of (3.6). Let us explain this in more detail. Follow the solution procedure presented above, but assume for a moment that $y_0 = 0$. The partial fraction expansion presented in (3.11) and the subsequent application of the inverse Laplace transform show that $y(t)$ is given as the sum of several functions of time that result from the fractions $\frac{1}{s}$ and $\frac{1}{s+a}$. Notice that the presence of $\frac{1}{s}$ in (3.11) is due to the fact that $U(s) = \frac{A}{s}$ appears in (3.10), which is because $u = A$ where A is a constant. Notice that the forced response $y_f(t)$ is a constant resulting from the term $\frac{kA}{a}\frac{1}{s}$ in (3.14), i.e., it arises from fraction $\frac{1}{s}$. This corroborates that $y_f(t)$ only depends on the function of time u.

On the other hand, fraction $\frac{1}{s+a}$ is due to terms $\dot{y} + ay$ as $\mathcal{L}\{\dot{y} + ay\} = (s+a)Y(s)$. According to (3.15), the consequence of this fraction is function e^{-at}, which constitutes the natural response $y_n(t)$. This corroborates that $y_n(t)$ is only determined by the differential equation structure (i.e., its nature). One important consequence of this fact is that $y_n(t)$ is always given by the function e^{-at} no matter what the value of u is. The reader can review the procedure presented above to solve this differential equation, thereby realizing that the initial condition y_0 always appears as a part of the natural response $y_n(t)$.

In control systems, y and u are known as the output and the input respectively. The polynomial resulting from the application of the Laplace transform in the differential equation, i.e., $s + a$, which arises from $\mathcal{L}\{\dot{y} + ay\} = (s + a)Y(s)$, is known as the characteristic polynomial. The solution in (3.16), or equivalently in (3.17), can evolve according to one of the following cases.

1. If $a > 0$, i.e., if the only root of the characteristic polynomial $s = -a$ is real and negative, then $\lim_{t \to \infty} y_n(t) = 0$ and $\lim_{t \to \infty} y(t) = y_f(t)$.
2. If $a < 0$, i.e., if the only root of the characteristic polynomial $s = -a$ is positive, then $y_n(t)$ and $y(t)$ grow without limit as time increases.
3. The case when $a = 0$, i.e., when the only root of the characteristic polynomial is zero $s = 0$, cannot be studied from results that have been obtained so far and it is studied as a special case in the next section. However, to include all the cases, let us talk about what is to be found in the next section. When the only root of the characteristic polynomial is at $s = -a = 0$, the natural response is constant $y_n(t) = y_0$, $\forall t \geq 0$ and the forced response is the integral of the input $y_f(t) = k \int_0^t u(t)dt$.

3.1.1   Graphical Study of the Solution

According to the previous section, if $a > 0$, then the natural response goes to zero over time: $\lim_{t \to \infty} y_n(t) = 0$; hence, when time is large enough the solution of the differential equation is equal to the forced response: $\lim_{t \to \infty} y(t) = y_f(t) = \frac{kA}{a}$. This means that the faster the natural response $y_n(t)$ tends toward zero (which occurs as $a > 0$ is larger), the faster the complete solution $y(t)$ reaches the forced response $y_f(t)$. Hence, the natural response can be seen as a means of transport, allowing the complete solution $y(t)$ to go from the initial value $y(0) = y_0$ to the forced response $y_f(t)$. Existence of such a means of transport is justified recalling that the solution of the differential equation in (3.6), i.e., $y(t)$, is a continuous function of time. This means that the solution cannot go from $y(t) = y_0$ to $y(t) = y_f(t) \neq y_0$ in a zero time interval.

The graphical representation of the solution in (3.17), when $y_0 = 0$ and $a > 0$, i.e.,

$$y(t) = \frac{kA}{a}\left(1 - e^{-at}\right), \tag{3.20}$$

is depicted in Fig. 3.1. An important parameter in first-order differential equations is *time constant* τ, which is defined as:

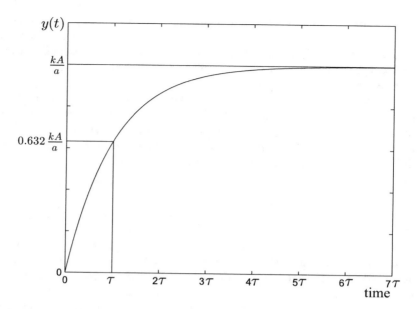

Fig. 3.1 Graphical representation of $y(t)$ in (3.20)

$$\tau = \frac{1}{a}. \tag{3.21}$$

This is indicative of how fast the natural response vanishes, i.e., how fast $y(t)$ approaches the forced response $y_f(t)$. Replacing (3.21) into (3.20), it is easy to verify that:

$$y(\tau) = 0.632\frac{kA}{a}. \tag{3.22}$$

Finally, it is important to stress that $y(t)$ grows without a limit if $a < 0$. If $a = 0$, then $y(t) = kAt$ also grows without a limit (see Sect. 3.2).

3.1.2   Transfer Function

Consider a zero initial condition, i.e., $y_0 = 0$, then (3.8) can be written as:

$$Y(s) = \frac{k}{s+a}U(s),$$

or, equivalently:

$$\frac{Y(s)}{U(s)} = G(s) = \frac{k}{s+a}. \tag{3.23}$$

The function $G(s)$ is known as the *transfer function* of the differential equation in (3.6). The polynomial in the denominator of $G(s)$, i.e., $s + a$, is known as the characteristic polynomial and its roots are known as the poles of $G(s)$. In this case, there is only one pole at $s = -a$. According to the discussion above, the following can be stated with respect to $y(t)$, i.e., the output of the transfer function $G(s)$:

- If pole located at $s = -a$ is negative, i.e., if $a > 0$, then the natural response $y_n(t)$ vanishes and the complete solution $y(t)$ approaches the forced response $y_f(t)$ as time increases. If $u(t) = A$ is a constant, then $y_f(t) = \frac{kA}{a}$.
- The faster $y(t)$ approaches $y_f(t)$, the farther to the left of the origin $s = 0$ is placed the pole at $s = -a$. This can be quantified using the time constant $\tau = \frac{1}{a}$. A large time constant implies a slow response whereas a small time constant implies a fast response.
- If $k = a > 0$, then it is said that the transfer function $G(s)$ has a unitary gain in a steady state because, according to the final value theorem (3.4), if $u(t) = A$ is a constant then:

$$\lim_{t \to \infty} y(t) = \lim_{s \to 0} sY(s) = \lim_{s \to 0} s \frac{k}{s+a} \frac{A}{s} = \frac{k}{a} A = A. \tag{3.24}$$

In general, the steady-state gain of the transfer function in (3.23) is computed as $\frac{k}{a}$.

- If the pole located at $s = -a$ is positive , i.e., if $a < 0$, then the complete solution $y(t)$ grows without limit. Notice that this implies that the pole is located on the right half of the plane s (see Fig. 3.2).
- The case when the pole is located at $s = a = 0$ cannot be studied from the above computations and it is analyzed as a special case in the next section. However, again to summarize all the possible cases, we present here results that are obtained in the next section. When the pole is located at $s = a = 0$, the natural response neither vanishes nor increases without limit and the forced response is given as the integral of the input $y_f(t) = k \int_0^t u(t)dt$.

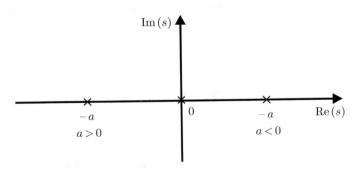

Fig. 3.2 Location of the poles of the transfer function in (3.23). It is usual to represent a pole on the s plane using a cross "×"

Fig. 3.3 A water level
system

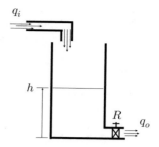

Example 3.2 Consider the tank containing water depicted in Fig. 3.3. Assume that
the tank section C is constant. Water enters the tank at a rate given by the input flow
q_i (m^3/s). Water leaves the tank at a rate given by the output flow q_o (m^3/s) through
a valve with hydraulic resistance R. The water level in the tank is represented
by h. The mathematical model describing this system is obtained using the mass
conservation law. As water is not compressible, this law can be stated in terms of
mass or volume.

Let ΔV_i and ΔV_o be the water volumes entering and leaving the tank respectively
during a time interval Δt. During this time interval, the water volume ΔV increases
inside the tank at a rate given as:

$$\frac{\Delta V}{\Delta t} = \frac{\Delta V_i}{\Delta t} - \frac{\Delta V_o}{\Delta t}. \tag{3.25}$$

On the other hand:

$$\Delta V = C \Delta h,$$
$$\Delta V_i = q_i \Delta t,$$
$$\Delta V_o = q_o \Delta t.$$

Replacing this in (3.25) and considering small increments of time such that $\Delta t \to$
dt and $\Delta h \to dh$:

$$C\frac{dh}{dt} = q_i - q_o. \tag{3.26}$$

Assuming that the water flow is laminar through the output valve [1], pp. 155, then:

$$q_o = \frac{h}{R}. \tag{3.27}$$

To understand the reason for this expression, consider the following situations.

a) Suppose that the opening of the output valve is kept without change, i.e., R remains constant, and the water level h increases. Everyday experience and the expression in (3.27) corroborate that the output flow q_o increases under this situation.

b) Suppose that the water level h is kept constant, because there is water entering the tank to exactly compensate for water leaving the tank. Then, slowly close the output valve, i.e., slowly increase R. Everyday experience and the expression in (3.27) corroborate that the output flow q_o decreases in this situation.

Substituting (3.27) in (3.26), it is found that:

$$C\frac{dh}{dt} = q_i - \frac{h}{R}.$$

Hence, the mathematical model of the water level system depicted in Fig. 3.3 is given as:

$$C\frac{dh}{dt} + \frac{1}{R}h = q_i.$$

Dividing by C, the following linear, ordinary, first-order differential equation with constant coefficients is finally obtained:

$$\frac{dh}{dt} + ah = kq_i,$$

$$a = \frac{1}{RC}, \quad k = \frac{1}{C}. \tag{3.28}$$

The reason why this differential equation is the mathematical model of the water level system is because, solving it, the evolution over time of the water level $h(t)$ can be known if the input flow $q_i(t)$, the tank section C, the hydraulic resistance R, and the initial water level are known.

Example 3.3 Consider the water level system depicted in Fig. 3.3. It was found in Example 3.2 that the mathematical model is given as:

$$\frac{dh}{dt} + ah = kq_i,$$

$$a = \frac{1}{RC} > 0, \quad k = \frac{1}{C}. \tag{3.29}$$

This differential equation can be written as in (3.6) if it is assumed that $y = h$, $u = q_i$ and $y_0 = h_0$. Hence, if $q_i = A$, the solution $h(t)$ is similar to that in (3.16), i.e.,:

$$h(t) = -\frac{kA}{a}e^{-at} + \frac{kA}{a} + h_0e^{-at}, \quad h_0 = h(0). \tag{3.30}$$

Let us analyze (3.30).

1. Assume that the tank is initially empty, i.e., $h_0 = 0$, and water enters the tank with a constant rate $q_i = A > 0$. Under these conditions, the water level $h(t)$ evolves as depicted in Fig. 3.1 by assuming that $h(t) = y(t)$. Notice that $a = \frac{1}{RC} > 0$. From this figure, the solution in (3.30), and using everyday experience, several cases can be studied.

   - If A is chosen to be larger, then the final water level $\lim_{t \to \infty} h(t) = \frac{kA}{a} = RA$ is also larger.
   - If the output valve is slowly closed, i.e., R is slowly increased, then the final water level $\lim_{t \to \infty} h(t) = \frac{kA}{a} = RA$ slowly increases.
   - The tank cross-section has no effect on the final water level, i.e., the final water level is the same in a thin tank and a thick tank.
   - If the product RC is larger, then a is smaller; hence, the system response is slower, i.e., more time is required for the level to reach the value $h(\tau) = 0.632\frac{kA}{a} = 0.632RA$ and, as consequence, also to reach the final water level $\lim_{t \to \infty} h(t) = \frac{kA}{a} = RA$. This is because the function e^{-at} tends toward zero more slowly when $a = \frac{1}{RC} > 0$ is smaller. Notice that a larger value of RC can be obtained by increasing the tank cross-section C, or decreasing the output valve opening, i.e., increasing R.

2. Suppose that no water is entering the tank, i.e., $q_i(t) = A = 0$, and the initial water level is not zero, i.e., $h_0 > 0$. From the solution in (3.30), and everyday experience can verify it, the water level decreases (recall that $a = \frac{1}{RC} > 0$) until the tank is empty, i.e., $\lim_{t \to \infty} h(t) = 0$ (see Fig. 3.4). This behavior is faster as RC is smaller, i.e., when a is larger, and slower as RC is larger, i.e., when a is smaller.

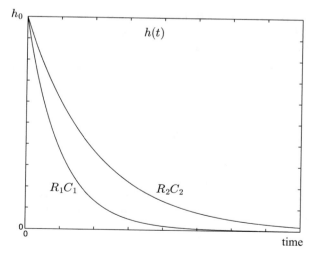

Fig. 3.4 Water level evolution when $h_0 > 0$ and $q_i(t) = A = 0$ ($R_1C_1 < R_2C_2$)

3. The case when $a < 0$ is not possible in an open-loop water level system because C and R cannot be negative. However, $a < 0$ may occur in closed-loop systems as a consequence of the interconnection of diverse components.
4. The case when $a = 0$ is studied in the next section.

3.2 An Integrator

Consider the following differential equation:

$$\dot{y} = ku, \tag{3.31}$$

where k is a real constant different from zero. Notice that this equation is obtained as a particular case of (3.6) when $a = 0$. By direct integration:

$$\int_{y(0)}^{y(t)} dy = \int_0^t ku \, dt,$$

$$y(t) = k \int_0^t u(t) \, dt + y(0), \tag{3.32}$$

$$y(t) = y_n(t) + y_f(t),$$

$$y_n(t) = y(0), \qquad y_f(t) = k \int_0^t u(t) \, dt.$$

In this case, $y_n(t)$ remains constant in time. Notice that, if $u = A$ is a constant, the forced response increases without a limit $y_f(t) = kAt$. The solution $y(t) = y(0) + kAt$ is depicted in Fig. 3.5 for $a)$ $y(0) \neq 0$ and $u = A > 0$ a constant, and $b)$ $y(0) \neq 0$ and $u = A = 0$. Physical systems represented by this class of equation are called *integrators* because, if $y(0) = 0$, the solution $y(t)$ is the integral of the excitation u.

Example 3.4 Consider the water level system depicted in Fig. 3.3. Suppose that the output valve is closed, i.e., $R \to \infty$. Then, the mathematical model in (3.28) becomes:

$$\frac{dh}{dt} = kq_i,$$

because $a = \frac{1}{RC} \to 0$ when $R \to \infty$. Then, the level evolution over time is described by (3.32), i.e.,:

$$h(t) = h(0) + k \int_0^t q_i(t) \, dt.$$

Fig. 3.5 Solution of
differential equation in (3.31).
(**a**) $y_0 \neq 0$ and $u = A > 0$ is
a constant. (**b**) $y_0 \neq 0$ and
$u = A = 0$

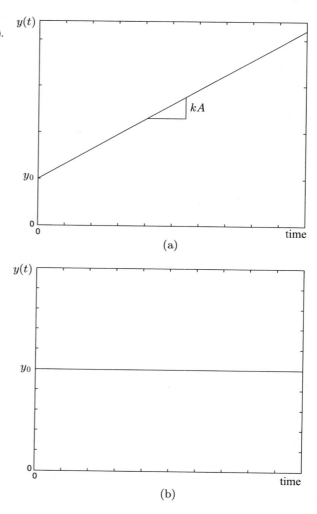

Hence, the water level system behaves in this case as an integrator. If $q_i = A > 0$,
then:

$$h(t) = h(0) + kAt. \qquad (3.33)$$

Figure 3.5 can be employed, assuming $h(t) = y(t)$ and $q_i(t) = u$, to graphically
depict the solution in (3.33) for the cases $i)$ $h(0) > 0$ and $q_i = A > 0$ is a constant,
and $ii)$ $h(0) > 0$ and $q_i = A = 0$. Notice that the water level remains constant if no
water enters the tank, i.e., when $A = 0$, and the level increases without a limit when
$A > 0$ is constant. The reader can resort to everyday experience to verify that these
two situations actually occur in a practical water level system.

Example 3.5 The differential equation in (3.31) can also be solved using partial fraction expansion, i.e., using the method employed in Sect. 3.1. From (3.2) and (3.31) it is found that:

$$sY(s) - y_0 = k\, U(s).$$

Solving for $Y(s)$ yields:

$$Y(s) = \frac{k}{s} U(s) + \frac{1}{s} y_0. \tag{3.34}$$

Assume that:

$$u = A, \qquad U(s) = \frac{A}{s},$$

where A is a constant. Hence, (3.34) can now be written as:

$$Y(s) = \frac{kA}{s^2} + \frac{1}{s} y_0. \tag{3.35}$$

According to the partial fraction expansion method [2], Ch. 4, [3], Ch. 7, it must be written:

$$\frac{kA}{s^2} = \frac{B}{s} + \frac{C}{s^2}, \tag{3.36}$$

where B and C are two constants. C is computed multiplying both sides of (3.36) by s^2 and evaluating at $s = 0$, i.e.,:

$$s^2 \frac{kA}{s^2}\bigg|_{s=0} = \frac{B}{s} s^2\bigg|_{s=0} + \frac{C}{s^2} s^2\bigg|_{s=0}.$$

Hence, $C = kA$ is found. B is computed multiplying both sides of (3.36) by s^2, differentiating once with respect to s and evaluating at $s = 0$:

$$\frac{d}{ds}\left(s^2 \frac{kA}{s^2}\right)\bigg|_{s=0} = \frac{d}{ds}\left(\frac{B}{s} s^2\right)\bigg|_{s=0} + \frac{d}{ds}\left(\frac{C}{s^2} s^2\right)\bigg|_{s=0}.$$

Hence, $B = 0$ is found. Thus, using these values and (3.36), the expression in (3.35) can be written as:

$$Y(s) = \frac{kA}{s^2} + \frac{1}{s} y_0. \tag{3.37}$$

From tabulated formulas, [4], Ch. 32, it is well known that:

$$\mathcal{L}\{t\} = \frac{1}{s^2}. \tag{3.38}$$

Using (3.38), it is found that the inverse Laplace transform of (3.37) is given as:

$$y(t) = kAt + y_0, \tag{3.39}$$

which is the solution of (3.31). This solution can be decomposed into two parts:

$$y(t) = y_n(t) + y_f(t), \tag{3.40}$$

$$y_n(t) = y_0,$$

$$y_f(t) = kAt,$$

where $y_n(t)$ and $y_f(t)$ are the natural and the forced responses respectively. As explained Sect. 3.1, the natural response is given by the Laplace transform of term $\dot{y}$ on the left-hand side of (3.31). As a consequence of this, the term $\frac{1}{s}y_0$ appears in (3.37), i.e., $y_n(t) = y_0$ in (3.40). The forced response $y_f(t)$ is produced by $u = A$. However, in this case, the forced response also receives the effect of the Laplace transform of the term $\dot{y}$ on the left-hand side of (3.31), as the combination of both of them results in $\frac{kA}{s^2}$ and, hence, $y_f(t) = kAt$.

The characteristic polynomial of the differential equation in (3.31) is s; hence, it only has one root at $s = 0$. Thus, the corresponding transfer function $G(s)$, when $y_0 = 0$:

$$\frac{Y(s)}{U(s)} = G(s) = \frac{k}{s}, \tag{3.41}$$

only has one pole at $s = 0$. Notice that the input $U(s) = \frac{A}{s}$ also has one pole at $s = 0$. The combination of these repeated poles in (3.35) produces the forced response $y_f(t) = kAt$, i.e., a first-degree polynomial of time, whereas the input $u = A$ is a zero-degree polynomial of time. Notice that the forced response corresponding to the differential equation in (3.6) is $y_f(t) = \frac{kA}{a}$, i.e., a zero-degree polynomial of time, whereas the input is $u = A$, i.e., a zero-degree polynomial of time. From these examples we arrive at the following important conclusion:

"The forced response of a first-order differential equation with a constant input, is also a constant if its characteristic polynomial has no root at $s = 0$."

In Fig. 3.2 the location, on the s plane of the pole of the transfer function in (3.41) is depicted, i.e., the pole at $s = 0$. The reader must learn to relate the pole location on the s plane to the corresponding time response $y(t)$. See the cases listed before Example 3.2.

Example 3.6 Consider the electric circuit depicted in Fig. 3.6. According to the Example 2.9, the mathematical model of this circuit is given as:

Fig. 3.6 A series RC circuit

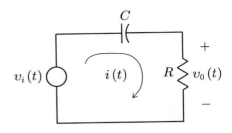

$$\dot{v}_0 + \frac{1}{RC}v_0 = \dot{v}_i.$$

Suppose that $v_i(t)$ is a step signal with amplitude A, i.e., $V_i(s) = \frac{A}{s}$. The solution $v_0(t)$ can be found to proceed as follows. Using the Laplace transform in the previous expression:

$$V_0(s)\left(s + \frac{1}{RC}\right) = sV_i(s) - v_i(0) + v_0(0),$$

$$V_0(s) = \frac{s}{s + \frac{1}{RC}}V_i(s) + \frac{v_0(0) - v_i(0)}{s + \frac{1}{RC}}.$$

Replacing $V_i(s) = \frac{A}{s}$ and using the inverse Laplace transform:

$$V_0(s) = \frac{s}{s + \frac{1}{RC}}\frac{A}{s} + \frac{v_0(0) - v_i(0)}{s + \frac{1}{RC}},$$

$$V_0(s) = \frac{A}{s + \frac{1}{RC}} + \frac{v_0(0) - v_i(0)}{s + \frac{1}{RC}},$$

$$v_0(t) = (A + v_0(0) - v_i(0))e^{-\frac{1}{RC}t}.$$

Applying KVL at $t = 0$, we have:

$$v_i(0) = v_0(0) + v_c(0),$$

where $v_c(0)$ represents the initial voltage at the capacitor. Then, we have:

$$v_0(t) = (A - v_c(0))e^{-\frac{1}{RC}t}. \tag{3.42}$$

In Fig. 3.7 this time response, when $v_i(t)$ is the square wave represented by the dashed line and the time constant is $RC = 0.1$[s], is represented. At $t = 0$ we have $v_c(0) = 0$ and $A = 1$. Thus:

$$v_0(t) = Ae^{-10t}. \tag{3.43}$$

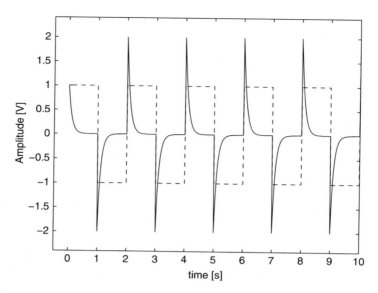

Fig. 3.7 Time response of the circuit in Fig. 3.6. Continuous: $v_0(t)$. Dashed: $v_i(t)$

This response is observed between $t = 0$ and $t = 1$[s]. Notice that $v_0(0) = A = 1$[V] is correctly predicted by (3.43). Also notice that $t = 1$[s] represents ten times the time constant, i.e., $RC = 0.1$[s]. Hence, $v_0(t)$ is very close to zero at $t = 1$[s]. Applying the KVL at $t = t_1 = 1$[s]:

$$v_i(t_1) = v_0(t_1) + v_c(t_1),$$

$$v_c(t_1) = A = 1[V], \tag{3.44}$$

as $v_i(t_1) = A = 1$[V] and $v_0(t_1) = 0$.

On the other hand, from $t = t_1 = 1$[s] to $t = 2$[s] we have that $A = -1$. Note that (3.42) is still valid in this time interval if $t = t - t_1$ is redefined and (3.42) is rewritten as:

$$v_0(t - t_1) = (A - v_c(t_1))e^{-\frac{1}{RC}(t-t_1)},$$

$$v_0(t - t_1) = -2e^{-10(t-t_1)}, \tag{3.45}$$

as $v_c(t_1) = 1$[V] (see (3.44)) and $A = -1$[V] at $t = t_1$[s]. In Fig. 3.7, it is observed that $v_0 = -2$[V], at $t = t_1$, is correctly predicted by (3.45). Also notice that, again, the time interval length 1[s] represents ten times the time constant, i.e., $RC = 0.1$[s]. Hence, $v_0(t - t_1)$ is very close to zero at $t = 2$[s]. The above procedure can be repeated to explain the results in Fig. 3.7 for any $0 \leq t \leq 10$.

Figure 3.7 has been obtained using the MATLAB/Simulink diagram presented in Fig. 3.8. The *signal generator block* is programmed to have a square wave form,

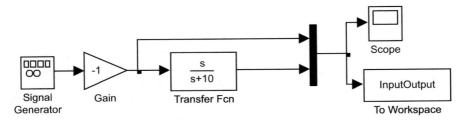

Fig. 3.8 MATLAB/Simulink diagram for the circuit in Fig. 3.6

unit amplitude, and 0.5[Hz] as the frequency. The negative unitary gain block is included because the signal delivered by the signal generator block starts with a negative semicircle. Once the simulation stops, the following MATLAB code is executed in an m-file to draw Fig. 3.7:

```
nn=length(InputOutput(:,1));
n=nn-1;
Ts=10/n;
t=0:Ts:10;
plot(t,InputOutput(:,1),'k--',t,InputOutput(:,2),'k-')
axis([-0.5 10 -2.5 2.5])
```

Notice that the transfer function of the circuit in Fig. 3.6 is:

$$\frac{V_0(s)}{V_i(s)} = G(s) = \frac{s}{s + \frac{1}{RC}}, \tag{3.46}$$

when $v_0(0) = 0$ and $v_i(0) = 0$, or equivalently, when $v_c(0) = 0$. Although this is a first-order transfer function, the main difference with respect to the transfer function defined in (3.23) is that a first-order polynomial appears at the numerator of $G(s) = \frac{s}{s + \frac{1}{RC}}$. It is concluded from Figs. 3.1 and 3.7 that the polynomial at the numerator of a transfer function affects time response too, despite the stability not being affected by the numerator of a transfer function.

In this respect, we can use (3.46), the initial value theorem in (3.5), and $V_i(s) = \frac{A}{s}$ to find:

$$v_0(0^+) = \lim_{t \to 0^+} v_0(t) = \lim_{s \to \infty} s V_0(s),$$

$$= \lim_{s \to \infty} s \frac{s}{s + \frac{1}{RC}} \frac{A}{s} = A.$$

The roots of the polynomial at the numerator of a transfer function are called the *transfer function zeros*. Thus, it is concluded that the discontinuity in the time response, when a step input is applied, is due to the fact that the transfer function has

Fig. 3.9 Phase-lead network

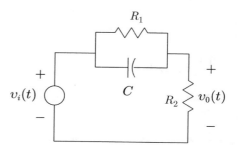

the same number of poles and zeros. In the next example, it is shown that a different location of a zero of a transfer function has different effects on the time response.

Example 3.7 Consider the electric circuit shown in Fig. 3.9. This circuit is also known as a *phase-lead network*. The mathematical model of this circuit was obtained in (2.49) and it is rewritten here for ease of reference:

$$V_0(s) = \frac{s+b}{s+a} V_i(s) - \frac{v_c(0)}{s+a}, \quad b = \frac{1}{R_1 C}, \quad a = \frac{1}{C}\frac{R_1+R_2}{R_1 R_2},$$

where $v_c(0)$ represents the initial voltage at the capacitor and $a > b > 0$. The solution $v_0(t)$ is computed in the following. Substituting $v_i(t) = A$, i.e., $V_i(s) = \frac{A}{s}$:

$$V_0(s) = \frac{(s+b)A}{s(s+a)} - \frac{v_c(0)}{s+a}. \tag{3.47}$$

As $a \neq 0$, the partial fraction expansion results in:

$$\frac{(s+b)A}{s(s+a)} = \frac{B}{s} + \frac{C}{s+a}. \tag{3.48}$$

Constant B is computed as:

$$s\frac{(s+b)A}{s(s+a)}\bigg|_{s=0} = s\frac{B}{s}\bigg|_{s=0} + s\frac{C}{s+a}\bigg|_{s=0},$$

$$B = \frac{bA}{a}. \tag{3.49}$$

Constant C is computed as:

$$(s+a)\frac{(s+b)A}{s(s+a)}\bigg|_{s=-a} = (s+a)\frac{B}{s}\bigg|_{s=-a} + (s+a)\frac{C}{s+a}\bigg|_{s=-a},$$

$$C = \frac{(b-a)A}{-a}. \tag{3.50}$$

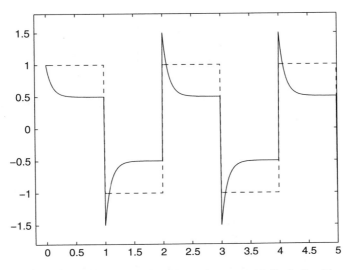

Fig. 3.10 Time response of the circuit in Fig. 3.9. Continuous: $v_0(t)$. Dashed: $v_i(t)$

Replacing (3.48), (3.49), (3.50), in (3.47), it is found that:

$$V_0(s) = \frac{bA}{a}\frac{1}{s} + \frac{(b-a)A}{-a}\frac{1}{s+a} - \frac{v_c(0)}{s+a}.$$

Hence, using the inverse Laplace transform:

$$v_0(t) = \frac{bA}{a} + \left(\frac{(b-a)A}{-a} - v_c(0)\right)e^{-at}. \tag{3.51}$$

This time response is depicted in Fig. 3.10 when $v_i(t)$ is the square wave shown with a dashed line, $b = 5$, $a = 10$ and $v_c(0) = 0$. At $t = 0$, $A = 1[\text{V}]$. Then, (3.51) becomes:

$$v_0(t) = \frac{1}{2} + \frac{1}{2}e^{-10t}.$$

Thus, $v_0(0) = 1[\text{V}]$ and $v_0(t_1) \approx \frac{1}{2}[\text{V}]$, for $t_1 = 1[\text{s}]$, are correctly predicted. Notice that a time interval of $1[\text{s}]$, between $t = 0$ and $t = t_1$, is ten times the time constant of $\frac{1}{a} = 0.1[\text{s}]$; hence, $\frac{1}{2}e^{-10t} \approx 0$ when $t = t_1 = 1[\text{s}]$. On the other hand, applying the KVL to the circuit in Fig. 3.9 it is found that;

$$v_i(t_1) = v_c(t_1) + v_0(t_1),$$

$$v_i(t_1) = A = 1, \quad v_0(t_1) = \frac{R_2}{R_1 + R_2} A,$$

$$v_c(t_1) = v_i(t_1) - v_0(t_1) = \frac{1}{2} = \frac{R_1}{R_1 + R_2}.$$

At $t = t_1 = 1$[s], $v_i(t)$ changes to $A = -1$. If $t = t - t_1$ is redefined, (3.51) is still valid if it is rewritten as:

$$v_0(t - t_1) = \frac{bA}{a} + \left(\frac{(b-a)A}{-a} - v_c(t_1) \right) e^{-a(t-t_1)},$$

$$v_0(t - t_1) = \frac{-1}{2} + \left(\frac{-1}{2} - \frac{R_1}{R_1 + R_2} \right) e^{-10(t-t_1)},$$

$$v_0(t - t_1) = \frac{-1}{2} - e^{-10(t-t_1)},$$

since $\frac{R_1}{R_1 + R_2} = \frac{1}{2}$. Thus, Fig. 3.10 corroborates that $v_0(t_1) = -1.5$[V], $t_1 = 1$[s], and $v_0 \approx -\frac{1}{2}$[V] at $t = 2$[s], because $-e^{-10(t-t_1)} \approx 0$ at $t = 2$[s], are correctly predicted again.

Figure 3.10 has been obtained using the MATLAB/Simulink simulation diagram presented in Fig. 3.11. The *signal generator block* is programmed to have a square wave form, unit amplitude, and 0.5[Hz] as the frequency. The negative, unit gain block is included because the signal delivered by the signal generator block starts with a negative semicircle. Once the simulation stops, the following MATLAB code is executed in an m-file to draw Fig. 3.10:

```
nn=length(InputOutput(:,1));
n=nn-1;
Ts=5/n;
t=0:Ts:5;
plot(t,InputOutput(:,1),'k--',t,InputOutput(:,2),'k-')
axis([-0.25 5 -2 2])
```

Notice that the transfer function of the circuit in Fig. 3.9, given as:

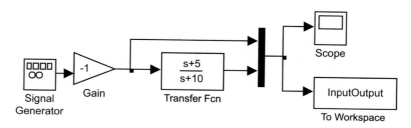

Fig. 3.11 MATLAB/Simulink simulation diagram for circuit in Fig. 3.9

$$\frac{V_0(s)}{V_i(s)} = G(s) = \frac{s+b}{s+a},\tag{3.52}$$

when $v_c(0) = 0$, has a zero at $s = -b < 0$, whereas the transfer function of the circuit in Fig. 3.6 has a zero at $s = 0$. The main difference produced by this fact is that the time response of the circuit in Fig. 3.6 reaches a zero value in a steady state whereas the steady-state response of the circuit in Fig. 3.9 is given as $\frac{b}{a}A \neq 0$.

Finally, we can use (3.52), the initial value theorem in (3.5), and $V_i(s) = \frac{A}{s}$ to find:

$$v_0(0^+) = \lim_{t \to 0^+} v_0(t) = \lim_{s \to \infty} sV_0(s),$$

$$= \lim_{s \to \infty} s\frac{s+b}{s+a}\frac{A}{s} = A.$$

Thus, it is concluded again that the discontinuity in time response when a step input is applied is due to the fact that the transfer function has the same number of poles and zeros.

3.3 Second-Order Differential Equation

Consider the following linear ordinary second-order differential equation with constant coefficients:

$$\ddot{y} + 2\zeta\omega_n\dot{y} + \omega_n^2 y = k\omega_n^2 u, \quad y(0) = y_0, \quad \dot{y}(0) = \dot{y}_0,\tag{3.53}$$

where $\omega_n > 0$ and k are real nonzero constants. Assume that $u = A$ is a real constant. Using (3.2) and $U(s) = A/s$, the following can be written:

$$s^2 Y(s) - s y_0 - \dot{y}_0 + 2\zeta\omega_n(sY(s) - y_0) + \omega_n^2 Y(s) = k\omega_n^2 U(s),$$

to obtain:

$$Y(s) = k\frac{\omega_n^2 A}{(s^2 + 2\zeta\omega_n s + \omega_n^2)s} + \frac{y_0(s + 2\zeta\omega_n) + \dot{y}_0}{s^2 + 2\zeta\omega_n s + \omega_n^2}.\tag{3.54}$$

First, suppose that all of the initial conditions are zero $\dot{y}_0 = 0$, $y_0 = 0$. Then:

$$Y(s) = k\frac{A\,\omega_n^2}{(s^2 + 2\zeta\omega_n s + \omega_n^2)s}.$$

Also, suppose that $0 \leq \zeta < 1$ is a real constant, then:

$$s^2 + 2\zeta\omega_n s + \omega_n^2 = (s - a)(s - \bar{a}), \tag{3.55}$$

$$a = \sigma + j\omega_d, \quad \bar{a} = \sigma - j\omega_d, \quad j = \sqrt{-1},$$

$$\omega_d = \omega_n\sqrt{1 - \zeta^2} > 0, \quad \sigma = -\zeta\omega_n \leq 0.$$

Notice that:

$$s^2 + 2\zeta\omega_n s + \omega_n^2 = (s - [\sigma + j\omega_d])(s - [\sigma - j\omega_d]) = (s - \sigma)^2 + \omega_d^2.$$

According to the partial fraction expansion method [2], Ch. 4, [3], Ch. 7, in this case the following can be written:

$$Y(s) = k\frac{A\omega_n^2}{(s^2 + 2\zeta\omega_n s + \omega_n^2)s} = k\frac{A\omega_n^2}{[(s - \sigma)^2 + \omega_d^2]s} = \frac{Bs + C}{(s - \sigma)^2 + \omega_d^2} + \frac{D}{s}, \tag{3.56}$$

where B, C, and D are constants to be computed. The inverse Laplace transform of (3.56) is given as:

$$y(t) = kA\left[1 - \frac{e^{-\zeta\omega_n t}}{\sqrt{1 - \zeta^2}} \sin(\omega_d t + \phi)\right], \tag{3.57}$$

$$\phi = \arctan\frac{\sqrt{1 - \zeta^2}}{\zeta}.$$

This expression constitutes the solution of (3.53) when all of the initial conditions are zero, i.e., when $\dot{y}_0 = 0$, $y_0 = 0$. The detailed procedure to find (3.57) from (3.56) is presented in the following. Readers who are not interested in these technical details may continue from (3.62) onward.

Multiplying both sides of (3.56) by the factor $(s - \sigma)^2 + \omega_d^2$ and evaluating at $s = a$ (see (3.55)) yields:

$$k\frac{A\omega_n^2}{[(s - \sigma)^2 + \omega_d^2]s}[(s - \sigma)^2 + \omega_d^2]\bigg|_{s=a} = \frac{Bs + C}{(s - \sigma)^2 + \omega_d^2}[(s - \sigma)^2 + \omega_d^2]\bigg|_{s=a}$$

$$+ \frac{D}{s}[(s - \sigma)^2 + \omega_d^2]\bigg|_{s=a},$$

to obtain:

$$k\frac{A\omega_n^2}{\sigma + j\omega_d} = B(\sigma + j\omega_d) + C.$$

Multiplying the left-hand term by $(\sigma - j\omega_d)/(\sigma - j\omega_d)$:

$$kA\omega_n^2\frac{(\sigma - j\omega_d)}{\sigma^2 + \omega_d^2} = B\sigma + C + jB\omega_d.$$

Equating the imaginary parts:

$$B = -\frac{kA\omega_n^2}{\sigma^2 + \omega_d^2}. \tag{3.58}$$

Equating the real parts:

$$B\sigma + C = \frac{kA\omega_n^2\sigma}{\sigma^2 + \omega_d^2}.$$

Replacing (3.58):

$$C = \frac{2kA\omega_n^2\sigma}{\sigma^2 + \omega_d^2}. \tag{3.59}$$

Using (3.58) and (3.59), the following can be written:

$$\frac{Bs + C}{(s - \sigma)^2 + \omega_d^2} = \frac{-\frac{kA\omega_n^2}{\sigma^2 + \omega_d^2}s + \frac{2kA\omega_n^2\sigma}{\sigma^2 + \omega_d^2}}{(s - \sigma)^2 + \omega_d^2},$$

$$= -\frac{kA\omega_n^2}{(\sigma^2 + \omega_d^2)}\frac{(s - \sigma)}{[(s - \sigma)^2 + \omega_d^2]} + \frac{kA\omega_n^2\sigma}{(\sigma^2 + \omega_d^2)}\frac{1}{[(s - \sigma)^2 + \omega_d^2]}.$$

Using the tabulated Laplace transform pairs [4], Ch. 32:

$$\mathcal{L}^{-1}\left\{\frac{s - \sigma}{(s - \sigma)^2 + \omega_d^2}\right\} = e^{\sigma t}\cos(\omega_d t),$$

$$\mathcal{L}^{-1}\left\{\frac{\omega_d}{(s - \sigma)^2 + \omega_d^2}\right\} = e^{\sigma t}\sin(\omega_d t),$$

the following is obtained:

$$\mathcal{L}^{-1}\left\{\frac{Bs + C}{(s - \sigma)^2 + \omega_d^2}\right\} = -\frac{kA\omega_n^2}{\sigma^2 + \omega_d^2}e^{\sigma t}\cos(\omega_d t) + \frac{kA\omega_n^2\sigma}{\omega_d(\sigma^2 + \omega_d^2)}e^{\sigma t}\sin(\omega_d t),$$

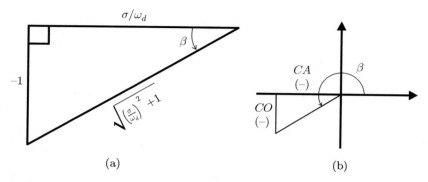

Fig. 3.12 Some important relations for the procedure in Sect. 3.3

$$= \frac{kA\omega_n^2}{\sigma^2 + \omega_d^2} e^{\sigma t} [-\cos(\omega_d t) + \frac{\sigma}{\omega_d} \sin(\omega_d t)],$$

$$= \frac{kA\omega_n^2}{\sigma^2 + \omega_d^2} e^{\sigma t} \quad [\sin \beta \cos(\omega_d t) + \cos \beta \sin(\omega_d t)] \sqrt{\left(\frac{\sigma}{\omega_d}\right)^2 + 1}.$$

Notice that some relations in Fig. 3.12a have been employed in the last step. On the other hand, it is possible to continue writing:

$$\sin \beta \cos(\omega_d t) + \cos \beta \sin(\omega_d t) = \frac{1}{2}[\sin(\beta - \omega_d t) + \sin(\beta + \omega_d t)] +$$

$$+ \frac{1}{2}[\sin(\omega_d t - \beta) + \sin(\omega_d t + \beta)],$$

$$= \sin(\omega_d t + \beta),$$

because $\sin(-x) = -\sin(x)$. Hence:

$$\mathcal{L}^{-1}\left\{\frac{Bs + C}{(s - \sigma)^2 + \omega_d^2}\right\} = kA\omega_n^2 \frac{\sqrt{\left(\frac{\sigma}{\omega_d}\right)^2 + 1}}{\sigma^2 + \omega_d^2} e^{\sigma t} \sin(\omega_d t + \beta).$$

On the other hand, from Fig. 3.12a:

$$\tan \beta = \frac{-1}{\sigma/\omega_d} = \frac{-1}{(-\zeta\omega_n)/(\omega_n\sqrt{1 - \zeta^2})} = \frac{-1}{\frac{-\zeta}{\sqrt{1-\zeta^2}}},$$

$$\beta = \pi + \arctan \frac{\sqrt{1 - \zeta^2}}{\zeta},$$

where signs of the adjacent and the opposite sides have been taken into account to conclude that β is an angle on the third quadrant (see Fig. 3.12b). This allows the following simplification [4], Ch. 5:

$$\sin(\omega_d t + \beta) = \sin(\omega_d t + \pi + \phi),$$

$$= -\sin(\omega_d t + \phi),$$

$$\phi = \arctan \frac{\sqrt{1-\zeta^2}}{\zeta},$$

and thus:

$$\mathcal{L}^{-1}\left\{\frac{Bs + C}{(s-\sigma)^2 + \omega_d^2}\right\} = -kA\omega_n^2 \frac{\sqrt{\left(\frac{\sigma}{\omega_d}\right)^2 + 1}}{\sigma^2 + \omega_d^2} e^{\sigma t} \sin(\omega_d t + \phi),$$

$$= -kA\omega_n^2 \frac{\sqrt{\sigma^2 + \omega_d^2}}{\omega_d(\sigma^2 + \omega_d^2)} e^{\sigma t} \sin(\omega_d t + \phi),$$

$$= -kA\omega_n^2 \frac{1}{\omega_d\sqrt{\sigma^2 + \omega_d^2}} e^{\sigma t} \sin(\omega_d t + \phi),$$

$$\mathcal{L}^{-1}\left\{\frac{Bs + C}{(s-\sigma)^2 + \omega_d^2}\right\} = -kA\frac{1}{\sqrt{1-\zeta^2}} e^{\sigma t} \sin(\omega_d t + \phi), \tag{3.60}$$

where $\sigma = -\zeta\omega_n$ and $\omega_d = \omega_n\sqrt{1-\zeta^2}$ have been employed. On the other hand, D in (3.56) is easily computed multiplying both sides of that expression by factor s and evaluating at $s = 0$:

$$D = \left.\frac{kA\,\omega_n^2}{s^2 + 2\zeta\omega_n s + \omega_n^2}\right|_{s=0} = kA. \tag{3.61}$$

Using (3.56), (3.60) and (3.61), the following solution is found:

$$y(t) = kA\left[1 - \frac{e^{-\zeta\omega_n t}}{\sqrt{1-\zeta^2}}\sin(\omega_d t + \phi)\right].$$

If it is assumed now that the initial conditions are not zero, then:

$$y(t) = kA\left[1 - \frac{e^{-\zeta\omega_n t}}{\sqrt{1-\zeta^2}}\sin(\omega_d t + \phi)\right] + p(t), \tag{3.62}$$

where:

$$p(t) = \mathcal{L}^{-1}\left\{\frac{y_0(s + 2\zeta\omega_n) + \dot{y}_0}{s^2 + 2\zeta\omega_n s + \omega_n^2}\right\}. \tag{3.63}$$

The inverse Laplace transform defining $p(t)$ may be computed using a similar procedure to that presented above. Moreover, the reader should realize that $p(t)$ is given by a function as that in (3.60) with coefficients depending on both y_0 and $\dot{y}_0$.

The solution given in (3.57) can be decomposed into two parts:

$$y(t) = y_n(t) + y_f(t), \tag{3.64}$$

$$y_n(t) = -kA\frac{e^{-\zeta\omega_n t}}{\sqrt{1 - \zeta^2}}\sin(\omega_d t + \phi), \tag{3.65}$$

$$y_f(t) = kA, \tag{3.66}$$

where $y_n(t)$ and $y_f(t)$ are the natural and the forced responses respectively. Notice that, according to (3.56) and (3.60), the natural response in (3.65) is due only to the characteristic polynomial $s^2 + 2\zeta\omega_n s + \omega_n^2 = (s - \sigma)^2 + \omega_d^2$, which results from applying the Laplace transform to terms $\ddot{y} + 2\zeta\omega_n\dot{y} + \omega_n^2 y$. Hence, no matter what the input u is, the natural response $y_n(t)$ is given as in (3.65). On the other hand, notice that the forced response in (3.66) is due to term D/s in (3.56), which is introduced by the constant input $u = A$, as $U(s) = A/s$. Also notice that, according to the paragraph after (3.63), when the initial conditions are different from zero they only affect the natural response in the sense that its coefficient depends on the initial conditions y_0 and $\dot{y}_0$, but the function of time is always given as $e^{-\zeta\omega_n t}\sin(\omega_d t + \phi)$ or its time derivative. Furthermore, $p(t)$ belongs to the natural response when the initial conditions are different from zero.

The reader may review the procedure presented above after (3.57) to corroborate that, in the case when $-1 < \zeta < 0$ with $k > 0$ and $\omega_n > 0$, the solution in (3.64) or in (3.57) becomes:

$$y(t) = y_n(t) + y_f(t), \tag{3.67}$$

$$y_n(t) = kA\frac{e^{-\zeta\omega_n t}}{\sqrt{1 - \zeta^2}}\sin\left(\omega_d t - \arctan\left(\frac{\sqrt{1 - \zeta^2}}{abs(\zeta)}\right)\right),$$

$$y_f(t) = kA,$$

where $abs(\cdot)$ stands for the absolute value function.

Employing (3.64), when $0 \le \zeta < 1$, and (3.67), when $-1 < \zeta < 0$, it is concluded that the solution of the differential equation in (3.53) has one of the following behaviors:

1. If $0 < \zeta < 1$, i.e., if both roots of the characteristic polynomial located at $s = -\zeta\omega_n \pm j\omega_d$ have negative real parts $-\zeta\omega_n < 0$, then $\lim_{t\to\infty} y_n(t) = 0$ and $\lim_{t\to\infty} y(t) = y_f(t)$.

2. If $-1 < \zeta < 0$, i.e., if both roots of the characteristic polynomial located at $s = -\zeta\omega_n \pm j\omega_d$ have positive real parts $-\zeta\omega_n > 0$, then $y_n(t)$ and $y(t)$ grow without a limit, describing oscillations, as time increases.

3. If $\zeta = 0$, i.e., if both roots of the characteristic polynomial located at $s = -\zeta\omega_n \pm j\omega_d$ have zero real parts $-\zeta\omega_n = 0$, then from (3.64) the following is obtained:

$$y(t) = y_n(t) + y_f(t), \tag{3.68}$$

$$y_n(t) = -kA\cos(\omega_n t), \quad y_f(t) = kA,$$

when the initial conditions are zero, i.e., $y_n(t)$ is an oscillatory function whose amplitude neither increases nor decreases. This means that although $y(t)$ does not grow without a limit, it will not reach the forced response $y_f(t)$ in a steady state.

4. The behavior obtained when ζ is out of the range $-1 < \zeta < 1$ is studied in the subsequent sections as the cases when the roots of the characteristic polynomial are real and repeated or real and different.

Finally, notice that, according to the above discussion and (3.62), (3.63), the solution of a second-order differential equation may present sustained oscillations ($\zeta = 0$) even if the input is zero ($u = 0$) if the initial conditions are different from zero. This explains the oscillations in the electronic circuits studied in Chap. 9 (see last paragraph in Sect. 9.3.1).

3.3.1 Graphical Study of the Solution

The graphical form of solution presented in (3.64) for the differential equation in (3.53) is studied in the following. Recall that it is assumed that $0 \le \zeta < 1$, $\omega_n > 0$, $k > 0$ and that the initial conditions are zero $y_0 = 0$, $\dot{y}_0 = 0$. Notice that the natural response tends toward zero as time increases if $0 < \zeta < 1$:

$$\lim_{t\to\infty} y_n(t) = \lim_{t\to\infty} \left(-kA\frac{e^{-\zeta\omega_n t}}{\sqrt{1-\zeta^2}}\sin(\omega_d t + \phi)\right) = 0, \tag{3.69}$$

because $\zeta\omega_n > 0$. Also recall that, in the case when the initial conditions are not zero, the natural response also tends toward zero as time increases. Hence, when time is large, the solution of the differential equation is equal to the forced response:

$$\lim_{t\to\infty} y(t) = y_f(t) = kA. \tag{3.70}$$

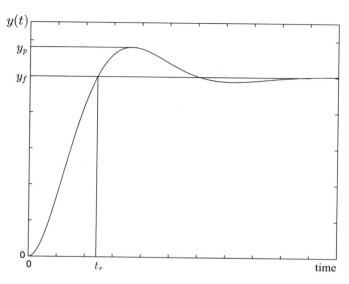

Fig. 3.13 $y(t)$ in (3.64)

This means that the faster the natural response vanishes $y_n(t)$, i.e., as $\zeta\omega_n > 0$ is larger, the faster the complete solution $y(t)$ reaches the forced response $y_f(t)$. As was noted for first-order differential equations, the natural response allows the complete solution $y(t)$ to go from the initial conditions (zero in this case) to the forced response $y_f(t)$.

Two important parameters of the response in (3.64) are the rise time t_r and overshoot M_p, which are shown in Fig. 3.13. Overshoot can be measured as:

$$M_p(\%) = \frac{y_p - y_f}{y_f} \times 100,$$

where $y_f = kA$ stands for the forced response. Recall that it is assumed that $y_0 = \dot{y}_0 = 0$. Carefully analyzing (3.64), it can be shown that [1], pp. 232:

$$t_r = \frac{1}{\omega_d}\left[\pi - \arctan\left(\frac{\sqrt{1 - \zeta^2}}{\zeta}\right)\right], \tag{3.71}$$

$$M_p(\%) = 100 \times e^{-\frac{\zeta}{\sqrt{1 - \zeta^2}}\pi}.$$

Figures 3.14 and 3.15 show how the solution in (3.64) is affected when the parameters ζ and ω_n change. Notice that overshoot M_p is affected only by ζ, whereas rise time t_r is mainly affected by ω_n, although ζ also has a small effect on t_r.

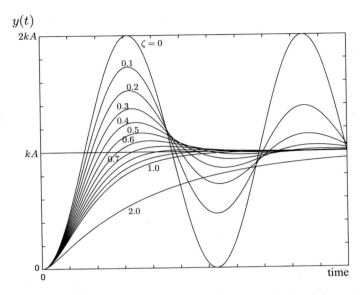

Fig. 3.14 The solution in (3.64) when different values for ζ are used and ω_n is kept constant

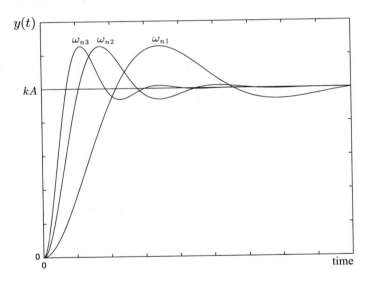

Fig. 3.15 The solution in (3.64) when different values of ω_n are used: $\omega_{n2} = 2\omega_{n1}$, $\omega_{n3} = 3\omega_{n1}$. Parameter ζ is kept constant

Finally, it is important to point out that, in the case when $u = A = 0$ but some of the initial conditions $\dot{y}_0$ or y_0 are different from zero, $y(t)$ behaves as the time function in (3.65) and is graphically represented as the oscillatory part without the constant component, i.e., it oscillates around $y = 0$, in any of the Figs. 3.13, 3.14 or 3.15. This is because of function $p(t)$ appearing in (3.62).

3.3.2 Transfer Function

If the initial conditions are zero $y_0 = 0$, $\dot{y}_0 = 0$, then (3.54) can be written as:

$$Y(s) = \frac{k\omega_n^2}{s^2 + 2\zeta\omega_n s + \omega_n^2} U(s),$$

or:

$$\frac{Y(s)}{U(s)} = G(s) = \frac{k\omega_n^2}{s^2 + 2\zeta\omega_n s + \omega_n^2}, \tag{3.72}$$

where $G(s)$ is the transfer function of the differential equation in (3.53). Notice that $G(s)$, defined in (3.72), has a second-degree characteristic polynomial, i.e., $G(s)$ has two poles. Hence, $G(s)$ is a second-order transfer function. According to (3.55) these poles are located at $s = a$ and $s = \bar{a}$, where:

$$a = \sigma + j\omega_d, \quad \bar{a} = \sigma - j\omega_d.$$

It is important to stress that these poles are complex conjugate only under the condition $-1 < \zeta < 1$. The reader can verify that these poles are real if ζ does not satisfy this condition. Such a case is studied in the subsequent sections.

According to the study in the previous sections, the following can be stated for the output of the transfer function $y(t)$:

- If both poles have a negative real part $\sigma = -\zeta\omega_n < 0$, i.e., $\zeta > 0$, then the natural response $y_n(t)$ vanishes and the complete response $y(t)$ reaches the forced response $y_f(t)$ as time increases. When the input is a constant $u(t) = A$, then the forced response is $y_f(t) = kA$.
- The natural response $y_n(t)$ vanishes faster as $\zeta\omega_n$ is larger. As is depicted in Fig. 3.16, the system response remains enveloped by two exponential functions whose vanishing rate is determined by $\zeta\omega_n$.
- The rise time t_r decreases if ω_n increases. This is corroborated by observing that, according to (3.55), if ω_n increases then ω_d also increases and, according to (3.71), t_r decreases.
- Overshoot M_p decreases if ζ increases.
- According to Fig. 3.17: $i)$ the rise time t_r decreases as the complex conjugate poles are located farther to the left of $s = 0$ because ω_n increases (see (3.71) and $\omega_d = \omega_n\sqrt{1 - \zeta^2}$), $ii)$ the parameter ζ increases and, hence, overshoot M_p decreases, as the angle θ is larger because $\zeta = \sin(\theta)$, $iii)$ $y_n(t)$ vanishes faster as the complex conjugate poles are located farther to the left of $s = 0$ because $\zeta\omega_n$ is larger.

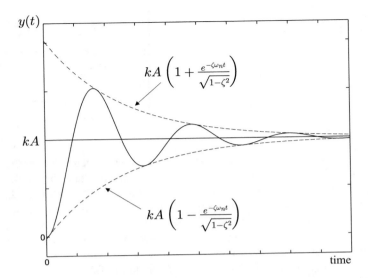

Fig. 3.16 The response of a second-order system is enveloped by two exponential functions when the input is constant and the initial conditions are zero

Fig. 3.17 One pole of $G(s)$ in (3.72) when $0 < \zeta < 1$

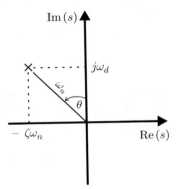

- If $k = 1$ then the transfer function $G(s)$ is said to have unitary gain in a steady state because, when $u = A$ is a constant, the final value theorem (3.4) can be used to find that:

$$\lim_{t \to \infty} y(t) = \lim_{s \to 0} sY(s) = \lim_{s \to 0} s \frac{k\omega_n^2}{s^2 + 2\zeta\omega_n s + \omega_n^2} \frac{A}{s} = \frac{k\omega_n^2}{\omega_n^2} A = A.$$

The constant k represents the steady-state gain of the transfer function in (3.72).
- If $-1 < \zeta < 0$, then the real part of the poles $\sigma = -\zeta\omega_n$ is positive; hence, the complete response $y(t)$ grows without limit as time increases. Notice that this implies that the complex conjugate poles are located on the right half of the plane s.

Fig. 3.18 A mass-spring-damper system

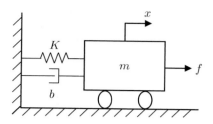

- The parameter ζ is known as the damping coefficient because it determines how oscillatory the response $y(t)$ is (see (3.71), for overshoot, and Fig. 3.14).
- The parameter ω_d is known as the damped natural frequency because, according to the previous discussion (see (3.64)), ω_d is the oscillation frequency in $y(t)$ when the damping ζ is different from zero. Notice that the frequency ω_d is the imaginary part of the poles of the transfer function in (3.72) (see (3.55)).
- The parameter ω_n is known as the undamped natural frequency because, according to (3.64) and (3.68), ω_n is the oscillation frequency when damping is zero.

Example 3.8 Consider the mass-spring-damper system depicted in Fig. 3.18. According to Example 2.2, Chap. 2, the mass position x is given by the following second-order differential equation:

$$m\ddot{x} = -Kx - b\dot{x} + f, \tag{3.73}$$

where K is the spring stiffness constant, b is the viscous friction coefficient, m is the body mass, and f is an applied external force. The expression in (3.73) can be written as:

$$\ddot{x} + \frac{b}{m}\dot{x} + \frac{K}{m}x = \frac{1}{m}f,$$

which can be rewritten as the expression in (3.53):

$$\ddot{x} + 2\zeta\omega_n\dot{x} + \omega_n^2 x = k\omega_n^2 f, \tag{3.74}$$

$$2\zeta\omega_n = \frac{b}{m}, \quad \omega_n^2 = \frac{K}{m}, \quad k = \frac{1}{K},$$

where x is the output y, whereas f is the input u. Hence, $x(t)$ is given as $y(t)$ in (3.64) if $0 \leq \zeta < 1$ when the initial conditions are zero. Notice that all constants in (3.73) are positive: the mass m, the viscous friction coefficient b, and the spring stiffness coefficient K only can be positive. According to (3.74), the damping coefficient is given as:

$$\zeta = \frac{b}{2\sqrt{mK}}. \tag{3.75}$$

If a constant force $f = A$ is applied, the results in Figs. 3.13, 3.14, and 3.15 can be employed to conclude the following.

- If there is no friction, i.e., if $b = 0$, then $\zeta = 0$ and the mass will oscillate forever around the position $x_f = kA = \frac{1}{K}A$. In the case when the external force is zero $f = 0$ but some of the initial conditions $\dot{x}_0$ or x_0 are different from zero, then the mass will oscillate again, though now around the position $x = 0$, as described by $p(t)$ appearing in (3.62).
- If the friction coefficient $b > 0$ increases, then $\zeta > 0$ also increases and the mass oscillations will disappear because the system will have more damping. It is shown in the subsequent sections that the poles are real and repeated when $\zeta = 1$. In all of these cases, the mass will stop moving at $x = x_f = kA = \frac{1}{K}A$. In the case when the external force is zero $f = 0$ but some initial conditions, $\dot{x}_0$ or x_0, are different from zero, the mass will stop at $x = 0$. This is again described by the function $p(t)$ appearing in (3.62). Notice that, according to (3.75), the damping ζ also increases if the mass m or the spring stiffness constant K decreases. To understand the reason for this, consider the opposite case: larger values of both m and K produce larger forces ($m\ddot{x}$ for the mass and Kx for the spring), which, hence, are less affected by the friction force ($b\dot{x}$) and thus, the mass can remain in movement for a longer period of time.
- As the rise time t_r inversely depends on the undamped natural frequency $\omega_n = \sqrt{\frac{K}{m}}$, then a more rigid spring (with a larger K) or a smaller mass produces faster responses (t_r smaller). On the other hand, advantage can be taken of the fact that the oscillation frequency is given by $\omega_d = \omega_n\sqrt{1 - \zeta^2}$ with $\omega_n = \sqrt{\frac{K}{m}}$ to adjust the rate of a mechanical clock based on a mass-spring-damper system: if the clock "lags," it is necessary to increase its oscillation frequency, or rate, which is achieved by applying tension to the spring to increase K.

The case when $-1 < \zeta < 0$ is not possible in this example because all of the constants b, m, and K are positive. However, the situation when $-1 < \zeta < 0$ commonly occurs in feedback systems as a result of the interconnection of diverse components.

3.4 Arbitrary-Order Differential Equations

In Exercise 3.6, it was shown that some differential equations can be written in terms of time derivatives of both the unknown function $y(t)$ and the excitation function $u(t)$. Moreover, the initial conditions of both variables can be combined to obtain the initial conditions in other circuit components. According to this reasoning, a linear ordinary differential equation with arbitrary order n and constant coefficients can always be written as:

$$y^{(n)} + a_{n-1}y^{(n-1)} + \cdots + a_1\dot{y} + a_0 y = b_0 u + b_1 \dot{u} + \cdots + b_m u^{(m)}, \qquad (3.76)$$

where $n \geq m$. If $n < m$ the differential equation has no physical meaning, i.e., there is no physical system in practice whose mathematical model can be represented by such a differential equation and, thus, it is not of any interest for engineers. Notice that all of the derivatives are related to time. Because of this feature, these differential equations are also called *dynamical systems*. The variable y is the unknown variable of the differential equation, whereas u is a function of time that is also called the excitation. The objective of solving a differential equation is to find a function $y(t)$ that satisfies the equality defined by the differential equation. To find $y(t)$, it is necessary to know $u(t)$, the real constants a_i, $i = 0, \ldots, n-1$, b_j, $j = 0, \ldots, m$, and the set of $n + m$ constants $y(0)$, $\dot{y}(0) \ldots y^{(n-1)}(0)$, $u(0)$, $\dot{u}(0), \ldots, u^{(m-1)}(0)$, which are known as the initial conditions of the problem, representing values that the unknown and the excitation, in addition to their time derivatives, have at $t = 0$.

Applying the Laplace transform (3.2) to (3.76), the following is found:

$$s^n Y(s) + a_{n-1} s^{n-1} Y(s) + \cdots + a_1 s Y(s) + a_0 Y(s) + P(s)$$

$$= b_0 U(s) + b_1 s U(s) + \cdots + b_m s^m U(s),$$

where $P(s)$ is a polynomial in s whose coefficients depend on the initial conditions $y(0)$, $\dot{y}(0), \ldots, y^{(n-1)}(0)$, $u(0)$, $\dot{u}(0), \ldots, u^{(m-1)}(0)$ and the coefficients of the differential equation. Hence, it is possible to write:

$$Y(s) = \frac{b_0 + b_1 s + \cdots + b_m s^m}{s^n + a_{n-1} s^{n-1} + \cdots + a_1 s + a_0} U(s) - \frac{P(s)}{s^n + a_{n-1} s^{n-1} + \cdots + a_1 s + a_0}.$$

Suppose that $u(t) = A$ is a constant, i.e., $U(s) = A/s$. Then:

$$Y(s) = \frac{B(s)}{N(s)} \frac{A}{s} - \frac{P(s)}{N(s)},$$

$$N(s) = s^n + a_{n-1} s^{n-1} + \cdots + a_1 s + a_0,$$

$$B(s) = b_0 + b_1 s + \cdots + b_m s^m.$$

Also suppose for the moment that all of the initial conditions are zero. This means that $P(s) = 0$ and it is possible to write:

$$Y(s) = \frac{B(s)}{N(s)} \frac{A}{s}.$$

According to the partial fraction expansion method [2], Ch. 4, [3], Ch. 7, the solution in time $y(t)$ depends on the roots of the characteristic polynomial $N(s)$, i.e., on poles of the following transfer function:

$$\frac{Y(s)}{U(s)} = G(s) = \frac{B(s)}{N(s)} \tag{3.77}$$

and factor $\frac{1}{s}$ introduced by $U(s) = A/s$. The variable $y(t)$ is known as the *output* and $u(t)$ is known as the *input* of the transfer function $G(s)$. The order of $G(s)$ is defined as the degree of the *characteristic polynomial* $N(s)$. As a n degree polynomial has n roots, then a n order transfer function has n poles (see paragraph after (3.23)). On the other hand, the roots of the polynomial $B(s)$ are known as the *zeros* of the transfer function $G(s)$. If $B(s)$ has the degree m then $G(s)$ has m zeros. Recall the condition $n \geq m$ imposed at the beginning of this section. It is assumed that $B(s)$ and $N(s)$ do not have common roots.

The solution for all of the possible cases for the roots of the polynomial $N(s)$ are studied in the next sections, i.e., all of the possible cases for the poles of $G(s)$. This provides the necessary information to understand how a system of arbitrary order n responds.

3.4.1 Real and Different Roots

Suppose that $N(s)$ has $k \leq n$ real and different roots, which are different from $s = 0$, i.e.,:

$$N(s) = N_0(s)(s - p_1)(s - p_2) \cdots (s - p_k),$$

where $p_i \neq 0$, $i = 1, \ldots, k$, with $p_j \neq p_i$ if $i \neq j$, stand for the roots of $N(s)$, whereas $N_0(s)$ is a polynomial containing the remaining $n - k$ roots of $N(s)$, where none of them is located at $s = 0$. According to the partial fraction expansion method [2], Ch. 4, [3], Ch. 7, in this case the following must be written:

$$Y(s) = \frac{B(s)}{N(s)} \frac{A}{s} = Q(s) + \frac{c_1}{s - p_1} + \frac{c_2}{s - p_2} + \cdots + \frac{c_k}{s - p_k} + \frac{f}{s}, \quad (3.78)$$

where c_i, $i = 1, \ldots, k$, and f are real constants to be computed, whereas $Q(s)$ represents all fractions corresponding to the roots of $N_0(s)$. Constant c_i can be computed as follows. Multiply both sides of (3.78) by the factor $(s - p_i)$ and evaluate the resulting expression at $s = p_i$ to obtain:

$$c_i = \left. \frac{B(s)}{N(s)} \frac{A}{s}(s - p_i) \right|_{s=p_i}, \quad i = 1, \ldots, n.$$

A similar procedure allows us to compute:

$$f = \left. \frac{B(s)A}{N(s)} \right|_{s=0}.$$

Using the tabulated inverse Laplace transform, [4], Ch.32:

$$\mathcal{L}^{-1} \left\{ \frac{c_i}{s - a} \right\} = c_i e^{at},$$

the following is finally obtained:

$$y(t) = q(t) + c_1\, e^{p_1 t} + c_2\, e^{p_2 t} + \cdots + c_k\, e^{p_k t} + f,$$

where $q(t) = \mathcal{L}^{-1}\{Q(s)\}$. Notice that the following can be written:

$$y(t) = y_n(t) + y_f(t),$$
$$y_n(t) = q(t) + c_1\, e^{p_1 t} + c_2\, e^{p_2 t} + \cdots + c_k\, e^{p_k t}, \qquad (3.79)$$
$$y_f(t) = f,$$

where the natural response $y_n(t)$ only depends on the roots of the characteristic polynomial $N(s)$, whereas the forced response $y_f(t)$ only depends on the fraction $\frac{1}{s}$ introduced by $U(s)$. Consider the following possibilities.

1. If $p_i < 0$ for all $i = 1, \ldots, k$, then $\lim_{t \to \infty}(c_1\, e^{p_1 t} + c_2\, e^{p_2 t} + \cdots + c_k\, e^{p_k t}) = 0$ and $\lim_{t \to \infty} y(t) = q(t) + y_f(t)$.
2. If $p_i > 0$ for at least one $i = 1, \ldots, k$, then $(c_1\, e^{p_1 t} + c_2\, e^{p_2 t} + \cdots + c_k\, e^{p_k t}) \to \infty$, as $t \to \infty$ and $\lim_{t \to \infty} y(t) = \infty$.

Example 3.9 Consider the following second-order differential equation also given in (3.53):

$$\ddot{y} + 2\zeta\omega_n\dot{y} + \omega_n^2 y = k\omega_n^2 u, \quad y(0) = y_0, \quad \dot{y}(0) = \dot{y}_0. \qquad (3.80)$$

If $\zeta > 1$, then the only two roots, p_1 and p_2, of the characteristic polynomial $N(s) = s^2 + 2\zeta\omega_n s + \omega_n^2$ are real and different:

$$p_1 = -\zeta\omega_n + \omega_n\sqrt{\zeta^2 - 1},$$

$$p_2 = -\zeta\omega_n - \omega_n\sqrt{\zeta^2 - 1},$$

$$p_1 \neq p_2.$$

Also, both roots are negative:

$$p_1 < 0, \quad p_2 < 0.$$

Although this is obvious for p_2, for p_1, a little observation is required: notice that $-\zeta\omega_n + \omega_n\sqrt{\zeta^2} = 0$, and since $-\zeta\omega_n + \omega_n\sqrt{\zeta^2} > -\zeta\omega_n + \omega_n\sqrt{\zeta^2 - 1}$, then $-\zeta\omega_n + \omega_n\sqrt{\zeta^2 - 1} < 0$. In this case $q(t) = 0$, because $Q(s) = 0$ because $N_0(s) = 1$, i.e., $N_0(s)$ has no roots. In Fig. 3.19 the solution $y(t)$ of the differential equation in (3.80) is shown, when $\zeta > 1$, $u = A$ and the initial conditions are zero. Notice that $y(t)$ does not present oscillations in this case. Recall that, when $-1 < \zeta < 1$, the oscillation frequency is equal to the imaginary part of both roots. Hence, when

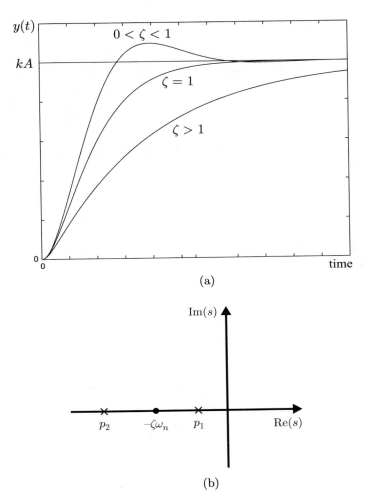

Fig. 3.19 Solution of the second-order differential equation in (3.80), when $\zeta > 1$, and in (3.85), when $\zeta = 1$, with $u = A$ and zero initial conditions. (a) Time response. (b) Location of poles when $\zeta > 1$

$\zeta > 1$, there is no oscillation, because the imaginary part of both roots is zero; thus, the oscillation frequency is also zero.

Example 3.10 Consider again the mass-spring-damper system in Example 3.8. Recall that (3.74) describes the mass movement. Assume that:

$$\zeta = \frac{b}{2\sqrt{mK}} > 1.$$

Then, according to the previous example, the roots p_1 and p_2 of the characteristic polynomial $s^2 + 2\zeta\omega_n s + \omega_n^2$ are real, different, and negative. Notice that this case stands when the viscous friction coefficient $b > 0$ is very large or when the spring stiffness constant $K > 0$ or mass m is very small. Also notice that one root $(p_1 = -\zeta\omega_n + \omega_n\sqrt{\zeta^2 - 1} < 0)$ approaches the origin as $\zeta > 1$ becomes larger. According to the discussion above in the present section, this means that the movement of the mass is slower as ζ grows, because it takes more time to function $e^{p_1 t}$ to vanish (despite the function $e^{p_2 t}$ vanishing faster because $p_2 = -\zeta\omega_n - \omega_n\sqrt{\zeta^2 - 1} < 0$ moves farther to the left of the origin). This explains why slower responses are depicted in Fig. 3.19, as $\zeta > 1$ is larger.

Example 3.11 Consider the following differential equation:

$$\ddot{y} + c\dot{y} + dy = eu, \quad y(0) = y_0, \quad \dot{y}(0) = \dot{y}_0,$$

with c, d and e some real constants. The roots p_1 and p_2 of the characteristic polynomial $N(s) = s^2 + cs + d$ are given as:

$$p_1 = -\frac{c}{2} + \frac{\sqrt{c^2 - 4d}}{2},$$

$$p_2 = -\frac{c}{2} - \frac{\sqrt{c^2 - 4d}}{2}.$$

The following cases have to be considered:

- If $d < 0$, then:

$$p_1 = -\frac{c}{2} + \frac{\sqrt{c^2 + 4\,abs(d)}}{2},$$

$$p_2 = -\frac{c}{2} - \frac{\sqrt{c^2 + 4\,abs(d)}}{2},$$

and both roots are real and different, with one positive and the other negative, no matter what the value of c. It is shown, in Example 7.5 of Chap. 7, that this case corresponds to a mass-spring-damper system with a negative spring stiffness constant and that this is obtained in practice when a simple pendulum works around its inverted configuration (also called *unstable*).
- If $c > 0$ and $d > 0$, the case studied in Sect. 3.3 is retrieved when $1 > \zeta > 0$, and the case studied in the present section is retrieved when $\zeta > 1$.
- If $c < 0$ and $d > 0$, with $abs(c)$ small and d large, the case studied in Sect. 3.3 is retrieved when $-1 < \zeta < 0$. The situation when $\zeta < -1$ is also obtained in this case when $abs(c)$ is large and d is small, but both roots are positive and different:

$$p_1 = -\zeta\omega_n + \omega_n\sqrt{\zeta^2 - 1} > 0,$$

$$p_2 = -\zeta \omega_n - \omega_n \sqrt{\zeta^2 - 1} > 0,$$

$$p_1 \neq p_2.$$

- The cases when $\zeta = -1$ and $\zeta = 1$ are studied in the next section.

3.4.2 Real and Repeated Roots

Suppose that $N(s)$ has one root that is repeated k times and it is different from zero, i.e.,:

$$N(s) = N_0(s)(s - p)^k,$$

with $p \neq 0$ and $N_0(s)$ a polynomial that contains the remaining $n - k$ roots in $N(s)$, and none of them is located at $s = 0$. According to the partial fraction expansion method [2], Ch. 4, [3], Ch. 7, the following must be written:

$$Y(s) = \frac{B(s)}{N(s)} \frac{A}{s} = Q(s) + \frac{c_k}{(s-p)^k} + \frac{c_{k-1}}{(s-p)^{k-1}} + \cdots + \frac{c_2}{(s-p)^2} + \frac{c_1}{s-p} + \frac{f}{s},$$

$$(3.81)$$

where c_i, $i = 1, \ldots, k$, and f represent real constants to be computed, whereas $Q(s)$ represents the fractions corresponding to roots of $N_0(s)$. One method of computing c_k is by multiplying both sides of (3.81) by the factor $(s - p)^k$ and evaluating at $s = p$ to obtain:

$$c_k = (s - p)^k \frac{B(s)A}{N(s)s}\bigg|_{s=p}.$$

On the other hand, the constants c_i, $i = 1, \ldots, k - 1$, can be computed by multiplying both sides of (3.81) by the factor $(s - p)^k$, differentiating $k - i$ times with regard to s and evaluating at $s = p$ to obtain:

$$c_i = \frac{1}{(k - i)!} \frac{d^{k-i}}{ds^{k-i}} \left(\frac{B(s)A}{N(s)s} \right)\bigg|_{s=p}, \quad i = 1, \ldots, k - 1.$$

Finally, the constant f is computed by multiplying both sides of (3.81) by the factor s and evaluating at $s = 0$ to find:

$$f = \frac{B(s)A}{N(s)}\bigg|_{s=0}.$$

Using the tabulated inverse Laplace transform [4], Ch. 32:

$$\mathcal{L}^{-1}\left\{\frac{1}{(s-a)^j}\right\} = \frac{t^{j-1}}{(j-1)!}\, e^{at}, \quad j = 1, 2, 3, \ldots, \quad 0! = 1,$$

the following is found:

$$y(t) = q(t) + c_k \frac{t^{k-1}}{(k-1)!}\, e^{pt} + c_{k-1}\frac{t^{k-2}}{(k-2)!}\, e^{pt} + \cdots$$
$$+ c_2 t\, e^{pt} + c_1\, e^{pt} + f,$$

where $q(t) = \mathcal{L}^{-1}\{Q(s)\}$. Notice that the following can be written:

$$y(t) = y_n(t) + y_f(t),$$

$$y_n(t) = q(t) + c_k \frac{t^{k-1}}{(k-1)!}\, e^{pt} + c_{k-1}\frac{t^{k-2}}{(k-2)!}\, e^{pt} + \cdots$$
$$+ c_2 t\, e^{pt} + c_1\, e^{pt},$$

$$y_f(t) = f. \tag{3.82}$$

It is remarked again that $y_n(t)$ only depends on the roots of the characteristic polynomial $N(s)$ and $y_f(t)$ only depends on the fraction $\frac{1}{s}$, introduced by $U(s)$. Consider the following possibilities:

1. If $p < 0$, it is useful to compute the following limit, where j is any positive integer number:

$$\lim_{t\to\infty} t^j\, e^{pt} = \lim_{t\to\infty} \frac{t^j}{e^{-pt}},$$

which represents an indetermination, because $-pt \to +\infty$. Then, L'Hopital's rule [12], pp. 303, can be employed:

$$\lim_{t\to\infty} t^j\, e^{pt} = \lim_{t\to\infty} \frac{t^j}{e^{-pt}} = \lim_{t\to\infty} \frac{\frac{dt^j}{dt}}{\frac{de^{-pt}}{dt}} = \lim_{t\to\infty} \frac{j\, t^{j-1}}{-p\, e^{-pt}}.$$

As an indetermination appears again, L'Hopital's rule can be applied several times to obtain:

$$\lim_{t\to\infty} t^j\, e^{pt} = \lim_{t\to\infty} \frac{t^j}{e^{-pt}} = \lim_{t\to\infty} \frac{\frac{dt^j}{dt}}{\frac{de^{-pt}}{dt}} = \lim_{t\to\infty} \frac{j\, t^{j-1}}{-p\, e^{-pt}},$$

$$= \lim_{t \to \infty} \frac{j(j-1) \, t^{j-2}}{(-p)^2 \, e^{-pt}} = \cdots = \lim_{t \to \infty} \frac{j!}{(-p)^j \, e^{-pt}} = 0.$$

Applying this result to the solution obtained above, it is concluded that:

$$\lim_{t \to \infty} \left(c_k \frac{t^{k-1}}{(k-1)!} \, e^{pt} + c_{k-1} \frac{t^{k-2}}{(k-2)!} \, e^{pt} + \cdots + c_2 t \, e^{pt} + c_1 \, e^{pt} \right) = 0,$$

and $\lim_{t \to \infty} y(t) = q(t) + y_f(t)$ for any $p < 0$, no matter how close to zero p is.

2. If $p > 0$, it is clear that:

$$\lim_{t \to \infty} \left(c_k \frac{t^{k-1}}{(k-1)!} \, e^{pt} + c_{k-1} \frac{t^{k-2}}{(k-2)!} \, e^{pt} + \cdots + c_2 t \, e^{pt} + c_1 \, e^{pt} \right) \to \infty,$$

as $t \to \infty$ and $\lim_{t \to \infty} y(t) = \infty$.

3. Finally, consider the case when the characteristic polynomial $N(s)$ has k real and repeated roots at $p = 0$, i.e., $N(s) = N_0(s)s^k$, where $N_0(s)$ is a polynomial containing the remaining $n - k$ roots of $N(s)$. Then, according to the partial fraction expansion method [2], Ch. 4, [3], Ch. 7, in this case the following must be written:

$$Y(s) = \frac{B(s)A}{N(s)s} = Q(s) + \frac{c_{k+1}}{s^{k+1}} + \frac{c_k}{s^k} + \frac{c_{k-1}}{s^{k-1}} + \cdots + \frac{c_2}{s^2} + \frac{c_1}{s},$$

where $Q(s)$ contains fractions corresponding to roots of $N_0(s)$. Using the tabulated Laplace transform [4], Ch. 32:

$$\mathcal{L}^{-1} \left\{ \frac{1}{s^j} \right\} = \frac{t^{j-1}}{(j-1)!}, \quad j = 1, 2, 3, \ldots, \quad 0! = 1, \tag{3.83}$$

yields:

$$y(t) = q(t) + c_{k+1} \frac{t^k}{k!} + c_k \frac{t^{k-1}}{(k-1)!} + \cdots + c_2 t + c_1, \tag{3.84}$$

where $q(t) = \mathcal{L}^{-1}\{Q(s)\}$. In this case:

$$y_n(t) = q(t) + c_k \frac{t^{k-1}}{(k-1)!} + \cdots + c_2 t + c_1,$$

$$y_f(t) = c_{k+1} \frac{t^k}{k!},$$

since the natural response only depends on the k real and repeated roots at $s = 0$ that $N(s)$ has, which, according to (3.83), only introduces terms included in $y_n(t)$. It is clear that $y_n(t) \to \infty$ and $y(t) \to \infty$ when $t \to \infty$ for $k \geq 2$ and that $y_n(t)$ is a constant if $k = 1$. On the other hand, the forced response $y_f(t)$ represents the integral of $u(t)$ iterated k times.

Example 3.12 Consider the following second-order differential equation, also given in (3.53):

$$\ddot{y} + 2\zeta\omega_n\dot{y} + \omega_n^2 y = k\omega_n^2 u, \quad y(0) = y_0, \quad \dot{y}(0) = \dot{y}_0. \tag{3.85}$$

If $\zeta = 1$, then both roots, p_1 y p_2, of the characteristic polynomial $N(s) = s^2 + 2\zeta\omega_n s + \omega_n^2$ are real, repeated, and negative:

$$p_1 = p_2 = -\zeta\omega_n.$$

The solution $y(t)$ of the differential equation in (3.85) is depicted in Fig. 3.19, when $\zeta = 1$, $u = A$ and the initial conditions are zero. Notice that $y(t)$ does not exhibit oscillations in this case because both roots have a zero imaginary part. This case, $\zeta = 1$, is at the edge between an oscillatory response $\zeta < 1$ and a response without oscillations $\zeta > 1$. In fact, the case $\zeta = 1$ represents the fastest response without oscillations because, when $\zeta > 1$, the response is slower as ζ increases.

Example 3.13 Consider again the mass-spring-damper system in Example 3.8. Recall that (3.74) describes the movement of mass. Assume that:

$$\zeta = \frac{b}{2\sqrt{mK}} = 1.$$

Then, in this case, the characteristic polynomial $s^2 + 2\zeta\omega_n s + \omega_n^2$ only has a real and negative root:

$$p = -\zeta\omega_n,$$

which is repeated twice. The mass position $x(t)$ evolves exactly as $y(t)$ in Fig. 3.19. This class of response may be very useful if a fast mass movement is required, but without oscillations.

Example 3.14 Suppose that there is neither a spring nor a damper in Fig. 3.18, i.e., assuming that $b = 0$ and $K = 0$, the differential equation in (3.73) becomes:

$$m\ddot{x} = f. \tag{3.86}$$

Notice that in this case the characteristic polynomial is s^2, which only has a real root at $p = 0$ repeated twice. Now suppose that all of the initial conditions are zero and that the mass receives a small disturbance force described by:

$$f = \begin{cases} \varepsilon_0, \ 0 \leq t \leq t_1 \\ 0, \quad \text{otherwise} \end{cases}, \tag{3.87}$$

where $\varepsilon_0 > 0$ and $t_1 > 0$ are small constant numbers. By direct integration of (3.86), it is found:

$$\dot{x}(t) = \frac{1}{m} \int_0^t f(\tau) d\tau + \dot{x}(0).$$

Again, integrating yields (assuming that $\dot{x}(0) = 0$):

$$x(t) = \frac{1}{m} \int_0^t \left\{ \int_0^r f(\tau) d\tau \right\} dr + x(0),$$

$$x(t) = \frac{1}{m} \int_0^t \left\{ \int_0^r f(\tau) d\tau \right\} dr, \quad x(0) = 0.$$

Replacing (3.87), the following is found:

$$x(t) = \begin{cases} \frac{1}{2m} \varepsilon_0 t^2, & 0 \leq t \leq t_1 \\ \frac{1}{2m} \varepsilon_0 t_1^2 + \frac{1}{m} \varepsilon_0 t_1 (t - t_1), & t > t_1 \end{cases}.$$

Notice that $x(t) \to \infty$ as $t \to \infty$ (see Fig. 3.20) despite $f = 0$ for $t > t_1$, i.e., despite the forced response being zero for all $t > t_1$. This means that the natural response $x_n(t) \to \infty$ when $t \to \infty$, which corroborates results obtained in the present section for real roots at zero, which are repeated at least twice. The reader may resort to everyday experience to verify that a slight hit (f in (3.87)) on the mass m is enough for the mass to begin moving to never stop ($x(t) \to \infty$) if no friction exists between the mass and the floor.

Example 3.15 Consider the second-order differential equation given in (3.53):

$$\ddot{y} + 2\zeta \omega_n \dot{y} + \omega_n^2 y = k \omega_n^2 u, \quad y(0) = y_0, \quad \dot{y}(0) = \dot{y}_0.$$

Fig. 3.20 Mass position in Fig. 3.18 when it is disturbed and neither a spring nor a damper exists

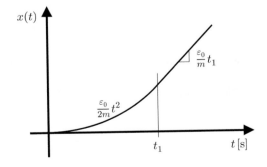

If $\zeta = -1$, then both roots, p_1 and p_2, of the characteristic polynomial $N(s) = s^2 + 2\zeta\omega_n s + \omega_n^2$ are real, repeated, and positive:

$$p_1 = p_2 = -\zeta\omega_n > 0.$$

3.4.3 Complex Conjugate and Nonrepeated Roots

Consider the following second-degree polynomial:

$$s^2 + bs + a = (s - p_1)(s - p_2), \quad p_1 = c_1 + jd_1, \quad p_2 = c_2 + jd_2,$$

where p_1 and p_2 are the complex, but not conjugate roots, i.e., c_1, c_2, d_1, d_2, are real numbers such that $c_1 \neq c_2$ if $d_1 = -d_2$ or $d_1 \neq -d_2$ if $c_1 = c_2$. We have that:

$$(s - p_1)(s - p_2) = s^2 - (p_1 + p_2)s + p_1 p_2,$$
$$b = p_1 + p_2 = c_1 + c_2 + j(d_1 + d_2),$$
$$a = p_1 p_2 = c_1 c_2 - d_1 d_2 + j(c_2 d_1 + c_1 d_2).$$

Note that, because of conditions $c_1 \neq c_2$ if $d_1 = -d_2$ or $d_1 \neq -d_2$ if $c_1 = c_2$, one of the coefficients a or b is a complex number, i.e., the polynomial $s^2 + bs + a$ has at least one complex coefficient. In the previous sections, we have seen that physical systems have characteristic polynomials whose coefficients are given only in terms of properties that can be quantified using real numbers, i.e., mass, inertia, friction coefficients, stiffness constant, electrical resistance, inductance, capacitance, etc. Thus, the characteristic polynomials of physical systems cannot have a complex root without including its corresponding complex conjugate. This means that in automatic control we are only interested in characteristic polynomials that have pairs of complex conjugate roots.

Suppose that $N(s)$ has k pairs of complex conjugate and nonrepeated roots, i.e., according to Sect. 3.3, the following can be written:

$$N(s) = N_0(s) \prod_{i=1}^{i=k} [(s - \sigma_i)^2 + \omega_{di}^2],$$

where $\sigma_i \neq \sigma_j$ or $\omega_{di} \neq \omega_{dj}$ if $i \neq j$, whereas $N_0(s)$ is a polynomial containing the remaining $n - 2k$ roots of $N(s)$, none of them located at $s = 0$. Then, according to the partial fraction expansion method [2], Ch. 4, [3], Ch. 7, in this case the following must be written:

$$Y(s) = \frac{B(s)}{N(s)}\frac{A}{s} = Q(s) + \sum_{i=1}^{i=k}\frac{F_i s + C_i}{[(s - \sigma_i)^2 + \omega_{di}^2]} + \frac{f}{s},$$

where $Q(s)$ contains fractions corresponding to roots of $N_0(s)$, f is a constant to be computed as in the previous sections, i.e.,:

$$f = \frac{B(s)A}{N(s)}\bigg|_{s=0},$$

whereas F_i and C_i, $i = 1, \ldots, k$, are constants to be computed as B and C were computed in Sect. 3.3. Hence, using (3.60) the following can be written:

$$y(t) = y_n(t) + y_f(t), \tag{3.88}$$

$$y_n(t) = q(t) + \sum_{i=1}^{i=k} \beta_i e^{-\zeta_i \omega_{ni} t} \sin(\omega_{di} t + \phi_i),$$

$$y_f(t) = f,$$

where β_i, ω_{ni}, ζ_i y ϕ_i, $i = 1, \ldots, k$, are real constants (see Sect. 3.3), whereas $q(t) = \mathcal{L}^{-1}\{Q(s)\}$. The evolution of the solution in (3.88) satisfies one of the following cases.

1. If $0 < \zeta_i < 1$ for all $i = 1, \ldots, k$, i.e., if the real parts of all of the complex conjugate roots are negative, $-\zeta_i \omega_{ni} < 0$ for all $i = 1, \ldots, k$, then $y(t)$ oscillates such that $\lim_{t \to \infty} \sum_{i=1}^{i=k} \beta_i e^{-\zeta_i \omega_{ni} t} \sin(\omega_{di} t + \phi_i) = 0$ and $\lim_{t \to \infty} y(t) = q(t) + y_f(t)$.
2. If $-1 < \zeta_i < 0$ for at least one $i = 1, \ldots, k$, i.e., if the real part is positive, $-\zeta_i \omega_{ni} > 0$ for at least one of the complex conjugate roots $i = 1, \ldots, k$, then $y(t)$ oscillates such that $y_n(t)$ and $y(t)$ grow without a limit as time increases.
3. If $\zeta_i = 0$ for all $i = 1, \ldots, k$, i.e., if the real parts of all the complex conjugate and nonrepeated roots are zero, $-\zeta_i \omega_{ni} = 0$ for all $i = 1, \ldots, k$, then $\sum_{i=1}^{i=k} \beta_i e^{-\zeta_i \omega_{ni} t} \sin(\omega_{di} t + \phi_i)$ does not disappear as time increases, but nor does it grow without a limit. Thus, although $y(t)$ may not grow without a limit (depending on the behavior of $q(t)$), it does not converge to $y_f(t)$. Notice that for this situation to stand, it is enough that $\zeta_i = 0$ for at least one i and $0 < \zeta_i < 1$ for all of the remaining i.

It is important to point out that, in the cases when $\zeta \geq 1$ or $\zeta \leq -1$, real roots are obtained and one of the cases in Sects. 3.4.1 or 3.4.2 is retrieved.

Example 3.16 The mathematical model of the mechanical system composed of two bodies and three springs depicted in Fig. 3.21 has been obtained in Example 2.4 in Chap. 2. From this result, the case when the spring at the left is not present can be studied. This is achieved by assuming that $K_1 = 0$, i.e.,:

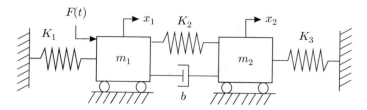

Fig. 3.21 A system with two bodies and three springs

$$\ddot{x}_1 + \frac{b}{m_1}(\dot{x}_1 - \dot{x}_2) + \frac{K_2}{m_1}(x_1 - x_2) = \frac{1}{m_1}F(t),$$

$$\ddot{x}_2 - \frac{b}{m_2}(\dot{x}_1 - \dot{x}_2) + \frac{K_3}{m_2}x_2 - \frac{K_2}{m_2}(x_1 - x_2) = 0.$$

Applying the Laplace transform (3.2) to each one of these differential equations and assuming that all of the initial conditions are zero, the following is found:

$$X_1(s) = \frac{\frac{1}{m_1}F(s) + \left(\frac{b}{m_1}s + \frac{K_2}{m_1}\right)X_2(s)}{s^2 + \frac{b}{m_1}s + \frac{K_2}{m_1}},$$

$$X_2(s) = \frac{\left(\frac{b}{m_2}s + \frac{K_2}{m_2}\right)X_1(s)}{s^2 + \frac{b}{m_2}s + \frac{K_2+K_3}{m_2}}.$$

Replacing the second of these equations in the first one, and rearranging, it is found that:

$$X_1(s) = \frac{\frac{1}{m_1}\left(s^2 + \frac{b}{m_2}s + \frac{K_2+K_3}{m_2}\right)F(s)}{\left(s^2 + \frac{b}{m_1}s + \frac{K_2}{m_1}\right)\left(s^2 + \frac{b}{m_2}s + \frac{K_2+K_3}{m_2}\right) - \left(\frac{b}{m_1}s + \frac{K_2}{m_1}\right)\left(\frac{b}{m_2}s + \frac{K_2}{m_2}\right)}.$$

To simplify the required algebra, assume that $b = 0$, then:

$$X_1(s) = \frac{\frac{1}{m_1}\left(s^2 + \frac{K_2+K_3}{m_2}\right)}{\left(s^2 + \frac{K_2}{m_1}\right)\left(s^2 + \frac{K_2+K_3}{m_2}\right) - \frac{K_2^2}{m_1 m_2}}F(s),$$

or:

$$X_1(s) = \frac{\frac{1}{m_1}\left(s^2 + \frac{K_2+K_3}{m_2}\right)}{s^4 + \left(\frac{K_2}{m_1} + \frac{K_2+K_3}{m_2}\right)s^2 + \frac{K_2 K_3}{m_1 m_2}}F(s). \tag{3.89}$$

Performing the indicated operation, it is found that:

$$s^4 + \left(\frac{K_2}{m_1} + \frac{K_2 + K_3}{m_2}\right)s^2 + \frac{K_2 K_3}{m_1 m_2} = (s^2 + c)(s^2 + e),$$

for some numbers c and e such that:

$$c + e = \frac{K_2}{m_1} + \frac{K_2 + K_3}{m_2}, \qquad ce = \frac{K_2 K_3}{m_1 m_2},$$

i.e.,:

$$c = \frac{\frac{K_2}{m_1} + \frac{K_2 + K_3}{m_2} + \sqrt{\left(\frac{K_2}{m_1} - \frac{K_2 + K_3}{m_2}\right)^2 + \frac{4K_2^2}{m_1 m_2}}}{2} > 0,$$

$$e = \frac{\frac{K_2}{m_1} + \frac{K_2 + K_3}{m_2} - \sqrt{\left(\frac{K_2}{m_1} - \frac{K_2 + K_3}{m_2}\right)^2 + \frac{4K_2^2}{m_1 m_2}}}{2} > 0.$$

Then, the following can be written:

$$X_1(s) = \frac{\frac{1}{m_1}\left(s^2 + \frac{K_2 + K_3}{m_2}\right)}{(s^2 + c)(s^2 + e)} F(s). \tag{3.90}$$

As e and c are positive and different, it is ensured that the characteristic polynomial in (3.90) has two pairs of complex conjugate and different roots (in fact, the roots are imaginary conjugate and different). Hence, mass m_1, whose position is represented by $x_1(t)$, moves by oscillating and it never stops. However, the amplitude of this oscillatory movement does not grow without a limit. These oscillations are performed at two angular frequencies defined by $\omega_{d1} = \omega_{n1} = \sqrt{c}$ and $\omega_{d2} = \omega_{n2} = \sqrt{e}$. Recall that the damping of these oscillations is zero.

3.4.4 Complex Conjugated and Repeated Roots

Assume that the characteristic polynomial $N(s)$ has a pair of complex conjugate roots that are repeated $k > 1$ times, i.e.,:

$$N(s) = N_0(s)(s^2 + 2\zeta\omega_n s + \omega_n^2)^k = N_0(s)[(s - \sigma)^2 + \omega_d^2]^k,$$

where $N_0(s)$ is a polynomial that includes the remaining $n - 2k$ roots of $N(s)$, without any of them located at $s = 0$. Then, the partial fraction expansion method establishes that in this case [2], Ch. 4:

$$Y(s) = \frac{B(s)}{N(s)} \frac{A}{s} = Q(s) + \frac{F_1 s + C_1}{[(s-\sigma)^2 + \omega_d^2]^k} + \frac{F_2 s + C_2}{[(s-\sigma)^2 + \omega_d^2]^{k-1}} + \cdots$$

$$+ \frac{F_{k-1} s + C_{k-1}}{[(s-\sigma)^2 + \omega_d^2]^2} + \frac{F_k s + C_k}{(s-\sigma)^2 + \omega_d^2} + \frac{f}{s},$$

where $Q(s)$ contains fractions corresponding to the roots of $N_0(s)$. In the case of real roots, it was found that when roots are not repeated, they introduce functions such as e^{pt}, where p is the corresponding root. When such a root is repeated j times, it was found that the functions introduced become $t^{j-1} e^{pt}$. In the case of complex conjugate and repeated roots, a similar situation occurs. In this case, any formal proof is not presented because of cumbersome notation, and only intuitive ideas are given to understand the reason for the resulting functions.

Using the inverse Laplace transform in (3.60), it was found that a complex conjugate nonrepeated root introduces the time function:

$$\mathcal{L}^{-1}\left\{ \frac{Bs + C}{(s-\sigma)^2 + \omega_d^2} \right\} = \beta e^{\sigma t} \sin(\omega_d t + \phi),$$

for some constants β and ϕ. Then, in the case of complex conjugated and repeated roots, $y(t)$ is given as:

$$y(t) = y_n(t) + y_f(t), \tag{3.91}$$

$$y_n(t) = q(t) + \beta_k t^{k-1} e^{\sigma t} \sin(\omega_d t + \phi_k) + \beta_{k-1} t^{k-2} e^{\sigma t} \sin(\omega_d t + \phi_{k-1})$$

$$+ \cdots + \beta_2 t e^{\sigma t} \sin(\omega_d t + \phi_2) + \beta_1 e^{\sigma t} \sin(\omega_d t + \phi_1),$$

$$y_f(t) = f,$$

where $q(t) = \mathcal{L}^{-1}\{Q(s)\}$. The solution in (3.91) has one of the following behaviors.

1. If the complex conjugate root has a positive real part $\sigma = -\zeta \omega_n > 0$, i.e., if $-1 < \zeta < 0$, it is easy to see that $y_n(t) \to \infty$ and $y(t) \to \infty$ as $t \to \infty$.
2. If the complex conjugate root has a negative real part $\sigma = -\zeta \omega_n < 0$, i.e., if $0 < \zeta < 1$, it is possible to proceed as in Sect. 3.4.2 to compute the following limit:

$$\lim_{t \to \infty} t^j e^{\sigma t} \sin(\omega_d t + \phi) = 0,$$

for any positive integer j and any real negative σ. Then, for complex conjugate and repeated roots with a negative real part:

$$\lim_{t \to \infty} \left(\beta_k t^{k-1} e^{\sigma t} \sin(\omega_d t + \phi_k) + \beta_{k-1} t^{k-2} e^{\sigma t} \sin(\omega_d t + \phi_{k-1}) \right.$$

$$\left. + \cdots + \beta_2 t e^{\sigma t} \sin(\omega_d t + \phi_2) + \beta_1 e^{\sigma t} \sin(\omega_d t + \phi_1) \right) = 0, \tag{3.92}$$

and $\lim_{t \to \infty} y(t) = q(t) + y_f(t)$.

3. If the root has a zero real part, i.e., $\sigma = 0$, then:

$$y(t) = y_n(t) + y_f(t),$$

$$y_n(t) = q(t) + \beta_k t^{k-1} \sin(\omega_d t + \phi_k) + \beta_{k-1} t^{k-2} \sin(\omega_d t + \phi_{k-1}) + \cdots$$
$$+ \beta_2 t \sin(\omega_d t + \phi_2) + \beta_1 \sin(\omega_d t + \phi_1).$$

$$y_f(t) = f.$$

Notice that, in this case:

$$\beta_k t^{k-1} \sin(\omega_d t + \phi_k) + \beta_{k-1} t^{k-2} \sin(\omega_d t + \phi_{k-1}) + \cdots$$
$$+ \beta_2 t \sin(\omega_d t + \phi_2) + \beta_1 \sin(\omega_d t + \phi_1) \rightarrow \infty, \qquad (3.93)$$

as time increases in the case when the root is repeated at least twice, i.e., if $k > 1$. But if this root is not repeated, i.e., if $k = 1$, then:

$$\beta_k t^{k-1} \sin(\omega_d t + \phi_k) + \beta_{k-1} t^{k-2} \sin(\omega_d t + \phi_{k-1}) + \cdots$$
$$+ \beta_2 t \sin(\omega_d t + \phi_2) + \beta_1 \sin(\omega_d t + \phi_1) = \beta_1 \sin(\omega_d t + \phi_1), \quad (3.94)$$

is an oscillatory function whose amplitude neither grows nor vanishes, i.e., although $y(t)$ does not grow without a limit (depending on $q(t)$) it will never reach $y_f(t)$ in a steady state.

3.4.5   Conclusions

In the previous sections it has been assumed that all of the initial conditions are zero, i.e., that:

$$-\frac{P(s)}{N(s)} = 0.$$

If some of the initial conditions are not zero, then the previous expression is not valid and the solutions $y(t)$ found in the previous sections have to be completed adding the function:

$$w(t) = \mathcal{L}^{-1}\left\{ -\frac{P(s)}{N(s)} \right\}.$$

As the denominator of:

$$W(s) = \mathcal{L}\{w(t)\} = -\frac{P(s)}{N(s)},$$

is still the characteristic polynomial $N(s)$, then, according to the partial fraction expansion method, $w(t)$ contains the same functions of time contained in the natural response $y_n(t)$ found in every case studied in the previous sections. This means that the complete natural response in every one of the cases studied in the previous sections must also include $w(t)$.

It is said that the differential equation in (3.76) or, equivalently, that transfer function $G(s) = \frac{B(s)}{N(s)}$, is:

- *Stable* if the natural response tends toward zero as time increases.
- *Unstable* if the natural response grows without a limit as time increases.
- *Marginally stable* if the natural response neither grows without limit nor tends toward zero as time increases.

Assume that all of the cases studied in the previous sections, i.e., regarding the roots of $N(s)$, appear simultaneously. Recall that the roots of $N(s)$ represent the poles of $G(s)$. The following can be concluded:

Conditions for the Stability of a Transfer Function

1. If all poles of $G(s)$ have negative real parts then $G(s)$ is stable.
2. If all poles of $G(s)$ have negative real parts, except for some nonrepeated poles having zero real parts then the transfer function $G(s)$ is marginally stable.
3. If at least one pole of $G(s)$ has a positive real part, then the transfer function $G(s)$ is unstable.
4. If there is at least one pole of $G(s)$ with a zero real part which is repeated at least twice, then $G(s)$ is unstable.

On the other hand, it has also been seen that $y_f(t)$ depends on the input $u(t)$ and that, in fact, both of them are represented by "similar" time functions if the characteristic polynomial has no root at $s = 0$, i.e., if $G(s)$ has no pole at $s = 0$. This means that in a control system, the input variable $u(t)$ can be used as the value it is desired that the output $y(t)$ reaches, i.e., $u(t)$ can be used to specify the desired value for $y(t)$. This is accomplished as follows. If the transfer function is stable, i.e., if $y_n(t) \to 0$, then $y(t) \to y_f(t)$; hence, the only thing that remains is to ensure that $y_f(t) = u$. The conditions to achieve this when $u = A$ is a constant are established in the following.

Conditions to Ensure $\lim_{t \to \infty} y(t) = A$

The complete response $y(t)$ reaches the desired output $u = A$ as time increases, i.e., $y_f(t) = A$ and $y_n(t) \to 0$, if $G(s)$ is stable and the coefficients of the independent terms of the polynomials $B(s)$ and $N(s)$ are equal, i.e., if $B(0) = N(0)$. This can be verified using the final value theorem (3.4):

$$\lim_{t \to \infty} y(t) = \lim_{s \to 0} s Y(s) = \lim_{s \to 0} s \frac{B(s)}{N(s)} \frac{A}{s} = \frac{B(0)}{N(0)} A = A. \tag{3.95}$$

Under these conditions, it is said that $G(s)$ is a transfer function with unitary gain in a steady state. When u is not a constant and it is desired that $\lim_{t \to \infty} y(t) = u(t)$,

$G(s)$ is also required to be stable, i.e., that $y_n(t) \rightarrow 0$, but some additional conditions must be satisfied. The determination of these conditions and how to satisfy them is the subject of the study presented in the subsequent chapters of this book (see Sect. 4.4).

Example 3.17 Consider the situation presented in Example 3.14. From this example, an experiment can be designed that allows us to know whether a system, a differential equation or a transfer function is stable or unstable:
  If the system is originally at "rest," apply a pulse disturbance to it and observe its evolution:

- *If the system is stable then it "moves" and after a while it comes back to the configuration where it was originally at "rest."*
- *If the system is unstable, then it "moves" more and more such that it goes far away from the configuration where it was originally at "rest."*

Example 3.18 Consider the mass-spring-damper system studied in Example 3.8, but now assume that the spring is not present. Replacing $K = 0$ in (3.73) yields:

$$m\ddot{x} + b\dot{x} = f.$$

Suppose that a force is applied according to:

$$f = k_p(x_d - x), \tag{3.96}$$

where k_p is a positive constant, x_d is a constant standing for the desired position, and x is the mass position. Combining these expressions yields:

$$m\ddot{x} + b\dot{x} = k_p(x_d - x),$$

and rearranging:

$$\ddot{x} + \frac{b}{m}\dot{x} + \frac{k_p}{m}x = \frac{k_p}{m}x_d.$$

Using the Laplace transform and assuming that all of the initial conditions are zero, the following is found:

$$G(s) = \frac{X(s)}{X_d(s)} = \frac{\frac{k_p}{m}}{s^2 + \frac{b}{m}s + \frac{k_p}{m}}.$$

As an exercise, the reader is to find the roots of the characteristic polynomial $s^2 + \frac{b}{m}s + \frac{k_p}{m}$ to corroborate that both roots have a negative real part if $\frac{b}{m} > 0$ and $\frac{k_p}{m} > 0$. Then, using (3.95), it is found that the position reached in a steady state:

$$\lim_{t \to \infty} x(t) = \lim_{s \to 0} sX(s) = \lim_{s \to 0} s \frac{\frac{k_p}{m}}{s^2 + \frac{b}{m}s + \frac{k_p}{m}} \frac{x_d}{s} = \frac{\frac{k_p}{m}}{\frac{k_p}{m}} x_d = x_d,$$

equals the desired position x_d. The reason for this result can be explained using, again, everyday experience: when $x = x_d$, the force produced according to (3.96) is $f = 0$; hence, the mass can stop and remain at that position. Moreover, if $x < x_d$, then $f > 0$ and the mass increases its position x approaching x_d (see Fig. 3.18 to recall that x increases in the same direction where f is positive). If $x > x_d$, then $f < 0$ and the mass position x decreases, approaching x_d again.

As an exercise, the reader is to verify that, in the case where a spring is present, i.e., when $K > 0$, then $\lim_{t \to \infty} x(t) \neq x_d$ if $x_d \neq 0$. Notice, however, that again this can be explained using everyday experience:

If $x = x_d$ then, according to (3.96), the external force applied to the mass is zero $f = 0$ and the friction force is also zero $-b\dot{x} = 0$, if it is assumed that the mass is at rest. However, if $x = x_d \neq 0$ then the force of the spring on the mass $-Kx$ is different from zero; hence, the mass will abandon the position $x = x_d \neq 0$. The mass will reach a position x such that the force of spring and the force f in (3.96) exactly cancel each other out, i.e., where $k_p(x_d - x) = Kx$.

3.5   Poles and Zeros in Higher-Order Systems

Systems of orders greater than two are known as higher-order systems. Contrary to first and second order systems, in higher-order systems it is not possible to perform a detailed graphical study of the solution $y(t)$. The main reason for this is the complexity of the expressions arising when the transfer function has three or more poles. Hence, it is important to approximate a higher-order system using a system with a smaller order. There are two ways of accomplishing this: i) The approximate cancellation of some poles with some zeros of the corresponding transfer function, and ii) Neglecting the effect of "fast" poles. Some examples are presented in the following.

3.5.1   Approximate Pole-Zero Cancellation and Reduced Order Models

Consider the following second-order system:

$$Y(s) = \frac{k(s - d)}{(s - p_1)(s - p_2)} U(s). \tag{3.97}$$

Suppose that $U(s) = A/s$, $p_1 \neq p_2$, $p_1 < 0$, $p_2 < 0$, $d < 0$, and $p_1 p_2 \approx -kd$ to render approximately the unitary gain in a steady state of the transfer function in (3.97). Using partial fraction expansion:

$$Y(s) = \frac{k(s - d)}{(s - p_1)(s - p_2)\,s} = \frac{A}{s} = \frac{B}{s - p_1} + \frac{C}{s - p_2} + \frac{D}{s}. \qquad (3.98)$$

Multiplying both sides by the factor $(s - p_1)$ and evaluating at $s = p_1$ yields:

$$B = \frac{k(s - d)A}{(s - p_2)s}\bigg|_{s=p_1} = \frac{k(p_1 - d)A}{(p_1 - p_2)p_1}.$$

Multiplying both sides of (3.98) by the factor $(s - p_2)$ and evaluating at $s = p_2$ yields:

$$C = \frac{k(s - d)A}{(s - p_1)s}\bigg|_{s=p_2} = \frac{k(p_2 - d)A}{(p_2 - p_1)p_2}.$$

Multiplying both sides of (3.98) by the factor s and evaluating at $s = 0$ yields:

$$D = \frac{k(s - d)A}{(s - p_1)(s - p_2)}\bigg|_{s=0} = -\frac{kd\,A}{p_1 p_2}.$$

If $p_1 \approx d < 0$ then, according to $p_1 p_2 \approx -kd$, we have $k \approx -p_2$ and, hence:

$$B \approx 0,$$

$$C \approx \frac{k(p_2 - d)A}{(p_2 - d)p_2} = \frac{kA}{p_2},$$

$$D \approx -\frac{kA}{p_2},$$

to finally obtain:

$$y(t) \approx \frac{kA}{p_2}e^{p_2 t} - \frac{kA}{p_2}.$$

The reader can verify that:

$$y(t) = \frac{kA}{p_2}e^{p_2 t} - \frac{kA}{p_2},$$

is the solution of:

$$Y(s) = \frac{k}{s - p_2}U(s), \qquad (3.99)$$

with $U(s) = A/s$ and $k = -p_2$. Thus, the following is concluded. If one pole and one zero of a transfer function are very close, then they approximately cancel each other out to obtain a reduced order transfer function. This means that (3.99) can be used instead of (3.97) to obtain very similar results. The advantages of using the model in (3.99) are: i) it is a model with a smaller order than (3.97), and ii) it has no zero. The advantage of using a reduced order model, which approximately describes a higher-order system is that the response of a reduced order system can often be handled as that of a first- or a second-order system. On the other hand, a zero in a transfer function modifies the time response in a manner that is not easy to quantify; hence, a transfer function that has no zeros is preferable.

It is important to say that the simplification obtained by the cancellation of one pole and one zero is useful only if the pole and the zero have negative real parts. This is because one pole with a positive real part, which is not exactly cancelled because of parameter uncertainty, has a dangerous effect that becomes larger as time increases.

3.5.2   Dominant Poles and Reduced Order Models

Consider the following transfer function:

$$Y(s) = \frac{k}{s + \frac{1}{RC}} \frac{d}{s + a} U(s), \tag{3.100}$$

where $U(s) = A/s$. Using partial fraction expansion, the following is found:

$$Y(s) = \frac{k}{s + \frac{1}{RC}} \frac{d}{s + a} \frac{A}{s} = \frac{B}{s + \frac{1}{RC}} + \frac{C}{s + a} + \frac{D}{s}. \tag{3.101}$$

Multiplying both sides by the factor $s + \frac{1}{RC}$ and evaluating at $s = -\frac{1}{RC}$:

$$B = k \frac{d}{s + a} \frac{A}{s} \bigg|_{s = -\frac{1}{RC}} = k \frac{d}{a - \frac{1}{RC}} \frac{A}{\left(-\frac{1}{RC}\right)}.$$

Multiplying both sides of (3.101) by the factor $s + a$ and evaluating at $s = -a$:

$$C = k \frac{d}{s + \frac{1}{RC}} \frac{A}{s} \bigg|_{s = -a} = k \frac{d}{-a + \frac{1}{RC}} \frac{A}{(-a)}.$$

Multiplying both sides of (3.101) by the factor s and evaluating at $s = 0$:

$$D = k \frac{d}{s + \frac{1}{RC}} \frac{A}{(s + a)} \bigg|_{s = 0} = k \frac{d}{\frac{1}{RC}} \frac{A}{a}.$$

If $a \gg \frac{1}{RC} > 0$, then:

$$B \approx -k \frac{d}{a} \frac{A}{\frac{1}{RC}},$$

$$C \approx k \frac{dA}{a^2} \ll D.$$

Hence, using (3.101) and neglecting C, the following is found:

$$y(t) \approx -k \frac{d}{a} \frac{A}{\frac{1}{RC}} e^{-\frac{1}{RC}t} + k \frac{d}{\frac{1}{RC}} \frac{A}{a}.$$

The reader can verify that:

$$y(t) = -k \frac{d}{a} \frac{A}{\frac{1}{RC}} e^{-\frac{1}{RC}t} + k \frac{d}{\frac{1}{RC}} \frac{A}{a},$$

is the solution of:

$$Y(s) = \frac{kd}{a(s + \frac{1}{RC})} U(s), \qquad (3.102)$$

with $U(s) = A/s$. Thus, (3.102) can be used instead of (3.100). The condition $a \gg \frac{1}{RC}$ is interpreted as "the pole at $s = -a$ is very fast compared with the pole at $s = -\frac{1}{RC}$". The pole at $s = -\frac{1}{RC}$ is known as the dominant pole because its effect is more important in the system response. The following is concluded. If one pole is much faster than the others, then a reduced order transfer function can be obtained by neglecting the fastest pole to only maintain the dominant (the slowest) poles. An accepted criterion is that the fastest poles (those to be neglected) have a real part that is five times the real parts of the dominant poles (those to be kept). Notice, however, that it is not just a matter of simply eliminating the factor $s + a$ at the denominator of the transfer function: the constant a still appears at the denominator of (3.102) because a is necessary to keep without change the steady-state gain of the transfer function in (3.100).

A simple way of obtaining (3.102) from (3.100) is the following:

$$Y(s) = \frac{k}{s + \frac{1}{RC}} \frac{d}{s + a} U(s) = \frac{k}{s + \frac{1}{RC}} \frac{d}{a(\frac{1}{a}s + 1)} U(s). \qquad (3.103)$$

If a is very large it can be assumed that $\frac{1}{a}s \ll 1$; hence:

$$Y(s) \approx \frac{kd}{a(s + \frac{1}{RC})} U(s),$$

which is the expression in (3.102).

Finally, it is important to say that order reduction, as presented above, is only valid if the neglected pole has a negative real part. If the neglected pole has a positive real part, then the contribution of this pole will grow to infinity as time increases, no matter how small C is. Thus, the contribution of such a pole cannot be neglected.

Example 3.19 According to Exercise 9 and Example 2.20 in Chap. 2, the mathematical model of a permanent magnet brushed DC motor is given as:

$$L_a \frac{di_a}{dt} = v - R_a i_a - k_e \omega,$$

$$J \frac{d\omega}{dt} = k_m i_a - B\omega, \tag{3.104}$$

where ω is the motor velocity (also see (10.9) and (10.10) in Chap. 10). Suppose that this motor actuates on a hydraulic mechanism, which, by centrifugal effect, produces a water flow, q_i, which is proportional to the motor velocity, i.e., $q_i = \gamma\omega$, where γ is a positive constant. Finally, this water flow enters the tank in Example 3.2 in the present chapter, whose mathematical model is given in (3.28), rewritten here for ease of reference:

$$\frac{dh}{dt} + ah = kq_i$$

$$a = \frac{1}{RC}, \quad k = \frac{1}{C}.$$

Using the Laplace transform (3.2) and assuming that all the initial conditions are zero, it is not difficult to verify that the corresponding equations become:

$$I(s) = \frac{1/L_a}{s + \frac{R_a}{L_a}}(V(s) - k_e\omega(s)), \tag{3.105}$$

$$\omega(s) = \frac{k_m/J}{s + \frac{B}{J}}I(s), \tag{3.106}$$

$$H(s) = \frac{1/C}{s + \frac{1}{RC}}Q_i(s). \tag{3.107}$$

Combining (3.105) and (3.106) yields:

$$\omega(s) = \frac{k_m/J}{s + \frac{B}{J}}\frac{1/L_a}{s + \frac{R_a}{L_a}}(V(s) - k_e\omega(s)).$$

According to (3.103), the following can be written:

$$\omega(s) = \frac{k_m/J}{\frac{B}{J}\left(\frac{J}{B}s + 1\right)}\frac{1/L_a}{\frac{R_a}{L_a}\left(\frac{L_a}{R_a}s + 1\right)}(V(s) - k_e\omega(s)).$$

In small motors, the inductance L_a and the viscous friction coefficient B are small; hence, $\frac{L_a}{R_a} \ll \frac{J}{B}$. Thus, it can be assumed that $\frac{L_a}{R_a}s \ll 1$ and the following can be written:

$$\omega(s) = \frac{1}{s + \frac{B}{J}} \frac{k_m}{J R_a}(V(s) - k_e\omega(s)).$$

Rearranging terms, this expression can be rewritten as:

$$\omega(s) = \frac{\frac{k_m}{JR}}{s + \left(\frac{B}{J} + \frac{k_m k_e}{J R_a}\right)} V(s).$$

Replacing this and $Q_i(s) = \gamma\omega(s)$ in (3.107) yields:

$$H(s) = \frac{\frac{1}{C}}{s + \frac{1}{RC}}\gamma\frac{\frac{k_m}{JR}}{s + \left(\frac{B}{J} + \frac{k_m k_e}{J R_a}\right)} V(s).$$

Proceeding again as in (3.103), the following can be written:

$$H(s) = \frac{\frac{\gamma}{C}}{\frac{1}{RC}(RCs + 1)} \frac{\frac{k_m}{JR}}{\left(\frac{B}{J} + \frac{k_m k_e}{J R_a}\right)\left(\frac{1}{\frac{B}{J} + \frac{k_m k_e}{J R_a}}s + 1\right)} V(s).$$

If the tank cross-section is large enough, the motor will reach its nominal velocity a long time before the water level in tank increases significantly. This can be stated by saying that "the tank time constant is very large compared with the motor time constant," i.e., $RC \gg \frac{1}{\frac{B}{J} + \frac{k_m k_e}{J R_a}}$. Then, it is possible to say that $\frac{1}{\frac{B}{J} + \frac{k_m k_e}{J R_a}}s \ll 1$; hence, the following approximation is valid:

$$H(s) = \frac{\frac{\gamma}{C}}{\frac{1}{RC}(RCs + 1)} \frac{\frac{k_m}{JR}}{\left(\frac{B}{J} + \frac{k_m k_e}{J R_a}\right)} V(s),$$

or:

$$H(s) = \frac{\frac{1}{C}}{s + \frac{1}{RC}} \frac{\frac{k_m}{JR}\gamma}{\left(\frac{B}{J} + \frac{k_m k_e}{J R_a}\right)} V(s).$$

Then, defining:

$$k_1 = \frac{\frac{k_m}{JR}\gamma}{\frac{B}{J} + \frac{k_m k_e}{J R_a}},$$

and comparing with (3.107), the expression in (3.139) is justified, i.e., that the water flow is proportional to the voltage applied to the motor through a constant k_1. This is possible if: 1) the motor electrical time constant is smaller than the motor mechanical time constant, i.e., if $\frac{L_a}{R_a} \ll \frac{J}{B}$, and 2) if the tank time constant is very large compared with the complete motor time constant, i.e., if $RC \gg \frac{1}{\frac{B}{J} + \frac{k_m k_e}{J R_a}}$.

Finally, it is important to say that this procedure is valid because $\frac{B}{J} + \frac{k_m k_e}{J R_a} > 0$ and $\frac{R_a}{L_a} > 0$, i.e., the neglected poles are negative.

3.5.3 Approximating the Transient Response of Higher-Order Systems

The simplicity of first- and second-order systems allows us to compute their exact solutions. However, in the case of systems represented by differential equations of an order greater than 2, also called higher-order systems, the complexity of the problem prohibits us from obtaining their exact solutions for control purposes. The traditional treatment of higher-order systems in classical control has been to approximate their response as a kind of extrapolation of responses obtained in first- and second-order systems.

According to the studies presented in Sects. 3.1 and 3.3, the poles of first- and second-order systems can be located on the s complex plane, as depicted in Fig. 3.22. Then, the distance of a pole to the origin, $s = 0$, is representative of the response time: the larger this distance, the shorter the response time. The angle between the imaginary axis and the location vector of the pole is representative of how damped the response is: the closer this angle is to zero, the more oscillatory the response; conversely, the closer this angle is to $90°$, the less oscillatory the response.

Fig. 3.22 Relative pole location on the s plane. $\zeta_1 < \zeta_2$ and $\omega_{n1} < \omega_{n2} < \omega_{n3}$

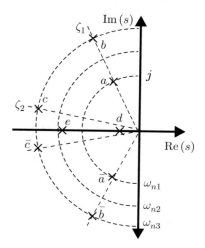

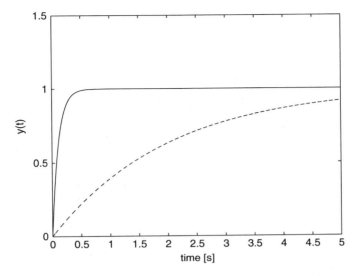

Fig. 3.23 Relative response of two systems with real poles (see Fig. 3.22). Dashed: pole at $s = d$. Continuous: pole at $s = e$

The responses of two first-order systems are compared in Fig. 3.23. The location of the respective real poles is depicted in Fig. 3.22. It is observed that a faster response if obtained, as the pole is located farther to the left of the origin.

The responses of three second-order systems are compared in Fig. 3.24. The location of the respective pair of complex conjugate poles is depicted in Fig. 3.22. It is observed that (i) a shorter response time is achieved as poles are located farther from the origin, and (ii) a less oscillatory response is achieved as the poles are located closer to the real axis.

The above results are applied to higher-order systems by considering the worst case, i.e., by observing the "slowest" pole (the closest pole to the origin) and the "least damped" pole (the closest pole to the imaginary axis).

3.6 The Case of Sinusoidal Excitations

Consider the n-order arbitrary transfer function:

$$\frac{Y(s)}{U(s)} = G(s) = \frac{E(s)}{N(s)}, \tag{3.108}$$

$$N(s) = s^n + a_{n-1}s^{n-1} + \cdots + a_1 s + a_0,$$

$$E(s) = b_0 + b_1 s + \cdots + b_m s^m,$$

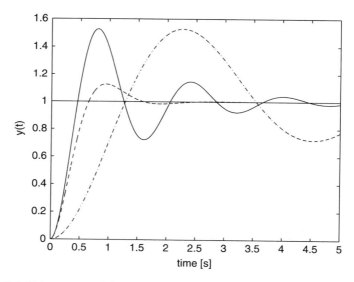

Fig. 3.24 Relative response of three systems with complex conjugate poles (see Fig. 3.22). Continuous: poles at $s = b$ and $s = \bar{b}$. Dash–dot: poles at $s = a$ and $s = \bar{a}$. Dashed: poles at $s = c$ and $s = \bar{c}$

where $n \geq m$ and $u(t) = A \sin(\omega t)$, i.e., $U(s) = A \frac{\omega}{s^2 + \omega^2}$. For the sake of simplicity, assume that $N(s)$ has no roots on the imaginary axis. Replacing $U(s)$ in (3.108) and using partial fraction expansion:

$$Y(s) = G(s) A \frac{\omega}{s^2 + \omega^2} = Y_n(s) + \frac{Cs + D}{s^2 + \omega^2}, \qquad (3.109)$$

where $Y_n(s) = \mathcal{L}\{y_n(t)\}$ is the natural response, i.e., those terms arising from the roots of $N(s)$. Recall that $\lim_{t \to \infty} y_n(t) = 0$ if $G(s)$ is stable. As $N(s)$ has been assumed not to have any roots on the imaginary axis, then $G(s)$ has no poles at $s = \pm j\omega$. Multiplying both sides of (3.109) by the factor $s^2 + \omega^2$ and evaluating at $s = j\omega$, it is found:

$$\omega A G(j\omega) = j\omega C + D.$$

Equating real and imaginary parts yields:

$$C = A \mathrm{Im}(G(j\omega)), \quad D = \omega A \mathrm{Re}(G(j\omega)),$$

where $G(j\omega) = \mathrm{Re}(G(j\omega)) + j\mathrm{Im}(G(j\omega))$. Hence:

$$\frac{Cs + D}{s^2 + \omega^2} = A \mathrm{Im}(G(j\omega)) \frac{s}{s^2 + \omega^2} + A \mathrm{Re}(G(j\omega)) \frac{\omega}{s^2 + \omega^2}.$$

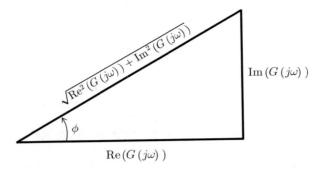

Fig. 3.25 Triangle defined by phase ϕ

Using the tabulated Laplace transforms:

$$\mathcal{L}\{\cos(\omega t)\} = \frac{s}{s^2 + \omega^2}, \quad \mathcal{L}\{\sin(\omega t)\} = \frac{\omega}{s^2 + \omega^2},$$

the following is found:

$$\mathcal{L}^{-1}\left\{\frac{Cs + D}{s^2 + \omega^2}\right\} = A[\mathrm{Im}(G(j\omega))\cos(\omega t) + \mathrm{Re}(G(j\omega))\sin(\omega t)].$$

Using Fig. 3.25 and the trigonometric identity, $\sin(\alpha)\cos(\beta) + \cos(\alpha)\sin(\beta) = \sin(\alpha + \beta)$, the following is obtained:

$$\mathcal{L}^{-1}\left\{\frac{Cs + D}{s^2 + \omega^2}\right\} = A|G(j\omega)|\ [\sin(\phi)\cos(\omega t) + \cos(\phi)\sin(\omega t)], \quad (3.110)$$

$$= B\ \sin(\omega t + \phi),$$

$$B = A|G(j\omega)|,$$

$$\phi = \arctan\left(\frac{\mathrm{Im}(G(j\omega))}{\mathrm{Re}(G(j\omega))}\right),$$

$$|G(j\omega)| = \sqrt{\mathrm{Re}^2(G(j\omega)) + \mathrm{Im}^2(G(j\omega))}.$$

Then, according to (3.109), the following is found:

$$y(t) = y_n(t) + y_f(t), \quad (3.111)$$

where the forced response is given as:

$$y_f(t) = B\ \sin(\omega t + \phi), \quad (3.112)$$

$$B = A|G(j\omega)|, \tag{3.113}$$

$$\phi = \arctan\left(\frac{\text{Im}(G(j\omega))}{\text{Re}(G(j\omega))}\right), \tag{3.114}$$

whereas $y_n(t) \to 0$ as $t \to \infty$ if $G(s)$ is stable. Hence, if the input $u(t)$ of an n−order linear system $G(s)$ is a sine function of time, then the output $y(t)$ is also, in a steady state, a sine function of time with the same frequency, but with an amplitude and a phase that are, in general, different from the amplitude and the phase of the signal at the input.

The ratio between the output and the input amplitudes at frequency ω, i.e., $\frac{B}{A} = |G(j\omega)|$, is known as the *magnitude of the transfer function* whereas the difference between the phase at the input and the phase at the output, ϕ, is known as the *phase of the transfer function*. Then, the following is concluded, which is an important result for the purposes of Chap. 6:

The frequency response is defined as the description of:

- How the ratio of the output amplitude and the input amplitude changes as the frequency of the signal at the input changes.
- How the difference between the phases of the signals at the input and the output changes as the frequency of the signal at the input changes.

Example 3.20 (Taken from [2], Ch. 4) Consider the spring-mass-damper system depicted in Fig. 3.18. Assume that there is no damper, i.e., $b = 0$, and the following external force $f = F_0 \sin(\omega t)$ is applied, where $\omega = \sqrt{\frac{K}{m}}$. It is desired to compute the mass position $x(t)$ if $x(0) = 0$ and $\dot{x}(0) = 0$. It is not necessary to compute the numerical value of constants appearing in $x(t)$.

Solution. The differential equation describing the situation in Fig. 3.18 when $b = 0$ is:

$$m\ddot{x} + Kx = f, \quad f = F_0 \sin(\omega t), \quad \omega = \sqrt{\frac{K}{m}}. \tag{3.115}$$

Applying the Laplace transform to (3.115) yields:

$$s^2 X(s) + \frac{K}{m} X(s) = \frac{1}{m} F(s).$$

Notice that this expression has the form:

$$s^2 X(s) + \omega_n^2 X(s) = \gamma \omega_n^2 F(s),$$

where $\omega_n = \sqrt{\frac{K}{m}} = \omega$, $\gamma = \frac{1}{K}$. As:

$$F(s) = \frac{F_0 \omega}{s^2 + \omega^2},$$

then:

$$X(s) = \frac{\gamma \omega^3 F_0}{(s^2 + \omega^2)^2}.$$

Using partial fraction expansion:

$$X(s) = \frac{As + B}{(s^2 + \omega^2)^2} + \frac{Cs + D}{s^2 + \omega^2}.$$

Thus, according to Sect. 3.4.4:

$$x(t) = \beta_1 t \sin(\omega t + \phi_1) + \beta_2 \sin(\omega t + \phi_2),$$

where $\beta_1, \beta_2, \phi_1, \phi_2$ are some real constants. The solution $x(t)$ is shown in Fig. 3.26 when $\omega = \sqrt{\frac{K}{m}} = 10[\text{rad/s}]$, $m = 1[\text{Kg}]$ and $F_0 = 1[\text{N}]$. This phenomenon is known as *resonance* and it means that the output reaches very large values despite the input being a bounded sine function of time. Notice that this may happen even if the characteristic polynomial has no roots with positive real parts nor located at $s = 0$. Thus, this is a different phenomenon to that described in Example 3.14. Also notice that resonance appears when a system is poorly damped ($\zeta \approx 0$) and the input is an oscillatory signal whose frequency is the same as (or very close to) the system natural frequency. As the mass position grows without a limit, resonance is a phenomenon that is considered to be dangerous.

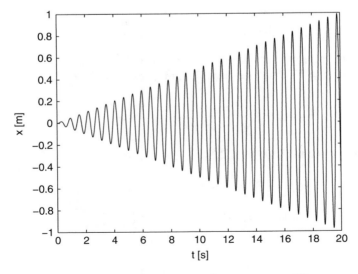

Fig. 3.26 Position in a mass-spring-damper system under resonance conditions

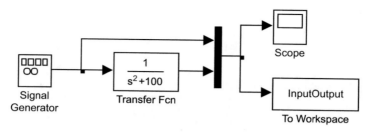

Fig. 3.27 MATLAB/Simulink simulation diagram for results in Fig. 3.26

Figure 3.26 has been obtained using the MATLAB/Simulink simulation diagram presented in Fig. 3.27. The *signal generator block* is programmed to have a sine wave form, unit amplitude and 10[rad/s] as frequency. Once the simulation stops, the following MATLAB code is executed in an m-file to draw Fig. 3.26:

```
nn=length(InputOutput(:,2));
n=nn-1;
Ts=20/n;
t=0:Ts:20;
plot(t,InputOutput(:,2),'k-');
axis([0 20 -1 1])
xlabel('t [s]')
ylabel('x [m]')
```

Example 3.21   Consider a linear system with transfer function:

$$\frac{Y(s)}{U(s)} = G(s) = \frac{10}{s+5}, \quad y(0) = 0.$$

Suppose that $u(t) = A\sin(5t + 90°)$, with $A = 1$. As $G(s)$ only has one pole at $s = -5$ and $\omega = 5$ in this case, according to (3.109), $Y_n(s) = \frac{h}{s+5}$ where:

$$h = (s+5)\frac{10}{s+5}\frac{\omega}{s^2+\omega^2}\Big|_{s=-5} = \frac{10 \times 5}{5^2 + 5^2} = 1.$$

Hence, according to (3.111), (3.112), (3.113), (3.114):

$$y(t) = e^{-5t} + B\,\sin(5t + 90° + \phi),$$

$$B = A|G(j\omega)|, \quad \phi = \arctan\left(\frac{\text{Im}(G(j\omega))}{\text{Re}(G(j\omega))}\right).$$

i.e.,:

$$G(j\omega) = \frac{10}{j\omega + 5}\left(\frac{-j\omega + 5}{-j\omega + 5}\right),$$

$$= \frac{10(-j\omega + 5)}{25 + \omega^2},$$

$$= \frac{10\sqrt{25 + \omega^2}}{\sqrt{25 + \omega^2}^2}\angle \arctan\left(\frac{-\omega}{5}\right),$$

$$|G(j\omega)| = \frac{10}{\sqrt{25 + \omega^2}}, \quad \phi = \arctan\left(\frac{-\omega}{5}\right).$$

Evaluating at the applied frequency, $\omega = 5$:

$$|G(j\omega)| = \frac{10}{\sqrt{25 + 25}} = \frac{2}{\sqrt{2}} = \sqrt{2}, \quad \phi = -45°.$$

Thus:

$$y(t) = e^{-5t} + \sqrt{2}\ \sin(5t + 45°).$$

This response is depicted in Fig. 3.28. Notice that the difference between $y(t)$ and $y_f(t)$ at $t = 0.2 = 1/5$[s] is equal to $1.38 - 1.015 = 0.3650 \approx e^{-5(0.2)}$, i.e., the natural response $y_n(t) = e^{-5t}$ evaluated at $t = 0.2$[s]. This fact corroborates the above results.

Figure 3.28 has been obtained using the MATLAB/Simulink diagram presented in Fig. 3.29. The *sine wave block* is programmed to be a time-based block, with unit amplitude, zero bias, 5[rad/s] as the frequency, and 1.57[rad], i.e., 90°, as the phase. The *sine wave 1 block* is programmed to be a time-based block, with 1.4142 as amplitude, zero bias, 5[rad/s] as the frequency, and 1.57/2[rad], i.e., 45°, as the phase. Once the simulation stops, the following MATLAB code is executed in an m-file to draw Fig. 3.28:

```
nn=length(InputOutput(:,1));
n=nn-1;
Ts=2/n;
t=0:Ts:2;
plot(t,InputOutput(:,1),'k-.',t,InputOutput(:,2),
'k-');
hold on
plot(t,InputOutput(:,3),'k--');
axis([0 2 -1.5 1.5])
xlabel('time [s]')
hold off
```

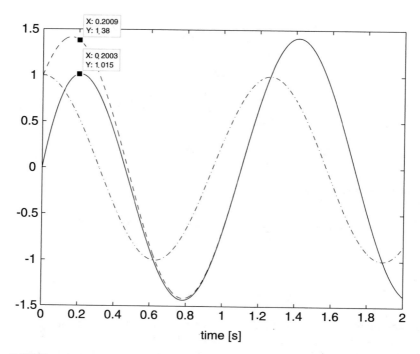

Fig. 3.28 Time response of a linear differential equation when excited with a sinusoidal function of time. Dash–dot: $u(t) = A \sin(5t+90°)$, $A = 1$. Dashed: $y_f(t) = \sqrt{2} \sin(5t+45°)$. Continuous: $y(t) = e^{-5t} + \sqrt{2} \sin(5t + 45°)$

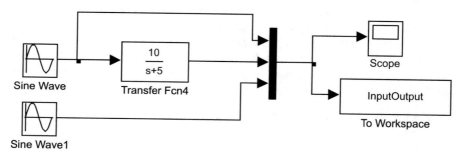

Fig. 3.29 MATLAB/Simulink diagram for the results in Fig. 3.28

3.7 The Superposition Principle

Every linear differential equation satisfies the superposition principle. Moreover, the fact that a differential equation satisfies the superposition principle is accepted as proof that the differential equation is linear. To simplify the exposition of ideas, the superposition principle is presented in the following only when all the initial conditions are zero.

Superposition Principle Consider the following n−order linear ordinary differential equation with constant coefficients:

$$y^{(n)} + a_{n-1}y^{(n-1)} + \cdots + a_1\dot{y} + a_0 y = b_0 u + b_1\dot{u} + \cdots + b_m u^{(m)}, \qquad (3.116)$$

where $n \geq m$. Assume that all of the initial conditions are zero. Also assume that $y_1(t)$ is the solution of (3.116) when $u_1(t)$ is applied at the input and $y_2(t)$ is the solution of (3.116) when $u_2(t)$ is applied at the input. Then, $\alpha_1 y_1(t) + \alpha_2 y_2(t)$, where α_1 and α_2 are arbitrary constants, is the solution of (3.116) when $\alpha_1 u_1(t) + \alpha_2 u_2(t)$ is applied at the input.

One way of verifying this result is presented in the following. As the differential equation in (3.116) is linear and all the initial conditions are zero, then it can be expressed in terms of the corresponding transfer function:

$$Y(s) = G(s)U(s). \qquad (3.117)$$

This means that it can be written as follows:

$$Y_1(s) = G(s)U_1(s),$$
$$Y_2(s) = G(s)U_2(s).$$

Adding these expressions yields:

$$\alpha_1 Y_1(s) + \alpha_2 Y_2(s) = \alpha_1 G(s)U_1(s) + \alpha_2 G(s)U_2(s),$$
$$= G(s)(\alpha_1 U_1(s) + \alpha_2 U_2(s)).$$

This is possible because α_1 and α_2 are constants. This expression corroborates that $\alpha_1 y_1(t) + \alpha_2 y_2(t)$ is the solution of (3.117); hence, the solution of (3.116) when $\alpha_1 u_1(t) + \alpha_2 u_2(t)$ is the input.

Example 3.22 Obtain the indicated voltage at impedance $Z_4(s)$ in Fig. 3.30a. This circuit can be simplified using the so-called *source transformation theorem*.

Theorem 3.1 (Source Transformation [5, 11], Chap. 9, pp. 61)

- When an impedance, $Z(s)$, and a voltage source, $V_{fv}(s)$, are series-connected between two terminals a and b, they can be replaced by a current source, $I_{fc}(s)$, parallel-connected to the same impedance, $Z(s)$. The magnitude of the current source is given as $I_{fc}(s) = V_{fv}(s)/Z(s)$.
- When an impedance, $Z(s)$, and a current source, $I_{fc}(s)$, are parallel-connected between two terminals a and b, they can be replaced by a voltage source, $V_{fv}(s)$, series-connected to the same impedance, $Z(s)$. The magnitude of the voltage source is given as $V_{fv}(s) = Z(s)I_{fc}(s)$.

Applying the first part of this theorem to voltage sources $V_1(s)$ and $V_2(s)$, circuit in Fig. 3.30b is obtained where:

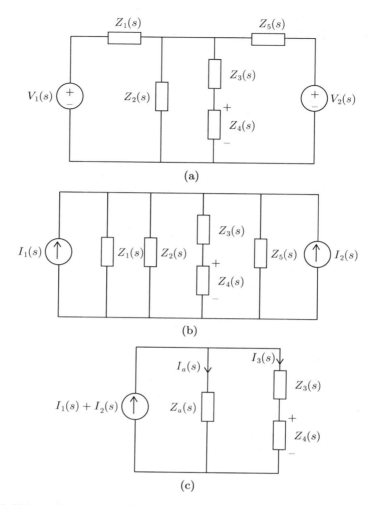

Fig. 3.30 Electric circuit studied in Example 3.22

$$I_1(s) = \frac{V_1(s)}{Z_1(s)}, \quad I_2(s) = \frac{V_2(s)}{Z_5(s)}. \tag{3.118}$$

It is clear that $Z_1(s)$, $Z_2(s)$ and $Z_5(s)$ are parallel-connected and its equivalent impedance is given as (see (2.69)):

$$Z_a(s) = \frac{1}{\frac{1}{Z_1(s)} + \frac{1}{Z_2(s)} + \frac{1}{Z_5(s)}}.$$

Using this and combining the current sources, the circuit in Fig. 3.30c is obtained. Now, another important result in network analysis is employed: the current divider.

Fact 3.1 (Current Divider [5], pp. 38) *When two parallel-connected impedances are parallel-connected to a current source, the current flowing through any of the impedances is given by the value of the current source multiplied by the opposite impedance to impedance where the current is to be computed and divided by the addition of the two impedances.*

Applying this result in addition to Ohm's Law to the circuit in Fig. 3.30c, the following expressions are found:

$$I_3(s) = \frac{Z_a(s)}{Z_a(s) + Z_3(s) + Z_4(s)}(I_1(s) + I_2(s)),$$

$$V_{z4}(s) = Z_4(s)I_3(s),$$

where $V_{z4}(s)$ is voltage at impedance $Z_4(s)$. Combining these expressions yields:

$$V_{z4}(s) = \frac{Z_a(s)Z_4(s)}{Z_a(s) + Z_3(s) + Z_4(s)}(I_1(s) + I_2(s)).$$

Finally, using (3.118), the following is found:

$$V_{z4}(s) = \frac{Z_a(s)Z_4(s)}{Z_a(s) + Z_3(s) + Z_4(s)}\left(\frac{V_1(s)}{Z_1(s)} + \frac{V_2(s)}{Z_5(s)}\right).$$

It is concluded that the voltage at $Z_4(s)$ can be obtained, as the addition of the voltages at $Z_4(s)$ due to each one of the voltage sources, i.e., $V_1(s)$ or $V_2(s)$, when the other source is put in short circuit (i.e., a zero value is assigned) and adding both results. This is exactly what the superposition principle establishes. It is interesting to say that the superposition principle has been established above in the present section, assuming that the different applied inputs are directly added, and, after that, they are applied to the system. However, this example shows that the superposition principle is valid in linear systems, even when the voltage sources, $V_1(s)$ and $V_2(s)$, cannot be added directly (see Fig. 3.30a).

Example 3.2 Computing of the voltage at impedance $Z_3(s)$ in the circuit shown in Fig. 3.31a is required. First, this circuit is simplified to obtain circuit in Fig. 3.31b. Then, the source transformation theorem can be employed (see the previous example) to obtain the circuit in Fig. 3.31c, where:

$$I_1(s) = \frac{V_1(s)}{Z_1(s)}. \tag{3.119}$$

As impedances $Z_1(s)$, $Z_2(s)$, and $Z_4(s)$ are parallel-connected, the equivalent impedance is given as (see (2.69)):

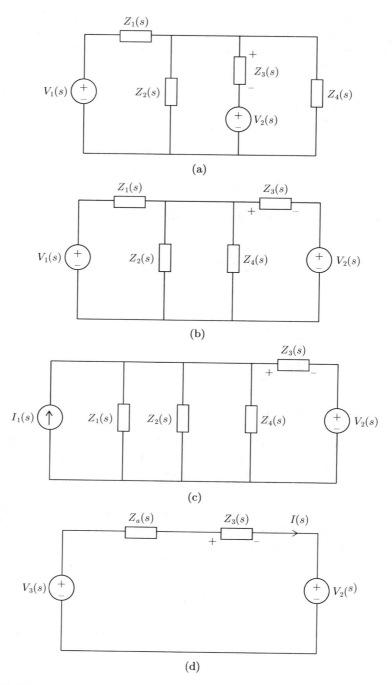

Fig. 3.31 Electric circuit studied in Example 3.2

$$Z_a(s) = \frac{1}{\frac{1}{Z_1(s)} + \frac{1}{Z_2(s)} + \frac{1}{Z_4(s)}}.$$

The equivalent impedance $Z_a(s)$ is parallel connected to the current source $I_1(s)$; hence, the second part of the source transformation theorem can be used (see the previous example) to obtain the circuit in Fig. 3.31d, where:

$$V_3(s) = Z_a(s)I_1(s),$$

$$I(s) = \frac{V_3(s) - V_2(s)}{Z_a(s) + Z_3(s)},$$

$$V_{z3}(s) = Z_3(s)I(s),$$

with $V_{z3}(s)$ voltage at the impedance $Z_3(s)$. Combining these expressions and using (3.119) yields:

$$V_{z3}(s) = \frac{Z_3(s)}{Z_a(s) + Z_3(s)}(V_3(s) - V_2(s)),$$

$$= \frac{Z_3(s)}{Z_a(s) + Z_3(s)}(Z_a(s)I_1(s) - V_2(s)),$$

$$= \frac{Z_3(s)}{Z_a(s) + Z_3(s)}\left(\frac{Z_a(s)}{Z_1(s)}V_1(s) - V_2(s)\right).$$

It is concluded, again, that the voltage at the impedance $Z_3(s)$ can be obtained as the addition of the voltages in $Z_3(s)$ because of each one of the voltage sources, i.e., $V_1(s)$ or $V_2(s)$, when the other source is put in a short circuit (i.e., its value is set to zero) and adding the two results. This example also shows that the superposition principle is valid in linear systems, even when the voltage sources $V_1(s)$ y $V_2(s)$ cannot be added directly (see Fig. 3.31a).

3.8 Controlling First- and Second-Order Systems

The simplicity of the first- and second-order systems allows us to design feedback controllers whose effects can be explained relying on the exact solution of the closed-loop system. This is shown in the present section by applying this methodology to some physical systems.

3.8.1   Proportional Control of Velocity in a DC Motor

According to Chap. 10, the mathematical model of a permanent magnet brushed DC motor is given in (10.9), (10.10), which are repeated here for ease of reference:

$$L \frac{di}{dt} = u - R\,i - n\,k_e\,\omega,$$

$$J\dot{\omega} = -b\,\omega + n\,k_m\,i - T_p. \tag{3.120}$$

The reader is encouraged to see Chap. 10 for an explanation of variables in this model. For the purposes of this example, it suffices to say that ω, u, i, T_p represent the load angular velocity, the applied voltage, the electric current through the armature circuit, and an external torque disturbance respectively, and all of the parameters are positive. If the motor is small, then it is widely accepted that $L \approx 0$ can be assumed, and the following is obtained:

$$0 = u - R\,i - n\,k_e\,\omega \quad \Rightarrow \quad i = \frac{u - n\,k_e\,\omega}{R}. \tag{3.121}$$

Replacing the electric current in (3.120) and rearranging:

$$J\dot{\omega} + \left(b + \frac{n^2\,k_m\,k_e}{R}\right)\omega = \frac{n\,k_m}{R}u - T_p. \tag{3.122}$$

Consider the following proportional velocity controller:

$$u = k_p(\omega_d - \omega), \tag{3.123}$$

where k_p is a constant known as the proportional gain, and ω_d represents the desired velocity given as a step signal of magnitude A, i.e., $\omega_d(s) = \frac{A}{s}$. The closed-loop system is obtained by replacing u, given by the proportional controller, in the previous equation, i.e.,:

$$J\dot{\omega} + \left(b + \frac{n^2\,k_m\,k_e}{R}\right)\omega = \frac{n\,k_m}{R}k_p(\omega_d - \omega) - T_p. \tag{3.124}$$

Rearranging terms yields:

$$\dot{\omega} + \left(\frac{b}{J} + \frac{n^2\,k_m\,k_e}{JR} + \frac{n\,k_m k_p}{JR}\right)\omega = \frac{n\,k_m k_p}{JR}\omega_d - \frac{1}{J}T_p.$$

This is a first-order system with two inputs. Then, we can apply superposition, i.e., define two signals ω_1 and ω_2 such that:

$$\omega(t) = \omega_1(t) + \omega_2(t),$$

$$\dot{\omega}_1 + \left(\frac{b}{J} + \frac{n^2 \, k_m \, k_e}{JR} + \frac{n \, k_m k_p}{JR}\right) \omega_1 = \frac{n \, k_m k_p}{JR} \omega_d,$$

$$\dot{\omega}_2 + \left(\frac{b}{J} + \frac{n^2 \, k_m \, k_e}{JR} + \frac{n \, k_m k_p}{JR}\right) \omega_2 = -\frac{1}{J} T_p.$$

First, analyze the equation for ω_1, applying the Laplace transform with zero initial conditions:

$$\omega_1(s) = G_1(s)\omega_d(s).$$

$$G_1(s) = \frac{\frac{n \, k_m k_p}{JR}}{s + \left(\frac{b}{J} + \frac{n^2 \, k_m \, k_e}{JR} + \frac{n \, k_m k_p}{JR}\right)}. \tag{3.125}$$

Notice that this transfer function is identical to:

$$\frac{b_2}{s + b_1}, \tag{3.126}$$

if we define:

$$b_1 = \frac{b}{J} + \frac{n^2 \, k_m \, k_e}{JR} + \frac{n \, k_m k_p}{JR} \neq 0, \quad b_2 = \frac{n \, k_m k_p}{JR} \neq 0.$$

Notice that this expression has the same structure as the transfer function in (3.23). Hence, if the parameters and variables b_1, b_2, ω_1 and ω_d are suitably associated with those in (3.23), the solution $\omega_1(t)$ is the same as in (3.16), i.e.,:

$$\omega_1(t) = -\frac{b_2 A}{b_1} e^{-b_1 t} + \frac{b_2 A}{b_1} + \omega_1(0) e^{-b_1 t},$$

Stability is ensured if, and only if, the only pole of $G_1(s)$, located at $s = -b_1$, is negative, i.e., $-b_1 < 0$, which implies that:

$$k_p > -\frac{JR}{n \, k_m}\left(\frac{b}{J} + \frac{n^2 \, k_m \, k_e}{JR}\right). \tag{3.127}$$

If this is the case, then the final value of the velocity can be computed using the final value theorem as:

$$\lim_{t \to \infty} \omega_1(t) = \lim_{s \to 0} s\omega_1(s),$$

$$= \lim_{s \to 0} s \frac{\frac{n \, k_m k_p}{JR}}{s + \left(\frac{b}{J} + \frac{n^2 \, k_m \, k_e}{JR} + \frac{n \, k_m k_p}{JR}\right)} \frac{A}{s} = \frac{b_2 A}{b_1}.$$

Recall that A represents the magnitude of the desired velocity. As it is desirable for the final value of the velocity to be close to A, it also desirable for $\frac{b_2 A}{b_1}$ to have the same sign as A, i.e., $b_2/b_1 > 0$. As $b_1 > 0$, because of the stability requirement, then $b_2 > 0$ must also be true. This is ensured if $k_p > 0$, which is in complete agreement with (3.127). Hence, we can conclude that:

$$\lim_{t \to \infty} \omega_1(t) = \frac{b_2 A}{b_1} = \frac{\frac{n \, k_m k_p}{JR}}{\frac{b}{J} + \frac{n^2 \, k_m \, k_e}{JR} + \frac{n \, k_m k_p}{JR}} A < A, \qquad (3.128)$$

which means that the final velocity remains below the desired velocity. This result may also be explained using everyday experience as follows. Assume for a moment that $\lim_{t \to \infty} \omega_1(t) = A$. Then, according to (3.123):

$$u = k_p(A - \omega) = 0,$$

the voltage applied at the motor terminals is zero. Hence, the motor tends to stop moving, i.e., ω_1 decreases to zero and cannot stay at the desired value A. This means that $\omega_1 < A$ will be obtained again. Hence, it is not possible to keep $\omega_1 = A$ forever, which justifies the result in (3.128).

The difference between the final velocity and A is known as the steady-state error e_{ss}. We remark that, according to (3.128), the steady-state error can be arbitrarily reduced by increasing $k_p > 0$, although a zero steady-state error will never be accomplished. On the other hand, if a specific steady-state error e_{ss} is required, the corresponding value for k_p can be computed as:

$$e_{ss} = A - \frac{\frac{n \, k_m k_p}{JR}}{\frac{b}{J} + \frac{n^2 \, k_m \, k_e}{JR} + \frac{n \, k_m k_p}{JR}} A,$$

$$= \frac{\frac{b}{J} + \frac{n^2 \, k_m \, k_e}{JR}}{\frac{b}{J} + \frac{n^2 \, k_m \, k_e}{JR} + \frac{n \, k_m k_p}{JR}} A,$$

$$k_p = \left[\left(\frac{b}{J} + \frac{n^2 \, k_m \, k_e}{JR} \right) \frac{A}{e_{ss}} - \left(\frac{b}{J} + \frac{n^2 \, k_m \, k_e}{JR} \right) \right] \frac{JR}{n \, k_m}.$$

Notice that to ensure $k_p > 0$, the steady-state error is required to be a fraction of the desired velocity, i.e., $e_{ss} = \rho A$ with $0 < \rho < 1$.

Now, consider the expression for ω_2. Using the Laplace transform:

$$\omega_2(s) = G_2(s) T_p(s),$$

$$G_2(s) = \frac{-\frac{1}{J}}{s + \left(\frac{b}{J} + \frac{n^2 \, k_m \, k_e}{JR} + \frac{n \, k_m k_p}{JR} \right)}.$$

Notice that $G_2(s)$ can be written as $G_2(s) = \frac{b_3}{s+b_1}$, with $b_3 = -\frac{1}{J}$, i.e., a transfer function such as that in (3.126). Thus, we may proceed as before to conclude that:

$$\lim_{t \to \infty} \omega_2(t) = \frac{-\frac{1}{J}}{\frac{b}{J} + \frac{n^2 k_m k_e}{JR} + \frac{n k_m k_p}{JR}} d, \quad T_p(s) = \frac{d}{s}.$$

The important observation in this expression is that the final value of ω_2 can be reduced by increasing $k_p > 0$. This is good news, as ω_2 is the velocity deviation due to external disturbance T_p; hence, it is desirable to keep ω_2 close or equal to zero, if possible. In this case, the deviation can be rendered arbitrarily small by increasing $k_p > 0$, but a zero deviation will never be achieved with a proportional controller.

As the problem at hand is a first-order system, the transient response is completely described by the time constant:

$$\tau = \frac{1}{b_1} = \frac{1}{\frac{b}{J} + \frac{n^2 k_m k_e}{JR} + \frac{n k_m k_p}{JR}}.$$

Recall that a faster response is obtained as τ is shorter. This means that a faster response is obtained as $k_p > 0$ is chosen to be larger. Moreover, if all the system parameters are known and a desired time constant T is specified, then the required proportional gain can be exactly computed as:

$$k_p = \frac{JR}{n k_m} \left[\frac{1}{\tau} - \left(\frac{b}{J} + \frac{n^2 k_m k_e}{JR} \right) \right].$$

This means that, to ensure $k_p > 0$, the closed-loop system is required to be faster than the open-loop system, i.e.:

$$\frac{1}{\tau} - \left(\frac{b}{J} + \frac{n^2 k_m k_e}{JR} \right) > 0.$$

The velocity responses of a DC motor, when using two different proportional gains $k_p = 12$ and $k_p = 40$, are shown in Fig. 3.32. This was performed using the MATLAB/Simulink diagram shown in Fig. 3.33 to represent the closed-loop system in (3.124). It is assumed that $J = 1$, $nk_m/R = 1$ and $b + \frac{n^2 k_m k_e}{R} = 7$. These values were selected merely to obtain representative, clearly appreciable results for the proportional control of velocity. Block ω_d is a step applied at $t = 0$ with a zero initial value and 2 as a final value. The external disturbance T_p is a step applied at $t = 0.5$ with a zero initial value and 7 as a final value. Notice that the use of a larger k_p results in a faster response, i.e., $T_2 < T_1$, a smaller steady-state error, i.e., a larger steady-state response β, and a smaller deviation δ because of a constant external torque disturbance.

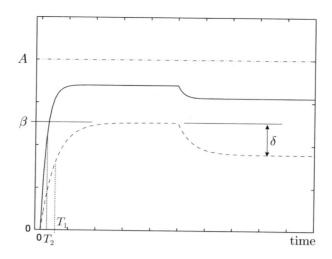

Fig. 3.32 Velocity response in the proportional control of velocity. Continuous: $k_p = 40$ is used. Dashed: $k_p = 12$ is used. Dash–dot: desired velocity

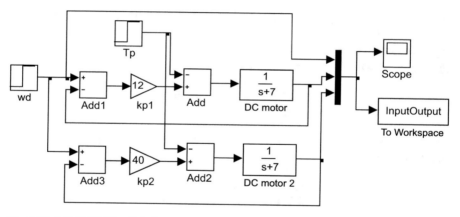

Fig. 3.33 MATLAB/Simulink diagram for the results in Fig. 3.32

The following is MATLAB code that is executed in an m-file to draw Fig. 3.32 after simulation in Fig. 3.33 stops:

```
nn=length(InputOutput(:,1));
n=nn-1;
Ts=1/n;
t=0:Ts:1;
plot(t,InputOutput(:,1),'k-.',t,InputOutput(:,3),
'k-');
hold on
```

```
plot(t,InputOutput(:,2),'k--');
axis([-0.02 1 0 2.5])
xlabel('time [s]')
hold off
```

3.8.2 Proportional Position Control Plus Velocity Feedback for a DC Motor

Consider the motor model in (3.122), but, now, expressed in terms of position θ, i.e., $\dot\theta = \omega$:

$$J\ddot\theta + \left(b + \frac{n^2 k_m k_e}{R}\right)\dot\theta = \frac{n k_m}{R}u - T_p. \tag{3.129}$$

Assume that the voltage applied at the motor terminals is computed according to the following expression:

$$u = k_p(\theta_d - \theta) - k_v\dot\theta,$$

where k_v is a constant known as the velocity gain, and θ_d represents the desired position, which is assumed to be given as a step signal with magnitude A, i.e., $\theta_d(s) = \frac{A}{s}$. This represents a controller known as *proportional position control plus velocity feedback*. The closed-loop equation is found by replacing u in (3.129) and rearranging terms, i.e.,:

$$J\ddot\theta + \left(b + \frac{n^2 k_m k_e}{R}\right)\dot\theta = \frac{n k_m}{R}\left(k_p(\theta_d - \theta) - k_v\dot\theta\right) - T_p, \tag{3.130}$$

$$\ddot\theta + \left(\frac{b}{J} + \frac{n^2 k_m k_e}{JR} + \frac{n k_m k_v}{JR}\right)\dot\theta + \frac{n k_m k_p}{JR}\theta = \frac{n k_m k_p}{JR}\theta_d - \frac{1}{J}T_p,$$

Using superposition and Laplace transform yields:

$$\theta(s) = \theta_1(s) + \theta_2(s),$$

$$\theta_1(s) = G_1(s)\theta_d(s), \quad \theta_2(s) = G_2(s)T_p(s),$$

$$G_1(s) = \frac{\frac{n k_m k_p}{JR}}{s^2 + \left(\frac{b}{J} + \frac{n^2 k_m k_e}{JR} + \frac{n k_m k_v}{JR}\right)s + \frac{n k_m k_p}{JR}}$$

$$G_2(s) = \frac{-\frac{1}{J}}{s^2 + \left(\frac{b}{J} + \frac{n^2 k_m k_e}{JR} + \frac{n k_m k_v}{JR}\right)s + \frac{n k_m k_p}{JR}}.$$

First, analyze the equation for θ_1. By equating:

$$s^2 + \left(\frac{b}{J} + \frac{n^2 k_m k_e}{JR} + \frac{n k_m k_v}{JR}\right) s + \frac{n k_m k_p}{JR} = s^2 + 2\zeta\omega_n s + \omega_n^2,$$

it is possible to write:

$$G_1(s) = \frac{\omega_n^2}{s^2 + 2\zeta\omega_n s + \omega_n^2}, \tag{3.131}$$

$$\omega_n = \sqrt{\frac{n k_m k_p}{JR}} \quad \zeta = \left(\frac{b}{J} + \frac{n^2 k_m k_e}{JR} + \frac{n k_m k_v}{JR}\right) \frac{1}{2\sqrt{\frac{n k_m k_p}{JR}}}.$$

Notice that, under these conditions, the inverse Laplace transform of $\theta_1(s) = G_1(s)\theta_d(s)$, is identical to the function presented in (3.57). Hence, stability is ensured if $k_p > 0$, to ensure that both ω_n and ζ are real numbers, and:

$$\frac{b}{J} + \frac{n^2 k_m k_e}{JR} + \frac{n k_m k_v}{JR} > 0,$$

to ensure that $\zeta > 0$. If this is the case, then we can use the final value theorem to compute the final position as:

$$\lim_{t\to\infty} \theta_1(t) = \lim_{s\to 0} s\theta_1(s),$$

$$= \lim_{s\to 0} s \frac{\omega_n^2}{s^2 + 2\zeta\omega_n s + \omega_n^2} \frac{A}{s} = A.$$

Thus, the motor position reaches the constant desired position in a steady state. On the other hand, according to (3.131), we conclude that:

- The closed-loop system is faster as $k_p > 0$ is chosen to be larger. This is because ω_n increases as $k_p > 0$ increases. However, as $k_p > 0$ increases the system becomes more oscillatory. This is because ζ decreases as k_p increases.
- The system response is less oscillatory as $k_v > 0$ is chosen to be larger. This is because ζ increases as k_v increases.

This means that the response can be rendered as fast and damped as desired merely by suitably selecting both k_p and k_v.

Now, consider the equation for ω_2. Notice that this variable represents the position deviation produced by an external torque disturbance. Hence, it is desirable for ω_2 to be close or equal to zero, if possible. Notice that $G_2(s)$ is stable if is $G_1(s)$ stable, because both transfer functions have the same characteristic polynomial. Thus, we can compute the final value of ω_2 using the final value theorem when the disturbance is a step signal, i.e., $T_p(s) = \frac{d}{s}$:

$$\lim_{t \to \infty} \omega_2(t) = \lim_{s \to 0} s\omega_2(s),$$

$$= \lim_{s \to 0} s \frac{-\frac{1}{J}}{s^2 + \left(\frac{b}{J} + \frac{n^2 k_m k_e}{JR} + \frac{n k_m k_v}{JR}\right) s + \frac{n k_m k_p}{JR}} \frac{d}{s},$$

$$= \frac{-\frac{1}{J}}{\frac{n k_m k_p}{JR}} d.$$

This means that the steady-state deviation is not zero. However, the good news is that this deviation can be arbitrarily reduced by increasing k_p. We conclude that a faster closed-loop system also results in smaller position deviations because of constant external torque disturbances. This is corroborated by simulations shown in Fig. 3.34, which were performed using the MATLAB/Simulink diagram in Fig. 3.35. There, the closed-loop system in (3.130) is simulated using $J = 1$, $b + \frac{n^2 k_m k_e}{R} = 7$, and $nk_m/R = 70$. These values were employed merely because they allow representative, clearly appreciable, results to be obtained. The block *theta_d* is a step applied at $t = 0$ with a zero initial value and 2 as a final value. The external disturbance T_p is a step applied at $t = 0.5[s]$ with a zero initial value and 7×70 as a final value. Once simulation in Fig. 3.35 stops, the following MATLAB code is executed using an m-file to draw Fig. 3.34:

```
nn=length(InputOutput(:,1));
n=nn-1;
Ts=1/n;
```

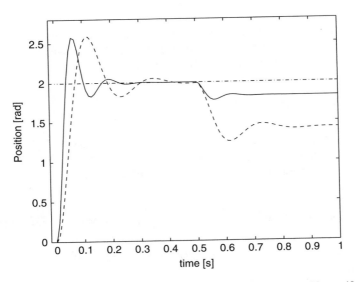

Fig. 3.34 Proportional control of the position with velocity feedback. Use of $k_p = 40$ and $k_v = 0.45$ (continuous) results in a faster response, but the same damping, than using $k_p = 12$ and $k_v = 0.2$ (dashed). See (3.132)

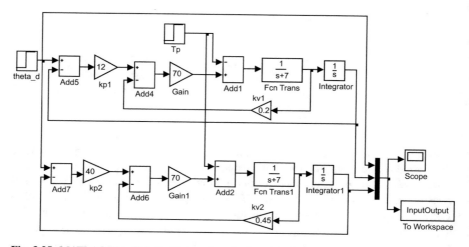

Fig. 3.35 MATLAB/Simulink diagram used to obtain results in Fig. 3.34

```
t=0:Ts:1;
plot(t,InputOutput(:,1),'k-.',t,InputOutput(:,3),
'k-');
hold on
plot(t,InputOutput(:,2),'k--');
axis([-0.02 1 0 2.75])
xlabel('time [s]')
ylabel('Position [rad]')
hold off
```

Notice that in the case where some t_r and M_p are specified, the use of expressions in (3.71) allow the required ω_n and ζ to be computed. These data and (3.131) allow the corresponding controller gains to be computed as:

$$k_p = \frac{J R}{n \, k_m} \omega_n^2, \quad k_v = \left[2\zeta\omega_n - \left(\frac{b}{J} + \frac{n^2 \, k_m \, k_e}{J R}\right)\right] \frac{J R}{n \, k_m}. \quad (3.132)$$

Using this procedure together with $t_r = 0.039$[s], $M_p = 29.9\%$, for the continuous line in Fig. 3.34, results in $k_p = 40$ and $k_v = 0.45$, whereas $t_r = 0.071$[s], $M_p = 29.9\%$, for the dashed line, results in $k_p = 12$ and $k_v = 0.2$.

3.8.3 Proportional–Derivative Position Control of a DC Motor

Consider the motor model in (3.129) in a closed loop with the following proportional–derivative (PD) controller:

$$u = k_p(\theta_d - \theta) + k_d \frac{d}{dt}(\theta_d - \theta),$$

i.e.,:

$$\ddot{\theta} + \left(\frac{b}{J} + \frac{n^2\,k_m\,k_e}{JR} + \frac{n\,k_m\,k_d}{JR}\right)\dot{\theta} + \frac{n\,k_m\,k_p}{JR}\theta = \frac{n\,k_m\,k_p}{JR}\theta_d + \frac{n\,k_m\,k_d}{JR}\dot{\theta}_d - \frac{1}{J}T_p,$$

where k_d is known as the derivative gain. Using superposition and the Laplace transform yields:

$$\theta(s) = \theta_1(s) + \theta_2(s),$$

$$\theta_1(s) = G_1(s)\theta_d(s), \quad \theta_2(s) = G_2(s)T_p(s),$$

$$G_1(s) = \frac{\frac{n\,k_m\,k_d}{JR}s + \frac{n\,k_m\,k_p}{JR}}{s^2 + \left(\frac{b}{J} + \frac{n^2\,k_m\,k_e}{JR} + \frac{n\,k_m\,k_d}{JR}\right)s + \frac{n\,k_m\,k_p}{JR}}$$

$$G_2(s) = \frac{-\frac{1}{J}}{s^2 + \left(\frac{b}{J} + \frac{n^2\,k_m\,k_e}{JR} + \frac{n\,k_m\,k_d}{JR}\right)s + \frac{n\,k_m\,k_p}{JR}}.$$

Notice that except for the term $\frac{n\,k_m\,k_d}{JR}s$ in the numerator of $G_1(s)$ and the use of k_d instead of k_v in the denominator, we have exactly retrieved the same closed-loop equations as in the previous example. Thus, we can proceed exactly as in the previous example to conclude:

- Expressions in (3.131) stand again with k_d instead of k_v.
- Stability is ensured if $k_p > 0$ and:

$$\frac{b}{J} + \frac{n^2\,k_m\,k_e}{JR} + \frac{n\,k_m\,k_d}{JR} > 0.$$

- If any external disturbance is not present, then the position reaches the constant desired position in a steady state.
- The closed-loop system is faster as $k_p > 0$ is chosen to be larger. This is because ω_n increases as $k_p > 0$ increases. However, as $k_p > 0$ increases, the system becomes more oscillatory. This is because ζ decreases as k_p increases.
- The system response is less oscillatory as $k_d > 0$ becomes larger. This is because ζ increases as k_d increases.
- The response can be rendered as fast and damped as desired merely by suitably selecting both k_p and k_d.
- If a step external torque disturbance $T_p(s) = \frac{d}{s}$ is present, then a position deviation exists in a steady state and it is given as:

$$\lim_{t\to\infty} \omega_2(t) = \frac{-\frac{1}{J}}{\frac{n\,k_m\,k_p}{JR}}d.$$

On the other hand, to study the transient response notice that the transfer function $G_1(s)$ can be rewritten as:

$$G_1(s) = \frac{\frac{k_d}{k_p}\omega_n^2 s + \omega_n^2}{s^2 + 2\zeta\omega_n s + \omega_n^2},$$

$$= \frac{k_d}{k_p}s\frac{\omega_n^2}{s^2 + 2\zeta\omega_n s + \omega_n^2} + \frac{\omega_n^2}{s^2 + 2\zeta\omega_n s + \omega_n^2},$$

where:

$$\omega_n = \sqrt{\frac{n\,k_m\,k_p}{JR}}, \quad \zeta = \left(\frac{b}{J} + \frac{n^2\,k_m\,k_e}{JR} + \frac{n\,k_m\,k_d}{JR}\right)\frac{1}{2\sqrt{\frac{n\,k_m\,k_p}{JR}}}, \quad (3.133)$$

have been used. Hence:

$$\theta_1(s) = \frac{k_d}{k_p}s\frac{\omega_n^2}{s^2 + 2\zeta\omega_n s + \omega_n^2}\theta_d(s) + \frac{\omega_n^2}{s^2 + 2\zeta\omega_n s + \omega_n^2}\theta_d(s). \quad (3.134)$$

As $\theta_d(s) = \frac{A}{s}$, notice that the inverse Laplace transform of the second term in this expression is the function presented in (3.57), whereas the inverse Laplace transform of the first term is the time derivative of the function in (3.57) multiplied by the factor $\frac{k_d}{k_p}$. This means that, although the shape of $\theta_1(t)$ is similar to that of the function in (3.57), there are some differences. For instance, formulas for the rise time and overshoot given in (3.71) are only approximate in the present control problem.

In the case in which some t_r and M_p are specified, use of the expressions in (3.71) allow only approximate values for the required ω_n and ζ to be computed. These data and (3.133) allow approximate values for the corresponding controller gains to be computed as:

$$k_p = \frac{JR}{n\,k_m}\omega_n^2, \quad k_d = \left[2\zeta\omega_n - \left(\frac{b}{J} + \frac{n^2\,k_m\,k_e}{JR}\right)\right]\frac{JR}{n\,k_m}.$$

Some simulation results are presented in Fig. 3.36 when $J = 1$, $b + \frac{n^2 k_m k_e}{R} = 7$, and $nk_m/R = 70$. The corresponding MATLAB/Simulink diagram is shown in Fig. 3.37. The block theta_d is a step applied at $t = 0$ with a zero initial value and 2 as a final value. The external disturbance T_p is a step applied at $t = 0.5[s]$ with a zero initial value and 7×70 as a final value. When the simulation in Fig. 3.37 stops, the following MATLAB code is executed in an m-file to draw Fig. 3.36:

```
nn=length(InputOutput(:,1));
n=nn-1;
Ts=1/n;
t=0:Ts:1;
plot(t,InputOutput(:,2),'k-.',t,InputOutput(:,3),
```

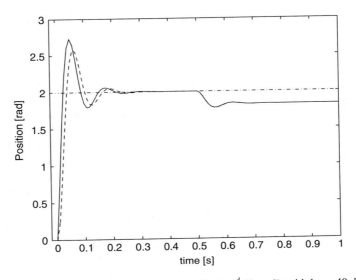

Fig. 3.36 Continuous: PD control law $u = k_p(\theta_d - \theta) + k_d \frac{d}{dt}(\theta_d - \theta)$, with $k_p = 40$, $k_d = 0.45$, is used. Dashed: control law $u = k_p(\theta_d - \theta) - k_v\dot{\theta}$, with $k_p = 40$, $k_v = 0.45$, is used. Dash–dot: desired position θ_d

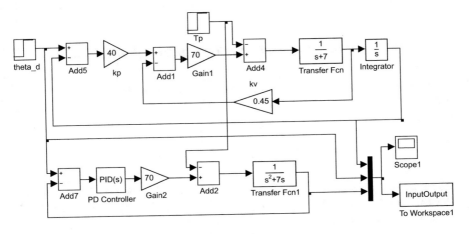

Fig. 3.37 MATLAB/Simulink diagram used to obtain results in Fig. 3.36

```
'k-');
hold on
plot(t,InputOutput(:,1),'k--');
axis([-0.02 1 0 3])
xlabel('time [s]')
ylabel('Position [rad]')
hold off
```

The dashed line represents the desired response, i.e., the inverse Laplace transform of:

$$\frac{\omega_n^2}{s^2 + 2\zeta\omega_n\, s + \omega_n^2}\theta_d(s), \tag{3.135}$$

which is identical to the response obtained with controller $u = k_p(\theta_d - \theta) - k_v\dot{\theta}$, introduced in the previous section, with $k_p = 40$, $k_v = 0.45$. The continuous line represents the inverse Laplace transform of the complete expression for $\theta_1(s)$ in (3.134), i.e., the response when a PD position controller is used with $k_p = 40$, $k_d = 0.45$. Notice that the continuous line in Fig. 3.36 has a smaller rise time and a larger overshoot than the dashed line. The reason for this is that the PD control law $u = k_p(\theta_d - \theta) + k_d\frac{d}{dt}(\theta_d - \theta)$ includes the term $k_d\frac{d\theta_d}{dt}$, which is not present in control law $u = k_p(\theta_d - \theta) - k_v\dot{\theta}$. Recall that θ_d is not a constant, but it is a step signal whose time derivative is very large at the time when such a step signal is applied. This contributes with a large voltage spike at this point in time, which results in a faster response, including a larger overshoot for the step response when a PD controller is employed.

Notice that a nonzero steady-state deviation results when a step external torque disturbance appears at $t = 0.5[s]$. We stress that this deviation is the same in both responses, the continuous and the dashed line, because the same k_p is used.

3.8.4 Proportional–Integral Velocity Control of a DC Motor

Consider again the velocity control problem in a permanent magnet brushed DC motor. Now assume that the following proportional–integral (PI) controller is employed:

$$u = k_p(\omega_d - \omega) + k_i \int_0^t (\omega_d - \omega)dr,$$

where k_i is a constant known as the integral gain, whereas ω_d is a step signal standing for the desired velocity, i.e., $\omega_d(s) = \frac{A}{s}$. The closed-loop system is obtained by replacing u, given above, in the model (3.122), i.e.,:

$$J\dot{\omega} + \left(b + \frac{n^2\,k_m\,k_e}{R}\right)\omega = \frac{n\,k_m}{R}\left(k_p(\omega_d - \omega) + k_i \int_0^t (\omega_d - \omega)dr\right) - T_p.$$

Differentiating once with respect to time and rearranging yields:

$$\ddot{\omega} + \left(\frac{b}{J} + \frac{n^2\,k_m\,k_e}{JR} + \frac{n\,k_m k_p}{JR}\right)\dot{\omega} + \frac{n\,k_m k_i}{JR}\omega = \frac{n\,k_m k_p}{JR}\dot{\omega}_d + \frac{n\,k_m k_i}{JR}\omega_d - \frac{1}{J}\dot{T}_p.$$

$$\tag{3.136}$$

This is a second-order system with two inputs. Then, superposition has to be used, i.e., to define two signals ω_1 and ω_2 such that:

$\omega(t) = \omega_1(t) + \omega_2(t)$,

$$\ddot{\omega}_1 + \left(\frac{b}{J} + \frac{n^2\,k_m\,k_e}{JR} + \frac{n\,k_m k_p}{JR}\right)\dot{\omega}_1 + \frac{n\,k_m k_i}{JR}\omega_1 = \frac{n\,k_m k_p}{JR}\dot{\omega}_d + \frac{n\,k_m k_i}{JR}\omega_d,$$

$$\ddot{\omega}_2 + \left(\frac{b}{J} + \frac{n^2\,k_m\,k_e}{JR} + \frac{n\,k_m k_p}{JR}\right)\dot{\omega}_2 + \frac{n\,k_m k_i}{JR}\omega_2 = -\frac{1}{J}\dot{T}_p.$$

First, analyze the equation for ω_1. Applying the Laplace transform with zero initial conditions:

$$\omega_1(s) = G_1(s)\omega_d(s).$$

$$G_1(s) = \frac{\frac{n\,k_m k_p}{JR}s + \frac{n\,k_m k_i}{JR}}{s^2 + \left(\frac{b}{J} + \frac{n^2\,k_m\,k_e}{JR} + \frac{n\,k_m k_p}{JR}\right)s + \frac{n\,k_m k_i}{JR}}. \tag{3.137}$$

Since the characteristic polynomial is second-degree, we proceed to equate:

$$s^2 + \left(\frac{b}{J} + \frac{n^2\,k_m\,k_e}{JR} + \frac{n\,k_m k_p}{JR}\right)s + \frac{n\,k_m k_i}{JR} = s^2 + 2\zeta\omega_n s + \omega_n^2,$$

to find:

$$\omega_n = \sqrt{\frac{n\,k_m k_i}{JR}} > 0, \; \zeta = \left(\frac{b}{J} + \frac{n^2\,k_m\,k_e}{JR} + \frac{n\,k_m k_p}{JR}\right)\frac{1}{2\sqrt{\frac{n\,k_m k_i}{JR}}}. \tag{3.138}$$

These expressions require $k_i > 0$ to ensure that ω_n and ζ are real numbers. On the other hand, $\zeta > 0$ is true if k_p is chosen to satisfy:

$$k_p > -\frac{JR}{n\,k_m}\left(\frac{b}{J} + \frac{n^2\,k_m\,k_e}{JR}\right).$$

Hence, $G_1(s)$ is a stable transfer function and, thus, the velocity final value can be computed using the final value theorem:

$$\lim_{t\to\infty} \omega_1(t) = \lim_{s\to 0} s\omega_1(s),$$

$$= \lim_{s\to 0} s\frac{\frac{n\,k_m k_p}{JR}s + \frac{n\,k_m k_i}{JR}}{s^2 + \left(\frac{b}{J} + \frac{n^2\,k_m\,k_e}{JR} + \frac{n\,k_m k_p}{JR}\right)s + \frac{n\,k_m k_i}{JR}}\frac{A}{s},$$

$$= \frac{\frac{n\,k_m k_i}{JR}}{\frac{n\,k_m k_i}{JR}}A = A.$$

Fig. 3.38 The integral $\int_0^t (\omega_d - \omega_1(r))dr$ is
represented by the shadowed
area under function
$e = \omega_d - \omega_1$. In this figure it
is assumed that
$\omega_d = A > \omega_1(0)$

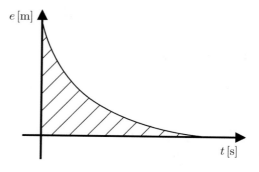

It is concluded that the desired velocity is reached in a steady state, an important
property of PI velocity control.

This result can be explained using everyday experience. As $\omega_1 = A$ is achieved
in a steady state, the error signal $e = A - \omega(t)$ is expected to behave as depicted
in Fig. 3.38. Recall that the integral operation $\int_0^t (\omega_d - \omega_1(r))dr$ represents the area
(positive) under the function shown in Fig. 3.38. This means that the above integral
operation remains constant at a positive value when $\omega_1 = \omega_d = A$ (because the
integrand becomes zero in these conditions). Then, according to $u = k_p(\omega_d - \omega) +
k_i \int_0^t (\omega_d - \omega)dr$, a positive voltage given by $u = k_i \int_0^t (\omega_d - \omega)dr$ continues to be
applied at the motor terminals. This allows the motor to stay rotating at the desired
velocity, i.e., the condition $\omega_1 = \omega_d = A$ holds forever.

On the other hand, to study the transient response, notice that the transfer function
in (3.137) can be rewritten as:

$$G_1(s) = \frac{\frac{k_p}{k_i}\omega_n^2 s + \omega_n^2}{s^2 + 2\zeta\omega_n s + \omega_n^2},$$

$$= \frac{k_p}{k_i} s \frac{\omega_n^2}{s^2 + 2\zeta\omega_n s + \omega_n^2} + \frac{\omega_n^2}{s^2 + 2\zeta\omega_n s + \omega_n^2},$$

where (3.138) has been used. Hence, according to (3.137):

$$\omega_1(s) = \frac{k_p}{k_i} s \frac{\omega_n^2}{s^2 + 2\zeta\omega_n s + \omega_n^2}\omega_d(s) + \frac{\omega_n^2}{s^2 + 2\zeta\omega_n s + \omega_n^2}\omega_d(s).$$

As $\omega_d(s) = \frac{A}{s}$, notice that the inverse Laplace transform of the second term in
this expression is the function presented in (3.57), whereas the inverse Laplace
transform of the first term is the time derivative of the function in (3.57) multiplied
by the factor $\frac{k_p}{k_i}$. This means that, although the shape of $\omega_1(t)$ is similar to that
of the function in (3.57), there are some differences. For instance, the rise time
and overshoot formulas given in (3.71) are only approximate in the present control
problem. This is a similar situation to that found in the study of the transient

response of PD position control of a DC motor. Despite this, the expressions in (3.138) are still valid and they can be used to conclude the following:

- A faster response is achieved as $k_i > 0$ is chosen to be larger. This is because ω_n increases as $k_i > 0$ is increased. However, a more oscillatory response is obtained as $k_i > 0$ is chosen to be larger. This is because ζ decreases as k_i increases.
- The oscillation can be reduced by increasing $k_p > 0$. This is because ζ increases as $k_p > 0$ increases.

These observations can be used as a tuning procedure to select both k_p and k_i for a PI velocity controller. Note that this is a trial and error procedure.

Finally, consider the expression for ω_2:

$$\ddot{\omega}_2 + \left(\frac{b}{J} + \frac{n^2 \, k_m \, k_e}{JR} + \frac{n \, k_m k_p}{JR}\right) \dot{\omega}_2 + \frac{n \, k_m k_i}{JR} \omega_2 = -\frac{1}{J}\dot{T}_p.$$

Using the Laplace transform yields:

$$\omega_2(s) = \frac{-\frac{1}{J}s}{s^2 + \left(\frac{b}{J} + \frac{n^2 \, k_m \, k_e}{JR} + \frac{n \, k_m k_p}{JR}\right) s + \frac{n \, k_m k_i}{JR}} T_p(s).$$

Then, the final value of ω_2 can be computed as:

$$\lim_{t \to \infty} \omega_2(t) = \lim_{s \to 0} s\omega_2(s),$$

$$= \lim_{s \to 0} s \frac{-\frac{1}{J}s}{s^2 + \left(\frac{b}{J} + \frac{n^2 \, k_m \, k_e}{JR} + \frac{n \, k_m k_p}{JR}\right) s + \frac{n \, k_m k_i}{JR}} \frac{d}{s} = 0,$$

when the external torque disturbance is a step signal, i.e., $T_p(s) = \frac{d}{s}$. Recall that it has already been ensured that the characteristic polynomial in this expression has two poles with a negative real part. Hence, the steady-state velocity deviation due to a constant external torque disturbance is zero, which is very good news.

The closed-loop response of velocity under PI control is shown in Fig. 3.39 where $J = 1$, $b + \frac{n^2 k_m k_e}{R} = 7$, and $nk_m/R = 70$ are used. These values are employed merely because they allow representative, clear, responses to be obtained. The corresponding MATLAB/Simulink diagram is shown in Fig. 3.40. The block ω_d is a step applied at $t = 0$ with a zero initial value and 2 as a final value. The external disturbance T_p is a step applied at $t = 0.5[s]$ with a zero initial value and 0.4×70 as a final value. Once the simulation stops, the following MATLAB code is executed in an m-file to draw Fig. 3.39:

```
nn=length(InputOutput(:,1));
n=nn-1;
Ts=1/n;
```

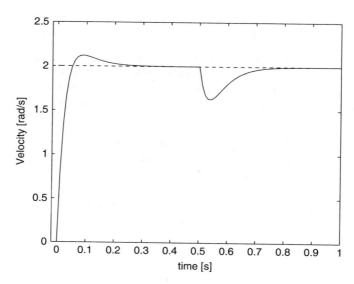

Fig. 3.39 PI velocity control when a constant external disturbance appears at $t = 0.5$[s]. The PI controller gains are $k_p = 0.7$ and $k_i = 10$. Continuous: velocity $\omega(t)$. Dashed: desired velocity ω_d

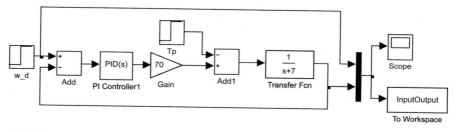

Fig. 3.40 MATLAB/Simulink diagram to obtain results in Fig. 3.39

```
t=0:Ts:1;
plot(t,InputOutput(:,1),'k--',t,InputOutput(:,2),
'k-');
axis([-0.02 1 0 2.5])
xlabel('time [s]')
ylabel('Velocity [rad/s]')
```

Notice that the velocity reaches its desired value before and after a constant external torque disturbance appears at $t = 0.5$[s].

3.8.5 Proportional, PI, and PID Control of First-Order Systems

Consider the water level system studied in Example 3.2 and depicted in Fig. 3.3. Suppose that a water pump is employed to produce a water flow q_i that is proportional to the voltage u applied at motor terminals actuating on the pump, i.e.,:

$$q_i = k_1 u, \tag{3.139}$$

where k_1 is a positive constant. Combining (3.139) and (3.28) it is obtained:

$$\frac{dh}{dt} + ah = kk_1 u.$$

Notice that this differential equation can also be represented by a transfer function with the form:

$$G(s) = \frac{Y(s)}{U(s)} = \frac{b_2}{s + b_1}. \tag{3.140}$$

Hence, identical conclusions are obtained as those for the proportional control of velocity, if a proportional level controller is used, or those for a proportional–integral velocity controller, if a PI level controller is used. Moreover, identical conclusions are valid for any first order plant with transfer function having the same structure as in (3.140).

On the other hand, first order systems represented by (3.140) do not need use of a controller with a derivative action. This can be explained as follows. Suppose that the following proportional–integral–derivative (PID) controller:

$$u = k_p e + k_d \frac{de}{dt} + k_i \int_0^t e(r) dr, \quad e = y_d - y$$

is used for plant in (3.140), i.e.,:

$$\dot{y} + b_1 y = b_2 \left(k_p e + k_d \frac{de}{dt} + k_i \int_0^t e(r) dr \right),$$

$$(1 + b_2 k_d) \dot{y} + (b_1 + b_2 k_p) y = b_2 k_p y_d + b_2 k_d \dot{y}_d + b_2 k_i \int_0^t e(r) dr,$$

and differentiating once with respect to time:

$$(1 + b_2 k_d) \ddot{y} + (b_1 + b_2 k_p) \dot{y} + b_2 k_i y = b_2 k_d \ddot{y}_d + b_2 k_p \dot{y}_d + b_2 k_i y_d.$$

The corresponding closed-loop transfer function is:

$$H(s) = \frac{Y(s)}{Y_d(s)} = \frac{b_2 k_d s^2 + b_2 k_p s + b_2 k_i}{(1 + b_2 k_d)s^2 + (b_1 + b_2 k_p)s + b_2 k_i}. \tag{3.141}$$

Notice that the characteristic polynomial is second degree. Thus, only two free parameters are necessary to assign the corresponding two roots at any location on the s complex plane, i.e., to achieve any desired transient response. However, three free parameters appear in this characteristic polynomial, i.e., k_p, k_d and k_i. Hence, one of them is not necessary, i.e., k_d.

On the other hand, the transfer function in (3.141) has polynomials at the numerator and the denominator having the same degree, i.e., it has the same number of poles and zeros. It has been shown in Examples 3.6 and 3.7 that time response of a system whose transfer function has the same number of poles and zeros is discontinuous when a step input is applied. It is important to remark that such a response is not possible, for instance, in a velocity control system because the velocity cannot be discontinuous. This is due to the fact that such a response would require a very large voltage to be applied at the motor terminals during a very short time interval. Mathematically, this is explained by the time derivative of the desired value y_d (a discontinuous, step signal) which is introduced by the derivative action of a controller. However, such large voltage spikes are not possible in practice. Thus, an unpredictable response is expected. Moreover, in the case of a velocity control system, the derivative action on velocity (equivalent to acceleration measurements, a very noisy signal) results in a strong amplification of noise, as velocity measurements are recognized to be noisy signals.

Thus, it is concluded that any performance improvement is not expected when using a PID controller with respect to performance achieved with a PI controller for first-order plants.

3.9   Case Study: A DC-to-DC High-Frequency Series Resonant Power Electronic Converter

In Sect. 2.5, Chap. 2 the mathematical model of a DC-to-DC high-frequency series resonant power electronic converter was obtained. This model is given in (2.142)–(2.144) and it is rewritten here for ease of reference:

$$L\frac{di}{dt} = -v - v_0 sign(i) + E(t), \tag{3.142}$$

$$C\frac{dv}{dt} = i, \tag{3.143}$$

$$C_0\frac{dv_0}{dt} = abs(i) - \frac{v_0}{R}, \tag{3.144}$$

where:

$$sign(i) = \begin{cases} +1, & i > 0 \\ -1, & i < 0 \end{cases}, \tag{3.145}$$

whereas $abs(i)$ stands for the absolute value of i. The operation of this circuit is studied in the present section through the solution of the mathematical model in (3.142)–(3.144). Performing a coordinate change for the variables involved is very useful. Hence, the following variables are defined (see Sect. 2.5, Chap. 2, for an explanation of the meaning of the parameters involved):

$$z_1 = \frac{i}{E}\sqrt{\frac{L}{C}}, \quad z_2 = \frac{v}{E}, \quad z_3 = \frac{v_0}{E}, \quad \tau = \frac{t}{\sqrt{LC}}. \tag{3.146}$$

Replacing this in (3.142), (3.143), (3.144) yields:

$$L\frac{d\left(E\sqrt{\frac{C}{L}}z_1\right)}{d\left(\tau\sqrt{LC}\right)} = -Ez_2 - Ez_3sign(z_1) + E(t),$$

$$C\frac{d\left(Ez_2\right)}{d\left(\tau\sqrt{LC}\right)} = z_1E\sqrt{\frac{C}{L}},$$

$$C_0\frac{d(Ez_3)}{d\left(\tau\sqrt{LC}\right)} = abs\left(E\sqrt{\frac{C}{L}}z_1\right) - \frac{Ez_3}{R}.$$

After some simplifications, the following is found:

$$\dot{z}_1 = -z_2 - z_3sign(z_1) + u, \tag{3.147}$$

$$\dot{z}_2 = z_1, \tag{3.148}$$

$$\alpha\dot{z}_3 = abs(z_1) - \frac{z_3}{Q}, \tag{3.149}$$

where the dot "·" represents the differentiation with regard to the normalized time τ whereas $\alpha = C_0/C$, $Q = R\sqrt{C/L}$ and u is a variable taking only the two possible values $+1$ (when $E(t) = +E$) and -1 (when $E(t) = -E$). The output capacitor C_0 is commonly chosen to be very large compared with the value of the resonant capacitor C, because this ensures that z_3, i.e., v_0, remains approximately constant during several cycles of the resonant current z_1, i.e., i. Hence, under this assumption (a constant z_3), (3.149) can be neglected and (3.147), (3.148), suffice to represent the evolution of the resonant variables z_1 and z_2, at least for several oscillations.

For operation at resonance, the transistors Q_1, Q_3 are turned on when $z_1 > 0$, i.e., when $i > 0$; hence, $u = +1$ because $E(t) = +E$ in this case, according to

Figs. 2.34a, 2.35 and 2.36. On the other hand, the transistors Q_2, Q_4 are turned on when $z_1 < 0$, i.e., when $i < 0$, and hence $u = -1$ because $E(t) = -E$ in this case. One way of achieving this is by designing $u = sign(z_1)$, where $sign(z_1) = +1$ if $z_1 > 0$ and $sign(z_1) = -1$ if $z_1 < 0$. This means that, according to (3.147) and (3.148), the evolution of the resonant variables can be described as:

$$\dot{z}_1 = -z_2 + v_e, \tag{3.150}$$

$$\dot{z}_2 = z_1, \tag{3.151}$$

$$v_e = \begin{cases} 1 - z_3, & Q_1, Q_3 \text{ turned-on} \\ -(1 - z_3), & Q_2, Q_4 \text{ turned-on} \end{cases},$$

where z_3 is assumed to be constant. Using the variable change $z_4 = z_2 - v_e$ in the previous equations yields $\dot{z}_4 = \dot{z}_2 = z_1$, because a constant v_e implies that $\dot{v}_e = 0$, and $\ddot{z}_4 = \dot{z}_1 = -z_2 + v_e = -z_4$, i.e.,:

$$\ddot{z}_4 + z_4 = 0. \tag{3.152}$$

This is a second-order linear ordinary differential equation with constant coefficients such as that defined in (3.53) with $\zeta = 0$ and $\omega_n = 1$. As the excitation, or input, of this differential equation is zero, then $A = 0$ and the complete solution is given by the natural response alone. Hence, according to Sect. 3.3, i.e., expressions in (3.62) and (3.63), the solution of (3.152) is given as:

$$z_4(\tau) = p(\tau) = \mathcal{L}^{-1}\left\{\frac{z_{40}(s + 2\zeta\omega_n) + \dot{z}_{40}}{s^2 + 2\zeta\omega_n s + \omega_n^2}\right\} = \mathcal{L}^{-1}\left\{\frac{z_{40}s + \dot{z}_{40}}{s^2 + 1}\right\},$$

$$= \mathcal{L}^{-1}\left\{\frac{z_{40}s}{s^2 + 1} + \frac{\dot{z}_{40}}{s^2 + 1}\right\} = z_{40}\cos(\tau) + \dot{z}_{40}\sin(\tau), \tag{3.153}$$

where the following Laplace transform pairs have been employed [4], Ch. 32:

$$\mathcal{L}^{-1}\left\{\frac{s}{s^2 + a^2}\right\} = \cos(a\tau),$$

$$\mathcal{L}^{-1}\left\{\frac{a}{s^2 + a^2}\right\} = \sin(a\tau),$$

whereas $z_{40} = z_4(0) = z_2(0) - v_e$ and $\dot{z}_{40} = \dot{z}_4(0) = z_1(0)$. Differentiating (3.153) once with respect to τ yields:

$$\dot{z}_4(\tau) = -z_{40}\sin(\tau) + \dot{z}_{40}\cos(\tau). \tag{3.154}$$

Then, using (3.153) and (3.154), it is easy to verify that:

$$z_4^2(\tau) + \dot{z}_4^2(\tau) = z_{40}^2 + \dot{z}_{40}^2. \tag{3.155}$$

If the solution of (3.152) is plotted on the phase plane $\dot{z}_4 - z_4$, then a circle is obtained whose center is located at origin $(z_4, \dot{z}_4) = (0, 0)$, and the radius is determined by the initial conditions, i.e., $\sqrt{z_{40}^2 + \dot{z}_{40}^2}$. Moreover, according to $z_4 = z_2 - v_e$ and $\dot{z}_4 = z_1$, it is concluded that the solution of (3.150), (3.151), when plotted on the phase plane $z_1 - z_2$, is a circle centered at $(z_2, z_1) = (v_e, 0)$, with a radius $\sqrt{(z_{20} - v_e)^2 + z_{10}^2}$ where $z_{20} = z_2(0)$ and $z_{10} = z_1(0)$. It is important to say that, according to (3.150) and (3.151), the values of z_1 and z_2 cannot be discontinuous because that would require infinite values for $\dot{z}_1$ and $\dot{z}_2$ respectively, which is not possible as the right-hand sides of (3.150) and (3.151) cannot take infinite values. This means that, when combining the solutions in both regions, i.e., when $z_1 > 0$ and when $z_1 < 0$, the initial conditions on each region must be equal to the final values reached in the previous region. Hence, Fig. 3.41 is obtained. The closed trajectory shown in this figure indicates that the resonant variables z_2, z_1, i.e., v and i, have sustained oscillations. It should be noted that this closed trajectory is clockwise-oriented because, according to (3.151), z_2 grows when $z_1 > 0$, i.e., if $\dot{z}_2 > 0$ then z_2 grows, and z_2 decreases when $z_1 < 0$. Notice that v_e can only take two different values, $1 - z_3$ when $z_1 > 0$, and $-(1 - z_3)$ when $z_1 < 0$. Thus, the only way in which the situation represented in Fig. 3.41 can be obtained is that $z_3 = 1$, i.e., that $v_e = 0$, in both regions. This means that the output voltage is equal to the power supply voltage, i.e., $v_0 = E$.

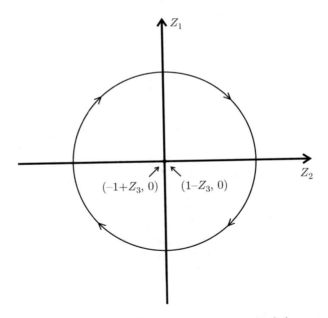

Fig. 3.41 Evolution of the resonant variables z_2 and z_1 when $u = sign(z_1)$

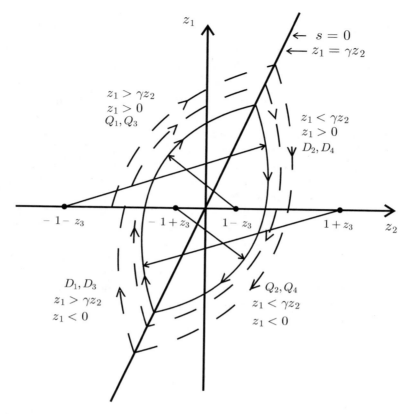

Fig. 3.42 Evolution of the resonant variables z_2 y z_1 when $u = sign(s)$ with $s = z_1 - \gamma z_2$

An important application of DC-to-DC high-frequency series resonant power electronic converters is to deliver output voltages that are smaller than the power supply voltage, i.e., $v_0 < E$ or $z_3 < 1$. Define a function $s = z_1 - \gamma z_2$, with γ a positive constant, and assign $u = sign(s)$. This defines the four regions shown in Fig. 3.42. Notice that the transistors Q_1, Q_3 are turned on when $u = +1$, i.e., above the straight line $z_1 = \gamma z_2$, but the electric current flows through them only when $z_1 > 0$ as, even when Q_1, Q_3 are turned on, the electric current flows through the diodes D_1, D_3 when $z_1 < 0$ (the electric current only flows in one direction through any of the transistors Q_1, Q_2, Q_3, Q_4). On the other hand, the transistors Q_2, Q_4 are turned on when $u = -1$, i.e., below the straight line $z_1 = \gamma z_2$, but the electric current flows through them only when $z_1 < 0$, as, despite Q_2, Q_4 being turned on, the electric current flows through diodes D_2, D_4 when $z_1 > 0$.

The resonant variables still evolve according to (3.150) and (3.151), but now v_e takes the following constant values:

$$
v_e = \begin{cases}
1 - z_3, & Q_1, Q_3 \text{ in saturation, } (z_1 > \gamma z_2, \ z_1 > 0) \\
-(1 - z_3), & Q_2, Q_4 \text{ in saturation, } (z_1 < \gamma z_2, \ z_1 < 0) \\
1 + z_3, & D_1, D_3 \text{ conduct, } (z_1 > \gamma z_2, \ z_1 < 0) \\
-(1 + z_3), & D_2, D_4 \text{ conduct, } (z_1 < \gamma z_2, \ z_1 > 0)
\end{cases}
$$

This means that, again, and according to arguments between (3.150) and the paragraph after (3.155), the resonant variables describe circles on the $z_2 - z_1$ plane. Radii of these circles are determined by the initial conditions in each region, which are equal to the final values in the previous region. The center of these circles is located at the following points, depending on the operation region:

$$
(z_2, z_1) = (1 - z_3, 0), \quad z_1 > \gamma z_2, \ z_1 > 0,
$$

$$
(z_2, z_1) = (-1 + z_3, 0), \quad z_1 < \gamma z_2, \ z_1 < 0,
$$

$$
(z_2, z_1) = (1 + z_3, 0), \quad z_1 > \gamma z_2, \ z_1 < 0,
$$

$$
(z_2, z_1) = (-1 - z_3, 0), \quad z_1 < \gamma z_2, \ z_1 > 0.
$$

Figure 3.42 depicts the corresponding situation when $\gamma = 3$. Notice that, again, the resonant variables z_2, z_1, i.e., v and i, describe sustained oscillations. To obtain the closed trajectory shown in Fig. 3.42, we may proceed as follows. Propose any $z_3 < 1$ and any point in the $z_2 - z_1$ plane as an initial condition. Using a pair of compasses, draw circles centered at the above indicated points according to the respective operation region. According to the previous discussion and (3.151), these circles are clockwise-oriented. The reader will realize that an helix appears similar to the dashed line in Fig. 3.42, which finally ends on a closed trajectory. The fact that this closed trajectory is reached from any initial point on the $z_2 - z_1$ plane is indicative that it is stable or attractive. It is possible to proceed analogously using some $z_3 > 1$ to verify that a closed trajectory is also reached in such a case. This is indicative that in a series resonant converter is not possible to deliver, in a steady state, an output voltage v_0 larger than the power supply voltage E, i.e., $z_3 > 1$ is not possible in a steady state. If an output voltage $v_0 > E$ is desired, then the parallel resonant converter shown in Fig. 2.34b is a suitable alternative.

On the other hand, arcs with a center at $(z_2, z_1) = (1 - z_3, 0)$ and $(z_2, z_1) = (-1 - z_3, 0)$, shown in Fig. 3.42, are drawn one after the other. Then, if both of them were centered at origin $(z_2, z_1) = (0, 0)$, they would together span a $180°$ angle. However, using basic geometry in Fig. 3.42 it is possible to conclude that arcs with the center at $(z_2, z_1) = (1 - z_3, 0)$ and $(z_2, z_1) = (-1 - z_3, 0)$ together describe an angle smaller that $180°$ despite completing a semicircle on both resonant variables z_2, z_1. Hence, the four arcs composing the whole closed trajectory in Fig. 3.42 together describe an angle smaller that $360°$ despite a complete cycle being described on both resonant variables. Notice that each one of these arcs is covered at a frequency that is still $\omega_n = 1$, but, according to the above description, less time is now required to complete an oscillation as the four arcs together describe a smaller angle at the same angular frequency. Thus, the resonant variables oscillate at a frequency ω_o, which is larger than $\omega_n = 1$, i.e., $\omega_o > \omega_n$. Using (3.146), it is possible to verify that:

$$\omega_{nt} = \frac{\omega_n}{\sqrt{LC}} = \frac{1}{\sqrt{LC}}, \quad \omega_{ot} = \frac{\omega_o}{\sqrt{LC}} > \omega_{nt},$$

where ω_{nt} and ω_{ot} stand respectively, for the resonance frequency and the operation frequency, expressed with respect to real time t. Hence, it is concluded that the circuit must operate at frequencies that are greater than the resonance frequency to deliver output voltages smaller than the power supply voltage $v_0 < E$, i.e., $z_3 < 1$. This is the reason why, when the transistors are activated according to $u = sign(s)$, the circuit in Figs. 2.34a and 2.35 is designated as DC-to-DC high-frequency series resonant power electronic converter. Further information on DC-to-DC resonant power electronic converters can be found in [8], where converters of this class (series and parallel) are designed, constructed, controlled, and experimentally tested, and [7], where the method used here to analyze series resonant converters drawing arcs of a circle is studied. On the other hand, the use of the phase plane $z_2 - z_1$ to study the operation of resonant power converters was introduced for the first time in [9] and [10].

3.10 Summary

The systems to be controlled in engineering, and every component in a closed-loop control system, are described by ordinary, linear differential equations with constant coefficients. Hence, for a control engineer, the study of differential equations must be oriented toward the understanding of the properties of the differential equations that shape the solution waveform.

The solution of an ordinary, linear differential equation with constant coefficients is given as the addition of the natural and the forced responses. The natural response only depends on the roots of the differential equation characteristic polynomial and it is always present, even when the excitation function (or input) is zero. On the other hand, the forced response depends on the excitation function (or input), i.e., the forced response is zero if the excitation function is zero. The forced response can be rendered to be equal to the excitation function or input. Furthermore, if the differential equation is stable, i.e., if the natural response tends toward zero, the complete response converges to the forced response. Hence, the input of a differential equation can be chosen to be equal to the way in which it is desired that the differential equation solution behaves.

Thus, the rationale behind control system design is the following. Given a system or plant to be controlled, i.e., a differential equation, choose a controller, i.e., another differential equation, such that when feedback connects the two differential equations, a new equivalent closed-loop differential equation is obtained with the following properties: 1) it is stable, 2) the forced response is equal to the input applied to the closed-loop system, i.e., the desired input of the closed-loop system, and 3) the natural response vanishes as fast and damped as desired.

The features listed in the previous paragraph are accomplished as follows:

1. Stability is determined exclusively by the roots of the differential equation characteristic polynomial or, equivalently, by the poles of the corresponding transfer function (see Sect. 3.4).
2. The steady-state response is achieved by rendering the forced response equal to the input applied to the closed-loop system. It is explained in Sect. 3.4 that this depends on the independent terms of polynomials at the numerator and denominator in the closed-loop transfer function for the case of constant inputs.
3. The transient response refers to the suitable shaping of the natural response waveform and is achieved by suitably assigning the closed-loop poles.

Controller design for arbitrary plants is studied in Chaps. 5, 6, 7 using the root locus method in addition to the frequency response and the state space approach. The main idea is to render the closed-loop system stable, so that the steady-state response is equal to the closed-loop system input and the transient response suitably shaped.

3.11 Review Questions

1. Why is a transfer function unstable if it has one pole with a zero real part that is repeated two or more times?
2. It has been said that the forced response has the same waveform as the input. Explain why this is not ensured if the transfer function has poles with a zero real part. Give an example of an input and a transfer function where this happens.
3. When an incandescent lamp is turned on it becomes hot. However, its temperature does not grow to infinity, but it stops at a certain value. Of the differential equations that have been studied in this chapter, which is the one that better describes the evolution in time of the incandescent lamp temperature? Why?
4. What is the relationship between a transfer function and a differential equation?
5. What are the conditions that determine the stability of a transfer function?
6. What are the necessary assumptions for a permanent magnet brushed DC motor to behave as an integrator or a double integrator?
7. How do you think the natural response waveform of a second-order differential equation whose characteristic polynomial has two real, repeated, and negative roots?
8. It has been explained that the time functions composing the natural response are determined by the transfer function poles, i.e., roots of the characteristic polynomial, but what is the effect of the transfer function zeros? What do you think determines the coefficients when using partial fraction expansion to apply the inverse Laplace transform? (see Sects. 3.1 and 3.5).

9. Indicate the zones of the s complex plane where the poles have to be located to achieve (i) fast but not oscillatory responses, (ii) fast and oscillatory responses, (iii) permanent oscillatory responses, and (iv) unstable responses.

3.12   Exercises

1. Consider the water level system in Example 3.2. The response in Fig. 3.43 is obtained when a constant water flow q_i is applied and the tank has a constant cross-section of $0.5[m^2]$. Find the values of both R and q_i.
2. Consider a proportional level control, i.e., when $q_i = k_1 k_p(h_d - h)$, with h_d the constant desired level. Using the results in the previous exercise to find the gain k_p, such that the steady-state error is less than or equal to 20% of the desired value and that the time constant is less than or equal to one third of the time constant in the previous exercise, i.e., the open-loop time constant. Find the maximal value of q_i that must be supplied. This is important for selecting the water pump to be employed. Suppose that the opening of the output valve is increased such that its hydraulic resistance decreases to 50%. Find the increment or the decrement, as a percentage, of the level steady-state error and the new time constant.

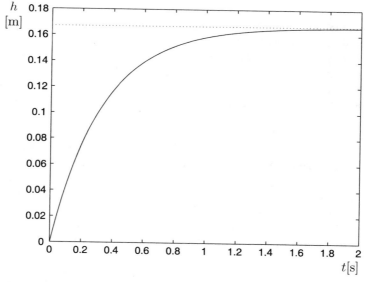

Fig. 3.43 Water level when a constant water flow is applied

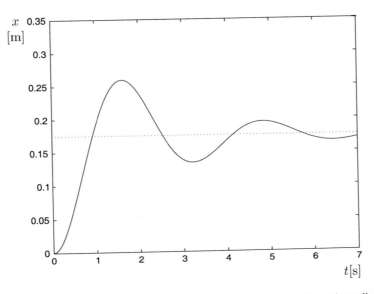

Fig. 3.44 Mass position in a mass-spring-damper system when a constant force is applied at the input. The dotted line stands for the final value of the mass position

3. Consider a proportional–integral level control. Show that oscillations in the water level appear if an adequate gain $k_i > 0$ is employed. Use your everyday experience to explain this phenomenon.

4. Given the following differential equation:

$$\ddot{y} + 127\dot{y} + 570y = 3990u, \quad y(0) = 0, \quad \dot{y}(0) = 0, \quad u = 1,$$

find values of ζ, ω_n and k in (3.53). With these values, write an expression for $y(t)$.

5. Consider the mass-spring-damper system described in Example 3.8. The response shown in Fig. 3.44 is obtained when the constant force $f=0.5[N]$ is applied. Find the values of b, K, and m.

6. Consider the following transfer functions:

$$Y_1(s) = G_1(s)U(s), \quad Y_2(s) = G_2(s)U(s),$$

$$G_1(s) = \frac{ab}{(s+a)(s+b)}, \quad G_2(s) = \frac{b}{s+b}, \quad a > 0, \ b > 0,$$

where $U(s) = A/s$, with A a constant. Perform simulations where a takes values that increase from small to large (with respect to b) and compare graphically $y_1(t)$ and $y_2(t)$. Explain what is observed.

7. According to Sect. 3.1, it is known that from any initial condition the following differential equation $\dot{y} + 7y = 2$ has a solution given as $y(t) = Ee^{-7t} + D$ where E and D are some constants that are not going to be computed in this exercise. Proceeding similarly, find the solution for each one of the following differential equations.

$$\ddot{y} + 4\dot{y} + y = 2,$$
$$-\ddot{y} + \dot{y} + 10y = 3t,$$
$$\ddot{y} + 7\dot{y} = 2\sin(t),$$
$$\ddot{y} + 3\dot{y} + 2y = 2e^{-t},$$
$$\ddot{y} + 2\dot{y} + 2y = 2\cos(2t) - 4\sin(2t),$$
$$\ddot{y} - 4y = 8t^2 - 4,$$
$$y^{(4)} + 11y^{(3)} + 28\ddot{y} = 10.$$

8. For each one of the following differential equations, find the value of $\lim_{t\to\infty} y(t)$. Notice that the same result can be obtained, no matter what the initial conditions are.

$$y^{(3)} + \ddot{y} - 2\dot{y} + y = 8,$$
$$\ddot{y} + \dot{y} - 10y = \cos(5t),$$
$$\ddot{y} + a\dot{y} + by = kA, \quad a > 0,\ b > 0,\ a, b, A, k \text{ are constants.}$$

9. Which one of the following differential equations has a solution with a smaller overshoot and which one has a solution with a smaller rise time?

$$3\ddot{y} + 3\dot{y} + 6y = 2A,$$
$$\ddot{y} + 4\dot{y} + 10y = 9A,$$

A is a constant.

10. Compute the poles of a second-order system with a 25% overshoot and 0.2[s] rise time.

11. Consider the proportional position control with velocity feedback of the following DC motor: $\ddot{\theta} + 0.4\dot{\theta} = 30u$, where θ is the angular position, whereas u is the applied voltage. Find the controller gains to achieve a 0.01[s] rise time and a 10% overshoot.

12. Consider the solution of a second-order differential equation when a step input is applied and assuming zero initial conditions, which is presented in (3.57). Find the maxima and minima of this response and solving for the first maximum, show that overshoot can be computed as:

$$M_p(\%) = 100 \times e^{-\frac{\zeta}{\sqrt{1-\zeta^2}}\pi}.$$

Find the times when the natural response is zero and solving for the first of these times, show that:

$$t_r = \frac{1}{\omega_d}\left[\pi - \arctan\left(\frac{\sqrt{1-\zeta^2}}{\zeta}\right)\right].$$

13. Consider the following expressions:

$$Y(s) = \frac{as+b}{s^2+c}U(s), \quad Z(s) = \frac{d}{s^2+c}U(s),$$

where $U(s) = A/s$ with A, a, b, c, d, are positive constants. Express $y(t)$ as a function of $z(t)$.

14. Consider the following transfer function:

$$G(s) = \frac{c}{s^2+bs+c}.$$

What must be done with b or c to achieve a faster system response?, What must be done with b or c to achieve a less oscillatory response?

15. Consider the solution $y(t)$ of an arbitrary n-order linear differential equation with constant coefficients.

- If the input is bounded, What are the conditions to ensure that $y(t)$ is also bounded for all time? i.e., to ensure that $y(t)$ does not become infinity.
- What are the conditions for the forced response to be the only component of $y(t)$ that remains as time increases?
- Assume that the input is zero. What are the different behaviors for $y(t)$ as time increases and under what conditions do they appear?
- Assume that the input is a polynomial of time. Show that the solution is not a polynomial with a larger degree than the polynomial at the input if all the roots of the characteristic polynomial have a negative real part.
- Assume that the input is given as the addition of several sine and cosine functions of time with different frequencies. Using the results in Sect. 3.6 and the superposition principle, Can you explain the following fundamental result for linear differential equations? [6, pp. 389]: "if the input is any periodic function with period T and all the roots of the characteristic polynomial have negative real parts, then as $t \to \infty$ the solution $y(t)$ is also a periodic function with period T, although with a waveform that may be different from that of the input." Recall the concept of the Fourier series.

16. Verify the following equations:

$$i) \quad \frac{5}{s^2+s-2} = \frac{-5/3}{s+2} + \frac{5/3}{s-1},$$

$ii)$ $\dfrac{2}{s(s+2)(s-1)} = \dfrac{1/3}{s+2} + \dfrac{2/3}{s-1} + \dfrac{-1}{s}$,

$iii)$ $\dfrac{1}{s^3 - s^2 - s + 1} = \dfrac{1/2}{(s-1)^2} + \dfrac{-1/4}{s-1} + \dfrac{1/4}{s+1}$,

$iv)$ $\dfrac{1}{s^4 - s^3 - s^2 + s} = \dfrac{1}{s} + \dfrac{1/2}{(s-1)^2} + \dfrac{-3/4}{s-1} + \dfrac{-1/4}{s+1}$,

$v)$ $\dfrac{2}{s^3 + 2s} = \dfrac{1}{s} + \dfrac{-s}{s^2 + 2}$,

$vi)$ $\dfrac{3}{s^4 - 3s^3} = \dfrac{-1}{s^3} + \dfrac{-1/3}{s^2} + \dfrac{-1/9}{s} + \dfrac{1/9}{s-3}$,

$vii)$ $\dfrac{8}{s(s+4)(s-4)} = \dfrac{-1/2}{s} + \dfrac{1/4}{s+4} + \dfrac{1/4}{s-4}$,

$viii)$ $\dfrac{8}{s^2 - 16} = \dfrac{1}{s-4} + \dfrac{-1}{s+4}$,

$ix)$ $\dfrac{4}{s^4 + s^3 + 2s^2} = \dfrac{2}{s^2} + \dfrac{-1}{s} + \dfrac{s-1}{s^2 + s + 2}$,

$x)$ $\dfrac{4}{s^5 + 3s^4 + 4s^3 + 4s^2 + 3s} = \dfrac{2}{(s+1)^3} + \dfrac{2}{(s+1)^2} + \dfrac{1}{s+1} + \dfrac{-s-1}{s^2 + 1}$,

$xi)$ $\dfrac{3s^2 + 3s - 2}{s^3 + 2s^2} = \dfrac{-1}{s^2} + \dfrac{2}{s} + \dfrac{1}{s+2}$,

$xii)$ $\dfrac{s+2}{s^2(s+1)^2} = \dfrac{2}{s^2} + \dfrac{-3}{s} + \dfrac{1}{(s+1)^2} + \dfrac{3}{s+1}$.

17. An induction oven is basically an inductor L with a heat-resistant material inside containing some metal to melt down. A capacitor C is series-connected to the inductor to improve the power factor. A high-frequency alternating voltage is applied at the circuit terminals that induces eddy currents in metal, producing heat and, hence, melting it down.
This device constitutes a series RLC circuit where the resistance R is represented by the equivalent electrical resistance of metal to melt down. The high-frequency alternating voltage applied to the circuit is given as a square wave described as $u = E \, sign(i)$, where i is the electric current through the circuit, E is a positive constant and $sign(i) = +1$ if $i \geq 0$ or $sign(i) = -1$ if $i < 0$.

• Assume that the initial electric current is zero. Show that, provided that the damping coefficient is less than unity, i.e., $0 < \zeta < 1$, the electric current through the inductor is zero periodically with period π/ω_d, where $\omega_d = \omega_n \sqrt{1 - \zeta^2}$, $\omega_n = \dfrac{1}{\sqrt{LC}}$ y $\zeta = \dfrac{R}{2L\omega_n}$.

- According to $u = E\ sign(i)$, the applied voltage, u, changes value when $i = 0$. Using this and $v_c(0)$ as the initial voltage at the capacitor, find an expression for the voltage at the capacitor that will be useful the next time that $i = 0$.
- Propose any values for R, L, and C such that $0 < \zeta < 1$. Assuming that the voltage at the capacitor and the current through the inductor are not discontinuous when the value of u changes (why?), write a computer program to use iteratively the formula in the previous item to compute voltage at the capacitor after u has changed several times. Verify that the voltage at the capacitor, when $i = 0$, converges to a constant.
- Perform a simulation of the series RLC circuit when $u = E\ sign(i)$, to verify the result in the previous item.

18. According to Example 3.19, when the armature inductance is negligible, the mathematical model of a DC motor is given as:

$$\omega(s) = \frac{\frac{k_m}{JR}}{s + \left(\frac{B}{J} + \frac{k_m k_e}{J R_a}\right)} V(s),$$

where $\omega(s)$ and $V(s)$ are the Laplace transforms of the motor velocity and voltage applied at the motor terminals respectively.

- Assume that the applied voltage is zero and that the initial velocity is not zero. Notice that applying a zero voltage at motor terminals is equivalent to putting a short circuit at the armature motor terminals and that a non-zero-induced voltage is present if the velocity is not zero. Depict the natural response under these conditions. What can be done to force the natural response to vanish faster? What happens with the electric current? Can you explain why the natural response vanishes faster if the armature resistance is closer to zero? Do you know the meaning of the term "dynamic braking"? Why does the time constant depend on motor inertia? What does this mean from the point of view of Newton's Laws of Mechanics?
- Assume that a voltage different from zero is applied when the initial velocity is zero. Depict the system response. Why does the final velocity not depend on the motor inertia J? Give an interpretation from the point of view of physics.

19. Consider the electric circuit in Fig. 3.45. This circuit is called a phase-lag network. The corresponding mathematical model is given in (2.146) and is rewritten here for ease of reference:

$$V_0(s) = \frac{R_2}{R_1 + R_2} \frac{s + \frac{1}{R_2 C}}{s + \frac{1}{(R_1 + R_2)C}} V_i(s) + \frac{R_1}{R_1 + R_2} \frac{1}{s + \frac{1}{(R_1 + R_2)C}} v_c(0),$$

Fig. 3.45 Phase-lag network

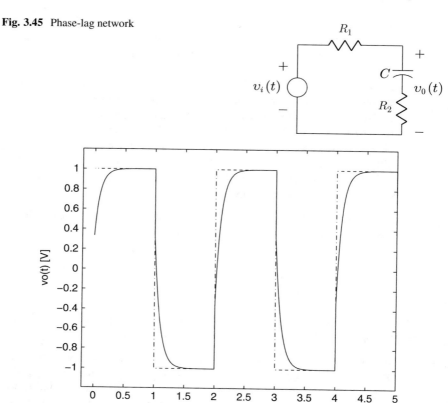

Fig. 3.46 Time response of the circuit in Fig. 3.45

where $v_c(0)$ is the initial voltage at the capacitor. Show that the solution is given as:

$$v_0(t) = A + \frac{R_1}{R_1 + R_2}(v_c(0) - A)e^{-at}, \quad a = \frac{1}{(R_1 + R_2)C},$$

and use this result to explain the response in Fig. 3.46 when $v_i(t)$ is the square wave represented by the dash-dot line, $v_c(0) = 0$, $R_1 = 2[\Omega]$, $R_2 = 1[\Omega]$ and $C = 1/33[F]$.

20. Repeat the previous exercise for all circuits in Fig. 2.39. Use MAT-LAB/Simulink to draw the time response and verify your results.

21. Consider the simple pendulum in Fig. 3.47 when the external torque $T(t) = 0$ and friction is negligible. The mathematical model is given as:

$$ml^2\ddot{\theta} + mgl\theta = 0,$$

Fig. 3.47 Simple pendulum

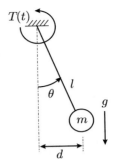

which is valid only when the pendulum remains close to the downward vertical configuration, i.e., θ is "small," $\theta \approx 0$ (see Example 7.5, Chap. 7). Find the required pendulum length l to produce oscillations with a period of 1 s.

References

1. K. Ogata, *Modern control engineering*, 4th edition, Prentice-Hall, Upper Saddle River, 2002.
2. E. Kreyszig, *Advanced engineering mathematics*, Vol. 1, 3rd. edition, John Wiley and Sons, New Jersey, 1972.
3. C. Ray Wylie, *Advanced engineering mathematics*, McGraw-Hill, 1975.
4. M. R. Spiegel, *Mathematical handbook of formulas and tables*, McGraw-Hill, Schaum series, New York, 1974.
5. W. H. Hayt and J. E. Kemmerly, *Engineering circuit analysis*, McGraw-Hill, 1971.
6. C.-T. Chen, *Linear system theory and design*, Holt, Rinehart, and Winston, New York, 1984.
7. V. Hernández Guzmán, A new stability analysis method for DC to DC series resonant converters, *Computación y Sistemas*, vol. 11, no. 1, pp. 14–25, 2007.
8. R. Silva-Ortigoza, Control of resonant power converters based on differential flatness: design and construction, *MSc. Thesis*, CINVESTAV-IPN (Zacatenco), México City, February, 2002.
9. R. Oruganti and F.C. Lee, Resonant power processors, Part I—State plane analysis, *IEEE Trans. Industry Applications*, vol. AI-21, no. 6, pp. 1453–1460, 1985.
10. R. Oruganti and F.C. Lee, Resonant power processors, Part II—Methods of control, *IEEE Trans. Industry Applications*, vol. AI-21, no. 6, pp. 1461–1471, 1985.
11. M.E. Van Valkenburg, *Network analysis*, Prentice-Hall, Upper-Saddle River, 1974.
12. J. Stewart, *Calculus concepts and contexts*, Brooks Cole Publishing and IPT Company, 1998.

Chapter 4
Stability Criteria and Steady-State Error

A general control system is made up of several interacting components : the plant to be controlled, the controller, the measurement systems, the actuators, etc. This is the reason why a control system may be very complex. Despite this, any closed-loop control system may be reduced to being represented by a single transfer function that relates the output to be controlled and the reference or the desired output. This closed-loop transfer function can be studied as in Chap. 3. In the present chapter, the concept of block diagrams (see [1] for a historical perspective) is introduced to represent how a closed-loop control system is constituted and how to manipulate them to obtain the closed-loop transfer function is also explained.

On the other hand, a control system is always designed such that: a) it is stable, b) the desired transient response specifications are satisfied, and c) the desired steady-state response specifications are satisfied. According to Chap. 3, a transfer function is stable if all its poles have a negative real part. However, checking this property analytically may be difficult and, because of that, it is necessary to find simple methods to solve this problem. Two alternatives are presented in this chapter: 1) *the rule of signs* to determine the sign of the real part of the roots of a polynomial and, 2) *Routh's stability criterion* (see [1] for a historical perspective).

The transient response of a control system depends on the location of poles (and zeros) of the closed-loop transfer function. There are two different methods of designing a controller, assigning poles and zeros of the closed-loop transfer function at suitable locations: the *root locus method* (see Chap. 5) and the *frequency response method* (see Chap. 6).

Finally, satisfying the desired specifications for the steady-state response is concerned with forcing the closed-loop system response to reach the reference or desired output once the natural response vanishes.[1] This is accomplished by

---

[1] When time is large enough or in a steady state.

© Springer International Publishing AG, part of Springer Nature 2019
V. M. Hernández-Guzmán, R. Silva-Ortigoza, *Automatic Control with Experiments*,
Advanced Textbooks in Control and Signal Processing,
https://doi.org/10.1007/978-3-319-75804-6_4

designing a controller that renders the closed-loop system forced response equal to, or at least close to, the reference or desired output. This subject is also studied in the present chapter.

4.1   Block Diagrams

Some examples are presented in the following to show how a block diagram can be simplified to obtain a single transfer function representing the whole block diagram. We also present some examples of block diagrams where a single output is affected by two different inputs.

Example 4.1 Consider two systems such that the input of one of them is the output of the other, as depicted in Fig. 4.1. It is said that these systems are cascade-connected. Notice that:

$$Y_1(s) = G_1(s)U_1(s), \quad Y_2(s) = G_2(s)U_2(s), \quad U_1(s) = Y_2(s),$$

to obtain:

$$Y_1(s) = G_1(s)G_2(s)U_2(s), \quad G(s) = G_1(s)G_2(s),$$
$$Y_1(s) = G(s)U_2(s).$$

This means that cascade-connected systems, such as those in Fig. 4.1, can be represented by a single transfer function $G(s)$ that can be computed as the product of the individual transfer functions of systems in the connection. The reader can verify that this is true no matter how many cascade-connected systems there are.

Example 4.2 Suppose that we have two parallel-connected systems, as depicted in Fig. 4.2. Notice that:

$$Y_1(s) = G_1(s)U(s), \quad Y_2(s) = G_2(s)U(s),$$

to write:

$$Y_1(s) + Y_2(s) = (G_1(s) + G_2(s))U(s), \quad G(s) = G_1(s) + G_2(s),$$
$$Y_1(s) + Y_2(s) = G(s)U(s).$$

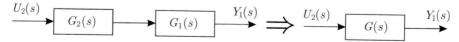

Fig. 4.1 Cascade-connected systems

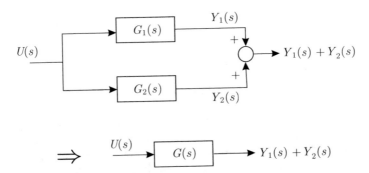

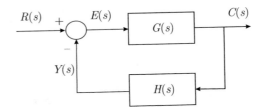

Fig. 4.2 Parallel-connected systems

Fig. 4.3 A closed-loop or feedback control system

This means that parallel-connected systems, such as those in Fig. 4.2, can be represented by a single transfer function $G(s)$ that is computed as the addition of the transfer functions of each single system in the parallel connection. The reader can verify that this is true no matter how many systems are in the parallel connection.

Example 4.3 Consider the closed-loop control system depicted in Fig. 4.3. Blocks $G(s)$ and $H(s)$ stand for the transfer functions of the several components of a control system. In fact, $G(s)$ and $H(s)$ can be the result of the combination of several transfer functions of several components of the control system, such as those in the previous examples. According to the definition of transfer function, from Fig. 4.3 it is found that:

$$C(s) = G(s)E(s), \quad E(s) = R(s) - Y(s), \quad Y(s) = H(s)C(s).$$

Replacing these expressions in each other, the following is obtained:

$$C(s) = G(s)[R(s) - Y(s)],$$
$$= G(s)[R(s) - H(s)C(s)],$$
$$= G(s)R(s) - G(s)H(s)C(s),$$
$$C(s)[1 + G(s)H(s)] = G(s)R(s),$$
$$C(s) = \frac{G(s)}{1 + G(s)H(s)}R(s),$$

where:

$$M(s) = \frac{C(s)}{R(s)} = \frac{G(s)}{1 + G(s)H(s)}, \tag{4.1}$$

is known as the closed-loop transfer function.

Example 4.4 Consider the block diagram shown in Fig. 4.4a. To simplify this block diagram, it is convenient to shift the subtraction point placed at the right of $U(s)$ to the point where $I^*(s)$ appears. This must be accomplished by keeping the definition of $I(s)$ in Fig. 4.4a without change. This means that, according to Fig. 4.4a:

$$I(s) = \frac{1}{Ls + R}[KA_p(I^*(s) - I(s)) - nk_e s\theta(s)].$$

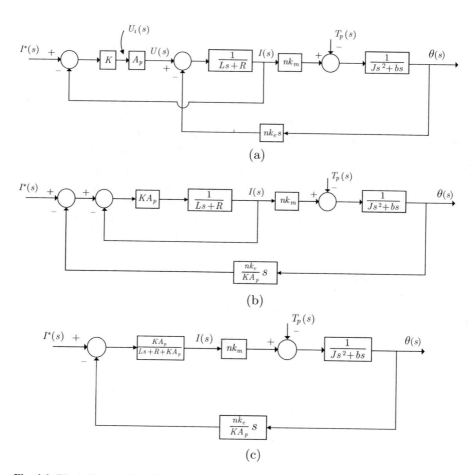

Fig. 4.4 Block diagram simplification

Notice that this expression is also valid in the block diagram of Fig. 4.4b; hence, this block diagram is equivalent to that in Fig. 4.4a. It is clear that the loop between the second subtraction point in Fig. 4.4b and $I(s)$ is identical to the closed-loop system shown in Fig. 4.3 with:

$$G(s) = \frac{KA_p}{Ls + R}, \quad H(s) = 1.$$

It is clear that the main result in Example 4.1 has been employed to compute the transfer function of two cascaded systems as the product of the individual transfer functions of the cascaded systems. Hence, using (4.1), the following transfer function is found, which must be placed before $I(s)$:

$$\frac{KA_p}{Ls + R + KA_p}.$$

This is shown in Fig. 4.4c. The block diagram in Fig. 4.4c has two inputs. To represent the output $\theta(s)$ as a function of both inputs, the superposition principle has to be employed (see Sect. 3.7), i.e.,

$$\theta(s) = G_1(s)I^*(s) + G_2(s)T_p(s). \tag{4.2}$$

$G_1(s)$ is the transfer function computed when $I^*(s)$ is the input and $\theta(s)$ is the output, and it is assumed that $T_p(s) = 0$ in the block diagram in Fig. 4.4c, i.e., when the block diagram is depicted as in Fig. 4.5a. Thus, using (4.1), the following is defined:

$$M(s) = \frac{G(s)}{1 + G(s)H(s)} = \frac{\theta(s)}{I^*(s)} = G_1(s),$$

with:

$$G(s) = \frac{KA_p n k_m}{(Ls + R + KA_p)(Js^2 + bs)}, \quad H(s) = \frac{nk_e s}{KA_p},$$

to find:

$$G_1(s) = \frac{nk_m}{\left[\left(\frac{sL + R}{KA_p} + 1\right)(sJ + b) + \frac{n^2 k_m k_e}{KA_p}\right]s}. \tag{4.3}$$

On the other hand, $G_2(s)$ is the transfer function obtained when $T_p(s)$ is the input, $\theta(s)$ is the output and it is assumed that $I^*(s) = 0$ in the block diagram in Fig. 4.4c, i.e., when block diagram is depicted as in Fig. 4.5b. Hence, using (4.1), the following is defined:

$$M(s) = \frac{-G(s)}{1 + G(s)H(s)} = \frac{\theta(s)}{T_p(s)} = G_2(s),$$

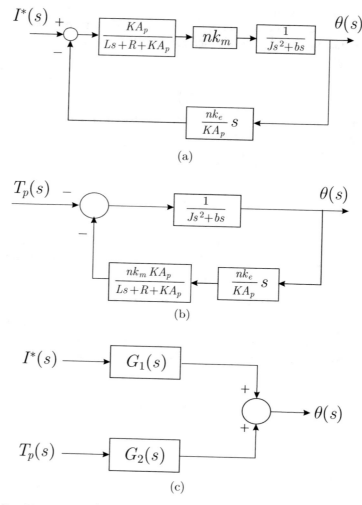

Fig. 4.5 Simplifying the block diagram in Fig. 4.4c. (**a**) Block diagram when $T_p(s) = 0$. (**b**) Block diagram when $I^*(s) = 0$. (**c**) Block diagram equivalent to any of the block diagrams in Fig. 4.4

with:

$$G(s) = \frac{1}{Js^2 + bs}, \quad H(s) = \frac{n^2 k_m k_e s}{Ls + R + KA_p},$$

to obtain:

$$G_2(s) = \frac{-\left(\frac{sL+R}{KA_p} + 1\right)}{\left[\left(\frac{sL+R}{KA_p} + 1\right)(sJ + b) + \frac{n^2 k_m k_e}{KA_p}\right]s}. \tag{4.4}$$

Thus, any of block diagrams in Figs. 4.4a, b or c can be represented as block diagram in Fig. 4.5c with $G_1(s)$ and $G_2(s)$ given in (4.3) and (4.4).

Example 4.5 Consider the closed-loop system shown in Fig. 4.6a. This block diagram represents a control scheme known as *two degrees of freedom control*. This is a closed-loop system with two inputs and one output. Hence, we can proceed as in the previous example to write, using the superposition principle:

$$C(s) = G_1(s)R(s) + G_2(s)D(s).$$

$G_1(s)$ is the transfer function obtained when using $R(s)$ as the input, $C(s)$ as the output and $D(s) = 0$, i.e., when the block diagram in Fig. 4.6b is employed. On the other hand, $G_2(s)$ is the transfer function obtained when using $D(s)$ as the input, $C(s)$ as the output and $R(s) = 0$, i.e., when the block diagram in Fig. 4.6c is employed.

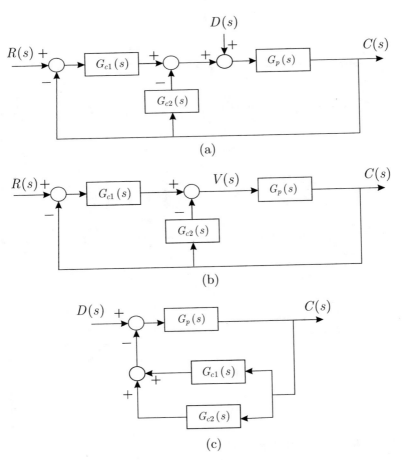

Fig. 4.6 A two-degree-of-freedom control system. (**a**) Closed-loop system. (**b**) Case when $D(s) = 0$. (**c**) Case when $R(s) = 0$

First, proceed to obtain $G_1(s)$. To simplify the block diagram in Fig. 4.6b notice that:

$$V(s) = G_{c1}(s)(R(s) - C(s)) - G_{c2}(s)C(s),$$
$$= G_{c1}(s)(R(s) - C(s)) + G_{c2}(s)(R(s) - C(s)) - G_{c2}(s)R(s),$$
$$= (G_{c1}(s) + G_{c2}(s))(R(s) - C(s)) - G_{c2}(s)R(s).$$

Then, the block diagrams in Figs. 4.7a, b, and c are successively obtained. Notice that, according to the Example 4.2, the following can be written:

$$F(s) = \left[1 - \frac{G_{c2}(s)}{G_{c1}(s) + G_{c2}(s)}\right] R(s),$$
$$= \frac{G_{c1}(s) + G_{c2}(s) - G_{c2}(s)}{G_{c1}(s) + G_{c2}(s)} R(s),$$
$$= \frac{G_{c1}(s)}{G_{c1}(s) + G_{c2}(s)} R(s). \tag{4.5}$$

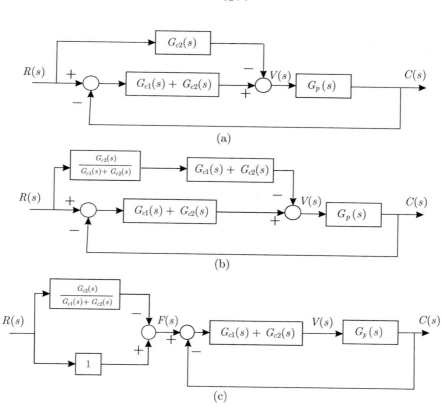

(a)

(b)

(c)

Fig. 4.7 Simplifying the block diagram in Fig. 4.6b

On the other hand, (4.1) can be used together with:

$$G(s) = (G_{c1}(s) + G_{c2}(s))G_p(s), \quad H(s) = 1,$$

to find:

$$C(s) = \frac{(G_{c1}(s) + G_{c2}(s))G_p(s)}{1 + (G_{c1}(s) + G_{c2}(s))G_p(s)} F(s). \tag{4.6}$$

Replacing (4.5) in (4.6), the following is found:

$$C(s) = \frac{G_{c1}(s)G_p(s)}{1 + (G_{c1}(s) + G_{c2}(s))G_p(s)} R(s),$$

hence:

$$G_1(s) = \frac{G_{c1}(s)G_p(s)}{1 + (G_{c1}(s) + G_{c2}(s))G_p(s)}. \tag{4.7}$$

On the other hand, from the block diagram in Fig. 4.6c and according to Example 4.2, the block diagram in Fig. 4.8 is obtained. Using (4.1) and:

$$G(s) = G_p(s), \quad H(s) = G_{c1}(s) + G_{c2}(s),$$

yields:

$$C(s) = \frac{G_p(s)}{1 + (G_{c1}(s) + G_{c2}(s))G_p(s)} D(s),$$

i.e.,

$$G_2(s) = \frac{G_p(s)}{1 + (G_{c1}(s) + G_{c2}(s))G_p(s)}. \tag{4.8}$$

Hence, the block diagram in Fig. 4.6a can be represented as the block diagram in Fig. 4.9 with $G_1(s)$ and $G_2(s)$ given in (4.7) and (4.8).

Fig. 4.8 Simplifying block diagram in Fig. 4.6c

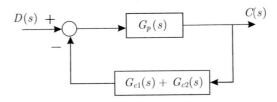

Fig. 4.9 Equivalent block
diagram to that in Fig. 4.6a

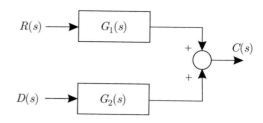

4.2 The Rule of Signs

According to Sect. 3.4, the stability of a transfer function is ensured if all its poles
have a negative real part, i.e., if all the roots of its characteristic polynomial have
a negative real part. However, sometimes, it is difficult to compute the exact value
of the poles and this is especially true when the polynomial has a degree that is
greater than or equal to 3. The reason for this is that the procedure to compute the
roots of third- or fourth-degree polynomials is complex. Moreover, for polynomials
with a degree greater than or equal to 5, analytical procedures do not exist. Even
for second-degree polynomials the analytical procedure may be tedious. Thus, it is
desirable to have a simple method for determining whether all roots of a polynomial
have a negative real part or whether there are some roots with a positive real part.
Some simple criteria for solving this problem are introduced in this section. These
criteria are based on the study of signs of the polynomial coefficients.

4.2.1 Second-Degree Polynomials

Criterion 4.1 *If all the coefficients of a second-degree polynomial have the same
sign, then all its roots have a negative real part.*

Proof Consider the following polynomial:

$$p(s) = s^2 + cs + d,$$

where $c > 0$, $d > 0$. If all the coefficients of the polynomial had a negative sign,
then this sign can be factorized to proceed as shown in the following, i.e., when all
the coefficients are positive. A second-degree polynomial $p(s)$ has two roots, which
can be computed using the general formula:

$$s_{1,2} = \frac{-c \pm \sqrt{c^2 - 4d}}{2}. \tag{4.9}$$

- Case (i): $c^2 - 4d < 0$.

In this case, both roots are complex conjugate with a negative real part because $c > 0$:

$$s_{1,2} = \frac{-c}{2} \pm \frac{\sqrt{4d - c^2}}{2} j.$$

- Case (ii): $c^2 - 4d > 0$.
  In this case,

$$c^2 > c^2 - 4d > 0,$$

because $d > 0$. Applying the square root on both sides yields:

$$c > \sqrt{c^2 - 4d},$$

because the square root is a strictly increasing function. Then:

$$c - \sqrt{c^2 - 4d} > 0.$$

According to (4.9), this means that both roots are real, different, and negative in this case:

$$s_1 = -\frac{c + \sqrt{c^2 - 4d}}{2} < 0,$$

$$s_2 = -\frac{c - \sqrt{c^2 - 4d}}{2} < 0.$$

- Case (iii): $c^2 - 4d = 0$.
  In this case, both roots are real, repeated, and negative:

$$s_{1,2} = \frac{-c}{2}.$$

Example 4.6 Notice that the polynomial $s^2 + s + 1$ satisfies the case (i); hence, its roots are complex conjugate with a negative real part. In fact, it is not difficult to verify that these roots are $-0.5 + j0.866$ and $-0.5 - j0.866$. On the other hand, the polynomial $s^2 + 2.5s + 1$ satisfies the case (ii); hence, its roots are real, different, and negative. It is not difficult to verify that these roots are -2 and -0.5. Finally, the polynomial $s^2 + 2s + 1$ satisfies the case (iii); hence, its roots are real, repeated, and negative. It is not difficult to verify that these roots are -1 and -1.

Criterion 4.2 *If some coefficients of a second-degree polynomial have a sign that is different to signs of the other coefficients, then at least one root has a positive real part.*

Proof Consider the following polynomial:

$$p(s) = s^2 + cs + d.$$

There are three possibilities:

1) $c < 0, d > 0$.
2) $c > 0, d < 0$.
3) $c < 0, d < 0$.

Both roots are computed using the general formula:

$$s_{1,2} = \frac{-c \pm \sqrt{c^2 - 4d}}{2}.$$

- Case (i): $c^2 - 4d < 0$.
  In this case, both roots are complex conjugate with a positive real part if 1) is satisfied. Cases 2) and 3) are not possible because $d < 0$ implies $c^2 - 4d > 0$.
- Case (ii): $c^2 - 4d > 0$.
  This case is possible for 2) and 3) and some small values for $d > 0$ in 1). For larger values of $d > 0$, i) is retrieved. Consider cases 2) and 3), then:

$$c^2 < c^2 - 4d.$$

Applying the square root on both sides of this inequality yields:

$$abs(c) < \sqrt{c^2 - 4d},$$

because the square root is a strictly increasing function. Then:

$$abs(c) - \sqrt{c^2 - 4d} < 0, \quad \text{or} \quad 0 < -abs(c) + \sqrt{c^2 - 4d}.$$

This means that both roots are real and different, with one of them positive:

$$s_1 = -\frac{c + \sqrt{c^2 - 4d}}{2} < 0,$$

$$s_2 = -\frac{c - \sqrt{c^2 - 4d}}{2} > 0.$$

Now consider case 1):

$$c^2 > c^2 - 4d.$$

Applying the square root on both sides of this inequality yields:

$$abs(c) > \sqrt{c^2 - 4d},$$

i.e.,

$$abs(c) - \sqrt{c^2 - 4d} > 0, \quad \text{or} \quad - abs(c) + \sqrt{c^2 - 4d} < 0.$$

As $c < 0$ and $d > 0$, then there are two positive real roots:

$$s_1 = -\frac{c + \sqrt{c^2 - 4d}}{2} > 0,$$

$$s_2 = -\frac{c - \sqrt{c^2 - 4d}}{2} > 0.$$

- Case (iii): $c^2 - 4d = 0$.
  In this case $d = c^2/4 > 0$, only if 1) is satisfied, i.e., $c < 0$. This implies that both roots are real, repeated, and positive:

$$s_{1,2} = \frac{-c}{2} > 0.$$

Example 4.7   Several second-degree polynomials and their corresponding roots are presented in what follows. It is left as an exercise for the reader to verify these results and to relate them to the cases analyzed above.

- $s^2 + 2s - 1$, roots: -2.4142 and 0.4142.
- $s^2 - 2.5s + 1$, roots: 2 and 0.5.
- $s^2 - s - 1$, roots: 1.618 and -0.618.
- $s^2 - s + 1$, roots: $0.5 + j0.866$ and $0.5 - j0.866$.
- $s^2 - 2s + 1$, roots: 1 and 1.

4.2.2   First-Degree Polynomials

In this case, is very easy to prove that the same conditions stand as for second-degree polynomials:

Criterion 4.3 *If both coefficients of a first-order polynomial have the same sign, then its only root is real and negative. If one coefficient has the opposite sign to the other, then the only root is real and positive.*

4.2.3 Polynomials with a Degree Greater Than or Equal to 3

Criterion 4.4 *If at least one coefficient has the opposite sign to the other coefficients, then there is at least one root with a positive real part.*

Proof Consider the following polynomial where $n \geq 3$:

$$p(s) = s^n + a_{n-1}s^{n-1} + \cdots + a_1 s + a_0 = (s - p_1)(s - p_2) \cdots (s - p_n). \tag{4.10}$$

According to the study of second-degree polynomials, it is known that the product $(s - p_i)(s - p_j) = s^2 + cs + d$ has $c < 0$ and/or $d < 0$ if at least one of the roots p_i or p_j has a positive real part. On the other hand, the first-degree polynomial $(s - p_k)$ has a negative coefficient if its root p_k is positive. Also notice that the coefficients $a_j, j = 0, \ldots, n-1$, on the left-hand side of (4.10) are obtained as the algebraic sum of the product of coefficients of polynomials of the form $s^2 + cs + d$ and $(s - p_k)$ existing on the right-hand side of (4.10). Then, the only way for a coefficient $a_j < 0$ to exist on the left-hand side of (4.10) is that at least one root with a positive real part exists on the right-hand side of (4.10), i.e., that some of the polynomials $s^2 + cs + d$ or $(s - p_k)$ have a negative coefficient.

Criterion 4.5 *Even when all the coefficients have the same sign, it is not a given that all roots have a negative real part.*

This can be explained using the same arguments as for the proof of the previous criterion, recalling that the product of two negative coefficients results in a positive number. Then, even when all the coefficients of the polynomial on the left-hand side of (4.10) are positive, there is the possibility that two negative coefficients (owing to roots with positive real parts) on the right-hand side of (4.10) multiply to give a positive coefficient on the left-hand side of (4.10).

Example 4.8 To illustrate what happens in a polynomial with a degree greater than 2, some polynomials and their roots are presented in the following. The reader can corroborate these results by checking that $s^3 + a_2 s^2 + a_1 s + a_0 = (s - p_1)(s - p_2)(s - p_3)$, where p_1, p_2, p_3 are the roots of the corresponding polynomial.

- $s^3 + s^2 + s + 1.5$, roots: $-1.2041, 0.102 + j1.1115$ and $0.102 - j1.1115$. There are two roots with a positive real part, despite all the polynomial coefficients having the same sign.
- $s^3 - s^2 + s + 1$, roots: $0.7718 + j1.1151, 0.7718 - j1.1151$ and -0.5437. There are two roots with a positive real part because one coefficient of the polynomial has the opposite sign to the other coefficients.
- $s^3 + s^2 + 1$, roots: $-1.4656, 0.2328 + j0.7926, 0.2328 - j0.7926$. There are two roots with a positive real part because one coefficient has a zero value, despite the other coefficients having the same sign.

- $s^3 + s^2 + 3s + 1$, roots: $-0.3194 + j1.6332$, $-0.3194 - j1.6332$ and -0.3611. All roots have a negative real part and all the polynomial coefficients have the same sign.

4.3 Routh's Stability Criterion

As explained in the previous section, although the rule of signs is very easy to apply, its main drawback appears when analyzing polynomials with a degree greater than or equal to 3: when all coefficients have the same sign, nothing can be concluded from the sign of the real part of the roots. The solution to this problem is provided by Routh's stability criterion, which, given a polynomial with an arbitrary degree, establishes necessary and sufficient conditions to ensure that all its roots have a negative real part.

Given an arbitrary polynomial, order its terms by descending powers of its variable:

$$a_n s^n + a_{n-1} s^{n-1} + \cdots + a_1 s + a_0. \qquad (4.11)$$

Routh's stability criterion is applied in two steps [2–5]:

1. Fill Table 4.1. Notice that the first two rows of the table are filled by direct substitution of the polynomial coefficients. Also notice that the last entries in the first two rows are zero, as that is the value of the coefficients of powers not appearing in the polynomial in (4.11). The entries of the remaining rows are computed using the following formulae:

$$b_1 = \frac{a_{n-1}a_{n-2} - a_n a_{n-3}}{a_{n-1}}, \qquad b_2 = \frac{a_{n-1}a_{n-4} - a_n a_{n-5}}{a_{n-1}},$$

$$b_3 = \frac{a_{n-1}a_{n-6} - a_n a_{n-7}}{a_{n-1}},$$

$$\vdots$$

$$c_1 = \frac{b_1 a_{n-3} - a_{n-1}b_2}{b_1}, \qquad c_2 = \frac{b_1 a_{n-5} - a_{n-1}b_3}{b_1}, \qquad c_3 = \frac{b_1 a_{n-7} - a_{n-1}b_4}{b_1},$$

$$\vdots$$

$$d_1 = \frac{c_1 b_2 - b_1 c_2}{c_1}, \qquad d_2 = \frac{c_1 b_3 - b_1 c_3}{c_1}, \qquad d_3 = \frac{c_1 b_4 - b_1 c_4}{c_1},$$

$$\vdots$$

Notice that, according to these formulae, the last entry in each row is zero and the table is triangular.

Table 4.1 Application of
Routh's stability criterion

s^n	a_n	a_{n-2}	a_{n-4}	a_{n-6}	$\ldots$	0
s^{n-1}	a_{n-1}	a_{n-3}	a_{n-5}	a_{n-7}	$\ldots$	0
s^{n-2}	b_1	b_2	b_3	b_4	$\ldots$	0
s^{n-3}	c_1	c_2	c_3	c_4	$\ldots$	0
s^{n-4}	d_1	d_2	d_3	d_4	$\ldots$	0
$\vdots$	$\vdots$	$\vdots$	$\vdots$	$\vdots$	$\ldots$	0
s^2	p_1	p_2	0			
s^1	q_1	0				
s^0	p_2					

Table 4.2 Application of
Routh's criterion to the
polynomial $s^2 + as + b$

s^2	1	b	0
s^1	a	0	
s^0	b		

2. Only analyze the first column in Table 4.1. The number of sign changes found from top to bottom in the first column of the table is equal to to the number of roots with a positive real part.

Several examples are presented in the following.

Example 4.9 Consider the following second-degree polynomial:

$$s^2 + as + b,$$

with a and b two real constants. The rule of signs (Sect. 4.2) can be employed to conclude that both roots have a negative real part if:

$$a > 0, \quad b > 0,$$

as the coefficient of s^2 is $+1$. Routh's criterion is now used to corroborate this result. First, Table 4.2 is filled. Routh's criterion establishes that the number of sign changes in the first column of the table is equal to the number of roots with a positive real part. Hence, if there are no sign changes, both roots have a negative real part. As the first entry of the first column is $+1$ then:

$$a > 0, \quad b > 0,$$

must be true to ensure that both roots have a negative real part. Thus, Routh's criterion and the rule of signs in Sect. 4.2 have the same conclusion, as expected.

Example 4.10 Consider the following polynomial:

$$s^3 + 5s^2 + 2s - 8.$$

Table 4.3 Application of Routh's criterion to the polynomial $s^3 + 5s^2 + 2s - 8$

s^3	1		2	0
s^2	5		-8	0
s^1	$\frac{5\times2-1\times(-8)}{5} = 3.6$		0	
s^0	-8			

Table 4.4 Application of Routh's criterion to the polynomial $s^3 + 1.8s^2 + 0.61s + 2.02$

s^3	1		0.61	0
s^2	1.8		2.02	0
s^1	$\frac{1.8\times0.61-1\times2.02}{1.8} = -0.512$		0	
s^0	2.02			

Because of coefficient -8, i.e., with a negative sign when all the other coefficients are positive, it is known from the rule of signs in Sect. 4.2 that there is at least one root with a positive real part. Routh's criterion is now used to corroborate this result. First, Table 4.3 is filled. Routh's criterion establishes that the number of sign changes in the first column is equal to the number of roots with a positive real part. Hence, it is concluded that the polynomial $s^3 + 5s^2 + 2s - 8$ has one root with a positive real part. In fact, the use of the MATLAB command:

```
roots([1 5 2 -8])
```

allows us to find that:

$$s = -2, \quad s = 1, \quad s = -4,$$

are the roots of the polynomial.

Example 4.11 Consider the following polynomial:

$$s^3 + 1.8s^2 + 0.61s + 2.02.$$

As all the coefficients have the same sign and the polynomial is third-degree, the rule of signs in Sect. 4.2 is not useful in this problem. In this case, Routh's stability criterion is a suitable alternative. First, Table 4.4 is filled. Routh's criterion establishes that the number of sign changes in the first column is equal to the number of roots with a positive real part. Hence, it is concluded that the polynomial $s^3 + 1.8s^2 + 0.61s + 2.02$ has two roots with a positive real part. In fact, use of the MATLAB command:

```
roots([1 1.8 0.61 2.02])
```

allows us to find that:

$$s = -2, \quad s = 0.1 + j, \quad s = 0.1 - j,$$

are the roots of the above polynomial.

Table 4.5 Application of
Routh's criterion to the
polynomial $s^3 + as^2 + bs + c$

s^3	1	b	0
s^2	a	c	0
s^1	$\frac{ab - 1 \times c}{a}$	0	
s^0	c		

Example 4.12 Consider the following polynomial:

$$s^3 + as^2 + bs + c,$$

where a, b and c are real unknown constants. This type of situation is very common
in control systems where coefficients of the characteristic polynomial depend on the
controller gains, which are not known and must be found such that closed-loop
stability is ensured. This is achieved by forcing all the roots of the closed-loop
characteristic polynomial to have a negative real part. However, the computation
of the roots of a third-degree polynomial using analytical methods is a very tedious
process; hence, that is not a practical solution. In this type of problem, Routh's
stability criterion is, again, very useful. First, fill Table 4.5. Routh's criterion
establishes that the number of sign changes in the first column of the table is equal
to the number of roots with a positive real part. Thus, as the first element in the first
column is $+1$, then if:

$$a > 0, \quad \frac{ab - c}{a} > 0 \quad c > 0, \tag{4.12}$$

it is ensured that all three roots have a negative real part. Notice that conditions
in (4.12) are equivalent to:

$$a > 0, \quad b > \frac{c}{a} > 0 \quad c > 0.$$

This means that, although the coefficients a and c are only required to be positive,
in the case of the coefficient b, this is not enough and a little more is demanded:
$b > \frac{c}{a} > 0$.

*Example 4.13 (Special Case: Only the Entry at the First Column of a Row Is Equal
to Zero)* Consider the following polynomial:

$$s^5 + 2s^4 + 3s^3 + 6s^2 + 5s + 3.$$

First fill Table 4.6. However, the procedure stops at the row corresponding to s^3
because the entry in the first column in that row is zero. As a consequence, some
divisions by zero would appear when continuing filling the remaining rows. In this
case, it is advised to replace the zero in the first column with a small $\varepsilon > 0$ to
continue filling the table as shown in Table 4.7.

Table 4.6 Application of Routh's criterion to the polynomial $s^5 + 2s^4 + 3s^3 + 6s^2 + 5s + 3$

s^5	1	3	5
s^4	2	6	3
s^3	$\frac{2\times3-1\times6}{2} = 0$	$\frac{2\times5-1\times3}{2} = 3.5$	0
s^2			
s^1			
s^0			

Table 4.7 Application of Routh's criterion to the polynomial $s^5 + 2s^4 + 3s^3 + 6s^2 + 5s + 3$ (cont.)

s^5	1	3	5
s^4	2	6	3
s^3	ε	3.5	0
s^2	$\frac{\varepsilon\times6-2\times3.5}{\varepsilon}$	$\frac{\varepsilon\times3-2\times0}{\varepsilon} = 3$	0
s^1	$\frac{42\varepsilon-49-6\varepsilon^2}{12\varepsilon-14}$	0	
s^0	3		

Now, determine the number of sign changes in the first column as $\varepsilon > 0$ tends toward zero. Notice that in this example there are two sign changes. This means that the polynomial under study has two roots with a positive real part. In fact, use of the MATLAB command:

```
roots([1 2 3 6 5 3])
```

allows us to find that the roots of the polynomial $s^5 + 2s^4 + 3s^3 + 6s^2 + 5s + 3$ are:

$$s = 0.3429 + j1.5083, \quad s = 0.3429 - j1.5083, \quad s = -1.6681,$$

$$s = -0.5088 + j0.7020, \quad s = -0.5088 - j0.7020.$$

If after $\varepsilon > 0$ is employed to fill the table, there were no sign changes in the first column, then the system would be marginally stable, i.e., there would be some roots on the imaginary axis.

Example 4.14 (Special Case: A Row Is Filled Exclusively with Zeros or a Row only Has One Entry, Which Is Zero) Consider the following polynomial:

$$s^3 + 3s^2 + s + 3. \tag{4.13}$$

First fill Table 4.8. The process stops at the row corresponding to s^1 because that row is filled exclusively with zeros. This is a special case that is indicative of three possibilities [3], pp. 336,[6], pp. 136:

1. There are real roots that are placed symmetrically with respect to origin.
2. There are imaginary roots that are placed symmetrically with respect to origin.
3. There are four complex conjugate roots placed at the vertices of a rectangle centered at the origin.

Table 4.8 Application of Routh's criterion to the polynomial $s^3 + 3s^2 + s + 3$

s^3	1		1	0
s^2	3		3	0
s^1	$\frac{3 \times 1 - 1 \times 3}{3} = 0$		0	
s^0				

Table 4.9 Application of Routh's criterion to the polynomial $s^3 + 3s^2 + s + 3$ (cont.)

s^3	1	1	0
s^2	3	3	0
s^1	6	0	
s^0	3		

To continue, use the entries in the row s^2 to form a new polynomial, i.e.,

$$P(s) = 3s^2 + 3,$$

compute its derivative:

$$\frac{dP(s)}{ds} = 6s,$$

and replace these coefficients in the row s^1, as shown in Table 4.9, to continue filling the table. As no sign changes exist in the first column, it is concluded that no roots exist with a positive real part; hence, only the case 2 above is possible: there are two imaginary roots that are placed symmetrically with respect to origin and the remaining root has a negative real part. Furthermore, another important feature of this special case is the following:

"Form a polynomial with entries of the row above the row filled exclusively with zeros. The roots of such a polynomial are also the roots of the polynomial in (4.13)."

This means that the roots of polynomial $P(s) = 3s^2 + 3$ are also roots of $s^3 + 3s^2 + s + 3$. These roots can be obtained as:

$$3s^2 + 3 = 0, \quad \Rightarrow s = \pm j.$$

Hence, the polynomial $s^3 + 3s^2 + s + 3$ has one root at $s = j$ and another at $s = -j$. In fact, use of the MATLAB command:

```
roots([1 3 1 3])
```

allows us to find that the roots of $s^3 + 3s^2 + s + 3$ are:

$$s = -3, \quad s = j, \quad s = -j.$$

Example 4.15 (Special Case: A Row Is Filled Exclusively with Zeros or a Row only Has One Entry, Which Is Zero) Consider the following polynomial:

$$s^4 + s^2 + 1.$$

Table 4.10 Application of Route's criterion to the polynomial $s^4 + s^2 + 1$

s^4	1	1	1	0
s^3	0	0	0	
s^2				
s^1				
s^0				

Table 4.11 Application of Route's criterion to the polynomial $s^4 + s^2 + 1$ (cont.)

s^4	1		1		1	0
s^3	4		2		0	
s^2	$\frac{4\times1-1\times2}{4} = 0.5$		$\frac{4\times1-1\times0}{4} = 1$			
s^1	$\frac{0.5\times2-4\times1}{0.5} = -1.5$	0				
s^0	1					

Fill Table 4.10. As there is a row filled exclusively with zeros, proceed as in the previous example. The polynomial $P(s) = s^4 + s^2 + 1$ is differentiated with respect to s:

$$\frac{dP(s)}{ds} = 4s^3 + 2s.$$

These coefficients are substituted in the row filled exclusively with zeros and continue filling Table 4.11. As two sign changes exist in the first column of this table, it is concluded that two roots with a positive real part exist. This means that only the following case is possible: there are four complex conjugate roots placed on vertices of a rectangle centered at the origin. In fact, use of the MATLAB command:

```
roots([1 0 1 0 1])
```

allows us to find that the roots of the polynomial $s^4 + s^2 + 1$ are:

$$s = -0.5 \pm j0.8660, \quad s = 0.5 \pm j0.8660.$$

Example 4.16 (Repeated Roots on the Imaginary Axis) When a characteristic polynomial has repeated roots on the imaginary axis, the transfer function is unstable. However, this class of instability is not detected by Route's criterion. For instance, consider the following polynomial:

$$s^4 + 2s^2 + 1 = [(s + j)(s - j)]^2.$$

Notice that there are two repeated roots on the imaginary axis. Fill Table 4.12. As there is a row exclusively filled with zeros, the polynomial $P(s) = s^4 + 2s^2 + 1$ is differentiated with respect to s:

$$\frac{dP(s)}{ds} = 4s^3 + 4s,$$

and continue filling Table 4.13.

Table 4.12 Application of
Routh's criterion to the
polynomial $s^4 + 2s^2 + 1$

s^4	1	2	1	0
s^3	0	0	0	
s^2				
s^1				
s^0				

Table 4.13 Application of
Routh's criterion to the
polynomial $s^4 + 2s^2 + 1$
(cont.)

s^4	1	2	1	0
s^3	4	4	0	
s^2	$\frac{4\times2-4\times1}{4} = 1$	$\frac{4\times1-1\times0}{4} = 1$	0	
s^1	0	0		
s^0				

Table 4.14 Application of
Routh's criterion to the
polynomial $s^4 + 2s^2 + 1$
(cont.)

s^4	1	2	1	0
s^3	4	4	0	
s^2	1	1	0	
s^1	2	0		
s^0	1			

As there is another row exclusively filled with zeros, the polynomial $P_1(s) = s^2 + 1$ is differentiated with respect to s:

$$\frac{dP_1(s)}{ds} = 2s,$$

and continue filling Table 4.14. Notice that no sign changes exist in the first column of Table 4.14; hence, the method correctly indicates that no roots exist with a positive real part. Also notice that the roots of $P_1(s) = s^2 + 1$ are $s = \pm j$, i.e., marginal stability is concluded. However, this is incorrect because the polynomial $s^4 + 2s^2 + 1$ has two imaginary roots, which are repeated twice, which implies instability. Thus, attention must be paid to these cases.

4.4 Steady-State Error

The solution of a differential equation is given as the addition of the natural response and the forced response. If the differential equation is stable or, equivalently, if the transfer function is stable then the natural response disappears as time increases and the complete solution reaches the forced response. Recall that the forced response depends on (or is similar to) the applied input. According to these ideas, in a closed-loop control system, we have the following. 1) The closed-loop system input is a signal that stands for the desired closed-loop output. This is why such a signal is known as the reference or desired output. 2) The controller is designed such that the closed-loop system is stable and the forced response is equal to or is close to the reference or desired output.

The properties that a closed-loop control system must possess to ensure that the forced response is equal to or close enough to the closed-loop system input are studied in this section. Notice that this is equivalent to requiring the system error, defined as the difference between the system output and the reference, to be zero or close to zero in a steady state, i.e., when time is large. Thus, the behavior of the steady-state error is studied in the following.

Consider the closed-loop system with unit feedback shown in Fig. 4.10. An important concept in the study of the steady-state error is the *system type*, which is defined next. It is assumed that any $n-$order open-loop transfer function can be written as:

$$G(s) = \frac{k(s - z_1)(s - z_2) \cdots (s - z_m)}{s^N (s - p_1)(s - p_2) \cdots (s - p_{n-N})},$$

where n is the number of open-loop poles and m is the number of open-loop zeros with $n > m$, $z_j \neq 0$, $j = 1, \ldots, m$ and $p_i \neq 0$, $i = 1, \ldots, n - N$. Hence, N is the number of poles that the open-loop transfer function has at the origin, i.e., after cancelling with any zeros that the open-loop transfer function might have at the origin. The *system type* is defined as N, i.e., it is the number of poles at the origin or, equivalently, the number of integrators that the open-loop transfer function possesses.

The error signal is given as the difference between the closed-loop system output and the reference or desired output:

$$E(s) = R(s) - C(s) \quad \Rightarrow C(s) = R(s) - E(s),$$

and, on the other hand:

$$C(s) = G(s)E(s) = R(s) - E(s),$$

hence:

$$E(s)[1 + G(s)] = R(s),$$

$$E(s) = \frac{1}{1 + G(s)} R(s). \tag{4.14}$$

Fig. 4.10 Closed-loop system with unitary feedback

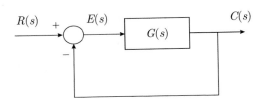

The steady-state error, e_{ss}, is obtained using (4.14) and the final value theorem in (3.4), i.e.,

$$e_{ss} = \lim_{t \to \infty} e(t) = \lim_{s \to 0} s E(s) = \lim_{s \to 0} \frac{s}{1 + G(s)} R(s). \qquad (4.15)$$

Notice that the steady-state error is the difference between the system output $c(t)$ and the reference, or desired output, $r(t)$, in a steady state. This explains why e_{ss} depends on the reference $R(s)$. Hence, to continue this study, it is necessary to know $R(s)$, which motivates the definition of the so-called *test signals*. A test signal is a function of time with two features: 1) It must make sense in real applications, and 2) It must be mathematically simple. Some important test signals, or test inputs, in classical control are defined in the following.

Consider the problem of a video camera tracking a target. The control objective in this problem is to force the video camera to aim at a target that moves very fast. Some situations arising in this problem are the following.

- **Step test signal.** Suppose that the target approaches the video camera directly along a constant direction represented by A. If video camera is at the beginning aiming in another direction and it is desired that it aims in the direction from which the target is approaching, then the reference or desired direction has the shape depicted in Fig. 4.11. This test signal is known as a *step* and it represents an abrupt change in the desired output. The system output is the direction in which the video camera is aiming. It is desired that the difference between these signals is zero or close to zero in a steady state. If $r(t)$ is a step:

$$r(t) = \begin{cases} 0, & t < 0 \\ A, & t \geq 0 \end{cases},$$

where A is a constant, then $R(s) = \mathcal{L}\{r(t)\} = \frac{A}{s}$ is a function that is simple enough to be mathematically handled.
- **Ramp test signal.** Suppose that the target passes in front of the video camera with a constant velocity A, approaching another point, and that the video camera is aiming at the target at the beginning. Then, the reference or desired direction has the shape shown in Fig. 4.12. This test signal is known as a *ramp* and it indicates

Fig. 4.11 Step test signal

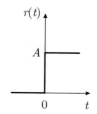

Fig. 4.12 Ramp test signal

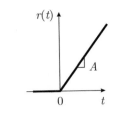

Fig. 4.13 Parabola test signal

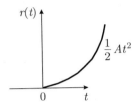

that the desired output changes at a constant rate A, i.e., the target velocity. If $r(t)$ is a ramp:

$$r(t) = \begin{cases} 0, & t < 0 \\ At, & t \geq 0 \end{cases},$$

where $A = \frac{dr(t)}{dt}$ is a constant standing for the rate of change of $r(t)$, then $R(s) = \mathcal{L}\{r(t)\} = \frac{A}{s^2}$ is a function that is simple enough to be mathematically handled.

- **Parabola test signal**. Suppose that the target passes in front of the video camera with a constant acceleration A, approaching another point, and that the video camera is aiming at the target at the beginning. Then, the reference or desired output has the shape shown in Fig. 4.13. This signal is known as a *parabola* and it indicates that the desired output changes with a constant acceleration A, i.e., the target acceleration. If $r(t)$ is a parabola:

$$r(t) = \begin{cases} 0, & t < 0 \\ \frac{1}{2}At^2, & t \geq 0 \end{cases},$$

where $A = \frac{d^2 r(t)}{dt^2}$ is a constant standing for the acceleration of $r(t)$, then $R(s) = \mathcal{L}\{r(t)\} = \frac{A}{s^3}$ is a function that is simple enough to be mathematically handled.

Any of these three situations may appear under the normal operating conditions of a recording video camera control system. Thus, the design of a controller must ensure that the steady-state error is zero or close to zero for any of these desired outputs. The conditions required to achieve this control objective are studied in the following.

4.4.1 Step Desired Output

Suppose that $R(s) = A/s$ is the step desired output. Using (4.15) the following is found:

$$e_{ss} = \lim_{s \to 0} \frac{s}{1 + G(s)} \frac{A}{s},$$

$$= \frac{A}{1 + k_p}, \quad k_p = \lim_{s \to 0} G(s),$$

where k_p is known as the *position constant* and is related to the system type. Hence, the following cases are considered:

- System type 0 ($N = 0$). In this case, the open-loop transfer function has the form:

$$G(s) = \frac{k(s - z_1)(s - z_2) \cdots (s - z_m)}{(s - p_1)(s - p_2) \cdots (s - p_n)},$$

i.e.,

$$k_p = \lim_{s \to 0} G(s) = \frac{k(-z_1)(-z_2) \cdots (-z_m)}{(-p_1)(-p_2) \cdots (-p_n)},$$

k_p is finite. This means that the steady-state error is constant and different from zero:

$$e_{ss} = \frac{A}{1 + k_p} \neq 0.$$

- System type greater than or equal to 1 ($N \geq 1$). In this case, the open-loop transfer function has the form:

$$G(s) = \frac{k(s - z_1)(s - z_2) \cdots (s - z_m)}{s^N(s - p_1)(s - p_2) \cdots (s - p_{n-N})},$$

i.e.,

$$k_p = \lim_{s \to 0} G(s) = \frac{k(-z_1)(-z_2) \cdots (-z_m)}{s^N(-p_1)(-p_2) \cdots (-p_{n-N})} \to \infty.$$

This means that the steady-state error is zero:

$$e_{ss} = \frac{A}{1 + k_p} = 0.$$

4.4.2 *Ramp Desired Output*

Suppose that $R(s) = A/s^2$ is the desired output. Using (4.15), the following is found:

$$e_{ss} = \lim_{s \to 0} \frac{s}{1 + G(s)} \frac{A}{s^2},$$

$$= \frac{A}{k_v}, \quad k_v = \lim_{s \to 0} sG(s),$$

where k_v is known as the *velocity constant* and it is related to the system type. Hence, consider the following cases:

- System type 0 ($N = 0$). In this case, the open-loop transfer function has the form:

$$G(s) = \frac{k(s - z_1)(s - z_2) \cdots (s - z_m)}{(s - p_1)(s - p_2) \cdots (s - p_n)},$$

i.e.,

$$k_v = \lim_{s \to 0} sG(s) = (0) \frac{k(-z_1)(-z_2) \cdots (-z_m)}{(-p_1)(-p_2) \cdots (-p_n)} = 0.$$

This means that the steady-state error grows without a limit:

$$e_{ss} = \frac{A}{k_v} \to \infty.$$

- System type 1 ($N = 1$). The open-loop transfer function has the form:

$$G(s) = \frac{k(s - z_1)(s - z_2) \cdots (s - z_m)}{s(s - p_1)(s - p_2) \cdots (s - p_{n-1})},$$

i.e.,

$$k_v = \lim_{s \to 0} sG(s) = \frac{k(-z_1)(-z_2) \cdots (-z_m)}{(-p_1)(-p_2) \cdots (-p_{n-1})},$$

k_v is finite and different from zero. This means that the steady-state error is constant and different from zero:

$$e_{ss} = \frac{A}{k_v} \neq 0.$$

- System type greater than or equal to 2 ($N \geq 2$). The open-loop transfer function can be written as:

$$G(s) = \frac{k(s - z_1)(s - z_2) \cdots (s - z_m)}{s^N(s - p_1)(s - p_2) \cdots (s - p_{n-N})},$$

i.e.,

$$k_v = \lim_{s \to 0} sG(s) = \frac{k(-z_1)(-z_2) \cdots (-z_m)}{s^{N-1}(-p_1)(-p_2) \cdots (-p_{n-N})} \to \infty,$$

because $N - 1 \geq 1$. This means that the steady-state error is zero:

$$e_{ss} = \frac{A}{k_v} = 0.$$

4.4.3   Parabola Desired Output

Suppose that $R(s) = A/s^3$ is the desired output. Using (4.15), the following is obtained:

$$e_{ss} = \lim_{s \to 0} \frac{s}{1 + G(s)} \frac{A}{s^3},$$

$$= \frac{A}{k_a}, \quad k_a = \lim_{s \to 0} s^2 G(s),$$

where k_a is known as the *acceleration constant* and is related to the system type. The open-loop transfer function is given as:

$$G(s) = \frac{k(s - z_1)(s - z_2) \cdots (s - z_m)}{s^N(s - p_1)(s - p_2) \cdots (s - p_{n-N})},$$

i.e.,

$$k_a = \lim_{s \to 0} s^2 G(s) = \frac{k(-z_1)(-z_2) \cdots (-z_m)}{s^{N-2}(-p_1)(-p_2) \cdots (-p_{n-N})}.$$

This means that:

• The steady-state error grows without limit (to infinity) when the system type is 0 or 1:

$$e_{ss} = \frac{A}{k_a} \to \infty, \quad N \leq 1.$$

• The steady-state error is constant and different from zero when the system type is 2:

$$e_{ss} = \frac{A}{k_a} \neq 0, \quad N = 2.$$

Fig. 4.14 Behavior of the output $c(t)$ when the reference $r(t)$ is a ramp. Notice that $e_{ss} = 0$ when the system type is greater than or equal to 2, $e_{ss} \neq 0$ when the system type is 1 and $e_{ss} \to \infty$ when the system type is 0

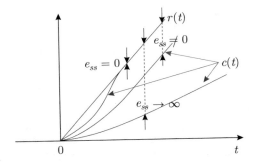

Table 4.15 Steady-state error

	Step	Ramp	Parabola
Type 0	$\frac{A}{1+k_p}$	∞	∞
Type 1	0	$\frac{A}{k_v}$	∞
Type 2	0	0	$\frac{A}{k_a}$
Type 3	0	0	0
⋮	⋮	⋮	⋮

- The steady-state error is zero when the system type is greater than or equal to 3:

$$e_{ss} = \frac{A}{k_a} = 0, \quad N \geq 3.$$

Notice that, in the case of the three references, or test signals, that have been considered, the steady-state error becomes zero if the number of open-loop integrators are suitably increased. This explains why some controllers include an integrator, i.e., proportional–integral (PI) or proportional–integral–derivative (PID) control. On the other hand, it is important to note that all the results presented above are true only if it is ensured that the closed-loop system is stable. This is because the final value theorem assumes that the natural response vanishes as time increases.

Finally, in Fig. 4.14, the behavior of the output is depicted $c(t)$ with respect to the reference $r(t)$, when the latter is a ramp for different system types. The results found above are summarized in Table 4.15.

Example 4.17 A velocity control system for a permanent magnet brushed DC motor is depicted in Fig. 4.15 (see Chap. 10). The motor model is given by the transfer function $G_m(s) = \frac{k}{s+a}$, with $k > 0$ and $a > 0$, whereas k_p is a positive constant standing for the transfer function of a proportional controller. This means that the electric current used as the control signal is computed as:

$$i^* = k_p(\omega_d - \omega),$$

where ω is the measured velocity, ω_d is the desired velocity and $I^*(s)$ is Laplace transform of i^*. The open-loop transfer function is:

$$G(s)H(s) = \frac{k_p k}{s + a}.$$

Fig. 4.15 Proportional
control of velocity

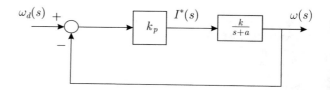

Notice that, as $a > 0$, the system type is 0 because the open-loop transfer function
has no pole at the origin. Hence, according to Sect. 4.4.1, if the desired velocity is
constant, i.e., a step signal, then the steady-state error $e_{ss} = \omega_d - \lim_{t\to\infty} \omega(t)$ is
constant and different from zero, i.e., a proportional velocity controller is not useful
for motor velocity to reach the desired velocity. See Fig. 3.32 for a simulation in
which these observations are corroborated. However, it is important to stress that the
steady-state error analysis that is presented in this chapter is useful for describing the
situation in Fig. 3.32 only as long as no external disturbance T_p is present. Notice
that, according to Sects. 4.4.2 and 4.4.3, the steady-state error would be greater if
the desired velocity was either a ramp or a parabola.

A PI controller solves this problem. A PI controller performs the following
operation on the velocity error:

$$i^* = k_p e + k_i \int_0^t e(r)dr, \quad e = \omega_d - \omega,$$

where k_p and k_i are positive constants. Use of the Laplace transform yields:

$$I^*(s) = k_p E(s) + k_i \frac{E(s)}{s},$$

$$= \left(k_p + \frac{k_i}{s}\right) E(s),$$

$$= \left(\frac{k_p s + k_i}{s}\right) E(s),$$

$$= k_p \left(\frac{s + \frac{k_i}{k_p}}{s}\right) E(s),$$

where $E(s)$ is the Laplace transform of the velocity error. Hence, the block diagram
in Fig. 4.16 is obtained. Notice that, now, the system type is 1 because the open-loop
transfer function:

$$G(s)H(s) = k_p \left(\frac{s + \frac{k_i}{k_p}}{s}\right) \frac{k}{s + a},$$

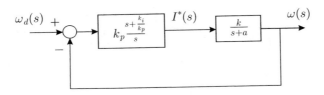

Fig. 4.16 Proportional–integral control of velocity

has one pole at $s = 0$ and, thus, according to Sect. 4.4.1, the steady-state error is zero, i.e., the motor velocity reaches the desired velocity when time is large enough if ω_d is constant, or a step signal. We conclude that the integral term of a PI controller is introduced to render the steady-state error zero. It must be stressed, however, that this result is absolutely true only if it is ensured that the closed-loop system is stable, i.e., if all the closed-loop poles have a negative real part.

Some simulation results are shown in Fig. 3.39 where these ideas are corroborated. Notice that in those simulations the steady-state error is still zero, even when a constant external disturbance is applied. However, it is important to say that this property is not described by the steady-state error analysis presented in this chapter. Recall that this analysis is only valid for a closed-loop system such as that represented in Fig. 4.10, i.e., when no external disturbance is present.

Finally, according to Sects. 4.4.2 and 4.4.3, a PI velocity controller is only useful for step desired outputs as the steady-state error is still different from zero if the desired velocity is either a ramp or a parabola.

Example 4.18 Consider the position control system of a permanent magnet brushed DC motor depicted in Fig. 4.17. Notice that the motor transfer function has one pole at $s = 0$, when the output to be controlled is in the position. Hence, the system type is 1 and the steady-state error is zero if the desired position is a constant, i.e., a step (see Sect. 4.4.1). It is important to realize that this result stands if any of the following controllers are employed (see Chap. 11): *a*) A PD controller, i.e., Fig. 4.18, *b*) A proportional position controller with velocity feedback, i.e., Fig. 4.19, or *c*) A lead-compensator, i.e., Fig. 4.20. The purpose of these controllers is to ensure that the closed-loop system is stable, introducing suitable damping and accomplishing a suitable response time. It is important to stress that closed-loop stability must be ensured to really achieve a zero steady-state error.

Some simulation results are presented in Figs. 3.34 and 3.36, which corroborate the above-mentioned ideas. Notice, however, that the steady-state error analysis presented in this chapter is not intended to explain the different from zero steady-state error observed in those simulations when a constant external disturbance appears. Recall that this analysis is only valid for a closed-loop system, such as that represented in Fig. 4.10, i.e., when no external disturbance is present.

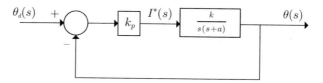

Fig. 4.17 Proportional control of the position

Fig. 4.18
Proportional–derivative
control of the position

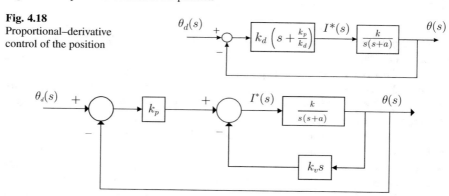

Fig. 4.19 Proportional control with velocity feedback

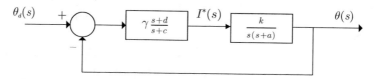

Fig. 4.20 Lead-compensator for position control

Example 4.19 Consider the control of a permanent magnet brushed DC motor when constant references are employed. How can it be explained that the proportional control of velocity cannot achieve a zero steady-state error, but a proportional control of the position does achieve a zero steady-state error?

The answer to this question is the following. When the error is zero in position control, then the commanded current $i^* = k_p(\theta_d - \theta)$ is zero; hence, the motor stops and the condition $\theta_d = \theta$ can stand forever.

On the other hand, when the error is zero in velocity control, then the commanded current $i^* = k_p(\omega_d - \omega)$ is zero; hence, the motor tends to stop. This means that the condition $\omega_d = \omega$ cannot stand forever. As a result the steady-state error is such that the difference between ω and ω_d is large enough to command an electric current $i^* = k_p(\omega_d - \omega)$, which maintains the motor rotating at that velocity. When a PI velocity controller is employed, then the integral of the error $\int_0^t (\omega_d - \omega(r))dr$ is constant when $\omega_d = \omega$. This constant value, multiplied by k_i is enough to command a suitable electric current to maintain the motor rotating at the desired velocity.

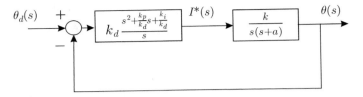

Fig. 4.21 Proportional–integral–derivative (PID) control of the position

Example 4.20 Consider the position control problem of a permanent magnet brushed DC motor where a PID controller is employed, i.e.,

$$i^* = k_p e + k_d \frac{de}{dt} + k_i \int_0^t e(r)dr, \quad e = \theta_d - \theta.$$

Use of the Laplace transform yields:

$$I^*(s) = k_p E(s) + k_d s E(s) + k_i \frac{E(s)}{s},$$

$$= \left(k_p + k_d s + \frac{k_i}{s}\right) E(s),$$

$$= k_d \frac{s^2 + \frac{k_p}{k_d}s + \frac{k_i}{k_d}}{s} E(s).$$

Hence, the block diagram in Fig. 4.21 is obtained. Notice that the motor has one pole at $s = 0$ and the controller has another pole at $s = 0$, i.e., the system type is 2 because the open-loop transfer function has two poles at the origin. This means that, according to Sects. 4.4.1 and 4.4.2, the position reaches the desired position θ_d, if this is either a step or a ramp. A steady-state error different from zero is obtained if the desired position is a parabola, see Sect. 4.4.3.

To verify the above conclusions, some simulations have been performed using the MATLAB/Simulink diagrams shown in Fig. 4.22. The top simulation diagram uses a ramp as the reference with a slope of 10 and an initial value of -10. The PID controller gains are $k_d = 1$, $k_p = 2$, $k_i = 1$. It is observed in Fig. 4.23 that the steady-state error is zero, which corroborates the steady-state error analysis presented above. The bottom simulation diagram in Fig. 4.22 uses a ramp with slope of 1 and a zero initial value, which passes through an integrator with an initial value of -10. This results in a parabola reference given as $\frac{1}{2}t^2 - 10$, i.e., $A = 1$. The PID controller gains are the same as in the previous simulation, i.e., $k_d = 1$, $k_p = 2$, $k_i = 1$. According to Sect. 4.4.3 and Figs. 4.21, 4.22:

$$e_{ss} = \frac{A}{k_a}, \quad k_a = \lim_{s \to 0} s^2 G(s), \quad G(s) = k_d \frac{s^2 + \frac{k_p}{k_d}s + \frac{k_i}{k_d}}{s} \frac{k}{s(s+a)},$$

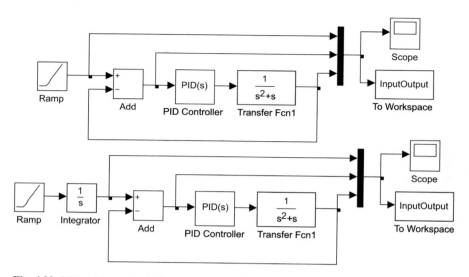

Fig. 4.22 MATLAB/Simulink diagrams used to obtain results in Figs. 4.23 and 4.24

i.e., $k_a = \frac{kk_i}{a}$, and $e_{ss} = \frac{a}{kk_i} = \frac{1}{1 \times 1} = 1$, because $A = 1$. The reader can verify in Fig. 4.24 that the system error is unity at $t = 20$, i.e., in a steady state. This corroborates the steady-state error analysis presented above.

Once simulations in Fig. 4.22 stop, the following MATLAB code is executed in an m-file to draw Figs. 4.23 and 4.24:

```
nn=length(InputOutput(:,1));
n=nn-1;
Ts=20/n;
t=0:Ts:20;
subplot(2,1,1)
plot(t,InputOutput(:,1),'k-.',t,InputOutput(:,3),
'k-');
axis([-1 20 -30 200])
subplot(2,1,2)
plot(t,InputOutput(:,2),'k-');
axis([-1 20 -12 8])
xlabel('time [s]')
```

On the other hand, consider the system depicted in Fig. 4.25. It represents the closed-loop control system of a ball and beam (see Chap. 14). Notice that the open-loop transfer function has two poles at $s = 0$. In this case, these two open-loop poles at the origin are part of the plant model, i.e., the plant naturally possesses those poles and it is not necessary to use a controller with an integral part to introduce them. Hence, the steady-state error is zero if the desired position is either a step or a ramp. However, if it is a parabola, then the steady-state error is different from zero.

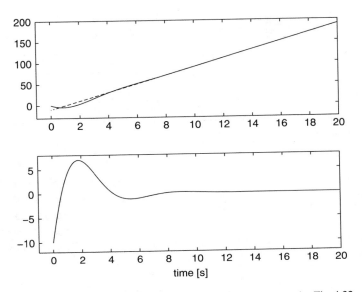

Fig. 4.23 Time response obtained from the top simulation diagram in Fig. 4.22, which is equivalent to the system in Fig. 4.21, when a ramp reference is commanded. Top subfigure: system response (continuous), ramp (dashed) Bottom subfigure: system error

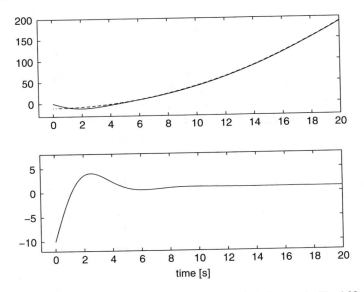

Fig. 4.24 Time response obtained from the bottom simulation diagram in Fig. 4.22, which is equivalent to the system in Fig. 4.21, when a parabola reference is commanded. Top subfigure: system response (continuous), parabola (dashed) Bottom subfigure: system error

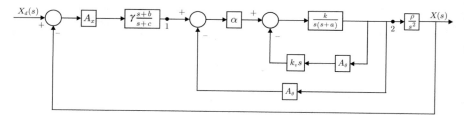

Fig. 4.25 Control of a ball and beam system

Example 4.21 Consider a permanent magnet brushed DC motor with transfer function $G_m(s) = \frac{\theta(s)}{I^*(s)} = \frac{k}{s(s+a)}$. Suppose that it is desired to achieve a zero steady-state error when the position reference θ_d is a parabola. In this case, the system type is required to be 3. As the plant has one pole at $s = 0$, it is necessary to use a controller introducing two poles at $s = 0$. Hence, the following controller is proposed:

$$i^* = k_p e + k_d \frac{de}{dt} + k_i \int_0^t e(r)dr + k_{ii} \int_0^t \int_0^r e(\tau)d\tau dr, \quad e = \theta_d - \theta,$$

which can be rewritten as:

$$I^*(s) = \frac{k_d s^3 + k_p s^2 + k_i s + k_{ii}}{s^2} E(s), \tag{4.16}$$

where $E(s)$ stands for the Laplace transform of the position error. This shows that the controller has two poles at $s = 0$ and, thus, it is useful for solving the problem. Notice that not only the double-integral term is included, with gain k_{ii}, but also the single-integral term with gain k_i. This is done for closed-loop stability reasons. If the single- integral term is not considered, then closed-loop instability is observed. This can be verified by the reader obtaining the closed-loop characteristic polynomial and applying Routh's stability criterion. As a general rule, if a controller has to introduce an integral term of order k, then the terms of integrals of order 1 to $k - 1$ must also be included.

Consider the simulation diagram in Fig. 4.26. This is a system with type 3, which is equivalent to a plant $G_m(s) = \frac{1}{s(s+1)}$ with a controller such as that in (4.16) where $k_d = 0.5$, $k_p = 3$, $k_i = 2$, $k_{ii} = 1$. A ramp with an unit slope and a zero initial value is passed through an integrator with -10 as initial value. This corresponds to a parabola reference given as $\frac{1}{2}t^2 - 10$, i.e., $A = 1$. In Fig. 4.27 the steady-state error is corroborated to be zero, which verifies the above predictions.

Once the simulation in Fig. 4.26 stops, the following MATLAB code is executed in an m-file to draw Fig. 4.27:

```
nn=length(InputOutput(:,1));
n=nn-1;
```

```
Ts=20/n;
t=0:Ts:20;
subplot(2,1,1)
plot(t,InputOutput(:,1),'k--',t,InputOutput(:,3),
'k-');
axis([-1 20 -30 200])
subplot(2,1,2)
plot(t,InputOutput(:,2),'k-');
axis([-1 20 -12 8])
xlabel('time [s]')
```

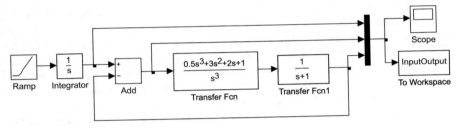

Fig. 4.26 MATLAB/Simulink diagram used to obtain results in Fig. 4.27

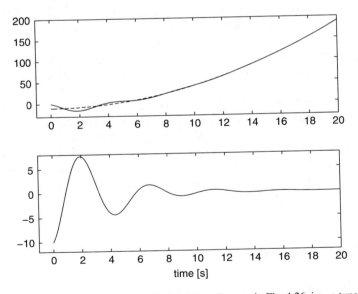

Fig. 4.27 Time response obtained from the simulation diagram in Fig. 4.26, i.e., a type 3 system, when a parabola reference is commanded. Top subfigure: system response (continuous), parabola (dashed) Bottom subfigure: system error

4.5 Summary

The basic preliminary tools for the analysis and design of arbitrary order control systems have been presented in this chapter. It must be understood that the response of any closed-loop control system, no matter how complex, is determined by the equivalent closed-loop transfer function.

The control system representation by means of block diagrams, and their simplification, are instrumental in obtaining the corresponding closed-loop transfer function. From this transfer function it is possible to determine: i) The closed-loop stability and the transient response, i.e., the closed-loop poles, and ii) The steady-state response.

It is explained in Chap. 3 that a necessary and sufficient condition for a transfer function to be stable is that all its poles have a negative real part. However, checking this condition through the exact computation of the poles is a complex problem. This is especially true when the numerical values of the characteristic polynomial coefficients are not known. This situation is common in control systems design as the characteristic polynomial coefficients depend on the controller gains, which are initially unknown. Moreover, it is desirable that the controller gains can be chosen within a range to render the design flexible. These features require the use of analytical tools, instead of numerical tools, useful to determine when a closed-loop control system is stable. This fact represents an important obstacle because the analytical formulas existing to compute the roots of higher-degree polynomials are complex. Furthermore, no analytical solution exists for polynomials with a degree greater than or equal to 5. This problem is successfully solved by Routh's stability criterion, which, however, only verifies whether all the polynomial roots have a negative real part, but does not compute the exact values of the roots. On the other hand, although the rule of signs presented in Sect. 4.2 has some limitations, it may be simpler to use than Routh's stability criterion in some applications. This is the reason for including such a methodology.

As a result of the study of the steady-state error, some criteria are established for suitable selection of a controller such that the steady-state response reaches the desired value; in other words, to render the forced response equal or very close to the close-loop system input, i.e., the desired response. Finally, the controller selection must be also performed such that the desired transient response is achieved through the suitable location of the closed-loop poles. The solution to this latter problem is presented in the next chapter.

4.6 Review Questions

1. What does system type mean?
2. How can the system type be increased in a control system?

3. How does Routh's stability criterion ensure that the real part of all roots of the characteristic polynomial is negative?
4. What are the advantages and the disadvantages or the rule of signs presented in Sect. 4.2?
5. Suppose that analysis of the steady-state error concludes that the output final value reaches the desired output value. Why is it still necessary for the closed-loop system to be stable?
6. Suppose that a controller possessing a five times iterated integral is required to render the steady-state error zero. Why is it also necessary to include terms with a four times integrated integral, a three times iterated integral, a two times iterated integral and a simple integral? Illustrate your answer using a permanent magnet brushed DC motor as a plant and Routh's stability criterion.
7. What is the relationship between the study of the steady-state error in this chapter and the requirement that the forced response reaches the closed-loop system input, i.e., the desired output?
8. The study of the steady-state error that has been presented in this chapter only considers test inputs such as step, ramp, and parabola. What would happen in applications, such as the video camera tracking control system, where the desired output is not exactly known in advance?

4.7 Exercises

1. Based on the knowledge of the system type required to ensure a zero steady-state error for references such as the step, the ramp and the parabola, show that the corresponding closed-loop transfer function (see Fig. 4.10) must possess the following features:

   - The terms independent of s must be equal in the polynomials at the numerator and the denominator to ensure a zero steady-state error when the reference is a step.
   - The terms independent of s in addition to the first-order terms in s must be equal in the polynomials at the numerator and the denominator to ensure a zero steady-state error when the reference is a ramp.
   - The terms independent of s in addition to the first-order and the second-order terms in s must be equal in the polynomials at the numerator and the denominator to ensure a zero steady-state error when the reference is a parabola.

   This means that the controller also has to suitably assign the closed-loop transfer function zeros, which supports the arguments presented in Example 4.21 (see (4.16)).

2. In Example 3.18, in Chap. 3, everyday experience-based arguments are presented to explain the steady-state error achieved by a proportional control of a mass-spring-damper system when x_d is constant. Now, explain this result using the system type as the argument.

3. Consider the mass-spring-damper system presented in Example 3.8, Chap. 3. If after a constant force A is applied, the mass reaches $\dot{x} = 0$ and $\ddot{x} = 0$, then replacing this in the corresponding differential equation, it is found that $x = \frac{A}{K}$ in a steady state. Now, use the final value theorem to compute the final value of x when $f = A$. What is the relationship among conditions $\dot{x} = 0$, $\ddot{x} = 0$, and $s \to 0$ in the final value theorem?

4. Consider the rotative mass-spring-damper system:

$$J\ddot{\theta} + f\dot{\theta} + K\theta = T,$$

where J is the inertia, f is the viscous friction coefficient, K is the spring stiffness constant, θ is the body angular position, and T the applied torque, which is given by the following PID controller:

$$T = k_p e + k_d \frac{de}{dt} + k_i \int_0^t e(r)dr, \quad e = \theta_d - \theta.$$

Use Routh's stability criterion to show that the following conditions:

$$J > 0 \quad k_i > 0, \quad K + k_p > 0, \quad \frac{f + k_d}{J} > \frac{k_i}{K + k_p} > 0,$$

ensure closed-loop stability. Notice that a large integral gain k_i tends to produce instability and this effect can be compensated for using large values for either k_p or k_d. Also notice that a small inertia J allows the use of larger integral gains before instability appears and something similar occurs if the stiffness constant K is large. Can you find an explanation for this from the point of view of physics (mechanics)?

5. Proceeding as in the Example 4.3 in this chapter, show that the closed-loop transfer function shown in Fig. 4.3 is:

$$M(s) = \frac{C(s)}{R(s)} = \frac{G(s)}{1 - G(s)H(s)},$$

when the system has positive feedback, i.e., when the feedback path adds (instead of subtracts) in Fig. 4.3.

6. Consider the mechanical system depicted in Fig. 4.28. In the Example 2.3, Chap. 2, it was found that the corresponding mathematical model is given as:

$$\frac{d^2 x_{m1}}{dt^2} + \frac{b_1}{m_1}\frac{dx_{m1}}{dt} + \frac{K}{m_1}(x_{m1} - x_{m2}) = \frac{1}{m_1}F(t),$$

$$\frac{d^2 x_{m2}}{dt^2} + \frac{b_2}{m_2}\frac{dx_{m2}}{dt} - \frac{K}{m_2}(x_{m1} - x_{m2}) = 0,$$

Fig. 4.28 Two bodies
connected through a spring

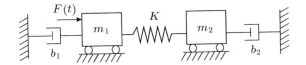

where x_{m1} is the position of m_1 and x_{m2} is the position of m_2. Apply the Laplace
transform to each one of these equations to express $X_{m1}(s)$ and $X_{m2}(s)$ as the
outputs of two transfer functions. Draw the corresponding block diagram suitably
connecting these two transfer functions according to their inputs and outputs.
Using the result in the previous exercise, simplify this block diagram to verify
that:

$$X_{m2}(s) = \frac{G_1(s)G_2(s)/m_1}{1 - G_1(s)G_2(s)K/m_1} F(s),$$

where $F(s)$ is the Laplace transform of $F(t)$ and:

$$G_1(s) = \frac{1}{s^2 + \frac{b_1}{m_1}s + \frac{K}{m_1}},$$

$$G_2(s) = \frac{\frac{K}{m_2}}{s^2 + \frac{b_2}{m_2}s + \frac{K}{m_2}}.$$

- Using this result, verify that the transfer function $\frac{X_{m2}(s)}{F(s)}$ has one pole at $s = 0$.
  What does this mean? Suppose that a constant force $F(t)$ is applied. What
  happens with position of mass m_2 under the effect of this force? What happens
  with the position of the mass m_1? It is suggested that the transfer function
  between $X_{m1}(s)$ and $X_{m2}(s)$ should be found.
- Use Routh's criterion to find the conditions for the transfer function $\frac{X_{m2}(s)}{F(s)}$ to
  be stable.

Notice that one may proceed similarly in the other examples including springs in
Chap. 2.

Some other exercises are proposed at the end of Chap. 5 whose solution involves
concepts and tools introduced in the present chapter.

References

1. S. Bennett, A brief history of automatic control, *IEEE Control Systems Magazine*, pp. 17–25,
   June 1996.
2. K. Ogata, *Modern control engineering*, 4th edition, Prentice-Hall, Upper Saddle river, 2002.

3. N. S. Nise, *Control systems engineering*, 5th. edition, John Wiley and Sons, New Jersey, 2008.
4. R. C. Dorf and R. H. Bishop, *Modern control systems*, 10th. edition, Pearson Prentice-Hall, 2005.
5. B. C. Kuo, *Automatic control systems*, Prentice-Hall, 1995.
6. G. H. Hostetter, C. J. Savant and R. T. Stefani, *Design of feedback control systems*, Holt, Rinehart and Winston, 1982.

Chapter 5
Time Response-Based Design

Consider the closed-loop system in Fig. 5.1. In Example 4.3, Sect. 4.1, it is shown that the closed-loop transfer function is given as:

$$M(s) = \frac{C(s)}{R(s)} = \frac{G(s)}{1 + G(s)H(s)}. \tag{5.1}$$

In Chap. 3 it is explained that the time response $c(t) = \mathcal{L}^{-1}\{C(s)\} = \mathcal{L}^{-1}\{M(s)R(s)\}$ is given as the addition of the natural response and the forced response: $c(t) = c_n(t) + c_f(t)$. The objectives of the control system in Fig. 5.1 are:

- The closed-loop control system is stable, i.e., $\lim_{t \to \infty} c_n(t) = 0$, which is ensured if all of poles of the closed-loop transfer function $M(s)$ have a negative real part. This is important because it ensures that $c(t) \to c_f(t)$ as time increases.
- The convergence $c_n(t) \to 0$ is fast.
- The forced response is equal to the desired output $c_f(t) = r(t) = \mathcal{L}^{-1}\{R(s)\}$.

The last item constitutes the specifications of the response in a steady state and the way to ensure this is studied in Sect. 4.4. The second item constitutes the transient response specifications and the way to satisfy them is studied in the current chapter. It must be stressed that ensuring the desired transient response specifications must also ensure closed-loop stability.

The design of a controller is performed by suitably assigning the poles (the roots of $1 + G(s)H(s)$) and the zeros (the roots of $G(s)$) of the closed-loop transfer function $M(s)$. In classical control, the plant poles and zeros are always assumed to be known, whereas the structure of the controller is proposed to satisfy the steady-state response specifications, for instance. However, the exact location of the poles and zeros of the controller must be selected to ensure that the closed-loop poles, i.e., the roots of $1 + G(s)H(s)$, are placed at the desired values and that the closed-

© Springer International Publishing AG, part of Springer Nature 2019
V. M. Hernández-Guzmán, R. Silva-Ortigoza, *Automatic Control with Experiments*,
Advanced Textbooks in Control and Signal Processing,
https://doi.org/10.1007/978-3-319-75804-6_5

Fig. 5.1 Closed-loop system

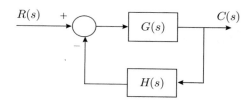

loop zeros, i.e., $G(s) = 0$, are also suitably located.[1] In this chapter, one of the main classical control techniques for control design is studied: *the root locus method* (see [1] for a historical perspective). This method assigns poles of the closed-loop transfer function $M(s)$ through a suitable modification of the *open-loop transfer function* $G(s)H(s)$. Moreover, it is also possible to place the zeros of the closed-loop transfer function $M(s)$ such that their effects on the closed-loop system response are small.

5.1 Drawing the Root Locus Diagram

Assuming that the poles and zeros of the plant are known, the root locus method is a graphical tool that is useful to locate the closed-loop poles by suitably selecting poles and zeros of a controller. Moreover, this can also be performed by cancelling some undesired (slow) closed-loop poles with some controller zeros. According to Sect. 3.5.1, this achieves system order reduction and the elimination of some zeros such that the closed-loop system transient response can be approximated by some of the simple cases studied in Sects. 3.1.1 and 3.3.1. This simplifies the task of designing a controller such that the closed-loop transfer function $M(s)$ has poles and zeros that ensure the desired transient response specifications.

The root locus diagram is represented by several curves whose points constitute the closed-loop system poles parameterized by the open loop-gain, k, which takes values from $k = 0$ to $k = +\infty$. According to (5.1), every closed-loop pole s, i.e., a point in the s plane belonging to the root locus, must satisfy $1 + G(s)H(s) = 0$. The root locus diagram is drawn by proposing points s on the complex plane s, which are checked to satisfy $1 + G(s)H(s) = 0$. To verify this condition in a simple manner, a set of rules are proposed, which are listed next. Following these rules, the root locus diagram is drawn using the poles and zeros of the open-loop transfer function $G(s)H(s)$ as data. This means that the closed-loop poles are determined by the open-loop poles and zeros.

It is assumed that the open-loop transfer function can be written as:

$$G(s)H(s) = k\frac{\prod_{j=1}^{m}(s - z_j)}{\prod_{i=1}^{n}(s - p_i)}, \quad n > m, \tag{5.2}$$

---

[1]Although the poles determine stability and transient response, zeros also have an effect on the final shape of the transient response (see Sect. 8.1, Chap. 8). This is the reason why the design based on the root locus is sometimes performed by trying to cancel one closed-loop pole with a closed-loop zero.

Fig. 5.2 Graphical
representation of factors
$s - z_j$ and $s - p_i$

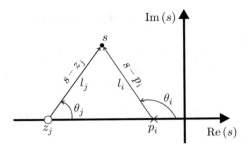

where z_j, $j = 1, \ldots, m$, are the open-loop zeros and p_i, $i = 1, \ldots, n$, are the open-loop poles. As z_j, p_i and s are, in general, complex numbers located on the s plane, the following can be written:

$$s - z_j = l_j \angle \theta_j,$$

$$s - p_i = l_i \angle \theta_i,$$

as depicted in Fig. 5.2. Then:

$$G(s)H(s) = k \frac{\prod_{j=1}^{m} l_j \angle \theta_j}{\prod_{i=1}^{n} l_i \angle \theta_i} = k \frac{\prod_{j=1}^{m} l_j}{\prod_{i=1}^{n} l_i} \angle \left(\sum_{j=1}^{m} \theta_j - \sum_{i=1}^{n} \theta_i \right). \qquad (5.3)$$

On the other hand, the closed-loop poles are those values of s satisfying the following:

$$1 + G(s)H(s) = 0. \qquad (5.4)$$

This means that closed-loop poles are those values of s satisfying:

$$G(s)H(s) = -1 = 1 \angle \pm (2q + 1)180°, \quad q = 0, 1, 2, \ldots \qquad (5.5)$$

Using (5.3) and (5.5) the so-called *magnitude condition* (5.6) and *angle condition* (5.7) are obtained, which must be satisfied by every closed-loop pole s, i.e., those points s belonging to root locus:

$$k \frac{\prod_{j=1}^{m} l_j}{\prod_{i=1}^{n} l_i} = 1, \qquad (5.6)$$

$$\sum_{j=1}^{m} \theta_j - \sum_{i=1}^{n} \theta_i = \pm (2q + 1)180°, \quad q = 0, 1, 2, \ldots \qquad (5.7)$$

From these conditions, the following rules are obtained [6–9].

5.1.1 Rules for Drawing the Root Locus Diagram

1. **The root locus starts ($k = 0$) at the open-loop poles.**
   Using (5.2) and (5.4):

$$1 + G(s)H(s) = 1 + k\frac{\prod_{j=1}^{m}(s - z_j)}{\prod_{i=1}^{n}(s - p_i)} = 0,$$

$$= \frac{\prod_{i=1}^{n}(s - p_i) + k\prod_{j=1}^{m}(s - z_j)}{\prod_{i=1}^{n}(s - p_i)} = 0,$$

$$= \prod_{i=1}^{n}(s - p_i) + k\prod_{j=1}^{m}(s - z_j) = 0, \qquad (5.8)$$

$$= \prod_{i=1}^{n}(s - p_i) = 0,$$

   if $k = 0$. This means that the closed-loop poles (those that satisfy $1 + G(s)H(s) = 0$) are identical to the open-loop poles ($s = p_i$, $i = 1, \ldots, n$) if $k = 0$.
2. **Root locus ends ($k \to \infty$) at the open-loop zeros.**
   From (5.8):

$$1 + G(s)H(s) = \prod_{i=1}^{n}(s - p_i) + k\prod_{j=1}^{m}(s - z_j) = 0,$$

$$\approx k\prod_{j=1}^{m}(s - z_j) = 0,$$

   because $\prod_{i=1}^{n}(s - p_i) \ll k\prod_{j=1}^{m}(s - z_j)$ if $k \to \infty$. This means that the closed-loop poles (those satisfying $1 + G(s)H(s) = 0$) are identical to the open-loop zeros ($s = z_j$, $j = 1, \ldots, m$) when $k \to \infty$.
3. **When $k \to \infty$ there are $n - m$ branches of the root locus that tend toward some point at the infinity of the plane s. This means that the open-loop transfer function $G(s)H(s)$ has $n - m$ zeros at infinity.**
   These branches can be identified by the angle of the asymptote where the root locus diagram approaches as $k \to \infty$. The angle that each asymptote forms with the positive real axis can be computed as:

$$\text{asymptote angle} = \frac{\pm 180°(2q + 1)}{n - m}, \qquad q = 0, 1, 2, \ldots \qquad (5.9)$$

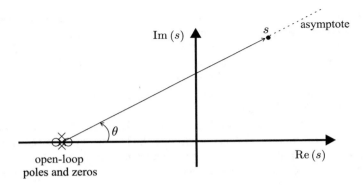

Fig. 5.3 Asymptote angle in rule 3

This formula is obtained as follows. Consider a point s that belongs to an asymptote and is located far away from the origin. Hence, all poles and zeros of $G(s)H(s)$, are observed as gathered in a single point of the plane s, as depicted in Fig. 5.3. Thus, the angle contributed by each open-loop pole and zero to the condition in (5.7) (see Fig. 5.3) is identical to the angle contributed by the others, i.e., representing such an angle by θ, the angle condition (5.7) can be written as:

$$\sum_{j=1}^{m} \theta_j - \sum_{i=1}^{n} \theta_i = (m - n)\theta = \pm(2q + 1)180°, \quad q = 0, 1, 2, \ldots$$

Solving this expression for θ yields:

$$\theta = \frac{\pm(2q + 1)180°}{n - m}, \quad q = 0, 1, 2, \ldots$$

which becomes (5.9) when θ =asymptote angle (see Fig. 5.3).

4. **Point σ_a where asymptotes intersect the real axis is computed as [5], Chap. 6:**

$$\sigma_a = \frac{\sum_i p_i - \sum_j z_j}{n - m}.$$

5. **Consider a point on the real axis of the plane s. Suppose that the number of real open-loop poles plus the number of real open-loop zeros on the right of such a point is equal to an odd number. Then such a point on the real axis belongs to the root locus. If this is not the case, then such a point does not belong to the root locus.**

Observe Fig. 5.4 and recall the angle condition (5.7). Notice that two open-loop complex conjugate poles or zeros produce angles that added to result in $0°$ or $360°$ and, hence, have no contribution to the angle condition (5.7). This means that, in this rule, only real open-loop poles and zeros must be considered in (5.7). On the other hand, notice that any real pole or zero located at the left of the test point s contributes with a zero angle. This means that, in this rule, in the expression (5.7), only real open-loop poles and zeros located on the right of the test point s must be considered. Notice that each real open-loop zero on the right of the point s contributes with a $+180°$ angle, whereas each real open-loop pole at the right of the same point contributes with a $-180°$ angle. Hence, the total angle contributed by all open-loop poles and zeros appearing on the right of the test point s is:

$$\sum_{j=1}^{m}\theta_j - \sum_{i=1}^{n}\theta_i = m_1(+180°) + n_1(-180°),$$

where m_1 and n_1 stand for the number of real open-loop zeros and poles respectively, on the right of the test point s. On the other hand, if the subtraction of two numbers is equal to an odd number, then the addition of the same numbers is also an odd number. Hence, if:

$$\sum_{j=1}^{m}\theta_j - \sum_{i=1}^{n}\theta_i = (m_1 - n_1) \times (+180°) = \pm(2q + 1)180°,$$

for some $q = 0, 1, 2, \ldots$, i.e., $m_1 - n_1$ is odd, then it is also true that:

$$\sum_{j=1}^{m}\theta_j - \sum_{i=1}^{n}\theta_j = (m_1 + n_1) \times (+180°) = \pm(2q + 1)180°,$$

for some $q = 0, 1, 2, \ldots$, i.e., $m_1 + n_1$ is odd. Notice that this last expression constitutes the angle condition (5.7) and, thus, the test point s in Fig. 5.4 is a closed-loop pole and belongs to the root locus.

6. **The root locus is symmetrical with respect to the real axis.**
   This is easily understood by recalling that all complex poles appear simultaneously with their complex conjugate pair (see first paragraph in Sect. 3.4.3).

7. **The departure angle from an open-loop complex pole is computed as:**
   Departure angle from a complex pole=
   $\pm(2q+1)180°+\sum$[angles of vectors from the open-loop zeros to the open-loop pole under study]
   $-\sum$[angles of vectors from the other open-loop poles to the open-loop pole under study]

$$\text{(5.10)}$$

Fig. 5.4 Open-loop poles
and zeros used to study rule 5

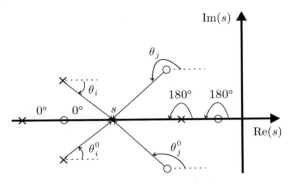

Fig. 5.5 Open-loop poles
and zeros in the study of
rule 7

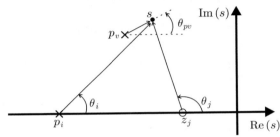

Consider Fig. 5.5. Assume that the point s belongs to the root locus and it is very close to open-loop pole at p_v. The angle θ_{pv} is the departure angle from the open-loop complex pole at p_v. According to the angle condition (5.7):

$$\sum_{j=1}^{m} \theta_j - \sum_{i=1}^{n} \theta_i =$$
$$= (\theta_{z1} + \theta_{z2} + \cdots + \theta_{zm}) - (\theta_{p1} + \theta_{p2} + \cdots + \theta_{pv} + \cdots + \theta_{pn})$$
$$= \pm(2q + 1)180°, \quad q = 0, 1, 2, \ldots$$

where θ_{zj} and θ_{pi} stand for the angles due to the open-loop zeros and poles respectively. Solving for θ_{pv}:

$$\theta_{pv} = \pm(2q + 1)180° + (\theta_{z1} + \theta_{z2} + \cdots + \theta_{zm}) - (\theta_{p1} + \theta_{p2} + \cdots + \theta_{pn}),$$

for $q = 0, 1, 2, \ldots$, which is equivalent to (5.10) because $s = p_v$ can be assumed as they are very close.

Fig. 5.6 Open-loop poles
and zeros in the study of
rule 8

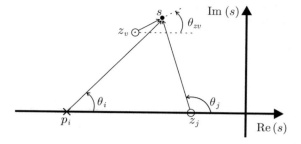

8. **The arrival angle to an open-loop complex zero is computed as:**
   Arrival angle to a complex zero$=$
   $\pm(2q + 1)180° - \sum$[angles of vectors from the other open-loop zeros to the
   open-loop zero under study]
   $+ \sum$[angles of vectors from the open-loop poles to the open-loop zero under
   study]

$$(5.11)$$

   Consider Fig. 5.6. Assume that the point s belongs to the root locus and is
very close to zero at z_v. The angle θ_{zv} is the arrival angle to the open-loop zero
at z_v. According to the angle condition (5.7):

$$\sum_{j=1}^{m} \theta_j - \sum_{i=1}^{n} \theta_i =$$
$$= (\theta_{z1} + \theta_{z2} + \cdots + \theta_{zv} + \cdots + \theta_{zm}) - (\theta_{p1} + \theta_{p2} + \cdots + \theta_{pn}),$$
$$= \pm(2q + 1)180°,$$

where $q = 0, 1, 2, \ldots$, whereas θ_{zj} and θ_{pi} stand for the angles due to the
open-loop zeros and poles respectively. Solving for θ_{zv}:

$$\theta_{zv} = \pm(2q + 1)180° - (\theta_{z1} + \theta_{z2} + \cdots + \theta_{zm}) + (\theta_{p1} + \theta_{p2} + \cdots + \theta_{pn}),$$

for $q = 0, 1, 2, \ldots$, which is equivalent to (5.11) because $s = z_v$ can be
assumed as they are very close.

9. **The open-loop gain, k, required for a point s, belonging to root locus, to
   actually be selected as a desired closed-loop pole is computed according to
   the magnitude condition (5.6):**

$$k = \frac{\prod_{i=1}^{n} l_i}{\prod_{j=1}^{m} l_j}.$$

10. The points where the root locus passes through the imaginary axis can be determined using Routh's criterion (see Sect. 4.3).

11. **If an additional open-loop pole located on the left half-plane is considered, "instability tends to increase" in the closed-loop system. This effect is stronger as the additional pole is placed closer to the origin.**
    Consider Fig. 5.7, where s represents a point that belongs to the root locus and the open-loop transfer function is assumed to have only the shown two poles without zeros. According to the angle condition (5.7):

    $$-(\theta_{p1} + \theta_{p2}) = -180°.$$

    The poles at p_1 and p_2 are retained in Fig. 5.8, but an additional pole is considered at p_3. The angle condition (5.7) now becomes:

    $$-(\theta_{p1} + \theta_{p2} + \theta_{p3}) = -180°.$$

    This means that the addition $\theta_{p1} + \theta_{p2}$ must be smaller in Fig. 5.8 with respect to Fig. 5.7, which is accomplished if point s in Fig. 5.7 moves to the right, as in Fig. 5.8, i.e., if the root locus bends toward the right in Fig. 5.8. It is easy to see that this effect is stronger, i.e., the root locus is further pushed to the right half-plane, as θ_{p3} is larger, i.e., as p_3 approaches the origin. This proves that and integral controller tends to produce instability in the closed-loop system.

12. **If an additional open-loop zero located on the left half-plane is considered, "stability tends to increase" in the closed-loop system. This effect is stronger as the additional zero is placed closer to the origin.**
    Consider again Fig. 5.7. According to the angle condition (5.7):

    $$-(\theta_{p1} + \theta_{p2}) = -180°.$$

    The poles at p_1 and p_2 are retained in Fig. 5.9, but an additional zero at z_1 is considered. The angle condition (5.7) now becomes:

    $$\theta_{z1} - (\theta_{p1} + \theta_{p2}) = -180°.$$

    This means that the addition $\theta_{p1} + \theta_{p2}$ must be larger in Fig. 5.9 than in Fig. 5.7, which is accomplished if point s in Fig. 5.7 moves to the left as in Fig. 5.9, i.e., if the root locus bends toward the left in Fig. 5.9. It is easy to see that this effect is stronger, i.e., the root locus is pulled to the left, as θ_{z1} is larger, i.e., as z_1 is closer to the origin. This proves that a derivative controller tends to improve the stability of a closed-loop system.

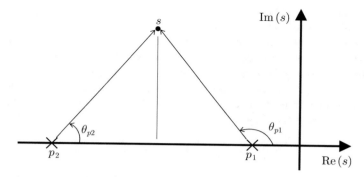

Fig. 5.7 A point belonging to the root locus for an open-loop system with two poles

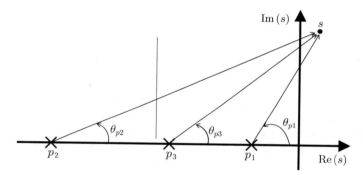

Fig. 5.8 The root locus is pushed toward the right when an additional open-loop pole is considered

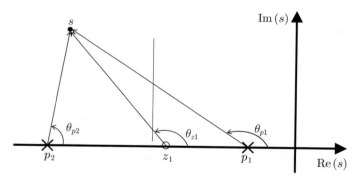

Fig. 5.9 The root locus is pulled toward the left when an additional open-loop zero is considered

5.2 Root Locus-Based Analysis and Design

5.2.1 Proportional Control of Position

According to Chap. 11, the permanent magnet brushed direct current (DC) motor model is given as:

$$\theta(s) = \frac{k}{s(s+a)} I^*(s), \qquad (5.12)$$

$$a = \frac{b}{J} > 0, \quad k = \frac{nk_m}{J} > 0,$$

where $\theta(s)$ and $I^*(s)$ stand for the position and the electric current commanded respectively. Suppose that the following proportional position controller is employed:

$$I^*(s) = k_p(\theta_d(s) - \theta(s)),$$

where $\theta_d(s)$ is the desired position and k_p is a constant known as the proportional gain. The closed-loop system can be represented as in Fig. 5.10, where it is concluded that the open-loop transfer function is given as:

$$G(s)H(s) = \frac{k_p k}{s(s+a)}. \qquad (5.13)$$

Notice that the system type is 1, i.e., the steady state error is zero when the desired position is a step. Hence, the only design problem that remains is to choose k_p such that the closed-loop poles are assigned to the desired locations. Use of the root locus method to solve this problem is shown in the following. The root locus method forces the gain k_p to take values from 0 to $+\infty$. First, $G(s)H(s)$ is rewritten as:

$$G(s)H(s) = \frac{k_p k}{l_1 l_2} \angle - (\theta_1 + \theta_2),$$

where the vectors $s - 0 = l_1 \angle \theta_1$ and $s - (-a) = l_2 \angle \theta_2$ have been defined (see Fig. 5.11). The fundamental conditions for drawing the root locus are the angle condition (5.7) and the magnitude condition (5.6) which are expressed respectively as:

$$-(\theta_1 + \theta_2) = \pm(2q + 1)180°, \quad q = 0, 1, 2, \ldots$$

$$\frac{k_p k}{l_1 l_2} = 1.$$

Fig. 5.10 Proportional
position control system

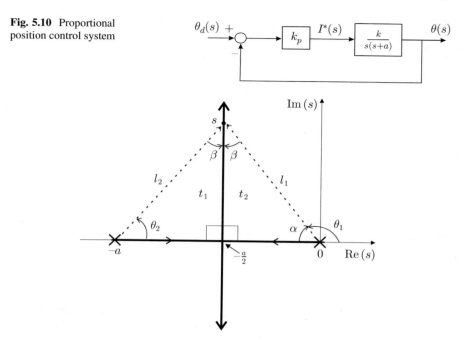

Fig. 5.11 Root locus for $G(s)H(s) = \frac{k_p k}{s(s+a)}$

Rule 5 indicates that, on the real axis, the root locus only exists between the points
$s = 0$ and $s = -a$. Furthermore, according to rule 1, the root locus begins ($k_p = 0$)
at $s = 0$ and $s = -a$. On the other hand, according to rules 2 and 3, the branches
starting at $s = 0$ and $s = -a$ must approach some open-loop zero (no zero exists in
this case) or some point at the infinity of the plane s as k_p tends to $+\infty$. Hence, the
root locus has to move away from points $s = 0$ and $s = -a$, on the real axis, as k_p
increases such that a breakaway point must exist somewhere between such points
and then two branches appear to be moving away to infinity in plane s.

On the other hand, according to the angle condition, $-(\theta_1 + \theta_2) = -180° =$
$-(\alpha + \theta_1)$ in Fig. 5.11, it is concluded that $\alpha = \theta_2$, i.e., both triangles t_1 and t_2 must
be identical for any closed-loop pole s. Hence, both branches referred above, which
are shown in Fig. 5.11, must be parallel to the imaginary axis. This means that the
breakaway point is located at the middle point between $s = 0$ and $s = -a$, i.e., at
$(0 - a)/2 = -a/2$. This can also be verified using rules 3 and 4. Finally, according
to rule 6, both branches are symmetrical with respect to the real axis.

The magnitude condition is employed when the exact value of k_p that renders a
specific point of the root locus an actual closed-loop pole is required. Notice, for
instance, that lengths l_1 and l_2 grow as k_p tends to $+\infty$ to satisfy the magnitude
condition:

$$\frac{k_p k}{l_1 l_2} = 1, \tag{5.14}$$

which means that the closed-loop poles corresponding to large values of k_p tend to some point at infinity on the s plane. Hence, it is concluded: (i) Both closed-loop poles are real, negative, different, and approach the point $s = -a/2$ when $k_p > 0$ is small; this means that the closed-loop system becomes faster because the slowest pole moves away from the origin, (ii) According to the magnitude condition, when:

$$k_p = \frac{l_1 l_2}{k} = \frac{a^2}{4k}, \quad \text{with } l_1 = l_2 = \frac{a}{2},$$

both closed-loop poles are real, repeated, negative, and located at $s = -\frac{a}{2}$, i.e., the fastest response without oscillations is obtained, (iii) As $k_p > \frac{a^2}{4k}$ increases, both poles move away from the real axis (one pole moves upward and the other downward) on the vertical line passing through $s = -\frac{a}{2}$; this means that the closed-loop system becomes faster (because ω_n increases; see Sect. 3.3) and more oscillatory (because the angle $90° - \alpha$ decreases and damping, given as $\zeta = \sin(90° - \alpha)$, decreases; see Sect. 3.3).

The above discussion shows that it is not possible to achieve a closed-loop system response that is simultaneously fast and well damped. This is a direct consequence of the fact that closed-loop poles cannot be assigned at any arbitrary location of the s plane, as they can only be located on the thick straight line shown in Fig. 5.11. In the next example, it is shown that the introduction of an additional zero in the open-loop transfer function of the present example allows the closed loop poles to be assigned at any point on the s plane.

Example 5.1 The root locus diagram can also be plotted using MATLAB. Given a closed-loop system as that in Fig. 5.1, it suffices to use the command:

```
rlocus(GH)
```

where GH stands for the open-loop transfer function $G(s)H(s)$. Consider, for instance, the closed-loop system in Fig. 5.10 when $k = 2$ and $a = 8$. The root locus diagram in Fig. 5.12 is plotted using the following commands:

```
k=2;
a=8;
gh=tf(k,[1 a 0]);
rlocus(gh);
rlocfind(gh)
```

The root locus in Fig. 5.12 is represented by the continuous lines. Notice that, contrary to the previous discussion, in the above commands the proportional gain k_p is not considered. This is because k_p is automatically increased from 0 to $+\infty$ by MATLAB when executing the command "rlocus()". This must be taken into account when defining the open-loop transfer function to be used as an argument of this MATLAB command. In Fig. 5.12, each open-loop pole is represented by a symbol "×", i.e., at $s = 0$ and $s = -a = -8$. Compare Figs. 5.12 and 5.11 to verify similarities between them. For instance, the horizontal and vertical lines in

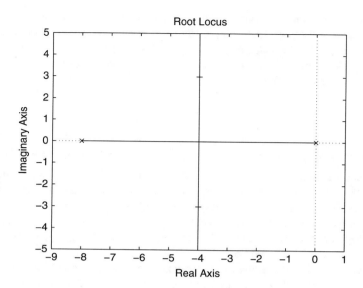

Fig. 5.12 Root locus for closed-loop system in Fig. 5.10

Fig. 5.12 intersect at $s = -4$, i.e., at $s = -a/2$, as predicted in Fig. 5.11. The root locus diagram is plotted by MATLAB when executing the command

```
rlocus(gh);
```

alone. However, the additional use of the command:

```
rlocfind(gh)
```

allows the user to select any desired point on the root locus. When selected, this point and all points on the root locus corresponding to the same gain k_p are marked with a symbol "+". For instance, in Fig. 5.12 the point $-4 + 3j$ was selected and, automatically, the point $-4 - 3j$ was also selected and marked with a "+" (only the horizontal line of this symbol is appreciable in Fig. 5.12 because of the particular geometry of this root locus). Also, MATLAB automatically sends the following text:

```
selected_point =
-3.9882 + 2.9969i
ans =
12.4907
```

This means that $k_p = 12.4907$ at $s = -4 \pm 3j$ and this can be verified using the magnitude condition in (5.14) as:

$$s - 0 = l_1 \angle \theta_1, \quad s - (-a) = l_2 \angle \theta_2,$$

$$l_1 = \sqrt{4^2 + 3^2} = 5, \quad l_2 = \sqrt{(8-4)^2 + 3^2} = 5,$$

$$k_p = \frac{l_1 l_2}{k} = \frac{25}{2} = 12.5.$$

When using the command "rlocus()", MATLAB automatically determines a maximal value of k_p to be considered to draw Fig. 5.12. However, the user can decide herself/himself the maximal value of k_p to be considered by using the command:

```
rlocus(gh,d);
```

where d is a vector containing all of the specific values of the gain k_p, which the user wants to be considered to plot the root locus.

Some closed-loop pole pairs are selected in Fig. 5.13. The time response of the closed-loop system Fig. 5.10 is presented in Fig. 5.14 when the corresponding closed-loop pole pairs are those indicated in Fig. 5.13. Recall that k_p increases as it passes from poles at "o" to poles at the square in Fig. 5.13. It is observed that the closed-loop system response becomes faster as k_p increases and overshoot is present once the closed-loop poles have imaginary parts that are different from zero. Moreover, overshoot is small if the imaginary parts are small. These observations corroborate the above root locus-based analysis.

The results in Figs. 5.13 and 5.14 were obtained by executing several times the following MATLAB code in an m-file:

```
k=2;
a=8;
kp=30; % 1 5 10 20 30
gh=tf(kp*k,[1 a 0]);
figure(1)
rlocus(gh);
hold on
rlocus(gh,34);
axis([-10 2 -9 9])
hold on
M=feedback(gh,1,-1);
v=pole(M);
figure(1)
plot(real(v(1)),imag(v(1)),'bs')
plot(real(v(2)),imag(v(2)),'bs')
figure(2)
step(M,'b:',5)
hold on
```

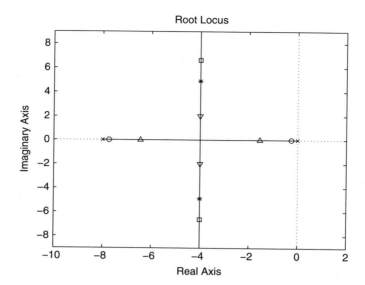

Fig. 5.13 Root locus in Fig. 5.12 when some closed-loop pole pairs are selected

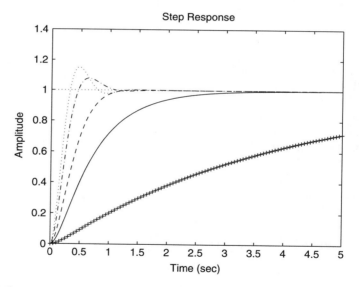

Fig. 5.14 Step response of the closed-loop system in Fig. 5.10, when the closed-loop poles are located as indicated in Fig. 5.13. The +line corresponds to poles at "o." The continuous line corresponds to poles at the triangle up. The dashed line corresponds to poles at the triangle down. The dash–dot line corresponds to poles at "*." The dotted line corresponds to poles at the square

5.2.2   *Proportional–Derivative Control of Position*

Consider again the motor model shown in (5.12), but now together with the following proportional–derivative (PD) controller:

$$i^* = k_p e + k_d \frac{de}{dt}, \quad e = \theta_d - \theta,$$

where k_p is the proportional gain and k_d is a constant known as the derivative gain. Use of the Laplace transform yields:

$$I^*(s) = k_p E(s) + k_d s E(s),$$
$$= (k_p + k_d s) E(s),$$
$$= k_d \left(s + \frac{k_p}{k_d} \right) E(s).$$

The corresponding closed-loop block diagram is shown in Fig. 5.15. The open-loop transfer function is given as:

$$G(s)H(s) = \frac{k_d k (s + c)}{s(s + a)}, \quad c = \frac{k_p}{k_d} > 0. \tag{5.15}$$

Notice that, now, k_d is the gain that varies from 0 to $+\infty$ to plot the root locus. First, $G(s)H(s)$ is rewritten as:

$$G(s)H(s) = \frac{k_d k \, l_3}{l_1 l_2} \angle \theta_3 - (\theta_1 + \theta_2), \tag{5.16}$$

where the vectors $s - 0 = l_1 \angle \theta_1$, $s - (-a) = l_2 \angle \theta_2$, and $s - (-c) = l_3 \angle \theta_3$ have been defined. The angle and the magnitude conditions are expressed respectively as:

$$\theta_3 - (\theta_1 + \theta_2) = \pm(2q + 1)180°, \quad q = 0, 1, 2, \ldots,$$
$$\frac{k_d k \, l_3}{l_1 l_2} = 1.$$

The root locus in this case can be obtained from the root locus obtained for the transfer function in (5.13) by simply taking into account that an additional zero at $s = -c$ has been included.

According to rule 12, the branches of the root locus in Fig. 5.11 will bend toward the left as a consequence of the additional zero. This can be verified using rule 3 to find that, now, the root locus only has one branch whose asymptote forms a $\pm 180°$ with the positive real axis. As shown in Fig. 5.16, the root locus has two different possibilities depending on the exact location of the zero at $s = -c$. If it is placed at the left of the pole at $s = -a$ (as in Fig. 5.16a) then, according to rule 5, the root

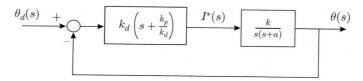

Fig. 5.15 Proportional–derivative (PD) control of position

Fig. 5.16 Root locus for
$G(s)H(s) = \frac{k_d k(s+c)}{s(s+a)}$. **(a)**
$c > a$. **(b)** $c < a$

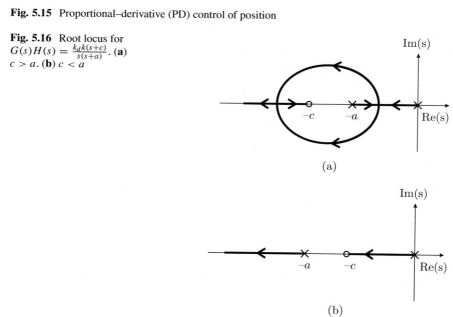

locus exists on two segments of the negative real axis: between the points $s = 0$ and $s = -a$, and to the left of zero at $s = -c$. Moreover, according to rules 2 and 3 one of the two root locus branches beginning ($k_d = 0$) at the open-loop poles located at $s = 0$ and $s = -a$ must tend toward the zero at $s = -c$ whereas the other branch must tend toward infinity in the s plane following the asymptote forming $\pm 180°$ with the positive real axis. Hence, a breakaway point must exist between points $s = 0$ and $s = -a$. Furthermore, as the root locus is symmetrical with respect to the real axis (rule 6), these branches must describe two semi circumferences toward the left of the breakaway point, which join again in a break-in point located on the negative real axis to the left of the zero at $s = -c$. After that, one branch approaches the zero at $s = -c$ and the other tends toward infinity along the negative real axis.

On the other hand, if the zero at $s = -c$ is placed between the open-loop poles at $s = 0$ and $s = -a$, then, according to rule 5, the root locus exists on two segments on the negative real axis located between points $s = -c$ and $s = 0$ and on the left of the pole at $s = -a$. Notice that, now, the branch beginning ($k_d = 0$) at $s = 0$ approaches the zero at $s = -c$, whereas the branch beginning ($k_d = 0$) at $s = -a$ tends toward infinity along the negative real axis. Also notice that this is possible without the necessity for any branch to exist outside the real axis, as is shown in Fig. 5.16b.

From the study of the resulting root locus, it is possible to realize that the closed-loop system is stable for any $k_p > 0$ and $k_d > 0$ because both possibilities for the root locus that have been presented in Fig. 5.16 show that the closed-loop poles are always located on the left half-plane s, i.e., the closed-loop poles have a negative real part. Finally, according to Sect. 3.8.3, there always exist gains k_p and k_d allowing any values to be chosen for both ω_n and ζ. This means that it is always possible to assign both closed-loop poles at any desired point on the left half-plane s.

Notice that the open-loop zero at $s = -c$ is also a zero of the closed-loop transfer function; hence, it also affects the closed-loop system transient response. This means that the transient response will not have the exact specifications computed using (3.71).

Example 5.2 The following MATLAB code is executed in an m-file to simulate the closed-loop system in Fig. 5.15:

```
clc
a=7;
k=70;
c=4; %7 4 10
Polo_des=-20;
l3=abs(Polo_des)-c;
l1=abs(Polo_des);
l2=abs(Polo_des)-a;
kd=l1*l2/(k*l3);
kp=c*kd;
Md=tf(20,[1 20]);
gm=tf(k,[1 a 0]);
PD=tf([kd kp],1);
M=feedback(gm*PD,1,-1);
v=pole(M)
%{
figure(1)
rlocus(gm*PD);
hold on
plot(real(v(1)),imag(v(1)),'k^');
plot(real(v(2)),imag(v(2)),'k^');
%}
figure(2)
step(Md*2,'r--',M*2,':',0.25)
hold on
```

It is assumed that $a = 7$ and $k = 70$. The desired response is that of a first-order system with time constant 0.05[s], i.e., with a real pole at $s = -20$. The idea is to propose different locations for the zero at $s = -c$ to observe when the desired response is accomplished. The derivative gain k_d is computed using the magnitude condition $\frac{k_d k \, l_3}{l_1 l_2} = 1$ where l_1, l_2, l_3 are the magnitude of the vectors defined in

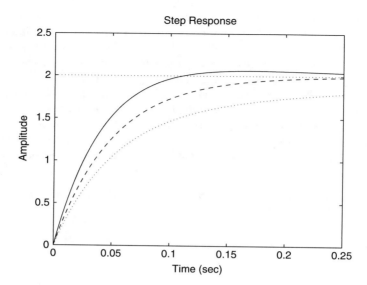

Fig. 5.17 Step response of the closed-loop system in Fig. 5.15, when a closed-loop first-order response with 0.05[s] time constant is desired. Dashed: desired response and closed-loop response when $c = a$. Dotted: closed-loop response when $c = 4$. Continuous: closed-loop response when $c = 10$

the paragraph after (5.16). Then $k_p = ck_d$ is computed. After that, it is possible to generate the closed-loop transfer function M and to simulate it together with the desired response. These results are shown in Fig. 5.17. Notice that the desired response is matched when $c = a$. The root locus diagram in this case is shown in Fig. 5.18 where the closed-loop poles are represented by a triangle up. One closed-loop pole is located at $s = -20$, as desired, and the other cancels with the zero at $s = -c$. This explains why the desired response is accomplished in this case.

When $c = 4$, the closed-loop poles are at $s = -20$, as desired, and $s = -3.25$. The latter pole is slow and does not cancel with zero at $s = -c = -4$; hence, its effects are important in the closed-loop transient response. This explains why, in this case, the response in Fig. 5.17 is slow. When $c = 10$, the closed-loop poles are located at $s = -20$ and $s = -13$, i.e., the pole at $s = -13$ cannot be cancelled by the zero at $s = -c$. This results in an overshoot, i.e., the desired response is not accomplished again. The reason for this overshoot is explained in Sect. 8.1.2, where it is stated that overshoot is unavoidable whenever all of the closed-loop poles are on the left of an open-loop zero despite all the closed-loop poles being real, i.e., as in the present case.

Example 5.3 The following MATLAB code is executed in an m-file to simulate the closed-loop system in Fig. 5.15:

```
clc
a=7;
```

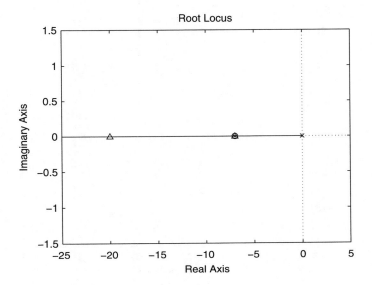

Fig. 5.18 Root locus corresponding to Fig. 5.15 when $c = a$

```
k=70;
tr=0.05;
Mp=14;%
z=sqrt( log(Mp/100)^2 /( log(Mp/100)^2 + pi^2 ) );
wn=1/(tr*sqrt(1-z^2))*(pi-atan(sqrt(1-z^2)/z));
s=-z*wn+wn*sqrt(1-z^2)*j
kd=(2*z*wn-a)/k;
kp=wn^2/k;
den1=conv([1 -s],[1 -real(s)+j*imag(s)]);
Md=tf(den1(3),den1);
gm=tf(k,[1 a 0]);
PD=tf([kd kp],1);
M=feedback(PD*gm,1,-1);
v=pole(M)
%%{
figure(1)
rlocus(gm*PD);
hold on
plot(real(v(1)),imag(v(1)),'k^');
plot(real(v(2)),imag(v(2)),'k^');
%}
figure(2)
step(Md*2,'r--',M*2,'-',0.25)
hold on
```

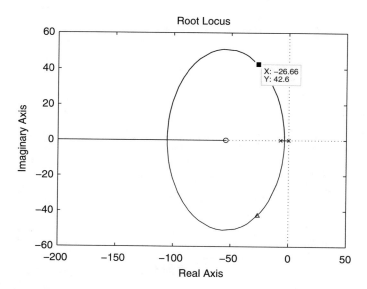

Fig. 5.19 Root locus diagram corresponding to Fig. 5.15 when $c > a$ and two closed-loop complex conjugate poles are specified (triangle up)

The desired specifications are a rise time of 0.05[s] and a 14% overshoot. This results in a pair of desired complex conjugate closed-loop poles located at $s = -26.66 \pm 42.59j$. The controller gains k_p and k_d were computed using:

$$k_d = \frac{2\zeta\omega_n - a}{k}, \quad k_p = \frac{\omega_n^2}{k},$$

(see Sect. 3.8.3). This allows the corresponding root locus diagram to be plotted in Fig. 5.19. Notice that this diagram is similar to that depicted in Fig. 5.16a because $c > a$ in this case. Although the desired closed-loop poles have been assigned, notice that the closed-loop zero at $s = -c$ cannot be cancelled; hence, it will modify the actual closed-loop response with respect to the desired one. As a matter of fact, in Fig. 5.20, the desired response, with a dashed line, and the actual closed-loop response, with a continuous line, are presented. We conclude that the effects of the zero at $s = -c$ are a shorter rise time and a larger overshoot.

5.2.3   Position Control Using a Lead Compensator

Consider again the DC motor model in (5.12) but, now, together with the following controller:

$$I^*(s) = \gamma \frac{s + d}{s + c}, \quad c > d > 0, \quad \gamma > 0,$$

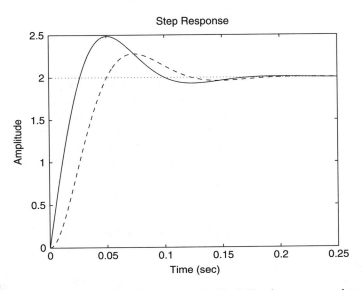

Fig. 5.20 Step response of the closed-loop system in Fig. 5.15, when $c > a$ and two closed-loop complex conjugate poles are specified. Dashed: desired response. Continuous: closed-loop response

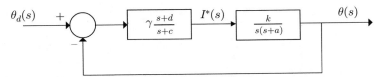

Fig. 5.21 Position control using a lead compensator

which is known as a lead compensator if $c > d$. This condition is introduced when it is desired to increase the damping in the system, i.e., when the closed-loop poles must be shifted to the left (see rules 11 and 12). The corresponding closed-loop block diagram is shown in Fig. 5.21. The open-loop transfer function is given as:

$$G(s)H(s) = \gamma \frac{k(s+d)}{s(s+a)(s+c)}. \tag{5.17}$$

If d is chosen as:

$$d = a, \tag{5.18}$$

then the close-loop transfer function is:

$$\frac{\theta(s)}{\theta_d(s)} = \frac{\gamma k}{s^2 + cs + \gamma k} = \frac{\omega_n^2}{s^2 + 2\zeta\omega_n s + \omega_n^2},$$

which means that:

$$c = 2\zeta\omega_n, \quad \gamma = \frac{\omega_n^2}{k}. \tag{5.19}$$

Hence, if the expressions in (3.71) are used to compute ζ and ω_n such that the desired rise time and overshoot are obtained, then (5.18) and (5.19) represent a simple tuning rule. Moreover, the formulas in (3.71) for overshoot and rise time are exact in this case; thus, the drawback that a PD controller has in this respect is eliminated. The transient response in this case is rather similar to that obtained with the controller in Sect. 3.8.2, i.e., proportional position control plus velocity feedback.

5.2.4 Proportional–Integral Control of Velocity

Although proportional–integral (PI) velocity control has been studied in Sect. 3.8.4, in this part, some ideas are presented that are clearer when employing block diagrams and the root locus. According to Chap. 10, when the velocity $\omega(s)$ is the output, the permanent magnet brushed DC motor model is:

$$\omega(s) = \frac{1}{s+a}[kI^*(s) - \frac{1}{J}T_p(s)],$$

$$a = \frac{b}{J} > 0, \quad k = \frac{nk_m}{J} > 0,$$

where the commanded electric current $I^*(s)$ is the input and $T_p(s)$ is an external torque disturbance. A PI velocity controller is given as:

$$i^* = k_p e + k_i \int_0^t e(r)dr, \quad e = \omega_d - \omega,$$

where ω_d is the desired velocity, k_p is the proportional gain and the constant k_i is known as the integral gain. Using the Laplace transform:

$$I^*(s) = k_p E(s) + k_i \frac{E(s)}{s},$$

$$= \left(k_p + \frac{k_i}{s}\right) E(s),$$

$$= \left(\frac{k_p s + k_i}{s}\right) E(s),$$

$$= k_p \left(\frac{s + \frac{k_i}{k_p}}{s}\right) E(s).$$

Hence, the closed-loop block diagram is depicted in Fig. 5.22a. As this control system has two inputs, the superposition principle can be employed (see Sect. 3.7) to write:

$$\omega(s) = G_1(s)\omega_d(s) + G_2(s)T_p(s),$$

where $G_1(s)$ is the transfer function obtained when using $\omega_d(s)$ as the input and $\omega(s)$ as the output and assuming $T_p(s) = 0$, i.e., using the block diagram in Fig. 5.22b, whereas $G_2(s)$ is the transfer function when $T_p(s)$ is the input and $\omega(s)$ is the output, assuming that $\omega_d(s) = 0$, i.e., when using the block diagram in Fig. 5.22c. It is important to say that, in the case when $T_p(s)$ is the input, $\omega(s)$ represents the velocity deviation produced by disturbance $T_p(s)$.

Now, the problem of choosing the PI controller gains such that the transient response of the closed-loop system satisfies the desired specifications is studied. It is observed in Fig. 5.22b, that the open-loop transfer function is:

$$G(s)H(s) = \frac{k_p k(s + c)}{s(s + a)}, \qquad c = \frac{k_i}{k_p}, \qquad (5.20)$$

Notice that the system type is 1, which ensures that $\omega(t) = \omega_d$ in a steady state if ω_d is a constant. This is one reason for using a PI controller in this example. Hence, the only problem that remains is to choose the controller gains k_p and k_i, such that the transient response satisfies the desired specifications. To this aim, notice that the open-loop transfer function shown in (5.20) is identical to the transfer function shown in (5.15), which corresponds to the PD control of position as only k_d has to be replaced by k_p. Thus, the root locus diagram corresponding to the PI velocity control is identical to both cases shown in Fig. 5.16 and the same conclusions are obtained:

i) There always exist some gains k_p and k_i allowing both closed-loop poles to be placed at any point on the left half-plane; hence, it is possible to tune the PI controller using a trial and error procedure (see Sect. 3.8.4).

ii) The open-loop zero located at $s = -c$ is also a zero of the closed-loop transfer function $G_1(s)$, i.e., it also affects the closed-loop transient response. This means that the transient response does not have the specifications designed using (3.71) to choose the closed-loop poles. This poses the following two possibilities.

1. The problem pointed out at $ii)$ can be eliminated if it is chosen:

$$c = a = \frac{k_i}{k_p}, \qquad (5.21)$$

because, in such a case and according to Fig. 5.22b and (5.1), the closed-loop transfer function is:

$$\frac{\omega(s)}{\omega_d(s)} = \frac{k_p k}{s + k_p k}.$$

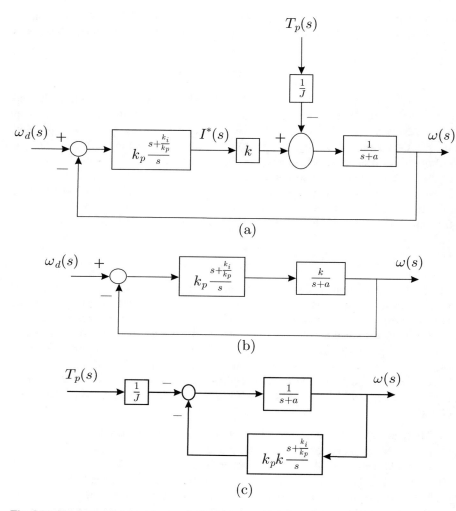

Fig. 5.22 Proportional–integral control of velocity. (**a**) The complete control system. (**b**) $T_p(s) = 0$. (**c**) $\omega_d(s) = 0$

This means that the closed-loop response is as that of a first-order system with a unit steady state gain and a time constant given as $\frac{1}{k_p k}$. Then, if a time constant τ is specified:

$$k_p k = \frac{1}{\tau}.\tag{5.22}$$

The conditions in (5.22) and (5.21) represent a simple tuning rule. Finally, the root locus diagram corresponding to this case ($c = a$, $G(s)H(s) = \frac{k_p k}{s}$) is shown

Fig. 5.23 Root locus for $c = a$, $G(s)H(s) = \frac{k_p k}{s}$

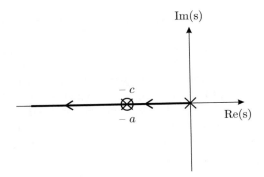

in Fig. 5.23. This can be easily verified using rules 3 and 5. As there is only one open-loop pole, there is only one closed-loop pole, which must be real and tends toward one zero at infinity (on an asymptote forming $-180°$ with the positive real axis) because there is no open-loop zero. Furthermore, the magnitude condition:

$$\frac{k_p k}{l_1} = 1,$$

where l_1 is the distance from the desired closed-loop pole to the origin, establishes that the desired closed-loop pole located at $s = -\frac{1}{\tau}$ is reached when:

$$k_p k = l_1, \quad l_1 = \frac{1}{\tau} \Rightarrow k_p k = \frac{1}{\tau},$$

which corroborates (5.22). However, this tuning rule has a problem when considering the closed-loop response when an external disturbance $T_p(s)$ appears. Consider the block diagram in Fig. 5.22c. If $c = a$ is chosen, then the closed-loop transfer function is:

$$\frac{\omega(s)}{T_p(s)} = \frac{-\frac{k}{J}s}{(s + k_p k)(s + a)}. \tag{5.23}$$

Using the final value theorem, the following is found:

$$\lim_{t \to \infty} \omega(t) = \lim_{s \to 0} s\omega(s),$$

$$= \lim_{s \to 0} s \frac{-\frac{k}{J}s}{(s + k_p k)(s + a)} \frac{t_d}{s} = 0,$$

i.e., that the velocity deviation produced by the external torque disturbance with a constant value t_d vanishes in a steady state. This is the reason why a PI velocity controller is chosen. However, there is a problem. The poles of the transfer function in (5.23) determine how fast this velocity deviation vanishes. These poles are located at $s = -k_p k$ and $s = -a$. Although one of these poles can be rendered as fast as desired just by choosing a suitable k_p, the other pole, at $s = -a$, cannot be modified and depends on the parameters of the DC motor. This has as a consequence that the velocity deviation due to the disturbance may approach zero very slowly, which is a serious drawback.

2. Trying to solve the above problem, we might abandon the tuning rule in (5.22) and (5.21). As we now have $c \neq a$, the closed-loop transfer function corresponding to the block diagram in Fig. 5.22c is:

$$\frac{\omega(s)}{T_p(s)} = \frac{-\frac{k}{J}s}{s^2 + (a + k_p k)s + k_p k c}. \tag{5.24}$$

Using the final value theorem, it can be easily verified that, in the case of a constant disturbance $T_p(s) = \frac{t_d}{s}$, the steady-state deviation is zero again: $\lim_{t \to \infty} \omega(t) = 0$, because the transfer function in (5.24) has a zero at $s = 0$. On the other hand, poles of the transfer function in (5.24), which determine how fast the velocity deviation vanishes, are identical to the poles of the transfer function $\frac{\omega(s)}{\omega_d(s)} = G_1(s)$; hence, they can be assigned using the root locus method from (5.20). As has been explained, the corresponding root locus is identical to both cases shown in Fig. 5.16 replacing k_d by k_p. It is important to stress that the transfer function in (5.24) has no zero at $s = -c$ shown in the root locus diagram in Fig. 5.16. This means that none of the closed-loop poles obtained using the root locus method can cancelled with the zero at $s = -c$ and the slowest pole will have the most significant effect on the time required for the velocity deviation due to the disturbance to vanish. With these ideas in mind, the following is concluded.

- To render small the effect of the zero at $s = -c$ on the transient response to a given velocity reference, one closed-loop pole must be placed close to the zero at $s = -c$. The other closed-loop pole (the fastest one) moves away to the left and tends toward infinity as the slow pole approaches to $s = -c$.
- This means that the fast pole determines the transient response, i.e., the time constant, to a given velocity reference. Then, if it is desired to fix some finite value for the time constant (such that the fast pole is placed at a finite point on the negative real axis), the slow pole is always relatively far from zero at $s = -c$. This means that the transient response is always affected by both poles and the zero at $s = -c$. Hence, if $c \neq a$, a tuning rule cannot be determined such that transient response to a given velocity reference satisfies the desired specifications. Notice that this is also true if the response to a reference is specified by two complex conjugate poles, as the zero at $s = -c$ cannot be cancelled; hence, it has the effect of modifying the transient response with respect to that specified by the two complex conjugate poles.

- It is more convenient to select $c > a$ because the slowest closed-loop pole (that approaching $s = -c$) is shifted further to the left (it is faster) compared with the case when $c < a$ is chosen.
- This also implies that, if it is desired that the velocity deviation due to a disturbance vanishes faster, we must choose $c \neq a$ with $c > 0$ larger. According to $c = k_i/k_p$, this means that a larger integral gain is obtained.

Thus, although it is possible to render shorter the time it takes to the velocity deviation due to disturbance to vanish, it is concluded that the controller gains k_p y k_i cannot be exactly computed to ensure that, simultaneously, the desired transient response specifications to a given velocity reference are accomplished and that the disturbance effects vanish as fast as desired. This statement is verified experimentally in Chap. 10 and, because of this, in that chapter a modified PI velocity controller is introduced, solving this situation.

Anticipating its experimental use in Chap. 10, a tuning rule is proposed next for the case when $c \neq a$. According to (5.20), the magnitude condition is:

$$\frac{k_p k l_1}{l_3 l_2} = 1,$$

where $s - (0) = l_2 \angle \theta_2$, $s - (-a) = l_3 \angle \theta_3$, and $s - (-c) = l_1 \angle \theta_1$. According to this tuning criterion, propose the slowest pole to be located at some known $s = -p_1$, to fix an upper limit on the time required to render negligible the disturbance effect. On the other hand, according to the items listed above, c is proposed to be close to p_1 such that $p_1 > c$. Then, using the magnitude condition above, the following tuning rule is obtained:

$$\frac{k_i}{k_p} = c, \quad c < p_1, \quad k_p = \frac{l_2 l_3}{l_1 k}, \tag{5.25}$$

$$l_1 = abs(-p_1 + c), \quad l_2 = abs(-p_1), \quad l_3 = abs(-p_1 + a).$$

According to the discussion above, the response to the reference of velocity is much faster than predicted by the pole at $s = -p_1$. For comparison purposes this is very important as some velocity controllers are designed in Chap. 10 accomplishing simultaneously transient responses to a velocity reference and to an external disturbance that are determined by a pole at $s = -p_1$. On the contrary, with the PI velocity controller studied in this section, if it is desired that the response to an external disturbance is faster (with a pole at $s = -p_1$), the response to a desired reference of velocity must be much faster.

Example 5.4 The MATLAB/Simulink diagram corresponding to Fig. 5.22a is shown in Fig. 5.24. It is assumed that $k = 70$ and $a = 7$. The desired velocity ω_d is a step command with 2 as the magnitude. The disturbance T_p is a step applied at $t = 2[s]$ with a magnitude 0.4×70. The PI controller block has k_p and k_i as proportional and integral gains.

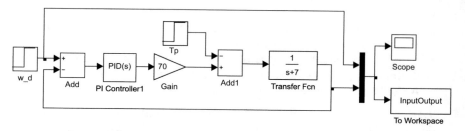

Fig. 5.24 MATLAB/Simulink diagram corresponding to the closed-loop system in Fig. 5.22a

The simulation in Fig. 5.24, was performed with the help of the following MATLAB code in an m-file:

```
a=7;
k=70;
Polo_des=-20;
c=15; %1 5 7 10 15
l1=abs(Polo_des)-c;
l2=abs(Polo_des);
l3=abs(Polo_des)-a;
kp=l2*l3/(k*l1);
ki=c*kp;
PI=tf(kp*[1 c],[1 0]);
gm=tf(k,[1 a]);
M=feedback(PI*gm,1,-1);
figure(1)
gd=tf(20,[1 20]);
step(M*2,'b:',gd*2,'r--',0.2)
hold on
%%{
nn=length(InputOutput(:,2));
n=nn-1;
Ts=10/n;
t=0:Ts:10;
figure(2)
plot(t,InputOutput(:,2),'k:');
axis([-0.5 10 0 2.5])
xlabel('t [s]')
ylabel('w [rad/s]')
hold on
%}
```

In this code, the desired closed-loop response is specified to be as that of a first-order system with a time constant $\tau = \frac{1}{20} = 0.05$[s] when a desired step velocity is commanded. This desired response is shown in Fig. 5.25 with a dashed line. The

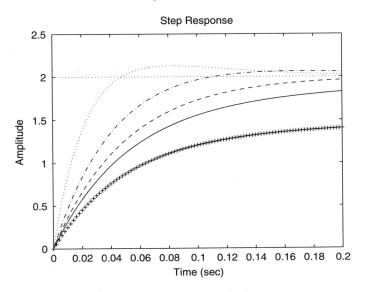

Fig. 5.25 Response to a step command in desired velocity $\omega_d(s)$

design criterion is to find k_p and k_i such that the desired closed-loop pole is assigned at $s = -20$, provided that a ratio $\frac{k_i}{k_p}$ is given. This allows to the system response to be analyzed when the integral gain is enlarged and one closed-loop pole is at the desired location.

First, the location of zero at $-\frac{k_i}{k_p} = -c$ is fixed by defining the constant c. The proportional gain k_p is computed using the magnitude condition $\frac{k_p k l_1}{l_3 l_2} = 1$, where l_1, l_2, l_3, are the magnitudes of vectors defined in the paragraph before (5.25), and $k_i = c k_p$ is also computed. Then, the closed-loop transfer function is obtained and the closed-loop system response as well as the desired response can be simulated using the command "step()". These signals are shown in Fig. 5.25. At this point, the simulation diagram in Fig. 5.24 is run and, with the help of the remaining code lines above, the corresponding results are shown in Fig. 5.26.

In Figs. 5.25 and 5.26 the following data are presented. (i) The line marked with "+" corresponds to $c = 1$, (ii) Continuous line: $c = 5$, (iii) Dashed line: $c = 7$, (iv) Dash–dot line: $c = 10$, and (v) Dotted line: $c = 15$. Notice that the disturbance effect is rejected very slowly when $c = 1$. In this case, the closed-loop poles are at $s = -20$ and $s = -0.6842$, which can be verified using the MATLAB command "pole(M)". Recall that the transfer function in (5.24) has no zero to cancel the pole at $s = -0.6842$.[2] This explains the slow rejection of the disturbance effects in this case. Also notice that the response to the velocity reference is also slow. This is because the effect of the pole at $s = -0.6842$ is not suitably cancelled by the zero at $s = -c = -1$.

---

[2]If it is assumed that this pole and zero at $s = 0$ cancel each other out, then a nonzero steady-state deviation would exist and the problem would be worse.

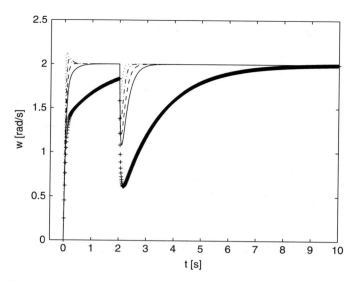

Fig. 5.26 Response to a step disturbance $T_p(s)$

Notice that the disturbance rejection and the response to the reference improve as c increases. Moreover, the system response exactly matches the desired response when $c = 7 = a$. However, the disturbance effects are still important in this case. When $c > 7 = a$, the disturbance rejection further improves, but the response to the reference exhibits overshoot, i.e., it is no longer a first-order response. According to Sect. 8.1.2, this behavior appears, and it is unavoidable, when the closed-loop poles are on the left of an open-loop zero. In this respect, notice that the closed-loop poles are at $s = -20$ and $s = -13$ when $c = 10$ and at $s = -20$ and $s = -39$ when $c = 15$.

The simulation results in this example corroborate the system behavior predicted in the above discussion, i.e., the classical PI control has limitations when it is required to satisfy simultaneously a specified response to a reference and a specified performance for disturbance rejection.

5.2.5   *Proportional–Integral–Derivative Control of Position*

Consider again the DC motor model but, now, assuming the presence of an external disturbance:

$$\theta(s) = \frac{1}{s(s+a)}[kI^*(s) - \frac{1}{J}T_p(s)],$$

together with the following proportional–integral–derivative controller:

$$i^* = k_p e + k_d \frac{de}{dt} + k_i \int_0^t e(r)dr, \quad e = \theta_d - \theta,$$

where θ_d is the desired position and the constants k_p, k_d and k_i are known as the proportional, derivative, and integral gains respectively. Use of the Laplace transform, yields:

$$I^*(s) = k_p E(s) + k_d s E(s) + k_i \frac{E(s)}{s},$$

$$= \left(k_p + k_d s + \frac{k_i}{s} \right) E(s),$$

$$= k_d \frac{s^2 + \frac{k_p}{k_d}s + \frac{k_i}{k_d}}{s} E(s),$$

and the corresponding block diagram is shown in Fig. 5.27a. As this is a system with two inputs, the superposition principle (see Sect. 3.7) can be used to write:

$$\theta(s) = G_1(s)\theta_d(s) + G_2(s)T_p(s),$$

where $G_1(s)$ is the closed-loop transfer function when $\theta_d(s)$ is the input and $\theta(s)$ is the output with $T_p(s) = 0$, i.e., when the block diagram in Fig. 5.27b is used to find:

$$\frac{\theta(s)}{\theta_d(s)} = G_1(s) = \frac{k_d k \left(s^2 + \frac{k_p}{k_d}s + \frac{k_i}{k_d} \right)}{s^3 + (a + k_d k)s^2 + k_p ks + k_i k}, \quad T_p(s) = 0. \quad (5.26)$$

On the other hand, $G_2(s)$ is the closed-loop transfer function when $T_p(s)$ is the input and $\theta(s)$ is the output with $\theta_d(s) = 0$, i.e., when the block diagram in Fig. 5.27c is employed to write:

$$\frac{\theta(s)}{T_p(s)} = G_2(s) = \frac{-\frac{k}{J}s}{s^3 + (a + k_d k)s^2 + k_p ks + k_i k}, \quad \theta_d(s) = 0. \quad (5.27)$$

It is stressed that when $T_p(s)$ is the input, then $\theta(s)$ stands for the position deviation (with respect to $\theta_d(s)$) produced by the external disturbance. Using the final value theorem, it is found that:

$$\lim_{t \to \infty} \theta(t) = \lim_{s \to 0} s\theta(s),$$

$$= \lim_{s \to 0} s \frac{-\frac{k}{J}s}{s^3 + (a + k_d k)s^2 + k_p ks + k_i k} \frac{t_d}{s} = 0,$$

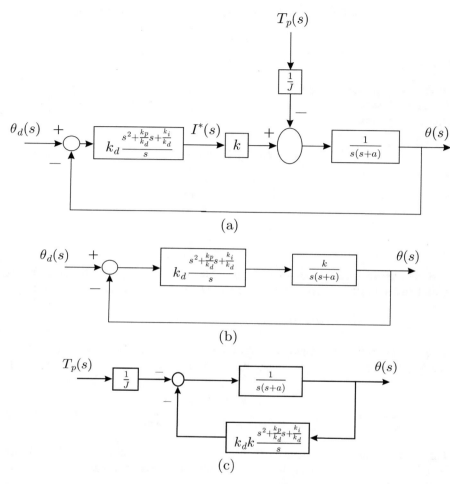

Fig. 5.27 Proportional–integral–derivative (PID) position control. (**a**) The complete control system. (**b**) $T_p(s) = 0$. (**c**) $\theta_d(s) = 0$

the deviation of position produced by the constant external torque disturbance $T_p(s) = \frac{t_d}{s}$ is zero in a steady state. Using the final value theorem again, it is not difficult to verify that, using (5.26), the final value of position θ is equal to the desired value θ_d when this is a constant. These are the main reasons for using a PID position controller.

The problem of choosing the controller gains for the PID controller such that the closed-loop response has the desired transient response specifications when a position reference θ_d is applied is studied in the following. To this aim, the three poles of the transfer function in (5.26) must be assigned at the desired locations on the left half-plane. This is accomplished by equating the characteristic polynomial of the transfer function in (5.26) to a polynomial with roots at the desired points $s = p_1, s = p_2, s = p_3$:

$$s^3 + (a + k_d k)s^2 + k_p ks + k_i k = (s - p_1)(s - p_2)(s - p_3).$$

It is clear that any values can be assigned to the coefficients of the characteristic polynomial by using suitable combinations of the controller gains k_p, k_d and k_i. This means that the three poles of the closed-loop system can be assigned at any desired location of the left half-plane. Any set of three poles satisfying this requirement can be written as:

$$p_1 = \sigma_1 + j\omega_1, \quad p_2 = \sigma_2 - j\omega_1, \quad p_3 < 0, \quad \sigma_1 < 0, \quad \sigma_2 < 0, \quad \omega_1 \geq 0.$$

Notice that $\sigma_1 = \sigma_2$ if $\omega_1 > 0$, i.e., when a pair of complex conjugate poles exists, but $\sigma_1 \neq \sigma_2$ is possible when $\omega_1 = 0$. Hence:

$$s^3 + (a + k_d k)s^2 + k_p ks + k_i k = (s - p_1)(s - p_2)(s - p_3),$$

$$s^3 - (p_1 + p_2 + p_3)s^2 + (p_1 p_2 + p_3 p_1 + p_3 p_2)s - p_1 p_2 p_3.$$

Equating the coefficients and using the above data, the following tuning rule is obtained:

$$k_d = \frac{-(\sigma_1 + \sigma_2 + p_3) - a}{k} > 0, \tag{5.28}$$

$$k_p = \frac{\sigma_1 \sigma_2 + \omega_1^2 + p_3 \sigma_1 + \sigma_2 p_3}{k} > 0,$$

$$k_i = \frac{-p_3(\sigma_1 \sigma_2 + \omega_1^2)}{k} > 0.$$

It is stressed that the three controller gains are positive. As a rule of thumb, if $-p_3 > 6|\sigma_1|$ then the real and imaginary parts of the poles at p_1 and p_2 can be computed using (3.71) such that the desired rise time t_r and overshoot $M_p(\%)$ are obtained. However, the system response presents some important differences with respect to these values because of the pair of zeros contained in the transfer function in (5.26). Although this is an important drawback of the tuning rule in (5.28), these values can be used as rough approximations of the required gains for the controller to perform fine adjustments afterward, by trial and error, until the desired transient response specifications are accomplished. To this aim, as the closed-loop system is third order, it is important to recall the stability rule obtained in Example 4.12 of Sect. 4.3. The possibility of adjusting the gains of a PID controller by trial and error (see Sect. 5.3) is one of the main reasons for the success of PID control in industry.

With the aim of finding a tuning rule allowing to exactly compute the gains of a PID controller, the root locus method is employed in the following. From the block diagram in Fig. 5.27b, it is concluded that the open-loop transfer function is given as:

$$G(s)H(s) = \frac{k_d k(s + \alpha)(s + \beta)}{s^2(s + a)}, \quad (s + \alpha)(s + \beta) = s^2 + \frac{k_p}{k_d}s + \frac{k_i}{k_d}, \quad (5.29)$$

for some constants α and β different from zero whose values are proposed as part of the design process. Then, from these values both k_p and k_i are computed. Notice that k_d is the gain that the method varies from 0 to $+\infty$ to plot the root locus diagram. It is clear that the system type is 2 (the motor is naturally provided with an integrator and another integrator is introduced by PID controller) and, hence, the steady-state error is zero if the reference of the position is constant.

Three possibilities for the corresponding root locus diagram are presented in Fig. 5.28. They depend on the values proposed for α and β. The reader is encouraged to use the rules introduced in Sect. 5.1.1 to corroborate these results. According to Fig. 5.28a, it is possible to design the control system such that two closed-loop poles are close to the zeros located at $s = -\alpha$ and $s = -\beta$. This would allow the closed-loop system to respond as a first-order system with one real and negative pole. One closed-loop pole may also be chosen to be close to the zero at $s = -\alpha$ such that the closed-loop system responds as a second-order system with complex conjugate poles and one real zero. Although the presence of this zero modifies the transient response, this is one way of specifying the transient response in terms of rise time and overshoot. This is the design criterion used in the following.

In cases like this one, the traditional root locus-based design procedures suggest performing the following three steps:

- Design a PD controller with transfer function:

$$k_d(s + \beta),$$

  such that some desired rise time and overshoot are accomplished.
- Given the open-loop transfer function designed in the previous step, introduce the following factor:

$$\frac{s + \alpha}{s},$$

  with α some positive value close to zero.
- Compute the PID controller gains using (5.29), i.e.,

$$k_p = (\alpha + \beta)k_d, \quad k_i = \alpha\beta k_d. \quad (5.30)$$

Now, the PD controller $k_d(s + \beta)$ is designed for the plant $\frac{k}{s(s+a)}$, i.e., when the closed-loop system has the form presented in Fig. 5.29 and the open-loop transfer function is:

$$G(s)H(s) = \frac{k_d k(s + \beta)}{s(s + a)}. \quad (5.31)$$

Fig. 5.28 Different possibilities for root locus diagrams of the PID control of position

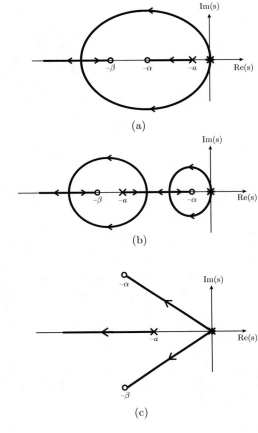

(a)

(b)

(c)

Fig. 5.29 Proportional–derivative control of the position

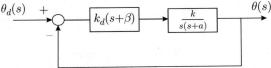

The closed-loop transfer function is, in this case:

$$\frac{\theta(s)}{\theta_d(s)} = \frac{k_d k(s + \beta)}{s^2 + (a + k_d k)s + k_d k\beta}.$$

Equating to a standard second-degree polynomial:

$$s^2 + (a + k_d k)s + k_d k\beta = s^2 + 2\zeta\omega_n s + \omega_n^2,$$

yields:

$$k_d = \frac{2\zeta\omega_n - a}{k}, \qquad \beta = \frac{\omega_n^2}{k_d k}. \tag{5.32}$$

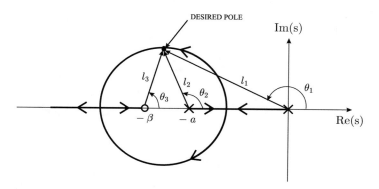

Fig. 5.30 Root locus diagram for the system in Fig. 5.29

The values for ζ and ω_n can be computed using (3.71) from the desired values for rise time and overshoot. The corresponding root locus diagram is shown in Fig. 5.30, which is identical to the case $c > a$ depicted in Fig. 5.16, because only in this case can the PD control of the position produce complex conjugate closed-loop poles. From (5.31), it is found that the angle and the magnitude conditions establish:

$$\theta_3 - (\theta_1 + \theta_2) = \pm 180°(2q + 1), \quad q = 1, 2, \dots \qquad \frac{k_d k l_3}{l_1 l_2} = 1, \qquad (5.33)$$

where $l_1, l_2, l_3, \theta_1, \theta_2$ and θ_3 are defined in Fig. 5.30. When the factor $\frac{s+\alpha}{s}$ is included in the open-loop transfer function shown in (5.31), the resulting open-loop transfer function becomes:

$$G(s)H(s) = \frac{k_d k(s + \beta)(s + \alpha)}{s^2(s + a)}. \qquad (5.34)$$

As $\alpha > 0$ is small, the corresponding root locus has the shape shown in Fig. 5.31. The reader is encouraged to follow the rules in Sect. 5.1.1 to verify this result. The angle and the magnitude conditions are established in this case from (5.34) as:

$$\theta_3 + \theta_5 - (2\theta_1 + \theta_2) = \pm 180°(2q + 1), \quad q = 1, 2, \dots \qquad \frac{k_d k l_3 l_5}{l_1^2 l_2} = 1, \qquad (5.35)$$

where $l_1, l_2, l_3, \theta_1, \theta_2$ and θ_3 are defined as in Fig. 5.30, whereas l_5 and θ_5 are defined in Fig. 5.31. The reason for choosing $\alpha > 0$ to be close to zero is to render l_1 and l_5 almost the same such that $l_5/l_1 \approx 1$. This also ensures that θ_1 and θ_5 are almost equal; hence, $\theta_5 - \theta_1 \approx 0$. Then, the angle and the magnitude conditions in (5.33) and (5.35) are almost identical. This ensures that the closed-loop poles

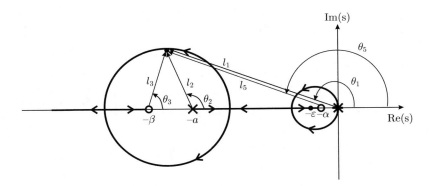

Fig. 5.31 Root locus for the open-loop transfer function shown in (5.34)

obtained when the open-loop transfer function is that presented in (5.31) are almost identical to the closed-loop poles obtained when the open-loop transfer function employed is that presented in (5.34). This ensures that the desired transient response specifications (rise time and overshoot) designed with the PD controller in (5.32) are also accomplished with the PID controller:

$$k_d \frac{(s+\alpha)(s+\beta)}{s} = k_d \frac{s^2 + \frac{k_p}{k_d}s + \frac{k_i}{k_d}}{s},$$

i.e., when the controller gains are chosen according to (5.30).

However, it is important to point out a drawback of this design approach: according to (5.30) a small α results in a small integral gain k_i. As a consequence, the deviation of the position produced by the external disturbance vanishes very slowly, which may be unacceptable in practice. This can also be explained from the root locus diagram in Fig. 5.31. Notice that a closed-loop pole (located, say, at $s = -\varepsilon$, $\varepsilon > 0$) approaches the zero at $s = -\alpha$, which means that both cancel each other out in the transfer function shown in (5.26), i.e., the effect of none of them is observed in the transient response to a reference of the position. However, the zero at $s = -\alpha$ does not appear in the transfer function shown in (5.27). But the pole at $s = -\varepsilon$ is still present in this transfer function; hence, it significantly affects the transient response: this slow pole (close to the origin) is responsible for a slow transient response when an external disturbance appears.

Despite this drawback, the traditional criteria for root locus-based design suggest proceeding as above when designing PID controllers. Moreover, it is interesting to say that these problems in traditional design methods remain without a solution despite the fact that they have been previously pointed out in some research works. See [10] for instance.

On the other hand, according to (5.30), a larger integral gain can be obtained (to achieve a faster disturbance rejection) by choosing a larger value for α. However, according to the previous discussion, this will result in a transient response to a

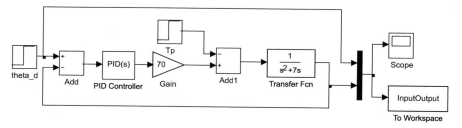

Fig. 5.32 MATLAB/Simulink diagram for the closed-loop system in Fig. 5.27a

reference of position that does not satisfy the desired specifications. Hence, it is concluded that there is no tuning rule allowing the gains of a PID controller for position to be computed exactly, simultaneously satisfying the desired transient response specifications to a given reference of the position and producing a satisfactory rejection of the effects of an external torque disturbance. These observations are experimentally verified in Chap. 11 where, given the described drawback, some new controllers solving these problems are designed and experimentally tested.

Finally, despite the above drawback, it is important to recall what was indicated just after (5.28): PID control is one of the most frequently employed controllers in industry because it can be tuned by trial and error (see Sect. 5.3). This has to be performed by taking into account the stability rule obtained in the Example 4.12, Sect. 4.3.

Example 5.5 The MATLAB/Simulink diagram corresponding to the block diagram in Fig. 5.27a is presented in Fig. 5.32. It is assumed that $a = 7$ and $k = 70$. The desired position θ_d is a step command applied at $t = 0$ with 2 as the magnitude, whereas the disturbance T_p is another step applied at $t = 0.5[\mathrm{s}]$ and 700 as the magnitude. The PID controller gains are k_p, k_d, and k_i, computed in the following MATLAB code executed in an m-file:

```
clc
a=7;
k=70;
tr=0.05;
Mp=14;%
z=sqrt( log(Mp/100)^2 /( log(Mp/100)^2 + pi^2 ) );
wn=1/(tr*sqrt(1-z^2))*(pi-atan(sqrt(1-z^2)/z));
s=-z*wn+wn*sqrt(1-z^2)*j
kd=(2*z*wn-a)/k;
beta=wn^2/(k*kd)
den1=conv([1 -s],[1 -real(s)+j*imag(s)]);
Md=tf(den1(3),den1);
gm=tf(k,[1 a 0]);
PD=tf(kd*[1 beta],1);
alpha=20;%0.1 1 10 20
```

```
PI=tf([1 alpha],[1 0]);
M_pid=feedback(PI*PD*gm,1,-1);
pole(M_pid)
figure(2)
step(Md*2,'r--',M_pid*2,'k--',0.25)
hold on
kp=(alpha+beta)*kd;
ki=alpha*beta*kd;
%%{
nn=length(InputOutput(:,2));
n=nn-1;
Ts=2/n;
t=0:Ts:2;
figure(3)
plot(t,InputOutput(:,2),'k--');
hold on
plot(t,InputOutput(:,1),'k:');
axis([0 1 0 3])
xlabel('t [s]')
ylabel('theta [rad]')
%}
```

The main idea is to propose different values for the design parameter α to see how the transient response is affected. Hence, the values for ζ and ω_n are first computed from the desired rise time $t_r = 0.05[s]$ and overshoot $M_p = 14\%$. This allows k_d and β to be computed using (5.32). Then, the desired time response, i.e., that dictated by the desired complex conjugate poles at $s = -z * wn \pm wn * sqrt(1 - z^2)j$ (the slowest dashed response in Fig. 5.33), and the actual closed-loop system response to the step reference are plotted on the same axes using the command "$step(Md * 2,'r - -', M\_pid * 2,'k - -', 0.25)$". The proportional and integral gains are computed using (5.30); hence, the simulation in Fig. 5.32 can be run. Once this is performed, the remaining commands in the code listed above can be used to draw Fig. 5.34.

Running the above code and the simulation in Fig. 5.32 several times, the responses are obtained when using different values for α. In Figs. 5.33 and 5.34: (*i*) Continuous lines stand for $\alpha = 0.1$, (*ii*) Dash–dot: $\alpha = 1$, (*iii*) Dotted: $\alpha = 10$, and the fastest dashed line: $\alpha = 20$.

It is observed that, when $\alpha = 0.1$, the response is different from the desired one, i.e., the slowest dashed line in Fig. 5.33. This is because of the closed-loop zero at $-\beta$, located at $s = -54.5225$ in this case, which deviates the system response from the desired one (See also Fig. 5.20. It is remarked that the desired closed-loop poles are located at $s = -\zeta\omega_n \pm j\omega_n\sqrt{1 - \zeta^2}$, as desired, in this case. Also notice that the disturbance rejection is so slow that it seems that a steady-state deviation might exist.

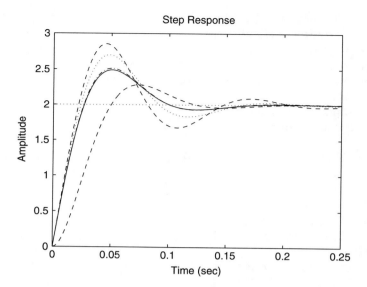

Fig. 5.33 Response to a step command in the desired position $\theta_d(s)$

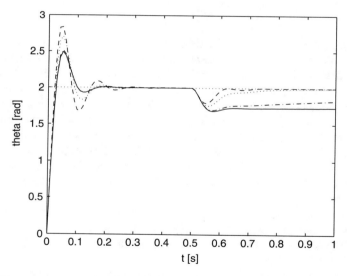

Fig. 5.34 Response to a step disturbance $T_p(s)$

As α is chosen to be larger, the performance observed in Fig. 5.34 improves with respect to the disturbance rejection, but the price to be paid is that the response to the position reference is increasingly different from the desired one, as observed in Fig. 5.33.

Thus, the simulation results in this example corroborate the predictions stated in the above discussion in the present section.

5.2.6 Assigning the Desired Closed-Loop Poles

An example is presented in this section to show how the root locus method can be used to assign the desired closed-loop poles. Consider the following transfer function:

$$G(s)H(s) = \frac{k}{(s - 35.7377)(s + 36.5040)}. \tag{5.36}$$

The gain k is varied by the method to take values from 0 to $+\infty$. First, $G(s)H(s)$ is rewritten as:

$$G(s)H(s) = \frac{k}{l_1 l_2} \angle - (\theta_1 + \theta_2),$$

where the vectors $s - 35.7377 = l_1 \angle \theta_1$ and $s - (-36.5040) = l_2 \angle \theta_2$ have been defined. The fundamental conditions in the root locus method are the angle condition (5.7) and the magnitude condition (5.6), which are expressed respectively as:

$$-(\theta_1 + \theta_2) = \pm(2q + 1)180°, \quad q = 0, 1, 2, \ldots$$

$$\frac{k}{l_1 l_2} = 1.$$

Rule 5 indicates that the root locus on the real axis only exists between the points $s = 35.7377$ and $s = -36.5040$. Moreover, according to rule 1, the root locus begins ($k = 0$) at $s = 35.7377$ and $s = -36.5040$. On the other hand, according to rules 2 and 3, when k tends to $+\infty$ the branches starting at $s = 35.7377$ and $s = -36.5040$ must end at some open-loop zero (there is no zero in this case) or at some point at infinity on the s plane. Hence, the root locus must move away from the points $s = 35.7377$ and $s = -36.5040$ on the real axis as k increases and a breakaway point must exist somewhere between such points. Then, the branches must move toward infinity on the s plane.

On the other hand, according to the angle condition, to $-(\theta_1 + \theta_2) = -180° = -(\alpha + \theta_1)$, and to Fig. 5.35, it can be observed that $\alpha = \theta_2$; hence, the triangles t_1 and t_2 must be identical for any closed-loop pole s. Then, both branches, which are also shown in Fig. 5.35, must be parallel to the imaginary axis. This means that the breakaway point on the real axis is located at the middle point between $s = 35.7377$ and $s = -36.5040$, i.e., at $(35.7377 - 36.5040)/2 = -0.7663$. This can also be verified using rules 3 and 4. Finally, according to rule 6, both branches are symmetrical with respect to the real axis.

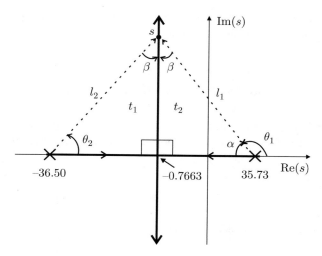

Fig. 5.35 Root locus for $G(s)H(s)$ in (5.36)

The magnitude condition is employed when it is required to know the exact value of k rendering a specific point on the root locus an actual closed-loop pole. Notice, for instance, that both l_1 and l_2 must increase to satisfy the magnitude condition as k grows to $+\infty$:

$$\frac{k}{l_1 l_2} = 1,$$

which means that the closed-loop poles corresponding to large values of k tend toward some points at infinity on the s plane.

Now suppose that an additional pole is included in the transfer function in (5.36) to obtain:

$$G(s)H(s) = \frac{116137\, k}{s^3 + 72.54s^2 - 1250s - 9.363 \times 10^4},$$

$$= \frac{116137\, k}{(s - 35.7377)(s + 36.5040)(s + 71.7721)}. \tag{5.37}$$

Again, k is the gain that the method varies from 0 to $+\infty$ to plot the root locus diagram. First, $G(s)H(s)$ is rewritten as:

$$G(s)H(s) = \frac{116137\, k}{l_1 l_2 l_3} \angle - (\theta_1 + \theta_2 + \theta_3),$$

where the vectors $s - 35.7377 = l_1 \angle \theta_1$, $s - (-36.5040) = l_2 \angle \theta_2$, and $s - (-71.7721) = l_3 \angle \theta_3$ have been defined. The angle condition (5.7) and the magnitude condition (5.6) are expressed respectively as:

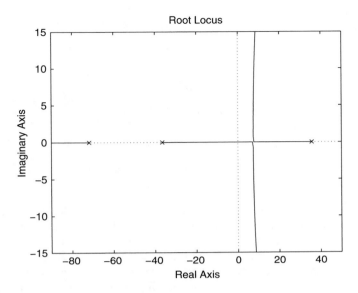

Fig. 5.36 Root locus diagram for $G(s)H(s)$ in (5.37)

$$-(\theta_1 + \theta_2 + \theta_3) = \pm(2q + 1)180°, \quad q = 0, 1, 2, \ldots$$

$$\frac{116137\,k}{l_1 l_2 l_3} = 1.$$

The root locus diagram corresponding to this case can be obtained from the root locus diagram obtained for the transfer function in (5.36) by simply taking into account the additional pole at $s = -71.7721$.

According to the rule 11, the new pole is responsible for both branches of the root locus shown in Fig. 5.35 bending to the right, as can be observed in Fig. 5.36. This can be corroborated using rule 3 to find that, now, three branches exist, which tend toward the asymptotes at $\pm 180°$ and $\pm 60°$ with respect to the positive real axis.

On the other hand, the breakaway point between the points $s = 35.7377$ and $s = -36.5040$, which was located at $(35.7377 - 36.5040)/2 = -0.7663$ in Fig. 5.35, is now shown in Fig. 5.36 shifted to the right of -0.7663. The reason for this is explained in Fig. 5.37 where s' and s stand for the points on the root locus which are very close to the breakaway point for the case of Figs. 5.35 and 5.36 respectively. As $-(\theta_1 + \theta_2 + \theta_3) = -180°$ must be satisfied in Fig. 5.36, whereas $-(\theta_1' + \theta_2') = -180°$ must be satisfied in Fig. 5.35, then both θ_1 and θ_2 must be smaller than θ_1' and θ_2'. This means that the point s must be shifted to the right of the point s'. Notice that this displacement of the breakaway point to the right of the point -0.7663 is greater as the angle θ_3 introduced by the pole at $s = -71.7721$ is greater, i.e., as this pole is placed farther to the right. In fact, it is observed in Fig. 5.36 that the breakaway point is located on the right half-plane.

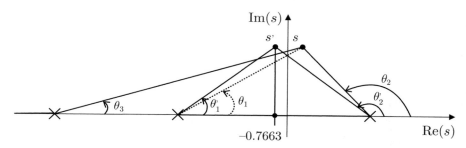

Fig. 5.37 An additional pole moves the breakaway point to the right

This description of the root locus diagram in Fig. 5.36 allows us to conclude that there is always at least one closed-loop pole with a positive real part, i.e., the closed-loop system is unstable for any positive value of k. As k can be interpreted as the gain of a proportional controller, it is concluded that it is not possible to render the closed-loop system stable using any proportional controller and, thus, another controller must be proposed. According to rule 12 it is required to use a controller introducing an open-loop zero because this bends the root locus to the left accomplishing closed-loop stability.

Consider the following open-loop transfer function:

$$G(s)H(s) = k\frac{116137(s + b)}{s^3 + 72.54s^2 - 1250s - 9.363 \times 10^4}, \tag{5.38}$$

$$= k\frac{116137(s + b)}{(s - 35.7377)(s + 36.5040)(s + 71.7721)},$$

with b a positive constant. Again, k is the gain that the method varies from 0 to $+\infty$ to draw the root locus diagram. First, $G(s)H(s)$ is rewritten as:

$$G(s)H(s) = \frac{116137\, k\, l_4}{l_1 l_2 l_3} \angle[\theta_4 - (\theta_1 + \theta_2 + \theta_3)], \tag{5.39}$$

where the vectors $s - 35.7377 = l_1 \angle\theta_1$, $s - (-36.5040) = l_2 \angle\theta_2$, $s - (-71.7721) = l_3 \angle\theta_3$, and $s - (-b) = l_4 \angle\theta_4$ have been defined. The angle condition (5.7) and the magnitude condition (5.6) are expressed respectively as:

$$\theta_4 - (\theta_1 + \theta_2 + \theta_3) = \pm(2q + 1)180°, \quad q = 0, 1, 2, \ldots \tag{5.40}$$

$$\frac{116137\, k\, l_4}{l_1 l_2 l_3} = 1. \tag{5.41}$$

Notice that, now, $n = 3$, $m = 1$, $p_1 = 35.7377$, $p_2 = -36.5040$, $p_3 = -71.7721$, and $z_1 = -b$. According to the rules 1, 2, and 3, one of the three branches starting ($k = 0$) at the points $s = 35.7377$, $s = -36.5040$, and $s = -71.7721$ (the open-loop poles) must end ($k = +\infty$) at the open-loop zero $s = -b$. Hence, there are two root locus branches tending toward some points at infinity of the s plane. According to rule 3, the angles of the asymptotes toward which these branches tend are:

$$\frac{\pm 180°(2 \times 0 + 1)}{n - m} = \frac{\pm 180°(2 \times 0 + 1)}{3 - 1} = \pm 90°.$$

Moreover, according to rule 4, the point where these asymptotes cross the real axis can be shifted by using the zero at $s = -b$:

$$\sigma_a = \frac{p_1 + p_2 + p_3 - (-b)}{n - m} = \frac{35.7377 - 36.5040 - 71.7721 + b}{2}. \quad (5.42)$$

Notice that b must be positive, as, if this is not the case, the root locus would be pushed to the right half-plane producing unstable closed-loop poles. Also notice that the asymptotes move to the left (increasing stability) as the zero is located closer to the origin, i.e., as b tends toward zero. This means that the closed-loop poles cannot be placed arbitrarily to the left.

According to rule 5, the root locus now exists on two segments of the real axis. The locations of these segments depend on the value of b as shown in Figs. 5.38a and b. Furthermore, it is shown in Fig. 5.38c that in the case when $b = 36.5040$ is selected, the open-loop zero at $s = -b$ cancels with the open-loop pole at $s = -36.5040$. This means that, in such a case, the root locus only exists in one segment on the real axis.

It is observed, in Fig. 5.38c, that only two closed-loop poles exist. It is observed, in Figs. 5.38a and b, that three closed-loop poles exist and one of them is on the right half-plane (closed-loop instability) if the gain k is too small. On the other hand, if k is too large, two complex conjugate poles exist with too large an imaginary part, i.e., producing a fast oscillation. Although in theory this can work because the three poles have negative real parts, a large k produces a number of practical problems such as noise amplification and power amplifier saturation, i.e., the control system may not work correctly in practice. Hence, a good design is that enabling the designer to locate as desired the closed-loop poles avoiding the use of values that are too large and values that are too small for the gain k.

Suppose that $s = -25 \pm j40$ are the desired closed-loop complex conjugate poles. A procedure that is useful for computing the exact values required for both b and k is presented in the following. According to Fig. 5.39 the following angles are computed:

$$\theta_3 = \arctan\left(\frac{40}{71.7721 - 25}\right),$$

$$\theta_2 = \arctan\left(\frac{40}{36.5040 - 25}\right),$$

$$\theta_1 = 180° - \arctan\left(\frac{40}{35.7377 + 25}\right).$$

Fig. 5.38 Root loci for
$G(s)H(s)$ defined in (5.38)
obtained by changing b. (**a**)
$b = 31.24 < 36.50$. (**b**)
$b > 36.50$. (**c**) $b = 36.50$

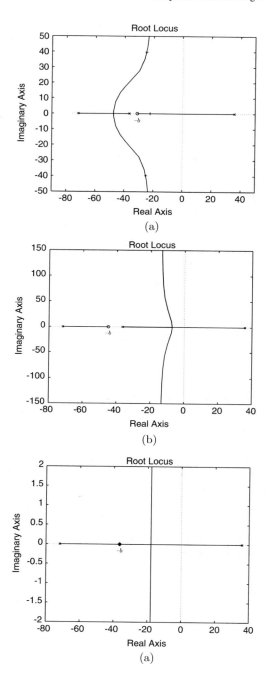

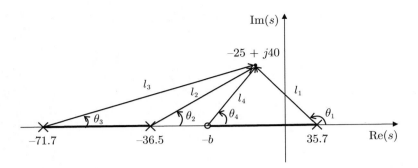

Fig. 5.39 Open-loop poles and zeros when trying to locate the desired closed-loop poles at $s = -25 \pm j40$

Using the angle condition (5.40), θ_4 is computed and, also, b:

$$\theta_4 = -180^0 + (\theta_1 + \theta_2 + \theta_3),$$

$$b = \frac{40}{\tan(\theta_4)} + 25 = 31.2463.$$

On the other hand, according to Fig. 5.39, the following lengths are computed:

$$l_4 = \sqrt{40^2 + (b - 25)^2},$$

$$l_3 = \sqrt{40^2 + (71.7721 - 25)^2},$$

$$l_2 = \sqrt{40^2 + (36.5040 - 25)^2},$$

$$l_1 = \sqrt{40^2 + (35.7377 + 25)^2},$$

and, finally, the magnitude condition (5.41) is employed to compute k:

$$k = \frac{l_1 l_2 l_3}{116137 \, l_4} = 0.0396.$$

The location of the third closed-loop pole can be found from the condition $1 + G(s)H(s) = 0$ using $G(s)H(s)$ given in (5.38) and $b = 31.2463$ and $k = 0.0396$, i.e.,

$$1 + 0.0396 \frac{116137(s + 31.2463)}{s^3 + 72.54s^2 - 1250s - 9.363 \times 10^4} = 0, \qquad (5.43)$$

hence:

$$s^3 + 72.54s^2 + (0.0396 \times 116137 - 1250)s$$
$$+ (0.0396 \times 116137 \times 31.2463 - 9.363 \times 10^4) = 0.$$

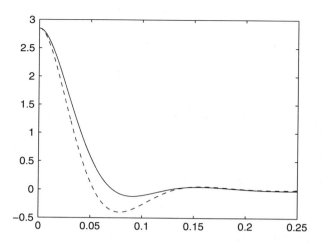

Fig. 5.40 Closed-loop response of the system (5.38) from an initial output that is different from zero. Continuous line: designed response. Dashed line: desired response. Vertical axis: y[m]. Horizontal axis: time in seconds

The roots of this polynomial are the closed-loop poles when $b = 31.2463$ and $k = 0.0396$. Using MATLAB, it is found that these roots are $-25.0037 + j39.9630$, $-25.0037 - j39.9630$ and -22.5326. These poles are also shown in Fig. 5.38a using symbols "+." Notice that the desired complex conjugate poles have been successfully assigned and the third pole, which is real, is also located on the left half-plane, i.e., closed-loop stability has been accomplished. Recall that the factor $k(s + b)$ represents a PD controller.

It is important to say that this control system is employed to regulate the output at a zero value. This implies that no matter what the system type, the desired output (i.e., zero) is reached. Hence, consideration with respect to the steady-state response is not necessary.

The closed-loop response (continuous line) is shown in Fig. 5.40 when the desired output is zero, $r = 0$ (see Fig. 5.1). In this simulation (5.38), $b = 31.2463$ and $k = 0.0396$ are employed. The dashed line represents the response of a system with the transfer function $\frac{\omega_n^2}{s^2 + 2\zeta\omega_n s + \omega_n^2}$ whose poles are located at $s = -25 \pm j40$. Thus, this system represents what is known as a reference model, because its response possesses the desired transient response specifications. Both responses start from an initial output equal to 2.85. Notice that some differences exist between these responses, which are due to the zero at $s = -31.2463$ and the closed-loop pole at $s = -22.5326$, because they are not close enough to completely cancel out their effects.

Finally, Figs. 5.36, 5.38 and 5.40 have been drawn using the following MATLAB code in an m-file:

```
clc
clear
```

```
gh=tf(116137,[1 72.54 -1250 -9.363e4]);
figure(1)
rlocus(gh)
axis([-90 50 -15 15])
b=31.24;
gh=tf(116137*[1 b],[1 72.54 -1250 -9.363e4]);
figure(2)
rlocus(gh)
axis([-90 50 -50 50])
b=45;
gh=tf(116137*[1 b],[1 72.54 -1250 -9.363e4]);
figure(3)
rlocus(gh)
axis([-80 40 -150 150])
b=36.5;
gh=tf(116137*[1 b],[1 72.54 -1250 -9.363e4]);
figure(4)
rlocus(gh)
axis([-80 40 -2 2])
b=31.2463;
k=0.0396;
gh=tf(116137*k*[1 b],[1 72.54 -1250 -9.363e4]);
M=feedback(gh,1,-1)
A=[0 1 0;
0 0 1;
-5.007e4 -3349 -72.54];
B=[0;
0;
1];
C=[1.437e5 4599 0];
Me=ss(A,B,C,0);
dend=conv([1 25+40*j],[1 25-40*j]);
Ad=[0 1;
-dend(3) -dend(2)];
Bd=[0;
dend(3)];
Cd=[1 0];
Md=ss(Ad,Bd,Cd,0);
figure(5)
initial(Me,'b-',[2.8/1.437e5 0 0])
hold on
initial(Md,'r--',[2.8 0])
```

5.2.7 Proportional–Integral–Derivative Control of an Unstable Plant

In this section, the use of the root locus method to select the gains of a PID controller for an open-loop unstable plant is presented. The open-loop transfer function is:

$$G(s)H(s) = \frac{11613700 \, k_d}{s^3 + 2739s^2 - 1250s - 3.536 \times 10^6} \left(\frac{s^2 + \frac{k_p}{k_d}s + \frac{k_i}{k_d}}{s} \right) \quad (5.44)$$

where k_d is the parameter that the method varies from 0 to $+\infty$ to draw the root locus, whereas $\frac{k_p}{k_d}$ and $\frac{k_i}{k_d}$ must be proposed. Notice that a PID controller is selected because it renders the system type 1, i.e., to ensure a zero steady-state error when the desired output is a constant. Notice that the PID controller introduces two open-loop zeros. There are four open-loop poles placed at:

$$s_1 = 35.9, \qquad s_2 = -2739.4, \qquad s_3 = -35.9, \qquad s_4 = 0.$$

Using rule 3 it is found that there are $n - m = 4 - 2 = 2$ branches of the root locus that tend toward infinity on the plane s following the asymptotes whose angles are given as:

$$\frac{\pm 180°}{2} = \pm 90°.$$

Moreover, according to rule 4, the point on the real axis where these asymptotes intersect is:

$$\sigma_a = \frac{35.9 - 2739.4 - 35.9 + \sigma_{z1} + \sigma_{z2}}{2},$$

where $-\sigma_{z1} < 0$ and $-\sigma_{z2} < 0$ are the real parts of both open-loop zeros introduced by the PID controller (these values have to be proposed). It is clear that σ_a moves to the left (which implies that the closed-loop system becomes more stable) if $-\sigma_{z1} < 0$ and $-\sigma_{z2} < 0$ are chosen to be close to zero. Notice that this is in agreement with rule 12. Thus, it is concluded that $\sigma_a < -1100$ is placed far to the left of the origin. Using these observations, in addition to rules 1, 2, 5, and 6, it is found that the root locus has the three possibilities depicted in Fig. 5.41.

Instead of assigning some desired closed-loop poles, the control objective is simply to render the closed-loop system stable. As previously stated, it is preferable to place both open-loop zeros close to the origin. This means that the possibility shown in Fig. 5.41a is to be used. On the other hand, two complex conjugate zeros would force two complex conjugate closed-loop poles to appear, which would result in a less damped closed-loop system. It is for this reason that the possibility in Fig. 5.41c is also avoided. Hence, searching for a root locus diagram such as that in

Fig. 5.41 Different possibilities for the root locus diagram when using (5.44) as the open-loop transfer function

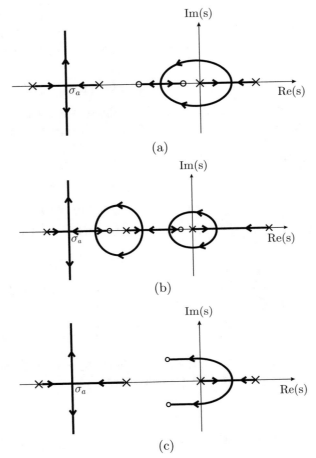

(a)

(b)

(c)

Fig. 5.41a, the following values are proposed:

$$\frac{k_p}{k_d} = 31.24, \quad \frac{k_i}{k_d} = \frac{31.24}{0.8}, \tag{5.45}$$

as this assigns two open-loop zeros at:

$$s_5 = -29.9355, \quad s_6 = -1.3045,$$

i.e., they are located between the open-loop poles at:

$$s_3 = -35.9, \quad s_4 = 0.$$

The facts that $\sigma_a < -1100$ is located far to the left of the origin and that the asymptotes form $\pm 90°$ angles, ensure that a minimal k_d exists for all closed-loop poles to be placed on the left half-plane. Using (5.45) and Routh's criterion, it is found that:

$$k_d > 0.01, \qquad (5.46)$$

ensures that all of the closed-loop poles are located on the left half-plane and, hence, the closed-loop system is stable. This is performed as follows. From the condition $1 + G(s)H(s) = 0$, i.e., using (5.44) and (5.45), the following characteristic polynomial is obtained:

$$s^4 + a_3 s^3 + a_2 s^2 + a_1 s + a_0 = 0, \qquad (5.47)$$

$$a_3 = 2739, \quad a_2 = 11613700\, k_d - 1250,$$

$$a_1 = 11613700 \times 31.24\, k_d - 3.536 \times 10^6,$$

$$a_0 = 11613700 \times \frac{31.28}{8}\, k_d.$$

To apply Routh's criterion, the Table 5.1 is filled. Closed-loop stability is ensured if no sign changes exist in the first column of Table 5.1, i.e., if:

$$\frac{a_3 a_2 - a_1}{a_3} > 0, \quad \frac{e a_1 - a_3 a_0}{e} > 0, \quad a_3 > 0, \quad a_0 > 0. \qquad (5.48)$$

Notice that the third condition in (5.48) is naturally satisfied and from the first and the last conditions it is found that:

$$k_d > -3.5695 \times 10^{-6}, \quad k_d > 0. \qquad (5.49)$$

From the second condition in (5.48) the following is found:

$$k_d^2 + \left(\frac{b_1}{b_2} - \frac{b_3}{b_4} - 2739^2 \frac{b_5}{b_2 b_4} \right) k_d - \frac{b_1 b_3}{b_2 b_4} > 0, \qquad (5.50)$$

$$b_1 = 112250, \quad b_2 = 3.1447 \times 10^{10}, \quad b_3 = 3.536 \times 10^6,$$

$$b_4 = 11613700 \times 31.24, \quad b_5 = 11613700 \frac{31.28}{8}.$$

Table 5.1 Applying Routh's criterion to the polynomial in (5.47)

s^4	1		a_2	a_0
s^3	a_3		a_1	0
s^2	$\frac{a_3 a_2 - a_1}{a_3} = e$		a_0	0
s^1	$\frac{e a_1 - a_3 a_0}{e}$		0	
s^0	a_0			

It is not difficult to find that the roots of the second-degree polynomial in (5.50) are $k_d = 0.01$ and $k_d = -3.5 \times 10^{-6}$. Moreover, it is possible to evaluate numerically to find that:

$$(k_d - 0.01)(k_d + 3.5 \times 10^{-6}) > 0, \quad \text{if } k_d < -3.5 \times 10^{-6},$$

$$(k_d - 0.01)(k_d + 3.5 \times 10^{-6}) < 0, \quad \text{if } -3.5 \times 10^{-6} < k_d < 0.01,$$

$$(k_d - 0.01)(k_d + 3.5 \times 10^{-6}) > 0, \quad \text{if } k_d > 0.01.$$

Thus, to simultaneously satisfy (5.49) and (5.50), i.e., to ensure closed-loop stability, (5.46) must be chosen.

Example 5.6 The root locus diagram for $G(s)H(s)$ given in (5.44) under the conditions in (5.45) is shown in Fig. 5.42. This has been obtained using the following MATLAB code in an m-file.

```
clc
clear
gh=tf(11613700*[1 31.24 31.24/0.8],[1 2739 -1250
  -3.536e6 0]);
k=0:0.00001:0.1;
figure(1)
rlocus(gh,k)
axis([-70 50 -10 10])
rlocfind(gh)
%%{
M=feedback(0.0226*gh,1,-1)
A=[0 1 0 0;
0 0 1 0;
0 0 0 1;
-1.025e007 -4.664e006 -2.612e005 -2739];
B=[0;
0;
0;
1];
C=[1.025e007 8.2e006 2.625e005 0];
Me=ss(A,B,C,0);
figure(2)
initial(Me,'b-',[1/1.025e007 0 0 0])
axis([-0.1 3 -0.8 1.2])
%}
```

Using the command "rlocfind(gh)", the closed-loop poles on the imaginary axis have been selected to find that $k_d = 0.01$, which corroborates the above findings, i.e., see (5.46). Using the command "pole(M)," all the closed-loop poles were found to be real, located at $s = -2640$, $s = -75.6$, $s = -20.1$, and $s = -2.6$ when

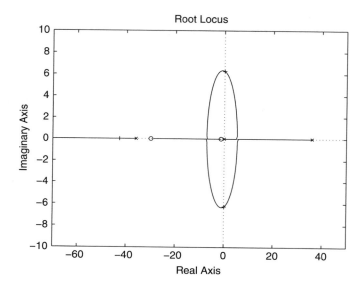

Fig. 5.42 Root locus diagram for $G(s)H(s)$ given in (5.44) under the conditions in (5.45)

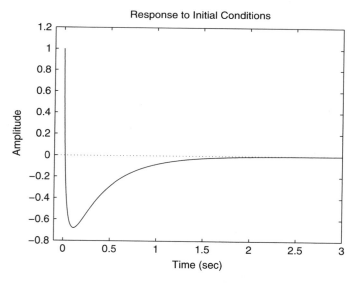

Fig. 5.43 Closed-loop time response when $G(s)H(s)$ is given in (5.44) under the conditions in (5.45) and $k_d = 0.0226$

$k_d = 0.0226$, i.e., the closed-loop system is stable. Notice that the open-loop zeros are located at $s = -29.93$ and $s = -1.3$, which was found using the command "zero(gh)." This means that all the closed-loop poles are located on the left of the open-loop zero at $s = -1.3$. Hence, according to Sect. 8.1.2, this means that overshoot is unavoidable despite the fact that all the closed-loop poles are real. This explains the large overshoot observed in Fig. 5.43, when a zero reference is commanded and the initial output is 1. Thus, in this example, it is not important to try to assign the closed-loop poles at some specific locations.

5.2.8 Control of a Ball and Beam System

In this section, a controller is designed for the ball and beam system that is built and experimentally tested in Chap. 14, Sect. 14.9. Assume that a proportional controller is proposed with gain γ. First, one proceeds to study the possibility that the closed-loop system might be rendered stable for some positive gain γ. The other system parameters are positive too, but they cannot be changed. The corresponding block diagram is shown in Fig. 5.44. The open-loop transfer function is:

$$\frac{X(s)}{X_d(s)} = \frac{\gamma A_x k\rho}{s^4 + as^3 + \gamma A_x k\rho},$$

$$A_x = 5.3750, \quad k = 16.6035, \quad a = 3.3132, \quad \rho = 5.$$

The stability of the closed-loop system is studied using Routh's criterion. Thus, Table 5.2 is filled using the closed-loop characteristic polynomial $s^4 + as^3 + \gamma A_x k\rho$. Notice that one entry at the first column is zero; hence, the method suggests replacing it by a small $\varepsilon > 0$. Also notice that under this condition, there are two sign changes in the first column of Table 5.2 and this cannot be modified by adjusting $\gamma > 0$. Then, it is concluded that no proportional controller exists to render the closed-loop system stable. Thus, another controller must be proposed.

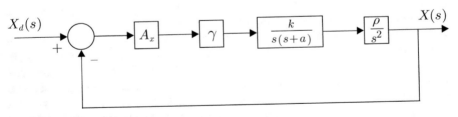

Fig. 5.44 Closed-loop system. Proportional control with gain γ

Table 5.2 Applying Routh's criterion to the polynomial $s^4 + as^3 + \gamma A_x k\rho$			
s^4	1	0	$\gamma A_x k\rho$
s^3	a	0	0
s^2	$0 \approx \varepsilon$	$\gamma A_x k\rho$	
s^1	$\frac{-a\gamma A_x k\rho}{\varepsilon}$	0	
s^0	$\gamma A_x k\rho$		

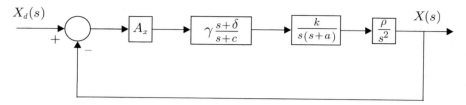

Fig. 5.45 Closed-loop system. Use of a lead compensator

According to rule 12 in Sect. 5.1.1, a PD controller can render closed-loop stable system an open-loop unstable system because a PD controller introduces an open-loop zero. However, a PD controller amplifies noise because of the derivative action. This is clearly shown in Chap. 6 where a PD controller is proved to be a high-pass filter and noise is a high-frequency signal. One way of maintaining the stabilizing properties of a PD controller, but reducing the effects of noise is to use a lead compensator, i.e., a controller with the transfer function $\gamma \frac{s+\delta}{s+c}$ with $c > \delta > 0$. It is for this reason that we study the possibility of stabilizing the ball and beam system using the block diagram in Fig. 5.45 in the following. The closed-loop transfer function is:

$$\frac{X(s)}{X_d(s)} = \frac{\gamma A_x k \rho (s + \delta)}{s^5 + (a + c)s^4 + acs^3 + \gamma A_x k \rho s + \gamma A_x k \rho \delta}.$$

The closed-loop stability is studied using Routh's criterion. Hence, the Table 5.3 is constructed using the closed-loop characteristic polynomial $s^5 + (a + c)s^4 + acs^3 + \gamma A_x k \rho s + \gamma A_x k \rho \delta$. As all the parameters are positive, there are at least two sign changes in the first column of Table 5.3. Then, it is concluded that there is no lead compensator rendering the closed-loop system stable; thus, another control strategy must be proposed. Carefully analyzing Table 5.3, it is concluded that the problem is originated by the first entry corresponding to row s^2, i.e., $\frac{-(a+c)e}{ac}$, which is negative. Notice that this is a consequence of the fact that the second entry in the row corresponding to s^4 is equal to zero. This is because s^2 has a zero coefficient in the polynomial $s^5 + (a+c)s^4 + acs^3 + \gamma A_x k \rho s + \gamma A_x k \rho \delta$. Thus, it is concluded that the closed-loop system may be rendered stable if the characteristic polynomial of the open-loop transfer function has a positive coefficient for s^2. Next, it is shown that this is possible if two internal loops are employed, as shown in Fig. 5.46. Notice that a lead compensator is still considered. In this case, the open-loop transfer function is given as:

$$G(s)H(s) = \frac{A_x \, \rho \, \gamma}{A_\theta} \frac{s+b}{s+c} \frac{\alpha k A_\theta}{s^2 + (a + k_v k A_\theta)s + \alpha k A_\theta} \frac{1}{s^2}, \qquad (5.51)$$

$$c > b > 0, \qquad \gamma > 0, \qquad A_\theta = 0.9167.$$

Table 5.3 Applying Routh's criterion to polynomial $s^5 + (a + c)s^4 + acs^3 + \gamma A_x kps + \gamma A_x kp\delta$

s^5	1	ac	$\gamma A_x kp$
s^4	$a + c$	0	$\gamma A_x kp\delta$
s^3	ac	$\frac{(a+c)\gamma A_x kp - \gamma A_x kp\delta}{a+c} = e$	
s^2	$\frac{-(a+c)e}{ac} = f$	$\gamma A_x kp\delta$	
s^1	$\frac{fe - ac\gamma A_x kp\delta}{f}$	0	
s^0	$\gamma A_x kp\delta$		

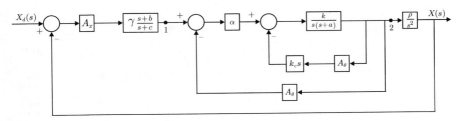

Fig. 5.46 Closed-loop system. Use of two internal loops and a lead compensator

Notice that the system type is 2, i.e., the steady state error is zero if the reference is a step or a ramp. The reader may obtain the closed-loop transfer function $\frac{X(s)}{X_d(s)}$ to verify that the corresponding characteristic polynomial has degree 5 and all of its coefficients are positive (including the coefficient of s^2), as expected. According to the previous discussion, this means that there exists the possibility of finding positive values for γ, c, b (with $c > b$), k_v and α such that the closed-loop system is stable. These values are found in the following using the root locus method.

Suppose that the roots of the polynomial $s^2 + (a + k_v k A_\theta)s + \alpha k A_\theta$ are complex conjugate, i.e., they can be written:

$$s^2 + (a + k_v k A_\theta)s + \alpha k A_\theta = (s + \sigma + j\omega)(s + \sigma - j\omega), \quad \sigma > 0, \quad \omega > 0.$$

According to rule 3, there are $n - m = 4$ root locus branches tending toward infinity on the plane s along four asymptotes whose angles are given as:

$$\frac{\pm 180°}{4} = \pm 45°,$$

$$\frac{\pm 180°(3)}{4} = \pm 3(45°) = \pm 135°.$$

Hence, using these results in addition to rules 1,2 and 5, it is concluded that the possibilities depicted in Fig. 5.50 exist. According to rule 4, the point where these asymptotes intersect with the real axis is given as:

$$\sigma_a = \frac{-2\sigma + (b - c)}{4}.$$

Fig. 5.47 Root locus
possibilities for a ball and
beam system

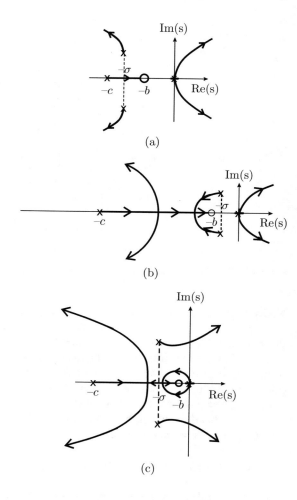

This means that the four branches are pulled to the left; hence, the closed-loop system becomes more stable, if:

- A greater $\sigma > 0$ is chosen, i.e., if the system:

$$\frac{\alpha k A_\theta}{s^2 + (a + k_v k A_\theta)s + \alpha k A_\theta},\tag{5.52}$$

  is damped enough, which is accomplished using a large enough value for $k_v > 0$.
- $b > 0$ Approaches to zero and $c > 0$ are large, i.e., if $c > b$. Note that this is in agreement with rules 11 and 12.

Following the second item above, the root locus shown in Fig. 5.47b is obtained. Notice that both branches starting at the poles located at $s = 0$ always remain on the right half-plane, i.e., the closed-loop system is unstable. The main reason for this behavior is that branches starting at the poles of the transfer function in (5.52) are pulled to the segment on the real axis between $s = -c$ and $s = -b$. This is because both open-loop complex conjugate poles are too close to the segment between $s = -c$ and $s = -b$. If these complex poles are placed far from such a segment, then the possibility exists for branches starting at $s = 0$ to be pulled to the segment between $s = -c$ and $s = -b$ to obtain the root locus diagram shown in Fig. 5.47c. This implies closed-loop stability for some of the small values of the loop gain. To achieve this, it is necessary to use large values for the coefficient $\alpha k A_\theta$ (because this increases the distance to the origin of the above-mentioned open-loop complex conjugate poles), i.e., using a large value for α.

According to the previous discussion, large values must be used for k_v and α. However, these values are limited in practice by the system noise content; hence, k_v and α must be obtained using experimental tests. In fact, noise is also an important reason not to select real poles for the transfer function in (5.52): real poles produce a very damped system that requires a large value for k_v. According to the experimental tests reported in Chap. 14 the following values were selected:

$$\alpha = 12, \quad k_v = 0.2.$$

Hence, the only values that remain to be determined are γ, c and b. Although these parameters can be determined such that the closed-loop poles are assigned at the desired locations (using the methodology presented in Sect. 5.2.6), in this case, it is preferred just to select them such that closed-loop stability is accomplished. To achieve this, we proceed as follows.

- b and c are proposed such that $c > b > 0$.
- MATLAB is employed to draw the root locus diagram.
- $\gamma > 0$ is chosen as that value placing all the closed-loop system poles on the left half-plane. This means that γ is the parameter used by the method to plot the root locus diagram.
- If such a $\gamma > 0$ does not exist, then we go back to the first step.

The root locus diagram shown in Fig. 5.48 is obtained following this procedure. It is observed that all the closed-loop poles, i.e., those indicated with a symbol "+," have a negative real part if:

$$\gamma = 1.2, \quad c = 20, \quad b = 2.5.$$

The simulation response of the closed-loop system when the reference is an unitary step is shown in Fig. 5.49. The graphical results in Figs. 5.48 and 5.49 were obtained by executing the following MATLAB code in an m-file:

```
Ax=5.375;
At=0.9167;
k=16.6035;
```

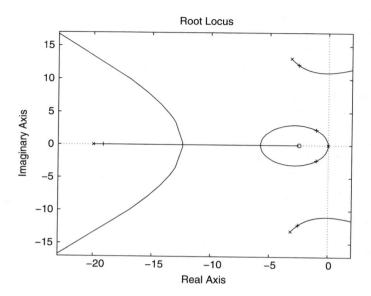

Fig. 5.48 Root locus diagram for the ball and beam system that has been designed

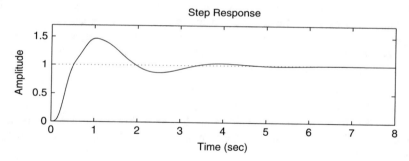

Fig. 5.49 The ball and beam closed-loop response

```
a=3.3132;
rho=5;
alpha=12;
kv=0.2;
c=20;
b=2.5;
gm=tf(k,[1 a 0]);
velFd=tf(kv*At*[1 0],1);
g1=feedback(gm,velFd,-1);
g2=feedback(alpha*g1,At,-1);
glead=tf([1 b],[1 c]);
```

```
gball=tf(rho,[1 0 0]);
gh=glead*g2*Ax*gball;
figure(1)
rlocus(gh);
axis([-23 2 -17 17])
gamma=rlocfind(gh)
M=feedback(gamma*gh,1,-1);
figure(2)
subplot(2,1,1)
step(M,8)
axis([0 8 0 1.7])
```

Finally, it is important to state that an additional step must be included:

- Once the values for γ, c and b are chosen, some experimental tests must be performed to verify that a good performance is obtained. If this is not the case, then we go back again to the first step in the procedure detailed above.

5.2.9 Assigning the Desired Closed-Loop Poles for a Ball and Beam System

Consider again the control problem in the previous section, but now using the following system parameters:

$$\rho = 4.8, \ a = 0.98, \ k = 10.729, \ k_v = 0.35, \ \alpha = 30, \ A_\theta = A_x = 1. \ (5.53)$$

Suppose[3] that the transient response specifications are stated by requiring a rise time $t_r = 1[s]$ and an overshoot $M_p = 25\%$. The use of this information and:

$$\zeta = \sqrt{\frac{\ln^2\left(\frac{Mp(\%)}{100}\right)}{\ln^2\left(\frac{Mp(\%)}{100}\right) + \pi^2}},$$

$$\omega_d = \frac{1}{t_r}\left[\pi - \arctan\left(\frac{\sqrt{1-\zeta^2}}{\zeta}\right)\right],$$

$$\omega_n = \frac{\omega_d}{\sqrt{1-\zeta^2}},$$

allows us to find that a complex closed-loop pole must be located at:

$$s = -\zeta\omega_n \pm j\omega_d = -0.8765 + j1.9864. \qquad (5.54)$$

---

[3] At this point, the reader is advised to see Example 8.2 in Chap. 8.

According to the closed-loop block diagram shown in Fig. 5.46, the open-loop transfer function in (5.51) becomes:

$$G(s)H(s) = \gamma \frac{s+b}{s+c} \frac{1545}{s^4 + 4.735s^3 + 321.9s^2}. \tag{5.55}$$

Using these numerical values, it is found that the open-loop poles are located at:

$$s_1 = 0, \ s_2 = 0, \ s_6 = -2.3676 + j17.7838, \ s_5 = -2.3676 - j17.7838.$$

Consider Fig. 5.50a. Proceed as in Sect. 5.2.6. The angle and the magnitude conditions establish that:

$$\theta_3 - (\theta_1 + \theta_2 + \theta_4 + \theta_5 + \theta_6) = -180°,$$

$$\frac{1545\gamma l_3}{l_1 l_2 l_4 l_5 l_6} = 1.$$

From these expressions, the following is found:

$$\theta_3 - \theta_4 = -180° + (\theta_1 + \theta_2 + \theta_5 + \theta_6), \tag{5.56}$$

$$\gamma = \frac{l_1 l_2 l_4 l_5 l_6}{1545 l_3}. \tag{5.57}$$

Also notice that:

$$\theta_3 = 180° - \arctan\left(\frac{1.9864}{0.8765 - b}\right), \tag{5.58}$$

$$\theta_4 = \arctan\left(\frac{1.9864}{c - 0.8765}\right),$$

$$\theta_1 = \theta_2 = 180° - \arctan\left(\frac{1.9864}{0.8765}\right), \tag{5.59}$$

$$\theta_5 = \arctan\left(\frac{1.9864 + 17.7838}{2.3676 - 0.8765}\right), \tag{5.60}$$

$$\theta_6 = -\arctan\left(\frac{(17.7838 - 1.9864)}{2.3676 - 0.8765}\right), \tag{5.61}$$

$$l_1 = l_2 = \sqrt{1.9864^2 + 0.8765^2},$$

$$l_3 = \sqrt{1.9864^2 + (0.8765 - b)^2},$$

$$l_4 = \sqrt{1.9864^2 + (c - 0.8765)^2},$$

$$l_5 = \sqrt{(17.7838 + 1.9864)^2 + (2.3676 - 0.8765)^2},$$

$$l_6 = \sqrt{(17.7838 - 1.9864)^2 + (2.3676 - 0.8765)^2}.$$

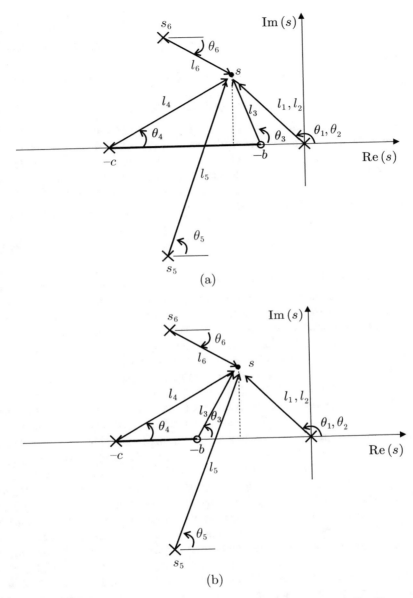

Fig. 5.50 Open-loop pole–zero contributions to the root locus diagram for a ball and beam system

From the second expression, the following is found:

$$c = \frac{1.9864}{\tan(\theta_4)} + 0.8765. \tag{5.62}$$

The following design procedure is proposed:

1. Compute $\theta_1, \theta_2, \theta_5, \theta_6$, using (5.59), (5.60), (5.61), and the difference $\theta_3 - \theta_4$ using (5.56).
2. Propose some b and compute θ_3 using (5.58). Using this and the value of the difference $\theta_3 - \theta_4$, compute θ_4.
3. Compute c and γ using (5.62) and (5.57) respectively.

Proposing $b = 0.7$, the above procedure yields $c = 2.8658$ and $\gamma = 1.3531$. The corresponding root locus diagram is depicted in Fig. 5.51 where the closed-loop poles are marked using the symbol "+". Notice that the desired closed-loop pole at $-0.8765 + j1.9864$ has been successfully assigned. In Fig. 5.52, we present the time response of the closed-loop system (continuous) when a unit step reference is applied. We also present, with a dashed line, the time response of the system:

$$M(s) = \frac{\omega_n^2}{s^2 + 2\zeta\omega_n s + \omega_n^2}, \tag{5.63}$$

where the values for ω_n and ζ ensure that the poles are identical to those defined in (5.54), i.e., the dashed line represents the desired response. An important difference between these time responses is observed. In particular, overshoot is almost twice its desired value. This difference may be explained by the fact that the closed-loop transfer function of the system in Fig. 5.46 has a zero[4] at $s = -b$. This feature deviates the time response from its desired behavior represented by the dashed line, i.e., the second-order system without any zero defined in (5.63).

Thus, to solve this problem it is proposed to use the block diagram in Fig. 5.53. Notice that the only difference with respect to the block diagram in Fig. 5.46 is that the lead-compensator $\gamma \frac{s+b}{s+c}$ is now placed in the feedback path. The reason for this is that the resulting closed-loop transfer function now has no zero.[5] Notice that the root locus diagram in this case is identical to that in Fig. 5.51 as the open-loop transfer function in this case $G(s)H(s)$ is identical to that in (5.55) because $b = 0.7, c = 2.8658, \gamma = 1.3531$ again. Recall that the zero at $s = -b$ is no longer closed-loop zero. The closed-loop time response for the block diagram in Fig. 5.53 is presented in Fig. 5.54 with a continuous line, whereas the desired response is represented by the dashed line. Notice that, again, an significant difference exists between the achieved response and the desired response. Analyzing the root locus diagram in Fig. 5.51, it is observed that a closed-loop real pole at ≈ -0.95 is very close to the poles at $-0.8765 \pm j1.9864$, which determine the transient response. Hence, the effects of this real pole are not negligible, and they are important in the transient response: the closed-loop system behaves as a third-order system instead of a second-order system. This fact explains the lag observed in the response achieved in Fig. 5.54.

---

[4]It is left as an exercise for the reader to verify this fact.
[5]Again, it is left as an exercise for the reader to verify this fact.

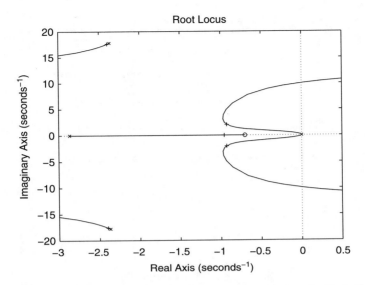

Fig. 5.51 Root locus diagram for the ball and beam system when using the block diagram in Fig. 5.46 and the controller gains $b = 0.7$, $c = 2.8658$, $\gamma = 1.3531$

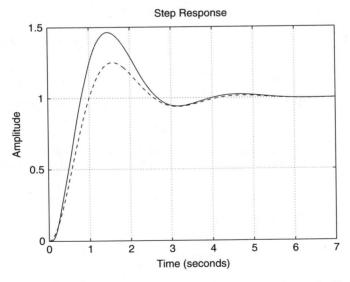

Fig. 5.52 Closed-loop time response of the ball and beam system when using the block diagram in Fig. 5.46 and the controller gains $b = 0.7$, $c = 2.8658$, $\gamma = 1.3531$ (continuous). Dashed: desired response

Hence, a remedy for this problem is to place this real pole further to the left. This may be accomplished by placing the zero at $s = -b$ further to the left as depicted

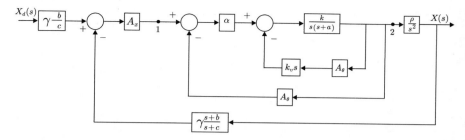

Fig. 5.53 Use of the lead compensator in the feedback path. The factor $\gamma \frac{b}{c}$ on the left is included to ensure that $\lim_{t \to \infty} x(t) = x_d$

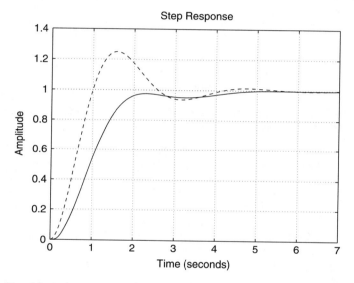

Fig. 5.54 Closed-loop time response of the ball and beam system when using the block diagram in Fig. 5.53 and the controller gains $b = 0.7$, $c = 2.8658$, $\gamma = 1.3531$ (continuous). Dashed: desired response

in Fig. 5.50b. Notice that in this case:

$$\theta_3 = \arctan\left(\frac{1.9864}{b - 0.8765}\right). \tag{5.64}$$

Thus, the design procedure proposed above is still valid, only requiring use of (5.64) instead of (5.58). Hence, proposing $b = 1.4$, the following is found: $c = 5.1138$ and $\gamma = 2.1868$. The root locus diagram for this case is shown in Fig. 5.55 where the closed-loop poles are marked with the symbol "+." It can be seen that the real pole is now placed at about -3.5, far enough from the poles in (5.54). Moreover, the time response achieved in this case (continuous) is shown in Fig. 5.56 and compared with

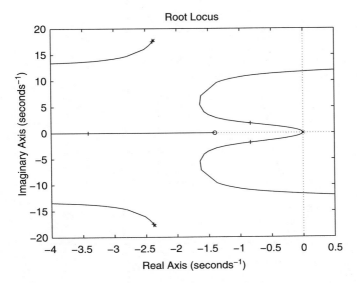

Fig. 5.55 Root locus for the ball and beam system according to the block diagram in Fig. 5.53 and controller gains $b = 1.4$, $c = 5.1138$, $\gamma = 2.1868$

the desired response (dashed). We realize that both responses are very similar in this case, which corroborates the correctness of the design. This controller is tested experimentally in Chap. 14.

Finally, to further improve the transient response, in Figs. 5.57 and 5.58, the root locus diagram and the time response respectively are shown when using the controller gains $b = 2$, $c = 11.7121$ and $\gamma = 4.6336$. Note that the real pole is now placed at about -10.5, i.e., it is far from the poles in (5.54). This results in a very small effect of this pole on the transient response, which, as shown in Fig. 5.58, is now very close to the desired response.

Except for Figs. 5.50 and 5.53, all the figures in this section have been drawn using the following MATLAB code in an m-file:

```
clc
clear
rho=4.8;
a=0.98;
k=10.729;
kv=0.35;
alpha=30;
Ao=1;
Ax=1;
Gdcm=tf(k,[1 a 0]);
Gfvel=tf(kv*Ao*[1 0],1);
Gfdbk1=feedback(Gdcm,Gfvel,-1);
```

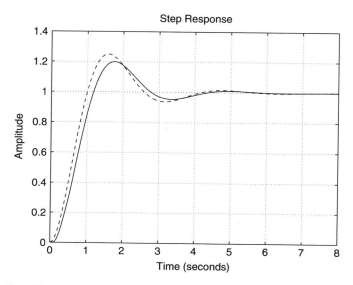

Fig. 5.56 Closed-loop time response of the ball and beam system according to the block diagram in Fig. 5.53 and controller gains $b = 1.4$, $c = 5.1138$, $\gamma = 2.1868$ (continuous). Dashed: desired response

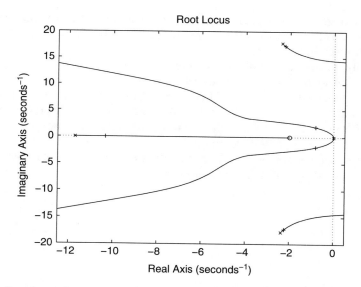

Fig. 5.57 Root locus diagram for the ball and beam system when using the block diagram in Fig. 5.53 and the controller gains $b = 2$, $c = 11.7121$ and $\gamma = 4.6336$

```
G2=alpha*Gfdbk1;
Gfdbk2=feedback(G2,Ao,-1);
```

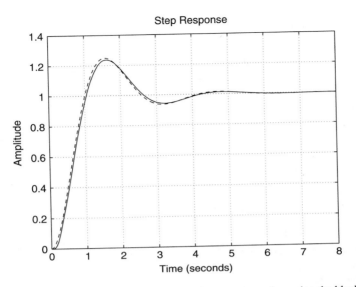

Fig. 5.58 Closed-loop time response of the ball and beam system when using the block diagram in Fig. 5.53 and the controller gains $b = 2$, $c = 11.7121$ and $\gamma = 4.6336$ (continuous). Dashed: desired response

```
Gball=tf(rho,[1 0 0]);
GH=Ax*Gfdbk2*Gball
pole(GH)
figure(1)
rlocus(GH)
axis([-5 2 -30 30])
tr=1; %s
Mp=25;%
z=sqrt(log(Mp/100)^2/(pi^2+log(Mp/100)^2));
wn=1/(tr*sqrt(1-z^2))*(pi-atan(sqrt(1-z^2)/z));
wd=wn*sqrt(1-z^2);
Res=-z*wn % -0.8765
Ims=wd % j 1.9864
th1=pi-atan(Ims/(-Res));
th5=atan((Ims+17.7838)/(2.3676+Res));
th6=-atan((17.7838-Ims)/(2.7636+Res));
th3menosth4=-pi+(2*th1+th5+th6);
b=2
%b=1.4
%b=0.7
%th3=pi-atan(Ims/(-Res-b));
th3=atan(Ims/(b+Res));
```

```
th4=-th3menosth4+th3;
c=Ims/tan(th4)-Res
l1=sqrt(Ims^2+Res^2);
l3=sqrt(Ims^2+(-Res-b)^2);
l4=sqrt(Ims^2+(c+Res)^2);
l5=sqrt((17.7838-Ims)^2+(2.3676+Res)^2);
l6=sqrt((17.7838+Ims)^2+(2.3676+Res)^2);
gamma=l1^2*l4*l5*l6/(l3*rho*alpha*k*Ax)
ctrl=tf(gamma*[1 b],[1 c]);
GHcomp=GH*ctrl;
gg=0:0.01:1;
figure(2)
rlocus(GHcomp) %,gg)
axis([-12.5 0.5 -20 20])
rlocfind(GHcomp)
M=feedback(GH,ctrl,-1);
M1=tf(wn^2,[1 2*z*wn wn^2])
figure(3)
step(M*0.7913,'b-',M1,'r--')
%step(M*0.5987,'b-',M1,'r--')
%step(M*0.3305,'b-',M1,'r--')
grid on
```

5.3 Case Study: Additional Notes on the PID Control of Position for a Permanent Magnet Brushed DC Motor

The PID control for position regulation in a DC motor is studied in Sect. 5.2.5. There, it was found that the motor position relates to the desired position and an external torque disturbance through two transfer functions, which, however, have the same characteristic polynomial:

$$s^3 + (a + k_d k)s^2 + k_p k s + k_i k. \qquad (5.65)$$

It was also found that the following closed-loop poles p_1, p_2, p_3 can be assigned:

$$p_1 = \sigma_1 + j\omega_1, \quad p_2 = \sigma_2 - j\omega_1, \quad p_3 < 0, \quad \sigma_1 < 0, \quad \sigma_2 < 0, \quad \omega_1 \geq 0.$$

Notice that $\sigma_1 = \sigma_2$ if $\omega_1 > 0$, i.e., when a pair of complex conjugate poles exists, but $\sigma_1 \neq \sigma_2$ is possible when $\omega_1 = 0$. Then, controller gains must be selected according to:

$$k_d = \frac{-(\sigma_1 + \sigma_2 + p_3) - a}{k} > 0, \qquad (5.66)$$

$$k_p = \frac{\sigma_1\sigma_2 + \omega_1^2 + p_3\sigma_1 + \sigma_2 p_3}{k} > 0, \tag{5.67}$$

$$k_i = \frac{-p_3(\sigma_1\sigma_2 + \omega_1^2)}{k} > 0. \tag{5.68}$$

On the other hand, in Example 4.12, Chap. 4, it was found that all of the roots of the following polynomial:

$$s^3 + as^2 + bs + c,$$

have a negative real part if and only if:

$$a > 0, \quad b > \frac{c}{a} > 0 \quad c > 0,$$

Applying these conditions to the characteristic polynomial in (5.65), yields:

$$a + k_d k > 0, \quad k_p k > \frac{k_i k}{a + k_d k}, \quad k_i k > 0. \tag{5.69}$$

Using the expressions in (5.66), (5.67), (5.68), (5.69), Fig. 5.59, the results in Sect. 3.3, and Example 3.9, the effects of the PID controller gains on the closed-loop system response can be studied.

- $k_p > 0$ and $k_d > 0$ are small, whereas $k_i > 0$ is large.

  - Suppose that p_1 and p_2 are complex conjugate. Then, $\omega_n^2 = \sigma_1\sigma_2 + \omega_1^2$ in Fig. 5.59. According to (5.68), the product $-p_3\omega_n^2$ is large, which implies a faster response because a real pole produces faster responses as p_3 moves to the left and a second-order system is faster as ω_n is larger. Note that k_i can be kept constant and, according to (5.67), $k_p > 0$ can be rendered small if both σ_1 and σ_2 approach zero, i.e., if $\theta > 0$ in Fig. 5.59 approaches zero and, hence, damping approaches zero (without affecting ω_n nor $-p_3$). According to (5.66), this also renders $k_d > 0$ small. Also notice that this is in agreement with the second inequality in (5.69), i.e., $k_p k > \frac{k_i k}{a + k_d k}$. If $k_i > 0$ is large and $k_p > 0$, $k_d > 0$ are small, then this condition tends not to be valid, i.e., the closed-loop system approaches instability and becomes more oscillatory.
  - Suppose that p_1 and p_2 are real and different, i.e., $\omega_1 = 0$. Thus, a more damped and, hence, slower response is obtained. According to (5.68), this results in a smaller $k_i > 0$. However, according to (5.66), (5.67), $k_p > 0$, and $k_d > 0$ can be rendered larger because $|\sigma_1|$ may become larger despite σ_2 approaching zero. Thus, from the second inequality in (5.69), it is concluded too that a more damped response is obtained.
    Notice that an increment in $k_d > 0$ has a greater effect on the increment of $|\sigma_1|$ than the effect that an increment in $k_p > 0$ has on the increment of $|\sigma_1|$ because this real part is affected by p_3 in (5.67), which is assumed to be large.

Fig. 5.59 Closed-loop pole
location for PID position
control in a DC motor

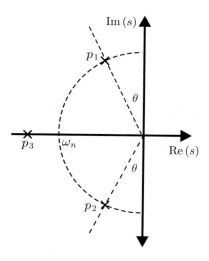

Thus, it is concluded that k_d has a greater effect on the damping than k_p, i.e., the slower response predicted above is more affected by k_d.

- $k_i > 0$ is small and $k_p > 0$ is large.

  This may be possible if p_3 is close to zero and p_1 and p_2, are complex conjugate. Thus, according to (5.67) and (5.66), $k_p > 0$ can be large and $k_d > 0$ small if $\omega_n^2 = \sigma_1\sigma_2 + \omega_1^2$ is large and σ_1 and σ_2, are close to zero. This means that the damping is small because $\theta > 0$ in Fig. 5.59 is close to zero. Thus, a more oscillatory but faster closed-loop system response is obtained.

  However, $|\sigma_1|$ and $|\sigma_2|$ may be rendered larger, i.e., to increase $k_d > 0$, without affecting ω_n if $\theta > 0$ in Fig. 5.59 is increased. Thus, a more damped but slower system response is obtained. Notice that $k_p > 0$ is less affected by the changes in $|\sigma_1|$ and $|\sigma_2|$ because they are affected by p_3 in (5.67), which is assumed to be close to zero.

According to the discussion above the following is concluded regarding the effects that the PID control gains have when used for position regulation in a DC motor:

- Larger positive values for the integral gain, k_i, result in a faster response that becomes more oscillatory.
- Large positive values of the proportional gain, k_p, result in either a) a slightly more damped response if $k_i > 0$ is large, or b) a more oscillatory response if $k_i > 0$ is small. In both cases, a fast response is expected.
- Larger positive values of the derivative gain, k_d, produce a more damped and slower response.

It is convenient to say that the transfer function between the position and its desired value ($G_1(s)$ in (5.26)) has two zeros whose effect on the transient response is not completely clear. Hence, it is possible that sometimes, some variations appear

Fig. 5.60 Simple pendulum
used as a load for a DC motor

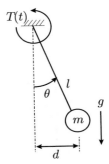

regarding the effects that have just been described for the gains of the PID controller.
Furthermore, the fact that the external disturbance is present because the desired
position is commanded in some mechanisms may favor these variations.

To illustrate the ideas above, some experimental results that were obtained
when controlling the position of a DC motor are presented in the following. The
experimental prototype is the same as that described in Chap. 11 with a simple
modification (that is not present in that chapter): the pendulum shown in Fig. 5.60
is fixed to the motor shaft. Hence, by means of the gravity effect g, a torque
disturbance is introduced that tries to deviate the motor position from its desired
value. In Fig. 5.60, $T(t)$ represents the torque generated by the motor and θ is the
controlled variable. The pendulum parameters l and m are not measured to show
that a PID controller can be tuned without any necessity of knowing the plant, if the
effects of the controller gains (listed above) are well understood.

The corresponding experimental results are presented in Figs. 5.61 and 5.62. The
desired position is set to $\theta_d = \pi/2$[rad] as this is where the torque disturbance due
to gravity has its greatest effect. A vertical line at $t = 0.13$[s] and a horizontal line
at $\theta = 1.8$[rad] are also shown to indicate the desired rise time and overshoot. Only
one parameter is adjusted each time to clearly appreciate its effect. Curves drawn in
Fig. 5.61 correspond to the following PID controller gains:

1. Upper figure:
   - Continuous: $k_p = 0.5$, $k_i = 0$, $k_d = 0$
   - Dash–dot: $k_p = 1$, $k_i = 0$, $k_d = 0$
   - Dotted: $k_p = 1$, $k_i = 0$, $k_d = 0.05$

2. Bottom figure:
   - Continuous: $k_p = 1$, $k_i = 2$, $k_d = 0.05$
   - Dash–dot: $k_p = 1$, $k_i = 5$, $k_d = 0.05$
   - Dotted: $k_p = 2$, $k_i = 5$, $k_d = 0.05$
   - Dashed: $k_p = 2$, $k_i = 5$, $k_d = 0.1$

As the closed-loop system is third-order when a PID controller is used and the motor
parameters are not known, it is difficult to know the controller gains that ensure

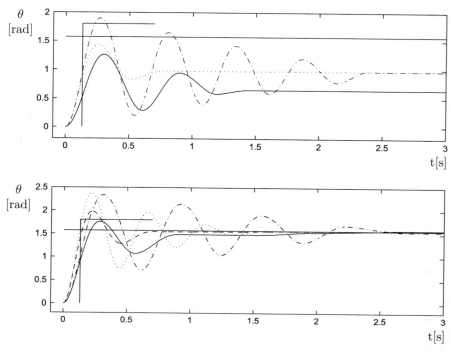

Fig. 5.61 The PID control of position for a DC motor with a pendulum as a load. Experimental results

stability. Thus, the best that can be done is to begin with a PD controller (assigning $k_i = 0$), which produces a second-order closed-loop system that is stable for any positive k_p with $k_d = 0$ (because any real mechanism possesses friction). Later, according to the transient response obtained and to the tuning rules listed above, the gains k_i and k_d are introduced such that stable responses are accomplished. It is for this reason that, in the upper part of Fig. 5.61 a PD controller is employed. Note that a steady-state error that is different from zero is obtained because of the torque disturbance introduced by gravity (recall that $k_i = 0$). The aim in this part of the experiment is to allow the system response time to approach its desired value ensuring stable behavior. Thus, the proportional gain is first increased to $k_p = 1$ and then the derivative gain is increased to $k_d = 0.05$. Then, an integral term can be introduced, which, although producing a zero steady-state error, increases the oscillations in the system response. This is shown at the bottom part of Fig. 5.61. Again, the idea is to allow the response time to approach its desired value. This is accomplished first by increasing the integral gain to $k_i = 5$. As this also increases the oscillation, $k_p = 2$ is employed to render the system response faster without appreciably increasing the oscillation. Finally, $k_d = 0.1$ is used to produce a well damped response. Notice that the increment on k_d increases the rise time a little.

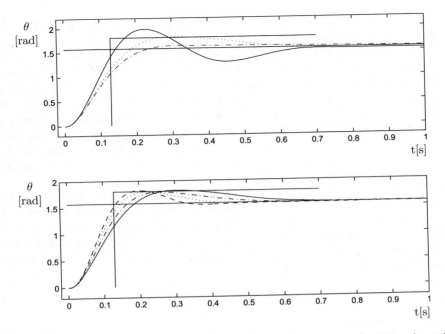

Fig. 5.62 The PID control of position for a DC motor with a pendulum as a load. Experimental results

Curves drawn in Fig. 5.62 correspond to the following PID controller gains:

1. Upper figure:

   - Continuous: $k_p = 2$, $k_i = 5$, $k_d = 0.1$
   - Dash–dot: $k_p = 2$, $k_i = 5$, $k_d = 0.2$
   - Dotted: $k_p = 2$, $k_i = 8$, $k_d = 0.2$

2. Bottom figure:

   - Continuous: $k_p = 2$, $k_i = 8$, $k_d = 0.2$
   - Dash–dot: $k_p = 2.5$, $k_i = 8$, $k_d = 0.2$
   - Dotted: $k_p = 3$, $k_i = 8$, $k_d = 0.2$
   - Dashed: $k_p = 3.5$, $k_i = 8$, $k_d = 0.2$

In the upper part of Fig. 5.62, it is observed that the response is even more damped and slower when the derivative gain is increased from $k_d = 0.1$ to $k_d = 0.2$. Because, at this moment, the response is slow and overshoot is small, $k_i = 8$ is employed to render the response faster and with a larger overshoot that is equal to the desired overshoot. To render the system faster, without appreciably affecting overshoot, the proportional gain is increased to $k_p = 3.5$ in the lower part of Fig. 5.62. This allows the desired rise time and overshoot to be accomplished.

It is important to stress the following. When connecting the pendulum to the motor shaft, the control system becomes nonlinear. This means that the control system behavior is not correctly predicted by the analysis presented above at certain operation regions. For instance, if the desired position is close to $\theta_d = \pm\pi$, a simple PD controller may be unstable if the proportional gain (positive) is not greater than a certain lower threshold. This result is obtained using a PD controller for the case when $x_1^* = \pm\pi$, $x_2^* = 0$ in Example 7.5 studied in Chap. 7 by finding the poles for such a linear approximation. The reader may also see the works reported in [2, 4], ch. 8, [3], ch. 7. This situation is worse for the case of a PID controller, but the problem does not appear if the desired position value is kept far from $\theta_d = \pm\pi$, as in the results presented in this section.

Finally, notice that the selection of the PID controller gains has been performed without requiring knowledge of the numerical value of any motor or pendulum parameter. However, the minimal knowledge that must be available is to verify that the motor can produce the required torque to perform the task. This must include the required torque to compensate for the gravity effect plus some additional torque to achieve the desired rise time.

5.4 Summary

The most general method of controller design using the time response approach is the root locus method. This means that the plants that can be controlled are of arbitrary order with any number of zeros as long as they are less than the number of poles. This method provides the necessary tools to determine, in a graphical way, the location of the closed-loop poles from the location of the open-loop poles and zeros. Some of the open-loop poles and zeros are due to the controller and the idea of the method is to select the location of the controller poles and zeros such that the desired closed-loop poles are assigned. The desired closed-loop poles are chosen from the knowledge of how they affect the corresponding transient response. The study presented in Chap. 3 is important for this. On the other hand, the controller structure is chosen from the knowledge of how the open-loop poles and zeros affect the closed-loop system steady- state response. The study presented in Chap. 4 is very important for this.

Although the root locus has been presented as a controller design method, it is also a powerful tool for control systems analysis. This means that it can be used to determine: (i) The relative stability of a control system, (ii) How the control system response changes as one of its parameters changes, (iii) What has to be done to modify the control system properties, etc. The use of the root locus method in these applications depends, to a large extent, on a good understanding of the method and the material presented in Chaps. 3 and 4.

5.5 Review Questions

1. When would you use each one of the following controllers?

   - Proportional.
   - Proportional–derivative.
   - Proportional–integral.
   - Proportional–integral–derivative.

2. Why do you think it is not advised to use the following controllers: (i) Derivative (alone), (ii) Integral (alone), and (iii) Derivative–integral (alone), i.e., without including a proportional part? Explain.

3. How do the requirements of the steady-state error determine the poles and/or zeros of a controller, i.e., the controller structure?

4. What is the main component that a controller must possess to improve the closed-loop system stability? What is the effect of this on the shape of the root locus diagram?

5. What is the main component that a controller must possess to improve the steady-state error? What is the effect of this on the shape of the root locus diagram?

6. Why does the root locus begin at the open-loop poles and end at the open-loop zeros? What do the words "begin" and "end" mean?

7. If the open-loop transfer function has no zeros, where does the root locus end?

8. It is often said that closed-loop system instability appears as the loop gain increases. However, this is not always true, because this depends on the properties of the plant to be controlled. Review the examples presented in this chapter and give an example of a plant requiring the loop gain to be large enough to render the closed-loop system stable.

9. What is a lead compensator and what is its main advantage?

10. Read Appendix F and Sect. 9.2 in Chap. 9. Use this information to explain how to practically implement a PID controller and a lead compensator using both software and analog electronics.

5.6 Exercises

1. Consider the control system in Fig. 5.63 where:

   $$k = 35.2671, \quad a = 3.5201, \quad c = 10.3672, \quad \gamma = 2.0465, \quad d = 3.8935.$$

   Verify that two closed-loop complex conjugate poles exist at $s = -4.8935 \pm j6.6766$ when $k_i = 0$ and $k_p = 1$. Use MATLAB to draw the root locus for the following values of k_i:

   $$k_i = 0.001, \ 0.01, \ 0.1, \ 1, \ 10,$$

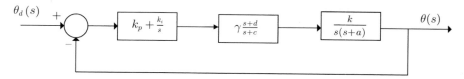

Fig. 5.63 A cascade connection of the PI control and lead compensator

Use k_p as the parameter the method varies from 0 to $+\infty$ to draw the root locus diagram. Select $k_p = 1$ and observe what happens with the closed-loop poles corresponding to $s = -4.8935 \pm j6.6766$. Perform simulations for each case when the reference is a unit step. Compare the obtained response as k_i grows with the response obtained when $k_p = 1$ and $k_i = 0$. Draw the corresponding root loci using the method presented in this chapter and explain what happens.

2. Consider the following plant:

$$Y(s) = H_1(s)U(s), \quad H_1(s) = \frac{k}{s(s+a)}, \quad k = 35.2671, \quad a = 3.5201.$$

a) Employ (3.71) in Chap. 3 to find the values for k_p and k_v such that the following transient response specifications are satisfied:

$$t_r = 0.33[s], \quad M_p(\%) = 10,$$

when y_d is a unit step and the following controller is used: $u(t) = k_p(y_d - y) - k_v\dot{y}$.

b) Assume that the input is given as:

$$u(t) = k_p(y_d - y) + k_d\frac{d(y_d - y)}{dt}, \quad \text{PD control.}$$

Obtain the closed-loop transfer function and use (3.71) in Chap. 3 to determine k_p and k_d such that the closed-loop poles are located at some points determining the following closed-loop transient response specifications:

$$t_r = 0.33[s], \quad M_p(\%) = 10.$$

Perform simulations corresponding to each one of the above items and compare the obtained responses. What is the reason for the differences between these responses? May zeros of a transfer function affect the transient response? Can these differences be explained using the root locus?

3. Consider a closed-loop system such as that shown in Fig. 5.1 with $H(s) = 1$ and $G(s) = \frac{1}{s^3+s^2+s+1}$.

- Using the rules presented in Sect. 5.1.1, draw the root locus diagram when using a proportional controller. Use Routh's criterion to determine the values of the proportional gain ensuring closed-loop stability.
- To stabilize the closed-loop system and to achieve a zero steady-state error when the reference is a step, design a PID controller proceeding as follows. (i) Propose $k_d s^2 + k_p s + k_i = k_d (s - z_1)(s - z_2)$, with $z_1 = -0.5 + 1.3j$, $z_2 = -0.5 - 1.3j$. Using the rules presented in Sect. 5.1.1, draw the root locus diagram to verify that no positive value exists for the derivative gain rendering the closed-loop system stable. (ii) Propose $k_d s^2 + k_p s + k_i = k_d (s - z_1)(s - z_2)$, with $z_1 = -0.25 + 1.3j$, $z_2 = -0.25 - 1.3j$. Using the rules presented in Sect. 5.1.1, draw the root locus diagram to verify that a range of positive values exists for the derivative gain k_d, rendering the closed-loop system stable. Use Routh's criterion to find the range of values for k_d, rendering the closed-loop system stable.
- How should the zeros of a PID controller be selected to improve the stability of the closed-loop system?

4. Verify that the transfer function of the circuit shown in Fig. 5.64 is:

$$\frac{V_o(s)}{V_i(s)} = \frac{s + a}{s + b}, \quad a = \frac{1}{R_1 C}, \quad b = \frac{1}{R_1 C} + \frac{1}{R_2 C}.$$

As $b > a$, this electric network can be used as a lead compensator.

5. Consider the following plant:

$$G(s) = \frac{4(s + 0.2)}{(s + 0.5)(s^2 - 0.2s + 0.3)}.$$

Use the root locus method to design a controller, ensuring closed-loop stability and a zero steady-state error when the reference is a step. It is suggested that the location of the poles and the zeros of the controller are proposed and then the open-loop gain is selected such that all the closed-loop poles have a negative real part.

Fig. 5.64 A lead compensator

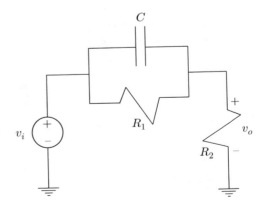

Fig. 5.65 A
mass-spring-damper system

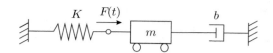

6. Consider the PD control of the position in Fig. 5.2.2. Prove that the value of the derivative gain k_d does not have any effect on the steady-state error when the reference is a step.

7. Consider the PID control of the position studied in Sect. 5.2.5. Prove that the values of the proportional k_p and the derivative k_d gains do not have any effect on the steady-state error, despite a constant disturbance being present if the reference is also constant.

   Assume that the integral part of the controller is not present, i.e., that $k_i = 0$. Prove that the value of the derivative gain k_d does not have any effect on the steady-state error.

8. Consider the mass-spring-damper system shown in Fig. 5.65, which has been modeled in Example 2.1, Chap. 2. In Example 3.18, Chap. 3, it is explained why a steady-state error that is different from zero exists when a proportional controller is employed and the desired position x_d is a constant that is different from zero. Recall that the force $F(t)$ applied to the mass is the control signal.

   Now suppose that a PID controller is used to regulate the position. Find the steady-state error when the desired position x_d is a constant that is different from zero. Use your everyday experience to explain your response, i.e., try to explain what happens with the force applied to the mass and the effect of the spring.

9. Consider the simple pendulum studied in Example 2.6, Chap. 2, which is shown in Fig. 5.66. Notice that the gravity exerts a torque that is different from zero on the pendulum if the angular position θ is different from zero. Suppose that it is desired to take the pendulum position θ to the constant value $\theta_d = 90°$. Note that torque $T(t)$ is the input for the pendulum. Using the experience in the previous exercise, state which controller you would employ to compute $T(t)$ such that θ reaches θ_d. Explain why. This problem is contrived to be solved without using a mathematical model; you merely need to understand the problem. In fact, the mathematical model is in this case nonlinear; hence, it cannot be analyzed using the control techniques studied in this book so far.

10. Consider an arbitrary plant $G(s)$ in cascade with the following controllers:

$$G_c(s) = \frac{s+a}{s+b}, \quad 0 < a < b,$$

$$G_d(s) = \frac{s+c}{s+d}, \quad c > d > 0.$$

$G_c(s)$ is known as a lead compensator, whereas $G_d(s)$ is known as a *lag compensator*.

Fig. 5.66 A simple
pendulum

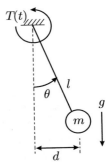

- What is the effect of each one of these compensators on the steady-state error when the reference is a step?
- What is the effect of each one of these compensators on the closed-loop stability?
- Which compensator is related to PI control and which to PD control? Explain why.
- How would you employ these compensators to construct a controller with similar properties to those of PID control? Explain.

11. In a ship, it is common to have a unique compass to indicate the course. However, it is important to know this information at several places in the ship. Hence, it is common to transmit this information to exhibit it using a needle instrument. The angular position of the needle is actuated using a DC motor by using the information provided by the compass as the desired position. Design a closed-loop control system such that the needle position tracks the orientation provided by the compass according to the following specifications:

- When a step change of 8° appears in the orientation provided by the compass, the system error decreases and remains at less than 1° in 0.3 s or less and overshoot is less than or equal to 25%.
- When the orientation provided by the compass changes as a ramp of 5° per second, the steady-state error must be 0.3° or less.
- There are no external disturbances.

Under which ship operating conditions may these situations appear? The transfer function between the voltage applied to the motor and the needle position in degrees is:

$$G(s) = \frac{45.84 \times 10^{-5}}{s(4.4 \times 10^{-9}s^2 + 308.5 \times 10^{-9}s + 3.3 \times 10^{-6})}.$$

Hint.
(i) On the basis of the specifications indicated, choose a controller to design.
(ii) Using the first specification, determine the zone in the time domain where

the closed-loop system response must lay. (iii) Recall that, according to Fig. 3.16, Chap. 3, the response of a second-order system remains between two exponential functions depending on ζ and ω_n. With this information and using the desired damping, determine the zone on the complex plane s where the dominant (complex conjugate) poles of the closed-loop system must be located. (iv) Use the root locus rules presented in this chapter to find the controller gains, ensuring that all the closed-loop poles are inside the desired zone in the plane s. (v) Verify that the second specification is satisfied and, if this is not the case, redesign the controller. (vi) Perform some simulations to corroborate that all the requirements are satisfied.

References

1. G. W. Evans, The story of Walter R. Evans and his textbook Control-Systems Dynamics, *IEEE Control Systems Magazine*, pp. 74–81, December 2004.
2. R. Kelly, V. Santibáñez, and A. Loría, *Control of robot manipulators in joint space*, Springer, London, 2005.
3. R. Kelly and V. Santibáñez, *Motion control of robot manipulators* (in Spanish), Pearson Prentice Hall, Madrid, 2003.
4. R. Kelly, PD control with desired gravity compensation of robotic manipulators: a review, *The International Journal of Robotics Research*, vo. 16, No. 5, pp. 660–672, 1997.
5. K. Ogata, *Modern control engineering*, 4th edition, Prentice-Hall, Upper Saddle River, 2002.
6. N. S. Nise, *Control systems engineering*, 5th edition, John Wiley and Sons, New Jersey, 2008.
7. R. C. Dorf and R. H. Bishop, *Modern control systems*, 10th edition, Pearson Prentice-Hall, 2005.
8. B. C. Kuo, *Automatic control systems*, Prentice-Hall, 1995.
9. G. H. Hostetter, C. J. Savant, and R. T. Stefani, *Design of feedback control systems*, Holt, Rinehart and Winston, 1982.
10. W. Ali and J. H. Burghart, Effects of lag controllers on the transient response, *IEE Proceedings-D, Control theory and applications*, vol. 138, no. 2, pp. 119–122, March 1991.

Chapter 6
Frequency Response-Based Design

In Chap. 5 a method for control systems design is presented, which is known as the *time response* method. The main feature of such an approach is to suitably locate the closed-loop poles to ensure that the transient response satisfies the desired specifications. Fundamental for that method is to know how the transient response is affected by the closed-loop poles if these are real, complex conjugate, repeated or different, etc.

In the present chapter, a new control system design method is introduced, which is known as the *frequency response* approach. Fundamental to this method is to know that when a sinusoidal function of time is applied at the input of a linear differential equation, then the output in a steady state is also a sinusoidal function of time, having the same frequency as the input, but, in general, having different amplitude and phase from the input (see Sect. 3.6). As a basis for this method, one must study how the amplitude and the phase at the output change as the frequency of the signal applied at the input changes. This is what term *frequency response* means. As shown in this chapter, the way in which the amplitude and the phase of the signal at the output change with the frequency depends on the location of the transfer function poles and zeros. The strategy of this method is to suitably modify the frequency response properties of the open-loop system to achieve the desired time response specifications for the output of the closed-loop system.

Finally, it is important to stress that control systems designed in this way are intended to respond to references that are not necessarily sinusoidal functions of time. Moreover, as shown in the subsequent sections, some methods that are useful to determine the closed-loop stability are also established. In this respect, it is important to say that these ideas can also be extended to unstable systems: although the natural response does not disappear in such a case, the forced response is still a sinusoidal function of time if the system input is too.

Every control design problem in classical control can be solved using one of two methods: the time response method or the frequency response method. This means that both methods provide the necessary tools for any control problem in

© Springer International Publishing AG, part of Springer Nature 2019
V. M. Hernández-Guzmán, R. Silva-Ortigoza, *Automatic Control with Experiments*,
Advanced Textbooks in Control and Signal Processing,
https://doi.org/10.1007/978-3-319-75804-6_6

classical control. However, as each of these methods analyzes the same problem from different points of view; each one of them provides information that can be complementary to information provided by the other method. Thus, it is important to understand both methods.

6.1 Frequency Response of Some Electric Circuits

In this section, we present some experimental results obtained when using some simple alternating current (AC) circuits. All of these circuits can be represented by the general circuit in Fig. 6.1a. We are interested in observing the output voltage $v_0(t)$ at the circuit element designated as z_2 when the voltage source delivers a signal given as:

$$v_i(t) = A \sin(\omega t). \tag{6.1}$$

When using a scope to measure $v_i(t)$ and $v_0(t)$, we observe a situation such as that depicted in Fig. 6.1b, i.e., the output voltage is given as:

$$v_0(t) = B \sin(\omega t + \phi),$$

$$\phi = \frac{360\, t_\phi}{T}\ [°], \quad T = \frac{2\pi}{\omega}.$$

Fig. 6.1 Input and output voltages in a general series of alternating current (AC) electric circuit

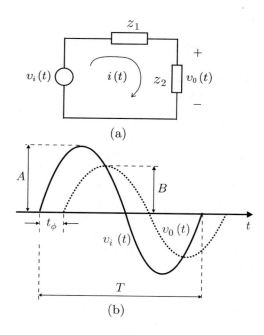

This means that both $v_i(t)$ and $v_0(t)$ are sinusoidal functions of time, having the same frequency. It is important to stress that, under the conditions shown in Fig. 6.1b, ϕ is defined as negative, i.e., the output voltage, $v_0(t)$, lags with respect to the input voltage, $v_i(t)$. Experiments to be performed consist of measuring the input and the output amplitudes, A and B, in addition to the time t_ϕ, to compute ϕ, for several frequencies ω. Then, the quantities B/A and ϕ are plotted as functions of the frequency ω. These plots are intended to explain the meaning of the term *frequency response*.

6.1.1 A Series Resistor–Capacitor Circuit: Output at the Capacitance

We first consider the AC circuit shown in Fig. 6.2a when $R = 1000$[Ohm] and $C = 0.1 \times 10^{-6}$[F]. Employing different values of frequency, ω, we obtain the measurements shown in Table 6.1, which are plotted in Fig. 6.3 (continuous line). The first of these plots represents B/A as a function of the frequency, whereas the second plot represents ϕ as a function of the frequency ω. The quantity B/A is called *magnitude* and ϕ is called *phase*.

From Fig. 6.3, we conclude that this circuit behaves as a *low-pass filter* and phase lag is produced. This means that the low-frequency signals (when ω is close to zero) are not attenuated or they are attenuated only a little, i.e., $B/A \approx 1$, whereas the high-frequency signals (when ω is large) are strongly attenuated, i.e., $B/A \approx 0$.

6.1.1.1 Model-Based Analysis

Using the tension divider in the circuit shown in Fig. 6.2a, the following is found:

$$V_0(s) = \frac{\frac{1}{sC}}{R + \frac{1}{sC}} V_i(s),$$

and rearranging:

$$\frac{V_0(s)}{V_i(s)} = G(s) = \frac{a}{s+a}, \quad a = \frac{1}{RC}. \tag{6.2}$$

Assume that $s = j\omega$, then:

$$G(j\omega) = \frac{a}{j\omega + a} = \frac{a}{j\omega + a}\frac{a - j\omega}{a - j\omega} = \frac{a(a - j\omega)}{\omega^2 + a^2},$$

$$|G(j\omega)| = \frac{a\sqrt{a^2 + \omega^2}}{\omega^2 + a^2} = \frac{a}{\sqrt{\omega^2 + a^2}}, \tag{6.3}$$

Fig. 6.2 Simple series of AC
circuits used in experiments

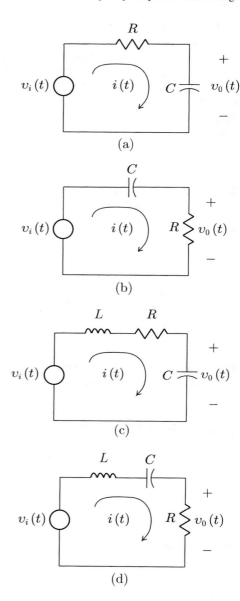

Table 6.1 Experimental data for the resistor–capacitor (RC) circuit in Fig. 6.2a

f [Hz]	$\omega = 2\pi f\,[\times 10^5\,\text{rad/s}]$	A [V]	B [V]	ϕ [°]
50	0.0031	4	4	0
60	0.0038	4	4	−2.3
70	0.0044	4	4	−2.5
80	0.0050	4	4	−2.8
90	0.0057	4	4	−3.8
100	0.0063	4	4	−4.3
200	0.0126	4	4	−8.5
300	0.0188	4	4	−12.9
400	0.0251	4	3.8	−15.8
500	0.0314	4	3.8	−19.8
600	0.0377	4	3.7	−23.7
700	0.0440	4	3.6	−27.7
800	0.0503	4	3.5	−31.6
900	0.0565	3.9	3.3	−35.6
1000	0.0628	3.9	3.2	−39.6
2000	0.1257	3.9	2.3	−57.6
3000	0.1885	3.9	1.7	−64.8
4000	0.2513	3.8	1.4	−74.8
5000	0.3142	3.8	1.1	−79.2
6000	0.3770	3.8	0.9	−73.4
7000	0.4398	3.8	0.8	−80.7
8000	0.5027	3.8	0.7	−74.7
9000	0.5655	3.8	0.6	−81
10000	0.6283	3.8	0.6	−79.8
20000	1.2566	3.8	0.3	−83.5
30000	1.8850	3.8	0.2	−82
40000	2.5133	3.9	0.1	−86
50000	3.1416	3.9	0.1	−87

$$\angle G(j\omega) = \arctan\left(\frac{\text{Im}(G(j\omega))}{\text{Re}(G(j\omega))}\right) = \arctan\left(\frac{-\omega}{a}\right). \qquad (6.4)$$

Thus, according to Sect. 3.6, if $v_i(t) = A\sin(\omega t)$, then $v_0(t) = B\sin(\omega t + \phi)$ where:

$$B = A\frac{a}{\sqrt{\omega^2 + a^2}}, \qquad (6.5)$$

$$\phi = \arctan\left(\frac{-\omega}{a}\right). \qquad (6.6)$$

At this point, it is convenient to recall that a transfer function is defined as the ratio of the system output and the system input, i.e., $G(s) = \frac{a}{s+a} = \frac{V_0(s)}{V_i(s)}$.

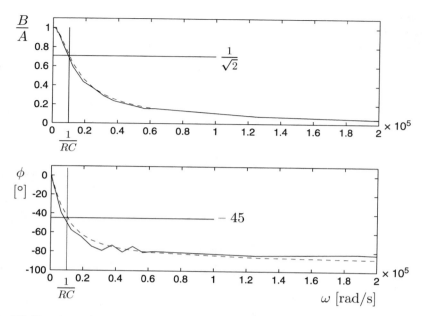

Fig. 6.3 Experimental (continuous line) and theoretical (dashed line) data for the circuit in Fig. 6.2a (see Table 6.1)

Hence, it is not a surprise that $|G(j\omega)|$ is given as the ratio of the output and the input amplitudes. On the other hand, if it assumed that $G(j\omega)$, $V_0(j\omega)$, $V_i(j\omega)$ are complex numbers, then the angle of $V_0(j\omega)$ must be given as the addition of the angles of $G(j\omega)$ and $V_i(j\omega)$ because $V_0(j\omega) = G(j\omega)V_i(j\omega)$. This means that the angle difference between $V_i(j\omega)$ and $V_0(j\omega)$ is equal to the angle of $G(j\omega)$.

In Fig. 6.3, the plot obtained (dashed line) using $|G(j\omega)|$ and ϕ defined in (6.3) and (6.4) is shown. The closeness observed in this figure between the continuous line and the dashed line is evidence that the analytical relationships obtained correctly represent the experimental situation. Finally, from (6.5), (6.6), we realize that:

$$\frac{B}{A} = \frac{1}{\sqrt{2}} = 0.7071,$$

$$\phi = -45°,$$

when $\omega = a = \frac{1}{RC}$, which is known as the *corner frequency*. Note that the phase of $\angle G(j\omega)$ takes values in the range $[-90°, 0°]$. This means that the corner frequency occurs when the phase is in the middle of its variation range and the output amplitude is equal to the root–mean–square (RMS) value of the input amplitude.

6.1.2 A Series RC Circuit: Output at the Resistance

Consider the AC circuit shown in Fig. 6.2b when $R = 1000$[Ohm] and $C = 0.1 \times 10^{-6}$[F]. Employing different frequency values, ω, we obtain the measurements shown in Table 6.2, which are plotted in Fig. 6.4 (continuous line). From Fig. 6.4, we conclude that this circuit behaves as a *high-pass filter* and a phase lead is produced. This means that the low-frequency signals (when ω is close to zero) are strongly attenuated, i.e., $B/A \approx 0$, whereas the high-frequency signals (when ω is large) are not attenuated or they are attenuated just a little, i.e., $B/A \approx 1$.

Table 6.2 Experimental data for the *RC* circuit in Fig. 6.2b

f [Hz]	$\omega = 2\pi f [\times 10^5 \text{rad/s}]$	A [V]	B [V]	ϕ [°]
50	0.0031	5.1	0.2	102.6
60	0.0038	5.1	0.24	97.2
70	0.0044	5.1	0.28	95.76
80	0.0050	5.1	0.32	97.92
90	0.0057	5.1	0.36	100.44
100	0.0063	5.1	0.4	90
200	0.0126	5.1	0.68	90.72
300	0.0188	5.1	1.0	84.24
400	0.0251	5.1	1.28	80.64
500	0.0314	5.1	1.56	79.2
600	0.0377	5.05	1.8	69.12
700	0.0440	5.1	2.4	70.56
800	0.0503	5.05	2.28	63.36
900	0.0565	5	2.7	58.32
1000	0.0628	5	2.77	60.48
2000	0.1257	4.96	3.88	41.76
3000	0.1885	4.96	4.32	30.24
4000	0.2513	4.6	3.92	24.76
5000	0.3142	4.88	4.44	18.72
6000	0.3770	4.92	4.72	16.41
7000	0.4398	4.92	4.76	13.1
8000	0.5027	4.92	4.76	12.67
9000	0.5655	4.84	4.56	11.01
10000	0.6283	4.92	4.72	9.36
20000	1.2566	4.92	4.84	0
30000	1.8850	4.92	4.92	0
40000	2.5133	4.92	4.92	0
50000	3.1416	5.2	5.2	0

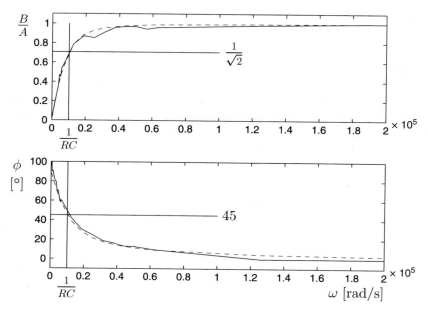

Fig. 6.4 Experimental (continuous) and theoretical (dashed) data for the circuit in Fig. 6.2b (see Table 6.2)

6.1.2.1   Model-Based Analysis

Using the tension divider in the circuit shown in Fig. 6.2b the following is found:

$$V_0(s) = \frac{R}{R + \frac{1}{sC}} V_i(s),$$

and rearranging:

$$\frac{V_0(s)}{V_i(s)} = G(s) = \frac{s}{s+a}, \quad a = \frac{1}{RC}.$$

Assume that $s = j\omega$, then:

$$G(j\omega) = \frac{j\omega}{j\omega + a} = \frac{j\omega}{j\omega + a} \frac{a - j\omega}{a - j\omega} = \frac{j\omega(a - j\omega)}{\omega^2 + a^2},$$

$$|G(j\omega)| = \frac{\omega\sqrt{\omega^2 + a^2}}{\omega^2 + a^2} = \frac{\omega}{\sqrt{\omega^2 + a^2}},$$

$$\angle G(j\omega) = 90° + \arctan\left(\frac{-\omega}{a}\right).$$

Thus, according to Sect. 3.6, if $v_i(t) = A\sin(\omega t)$, then $v_0(t) = B\sin(\omega t + \phi)$ where:

$$B = A\frac{\omega}{\sqrt{\omega^2 + a^2}},$$

$$\phi = 90° + \arctan\left(\frac{-\omega}{a}\right).$$

In Fig. 6.4, the plot obtained (dashed line) using the expressions for $|G(j\omega)| = \frac{B}{A}$ and ϕ that have just been defined is shown. The closeness observed between the continuous line and the dashed line is evidence that the analytical relationships obtained correctly represent the experimental situation. Finally, note that:

$$\frac{B}{A} = \frac{1}{\sqrt{2}} = 0.7071,$$

$$\phi = +45°,$$

when $\omega = a = \frac{1}{RC}$, which is known, again, as the corner frequency. Also note that the phase of $\angle G(j\omega)$ takes values in the range $[0°, +90°]$. This means that the corner frequency occurs when the phase is at the middle of its variation range and the output amplitude is equal to the RMS value of the input amplitude.

6.1.3 A Series RLC Circuit: Output at the Capacitance

Consider the AC circuit shown in Fig. 6.2c when $R = 100[\text{Ohm}]$, $C = 0.1 \times 10^{-6}[\text{F}]$ and $L = 14.262[\text{mH}]$. Employing different values of frequency, ω, we obtain the measurements shown in Table 6.3, which are plotted in Fig. 6.5 (continuous line). From Fig. 6.5, we conclude that this circuit behaves as a *low-pass filter* and phase lag is produced. This means that the low-frequency signals (when ω is close to zero) are not attenuated or they are attenuated just a little, i.e., $B/A \approx 1$, whereas the high-frequency signals (when ω is large) are strongly attenuated, i.e., $B/A \approx 0$. Note that $B/A > 1$ for a range of frequencies such that B/A reaches a maximum. This maximum is known as the *resonance peak*.

6.1.3.1 Model-Based Analysis

Using the tension divider in the circuit shown in Fig. 6.2c yields:

$$V_0(s) = \frac{\frac{1}{sC}}{R + sL + \frac{1}{sC}}V_i(s),$$

and rearranging:

Table 6.3 Experimental data for the resistor–inductor–capacitor (*RLC*) circuit in Fig. 6.2c

f [Hz]	$\omega = 2\pi f [\times 10^5 \text{rad/s}]$	A [V]	B [V]	ϕ [°]
100	0.0063	0.4	0.4	0
500	0.0314	0.38	0.39	0
800	0.0503	0.36	0.38	0
900	0.0565	0.35	0.38	0
1000	0.0628	0.34	0.38	0
2000	0.1257	0.3	0.38	−5.4
3000	0.1885	0.29	0.5	−18.9
4000	0.2513	0.28	0.85	−54
4293	0.2697	0.28	0.925	−77.27
5000	0.3142	0.28	0.625	−126
6000	0.3770	0.28	0.31	−151.2
7000	0.4398	0.28	0.22	−157.5
8000	0.5027	0.28	0.16	−165.6
10000	0.6283	0.28	0.08	−171
20000	1.2566	0.27	0.02	−171
40000	2.5133	0.27	0.004	−187.2

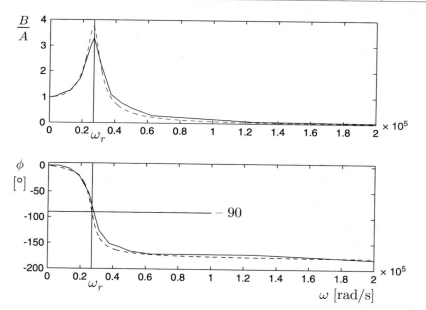

Fig. 6.5 Experimental (continuous) and theoretical (dashed) data for the circuit in Fig. 6.2c (see Table 6.3)

$$\frac{V_0(s)}{V_i(s)} = G(s) = \frac{\omega_n^2}{s^2 + 2\zeta\omega_n s + \omega_n^2}, \quad \omega_n^2 = \frac{1}{LC}, \quad 2\zeta\omega_n = \frac{R}{L}.$$

Assume that $s = j\omega$, then:

$$G(j\omega) = \frac{\omega_n^2}{(j\omega)^2 + j2\zeta\omega_n\omega + \omega_n^2},$$

$$= \frac{\omega_n^2}{\omega_n^2 - \omega^2 + j2\zeta\omega_n\omega}\left(\frac{\omega_n^2 - \omega^2 - j2\zeta\omega_n\omega}{\omega_n^2 - \omega^2 - j2\zeta\omega_n\omega}\right),$$

$$= \frac{\omega_n^2(\omega_n^2 - \omega^2 - j2\zeta\omega_n\omega)}{(\omega_n^2 - \omega^2)^2 + 4(\zeta\omega_n\omega)^2},$$

$$|G(j\omega)| = \frac{\omega_n^2\sqrt{(\omega_n^2 - \omega^2)^2 + 4(\zeta\omega_n\omega)^2}}{(\omega_n^2 - \omega^2)^2 + 4(\zeta\omega_n\omega)^2} = \frac{\omega_n^2}{\sqrt{(\omega_n^2 - \omega^2)^2 + 4(\zeta\omega_n\omega)^2}},$$

$$\angle G(j\omega) = \arctan\left(\frac{-2\zeta\omega_n\omega}{\omega_n^2 - \omega^2}\right).$$

Thus, according to Sect. 3.6, if $v_i(t) = A\sin(\omega t)$, then $v_0(t) = B\sin(\omega t + \phi)$ where:

$$B = A\frac{\omega_n^2}{\sqrt{(\omega_n^2 - \omega^2)^2 + 4(\zeta\omega_n\omega)^2}},$$

$$\phi = \arctan\left(\frac{-2\zeta\omega_n\omega}{\omega_n^2 - \omega^2}\right).$$

In Fig. 6.5, the plot obtained (dashed line) using the expressions for $|G(j\omega)| = \frac{B}{A}$ and ϕ that have just been defined is shown. The closeness observed between the continuous and the dashed lines is evidence that the analytical relationships obtained correctly represent the experimental situation. Note that:

$$\frac{B}{A} = \frac{\omega_n^2}{\sqrt{4(\zeta\omega_n^2)^2}} = \frac{1}{2\zeta} = \frac{\omega_n}{\frac{R}{L}} = \frac{1}{R}\sqrt{\frac{L}{C}},$$

$$\phi = -90°,$$

when $\omega = \omega_n = \sqrt{\frac{1}{LC}}$, which is known as the corner frequency. We note, however, that the amplitude maximum occurs when $\omega = \omega_r$, which is known as the resonance frequency satisfying $\omega_r < \omega_n$ and $\omega_r = \omega_n$ only when $\zeta = 0$. Note that, according to Table 6.3, B/A reaches its maximum when $f = 4293[\text{Hz}]$, i.e., when $\omega = 0.2697 \times 10^5[\text{rad/s}] = \omega_r \approx \omega_n$. Thus, we can compute an approximate value for inductance from these experimental data as:

$$L = \frac{1}{\omega_n^2 C} \approx \frac{1}{(0.2697 \times 10^5)^2(0.1 \times 10^{-6})} = 13.7 \times 10^{-3}[\text{H}], \qquad (6.7)$$

value, which is very close to $L = 14.262[\text{mH}]$ obtained using an inductance–capacitance–resistance (LCR) meter. On the other hand, note that the phase of $G(j\omega)$ takes values in the range $[-180°, 0°]$. This means that the corner frequency occurs when the phase is at the middle of its variation range and the output amplitude tends toward ∞ as the circuit resistance tends toward zero, i.e., as the system damping tends toward zero.

6.1.4 A Series RLC Circuit: Output at the Resistance

Consider the AC circuit shown in Fig. 6.2d when $R = 100[\text{Ohm}]$, $C = 0.1 \times 10^{-6}[\text{F}]$ and $L = 14.262[\text{mH}]$. Employing different values of frequency, ω, we obtain the measurements shown in Table 6.4, which are plotted in Fig. 6.6 (continuous line). From Fig. 6.6, we conclude that this circuit behaves as a *bandpass filter* and phase lag in addition to phase lead are produced for different values of frequencies. This means that both low-frequency and high-frequency signals (when ω is close to zero and ω is large) are strongly attenuated, i.e., $B/A \approx 0$, whereas the frequencies around ω_r are not attenuated or they are attenuated just a little, i.e., $B/A \approx 1$.

Table 6.4 Experimental data for the RLC circuit in Fig. 6.2d

f [Hz]	$\omega = 2\pi f [\times 10^5 \text{rad/s}]$	A [V]	B [V]	ϕ [°]
100	0.0063	0.4	0.004	72
500	0.0314	0.38	0.012	99
800	0.0503	0.35	0.018	100.8
900	0.0565	0.34	0.018	97.2
1000	0.0628	0.34	0.024	90
2000	0.1257	0.3	0.04	79.2
3000	0.1885	0.29	0.09	64.8
4000	0.2513	0.29	0.22	28.8
4497	0.2826	0.29	0.26	0
5000	0.3142	0.26	0.2	−36
6000	0.3770	0.26	0.11	−59.4
7000	0.4398	0.26	0.08	−75.6
8000	0.5027	0.26	0.07	−79.2
10000	0.6283	0.26	0.045	−81
20000	1.2566	0.26	0.02	−79.2
40000	2.5133	0.26	0.01	−72

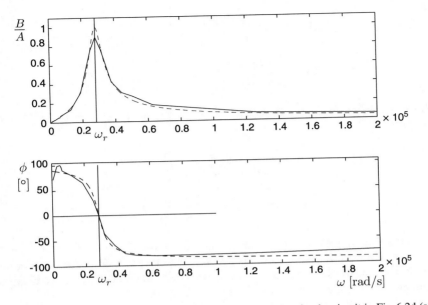

Fig. 6.6 Experimental (continuous) and theoretical (dashed) data for the circuit in Fig. 6.2d (see Table 6.4)

6.1.4.1 Model-Based Analysis

Using the tension divider in the circuit shown in Fig. 6.2d yields:

$$V_0(s) = \frac{R}{R + sL + \frac{1}{sC}} V_i(s),$$

and rearranging:

$$\frac{V_0(s)}{V_i(s)} = G(s) = \frac{2\zeta\omega_n s}{s^2 + 2\zeta\omega_n s + \omega_n^2}, \quad \omega_n^2 = \frac{1}{LC}, \quad 2\zeta\omega_n = \frac{R}{L}.$$

Assume that $s = j\omega$, then:

$$G(j\omega) = \frac{j2\zeta\omega_n\omega}{(j\omega)^2 + j2\zeta\omega_n\omega + \omega_n^2},$$

$$= \frac{j2\zeta\omega_n\omega}{\omega_n^2 - \omega^2 + j2\zeta\omega_n\omega}\left(\frac{\omega_n^2 - \omega^2 - j2\zeta\omega_n\omega}{\omega_n^2 - \omega^2 - j2\zeta\omega_n\omega}\right),$$

$$= \frac{j2\zeta\omega_n\omega(\omega_n^2 - \omega^2 - j2\zeta\omega_n\omega)}{(\omega_n^2 - \omega^2)^2 + 4(\zeta\omega_n\omega)^2},$$

$$|G(j\omega)| = \frac{2\zeta\omega_n\omega\sqrt{(\omega_n^2 - \omega^2)^2 + 4(\zeta\omega_n\omega)^2}}{(\omega_n^2 - \omega^2)^2 + 4(\zeta\omega_n\omega)^2} = \frac{2\zeta\omega_n\omega}{\sqrt{(\omega_n^2 - \omega^2)^2 + 4(\zeta\omega_n\omega)^2}},$$

$$\angle G(j\omega) = 90° + \arctan\left(\frac{-2\zeta\omega_n\omega}{\omega_n^2 - \omega^2}\right).$$

Thus, according to Sect. 3.6, if $v_i(t) = A\sin(\omega t)$, then $v_0(t) = B\sin(\omega t + \phi)$ where:

$$B = A\frac{2\zeta\omega_n\omega}{\sqrt{(\omega_n^2 - \omega^2)^2 + 4(\zeta\omega_n\omega)^2}},$$

$$\phi = 90° + \arctan\left(\frac{-2\zeta\omega_n\omega}{\omega_n^2 - \omega^2}\right).$$

In Fig. 6.6, the plot obtained (dashed line) using the expressions for $|G(j\omega)| = \frac{B}{A}$ and ϕ that have just been defined is shown. The closeness observed between the continuous and the dashed lines is evidence that the analytical relationships obtained correctly represent the experimental situation. Note that:

$$\frac{B}{A} = \frac{2\zeta\omega_n\omega}{\sqrt{4(\zeta\omega_n\omega)^2}} = 1,$$

$$\phi = 0°,$$

when $\omega = \omega_n = \sqrt{\frac{1}{LC}}$, which is known as the band center. We note that the magnitude at this maximum does not depend on any circuit parameter and $\omega_r = \omega_n$ in this case. According to Table 6.4, the maximum value of B/A is obtained experimentally when $f = 4497[\text{Hz}]$, i.e., when $\omega = 0.2826 \times 10^5[\text{rad/s}] = \omega_n$. Hence, we can compute the value of the inductance from the experimental data as:

$$L = \frac{1}{\omega_n^2 C} = \frac{1}{(0.2826 \times 10^5)^2(0.1 \times 10^{-6})} = 12.5 \times 10^{-3}[\text{H}], \qquad (6.8)$$

value, which is very close to $L = 14.262[\text{mH}]$ obtained using an LCR meter. On the other hand, the phase of $G(j\omega)$ takes values in the range $[-90°, 90°]$, whereas the band center occurs when the phase is zero, i.e., it is at the middle of the phase variation range.

6.2   The Relationship Between Frequency Response and Time Response

According to *Fourier series*, a periodic function of time can be represented as the addition of many different frequency sine or cosine functions of time. For instance, consider the function $f(t)$ presented in Fig. 6.7. Its Fourier series expansion is given as [1], pp. 457:

$$f(t) = \frac{4k}{\pi}\left(\sin(t) + \frac{1}{3}\sin(3t) + \frac{1}{5}\sin(5t) + \cdots\right). \tag{6.9}$$

The frequencies $\omega = 1, 3, 5, \ldots$, are known as the *frequency components* of $f(t)$. For any periodic function $f(t)$, these frequency components are integer multiples of a fundamental frequency ω_0, in this case $\omega_0 = 1$, which can be computed as $\omega_0 = 2\pi/T$, where T is known as the period of $f(t)$. In this case, $T = 2\pi$. The amplitudes $\frac{4k}{\pi}, \frac{4k}{\pi}\frac{1}{3}, \frac{4k}{\pi}\frac{1}{5}, \ldots$, represent the contribution of each one of the frequency components to $f(t)$. Roughly speaking, these amplitudes determine how large, or significant, the influence of each frequency component is, to determine the waveform of $f(t)$ in the time domain.

Consider Fig. 6.8 where two different frequency sine functions are shown. Note that a higher-frequency signal presents faster variations in time. This means that, if $f(t)$ has components of higher frequencies then $f(t)$ contains faster variations. For instance, consider the function $f(t)$ shown in Fig. 6.7. Its Fourier series expansion is given in (6.9). In Fig. 6.9, several approximations are shown for this function, which include different frequency components [1], pp. 458. Note that the greater the content in high frequencies the faster the variations that this function has in the time domain. Furthermore, as higher-frequency components are included when trying

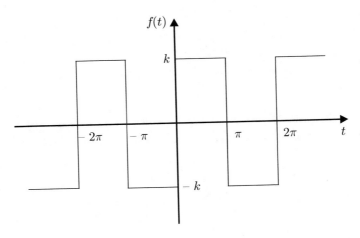

Fig. 6.7 A periodic function of time

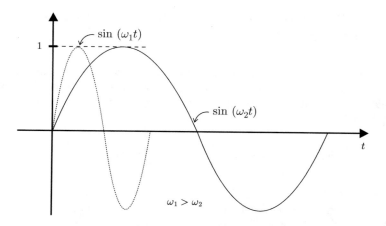

Fig. 6.8 Comparison of two different frequency sine functions of time

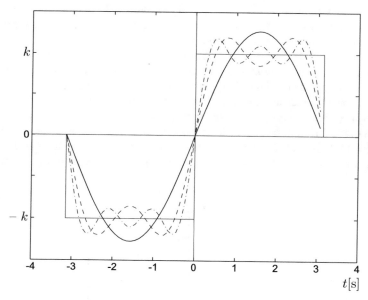

Fig. 6.9 Different approximations of $f(t)$ shown in Fig. 6.7. Continuous $f_1(t) = \frac{4k}{\pi} \sin(t)$. Dashed $f_2(t) = \frac{4k}{\pi} \left(\sin(t) + \frac{1}{3} \sin(3t) \right)$. Dash–dot $f_3(t) = \frac{4k}{\pi} \left(\sin(t) + \frac{1}{3} \sin(3t) + \frac{1}{5} \sin(5t) \right)$

to complete the corresponding Fourier series, the resulting function approaches the abrupt discontinuities that $f(t)$ has at specific points. According to these ideas, the following is concluded.

High-Frequency Components The larger the amplitude of the high-frequency components, the faster the variations of the function in time. Very high-frequency components imply discontinuity in time.

Noise is an undesirable fast-changing signal. This means that the noise has high-frequency components.

Zero-Frequency Component A zero-frequency signal is a constant signal. This can be concluded from:

$$A_u \cos(\omega t) = A_u = \text{constant}, \quad \text{if } \omega = 0.$$

Hence, using the terminology of electrical and electronic engineering, a signal $f(t)$ containing a zero frequency component is a signal containing a direct current (DC) component. This means that, if $f(t)$ varies in time these variations are performed around a constant value that is different from zero.

Finally, let us say that when the input is not periodic, which is when a step signal is applied (see Fig. 6.10), the Fourier transform [4] allows us to represent such a signal as the addition of an infinite number of sinusoidal signals with different frequencies. In this case, the frequency components are not integer multiples of a fundamental frequency, but they take continuous values from $\omega = -\infty$ to $\omega = +\infty$.

6.2.1   Relationship Between Time Response and Frequency Response

A fundamental property of the arbitrary order linear system given in (3.108) is superposition (see Sect. 3.7, Chap. 3). This means that when the input signal can be represented as the addition of several signals with different frequencies, then

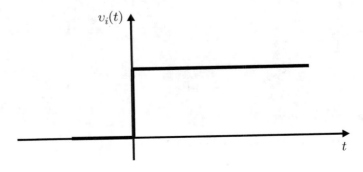

Fig. 6.10 Step input

the system response can be obtained by computing the response to each signal composing the input and, finally, adding all these responses. According to this and the above discussion, we arrive at the following conclusions, which are valid if $G(s)$ is stable.

- If $u(t)$ is a step with magnitude A, i.e., the zero frequency component has amplitude A, then the corresponding output $y(t)$, in a steady state, is $\lim_{t\to\infty} y(t) = AG(j0)$ (see (3.108)), i.e., $y(\infty)$ is equal to the system response to the zero frequency component of the input. For instance, in the circuits studied in Sects. 6.1.1, 6.1.2, 6.1.3, 6.1.4, we have $|G(j0)| = |G(j\omega)|_{\omega=0} = 1$, $|G(j0)| = |G(j\omega)|_{\omega=0} = 0$, $|G(j0)| = |G(j\omega)|_{\omega=0} = 1$ and $|G(j0)| = |G(j\omega)|_{\omega=0} = 0$ respectively; hence, $\lim_{t\to\infty} y(t) = A$, $\lim_{t\to\infty} y(t) = 0$, $\lim_{t\to\infty} y(t) = A$, $\lim_{t\to\infty} y(t) = 0$, respectively. See Figs. 6.11, 6.12, 6.13 and 6.14, where a step signal is applied experimentally at the input of each one of these circuits.

It is interesting to realize that, as shown in Sect. 3.4, if $u(t)$ is a step with amplitude A, then the final value of the output, in a steady state, is computed using the final value theorem, i.e.,:

$$\lim_{t\to\infty} y(t) = \lim_{s\to 0} sY(s) = \lim_{s\to 0} sG(s)\frac{A}{s} = G(0)A,$$

where it is important to see that $\omega \to 0$ if $s = \sigma + j\omega \to 0$.

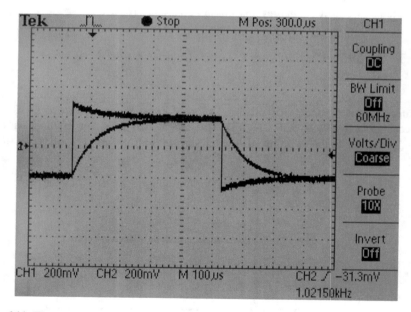

Fig. 6.11 Time response of the circuit in Fig. 6.2a when $v_i(t)$ is a step signal

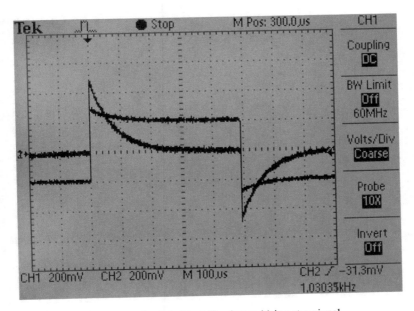

Fig. 6.12   Time response of the circuit in Fig. 6.2b when $v_i(t)$ is a step signal

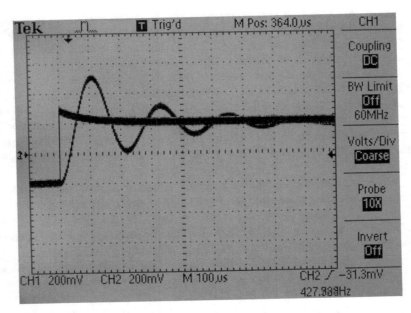

Fig. 6.13   Time response of the circuit in Fig. 6.2c when $v_i(t)$ is a step signal

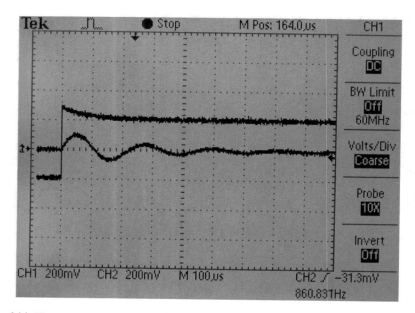

Fig. 6.14 Time response of the circuit in Fig. 6.2d when $v_i(t)$ is a step signal

- Suppose that $u(t)$ is a step with amplitude A. In this case, $u(t)$ possesses a discontinuity at $t = 0$. As discussed in the previous section, the high-frequency components produce such a discontinuity; hence, it is concluded that $u(t)$ possesses a large content of high-frequency components.

  In the case of the circuits studied in Sects. 6.1.1, 6.1.3, and 6.1.4, the input high-frequency components, $u(t) = v_i(t)$, are strongly attenuated, i.e., they have only a very small effect on the output $y(t) = v_0(t)$, because $|G(j\omega)| \to 0$ as $\omega \to \infty$ (see Figs. 6.3, 6.5 and 6.6). Then, the output $y(t) = v_0(t)$ does not present any discontinuity at $t = 0$ (see Figs. 6.11, 6.13, and 6.14). Note that this also means that the effect of noise on the output is also strongly attenuated in these cases.

  On the other hand, in the case of the circuit studied in Sect. 6.1.2, the high-frequency components at the input, $u(t) = v_i(t)$, are not attenuated and they appear identical at the output $y(t) = v_0(t)$, because $|G(j\omega)| \to 1$ and $\angle G(j\omega) \to 0$ as $\omega \to \infty$ (see Fig. 6.4). Then, the output $y(t) = v_0(t)$ presents the same discontinuity that the input $u(t)$ possesses at $t = 0$ (see Fig. 6.12).

  Finally, we note that the above ideas on the relationship between the discontinuity of the function in time and the high-frequency components is also supported by the initial value theorem (see (3.5)). Let us apply this theorem to circuits studied experimentally in Sect. 6.1. We recall that $v_i(t)$ is a step function with the magnitude A. In the case of the electric circuit studied in Sect. 6.1.1, we have:

$$v_0(0^+) = \lim_{s \to \infty} s \frac{a}{s+a} \frac{A}{s} = 0.$$

For the circuit in Sect. 6.1.3, we have:

$$v_0(0^+) = \lim_{s \to \infty} s \frac{\omega_n^2}{s^2 + 2\zeta\omega_n s + \omega_n^2} \frac{A}{s} = 0.$$

Analogously, for the circuit in Sect. 6.1.4:

$$v_0(0^+) = \lim_{s \to \infty} s \frac{2\zeta\omega_n s}{s^2 + 2\zeta\omega_n s + \omega_n^2} \frac{A}{s} = 0.$$

Finally, for the electric circuit in Sect. 6.1.2:

$$v_0(0^+) = \lim_{s \to \infty} s \frac{s}{s + a} \frac{A}{s} = A.$$

We stress that all signals are assumed to be zero for all $t < 0$ when using the Laplace transform and that all initial conditions are assumed to be zero because a transfer function representation of systems is considered. Hence, $v_0(0^+) = 0$ implies that $v_0(t)$ is continuous at $t = 0$ and $v_0(0^+) = A$ implies that $v_0(t)$ has a discontinuity of magnitude A, i.e., as large as the step applied at $v_i(t)$, at $t = 0$. These results are corroborated experimentally in Figs. 6.11, 6.12, 6.13, and 6.14.

- A system becomes faster if it does not attenuate the high-frequency components. However, if too high-frequency components are not attenuated, some problems appear because of noise amplification. Hence, the frequency components that are not to be attenuated are called the *intermediate frequencies*.

For instance, in the circuit studied in Sect. 6.1.1, this is achieved if the parameter $a = \frac{1}{RC} > 0$ is increased (see (6.2)). This can be understood by observing Fig. 6.15 where it is shown that the value of $|G(j\omega)|$ can be increased for larger frequencies if a larger value of a is employed. This is corroborated by

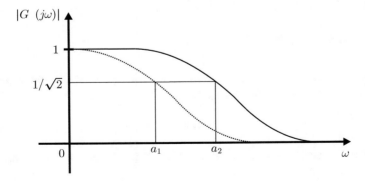

Fig. 6.15 A larger value of a in $\frac{a}{s+a}$ allows a greater effect of the intermediate frequencies and, hence, a faster time response is accomplished

what was exposed in Sect. 3.1.1, Chap. 3, where it is explained that a system with the transfer function $G(s) = a/(s + a)$ is faster as a is increased.

Using similar arguments, we conclude that in the case of the electric circuit studied in Sect. 6.1.3, a faster response is obtained if $\omega_n = \frac{1}{\sqrt{LC}}$ is increased. Moreover, the presence of a resonance peak in Fig. 6.5 implies that a frequency component at $\omega = \omega_r \approx \omega_n$ is significantly amplified. This is observed as a clear oscillation in Fig. 6.13 whose frequency is equal to $\omega_r \approx \omega_n$. Hence, if the frequency of this oscillation increases, then the rise time decreases (a faster response is obtained). This corroborates the notion that a faster response is obtained if $\omega_r \approx \omega_n$ is increased. According to these ideas, it is obvious that a faster response is also produced in the electric circuit studied in Sect. 6.1.4 if ω_n is increased.

In arbitrary $n-$order systems $G(s)$, the *bandwidth*, ω_{BW}, is defined as the frequency where $|G(j\omega_{BW})| = \frac{1}{\sqrt{2}}|G(j0)| = 0.7071|G(j0)|$, i.e., the frequency where the system magnitude is 0.7071 times the magnitude at $\omega = 0$. For systems with a transfer function:

$$\frac{a}{s+a},$$

$\omega_{BW} = a$ and for systems with a transfer function:

$$\frac{\omega_n}{s^2 + 2\zeta\omega_n s + \omega_n^2},$$

$\omega_{BW} \approx \omega_n$, although the exact value depends on system damping ζ. Thus, in arbitrary $n-$order systems $G(s)$, a faster response is obtained as the bandwidth is rendered larger.

6.3   Common Graphical Representations

6.3.1   Bode Diagrams

Let k be a positive number. The corresponding value in *decibels* (dB) is computed using the following operation (see Appendix C):

$$k_{\mathrm{dB}} = 20\log(k),$$

where $\log(k)$ stands for the decimal logarithm of k. The decimal logarithm of k is defined as follows. If:

$$10^z = k,$$

then $z = \log(k)$. Some properties of the decimal logarithm are as follows:

$$\log(xy) = \log(x) + \log(y), \quad \log(x/y) = \log(x) - \log(y), \quad (6.10)$$

$$\log(x^m) = m \log(x), \quad \log(1) = 0,$$

$$\text{If } x \to 0 \quad \text{then } \log(x) \to -\infty.$$

Bode diagrams are drawn on semilogarithmic axes: the horizontal axis (ω, in [rad/s]) has a logarithmic scale whereas the vertical axis has a linear scale, in decibels (dB) for the magnitude Bode diagram or in degrees for the phase Bode diagram. An important concept in Bode diagrams is *decade* or *dec*. In Bode diagrams, it is very important to analyze the frequency ω by decades. One decade represents an increment of ten times in frequency. For instance, if $\omega_1 = 2.5$ and $\omega_2 = 25$, then it is said that one decade exists between ω_1 and ω_2.

The arbitrary n−order transfer function defined in (3.108) can be written as:

$$G(s) = \frac{b_m \prod_{j=1}^{j=m} (s - z_j)}{\prod_{i=1}^{i=n} (s - p_i)}, \quad (6.11)$$

where z_j, $j = 1, \ldots, m$, are the zeros of $G(s)$ and p_i, $i = 1, \ldots, n$, are the poles of $G(s)$. Because of the first logarithm property in (6.10), taking into account the fact that poles and zeros can be real or complex conjugate, and assuming that $G(s)$ is a minimum phase transfer function (see Sect. 6.5), then Bode diagrams of $G(s)$ can always be constructed as the addition of Bode diagrams of the following first- and second-order basic factors[1]:

$$s, \quad \frac{1}{s}, \quad \frac{s+a}{a}, \quad \frac{a}{s+a}, \quad (6.12)$$

$$\frac{s^2 + 2\zeta\omega_n s + \omega_n^2}{\omega_n^2}, \quad \frac{\omega_n^2}{s^2 + 2\zeta\omega_n s + \omega_n^2}.$$

For this reason, in Figs. 6.16 and 6.17, Bode diagrams of these factors are presented. Recall that the magnitude Bode diagram represents the quantity:

$$20 \log (|G(j\omega)|),$$

as a function of the frequency ω. The phase Bode diagram represents the quantity:

$$\phi = \text{atan} \left(\frac{\text{Im}(G(j\omega))}{\text{Re}(G(j\omega))} \right),$$

---

[1]This is further explained later

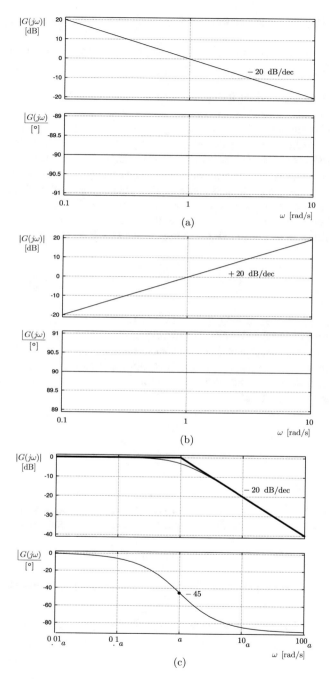

Fig. 6.16 Bode diagrams of first- and second-order factors in (6.12). (**a**) $G(s) = \frac{1}{s}$. (**b**) $G(s) = s$. (**c**) $G(s) = \frac{a}{s+a}$

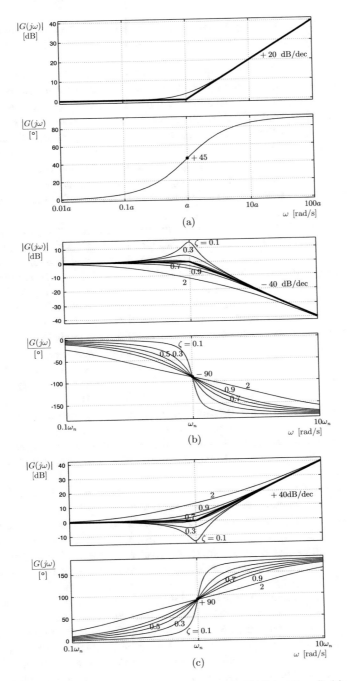

Fig. 6.17 Bode diagrams of first- and second-order factors in (6.12) (continued). (**a**) $G(s) = \frac{s+a}{a}$.
(**b**) $G(s) = \frac{\omega_n^2}{s^2+2\zeta\omega_n s+\omega_n^2}$. (**c**) $G(s) = \frac{s^2+2\zeta\omega_n s+\omega_n^2}{\omega_n^2}$

as a function of the frequency ω. The reader is encouraged to see Appendix B to find out about the detailed procedure followed to obtain the plots shown in Figs. 6.16 and 6.17.

The thick lines in the Bode diagrams presented in Figs. 6.16 and 6.17 are known as the asymptotes, whereas the thin lines represent the exact value of the functions represented. The asymptotes are merely approximations, which, however, are very useful for recognizing the fundamental shape of the corresponding Bode diagrams. Note that in some of the magnitude Bode diagrams, the asymptotes form a corner. The frequency where this corner appears is known as the *corner frequency* and it is directly related to the zero or the pole of the first- or the second-order factor that the diagram represents. Recall that, according to Sect. 3.3.1, ω_n is the distance from the corresponding root (pole or zero) to the origin and the roots of a second-order factor are given as $s = -\zeta\omega_n \pm j\omega_n\sqrt{1 - \zeta^2}$.

According to the asymptotes, the magnitude of all the factors in (6.12), except for s and $1/s$, are equal to 0[dB] for frequencies that are less than or equal to the corner frequency but, for frequencies that are greater than the corner frequency, the magnitude increases (zeros) or decreases (poles) with a slope of ± 20[dB/dec] for the first-order factors and ± 40(dB/dec) for the second-order factors. Note that in the second-order factors, the behavior of the exact curve around the corner frequency strongly depends on the damping coefficient ζ. Moreover, if $\zeta < 0.7071$, there is a maximum in the magnitude whose value increases to infinity as ζ tends toward zero. This maximum of magnitude is known as the *resonance peak* and is represented by M_r.

In the case of the phase Bode diagram we note the following. Each first-order factor has a phase that varies from $0°$ to $+90°$ (zeros) or to $-90°$ (poles). Each second-order factor has a phase that varies from $0°$ to $+180°$ (zeros) or to $-180°$ (poles). In all the cases, when the frequency equals the corner frequency the phase is at the middle of the corresponding range. Again, in the case of the second-order factors, around the corner frequency, the phase curve strongly depends on the damping coefficient ζ: the phase changes faster between $0°$ and $\pm 180°$ as ζ tends toward zero.

It is important to observe that in the case when the factor in (6.12) contains zeros, then the corresponding Bode diagrams show that such a factor behaves as a high-pass filter and that a *phase lead* is produced, i.e., the factor contributes with a positive phase. This observation is important because it means that in a transfer function the zeros tend to increase the effects of noise in a control system. This implies that the use of an excessively large derivative gain in a proportional–derivative (PD) or proportional–integral–derivative (PID) controller may result in performance deterioration. To understand this, note that a PD controller introduces one open-loop zero, whereas a PID controller introduces two open-loop zeros. On the other hand, as discussed in Sect. 5.1.1, an open-loop zero tends to improve the closed-loop stability. Hence, a trade-off must be established between stability and performance deterioration due to noise, when designing a control system. In Sect. 6.5 it is explained that the stabilizing effect of an open-loop zero is due to the phase lead contributing to the frequency response of the open-loop transfer function.

On the other hand, if the factor in (6.12) contains poles, then the corresponding Bode diagrams show that such a factor behaves as a low-pass filter and that a *phase lag* is produced, i.e., the factor contributes with a negative phase. This means that poles in a transfer function reduce the effects of noise. On the other hand, as discussed in Sect. 5.1.1, an open-loop pole tends to deteriorate the closed-loop stability. In Sect. 6.5, it is explained that this effect is due to the phase lag contributed by a pole in the frequency response of the open-loop transfer function.

The frequency response plots of the arbitrary $n-$order minimum phase transfer function defined in (6.11) are obtained as the combination of the factors in (6.12). In the case of Bode diagrams, this combination process of first- and second-order factors is very simple, as explained in the following. Note that the arbitrary $n-$order transfer function, $G(s)$, given in (6.11) can be written as:

$$G(s) = \beta \prod_{k=1}^{k=r} G_k(s),$$

for some real constant β and some integer positive number r, where $G_k(s)$, for $k = 1, \ldots, r$, has the form of one of the factors in (6.12). Then:

$$|G(j\omega)| = \left| \beta \prod_{k=1}^{k=r} G_k(j\omega) \right| = |\beta| \prod_{k=1}^{k=r} |G_k(j\omega)|.$$

According to this, the decibels definition $|G(j\omega)|_{dB} = 20 \log(|G(j\omega)|)$ and the property $\log(xy) = \log(x) + \log(y)$, shown in (6.10), it is possible to write:

$$|G(j\omega)|_{dB} = 20 \log(|\beta|) + \sum_{k=1}^{k=r} 20 \log(|G_k(j\omega)|).$$

This means that the magnitude Bode diagram (the phase Bode diagram) of an arbitrary $n-$order transfer function, $G(s)$, can always be obtained as the addition of the magnitude Bode diagrams (the phase Bode diagrams) of every first- and second-order factor of $G(s)$.[2]

6.3.2 Polar Plots

In Fig. 6.18 a complex number G is defined as follows:

---

[2]Recall that given three complex numbers z, x, y such that $z = xy$, the angle of z is given as the addition of the angles of x and y.

Fig. 6.18 A complex number

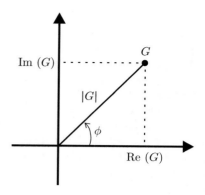

$$G = \frac{B}{A}\angle\phi = \mathrm{Re}(G) + j\mathrm{Im}(G),$$

$$|G| = \frac{B}{A}, \quad \mathrm{Re}(G) = \frac{B}{A}\cos(\phi), \quad \mathrm{Im}(G) = \frac{B}{A}\sin(\phi).$$

Given a transfer function $G(s)$, the quantity $G(j\omega)$ is a complex number that can be represented as in Fig. 6.18. The polar plot of $G(j\omega)$ is the curve obtained in the plane shown in Fig. 6.18 as ω goes from $-\infty$ to $+\infty$. In polar plots, the magnitude is no longer represented using decibels.

Note that the polar plot of the arbitrary $n-$order transfer function shown in (6.11) cannot be obtained as the *simple* combination of the polar plots of the first- and second-order factors shown in (6.12). To simplify this process, it is suggested to first obtain the corresponding Bode diagrams and, from them, obtain the required information to draw the corresponding polar plots for positive frequencies, i.e., when ω goes from 0 to $+\infty$. The polar plots for negative frequencies, i.e., when ω goes from $-\infty$ to 0, are obtained from polar plots for positive frequencies just by projecting, as in a mirror, the part of the curve on the upper half-plane to the bottom half-plane and vice versa. This is the procedure that has been followed to obtain the polar plots shown in Figs. 6.19 and 6.20, from the Bode diagrams in Figs. 6.16 and 6.17. The arrows in the polar plots indicate the sense where frequency increases from $\omega = -\infty$, passing through $\omega = 0$, to $\omega = +\infty$. Some examples of where these ideas are applied are presented in the next section.

Example 6.1 (Resonance) Consider the mass-spring-damper system depicted in Fig. 6.21. The transfer function in this case is the same as in Example 3.8 Chap. 3, i.e.,:

$$X(s) = G(s)F(s), \quad G(s) = \frac{k\omega_n^2}{s^2 + 2\zeta\omega_n s + \omega_n^2},$$

$$\omega_n = \sqrt{\frac{K}{m}}, \quad \zeta = \frac{b}{2\sqrt{mK}}, \quad k = \frac{1}{K},$$

Fig. 6.19 Polar plots of the
first- and second-order factors
in (6.12). (**a**) $G(s) = \frac{1}{s}$. (**b**)
$G(s) = s$. (**c**) $G(s) = \frac{a}{s+a}$

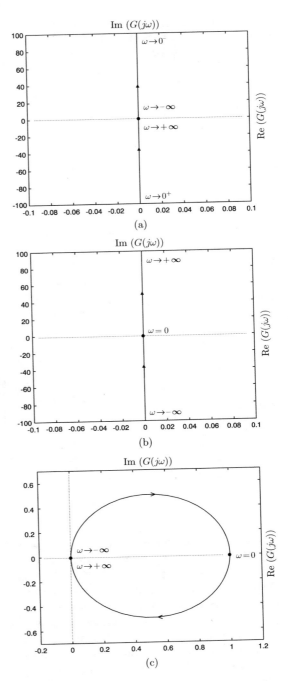

Fig. 6.20 Polar plots of the first- and second-order factors in (6.12) (continued). (**a**) $G(s) = \frac{s+a}{a}$. (**b**) $G(s) = \frac{\omega_n^2}{s^2 + 2\zeta\omega_n s + \omega_n^2}$

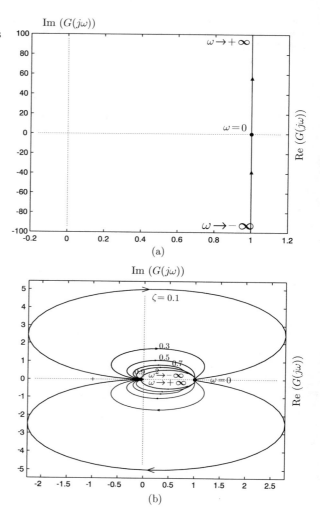

where $x = 0$ is defined as the position when the system is at rest with $f(t) = 0$ under the effect of the spring and weight of the mass. The situation shown in Fig. 6.21 is equivalent to that of a car at rest on the floor: m stands for the car mass, K is the suspension system stiffness, and b represents the effect of the dampers. Everyday experience teaches us that, although heavy, a small car can be flipped over by a small group of people if they apply a force as follows. The car is pushed up and then it is allowed to move down. After bouncing on the floor it is pushed up again when the car is moving upward. Although the applied force is always too small to flip over the car, after applying the same force several times the car is finally flipped over. This is a clear example of the resonance effects and this situation is analyzed in the following.

Fig. 6.21 A car suspension
system

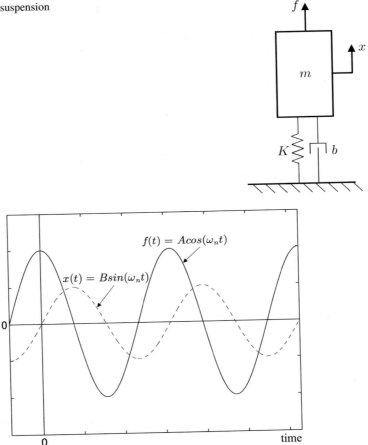

Fig. 6.22 The applied force and the resulting position of the body in Fig. 6.21 under resonance
conditions

The force applied by the people has a periodic form with a cosine fundamental
component. When this force is positive, the car is pushed upward. When it is
negative the car is allowed to fall down. The fact that this force is applied only
when the car moves upward implies that this cosine component has a frequency ω
that is equal to the car's natural oscillation frequency ω_n. If it is assumed that there is
no damper, $b = 0$, then $\zeta = 0$. According to Fig. 6.17b, under these conditions, the
output $x(t)$ (the car position) lags 90° with respect to the applied force. This means
that $x(t) = B \sin(\omega t)$ if $f(t) = A \cos(\omega t)$, for some positive constant B, if $\omega = \omega_n$.
Then, according to Fig. 6.22, a maximal force is applied (f has a crest) exactly when
the car is moving upward (x is passing from a trough to a crest), which is in complete
agreement with the intuitive application of the force by people. Moreover, if $\omega = \omega_n$

and $\zeta = 0$ the oscillation amplitude B of the motion described by the car will grow without limit, even when the amplitude of the force applied at the input A is small ($B = |G(\omega_n)|A$ and $|G(\omega_n)| \to \infty$). This explains why the car can be flipped over.

An explanation from the point of view of physics is presented next to help to understand why this happens. The system energy is given as the addition of the kinetic energy and the potential energy (only due to the spring):

$$E = E_c + E_p,$$

$$E_c = \frac{1}{2}m\dot{x}^2, \quad E_p = \frac{1}{2}Kx^2.$$

The gravity effect is not taken into account because it has been assumed that $x = 0$ is the position when the body is at rest when $f = 0$ under the effects of the gravity and the spring. As time increases, the variation of this energy is given as:

$$\dot{E} = \dot{E}_c + \dot{E}_p = m\dot{x}\ddot{x} + Kx\dot{x}.$$

Using $\ddot{x}$ from the system model, i.e.,:

$$\ddot{x} = -\frac{b}{m}\dot{x} - \frac{K}{m}x + \frac{1}{m}f,$$

yields:

$$\dot{E} = -b\dot{x}^2 + \dot{x}f.$$

Consider two cases:

1. $\zeta = 0$, $f = A\cos(\omega_n t)$. In this case $b = 0$ and, according to Fig. 6.17b, the output $x(t)$ lags $90°$ with respect to the applied force, i.e., $x(t) = B\sin(\omega_n t)$ and $\dot{x} = \omega_n B\cos(\omega_n t)$ for some positive B; hence:

$$\dot{E} = \dot{x}f = \omega_n AB\cos^2(\omega_n t).$$

The mean energy delivered to the mechanical system with each oscillation period $T = 2\pi/\omega_n$, is given as:

$$E = \frac{\omega_n AB}{T}\int_0^T \cos^2(\omega_n t)dt > 0.$$

Note that $\cos^2(\omega_n t) \geq 0$ for all t. On the other hand, when considering that $b = 0$, it is assumed that there is no energy dissipation; hence, that all the energy supplied to the system is stored as the total energy of the mass $E = E_c + E_p$. Thus, this energy increases without limit as time increases. As $x(t)$ and $\dot{x}(t)$ are sine functions of time, then the amplitude of these oscillations also grows without limit, as described by Fig. 6.17b.

2. $\zeta > 0$, $f = A\cos(\omega_n t)$. In this case $b > 0$ and, according to Fig. 6.17b, the output $x(t)$ converges (after the natural response becomes negligible) to a function that lags at an angle 90° with respect to the applied force, i.e., $x(t) = B\sin(\omega_n t)$ y $\dot{x} = \omega_n B\cos(\omega_n t)$ for some positive B and, then:

$$\dot{E} = -b\dot{x}^2 + \dot{x}f = -b\omega_n^2 B^2 \cos^2(\omega_n t) + \omega_n AB\cos^2(\omega_n t).$$

Thus, the mean stored energy in the mechanical system during each oscillation period is given as the subtraction of the mean supplied energy and the mean dissipated energy, i.e.,:

$$E = \frac{1}{T}\int_0^T [-b\omega_n^2 B^2 \cos^2(\omega_n t) + \omega_n AB\cos^2(\omega_n t)]dt,$$

$$= \frac{-b\omega_n^2 B^2 + \omega_n AB}{T}\int_0^T \cos^2(\omega_n t)dt.$$

As long as the amplitude position B is small, $-b\omega_n^2 B^2 + \omega_n AB > 0$ is true; hence, the energy E and the amplitude B increase. Then, B increases until $-b\omega_n^2 B^2 + \omega_n AB < 0$ because A is constant. At that moment, the energy stops increasing and a steady state is reached when $b\omega_n^2 B^2 = \omega_n AB$, i.e., the position amplitude B remains constant at a finite value given as:

$$B = \frac{A}{b\omega_n} = \frac{A}{2\zeta\sqrt{mK}\sqrt{\frac{K}{m}}} = \frac{A}{2\zeta K} = \frac{kA}{2\zeta}. \tag{6.13}$$

Note that the position amplitude B grows without limit as $\zeta \to 0$, which is in agreement with Fig. 6.17b and, at the same time, explains why the resonance peak decreases as ζ increases. In this respect, note that the term $-b\dot{x}^2 \le 0$ if $b > 0$, i.e., if $\zeta > 0$. This represents the dissipated energy caused by friction. Thus, friction avoids the energy and the oscillation amplitude grows without limit. It can be stated that the greater the friction that exists, the greater the percentage of the supplied energy to the mass through the force f converts into heat and the smaller the amount of energy is stored in the mass, i.e., the smaller the oscillation amplitude results.

Finally, note that $|G(j\omega)|_{\omega=\omega_n} = \frac{k}{2\zeta}$, which is in agreement with (6.13). It is also important to recall that the resonance peak occurs at a frequency that is slightly different from ω_n when $\zeta > 0$. However, the magnitude of the transfer function at $\omega = \omega_n$ also grows as $\zeta \to 0$.

Example 6.2 Consider the following transfer function:

$$G_d(s) = \frac{s + \frac{1}{T}}{s + \frac{1}{\xi T}}, \quad 0 < \xi < 1, \quad T > 0. \tag{6.14}$$

Fig. 6.23 Bode diagrams for
Example 6.2 (see (6.15)): *(i)*
$0 < \frac{1}{T} < 1$, *(ii)* $\frac{s+\frac{1}{T}}{\frac{1}{T}}$, *(iii)*
$\frac{\frac{1}{\xi T}}{s+\frac{1}{\xi T}}$, *(iv)* $G_d(s)$

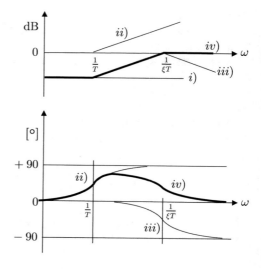

According to the properties in (6.10), the magnitude Bode diagram (in dB) of $G_d(s)$
in (6.14) is obtained as the addition of the magnitude Bode diagrams of two first-
order factors:

$$G_d(s) = \frac{\frac{1}{T}}{\frac{1}{\xi T}} \frac{s+\frac{1}{T}}{\frac{1}{T}} \frac{\frac{1}{\xi T}}{s+\frac{1}{\xi T}}. \tag{6.15}$$

The pole has a corner frequency at $\omega = \frac{1}{\xi T}$ that is greater than the corner frequency
at $\omega = \frac{1}{T}$ of the zero, because $0 < \xi < 1$. This fact determines that $G_d(s)$
contributes a phase lead and possesses properties of a high-pass filter: the low
frequencies are attenuated, but not the high frequencies. The maximal phase lead
produced and the attenuation at low frequencies only depend on the constant ξ, i.e.,
on the distance between the pole and the zero corner frequencies. These observations
can be verified in Fig. 6.23, where the Bode diagrams of $G_d(s)$ and the factors
composing it are sketched.

The software MATLAB can also be used to plot Bode diagrams using the
command "bode()." For instance, the Bode diagrams of $G_d(s)$ in (6.14) that are
presented in Fig. 6.24 were drawn by executing the following MATLAB code in an
m-file several times:

```
xi=0.7;%0.14 0.2 0.4 0.7 0.9
T=4;
gd=tf([1 1/T],[1 1/(T*xi)]);
figure(2)
bode(gd)
grid on
hold on
```

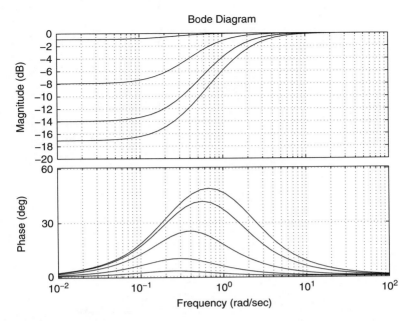

Fig. 6.24 Bode diagrams for $G_d(s)$ in (6.14), when $T = 4$ and ξ takes the values $0.14, 0.2, 0.4, 0.7, and 0.9$. In the magnitude plot, ξ increases from bottom to top. In the phase plot, ξ increases from top to bottom

Note that $T = 4$ and several values of $0 < \xi < 1$ were used. The reader can verify that the first corner frequency appears at $\omega = \frac{1}{T} = 0.25[\text{rad/s}]$ and the second corner frequency appears at $\omega = \frac{1}{\xi T} > 0.25[\text{rad/s}]$. Also note that the high-frequency gain is unitary, whereas the low-frequency gain $\frac{\frac{1}{T}}{\frac{1}{\xi T}} = \xi$ decreases with ξ. This corroborates Fig. 6.23.

On the other hand, the transfer function in (6.14) is very useful for designing a special class of controllers known as lead compensators. This application, however, is rendered easier if the following procedure is employed:

$$G_d(Ts) = \frac{Ts + 1}{Ts + \frac{1}{\xi}}.$$

The use of the variable change $z = Ts$ yields:

$$G_d(Ts) = G_d(z) = \xi \frac{z + 1}{\frac{1}{\xi}} \cdot \frac{\frac{1}{\xi}}{z + \frac{1}{\xi}}. \tag{6.16}$$

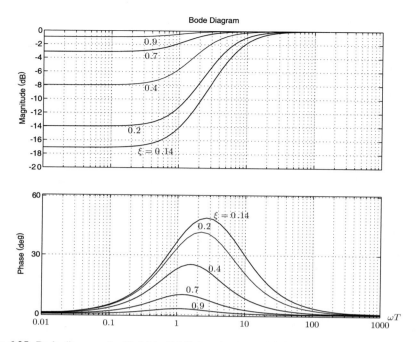

Fig. 6.25 Bode diagrams for $G_d(z)$ in (6.16), i.e., the normalized version of $G_d(s)$ in (6.14)

This allows us to draw a family of normalized Bode diagrams without needing to know T, which is important during the design stage. In such Bode diagrams, the horizontal axis, i.e., the frequency axis, represents the imaginary part of the variable $z = Ts = T(\sigma + j\omega)$, where σ stands for the real part of the Laplace variable s. This means that the frequency axis stands for the variable $T\omega$. Hence, the first corner frequency appears at $T\omega = 1$ and the second corner frequency appears at $T\omega = \frac{1}{\xi}$. These facts can be observed in Fig. 6.25, where the Bode diagrams of $G_d(z)$ in (6.16) are presented. Note that $G_d(z)$ is the normalized version of $G_d(s)$ in (6.14). The Bode diagrams in Fig. 6.25 were drawn by executing the following MATLAB code in an m-file several times:

```
xi=0.7;%0.14 0.2 0.4 0.7 0.9
gdn=tf([1 1],[1 1/xi]);
figure(1)
bode(gdn)
grid on
hold on
```

The polar plot of $G_d(s)$ can be drawn easily, employing the information provided by the Bode diagrams in Figs. 6.23 or 6.24. Recall that the magnitude Bode diagram is in dB, but not the polar plot. The magnitude and the phase for zero and infinity frequencies must be observed to draw some sketches of the corresponding behaviors

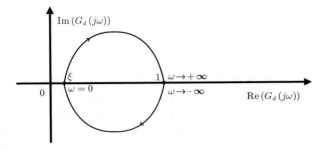

Fig. 6.26 Polar plot of $G_d(s)$ in (6.14)

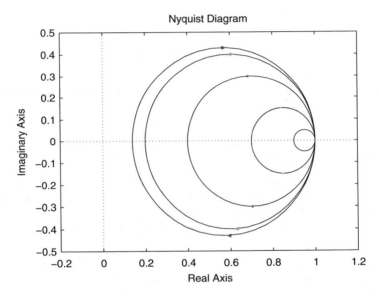

Fig. 6.27 Polar plot of $G_d(s)$ in (6.14) when using the numerical values in Fig. 6.24

in the polar plots. Then, sketches representing the behavior at the intermediate frequencies are drawn that satisfy the previously drawn sketches for zero and infinity frequencies. Finally, the polar plots for negative frequencies, i.e., from $\omega = -\infty$ to $\omega = 0$, are symmetrical images with respect to the real axis of the plots for positive frequencies. Compare the polar plot shown in Fig. 6.26 and the Bode diagrams in Fig. 6.23.

Polar plots are drawn in MATLAB using the command "nyquist()." The polar plot corresponding to the Bode diagrams in Fig. 6.24 is shown in Fig. 6.27. This was achieved by executing the following MATLAB code in an m-file several times:

```
xi=0.9;%0.14 0.2 0.4 0.7 0.9
T=4;
gd=tf([1 1/T],[1 1/(T*xi)]);
```

```
figure(3)
nyquist(gd)
axis([-0.2 1.2 -0.5 0.5])
hold on
```

It is observed in Fig. 6.27 that, for each different value of ξ, the polar plot is a circle crossing the positive real axis at two points: $+1$, i.e., the high-frequency gain, and ξ, i.e., the low-frequency gain. This means that the external-most circle corresponds to $\xi = 0.14$ and the internal-most circle corresponds to $\xi = 0.9$. This corroborates the plot in Fig. 6.26. Note the arrow senses indicating that the frequency ω increases from $-\infty$ to $+\infty$.

Example 6.3 Consider the transfer function:

$$G(s)H(s) = \frac{A_x \gamma k}{s+a} \frac{\rho}{s^3}, \tag{6.17}$$

where the constants A_x, γ, k, a, ρ are all positive. Writing this transfer function in terms of its fundamental factors, yields:

$$G(s)H(s) = \frac{\rho A_x \gamma k}{a} \frac{a}{s+a} \frac{1}{s^3}.$$

According to the properties in (6.10), the magnitude Bode diagram of the factor $\frac{1}{s^3}$ is a straight line with a slope of $-60[dB/dec]$. The Bode diagrams of $G(s)H(s)$ and its components are shown in Fig. 6.28. As explained in the previous example, the polar plot in Fig. 6.29 can be easily obtained from the Bode diagrams in Fig. 6.28.

Fig. 6.28 Bode diagrams of $G(s)H(s)$ defined in (6.17): (i) $\frac{\rho A_x \gamma k}{a}$, (ii) $\frac{1}{s^3}$, (iii) $\frac{a}{s+a}$, iv) $G(s)H(s)$

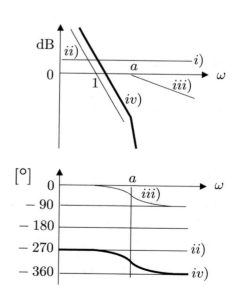

Fig. 6.29 Polar plot of $G(s)H(s)$ in (6.17)

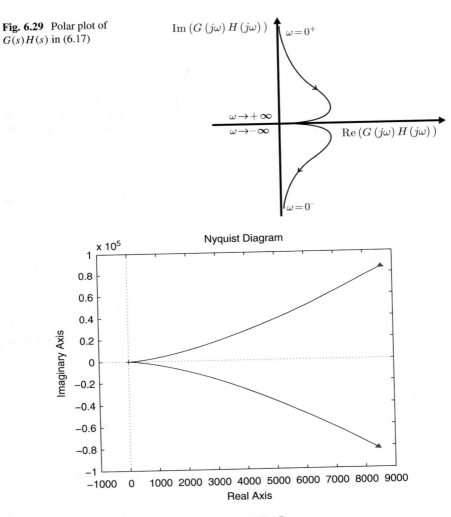

Fig. 6.30 Polar plot of $G(s)H(s)$ in (6.17) using MATLAB

To verify these results, the polar plot of $G(s)H(s)$ defined in (6.17) is presented in Fig. 6.30 when using:

$$k = 10.7219, \quad a = 0.98, \quad A_x = 1, \quad \rho = 4.7, \quad \gamma = 1.64.$$

and executing the following MATLAB code in an m-file:

```
k=10.7219;
a=0.98;
Ax=1;
rho=4.7;
```

```
gamma=1.6406;
gh=tf(Ax*gamma*k*rho,[1 a 0 0 0])
figure(1)
nyquist(gh,{0.1,10})
```

It is important to learn to correctly interpret the polar plot in Fig. 6.30. Note that neither infinite nor zero frequencies can be handled numerically by MATLAB in this example because of the singularities appearing at these frequencies. Hence, the polar plot in Fig. 6.30 is just a small part of the polar plot in Fig. 6.29, i.e., a small part around the positive real axis close to the origin. Observe the sense of the arrows in both figures.

Example 6.4 Following similar procedures to those described in the previous examples, the Bode diagrams and the polar plots of the following transfer functions can be obtained:

$$G(s)H(s) = A_x \, \gamma \frac{s+b}{s+c} \frac{k}{s(s+a)} \frac{\rho}{s^2}, \tag{6.18}$$

$$G(s)H(s) = \frac{A_x \, \gamma}{A_\theta} \frac{s+b}{s+c} \frac{\alpha k A_\theta}{s^2 + (a + k_v k A_\theta)s + \alpha k A_\theta} \frac{\rho}{s^2}, \tag{6.19}$$

where A_x, A_θ, γ, k, a, ρ, b, c, α, k_v are positive constants. The corresponding plots are shown in Figs. 6.31, 6.32, 6.33 and 6.34.

Fig. 6.31 Bode diagram of $G(s)H(s)$ in (6.18): (*i*) $\frac{A_x \, \gamma bk\rho}{ac}$, (*ii*) $\frac{1}{s^3}$, (*iii*) $\frac{a}{s+a}$, (*iv*) $\frac{c}{s+c}$, (*v*) $\frac{s+b}{b}$, (*vi*) $G(s)H(s)$

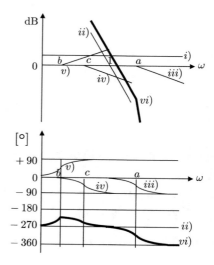

Fig. 6.32 Polar plot of $G(s)H(s)$ in (6.18)

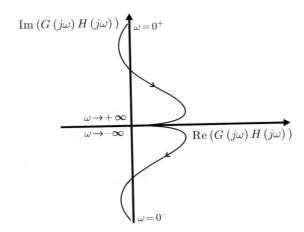

Fig. 6.33 Bode diagrams of $G(s)H(s)$ in (6.19): (i) $\frac{A_x \gamma b\rho}{A_\theta c}$, (ii) $\frac{1}{s^2}$, (iii) $\frac{\alpha k A_\theta}{s^2+(a+k_v k A_\theta)s+\alpha k A_\theta}$, (iv) $\frac{c}{s+c}$, (v) $\frac{s+b}{b}$, (vi) $G(s)H(s)$

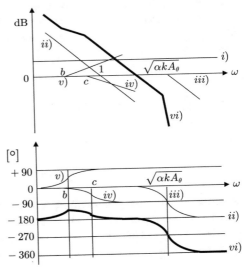

6.4 Nyquist Stability Criterion

Routh's stability criterion is a tool useful for determining whether a system has some of its poles on the right half plane. If this is not the case, then all poles are located in the left half-plane, which ensures system stability. Routh's criterion must be applied to the characteristic polynomial of the system under study and it is introduced as the stability criterion for control systems when analyzed from the point of view of time response. The reader is referred to Chap. 4 in the present book, or [5, 6], for further information on Routh's criterion.

Now, a new stability criterion is introduced for closed-loop control systems, which is established in terms of its open-loop frequency response. This feature

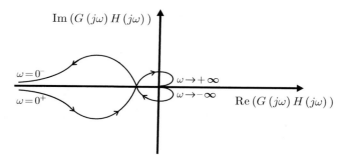

Fig. 6.34 Polar plot of $G(s)H(s)$ in (6.19)

must be correctly understood because students often become confused asking the following question: how is it possible to determine *closed-loop stability* from the study of the *open-loop* frequency response? The answer to this question is given in the present section and the tool solving this problem is known as the *Nyquist stability criterion*.

6.4.1   Contours Around Poles and Zeros

Consider the following transfer function:

$$G(s) = s - a,$$

where a is any constant number, real or complex. Note that this system only possesses one zero at $s = a$. Consider a clockwise contour Γ on the plane s, around the zero at $s = a$ (see Fig. 6.35). When $G(s)$ is evaluated on Γ, it means that s in $G(s)$ successively takes all values of s on Γ. Evaluate $G(s)$ on Γ:

$$G(s) = l\angle\theta, \quad l = |s - a|, \quad \theta = \angle(s - a).$$

Note that $s - a$ can be represented as a vector from a to s. It is easy to observe that the angle θ has a negative increment (clockwise) of -2π [rad] each time the contour Γ is completed. On the other hand, as Γ does not pass through $s = a$, then $l > 0$ on the whole contour. This means that each time the clockwise contour Γ is completed, the vector $l\angle\theta$ describes one clockwise contour around the origin. As $G(s) = l\angle\theta$, it is said that one clockwise contour has been completed around the origin of the plane $G(s)$, i.e., around $G(s) = 0$ (see Fig. 6.36).

Suppose that $G(s)$ is composed only of m zeros. It is not difficult to see from the above discussion, that m clockwise contours around the origin of the plane $G(s)$ are completed each time the clockwise contour Γ is completed if this contour encircles the m zeros of $G(s)$.

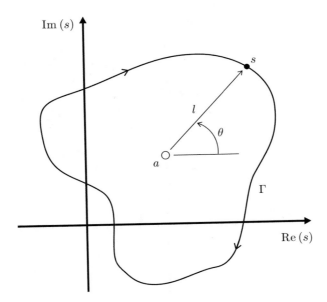

Fig. 6.35 A contour encircling a zero

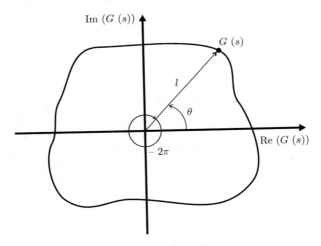

Fig. 6.36 A clockwise contour around the origin of the plane $G(s)$

Now consider the following transfer function:

$$G(s) = \frac{1}{s - a},$$

Fig. 6.37 A contour encircling a pole

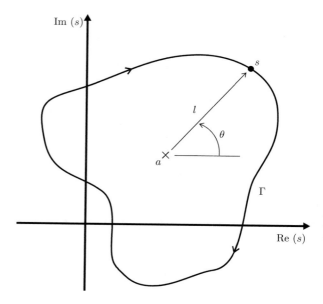

where a is any constant number, real or complex. Note that this system only possesses one pole at $s = a$. Consider a clockwise contour Γ on the plane s, encircling the pole at $s = a$ (see Fig. 6.37). Evaluate $G(s)$ on Γ:

$$G(s) = \frac{1}{l\angle\theta} = \frac{1}{l}\angle - \theta, \quad l = |s - a|, \quad \theta = \angle(s - a).$$

Note, again, that $s - a$ can be represented by a vector from a to s. It is easy to realize that each time the contour Γ is completed, the angle $-\theta$ has a positive increment (counterclockwise) of $+2\pi$ [rad]. On the other hand, as Γ does not pass through $s = a$ then $1/l > 0$ (it is finite) on the whole contour. This means that each time the clockwise contour Γ is completed, the vector $\frac{1}{l}\angle - \theta$ describes a counterclockwise contour around the origin. As $G(s) = \frac{1}{l}\angle - \theta$, it is said that a counterclockwise contour has been completed around the origin of the plane $G(s)$, i.e., around $G(s) = 0$ (see Fig. 6.38).

Suppose that $G(s)$ is only composed of n poles. From the above discussion, it is not difficult to realize that when Γ encircles all these poles, then each time the clockwise contour Γ is completed, n counterclockwise contours are completed around the origin of the plane $G(s)$.

6.4.2 Nyquist Path

Consider the transfer function of the closed-loop system in Fig. 6.39:

$$\frac{C(s)}{R(s)} = \frac{G(s)}{1 + G(s)H(s)}.$$

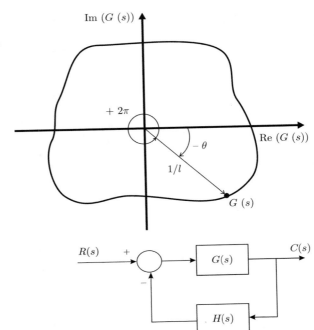

Fig. 6.38 A counterclockwise contour around the origin of the plane $G(s)$

Fig. 6.39 Closed-loop system

Closed-loop stability is ensured if this transfer function is proven not to have any poles on the right half-plane. The main idea is to investigate the existence of such poles using a contour (path), as in the previous section. Suppose that a clockwise closed path is employed such as that shown in Fig. 6.40. Note that this path encloses any pole or zero that the transfer function $\frac{G(s)}{1+G(s)H(s)}$ might have on the right half-plane. This closed path is known as the *Nyquist path*. Suppose that the transfer function $\frac{G(s)}{1+G(s)H(s)}$ has p poles and z zeros on the right half-plane. According to the previous section, if the Nyquist path is completed once, the number of clockwise rotations of the contour around the origin of the plane $\frac{G(s)}{1+G(s)H(s)}$ is $z - p$, whereas a negative result indicates counterclockwise rotations. Suppose that such a number of rotations is z, i.e., $z - p = z$. This would imply that $p = 0$, i.e., that any closed-loop pole is not placed on the right half-plane; hence, closed-loop stability would be ensured. This is the fundamental idea behind the Nyquist stability criterion presented next. However, some details have to be considered first.

Given a closed-loop system, the open-loop transfer function $G(s)H(s)$ is known (the plant and the controller), but the closed-loop transfer function $\frac{G(s)}{1+G(s)H(s)}$ is unknown. This means that, despite the poles and zeros of $G(s)H(s)$ being known, the poles of $\frac{G(s)}{1+G(s)H(s)}$ are unknown. If the poles of $\frac{G(s)}{1+G(s)H(s)}$ were known, then the stability problem would be solved. Hence, one problem to be solved by the Nyquist criterion is to determine closed-loop stability only from the knowledge of the open-loop transfer function. This means that the function $\frac{G(s)}{1+G(s)H(s)}$ cannot be evaluated on the Nyquist path as described in the previous paragraphs. This implies

Fig. 6.40 Nyquist path

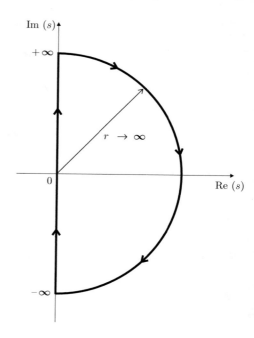

that the number of contours completed around the origin of the plane $\frac{G(s)}{1+G(s)H(s)}$ cannot be known; hence, such a method has to be modified. This is performed in the next sections.

6.4.3 Poles and Zeros

Consider the transfer function of the closed-loop system in Fig. 6.39:

$$\frac{C(s)}{R(s)} = \frac{G(s)}{1 + G(s)H(s)}.$$

- The closed-loop poles are the zeros of $1 + G(s)H(s)$. Recall that the closed-loop poles are those values of s satisfying $1 + G(s)H(s) = 0$.
- The poles of $G(s)H(s)$ (the open-loop poles) are also poles of $1 + G(s)H(s)$. Note that $G(s)H(s) \to \infty$ implies that $1 + G(s)H(s) \to \infty$.

6.4.4 Nyquist Criterion: A Special Case

Suppose that $G(s)H(s)$ has P poles on the right half-plane (open-loop unstable poles), i.e., according to the previous section, $1 + G(s)H(s)$ has P poles on the right half-plane. Also suppose that the function $1 + G(s)H(s)$ has Z zeros on the right half-plane (closed-loop unstable poles). Suppose that, once the Nyquist path is completed, N contours around the origin of the plane $1 + G(s)H(s)$ are completed, i.e., $N = Z - P$. If $N = -P < 0$, i.e., if P counterclockwise contours are completed around the origin of the plane $1 + G(s)H(s)$, then the number of closed-loop unstable poles is zero, $Z = 0$, and closed-loop stability is ensured.

The origin of the plane $1 + G(s)H(s)$ satisfies the condition $1 + G(s)H(s) = 0$, which is equivalent to $G(s)H(s) = -1$. Hence, the origin of the plane $1+G(s)H(s)$ corresponds to the point $(-1, j0)$ on the plane $G(s)H(s)$. This means that the number N of contours around the origin of the plane $1+G(s)H(s)$ is identical to the number of contours around the point $(-1, j0)$ on the plane $G(s)H(s)$. This is very convenient because it only requires us to know $G(s)H(s)$, instead of $1+G(s)H(s)$,.

Finally, if the number of poles of $G(s)H(s)$ is greater than or equal to its number of zeros, then $\lim_{s \to \infty} G(s)H(s) =$ constant. This means that the function $G(s)H(s)$ remains at a single point along the complete semicircular part of the Nyquist path, which has an infinite radius. Then, the very interesting part of $G(s)H(s)$ is that obtained when the Nyquist path is on the imaginary axis, from $j\omega = -j\infty$ to $j\omega = +j\infty$. Note that along this part of the Nyquist path, $G(s)H(s)$ becomes $G(j\omega)H(j\omega)$, which represents the polar plot of the open-loop transfer function. The above discussion allows us to establish the following result.

Nyquist Stability Criterion *Special case when $G(s)H(s)$ has neither poles nor zeros on the imaginary axis.*

Consider the closed-loop system shown in Fig. 6.39. If the open-loop transfer function $G(s)H(s)$ has P poles on the right half-plane s and $\lim_{s \to \infty} G(s)H(s)$ = constant, then closed-loop stability is ensured if the polar plot of $G(j\omega)H(j\omega)$ (when ω passes from $-\infty$ to $+\infty$) describes P counter clockwise contours around the point $(-1, j0)$.

The key to remembering the Nyquist criterion is the following formula:

$$N = Z - P$$

where P represents the number of open-loop poles with a positive real part, Z stands for the number of unstable closed-loop poles and N represents the number of contours that the polar plot of $G(j\omega)H(j\omega)$ describes around the point $(-1, j0)$, when ω passes from $-\infty$ to $+\infty$. A positive N stands for clockwise contours, whereas a negative N represents counterclockwise contours. Closed-loop stability is ensured when $Z = 0$, i.e., when $N = -P$.

6.4.5   Nyquist Criterion: The General Case

The trick of the closed path Γ introduced in Sect. 6.4.1 is valid only if such a closed path completely encircles the corresponding poles and zeros. Hence, it cannot be applied when one pole or zero is located on the path. Note that in the case when the open-loop transfer function $G(s)H(s)$ has poles or zeros on the imaginary axis the Nyquist path passes through such poles or zeros; hence, the criterion established in Sect. 6.4.4 cannot be applied. This problem is solved by defining a new path known as the *modified Nyquist path*, which is shown in Fig. 6.41. This path is identical to the Nyquist path introduced in Fig. 6.40 and the former only differs from the latter at points containing poles or zeros of $G(s)H(s)$ on the imaginary axis. The main idea is to contour such points using a semicircular path with a very small radius $\varepsilon \to 0$, which describes an angle from $-90°$ to $+90°$. The reason to include a very small radius is to ensure that the modified Nyquist path completely encloses any possible pole or zero (unknown) that could be located on the right half-plane close to the imaginary axis. Thus, the general form of the Nyquist stability criterion can be stated as follows.

Nyquist Stability Criterion *General case when $G(s)H(s)$ may have poles or zeros on the imaginary axis.*

Consider the closed-loop system shown in Fig. 6.39. If the open-loop transfer function $G(s)H(s)$ has P poles on the right half-plane s, then closed-loop stability

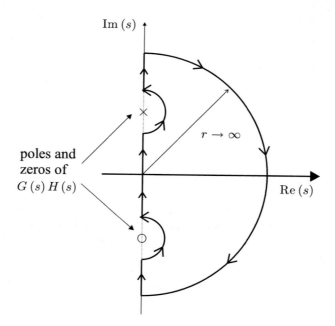

Fig. 6.41 Modified Nyquist path

is ensured if the polar plot of $G(s)H(s)$ describes P counterclockwise contours around the point $(-1, j0)$ each time the modified Nyquist path is completed clockwise.

It is important to stress that in this case, $G(s)H(s) \rightarrow 0$ is also required to be constant when $s \rightarrow \infty$, i.e., that $G(s)H(s)$ has a number of poles that is greater than or equal to its number of zeros. Moreover, the relation $N = Z - P$, with Z the number of unstable closed-loop poles, P the number of open-loop poles with a positive real part and N the number of contours around the point $(-1, j0)$, is still valid.

6.5 Stability Margins

6.5.1 Minimum Phase Systems

A system is said to be *minimum phase* if all its poles and zeros have a negative or zero real part and its gain is positive. If this is not the case, the system is said to be *nonminimum phase*. The Nyquist criterion is valid for closed-loop systems whose open-loop transfer function is either minimum phase or nonminimum phase, i.e., it is a result that can be applied to any case. However, the Nyquist stability criterion is simplified if the system has a minimum phase open-loop transfer function, as explained in the following.

If the open-loop system is minimum phase then $P = 0$; hence, closed-loop stability is ensured if $N = 0$, then the Nyquist stability criterion is simplified as follows:

Nyquist Stability Criterion for Open-Loop Minimum Phase Systems [10, 11]
Consider that the frequency ω only takes values from 0 to $+\infty$. Draw the part of the polar plot of $G(s)H(s)$ corresponding to these frequencies. Suppose that this plot represents a path where you walk from $\omega = 0$ to $\omega \rightarrow +\infty$. If the point $(-1, j0)$ lies to the left of this path, then the closed-loop system in Fig. 6.39 is stable. If such a point lies to the right, then the closed-loop system is unstable.

If the polar plot passes through the point $(-1, j0)$, it is said that the closed-loop system is marginally stable, which means that there are some closed-loop poles with a zero real part. It can be said that the closed-loop system is more stable, as the point $(-1, j0)$ is farther from the polar plot of $G(s)H(s)|_{s=j\omega}$, but always to the left. The stability margin is a concept introduced to measure the distance existing between this point and the polar plot, i.e., the *distance* of a system to instability. There are two ways of measuring this distance: the phase margin and the gain margin.

- *Phase margin* is the amount of phase lag required to be added at the crossover frequency to take the closed-loop system to the edge of instability. The *crossover frequency*, ω_1, is such that $|G(j\omega_1)H(j\omega_1)| = 1$. Let K_f and ϕ represent the phase margin and the phase of $G(j\omega_1)H(j\omega_1)$ respectively, then:

$$K_f = 180° + \phi.$$

As shown in Fig. 6.42, a negative phase margin implies closed-loop instability, whereas a positive phase margin implies closed-loop stability. According to these ideas, one additional open-loop zero tends to improve closed-loop stability, because it contributes a positive phase, i.e., it increases K_f or renders it positive. Note that, according to the plots in Figs. 6.16 and 6.17, the phase lead contributed by a zero is larger for all frequencies as the corner frequency is smaller (which is numerically equal to the absolute value of the zero location). Hence, the stabilizing effect of an open-loop zero increases as the zero is located closer to the origin (but on the left half-plane).

On the other hand, an additional open-loop pole tends to render the closed-loop system less stable because it contributes a negative phase, i.e., it decreases K_f or renders it negative. Note that, according to the plots in Figs. 6.16 and 6.17, the phase lag due to a pole is larger for all frequencies, as the corner frequency is smaller (which is numerically equal to the absolute value of the pole location). Hence, the destabilizing effect of an open-loop pole increases, as this pole is located closer to the origin (but on the left half-plane).

- *Gain margin.* Let K_g represent the gain margin, then:

$$K_g = \frac{1}{|G(j\omega_2)H(j\omega_2)|},$$

where ω_2 represents the frequency, where the phase of $G(j\omega_2)H(j\omega_2)$ is $-180°$. It is observed in Fig. 6.42 that $K_g > 1$ ($20\log(K_g) > 0$[dB]) implies closed-loop stability, whereas $K_g < 1$ ($20\log(K_g) < 0$[dB]) implies closed-loop instability.

It is important to stress that the phase and gain margins are defined on the frequency response plots of the *open-loop transfer function* and they give information on the stability and the transient response of the *closed-loop system*.

6.5.2   Time Delay Systems

Consider the water level system shown in Fig. 6.43. Note that a considerable distance x exists between the tank and the position of the input valve whose opening determines the input water flow q_i. This means that if pipes have a cross-section area A, then:

$$T = \frac{xA}{q_i},$$

is the time it takes for water to travel from the input valve to the tank. Hence, if the input valve opening changes a little, then it takes time T for the corresponding change in q_i to arrive to the tank. The time T is known as *time delay*. From a mathematical point of view, this fact is expressed by writing $q_i = q_i(t - T)$. The Laplace transform of a delayed function $q_i(t - T)$ is given by the *second translation theorem*:

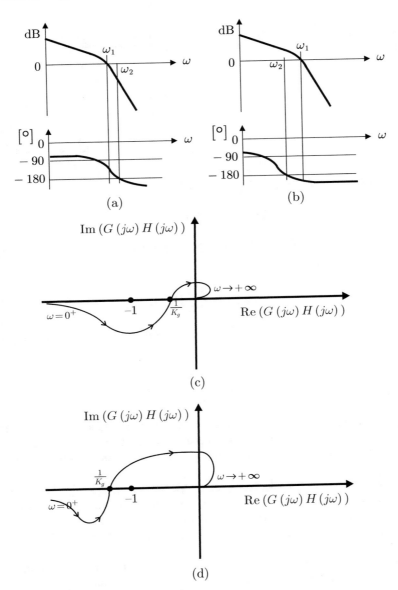

Fig. 6.42 Frequency response of $G(j\omega)H(j\omega)$. The phase and gain margins are determined from the frequency response of the open-loop transfer function. (**a**) Closed-loop stability. (**b**) Closed-loop instability. (**c**) Closed-loop stability. (**d**) Closed-loop instability

Fig. 6.43 A water level
system with time delay at the
input

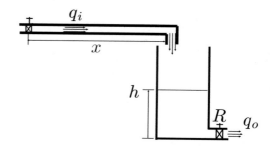

Theorem 6.1 ([2]) *If $F(s) = \mathcal{L}\{f(t)\}$, then:*

$$\mathcal{L}\{f(t - T)\} = e^{-Ts} F(s).$$

According to Sect. 3.1.2 and (3.28), the mathematical model of the undelayed water level system is:

$$H(s) = \frac{k}{s + a} Q_i(s),$$

thus, according to the theorem above, the model of the water level system when the input is delayed becomes:

$$H(s) = \frac{k\,e^{-Ts}}{s + a} Q_i(s).$$

Note that, from the mathematical point of view, this model stands if either the input or the output is the delayed variable. It is concluded that the time delay is considered in a system just by including the factor e^{-Ts} in the corresponding transfer function [9].

On the other hand, a time delay may also appear when using a digital computer to implement a continuous time controller. In such a case, the time delay T represents the time required to perform the computations involved in the control law. Thus, in such a case, it suffices again to include the factor e^{-Ts} in the plant transfer function [3], as the delay appears at the plant input.

Consider the open-loop transfer function of a system that includes a time delay T as a result of the digital implementation of a continuous time controller, i.e.,:

$$e^{-Ts} G(s)H(s).$$

To analyze the closed-loop stability, consider the magnitude and the phase of this open-loop transfer function. Note that:

$$\left| e^{-Ts} \right|_{s=j\omega} = \left| e^{-jT\omega} \right| = |\cos(\omega T) - j\sin(\omega T)| = 1.$$

and:

$$\angle\,e^{-Ts}\Big|_{s=j\omega} = \angle e^{-jT\omega} = \angle[\cos(\omega T) - j\,\sin(\omega T)] = \arctan\left(\frac{-\sin(\omega T)}{\cos(\omega T)}\right)$$

$$= -\omega T,$$

because $\tan(\theta) = \frac{\sin(\theta)}{\cos(\theta)}$. Suppose that an open-loop transfer function has been designed to ensure closed-loop stability, i.e., to satisfy suitable stability margins. Also, suppose that the controller is to be implemented using a digital computer. To analyze the stability effects that the time delay, because of the real-time computations required, has on the closed-loop system, consider the following. The time delay has no effect on the magnitude of the open-loop transfer function, but an additional phase lag of $-\omega T$ [rad] is included. Note that this phase lag has the effect of reducing the phase margin; thus, attention must be paid to designing a considerable phase margin rendering the closed-loop system robust with respect to the digital computer-induced time delay.

6.6   The Relationship Between Frequency Response and Time Response Revisited

As stated in Chap. 5, the time response of a control system is composed of two parts: the transient response and the steady-state response. It is important to know how the characteristics of these time responses relate to the characteristics of the frequency response, to know what is to be modified in the latter to suitably design the former. The most important points of this relationship are described in the following.

A closed-loop system with the transfer function:

$$\frac{C(s)}{R(s)} = M(s) = \frac{G(s)}{1 + G(s)H(s)},$$

is shown in Fig. 6.44. In the following section, it is described how the frequency response of the closed-loop transfer function, $M(s)$, determines the properties of the time response $c(t)$. In the subsequent section, the way in which the frequency response of the open-loop transfer function, $G(s)H(s)$, determines the properties of the time response $c(t)$ is described.

Fig. 6.44 A control system with the closed-loop transfer function $M(s)$ and the open-loop transfer function $G(s)H(s)$

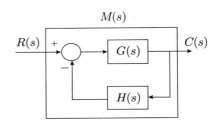

Fig. 6.45 Effects of the
closed-loop frequency
response characteristics on
the close-loop time response

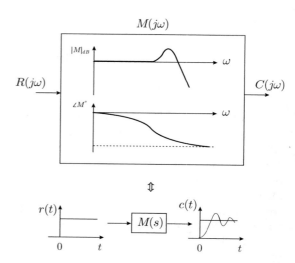

6.6.1   Closed-Loop Frequency Response and Closed-Loop Time Response

The following discussion is based on Sect. 6.2.1 and it refers to the situation illustrated in Fig. 6.45.

1. The resonant peak height M_r is inversely related to the damping coefficient, ζ: the larger M_r, the smaller ζ (see Fig. 6.17b). Recall that a smaller damping ζ produces a larger overshoot, $M_p(\%)$, in the time response when a step input is applied. Hence, the overshoot is larger (and the system is more oscillatory), as M_r is larger. Roughly speaking, when $1.0 < M_r < 1.4$ ($0[\mathrm{dB}] < M_r < 3[\mathrm{dB}]$), then damping is in the range $0.7 > \zeta > 0.4$ [5] pp. 571, [6] pp. 637.
2. The corner frequency indicates how fast the time response is [5] pp. 573, [6] pp. 638. Recall that in a second-order system, the corner frequency is ω_n, which, when increased, reduces the rise time (see Sect. 3.3.1). In the case of a first-order system the corner frequency is given as the absolute value of the system pole. According to Sect. 3.1.1, the response time (the time constant) reduces if the pole absolute value increases. In the case of an arbitrary $n-$order system, the bandwidth is defined to describe how fast the time response is in terms of the frequency response. The bandwidth is the frequency where the transfer function magnitude reduces $-3[\mathrm{dB}]$ with respect to its magnitude at zero frequency. The time response is faster as the bandwidth is larger.
3. The output in a steady state, $c(\infty)$, when a step with magnitude A is applied at the input, is determined by the system response at zero frequency $M(j\omega)|_{\omega=0}$:

$$\lim_{t\to\infty} c(t) = \lim_{s\to 0} sM(s)\frac{A}{s} = M(0)A.$$

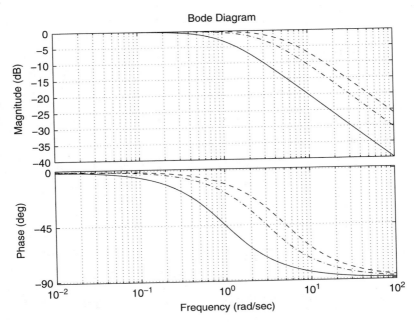

Fig. 6.46 Bode diagrams of the transfer function $M(s) = \frac{a}{s+a}$. Continuous: $a = 1$. Dash–dot: $a = 3$. Dashed: $a = 5$

Example 6.5 To illustrate the above-mentioned ideas, in Figs. 6.46 and 6.47 the Bode diagrams and the time response to a step input of the first-order transfer function $M(s) = \frac{a}{s+a}$ are presented. It is observed that, as predicted above, the time response becomes faster as the bandwidth becomes larger.

In Figs. 6.48 and 6.49 the Bode diagrams and the time response to a step input of the transfer function $M(s) = \frac{\omega_n^2}{s^2+2\zeta\omega_n s+\omega_n^2}$ are presented, when $\omega_n = 5$ is kept constant and different values for ζ are employed. Note that overshoot in the time response becomes larger as the resonance peak also becomes larger, as predicted above.

In Figs. 6.50 and 6.51 the Bode diagrams and the time response to a step input of the transfer function $M(s) = \frac{\omega_n^2}{s^2+2\zeta\omega_n s+\omega_n^2}$ are presented, when $\zeta = 0.4$ is kept constant and different values for ω_n are employed. Note that the time response becomes faster as the bandwidth becomes larger, as predicted above.

These results were obtained by executing the following MATLAB code in an m-file:

```
%{
z=0.7;% 0.1 0.4 0.7
wn=5;
M=tf(wn^2,[1 2*z*wn wn^2]);
```

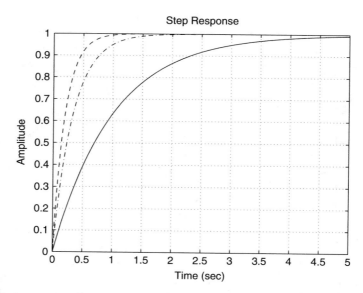

Fig. 6.47 Response to an unit step input of the transfer function $M(s) = \frac{a}{s+a}$. Continuous: $a = 1$. Dash–dot: $a = 3$. Dashed: $a = 5$

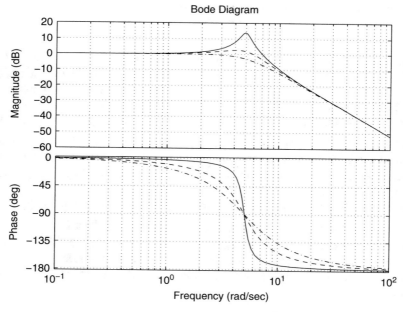

Fig. 6.48 Bode diagrams of the transfer function $M(s) = \frac{\omega_n^2}{s^2+2\zeta\omega_n s+\omega_n^2}$ when $\omega_n = 5$. Continuous: $\zeta = 0.1$. Dashed: $\zeta = 0.4$. Dash–dot: $\zeta = 0.7$

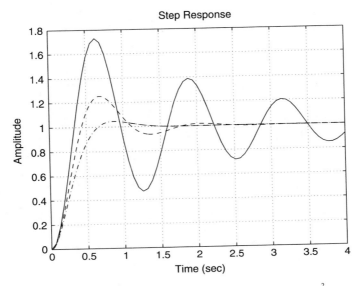

Fig. 6.49 Response to a unit step input of the transfer function $M(s) = \frac{\omega_n^2}{s^2+2\zeta\omega_n s+\omega_n^2}$ when $\omega_n = 5$. Continuous: $\zeta = 0.1$. Dashed: $\zeta = 0.4$. Dash–dot: $\zeta = 0.7$

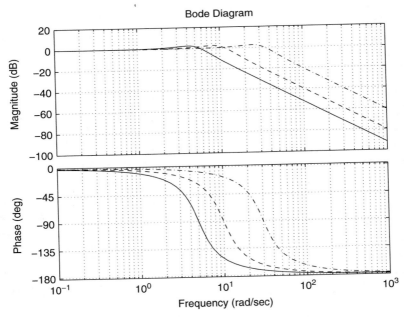

Fig. 6.50 Bode diagrams of the transfer function $M(s) = \frac{\omega_n^2}{s^2+2\zeta\omega_n s+\omega_n^2}$ when $\zeta = 0.4$. Continuous: $\omega_n = 5$. Dashed: $\omega_n = 10$. Dash–dot: $\omega_n = 30$

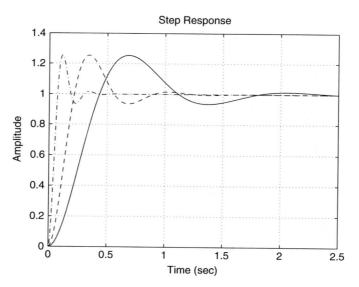

Fig. 6.51 Response to an unit step input of the transfer function $M(s) = \frac{\omega_n^2}{s^2+2\zeta\omega_n s+\omega_n^2}$ when $\zeta = 0.4$. Continuous: $\omega_n = 5$. Dashed: $\omega_n = 10$. Dash–dot: $\omega_n = 30$

```
figure(1)
bode(M,'k-.')
grid on
hold on
figure(2)
step(M,'k-.',4)
grid on
hold on
%}
%{
figure(3)
z=0.4;
wn=30; % 5 10 30
M=tf(wn^2,[1 2*z*wn wn^2]);
bode(M,'k-.')
grid on
hold on
figure(4)
step(M,'k-.',2.5)
grid on
hold on
%}
%%{
```

```
figure(5)
a=3;% 1 5 3
M=tf(a,[1 a]);
bode(M,'k-.')
grid on
hold on
figure(6)
step(M,'k-.',5)
grid on
hold on
%}
```

Note that this code can be commented on in blocks. Enabling the corresponding blocks and executing several times, the results above are obtained.

6.6.2 Open-Loop Frequency Response and Closed-Loop Time Response

There is also a relationship between the frequency response of the open-loop transfer function $G(s)H(s)$ and the time response of the closed-loop system $\frac{C(s)}{R(s)} = M(s)$ (see Fig. 6.52). This relationship is very important for designing feedback control systems.

1. The phase margin K_f (measured on the system open-loop frequency response characteristics) is related to the damping coefficient of the closed-loop system according to the following table [5] pp. 570,[6] pp. 648:

    Note that $\zeta \approx \frac{K_f}{100}$ in the range $0 \leq \zeta \leq 0.6$. It must be stressed that this relationship is computed for a second-order system without zeros. Although differences are expected for higher order systems with zeros with respect to these values, it is suggested to use these data as directives for design.

2. The larger the crossover frequency (measured on the open-loop frequency response), the faster the closed-loop system time response [5], pp. 571. This can be explained as follows. The crossover frequency is that where $|G(j\omega)H(j\omega)| = 0[\text{dB}]$. Generally, this magnitude increases to $+\infty[\text{dB}]$ as frequency decreases to 0, and this magnitude decreases to $-\infty[\text{dB}]$ as frequency increases to $+\infty$. Hence, the closed-loop bandwidth is close to the crossover frequency. Thus, a larger open-loop crossover frequency implies a larger closed-loop bandwidth, i.e., as a faster closed-loop time response.

3. The steady-state error (closed-loop time response) is obtained from the open-loop frequency response at zero frequency. For instance, if the reference input is $R(s) = A/s$, then (see Sect. 4.4.1, Chap. 4):

Fig. 6.52 Effects of the
open-loop frequency response
characteristics on the
close-loop time response

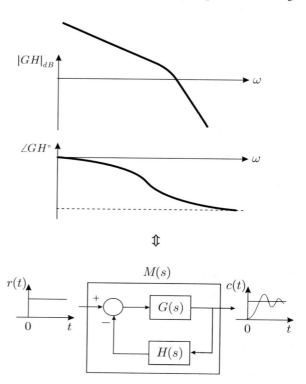

$$e_{ss} = \lim_{t \to \infty} [r(t) - y(t)] = \frac{A}{1 + k_p}, \quad k_p = G(0)H(0) = G(j\omega)H(j\omega)|_{\omega=0}.$$

Example 6.6 To illustrate the above-mentioned ideas, in Figs. 6.53 and 6.54 the Bode diagrams of the open-loop transfer function $G(s)H(s) = \frac{k}{s(s+3)(s+5)}$ are presented in Fig. 6.52 and the closed-loop time response of $M(s)$ in Fig. 6.52 when a unit step reference is applied. Note that the phase plot remains the same for all values of k in Fig. 6.53. Hence, it is easy to realize that the phase margin decreases as k increases. Moreover, the phase margin becomes slightly negative for $k = 145$. This explains why the time response in Fig. 6.54 becomes more oscillatory and, eventually, unstable as k increases.

On the other hand, it is clear from Fig. 6.53 that the crossover frequency increases as k increases. As a consequence, the closed-loop time response in Fig. 6.54 becomes faster as k increases. The graphical results above were obtained using the following MATLAB code executed several times in an m-file:

```
k=145; % 15 55 95 145
denom=conv([1 0],[1 3]);
denom=conv([1 5],denom);
g=tf(k,denom);
```

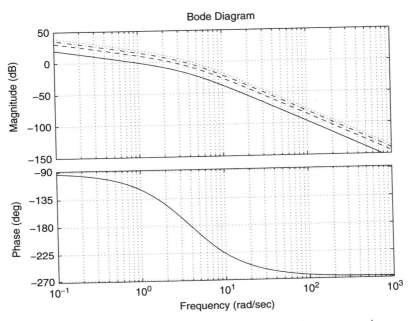

Fig. 6.53 Bode diagrams of the open-loop transfer function $G(s)H(s) = \frac{k}{s(s+3)(s+5)}$ in the closed-loop system shown in Fig. 6.52. Continuous: $k = 15$. Dashed: $k = 55$. Dash–dot: $k = 95$. Dotted: $k = 145$

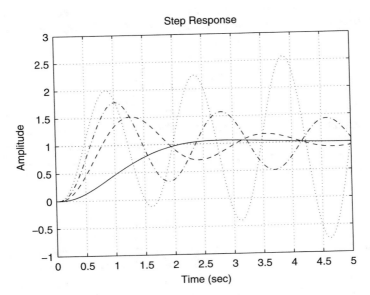

Fig. 6.54 Time response to a unit step reference of the closed-loop system $M(s)$ in Fig. 6.52. Continuous: $k = 15$. Dashed: $k = 55$. Dash–dot: $k = 95$. Dotted: $k = 145$

```
M=feedback(g,1,-1);
figure(1)
bode(g,'k:')
grid on
hold on
figure(2)
step(M,'k:',5)
grid on
hold on
```

6.7 Analysis and Design Examples

6.7.1 PD Position Control of a DC Motor

Consider the DC motor model presented in (5.12), Chap. 5, which is rewritten here for ease of reference:

$$\theta(s) = \frac{k}{s(s+a)} I^*(s),$$ (6.20)

$$a = \frac{b}{J} > 0, \quad k = \frac{nk_m}{J} > 0,$$

where $\theta(s)$ and $I^*(s)$ represent the position (output) and the commanded electric current (input) respectively. The Bode diagrams of the following transfer function:

$$G(s) = \frac{k}{s(s+a)},$$ (6.21)

are sketched in Fig. 6.55. Based on this result, the polar plot in Fig. 6.56 is obtained when mapping the modified Nyquist path shown in Fig. 6.57. Note that $G(s)$ has a single pole at $s = 0$ and this is the reason for the semicircle with infinite radius appearing in Fig. 6.56. This is explained as follows. The variable change $s = \varepsilon \angle \phi$ is first performed to rewrite (6.21) as:

$$G(s) = \frac{k}{\varepsilon \angle \phi (\varepsilon \angle \phi + a)} = \frac{k}{\varepsilon \angle \phi (a)} = \frac{k}{\varepsilon a} \angle -\phi.$$

Recalling that $\varepsilon \to 0$ and ϕ passes from $-90°$ to $+90°$ as the contour around $s = 0$ is performed (shown in Fig. 6.57), then the clockwise-oriented semicircle with infinite radius in Fig. 6.56 is obtained. As the phase of $G(j\omega)$ asymptotically tends toward $-180°$ as $\omega \to +\infty$, it is clear that the polar plot in Fig. 6.56 will never encircle the point $(-1, j0)$ for any positive value of k and a. This means that $N = 0$. Note that $P = 0$ because $G(s)$ has no poles with a positive real part.

Fig. 6.55 Bode diagrams of $G(s)$ in (6.21). $i)\frac{k}{a}$, $ii)\frac{1}{s}$, $iii)\frac{a}{s+a}$, $iv)\frac{k}{s(s+a)}$

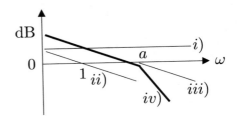

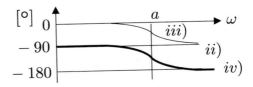

Fig. 6.56 Polar plot of $G(s)$ in (6.21)

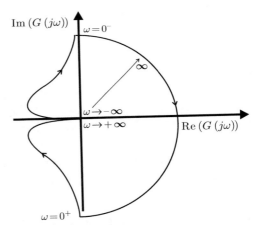

Suppose that a loop is closed around $G(s)$ including a positive gain k_p, to obtain a proportional position control system. The gain $k_p > 0$ has the following effects on the open-loop frequency response:

- The phase for a given frequency does not change with k_p.
- The magnitude increases proportionally to k_p.

Thus, every point belonging to the polar plot of the open-loop transfer function becomes farther from the origin, but keeps the same phase, as $k_p > 0$ increases. This means that the closed-loop system is stable for any positive k_p because $Z = N + P = 0$, but the polar plot in Fig. 6.56 becomes closer to the point $(-1, j0)$ as $k_p > 0$ is chosen to be larger, i.e., the phase margin decreases. As a consequence, although the time response becomes faster, it also becomes more oscillatory (as concluded in Sect. 5.2.1). To achieve a time response that is as fast as desired and, simultaneously, as damped as desired, a PD position controller is designed in the following. Note

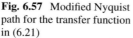

Fig. 6.57 Modified Nyquist
path for the transfer function
in (6.21)

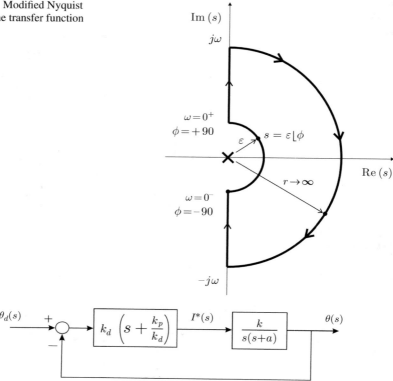

Fig. 6.58 Proportional–derivative position control system

that, according to (6.21), the plant system type is 1; hence, an additional integrator
is not required to achieve a zero steady-state error when the desired position is a
constant. This justifies the use of a PD controller instead of a PID controller.

Consider the DC motor model presented in (6.20) together with the following PD
controller:

$$i^* = k_p e + k_d \frac{de}{dt}, \quad e = \theta_d - \theta.$$

Use of the Laplace transform yields:

$$I^*(s) = (k_p + k_d s)E(s), \quad E(s) = \theta_d(s) - \theta(s).$$

The closed-loop block diagram is shown in Fig. 6.58. The open-loop transfer
function is:

$$G(s)H(s) = k_d \left(s + \frac{k_p}{k_d} \right) \frac{k}{s(s+a)}. \tag{6.22}$$

The PD controller transfer function can be rewritten as:

$$\frac{I^*(s)}{E(s)} = k_d(s + b), \quad b = \frac{k_p}{k_d}, \tag{6.23}$$

$$= k_d b \frac{s + b}{b} = k_d b \left(\frac{1}{b}s + 1\right).$$

Introducing the variable change $z = \frac{1}{b}s$ yields:

$$\frac{I^*(s)}{E(s)} = k_d b(z + 1).$$

The Bode diagrams of the factor $z+1$ are shown in Fig. 6.59. Note that the horizontal axes of these Bode diagrams represent the variable $\frac{1}{b}\omega$ (see Example 6.2). Hence, the open-loop transfer function, given in (6.22), can be rewritten as:

$$G(s)H(s) = k_d b(z + 1)\frac{k}{s(s + a)}. \tag{6.24}$$

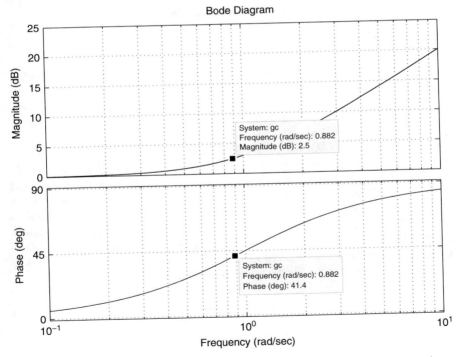

Fig. 6.59 Bode diagrams of the factor $z + 1$, where $z = \frac{1}{b}s$. The horizontal axis stands for $\frac{1}{b}\omega$

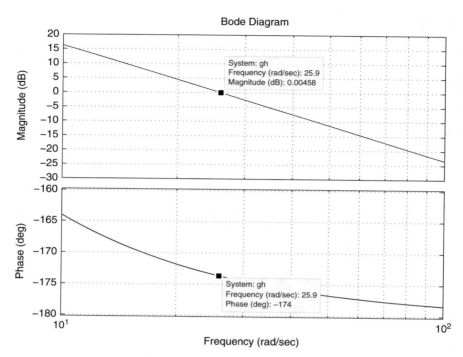

Fig. 6.60 Bode diagrams of factor $\frac{k}{s(s+a)}$, where $k = 675.4471$ and $a = 2.8681$

Consider the DC motor experimentally studied in Chap. 11. The numerical values of this motor are:

$$k = 675.4471, \quad a = 2.8681.$$

Using these data, the Bode diagrams of the factor $\frac{k}{s(s+a)}$ are obtained and they are shown in Fig. 6.60. It is observed that the crossover frequency is $\omega_1 = 25.9[\text{rad/s}]$ and the phase at this frequency is $-174°$, which corresponds to a phase margin of $+6°$. As this phase margin is small, the time response is very oscillatory when using a proportional controller with $k_p = 1$, as shown in Fig. 6.61. Note that the rise time is $t_r = 0.0628[\text{s}]$ and overshoot is $M_p = 83\%$. A PD controller is designed in the following such that the close-loop time response has the same rise time $t_r = 0.0628[\text{s}]$, but overshoot is reduced to $M_p = 20\%$. To maintain the same response time, it is proposed to maintain the same crossover frequency, i.e.,:

$$\omega_1 = 25.9[\text{rad/s}]. \tag{6.25}$$

On the other hand, using the desired overshoot $M_p = 20\%$ and:

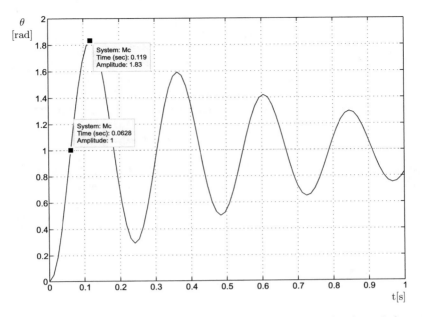

Fig. 6.61 Time response of the closed-loop system shown in Fig. 6.58, when $k_p = 1$, $k_d = 0$, $k = 675.4471$ and $a = 2.8681$

Table 6.5 Relationship between the phase margin (open-loop) and the damping coefficient of a closed-loop second-order system without zeros

$K_f[°]$	0	11	23	33	43	52	59	65	70	74	76
ζ	0	0.1	0.2	0.3	0.4	0.5	0.6	0.7	0.8	0.9	1

$$\zeta = \sqrt{\frac{\ln^2\left(\frac{Mp(\%)}{100}\right)}{\ln^2\left(\frac{Mp(\%)}{100}\right) + \pi^2}},$$

$\zeta = 0.4559$ is obtained. Using linear interpolation in Table 6.5, it is found that this damping corresponds to the phase margin:

$$K_f = +47.5°. \tag{6.26}$$

As the phase margin of the uncompensated motor model is $+6°$, then the controller is required to contribute the phase lead $+47.5° - 6° = 41.5°$. In Fig. 6.59, it is observed that this happens when $\frac{1}{b}\omega = 0.882$. Hence, the parameter b is found by forcing this to occur when $\omega = \omega_1 = 25.9$[rad/s], i.e.,:

$$b = \frac{\omega}{0.882}\bigg|_{\omega=25.9} = 29.3984.$$

On the other hand, as it is desired to maintain the crossover frequency of the uncompensated plant, i.e., $\omega = \omega_1 = 25.9$[rad/s], it is necessary for the complete controller to contribute a 0[dB] magnitude when $\omega = \omega_1$. This is accomplished by imposing the following condition on the magnitude of the complete controller:

$$|k_d b(z+1)|_{\text{dB}} = 20\log(k_d b) + |z+1|_{\text{dB}} = 0\text{[dB]},$$

where, according to Fig. 6.59, $|z+1|_{\text{dB}} = 2.5$[dB]. Hence, solving for k_d and performing the corresponding computations the following is found:

$$k_d = \frac{1}{b}10^{\frac{-2.5}{20}} = 0.0255.$$

Finally, using (6.23):

$$k_p = b k_d = 0.7499.$$

The Bode diagrams of the compensated open-loop transfer function (see (6.22)) are shown in Fig. 6.62. It can be verified that the crossover frequency is $\omega = 25.9$[rad/s] and the phase at this frequency is $-132°$, which corresponds to a phase margin of $+48°$. This means that the frequency response specifications, established in (6.25) and (6.26), have been accomplished. On the other hand, in Fig. 6.63, the time response of the corresponding closed-loop system is shown, when the desired position is a unit step. It is observed that the rise time is 0.0612[s] and overshoot is 30%. Although these values are not identical to $t_r = 0.0628$[s] and $M_p = 20\%$, which were specified at the beginning, they are relatively close. This difference is attributed to the zero present in the closed-loop transfer function (contributed by the PD controller) as it has an important effect on the transient response, which has not been precisely quantified. Recall that Table 6.5, which has been employed to determine the required phase margin, is precise only for second-order closed-loop systems with no zeros.

If it is desired to exactly satisfy the proposed time response specifications, the reader may proceed to redesign: propose a little larger phase margin to reduce overshoot and repeat the procedure. However, instead of this, in the following section, it is explained how to proceed to achieve a different desired rise time.

Figures 6.59 to 6.63 were drawn using the following MATLAB code in an m-file:

```
gc=tf([1 1],1);
figure(1)
bode(gc,{0.1,10})
grid on
k=675.4471;
```

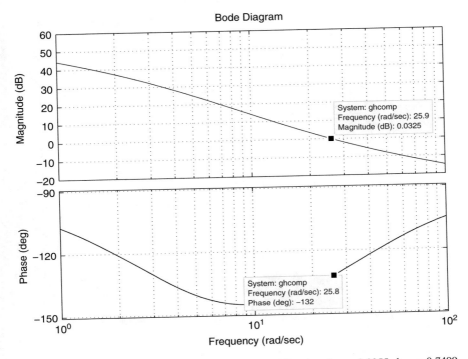

Fig. 6.62 Bode diagrams of the transfer function in (6.22), when $k_d = 0.0255$, $k_p = 0.7499$, $k = 675.4471$ and $a = 2.8681$

```
a=2.8681;
g=tf(k,[1 a 0]);
figure(2)
bode(g,{10,100})
grid on
M=feedback(g,1,-1);
figure(3)
step(M,1)
grid on
kp=0.7499;
kd=0.0255;
PD=tf([kd kp],1);
figure(4)
bode(PD*g,{1,100})
grid on
M=feedback(PD*g,1,-1);
figure(5)
step(M)
grid on
```

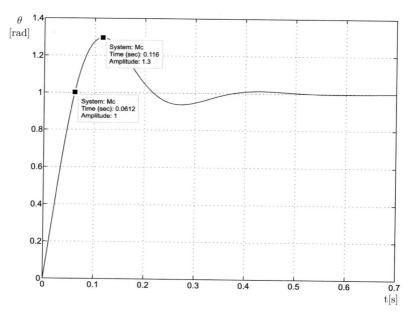

Fig. 6.63 Time response of the closed-loop system shown in Fig. 6.58, when $k_d = 0.0255$, $k_p = 0.7499$, $k = 675.4471$ and $a = 2.8681$

6.7.2   Redesign of the PD Position Control for a DC Motor

Consider again the PD position control of a DC motor. The corresponding closed-loop block diagram is shown in Fig. 6.58 and the open-loop transfer function is shown in (6.22). This expression is rewritten here for ease of reference:

$$G(s)H(s) = k_d(s+b)\frac{k}{s(s+a)}, \quad b = \frac{k_p}{k_d}. \tag{6.27}$$

In Sect. 6.7.1, it was shown that using:

$$k = 675.4471, \quad a = 2.8681,$$

and only the factor $\frac{k}{s(s+a)}$, the crossover frequency is $\omega_1 = 25.9[\text{rad/s}]$. Moreover, the closed-loop rise time is $t_r = 0.0628[\text{s}]$ and overshoot is $M_p = 83\%$ when a proportional controller with unitary gain $k_p = 1$ is employed. In the following, a PD controller is designed to accomplish half of the above rise time and an overshoot of 20%, i.e.,

$$t_r = \frac{0.0628}{2} = 0.0314[\text{s}], \quad M_p = 20\%. \tag{6.28}$$

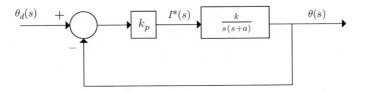

Fig. 6.64 Proportional position control

Fig. 6.65 Bode diagrams of $G(s)$ in (6.29). $i)\frac{k_pk}{a}$, $ii)\frac{1}{s}$, $iii)\frac{a}{s+a}$, $iv)\frac{k_pk}{s(s+a)}$

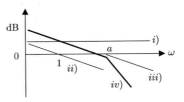

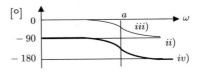

It has been already stated that reducing the rise time, i.e., rendering the time response faster, requires the crossover frequency ω_1 to be increased. However, a mathematical expression has not yet been defined relating the rise time to the crossover frequency. Next, it is explained how to select ω_1 to obtain a particular t_r.

First, consider the closed-loop system in Fig. 6.64. The open-loop transfer function is:

$$G(s) = \frac{k_pk}{s(s+a)}, \tag{6.29}$$

whereas the closed-loop transfer function is:

$$\frac{\theta(s)}{\theta_d(s)} = \frac{G(s)}{1+G(s)} = \frac{\omega_n^2}{s^2 + 2\zeta\omega_n s + \omega_n^2}, \quad \omega_n^2 = k_pk, \quad 2\zeta\omega_n = a. \tag{6.30}$$

The Bode diagrams of (6.29) are depicted in Fig. 6.65, whereas the Bode diagrams of (6.30) are depicted in Fig. 6.17b. According to the latter figure, the transfer function in (6.30) has a unit magnitude when $\omega \to 0$. This can be explained from the expression $\frac{G(s)}{1+G(s)}$ and Fig. 6.65, noting that $|G(j\omega)| \to +\infty$ when $\omega \to 0$. This means that the closed-loop transfer function $\left|\frac{G(j\omega)}{1+G(j\omega)}\right| \to 1$ when $\omega \to 0$.

On the other hand, according to Fig. 6.17b, the magnitude of the transfer function in (6.30) tends toward zero when $\omega \to +\infty$. Again, this can be explained using the expression $\frac{G(s)}{1+G(s)}$ and Fig. 6.65, noting that $|G(j\omega)| \to 0$ when $\omega \to +\infty$. This means that the closed-loop transfer function $\left|\frac{G(j\omega)}{1+G(j\omega)}\right| \to 0$ when $\omega \to +\infty$.

According to these ideas, the corner frequency that the transfer function in (6.30) has at $\omega = \omega_n$ (see Fig. 6.17b) must occur approximately at the crossover frequency in Fig. 6.65 (when $|G(j\omega)| = 1$; hence, $\left|\frac{G(j\omega)}{1+G(j\omega)}\right| = 0.5 \approx 0.7071$, which corresponds to $-3[\text{dB}]$). This observation allows us to conclude that ω_n in Fig. 6.17b increases if ω_1, measured in Fig. 6.65, also increases.

On the other hand, according to:

$$t_r = \frac{1}{\omega_n\sqrt{1-\zeta^2}}\left[\pi - \arctan\left(\frac{\sqrt{1-\zeta^2}}{\zeta}\right)\right], \qquad \omega_d = \omega_n\sqrt{1-\zeta^2},$$

if it is desired to reduce the rise time by half (neglecting the effect of the damping variation) then ω_n must be increased twice. This means that, according to the above discussion, the crossover frequency ω_1 must also be increased twice.

Hence, if $t_r = 0.0628[\text{s}]$ when $\omega_1 = 25.9[\text{rad/s}]$, the following:

$$\omega_1 = 51.8[\text{rad/s}], \tag{6.31}$$

must be used to accomplish $t_r = 0.0314[\text{s}]$. On the other hand, it was found in Sect. 6.7.1 that to obtain an overshoot of 20%, the follow phase margin must be set:

$$K_f = +47.5°. \tag{6.32}$$

The Bode diagrams of $\frac{k}{s(s+a)}$ when $k = 675.4471$ and $a = 2.8681$ are shown in Fig. 6.66. It is observed that the magnitude and the phase are $-12[\text{dB}]$ and $-177°$ when $\omega = 51.8[\text{rad/s}]$. This corresponds to a phase margin of $+3°$. As the desired phase margin is $+47.5°$, then the controller is required to contribute the positive phase $47.5° - 3° = 44.5°$ when $\omega = 51.8[\text{rad/s}]$. Recall that, according to (6.24), the open-loop transfer function of the compensated system is given as:

$$G(s)H(s) = k_d b(z+1)\frac{k}{s(s+a)},$$

where $k_d b(z+1) = k_d s + k_p$, with $b = \frac{k_p}{k_d}$ and $z = \frac{1}{b}s$, represents the PD controller. The Bode diagrams of the factor $z + 1$ are shown in Fig. 6.67. Recall that the horizontal axis in these diagrams corresponds to $\frac{1}{b}\omega$ (see Example 6.2). The required phase lead $+44.5°$ appears when $\frac{1}{b}\omega = 0.982$. As this phase lead must occur when the frequency is equal to the desired crossover frequency $\omega = 51.8[\text{rad/s}]$, then:

$$b = \left.\frac{\omega}{0.982}\right|_{\omega=51.8} = 52.8033.$$

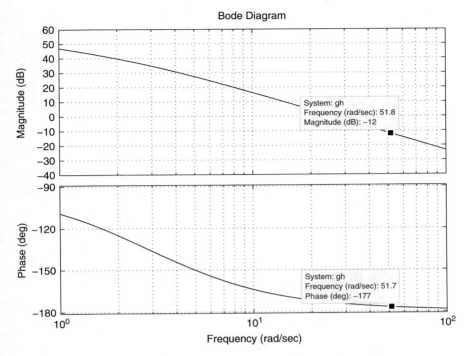

Fig. 6.66 Bode diagrams of $\frac{k}{s(s+a)}$ when $k = 675.4471$ and $a = 2.8681$

On the other hand, the crossover frequency is desired to be $\omega = 51.8[\text{rad/s}]$. This requires the controller to contribute, at this frequency, the same magnitude, but with the opposite sign that the factor $\frac{k}{s(s+a)}$ contributes at this frequency $(-12[\text{dB}])$. This is to render $0[\text{dB}]$ the magnitude of the compensated system at the new crossover frequency. This implies that:

$$|k_d b(z + 1)|_{\text{dB}} = 20 \log(k_d b) + |z + 1|_{\text{dB}} = 12[\text{dB}],$$

where, according to Fig. 6.67, $|z + 1|_{\text{dB}} = 2.95[\text{dB}]$ must be used. Thus, solving for k_d and performing the corresponding computations, it is found that:

$$k_d = \frac{1}{b} 10^{\frac{12-2.95}{20}} = 0.0537.$$

Finally, the use of $b = \frac{k_p}{k_d}$ yields:

$$k_p = b k_d = 2.8347.$$

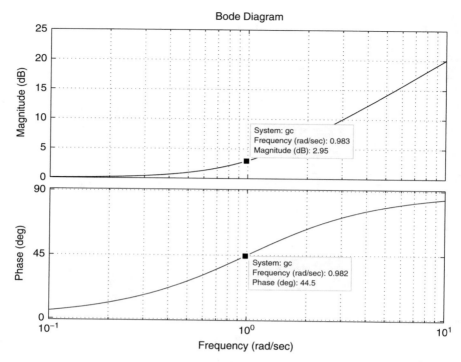

Fig. 6.67 Bode diagrams of the factor $z + 1$. The horizontal axis represents $\frac{1}{b}\omega$

The Bode diagrams of the open-loop transfer function of the compensated system shown in (6.27) are presented in Fig. 6.68. There, it can be verified that the crossover frequency is $\omega = 51.8[\text{rad/s}]$ and the phase at this frequency is $-132°$, which corresponds to a phase margin of $+48°$. This means that the desired frequency response characteristics established in (6.31), (6.32), have been accomplished. On the other hand, the time response of the corresponding closed-loop system is shown in Fig. 6.69 when the reference of the position is a unit step. It can be observed that the rise time is 0.03[s] and overshoot is 31%. Although these values are not exactly equal to $t_r = 0.0314[\text{s}]$ and $M_p = 20\%$, they are considered to be close enough. The reason for the difference in overshoot and the way of reducing it has been already explained in Sect. 6.7.1.

Figures 6.66 to 6.69 were drawn using the following MATLAB code executed in an m-file:

```
k=675.4471;
a=2.8681;
g=tf(k,[1 a 0]);
figure(1)
bode(g,{1,100})
```

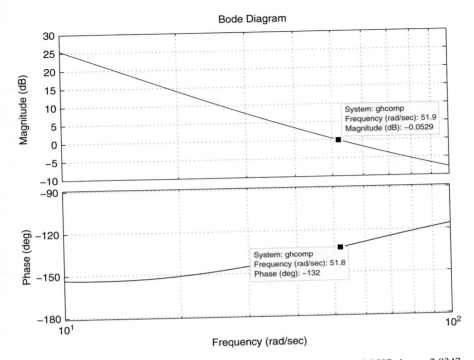

Fig. 6.68 Bode diagrams of the transfer function in (6.27), when $k_d = 0.0537$, $k_p = 2.8347$, $k = 675.4471$ and $a = 2.8681$

```
grid on
gc=tf([1 1],1);
figure(2)
bode(gc,{0.1,10})
grid on
kp=2.8347;
kd=0.0537;
PD=tf([kd kp],1);
figure(3)
bode(PD*g,{10,100})
grid on
M=feedback(PD*g,1,-1);
figure(4)
step(M)
grid on
```

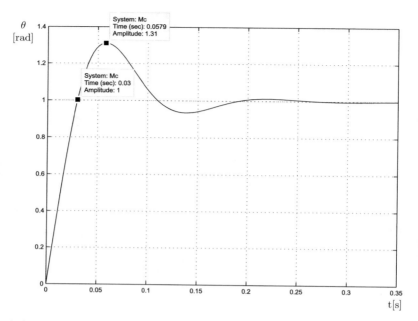

Fig. 6.69 Time response of the closed-loop system in Fig. 6.58, when $k_d = 0.0537$, $k_p = 2.8347$, $k = 675.4471$ and $a = 2.8681$

6.7.3   PID Position Control of a DC Motor

Consider the DC motor model presented in (5.12), Chap. 5, which is rewritten here for ease of reference:

$$\theta(s) = \frac{k}{s(s+a)} I^*(s),\tag{6.33}$$

$$a = \frac{b}{J} > 0, \quad k = \frac{nk_m}{J} > 0,$$

where $\theta(s)$ and $I^*(s)$ stand for the position (output) and the commanded electric current (input) respectively. In this section, the following PID position controller is designed:

$$i^* = k_p e + k_d \frac{de}{dt} + k_i \int_0^t e(r)dr, \quad e = \theta_d - \theta,$$

where θ_d is the desired position and the constants k_p, k_d, and k_i are known as the proportional, derivative, and integral gains, respectively. Use of the Laplace transform yields:

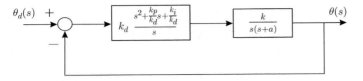

Fig. 6.70 Closed-loop system for PID position control of a DC motor

$$I^*(s) = k_p E(s) + k_d s E(s) + k_i \frac{E(s)}{s},$$

$$= \left(k_p + k_d s + \frac{k_i}{s} \right) E(s),$$

$$= k_d \frac{s^2 + \frac{k_p}{k_d} s + \frac{k_i}{k_d}}{s} E(s).$$

Then, the closed-loop system is represented by the block diagram in Fig. 6.70. The open-loop transfer function is:

$$G(s)H(s) = k_d \frac{s^2 + \frac{k_p}{k_d} s + \frac{k_i}{k_d}}{s} \frac{k}{s(s+a)}. \tag{6.34}$$

Note that the system type is 2; hence, a zero steady-state error is ensured if the desired position is either a step or a ramp. Moreover, as explained in Sects. 5.2.5 and 5.3, if the desired position is a step and a constant external torque disturbance appears, a zero steady-state error is still ensured. These are the reasons for designing a PID position controller for a DC motor.

Defining:

$$s^2 + \frac{k_p}{k_d} s + \frac{k_i}{k_d} = (s + z_1)(s + z_2) = s^2 + (z_1 + z_2)s + z_1 z_2,$$

it is concluded that:

$$\frac{k_p}{k_d} = z_1 + z_2, \qquad \frac{k_i}{k_d} = z_1 z_2. \tag{6.35}$$

In Fig. 6.71, the polar plot of (6.34) is sketched, when mapping the modified Nyquist path shown in Fig. 6.57. Note that there are two poles at $s = 0$; hence, a contour must be described around the origin of the plane s, as explained in Sect. 6.7.1. This is the reason for the circumference with the infinite radius in Fig. 6.71. Furthermore, it is not difficult to realize that, for $0 < \omega < +\infty$, the polar plot of the factor $\frac{k}{s^2(s+a)}$ is similar to that shown in Fig. 6.56, but rotated 90° clockwise (counterclockwise

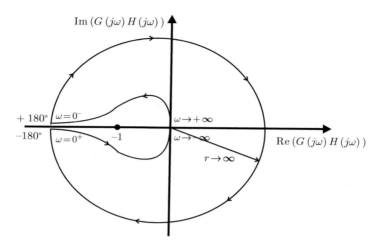

Fig. 6.71 Polar plot of (6.34) when mapping the modified Nyquist path shown in Fig. 6.57

for $-\infty < \omega < 0$). Hence, the phase lead contributed by the second-order factor $s^2 + \frac{k_p}{k_d}s + \frac{k_i}{k_d}$ must modify the polar plot as in Fig. 6.71. This achieves closed-loop stability because $P = 0$, $N = 0$, and $Z = N + P = 0$.

Recall that the design based on frequency response relies on satisfying two conditions: the desired crossover frequency and the desired phase margin. As has been seen in the previous sections, these conditions are satisfied by suitably selecting two controller parameters. In the present problem, a PID controller must be designed that possesses three parameters or gains. Hence, the problem can be solved by proposing an arbitrary value for one of these gains, whereas the other two must be found by satisfying the crossover frequency condition and the phase margin condition. Thus, the following design criterion is proposed:

$$z_1 = a, \tag{6.36}$$

because this simplifies the open-loop transfer function shown in Fig. 6.34 as follows:

$$G(s)H(s) = k_d(s + z_2)\frac{k}{s^2}. \tag{6.37}$$

It must be stressed that the cancellation of factors $s + z_1$ and $s + a$ is valid because both of them represent poles and zeros with a negative real part because $a > 0$ and $z_1 > 0$. Performing some additional algebra in (6.37) the following is found:

$$G(s)H(s) = k_d z_2 \frac{s + z_2}{z_2}\frac{k}{s^2} = k_d z_2 \left(\frac{1}{z_2}s + 1\right)\frac{k}{s^2} = k_d z_2(z + 1)\frac{k}{s^2}, \tag{6.38}$$

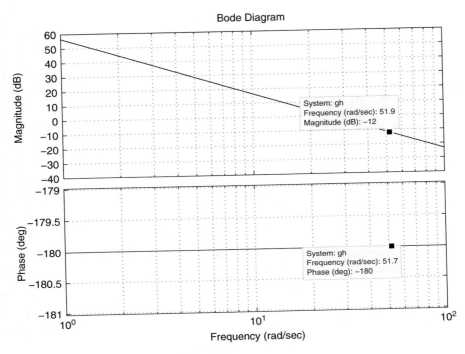

Fig. 6.72 Bode diagrams of the factor $\frac{k}{s^2}$ when $k = 675.4471$

where $z = \frac{1}{z_2}s$ has been defined. The factor $\frac{k}{s^2}$ now represents the uncompensated plant whose Bode diagrams are shown in Fig. 6.72 when employing the numerical values of the DC motor experimentally controlled in Chap. 11, i.e.,:

$$k = 675.4471.$$

It is observed that the magnitude and the phase are $-12[\mathrm{dB}]$ and $-180°$ respectively when the frequency is 51.8[rad/s]. According to Sect. 6.7.2, use of the following crossover frequency is proposed:

$$\omega_1 = 51.8[\mathrm{rad/s}], \tag{6.39}$$

and the following phase margin:

$$K_f = +47.5°, \tag{6.40}$$

to achieve a rise time $t_r = 0.0314[\mathrm{s}]$ and an overshoot 20%. Hence, the controller defined by $k_d z_2 (z + 1)$ must contribute a $+47.5°$ phase lead and $+12[\mathrm{dB}]$ as magnitude when the frequency is 51.8[rad/s]. It is observed in Fig. 6.73 that the

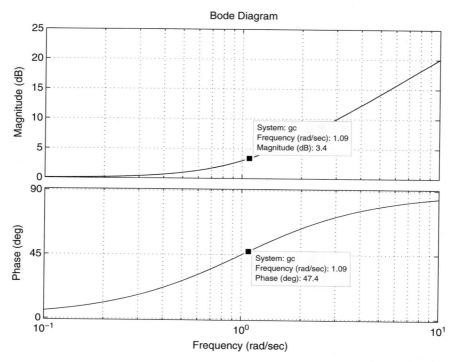

Fig. 6.73 Bode diagrams of the factor $z + 1$. The horizontal axis represents $\frac{1}{z_2}\omega$

factor $z + 1$ contributes a $+47.4°$ phase when $\frac{1}{z_2}\omega = 1.09$ (see Example 6.2). As this must occur when $\omega = 51.8[\text{rad/s}]$, then:

$$z_2 = \left.\frac{\omega}{1.09}\right|_{\omega=51.8} = 47.5229.$$

On the other hand, it is desired that the crossover frequency is $\omega = 51.8[\text{rad/s}]$. This requires the controller to contribute, at this frequency, the same magnitude but with opposite sign to the magnitude contributed by the factor $\frac{k}{s^2}$ at this frequency, i.e., $-12[\text{dB}]$, to render $0[\text{dB}]$ the magnitude of the compensated system at the new crossover frequency. Thus:

$$|k_d z_2(z + 1)|_{\text{dB}} = 20\log(k_d z_2) + |z + 1|_{\text{dB}} = 12[\text{dB}],$$

where, according to Fig. 6.73, $|z + 1|_{\text{dB}} = 3.4[\text{dB}]$ must be used. Hence, solving for k_d and performing the corresponding computations the following is found:

$$k_d = \frac{1}{z_2}10^{\frac{12-3.4}{20}} = 0.0566.$$

Finally, use of (6.35), (6.36) and $a = 2.8681$ (see Chap. 11) yields:

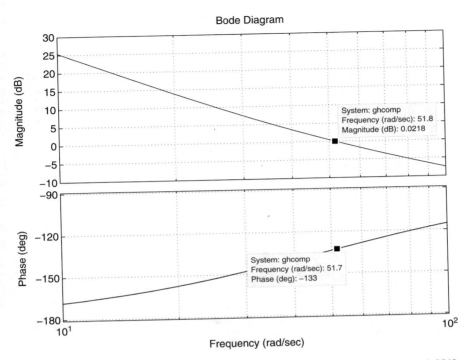

Fig. 6.74 Bode diagrams of the transfer function in (6.34), when $k_d = 0.0566$, $k_p = 2.8540$, $k_i = 7.7196$ $k = 675.4471$ and $a = 2.8681$

$$k_p = k_d(z_1 + z_2) = 2.8540, \quad k_i = k_d z_1 z_2 = 7.7196.$$

The Bode diagrams of the compensated system open-loop transfer function shown in (6.34) are presented in Fig. 6.74. Here, it can be verified that the crossover frequency is $\omega = 51.8[\text{rad/s}]$ and the phase at this frequency is $-133°$, which corresponds to a phase margin of $+47°$. This means that the frequency response specifications established in (6.39) and (6.40) are satisfied. On the other hand, in Fig. 6.75 the corresponding closed-loop time response is shown when the reference of the position is a unit step. There, a rise time of $0.0291[\text{s}]$ and an overshoot of 33% are measured. Although these values are not identical to $t_r = 0.0314[\text{s}]$ and $M_p = 20\%$, they are close enough. The reason for the difference between the desired overshoot and the achieved overshoot in addition to the way in which to improve it have been explained in Sect. 6.7.1.

Figures 6.72 to 6.75 were drawn executing the following MATLAB code in an m-file:

```
k=675.4471;
a=2.8681;
z1=a;
```

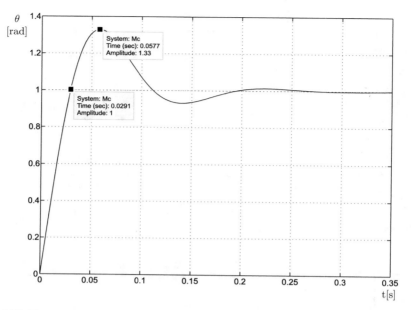

Fig. 6.75 Time response of the closed-loop system shown in Fig. 6.70, when $k_d = 0.0566$, $k_p = 2.8540$, $k_i = 7.7196$ $k = 675.4471$ and $a = 2.8681$

```
g=tf(k,[1 0 0]);
figure(1)
bode(g,{1,100})
grid on
gc=tf([1 1],1);
figure(2)
bode(gc,{0.1,10})
grid on
kp=2.854;
ki=7.7196;
kd=0.0566;
PID=tf([kd kp ki],[1 0]);
g=tf(k,[1 a 0]);
figure(3)
bode(PID*g,{10,100})
grid on
M=feedback(PID*g,1,-1);
figure(4)
step(M)
grid on
```

6.7.4 PI Velocity Control of a DC Motor

Consider the PI velocity control closed-loop system shown in (3.136), Sect. 3.8.4:

$$\ddot{\omega} + \left(\frac{b}{J} + \frac{n^2\, k_m\, k_e}{J\,R} + \frac{n\, k_m k_p}{J\,R} \right) \dot{\omega} + \frac{n\, k_m k_i}{J\,R}\omega = \frac{n\, k_m k_p}{J\,R}\dot{\omega}_d + \frac{n\, k_m k_i}{J\,R}\omega_d - \frac{1}{J}\dot{T}_p.$$

$$(6.41)$$

Define:

$$\alpha = \frac{b}{J} + \frac{n^2\, k_m\, k_e}{J\,R}, \qquad \beta = \frac{n\, k_m}{J\,R}.$$

Hence, (6.41) can be written as:

$$\ddot{\omega} + (\alpha + \beta k_p)\,\dot{\omega} + \beta k_i \omega = \beta k_p\,\dot{\omega}_d + \beta k_i \omega_d - \frac{1}{J}\dot{T}_p.$$

Suppose that ω_d and T_p are no longer step signals, but they are more elaborate functions of time. Adding and subtracting some terms in the previous expression, the following is found:

$$\frac{d^2}{dt^2}(\omega - \omega_d) + (\alpha + \beta k_p)\,\frac{d}{dt}(\omega - \omega_d) + \beta k_i(\omega - \omega_d) = -\ddot{\omega}_d - \alpha\,\dot{\omega}_d - \frac{1}{J}\dot{T}_p.$$

Defining the tracking error as $e = \omega - \omega_d$ yields:

$$\ddot{e} + (\alpha + \beta k_p)\,\dot{e} + \beta k_i e = -\ddot{\omega}_d - \alpha\,\dot{\omega}_d - \frac{1}{J}\dot{T}_p.$$

Using the Laplace transform, assuming all the initial conditions to be zero, and using superposition, the following is found:

$$E(s) = E_1(s) + E_2(s),$$

$$E_1(s) = G_1(s)\omega_d(s), \qquad G_1(s) = \frac{-s(s + \alpha)}{s^2 + (\alpha + \beta k_p)s + \beta k_i}, \qquad (6.42)$$

$$E_2(s) = G_2(s)T_p(s), \qquad G_2(s) = \frac{-\frac{1}{J}s}{s^2 + (\alpha + \beta k_p)s + \beta k_i}. \qquad (6.43)$$

Note that $G_1(s)$ and $G_2(s)$ are stable transfer functions if $\alpha + \beta k_p > 0$ and $\beta k_i > 0$. The magnitude Bode diagram of $G_1(s)$ is depicted in Fig. 6.76. Note that $\lim_{s \to \infty} G_1(s) = 1$; hence, the high-frequency magnitude of this Bode plot is 0[dB]. On the other hand, $G_1(s)$ has a zero at $s = 0$, thus, a +20[dB/dec] slope is observed as $\omega \to 0$. This slope increases to +40[dB/dec] when the zero at $s = -\alpha$

Fig. 6.76 Magnitude Bode diagram of $G_1(s)$ defined in (6.42)

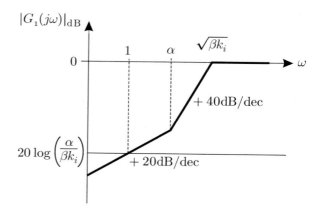

appears. As a complex conjugate pole pair appears at $\omega = \sqrt{\beta k_i}$, contributing with an additional slope of -40[dB/dec], any increment of the magnitude Bode plot is canceled out, remaining constant for $\omega \geq \sqrt{\beta k_i}$. Finally, by writing:

$$G_1(s) = \frac{-s(s + \alpha)}{s^2 + (\alpha + \beta k_p)s + \beta k_i} = -\frac{\alpha}{\beta k_i}s\frac{s + \alpha}{\alpha}\frac{\beta k_i}{s^2 + (\alpha + \beta k_p)s + \beta k_i},$$

the constant magnitude component $\frac{\alpha}{\beta k_i}$ is found.

From this Bode diagram, we conclude that the tracking error e is small if the integral gain is chosen to be large. We also see that, although this is valid for slow changing velocity profiles $\omega_d(t)$, i.e., for desired velocities with low-frequency components, choosing larger integral gains allows us to track velocity profiles with faster changes, i.e., desired velocities with higher-frequency components. This is because the corner frequency $\sqrt{\beta k_i}$ is shifted to the right. Moreover, to avoid a large resonant peak, the proportional gain must be chosen to be larger as the integral gain is increased.

Thus, we conclude that $\omega(t) \approx \omega_d(t)$ is accomplished as time grows, for an arbitrarily fast changing velocity profile $\omega_d(t)$, if k_p and k_i are chosen to be large enough. Note that a 0[dB] magnitude for high frequencies in Fig. 6.76 means that $|\omega - \omega_d| = |\omega_d|$, i.e., that $\omega(t) = 0$. This means that the motor remains stopped because it cannot respond when very fast changing velocity profiles are commanded. Note that velocity might be small, but it changes very fast.

The magnitude Bode diagram of $G_2(s)$ is depicted in Fig. 6.77. We observe that $G_2(s)$ behaves as a band-pass filter, attenuating small and large frequencies. This feature is due to zero at $s = 0$ and the second-order low-pass component that we find by decomposing $G_2(s)$ as:

$$G_2(s) = \frac{-\frac{1}{J}s}{s^2 + (\alpha + \beta k_p)s + \beta k_i} = -\frac{1}{J\beta k_i}s\frac{\beta k_i}{s^2 + (\alpha + \beta k_p)s + \beta k_i}.$$

Fig. 6.77 Magnitude Bode diagram of $G_2(s)$ defined in (6.43)

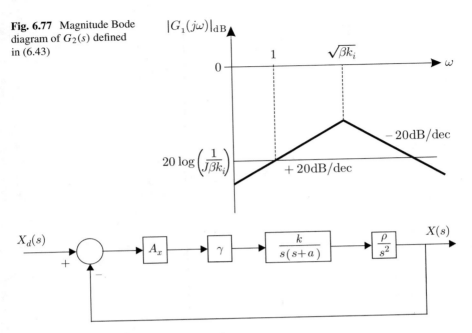

Fig. 6.78 Closed-loop system. First proposal

Note that, because of the factor $\frac{1}{J\beta k_i}$, attenuation for all frequencies is achieved as k_i is chosen to be large. However, k_p must also be large as k_i is large to avoid a large resonant peak. Thus, the effect of fast-changing external torque disturbances can be attenuated by using large values for both k_p and k_i.

6.7.5 A Ball and Beam System

This example constitutes an application to a popular prototype known as *ball and beam*. This prototype is built and experimentally controlled in Chap. 14.

6.7.5.1 Stability Analysis

Consider the block diagram in Fig. 6.78, where A_x, a, ρ, γ, and k are positive constants. The open-loop transfer function is given as:

$$G(s)H(s) = \frac{\rho A_x \gamma k}{a} \frac{a}{s+a} \frac{1}{s^3}. \tag{6.44}$$

Fig. 6.79 Modified Nyquist
path used in Sect. 6.7.5.1

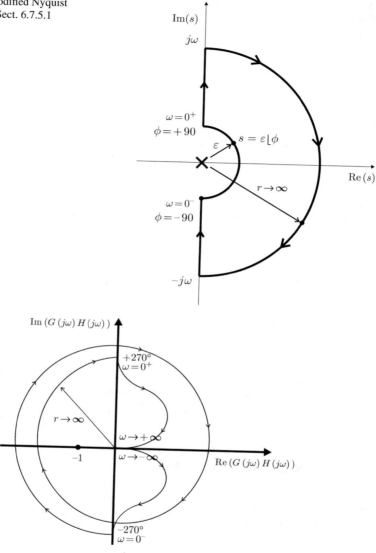

Fig. 6.80 Polar plot of $G(s)H(s)$ in (6.44) along the modified Nyquist path shown in Fig. 6.79

This open-loop transfer function has three poles at the origin, i.e., on the imaginary axis of the complex plane s. Hence, the modified Nyquist path shown in Fig. 6.79 must be used. The polar plot of $G(s)H(s)$, given in (6.44), obtained along the modified Nyquist trajectory is shown in Fig. 6.80. This is explained as follows.

The polar plot of (6.44) is depicted in Fig. 6.29. To take into account the contour around $s = 0$, included in the modified Nyquist path, proceed as follows. First, perform the variable change $s = \varepsilon \angle \phi$ in the transfer function in (6.44) to obtain:

$$G(s)H(s) = \frac{\rho A_x \gamma k}{\varepsilon \angle \phi + a} \frac{1}{\varepsilon^3} \angle - 3\phi.$$

As $\varepsilon \to 0$ is a very small constant value, it can be assumed that $a \gg \varepsilon$ to write:

$$G(s)H(s) = \frac{\rho A_x \gamma k}{a \varepsilon^3} \angle - 3\phi = \beta_1 \angle - 3\phi,$$

where $\beta_1 = \frac{\rho A_x \gamma k}{a \varepsilon^3} \to +\infty$ is a very large constant because $\varepsilon \to 0$. On the other hand, ϕ varies from $-90°$ (when $\omega = 0^-$) to $+90°$ (when $\omega = 0^+$). Then:

$$G(s)H(s) = \beta_1 \angle - 3\phi = \begin{cases} \beta_1 \angle + 270°, & \text{when } \omega = 0^- \\ \beta_1 \angle - 270°, & \text{when } \omega = 0^+ \end{cases}. \tag{6.45}$$

This explains the two circles with infinite radius appearing in Fig. 6.80, which, according to (6.45), are clockwise-oriented as ω passes from $\omega = 0^-$ to $\omega = 0^+$.

Note that $P = 0$ because $G(s)H(s)$, given in (6.44), has no zeros with a positive real part, and $N = 2$ because the polar plot in Fig. 6.80 encircles the point $(-1, j0)$ twice clockwise each time that the modified Nyquist path is completed once in the indicated orientation. Hence, the application of the Nyquist stability criterion yields $Z = N + P = 2$, which indicates that there are two closed-loop poles on the right half-plane and, thus, the closed-loop system in Fig. 6.78 is unstable. Moreover, as the circumferences in Fig. 6.80 have infinite radius, this result does not change if the following gain is chosen to be small:

$$\frac{\rho A_x \gamma k}{a}, \tag{6.46}$$

selecting, for instance, γ to be small. This is because β_1 tends toward infinity for any finite value in (6.46) as $\varepsilon \to 0$; hence, $P = 0$, $N = 2$, and $Z = 2$ still stand.

Now consider the block diagram in Fig. 6.81, where δ and c are positive constants. This block diagram is motivated by the intention to achieve closed-loop stability using a PD controller with the transfer function $(\gamma s + \gamma \delta)$. However, instead

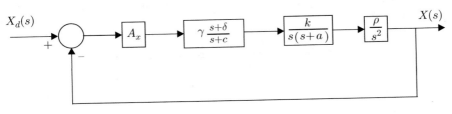

Fig. 6.81 Closed-loop system. Second proposal

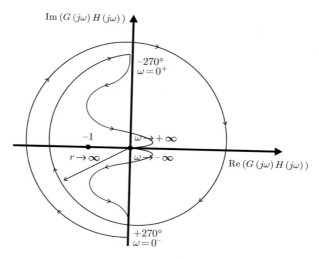

Fig. 6.82 Polar plot of $G(s)H(s)$ in (6.47) obtained along the modified Nyquist path shown in Fig. 6.79

of a PD controller a phase lead compensator with a transfer function $\gamma \frac{s+\delta}{s+c}$, where $\delta < c$, is proposed, as this class of controller attenuates the noise amplified by a PD controller. The open-loop transfer function is given in this case as:

$$G(s)H(s) = \frac{\rho A_x \gamma \delta k}{ac} \frac{s+\delta}{\delta} \frac{c}{s+c} \frac{a}{s+a} \frac{1}{s^3}. \tag{6.47}$$

This open-loop transfer function also has three poles at the origin, i.e., on the imaginary axis of the complex plane s; hence, the modified Nyquist path shown in Fig. 6.79 must be considered. The polar plot of $G(s)H(s)$, given in (6.47), obtained along the modified Nyquist path is shown in Fig. 6.82. This is explained as follows.

The polar plot of (6.47) is depicted in Fig. 6.32. To take into account the contour around $s = 0$ included in the modified Nyquist path, proceed as follows. First, perform the variable change $s = \varepsilon \angle \phi$ in the transfer function (6.47) to obtain:

$$G(s)H(s) = \rho A_x \gamma k \frac{\varepsilon \angle \phi + \delta}{(\varepsilon \angle \phi + c)(\varepsilon \angle \phi + a)} \frac{1}{\varepsilon^3} \angle - 3\phi.$$

As $\varepsilon \to 0$ is a very small constant value, it can be assumed that $a \gg \varepsilon$, $c \gg \varepsilon$ and $\delta \gg \varepsilon$ to write:

$$G(s)H(s) = \frac{\rho A_x \gamma k \delta}{ac\varepsilon^3} \angle - 3\phi = \beta_2 \angle - 3\phi,$$

where $\beta_2 = \frac{\rho A_x \gamma k \delta}{ac\varepsilon^3} \to +\infty$ is a very large constant because $\varepsilon \to 0$. On the other hand, ϕ varies from $-90°$ (when $\omega = 0^-$) to $+90°$ (when $\omega = 0^+$). Then:

$$G(s)H(s) = \beta_2 \angle - 3\phi = \begin{cases} \beta_2 \angle + 270°, & \text{when } \omega = 0^- \\ \beta_2 \angle - 270°, & \text{when } \omega = 0^+ \end{cases} \quad (6.48)$$

This explains the presence of the circles with infinite radius in Fig. 6.82, which, according to (6.48), are clockwise-oriented as ω passes from $\omega = 0^-$ to $\omega = 0^+$.

Note that, again, $P = 0$, $N = 2$ and $Z = N + P = 2$. Hence, there are two closed-loop poles with a positive real part and the closed-loop system is unstable. Furthermore, recalling the reasoning in the previous example, it is concluded that the instability cannot be eliminated by using small values for the constant:

$$\frac{\rho A_x \gamma k \delta}{ac}. \quad (6.49)$$

Hence, the use of a phase lead compensator is not enough to achieve closed-loop stability. Careful reasoning must be performed. Recall that, in Sect. 6.5, it was shown that the introduction of an open-loop pole (with a negative real part) results in a closed-loop system that approaches =instability. Moreover, this effect increases as that pole is placed closer to the origin. Now, observe that the transfer functions in (6.44) and (6.47) have three poles at the origin. Thus, it is reasonable to wonder whether the instability problems that we have found are due to the fact that the open-loop transfer function has too many poles at the origin.

Consider the block diagram in Fig. 6.83, where α, A_θ, k_v, b and c are positive constants. The internal loops around the transfer function $\frac{k}{s(s+a)}$ are intended to move one of poles at the origin to the left. Also note that the lead compensator is given as:

$$G_c(s) = \gamma \frac{s+b}{s+c}, \quad 0 < b < c. \quad (6.50)$$

The transfer function between points 1 and 2 in Fig. 6.83 is given as:

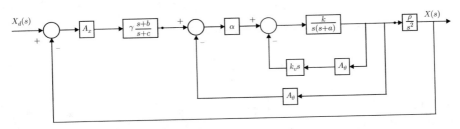

Fig. 6.83 Closed-loop system. Third proposal

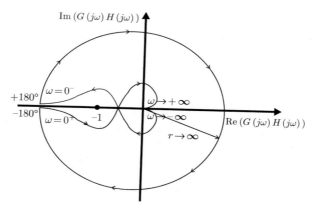

Fig. 6.84 Polar plot of $G(s)H(s)$ in (6.51) obtained on the modified Nyquist path shown in Fig. 6.79

$$G_m(s) = \frac{\alpha k}{s^2 + (a + k_v k A_\theta)s + \alpha k A_\theta}.$$

Hence, the open-loop transfer function is:

$$G(s)H(s) = \frac{A_x\,\rho\,\gamma b}{c A_\theta}\,\frac{s+b}{b}\,\frac{c}{s+c}\,\frac{\alpha k A_\theta}{s^2 + (a + k_v k A_\theta)s + \alpha k A_\theta}\,\frac{1}{s^2}. \quad (6.51)$$

Note that this open-loop transfer function has two poles at the origin, i.e., on the imaginary axis of the complex plane s; hence, the modified Nyquist path shown in Fig. 6.79 must be considered. In Fig. 6.84 the polar plot of $G(s)H(s)$ is shown, given in (6.51), obtained on the modified Nyquist path. This is explained as follows.

The polar plot of (6.51) is shown in Fig. 6.34. To take into account the contour around $s = 0$ included in the modified Nyquist path, proceed as follows. First, perform the variable change $s = \varepsilon \angle \phi$ in the transfer function (6.51) to obtain:

$$G(s)H(s) = A_x\,\rho\,\gamma\,\frac{\varepsilon\angle\phi + b}{\varepsilon\angle\phi + c}\,\frac{\alpha k}{\varepsilon^2\angle 2\phi + (a + k_v k A_\theta)\varepsilon\angle\phi + \alpha k A_\theta}\,\frac{1}{\varepsilon^2}\angle - 2\phi.$$

As $\varepsilon \to 0$ is a very small constant, it can be written:

$$G(s)H(s) = \frac{A_x\,\rho\,\gamma\,b}{c A_\theta \varepsilon^2}\,\angle - 2\phi,$$

where $\beta_3 = \frac{A_x\,\rho\,\gamma\,b}{c A_\theta \varepsilon^2} \to +\infty$ is a very large constant because $\varepsilon \to 0$. On the other hand, ϕ varies from $-90°$ (when $\omega = 0^-$) to $+90°$ (when $\omega = 0^+$). Then:

$$G(s)H(s) = \beta_3 \angle -2\phi = \begin{cases} \beta_3 \angle +180°, & \text{when } \omega = 0^- \\ \beta_3 \angle -180°, & \text{when } \omega = 0^+ \end{cases}. \qquad (6.52)$$

This explains the presence of the circumference with an infinite radius in Fig. 6.84, which, according to (6.52), is clockwise-oriented as ω passes from $\omega = 0^-$ to $\omega = 0^+$. Note that, now, $P = 0$, $N = 0$ and $Z = N + P = 0$. Thus, there is no closed-loop pole with a positive real part and the closed-loop system is stable.

Finally, note that the system type is 2, which ensures a zero steady-state error if either a step or a ramp reference is employed and any external disturbance is not assumed to appear. This explains why a controller with integral action, i.e., neither a PI nor a PID controller, is not considered at this point. See Sect. 14.5, Chap. 14, for a case where a PI controller is included in a ball and beam control system.

6.7.5.2 Design Using Bode Diagrams

Consider the following numerical values:

$$k = 10.7219, \quad a = 0.98, \quad A_\theta = 1, \qquad (6.53)$$

$$A_x = 1, \quad \rho = 4.7, \quad \alpha = 30, \quad k_v = 0.35.$$

The constants α and k_v are gains of the controller to be designed; hence, a criterion is required to select them. However, this criterion depends on some constraints present in practice, i.e., it is defined on the basis of the actual experimental prototype to be controlled. In Chap. 14, this criterion is defined and it is explained how the above values for α and k_v have been selected. The Bode diagrams of the following transfer function are shown in Fig. 6.85:

$$\frac{A_x \rho}{A_\theta} \frac{\alpha k A_\theta}{s^2 + (a + k_v k A_\theta)s + \alpha k A_\theta} \frac{1}{s^2}, \qquad (6.54)$$

which have been obtained using the values in (6.53). The phase is about $-182°$ when the magnitude is 0[dB] (when $\omega \approx 2.1[\text{rad/s}]$). This means that the phase margin is negative; hence, the closed-loop system obtained with (6.54) as the open-loop transfer function would be unstable. To stabilize the closed-loop system, the compensator shown in (6.50) is included, i.e., the open-loop transfer function of the compensated system is that presented in (6.51).

The compensator in (6.50) is designed by requiring $M_p(\%) = 20\%$. Then, using:

$$\zeta = \sqrt{\frac{\ln^2\left(\frac{M_p(\%)}{100}\right)}{\ln^2\left(\frac{M_p(\%)}{100}\right) + \pi^2}},$$

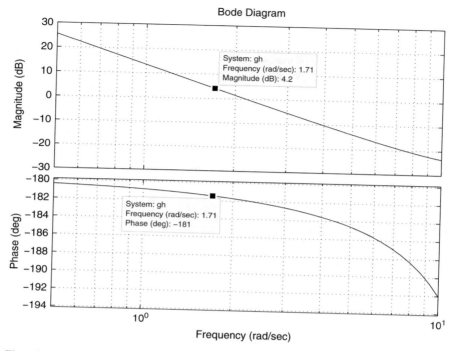

Fig. 6.85 Bode diagrams of the transfer function in (6.54)

$\zeta = 0.4559$ is found. Using Table 6.5 and linear interpolation, it is found that the following phase margin is required:

$$K_f = +47.5°.$$

After several essays, it is found that a suitable crossover frequency[3] is $\omega_1 = 1.71[\text{rad/s}]$. Note that:

- The uncompensated system has a phase of $-181°$ and a magnitude of $4.2[\text{dB}]$ at $\omega = 1.71[\text{rad/s}]$. This can be seen in Fig. 6.85.
- The compensator in (6.50) must contribute a $181° - 132.5° = 48.5°$ phase lead when $\omega = 1.71[\text{rad/s}]$, to achieve a $+47.5°$ phase margin, and $-4.2[\text{dB}]$ to render $0[\text{dB}]$ the magnitude of the compensated system when $\omega = 1.71[\text{rad/s}]$.

The Bode diagrams of the following transfer function are shown in Fig. 6.86 where the horizontal axis stands for $T\omega$ (see Sect. 6.2):

---

[3]The frequency where the compensated open-loop system must have a magnitude of $0[\text{dB}]$.

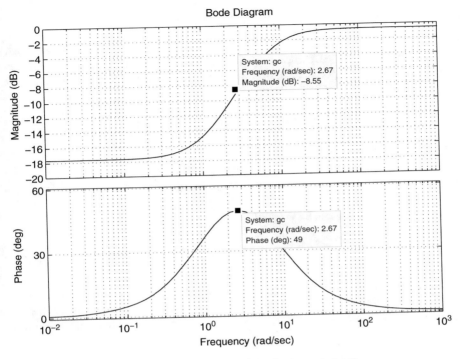

Fig. 6.86 Bode diagrams of $G_d(Ts)$ in (6.55). Horizontal axis stands for $T\omega$

$$G_d(Ts) = \frac{Ts + 1}{Ts + \frac{1}{\xi}}, \quad \xi = 0.14. \tag{6.55}$$

$\xi = 0.14$ is chosen because it is observed in Fig. 6.86 that this value forces the compensator in (6.55) to produce a maximal phase lead of $49° \approx 48.5°$. This occurs when $\omega T = 2.67$, which also contributes a magnitude of -8.55[dB]. The value for T is computed as:

$$T = \frac{2.67}{\omega_1} = \frac{2.67}{1.71} = 1.5614,$$

because $\omega = \omega_1 = 1.71$[rad/s] is the frequency where a $48.5°$ phase lead must be produced. Hence, (6.55) becomes:

$$G_d(s) = \frac{s + 0.6404}{s + 4.5746}.$$

Note that the compensator in (6.50) can be written as:

$$G_c(s) = \gamma G_d(s),   \tag{6.56}$$

if $b = 0.6404$ and $c = 4.5746$ are selected. The compensator $G_c(s)$ must contribute a magnitude of -4.2[dB] when $\omega = 1.71$[rad/s]. According to (6.56), this number is given as the sum, in decibels, of the magnitude of the transfer function $G_d(s)$ at $\omega = 1.71$[rad/s], i.e., -8.55[dB], and the gain γ in decibels:

$$-8.55[\text{dB}] + 20\log(\gamma) = -4.2[\text{dB}].$$

Solving for γ:

$$\gamma = 10^{\frac{8.55-4.2}{20}} = 1.6406.$$

Finally, the following can be written:

$$G_c(s) = 1.6406\frac{s + 0.6404}{s + 4.5746}.   \tag{6.57}$$

The Bode plots of the compensated system (6.51) using the controller (6.57) are presented in Fig. 6.87. Note that the crossover frequency is about $\omega = 1.71$[rad/s] and the phase margin is about $47.5°$, which corroborates that the design specifica-

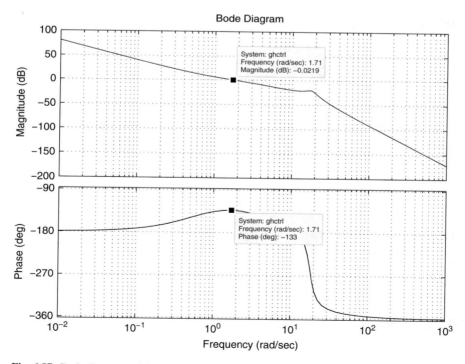

Fig. 6.87 Bode diagrams of the compensated ball and beam system

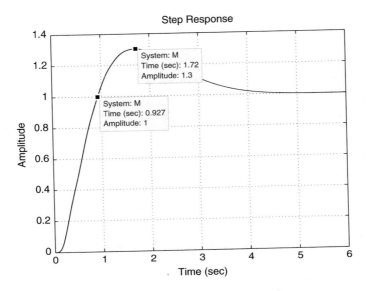

Fig. 6.88 Time response $x(t)$ of system in Fig. 6.83 when $x_d = 1[m]$

tions are satisfied. In Fig. 6.88 the time response of the closed-loop system shown in Fig. 6.83 is presented when a step reference is applied and (6.57), (6.53), are employed. Note that the overshoot is $M_p(\%) = 30\%$ instead of the desired value of 20%.

Recall that Table 6.5 has been employed to find the required phase margin to achieve a 20% overshoot and that this table is exact only for second-order closed-loop systems without zeros. Note that the closed-loop system that has been designed, see Fig. 6.83, is fifth order with a zero. This explains the difference between the desired 20% overshoot and the achieved 30% overshoot. On the other hand, the rise time is $t_r = 0.927[s]$. Also note that the system reaches a steady state in about 5[s], which is a good time for the ball and beam system that we present in Chap. 14.

Figures 6.85 to 6.88 are drawn using the following MATLAB code in an m-file:

```
k=10.7219;
a=0.98;
Ao=1;
Ax=1;
rho=4.7;
alpha=30;
kv=0.35;
g=tf(Ax*alpha*rho*k,[1 a+kv*k*Ao alpha*k*Ao 0 0]);
figure(1)
bode(g,{0.5,10})
grid on
xi=0.14;
```

```
gd=tf([1 1],[1 1/xi]);
figure(2)
bode(gd)
grid on
gc=tf(1.6406*[1 0.6404],[1 4.5746]);
figure(3)
bode(gc*g)
grid on
M=feedback(gc*g,1,-1);
figure(4)
step(M)
grid on
```

6.7.6 Analysis of a Non-minimum Phase System

Consider the following transfer function:

$$G(s)H(s) = \frac{116137\,k}{s^3 + 72.54s^2 - 1250s - 9.363 \times 10^4},$$ (6.58)

$$= \frac{116137\,k}{(s - 35.7377)(s + 36.5040)(s + 71.7721)},$$

$$= \frac{116137\,k}{(35.7377)(36.5040)(71.7721)}\frac{35.7377}{(s - 35.7377)}\frac{36.5040}{(s + 36.5040)}\frac{71.7721}{(s + 71.7721)}.$$

For this example, it is important to find the frequency response characteristics of the following transfer function:

$$G_1(s) = \frac{a}{s - a}, \quad a > 0.$$

Use of the variable change $s = j\omega$, yields:

$$G_1(j\omega) = \frac{a}{j\omega - a},$$

$$= \frac{a}{j\omega - a}\frac{-j\omega - a}{-j\omega - a},$$

$$= \frac{a(-j\omega - a)}{\omega^2 + a^2},$$

$$= \frac{a\sqrt{\omega^2 + a^2}}{\omega^2 + a^2} \angle \arctan\left(\frac{-\omega}{-a}\right),$$

$$= \frac{a}{\sqrt{\omega^2 + a^2}} \angle \arctan\left(\frac{-\omega}{-a}\right),$$

$$|G_1(j\omega)| = \frac{a}{\sqrt{\omega^2 + a^2}}, \quad \angle G_1(j\omega) = \arctan\left(\frac{-\omega}{-a}\right).$$

Thus:

$$\text{if } \omega = 0: |G_1(j\omega)| = 1, \quad \angle G_1(j\omega) = -180°,$$

$$\text{if } \omega \to \infty: |G_1(j\omega)| \to 0, \quad \angle G_1(j\omega) \to -90°,$$

$$\text{if } \omega = a: |G_1(j\omega)| = \frac{1}{\sqrt{2}}, \quad \angle G_1(j\omega) = -135°.$$

The Bode diagrams in Fig. 6.89 are obtained using this information and the results in Sect. 6.3. The polar plot in Fig. 6.90 is obtained using these Bode diagrams. It is stressed that it is also correct to say that:

$$\text{if } \omega = 0: |G_1(j\omega)| = 1, \quad \angle G_1(j\omega) = +180°,$$

$$\text{if } \omega \to \infty: |G_1(j\omega)| \to 0, \quad \angle G_1(j\omega) \to +270°,$$

$$\text{if } \omega = a: |G_1(j\omega)| = \frac{1}{\sqrt{2}}, \quad \angle G_1(j\omega) = +225°.$$

Fig. 6.89 Bode diagrams of $G(s)H(s)$ in (6.58):
$i)$ $\frac{116137\,k}{(35.7377)(36.5040)(71.7721)}$,
$ii)$ $\frac{35.7377}{(s-35.7377)}$,
$iii)$ $\frac{36.5040}{(s+36.5040)}$,
$iv)$ $\frac{71.7721}{(s+71.7721)}$, $v)$ $G(s)H(s)$

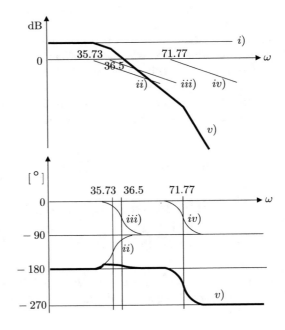

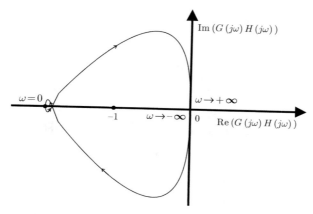

Fig. 6.90 Polar plot of $G(s)H(s)$ in (6.58)

As a consequence, in this case, the phase Bode diagram shifts upward 360° with respect to the phase Bode diagram shown in Fig. 6.89. However, the reader can verify that the polar plot in both cases is identical to that shown in Fig. 6.90. Thus, the analysis presented in the following to study the effect of a compensator is valid for both cases. The Bode diagrams in Fig. 6.89 and the polar plot in Fig. 6.90 can also be drawn using MATLAB, as shown in Figs. 6.91 and 6.92. Note that the phase lead due to the pole at $s = 35.73$, which is shown in Fig. 6.89, cannot be appreciated in Fig. 6.91 owing to the presence of the pole at $s = -36.50$.

From Figs. 6.90 and 6.92, note that the distance from any point on the polar plot to the origin is larger as k is larger. Hence, for small k, $N = 0$, and for large k, $N = 1$. The number of open-loop poles with a positive real part is $P = 1$. Then, according to the Nyquist criterion the number of unstable closed-loop poles is $Z = P + N > 0$. This means that the closed-loop system is unstable for any $k > 0$. In the following, it is shown that this problem can be solved if a suitable phase lead is included, i.e., if one zero is included in the open-loop transfer function.

Consider the following open-loop transfer function:

$$G(s)H(s) = k \frac{116137(s + b)}{s^3 + 72.54s^2 - 1250s - 9.363 \times 10^4},$$

$$= k \frac{116137(s + b)}{(s - 35.7377)(s + 36.5040)(s + 71.7721)},$$

$$= \frac{116137\,k\,b}{(35.7377)(36.5040)(71.7721)} \frac{35.7377}{(s - 35.7377)} \frac{36.5040}{(s + 36.5040)} \frac{71.7721}{(s + 71.7721)} \frac{s + b}{b},$$

(6.59)

where b is some positive constant. If $b < 35.7377$ is chosen, then the Bode diagrams in Fig. 6.93 are obtained in addition to the polar plot in Fig. 6.94. This means that,

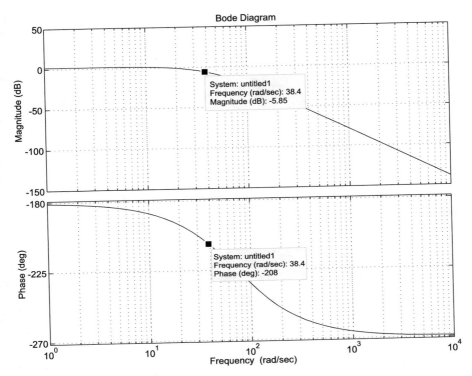

Fig. 6.91 Bode diagrams of $\frac{116137}{s^3+72.54s^2-1250s-9.363\times10^4}$

if k is large enough, then $N = -1$ and, as $P = 1$, then the closed-loop system is stable because $Z = P + N = 0$.

The Bode diagrams of the following transfer function are shown in Fig. 6.91:

$$\frac{116137}{s^3 + 72.54s^2 - 1250s - 9.363 \times 10^4}.$$

Note that the phase margin[4] being negative; hence, the closed-loop system is unstable. As explained above, closed-loop stability can be achieved if sufficient phase lead is produced by including the factor:

$$k(s + b). \tag{6.60}$$

---

[4] According to Sect. 6.5, the phase margin and the gain margin are defined for minimum phase systems and, although the system studied here is non-minimum phase, (because of the unstable pole at $s = 35.7377$), the term *phase margin* is used to indicate the comparison of the system phase with respect to the fundamental phase $-180°$.

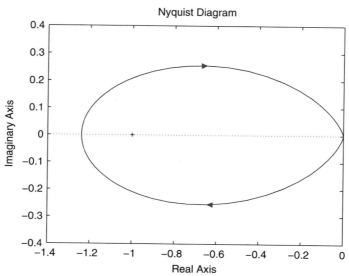

Fig. 6.92 Polar plot drawn with MATLAB for $G(s)H(s)$ in (6.58)

Fig. 6.93 Bode diagrams of $G(s)H(s)$ in (6.59):
$i)$ $\frac{116137\,k\,b}{(35.7377)(36.5040)(71.7721)}$,
$ii)$ $\frac{35.7377}{(s-35.7377)}$,
$iii)$ $\frac{36.5040}{(s+36.5040)}$,
$iv)$ $\frac{71.7721}{(s+71.7721)}$, $v)$ $G(s)H(s)$,
$vi)$ $\frac{s+b}{b}$, $b < 35.7377$

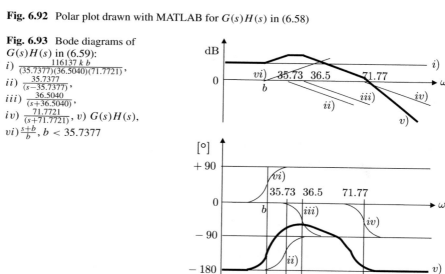

Also, the closed-loop time response can be rendered faster if a large enough crossover frequency is accomplished. These features can be achieved if k and b are suitably selected. Using $b = 31.2463$ and $k = 0.0396$ computed in Sect. 5.2.6, the following is obtained. For $\omega = 38.5$[rad/s], the magnitude and the phase are -5.85[dB] and $-208°$ respectively in Fig. 6.91. It is shown in Fig. 6.95 that, at the same frequency, the magnitude and the phase are 5.87[dB] and $50.9°$ for the factor in (6.60). It is shown in Fig. 6.96 that this allows the transfer function $G(s)H(s)$

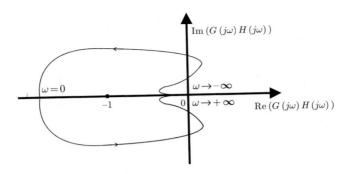

Fig. 6.94 Polar plot of $G(s)H(s)$ in (6.59)

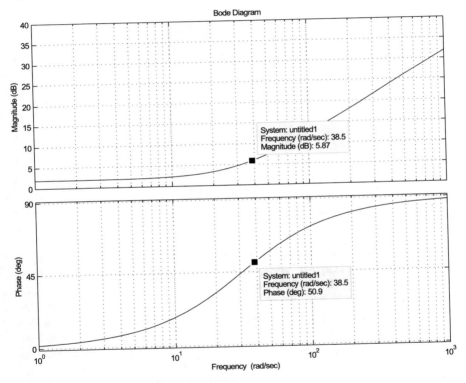

Fig. 6.95 Bode diagrams of $k(s + b)$, $b = 31.2463$, $k = 0.0396$

in (6.59) to have a 0[dB] magnitude and a $-157°$ phase when $\omega = 38.5$[rad/s]. The polar plot of $G(s)H(s)$ in (6.59) when $b = 31.2463$, $k = 0.0396$, is shown in Fig. 6.97. Note that the slight difference with respect to Fig. 6.94 is due to the fact that the phase in Fig. 6.96 is always less than $-90°$. It is noted that a $23°$ phase margin is achieved. According to Sect. 6.6.2, if the system were second-order

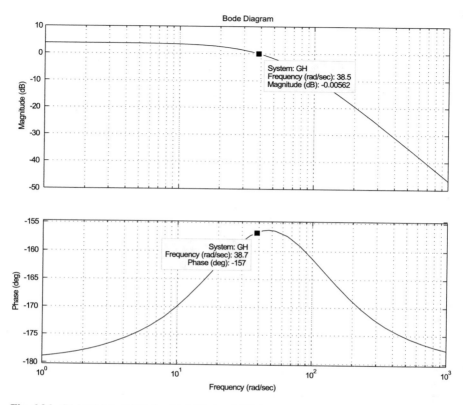

Fig. 6.96 $G(s)H(s)$ in (6.59), $b = 31.2463$, $k = 0.0396$

without zeros this would correspond to a closed-loop damping factor of 0.2, i.e., a large overshoot should appear. However, in Fig. 6.98 the closed-loop time response is shown when $G(s)H(s)$ in (6.59) with $b = 31.2463$, $k = 0.0396$, is placed in the direct path, unit-negative feedback is used and the reference is a unit step. Note that only a small overshoot is observed. This can be explained by recalling that $G(s)H(s)$ has no pole at $s = 0$ in this case. Hence, attention must be paid to these cases. Moreover, in Fig. 6.99, the Bode diagrams of the corresponding closed-loop transfer function are shown. It is observed that no resonant peak is observed; hence, it is concluded that only a small overshoot might appear, as verified in Fig. 6.98. Finally, the steady-state error is computed as $e_{ss} = \frac{A}{1+k_p}$, where $A = 1$ and $k_p = G(j\omega)H(j\omega)|_{\omega=0} \approx -1.5171$, because $|G(j\omega)H(j\omega)|_{\omega=0} \approx 3.62[\text{dB}]$, and $\angle G(j\omega)H(j\omega)|_{\omega=0} = -180°$, as observed in Fig. 6.96. Thus, $e_{ss} = -1.934$, which explains the final response of ≈ 2.87 in Fig. 6.98. A steady-state error that is different from zero is the reason why a zero reference was used in Sect. 5.2.6.

All of the figures in this example that have been drawn using MATLAB were executed using the following code in an m-file:

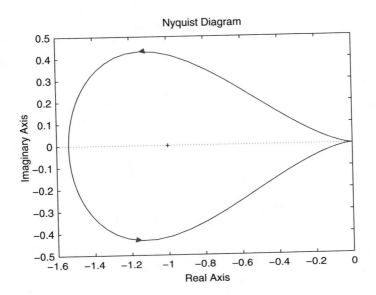

Fig. 6.97 Polar plot of $G(s)H(s)$ in (6.59) when $b = 31.2463$, $k = 0.0396$

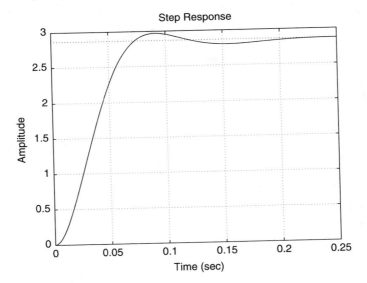

Fig. 6.98 Closed-loop time response when the reference is an unit step

```
gh=tf(116137,[1 72.54 -1250 -9.363e4]);
figure(1)
nyquist(gh)
figure(2)
```

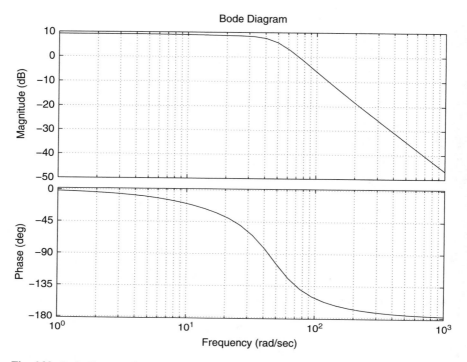

Fig. 6.99 Bode diagram of the closed-loop transfer function

```
bode(gh)
grid on
gc=tf(0.0396*[1 31.2463],1);
figure(3)
bode(gc)
grid on
figure(4)
bode(gh*gc)
grid on
figure(5)
nyquist(gh*gc)
M=feedback(gc*gh,1,-1)
figure(6)
step(M)
grid on
figure(7)
bode(M)
grid on
```

6.8 Case Study: PID Control of an Unstable Plant

Consider the following plant:

$$G(s) = \frac{11613700}{s^3 + 2739s^2 - 1250s - 3.536 \times 10^6}.$$

The characteristic polynomial of this transfer function can be written as $s^3 + 2739s^2 - 1250s - 3.536 \times 10^6 = (s + 35.9)(s - 35.9)(s + 2739)$. This plant is unstable because it has a pole with a positive real part located at $s = 35.9$. Thus, a controller with a derivative action must be proposed. On the other hand, this plant has no pole at $s = 0$. This means that, to achieve a zero steady-state error when the reference is a constant, a controller with an integral action must be proposed to render the system type 1. Thus, the following PID controller is proposed (see Sect. 6.7.3):

$$k_d \frac{s^2 + \frac{k_p}{k_d}s + \frac{k_i}{k_d}}{s} = k_d(s + z_1)(s + z_2)\frac{1}{s} = k_d(s^2 + (z_1 + z_2)s + z_1 z_2)\frac{1}{s},$$

$$\frac{k_p}{k_d} = z_1 + z_2, \quad \frac{k_i}{k_d} = z_1 z_2. \tag{6.61}$$

The open-loop transfer function becomes:

$$G(s)H(s) = k_d \frac{(s + z_1)(s + z_2)}{s} \frac{k}{s^3 + 2739s^2 - 1250s - 3.536 \times 10^6},$$

$$= k_d \frac{(s + z_1)(s + z_2)}{s} \frac{k}{(s + 35.9)(s - 35.9)(s + 2739)}, \tag{6.62}$$

$$k = 11613700.$$

A PID controller must be designed such that the closed-loop system is stable. As explained in Sect. 6.7.3, we can propose the following:

$$z_1 = 35.9, \tag{6.63}$$

and proceed to compute the two remaining PID controller gains such that the crossover frequency and the phase margin conditions are satisfied. Thus, the open-loop transfer function simplifies to:

$$G(s)H(s) = k_d(s + z_2)\frac{k}{s(s - 35.9)(s + 2739)},$$

or:

$$G(s)H(s) = k_d z_2(z + 1)\frac{k}{s(s - 35.9)(s + 2739)}, \tag{6.64}$$

where $z = \frac{1}{z_2}s$ has been defined. It is important to stress that the cancellation of the terms $s + z_1$ and $s + 35.9$ is valid because these terms represent poles and zeros with a negative real part as $35.9 > 0$ and $z_1 > 0$. In Fig. 6.100, the Bode diagrams of the transfer function:

$$G_0(s) = \frac{k}{s(s - 35.9)(s + 2739)}, \tag{6.65}$$

are depicted. There, the results in Sect. 6.7.6 are employed to obtain the Bode diagrams of a nonminimum phase factor of the form $\frac{a}{s-a}$ with $a > 0$. Using Fig. 6.100, the polar plot in Fig. 6.101 is obtained. It is important to stress that the transfer function in (6.65) has one pole at the origin; hence, the modified Nyquist path shown in Fig. 6.57 must be employed. According to this, a contour around $s = 0$ must be performed as follows. The variable change $s = \varepsilon \angle \phi$ must be used in (6.65):

$$G_0(\varepsilon \angle \phi) = \frac{k}{\varepsilon \angle \phi (\varepsilon \angle \phi - 35.9)(\varepsilon \angle \phi + 2739)} = \frac{k}{\varepsilon \angle \phi (-35.9)(2739)},$$

$$= \frac{k}{\varepsilon(-35.9)(2739)} \angle - \phi = \frac{k}{\varepsilon(35.9)(2739)} \angle (-\phi - 180°),$$

Fig. 6.100 Bode diagrams of (6.65). $i) \frac{k}{(35.9)(2739)}$, $ii) \frac{1}{s}$, $iii) \frac{35.9}{s-35.9}$, $iv) \frac{2739}{s+2739}$, $v) \frac{k}{s(s-35.9)(s+2739)}$

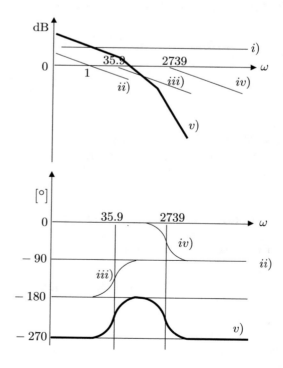

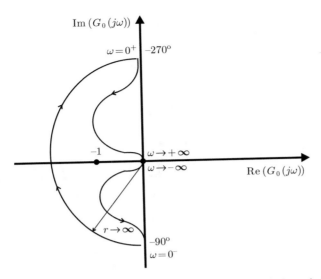

Fig. 6.101 Polar plot of (6.65) when mapping the modified Nyquist path shown in Fig. 6.57

recalling that $\varepsilon \to 0$. This explains the infinite radius circumference passing from $-90°$, when $\omega = 0^-$, to $-270°$, when $\omega = 0^+$ (see Fig. 6.57). It is clear in Fig. 6.101 that $N = 1$ and, as $P = 1$ (see (6.65)), then $Z = N + P = 2$, and the closed-loop system would be unstable. Because of this, to achieve closed-loop stability using the open-loop transfer function (6.64) (with a PID controller), the factor $k_d z_2(z + 1)$ must contribute enough phase lead to obtain a polar plot such as that shown in Fig. 6.102. In such a case $N = -1$ and $P = 1$, i.e., $Z = N + P = 0$ and the closed-loop system would be stable. The Bode diagrams of (6.65) are shown in Fig. 6.103. The magnitude and the phase are observed to be 0[dB] and $-212°$ respectively, when the frequency is 60[rad/s]. It is proposed to maintain the same crossover frequency:

$$\omega_1 = 60[\text{rad/s}], \tag{6.66}$$

and to obtain the following phase margin[5]:

$$K_f = +20°. \tag{6.67}$$

As the plant to control is naturally unstable, the main objective is to ensure closed-loop stability without defining any other specification, such as rise time or overshoot.

---

[5] Although the phase margin was introduced in Sect. 6.5 for minimum phase systems, this concept is also applicable in this example

Fig. 6.102 Polar plot of (6.64) when mapping the modified Nyquist path shown in Fig. 6.57

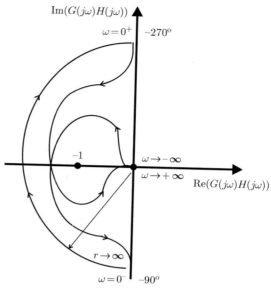

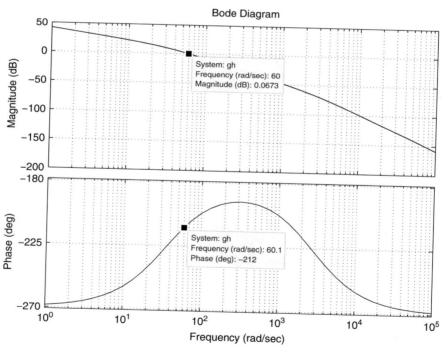

Fig. 6.103 Bode diagrams of (6.65)

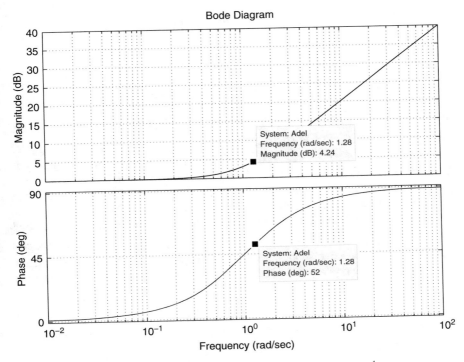

Fig. 6.104 Bode diagrams of the factor $z + 1$. The horizontal axis represents $\frac{1}{z_2}\omega$

Hence, the controller defined by $k_d z_2(z + 1)$ must contribute $+52°$ phase lead and 0[dB] magnitude when the frequency is 60[rad/s]. It is observed in Fig. 6.104 that the factor $z + 1$ contributes $+52°$ phase when $\frac{1}{z_2}\omega = 1.28$ (see Example 6.2). As this must occur when $\omega = 60$[rad/s], then:

$$z_2 = \left.\frac{\omega}{1.28}\right|_{\omega=60} = 46.875.$$

On the other hand, the desired crossover frequency is $\omega = 60$[rad/s]. This requires the controller to contribute, at this frequency, a magnitude of 0[dB] to achieve a magnitude 0[dB] for the compensated system at the crossover frequency. This implies that:

$$|k_d z_2(z + 1)|_{dB} = 20\log(k_d z_2) + |z + 1|_{dB} = 0[dB],$$

where, according to Fig. 6.104, $|z + 1|_{dB} = 4.24$[dB] must be used. Hence, solving for k_d and performing the corresponding computations:

$$k_d = \frac{1}{z_2} 10^{\frac{-4.24}{20}} = 0.0131.$$

Finally, using (6.61) and (6.63) the following is found:

$$k_p = k_d(z_1 + z_2) = 1.0838, \quad k_i = k_d z_1 z_2 = 22.0341.$$

The Bode diagrams of the open-loop transfer function of the compensated system, given in (6.62), i.e., $G(s)H(s) = \left(k_p + k_d s + \frac{k_i}{s}\right) \times \frac{k}{s^3 + 2739s^2 - 1250s - 3.536 \times 10^6})$ are shown in Fig. 6.105. There, it can be verified that the crossover frequency is $\omega = 60$[rad/s] and the phase at this frequency is $-160°$, which corresponds to a phase margin $+20°$. Thus, the desired frequency response specifications established in (6.66) and (6.67) have been accomplished. Finally, in Fig. 6.106, the time response is shown when the reference is a unit step and a loop is closed around $G(s)H(s) = \left(k_p + k_d s + \frac{k_i}{s}\right) \frac{k}{s^3 + 2739s^2 - 1250s - 3.536 \times 10^6}$. It is observed that the rise time is 0.0166[s] and overshoot is 89%. As stated before, no effort has been made to achieve particular values for either the rise time or the overshoot, i.e., the only

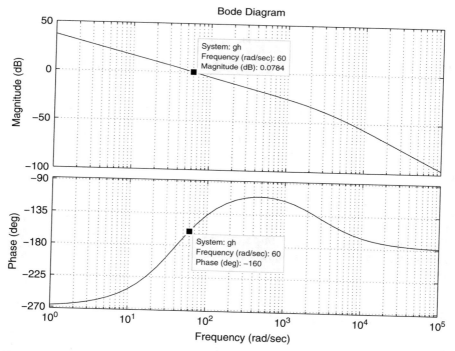

Fig. 6.105 Bode diagrams of the transfer function in (6.62), i.e., $G(s)H(s) = \left(k_p + k_d s + \frac{k_i}{s}\right) \frac{k}{s^3 + 2739s^2 - 1250s - 3.536 \times 10^6}$ with $k_d = 0.0131, k_p = 1.0838, k_i = 22.0341$

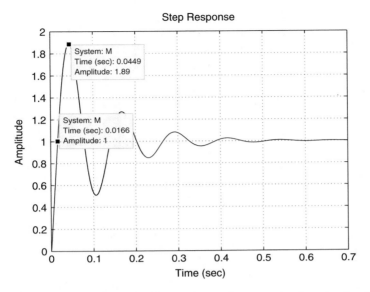

Fig. 6.106 Time response when the reference is a unit step and a loop is closed around $G(s)H(s) = \left(k_p + k_d s + \frac{k_i}{s}\right)\frac{k}{s^3+2739s^2-1250s-3.536\times10^6}$ with $k_d = 0.0131$, $k_p = 1.0838$, $k_i = 22.0341$

objective has been to ensure closed-loop stability. However, a way of improving the performance is presented in the following. At this point, the reader is advised to see Example 8.4, in Chap. 8.

The Bode diagrams shown in Fig. 6.103 are depicted again in Fig. 6.107. However, this time $\omega_1 = 344[\text{rad/s}]$ is selected as the new desired crossover frequency because the phase at this frequency, i.e., $-193°$, is maximal. Hence, it is expected that using the same phase lead as before, i.e., $52°$, a larger phase margin is accomplished. Thus, using the same data shown in Fig. 6.104 the following is found that:

$$z_2 = \frac{\omega}{1.28}\bigg|_{\omega=344} = 268.75$$

On the other hand, to render $\omega = 344[\text{rad/s}]$ to be the new crossover frequency, the magnitude of the controller is required to be $+29[\text{dB}]$ when $\omega = 344[\text{rad/s}]$. This means that

$$|k_d z_2(z+1)|_{dB} = 20\log(k_d z_2) + |z+1|_{dB} = 29[\text{dB}] \tag{6.68}$$

where, according to Fig. 6.104, $|z+1|_{dB} = 4.24[\text{dB}]$ must be used. Thus:

$$k_d = \frac{1}{z_2}10^{\frac{29-4.24}{20}} = 0.0644.$$

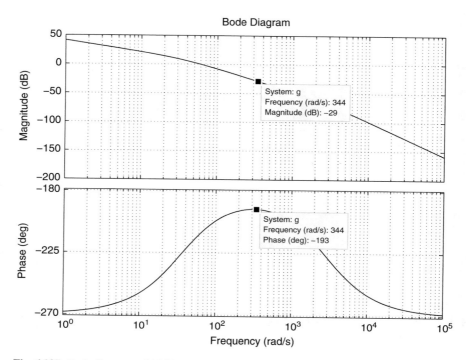

Fig. 6.107 Bode diagrams of (6.65)

Finally, using (6.61) and (6.63), the following is found:

$$k_p = k_d(z_1 + z_2) = 19.6089, \quad k_i = k_d z_1 z_2 = 621.0041.$$

The Bode diagrams of $G(s)H(s) = \left(k_p + k_d s + \frac{k_i}{s}\right) \frac{k}{s^3 + 2739s^2 - 1250s - 3.536 \times 10^6}$, when $k_d = 0.0644$, $k_p = 19.6089$, $k_i = 621.0041$, are depicted in Fig. 6.108 where it is verified that a crossover frequency of 344[rad/s] and a phase margin of $+39°$ are accomplished. Finally, the loop is closed around this open-loop transfer function and a step reference is applied. The corresponding time response is depicted in Fig. 6.109, where a 44% overshoot and a 0.004[s] rise time are observed. Thus, the performance of the closed-loop system has been improved, as expected, by rendering it faster and more damped.

Figures 6.103 to 6.109 were drawn by executing the following MATLAB code in an m-file:

```
k=11613700;
z1=35.9;
den1=conv([1 2739],[1 -39.5 0]);
g0=tf(k,den1);
figure(1)
```

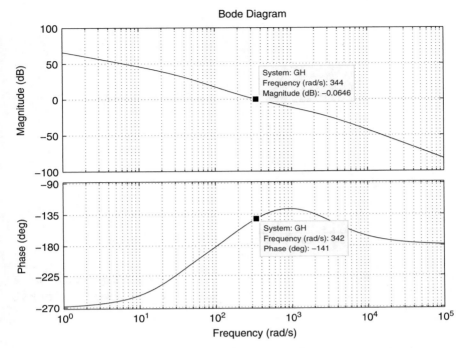

Fig. 6.108 Bode diagrams of $G(s)H(s) = \left(k_p + k_d s + \frac{k_i}{s}\right) \frac{k}{s^3+2739s^2-1250s-3.536\times10^6}$, when $k_d = 0.0644$, $k_p = 19.6089$, $k_i = 621.0041$

```
bode(g0,{1,1e5})
grid on
gc=tf([1 1],1);
figure(2)
bode(gc,{0.01,100})
grid on
kp=1.0838;
ki=22.0341;
kd=0.0131;
PID=tf([kd kp ki],[1 0]);
g=tf(k,[1 2739 -1250 -3.536e6]);
figure(3)
bode(PID*g,{1,1e5})
grid on
M=feedback(PID*g,1,-1);
figure(4)
step(M)
grid on
```

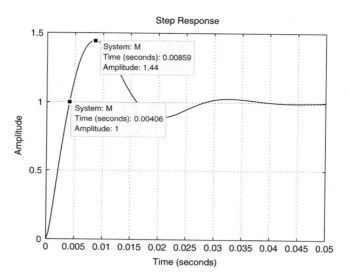

Fig. 6.109 Time response when the reference is an unit step and a loop is closed around $G(s)H(s) = \left(k_p + k_d s + \frac{k_i}{s}\right) \frac{k}{s^3 + 2739s^2 - 1250s - 3.536 \times 10^6}$, with $k_d = 0.0644$, $k_p = 19.6089$, $k_i = 621.0041$

```
kp=19.6089;
ki=621.0041;
kd=0.0644;
PID=tf([kd kp ki],[1 0]);
figure(5)
bode(PID*g,{1,1e5})
grid on
M=feedback(PID*g,1,-1);
figure(6)
step(M)
grid on
```

6.9   Summary

A fundamental result in the solution of linear differential equations is as follows. Given an arbitrary linear differential equation, if a sinusoidal function of time is applied at the input, then the forced response is also a sinusoidal function of time. Although the amplitudes and the phases of these sinusoidal functions of time may be different, they both have the same frequency. The relationship between these amplitudes and phases is given as functions of the frequency. These functions of the frequency are known as the system frequency response.

The fundamental property of the frequency response that is exploited by control theory is the fact that the above-cited functions of the frequency are determined by the poles and zeros of the corresponding transfer function. In open-loop systems, this allows us to interpret linear differential equations, and linear control systems, as filters. This means that both the response time and oscillations in a control system depend on the frequency content of the signal applied at the system input and how the control system amplifies or attenuates those frequency components: (*i*) The high-frequency components favor a fast response; (*ii*) If a band of frequencies is amplified, i.e., if there is a resonance peak, then the time response is oscillatory; (*iii*) The steady-state response depends on how the zero-frequency components are amplified or attenuated; (*iv*) The effects of the noise are small if the high-frequency components are attenuated.

An important feature of the frequency response approach to control system design is that it relates the open-loop frequency response characteristics, i.e., crossover frequency and stability margins, to the closed-loop time response characteristics such as rise time, time constant, and overshoot. Thus, a controller can be designed to suitably modify the open-loop system frequency response to accomplish the desired closed-loop time response specifications.

6.10 Review Questions

1. What do you understand by frequency response? What do the magnitude and the phase of a transfer function represent?
2. Transfer functions are composed of poles and zeros, Which of these components act as high-pass filters and which as low-pass filters? Which of them contribute with a positive phase and which contribute with a negative phase?
3. Why do open-loop zeros improve closed-loop stability? How do open-loop poles contribute to closed-loop stability? What is the advantage of adding poles in the open-loop transfer function?
4. What is the relationship between the Nyquist stability criterion and the stability margins?
5. Given an open-loop system, why does a larger corner frequency contribute to a faster time response?
6. How does the crossover frequency have to be modified to render the closed-loop system time response faster?
7. How does the phase margin affect the closed-loop system time response?
8. How does the Nyquist stability criterion have to be modified to be applied when some open-loop poles and/or zeros are located on the imaginary axis? What is the reason for this?
9. What is a non-minimum phase system?
10. What are the Nyquist path and the modified Nyquist path?

6.11 Exercises

1. Investigate how to compute the steady-state error for step, ramp, and parabola references from the open-loop system Bode diagrams.
2. Consider the integrator system:

$$y(t) = k \int_0^t u \, dt,$$

and the differentiator system:

$$y(t) = k\dot{u}(t).$$

a) In both cases, assume that $u = A\cos(\omega t)$, with A a constant, and compute $y(t)$. Draw a plot showing how the amplitude of $y(t)$ changes with respect to frequency and compare with the Bode diagrams in Figs. 6.16a and b.

b) What is the effect of a differentiator when u contains noise (i.e., high-frequency sinusoidal functions of time)? According to this, state what you think would happen in the following cases: What is the effect of a controller possessing a derivative action in a position control system containing high levels of noise? What is the effect of noise in a control system possessing only an integral action?

c) Consider the case of a pneumatic piston. The air input flow is u and the position of the piston is y. Suppose that air neither compresses nor expands inside the piston chamber. If the air flow leaves the piston chamber, then $u < 0$; if the air flow enters the piston chamber, then $u > 0$. Under these conditions, the system behaves as an integrator. Assume that an input of the form $u = A\cos(\omega t)$ is applied, with a value for ω that successively approaches zero. Using your everyday experience, explain what happens with the piston position and compare it with the Bode diagrams of an integrator and the plots obtained in the first item of this exercise. What does the term *infinite gain at zero frequency* mean? Although a permanent magnet brushed DC motor is not exactly an integrator, use your daily experience to investigate whether a similar behavior is obtained when y is the rotor angular position and $u = A\cos(\omega t)$ is the applied voltage. From the point of view of the motor model, how can this behavior be explained?

3. Consider the following system:

$$Y(s) = \frac{a}{s+a}U(s), \quad a > 0.$$

Suppose that $u(t)$ is a square wave, with a period equal to 2, which is equal to unity during the first half-period and zero during the second half-period.

Perform some simulations employing different values for a from almost zero to very large values. Observe the resulting waveform of $y(t)$ in each case. Why do you think $y(t)$ is very different from $u(t)$ when a is very close to zero? Why do you think $y(t)$ looks very similar to $u(t)$ as a becomes too large? Can you explain this behavior from the point of view of the frequency response of the transfer function $\frac{a}{s+a}$?

4. There is a fundamental result in linear systems theory stating that given a periodic function of time $u(t)$ and a linear system $Y(s) = G(s)U(s)$ with all poles of $G(s)$ having a strictly negative real part, then $y(t)$, in a steady state, is a periodic function of time with the same period as $u(t)$, but, possibly, with a different waveform [7], pp. 389 (see also Exercise 15, Chap. 3). Does the example in Exercise 3, in this chapter, suggest to you some explanation of why $u(t)$ and $y(t)$ may not have the same waveform? What is the main property that determines that the waveforms of $u(t)$ and $y(t)$ are similar or different?

5. Consider the following differential equation:

$$\ddot{y} + \omega_n^2 y = \omega_n^2 u.$$

The computation of $y(t)$ when $t \to \infty$ and $u(t) = A\sin(\omega t)$ is simplified using the Bode diagrams in Fig. 6.17b. With this information, draw some plots to represent the steady-state value of $y(t)$ for several values of ω from some $\omega = \omega_1 < \omega_n$ to some $\omega = \omega_2 > \omega_n$.

From this result, explain what happens when the same experiment is performed, but, now, employing the following differential equation:

$$\ddot{y} + 2\zeta\omega_n\dot{y} + \omega_n^2 y = \omega_n^2 u,$$

with $\omega_n > 0$ and $\zeta > 0$. What happens as ζ grows? Why is resonance considered to be dangerous in many applications?

6. Consider a closed-loop system such as that in Fig. 4.10. The corresponding expression for the error is given as $E(s) = \frac{1}{1+G(s)} R(s)$ where $E(s) = R(s) - C(s)$. If the desired reference is a sinusoidal function of time $r(t) = A\sin(\omega t)$, what can be done to render $\lim_{t\to\infty} e(t)$ increasingly smaller? Is it possible to accomplish $\lim_{t\to\infty} e(t) = 0$? If the answer is affirmative, under what conditions? Verify your answer by performing simulations.

7. Consider the parallel resistor–inductor–capacitor (RLC) circuit shown in Fig. 6.110. Show that the transfer function is given as:

$$\frac{V(s)}{I(s)} = \frac{k\omega_n^2 s}{s^2 + 2\zeta\omega_n s + \omega_n^2}, \tag{6.69}$$

where $\omega_n = \frac{1}{\sqrt{LC}}$, $\zeta = \frac{1}{2R}\sqrt{\frac{L}{C}}$, $k = L$. Plot the corresponding Bode diagrams when ζ is close to zero (how can this be possible in practice?) and verify that

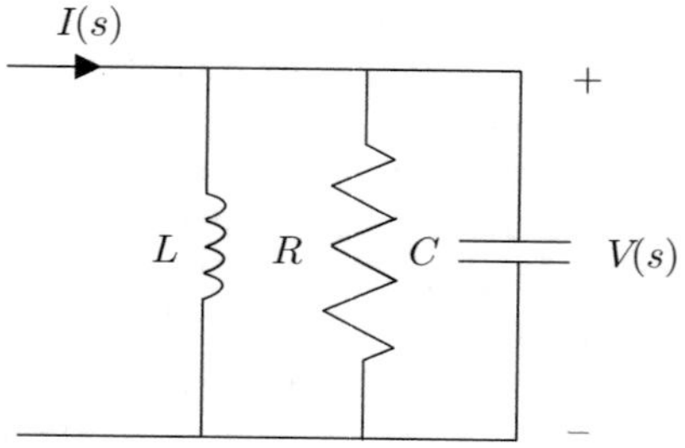

Fig. 6.110 Parallel
resistor–inductor–capacitor
(*RLC*) circuit

Fig. 6.111 Basic amplitude
modulation (AM)
radio-receiver

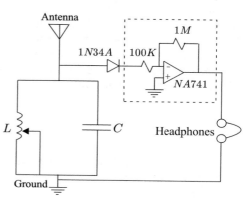

this is a band-pass filter around some frequency $\omega \approx \omega_n$ (in fact $\omega \rightarrow \omega_n$ as $\zeta \rightarrow 0$). How can ω_n be changed in practice?

An interesting and important application of this exercise is that of the amplitude modulation (AM) radio-receiver shown in Fig. 6.111 [8], pp. 94–99. Note that the antenna collects electric signals carried by the electromagnetic waves radiated by commercial AM broadcast stations. The antenna is merely a long wire. The longer the antenna, the stronger the electric signals at the radio-receiver input. Ground connection is performed using a copper water pipe.

The germanium diode 1N34A is the AM detector, i.e., it extracts the information carried by the electromagnetic waves. It is important that this diode is made of germanium because it is a semiconductor material with a smaller potential barrier than silicon. This renders it easier for the weak received signals to pass through the diode. Different broadcast AM stations can be tuned merely by changing the value of inductance L.

Can the reader explain how the tuning circuit works using frequency response arguments? Why do only the inductance L and the capacitance C appear, but not resistance R in Fig. 6.111? What signals in the radio-receiver have to be considered as $V(s)$ and $I(s)$ in (6.69)?

8. Consider the *RC* circuit in Fig. 6.112. If $v_i(t)$ is a step, find $v_o(t)$ and plot both of these variables. Is signal $v_o(t)$ continuous or discontinuous at $t = 0$? Obtain the transfer function $\frac{V_o(s)}{V_i(s)}$, plot the corresponding Bode diagrams, and state

Fig. 6.112 *RC* circuit

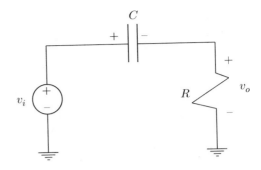

whether this is a high-pass or a low-pass filter. According to these results, how can the reader explain the continuity (or discontinuity) of $v_o(t)$ at $t = 0$ when $v_i(t)$ is a step?

9. It has been explained in this chapter how to plot Bode diagrams of a known transfer function. However, the inverse process is also possible: given some Bode diagrams, it is possible to obtain the corresponding transfer function. In this class of problems, the Bode diagrams are obtained experimentally and the process of finding the corresponding transfer function is called *system identification*. Some Bode diagrams that have been plotted from experimental data are shown in Fig. 6.113. Find the corresponding transfer function. Verify your answer by plotting the Bode diagrams of the identified transfer function and comparing them with those in Fig. 6.113.

10. In Exercise 7, Chap. 2, it is explained how to obtain the mathematical model of an analog DC voltmeter. If it is assumed that the inductance of the voltmeter is negligible ($L \approx 0$), the model reduces to that of a second-order rotative mass-spring-damper system. With this information, perform the following experiment.

    - Connect the analog voltmeter in DC mode.
    - Use a signal generator to apply a sinusoidal wave form. This signal must contain a constant DC component. The value of this DC component must be the same as half the scale of the DC voltmeter that has been selected. Furthermore, the maximal and the minimal values of the signal delivered must not exceed the range of the scale of the DC voltmeter that has been selected.
    - Connect the terminals of the DC voltmeter at the signal generator output. Apply a sinusoidal signal of frequency ω and measure v_{min} and v_{max} as the minimal and maximal values reached by the voltmeter needle. Employ a scope to measure the peak-to-peak voltage delivered by the signal generator and designate it as A. Use different values for the frequency ω and complete the following table (Table 6.6):
    - From these data, draw the Bode diagram of the corresponding magnitude and obtain the numerical values of the transfer function $G(s) = \frac{V_o(s)}{V_i(s)}$, where

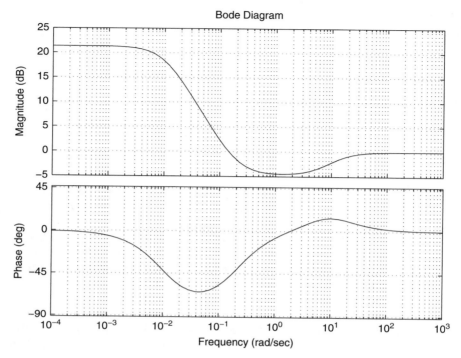

Fig. 6.113 Bode diagrams of a plant to identify

Table 6.6 Frequency response data of a DC voltmeter

ω, [rad/s]	$B = v_{max} - v_{min}$, [Volts]	A, [Volts]	$\frac{B}{A}$

$V_i(s)$ and $V_o(s)$ are the Laplace transforms of voltages delivered by the generator and measured by the DC voltmeter respectively. Note that this transfer function must have a unit gain at zero frequency because this is the frequency of a DC signal, i.e., when a DC voltmeter correctly measures voltage. Also note that the value of ω_n gives information on what is the maximal frequency where the DC voltmeter delivers a correct measurement. This part of the result is highly dependent on the ability of the scope and the DC voltmeter to deliver the same voltage measurements at low frequencies.

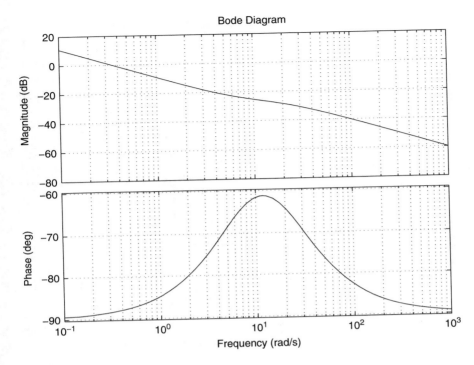

Fig. 6.114 Bode diagrams

11. Draw the Bode diagrams and the polar plots of the following systems:

$$G(s) = \frac{0.5(s + 0.8)}{(s + 0.1)(s + 20)}, \quad G(s) = \frac{0.5(s - 0.8)}{(s + 0.1)(s + 20)}.$$

12. The Bode diagrams of a plant are depicted in Fig. 6.114. Find the corresponding transfer function. Assume that a sine function with 5 as amplitude and 10[rad/s] as frequency is applied at the input of this system. State the amplitude and phase of the output signal. Now assume that the above plots are the Bode diagrams of an open-loop transfer function $G(s)$; hence, the same input signal that was used before is now used as the reference of the closed-loop system $\frac{G(s)}{1+G(s)}$. State the amplitude and the phase at the output.

13. Sketch the Bode diagrams for small frequencies, corresponding to systems type 0, type 1, and type 2. Perform the same with polar plots.

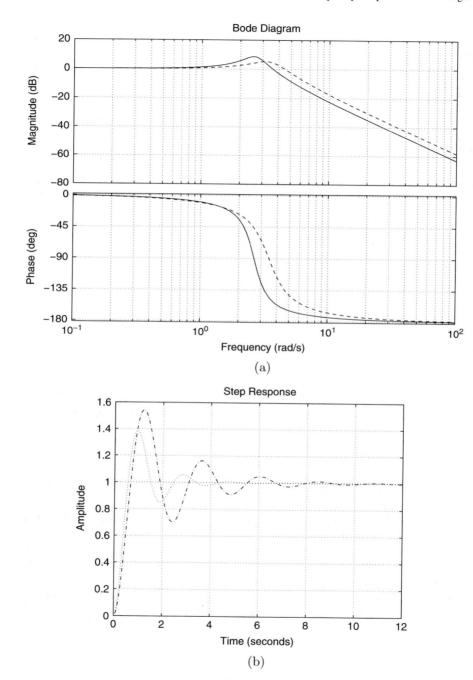

Fig. 6.115 (**a**) Bode diagrams and (**b**) time responses of two different systems

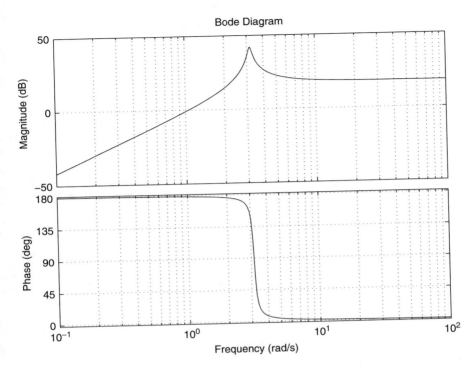

Fig. 6.116 Bode diagrams

14. The Bode diagrams and the time responses, when a unitary step input is applied, are depicted in Fig. 6.115 for two different open-loop systems. State which one of the step responses corresponds to each one of the Bode plots.

15. Of the Bode diagrams in Fig. 6.116, state which is the corresponding transfer function and sketch the corresponding time response when the input is a step.

16. The Bode diagrams of two different open-loop systems are depicted in Fig. 6.117a. The time responses in Fig. 6.117b are obtained when the loop is closed, with unit feedback around each one of the systems in Fig. 6.117a, and a step is applied as a reference. State which time response corresponds to each Bode diagram.

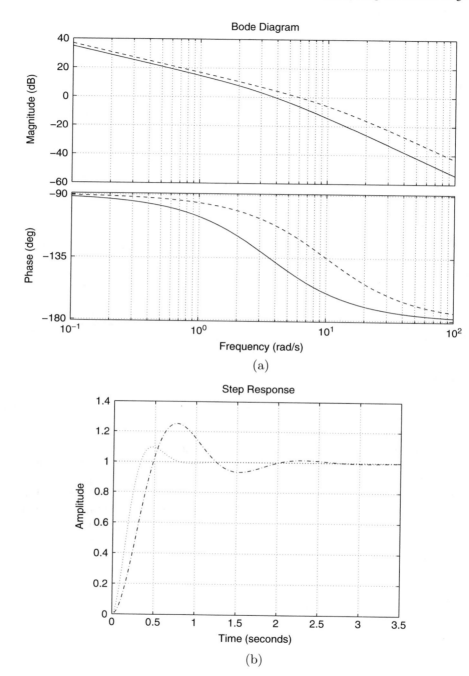

Fig. 6.117 (a) Open-loop Bode diagrams and (b) closed-loop time response to a step input of two different systems

References

1. E. Kreyszig, *Advanced engineering mathematics*, Vol. 1, 3rd. edition, John Wiley and Sons, New Jersey, 1972.
2. C. Ray Wylie, *Advanced engineering mathematics*, McGraw-Hill, New York, 1975.
3. G. F. Franklin, J. D. Powell, and M. L. Workman, *Digital control of dynamic systems*, Second edition, Addison-Wesley, USA, 1994.
4. H. P. Hsu, *Fourier analysis*, Simon & Schuster, New York, 1970.
5. K. Ogata, *Modern control engineering*, 4th edition, Prentice-Hall, Upper Saddle River, 2002.
6. N. S. Nise, *Control systems engineering*, 5th. edition, John Wiley and Sons, New Jersey, 2008.
7. C.-T. Chen, *Linear system theory and design*, Holt, Rinehart, and Winston, New York, 1984.
8. The Thomas Alva Edison Foundation, *The Thomas Edison book of easy and incredible experiments*, John Wiley and Sons, New Jersey, 1988.
9. R. C. Dorf and R. H. Bishop, *Modern control systems*, 10th edition, Pearson Prentice-Hall, 2005.
10. B. C. Kuo, *Automatic control systems*, Prentice-Hall, 1995.
11. G. H. Hostetter, C. J. Savant, and R. T. Stefani, *Design of feedback control systems*, Holt, Rinehart, and Winston, 1982.

Chapter 7
The State Variables Approach

The design methods presented in Chaps. 5 and 6 constitute what is now known as classical control. One of the features of classical control is that it relies on the so-called input–output approach, i.e., the use of a transfer function. This means that a transfer function is like a black box receiving an input and producing an output, i.e., nothing is known about what happens inside the black box. On the contrary, the state variable approach allows us to study what happens inside the system.

The study of the state variable approach is a complex subject involving the use of advanced mathematical tools. However, the aim in this chapter is not to present a detailed formal exposition of this approach, but merely to present the basic concepts allowing its application to control simple plants. Thus, only controllable and observable, single-input single-output plants are considered.

7.1 Definition of State Variables

Consider the following linear, constant coefficients differential equations:

$$L \frac{di}{dt} = u - R\,i - n\,k_e\,\dot{\theta}, \tag{7.1}$$

$$J\ddot{\theta} = -b\,\dot{\theta} + n\,k_m\,i. \tag{7.2}$$

Note that each one of these differential equations has an effect on the other. It is said that these differential equations must be solved simultaneously. One useful way of studying this class of differential equations is the state variables approach. The state variables are defined as follows.

© Springer International Publishing AG, part of Springer Nature 2019
V. M. Hernández-Guzmán, R. Silva-Ortigoza, *Automatic Control with Experiments*,
Advanced Textbooks in Control and Signal Processing,
https://doi.org/10.1007/978-3-319-75804-6_7

Definition 7.1 ([1], pp. 83) If the input $u(t)$ is known for all $t \geq t_0$, the state variables are the set of variables the knowledge of which at $t = t_0$ allows us to compute the solution of the differential equations for all $t \geq t_0$.

Note that this definition of the state variables is ambiguous. Although this may seem to be a drawback, it becomes an advantage because this allows us to select the state variables that are most convenient. According to this, although several criteria are proposed in the literature to select the state variables, it must be understood that no matter how they are selected they must be useful in solving the corresponding differential equations at any future time. In this chapter, the state variables are selected using the following criterion.

Criterion 7.1 *Identify the unknown variable in each one of the differential equations involved. Identify the order of each differential equation and designate it as r. Select the state variables as the unknown variable and its first r − 1 time derivatives in each one of the differential equations involved. It is also possible to select some of these variables multiplied by some nonzero constant.*

It is common to use n to designate the number of state variables. On the other hand, *the state* is a vector whose components are given by each one of the state variables. Hence, the state is a vector with n components. According to this criterion, the state variables selected for equations (7.1), (7.2) are i, θ, $\dot{\theta}$, i.e., $n = 3$ and the state is given as:

$$x = \begin{bmatrix} x_1 \\ x_2 \\ x_3 \end{bmatrix} = \begin{bmatrix} i \\ \theta \\ \dot{\theta} \end{bmatrix}.$$

Using this nomenclature, the differential equations in (7.1), (7.2), can be written as:

$$L\,\dot{x}_1 = u - R\,x_1 - n\,k_e\,x_3,$$
$$J\dot{x}_3 = -b\,x_3 + n\,k_m\,x_1,$$

i.e.,:

$$\frac{d}{dt} \begin{bmatrix} x_1 \\ x_2 \\ x_3 \end{bmatrix} = \begin{bmatrix} (u - R\,x_1 - n\,k_e\,x_3)/L \\ x_3 \\ (-b\,x_3 + n\,k_m\,x_1)/J \end{bmatrix},$$

, which can be rewritten in a compact form as:

$$\dot{x} = Ax + Bu,$$

$$A = \begin{bmatrix} -\frac{R}{L} & 0 & -\frac{n\,k_e}{L} \\ 0 & 0 & 1 \\ \frac{n\,k_m}{J} & 0 & -\frac{b}{J} \end{bmatrix}, \qquad B = \begin{bmatrix} \frac{1}{L} \\ 0 \\ 0 \end{bmatrix}.$$

On the other hand, it is common to use y to represent the output to be controlled. It must be stressed that the state x consists variables that are internal to the system; hence, in general, they cannot be measured. On the contrary, the output y is a variable that can always be measured. If the output is defined as the position, i.e., if $y = \theta = x_2$, then:

$$y = Cx, \quad C = [0\ 1\ 0].$$

The set of equations:

$$\dot{x} = Ax + Bu, \tag{7.3}$$

$$y = Cx, \tag{7.4}$$

is known as a linear, time invariant, dynamical equation (A, B, and C are constant matrices and vectors). The expression in (7.3) is known as the *state equation* and the expression in (7.4) is known as the *output equation*.

Although the state equation in (7.3) has been obtained from a single input differential equation, i.e., u is an scalar, it is very easy to extend (7.3) to the case where there are p inputs: it suffices to define u as a vector with p components, each one of them representing a different input, and B must be defined as a matrix with n rows and p columns. On the other hand, it is also very easy to extend the output equation in (7.4) to the case where there are q outputs: merely define y as a vector with q components, each one representing a different output, and C must be defined as a matrix with q rows and n columns.

Note that a state equation is composed of n first-order differential equations. A property of (7.3), (7.4), i.e., of a linear, time invariant dynamical equation, is that the right-hand side of each one of the differential equations involved in (7.3) is given as a linear combination of the state variables and the input using constant coefficients. Also note that the right-side of (7.4) is formed as the linear combination of the state variables using constant coefficients.

A fundamental property of linear, time invariant dynamical equations is that they satisfy the *superposition principle*, which is reasonable as this class of state equations is obtained, as explained above, from linear, constant coefficients differential equations, which, according to Sect. 3.7, also satisfy the superposition principle.

Example 7.1 Consider the mechanical system shown in Fig. 7.1. The corresponding mathematical model was obtained in Example 2.4, Chap. 2, and it is rewritten here for ease of reference:

$$\ddot{x}_1 + \frac{b}{m_1}(\dot{x}_1 - \dot{x}_2) + \frac{K_1}{m_1}x_1 + \frac{K_2}{m_1}(x_1 - x_2) = \frac{1}{m_1}F(t), \tag{7.5}$$

$$\ddot{x}_2 - \frac{b}{m_2}(\dot{x}_1 - \dot{x}_2) + \frac{K_3}{m_2}x_2 - \frac{K_2}{m_2}(x_1 - x_2) = 0. \tag{7.6}$$

Fig. 7.1 Mechanical system

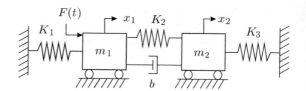

Note that there are two second-order differential equations. This means that only four state variables exist: the unknown variable in each one of these differential equations, i.e., x_1 and x_2, and the first time derivative of each one of these unknown variables, i.e., $\dot{x}_1$ and $\dot{x}_2$:

$$x = \begin{bmatrix} x_1 \\ x_2 \\ x_3 \\ x_4 \end{bmatrix} = \begin{bmatrix} x_1 \\ x_2 \\ \dot{x}_1 \\ \dot{x}_2 \end{bmatrix}, \quad u = F(t).$$

Using this nomenclature, the differential equations in (7.5), (7.6), can be rewritten as:

$$\ddot{x}_1 = \dot{x}_3 = -\frac{b}{m_1}(x_3 - x_4) - \frac{K_1}{m_1}x_1 - \frac{K_2}{m_1}(x_1 - x_2) + \frac{1}{m_1}u,$$

$$\ddot{x}_2 = \dot{x}_4 = \frac{b}{m_2}(x_3 - x_4) - \frac{K_3}{m_2}x_2 + \frac{K_2}{m_2}(x_1 - x_2),$$

i.e.,:

$$\frac{d}{dt}\begin{bmatrix} x_1 \\ x_2 \\ x_3 \\ x_4 \end{bmatrix} = \begin{bmatrix} x_3 \\ x_4 \\ -\frac{b}{m_1}(x_3 - x_4) - \frac{K_1}{m_1}x_1 - \frac{K_2}{m_1}(x_1 - x_2) + \frac{1}{m_1}u \\ \frac{b}{m_2}(x_3 - x_4) - \frac{K_3}{m_2}x_2 + \frac{K_2}{m_2}(x_1 - x_2) \end{bmatrix}. \qquad (7.7)$$

Hence, the following state equation is obtained:

$$\dot{x} = Ax + Bu,$$

$$A = \begin{bmatrix} 0 & 0 & 1 & 0 \\ 0 & 0 & 0 & 1 \\ -\frac{K_1}{m_1} - \frac{K_2}{m_1} & \frac{K_2}{m_1} & -\frac{b}{m_1} & \frac{b}{m_1} \\ \frac{K_2}{m_2} & -\frac{K_2}{m_2} - \frac{K_3}{m_2} & \frac{b}{m_2} & -\frac{b}{m_2} \end{bmatrix}, \quad B = \begin{bmatrix} 0 \\ 0 \\ \frac{1}{m_1} \\ 0 \end{bmatrix}. \qquad (7.8)$$

On the other hand, the output equation depends on the variable to be controlled. For instance, if it is desired to control the position of mass 1, then:

$$y = Cx = x_1, \quad C = [1\ 0\ 0\ 0],$$

but, if it is desired to control position of mass 2, then:

$$y = Cx = x_2, \quad C = [0\ 1\ 0\ 0].$$

Example 7.2 A mechanism known as ball and beam is studied in Chap. 14. In that chapter, the corresponding mathematical model is obtained and presented in (14.8). This model is rewritten here for ease of reference:

$$\frac{X(s)}{\theta(s)} = \frac{\rho}{s^2}, \quad \theta(s) = \frac{k}{s(s+a)} I^*(s).$$

Using the inverse Laplace transform, the corresponding differential equations can be obtained, i.e.,:

$$\ddot{x} = \rho\theta, \quad \ddot{\theta} + a\dot{\theta} = ki^*. \tag{7.9}$$

As there are two second-order differential equations, then there are four state variables: the unknown variable in each differential equation, i.e., x and θ, and the first time derivative of each unknown variable, i.e., $\dot{x}$ and $\dot{\theta}$:

$$z = \begin{bmatrix} z_1 \\ z_2 \\ z_3 \\ z_4 \end{bmatrix} = \begin{bmatrix} x \\ \dot{x} \\ \theta \\ \dot{\theta} \end{bmatrix}, \quad u = i^*.$$

Using this nomenclature, the differential equations in (7.9) can be rewritten as:

$$\ddot{x} = \dot{z}_2 = \rho z_3, \quad \ddot{\theta} = \dot{z}_4 = -az_4 + ku,$$

i.e.,:

$$\frac{d}{dt} \begin{bmatrix} z_1 \\ z_2 \\ z_3 \\ z_4 \end{bmatrix} = \begin{bmatrix} z_2 \\ \rho z_3 \\ z_4 \\ -az_4 + ku \end{bmatrix}. \tag{7.10}$$

Hence, the following state equation is obtained:

$$\dot{z} = Az + Bu$$

$$A = \begin{bmatrix} 0 & 1 & 0 & 0 \\ 0 & 0 & \rho & 0 \\ 0 & 0 & 0 & 1 \\ 0 & 0 & 0 & -a \end{bmatrix}, \quad B = \begin{bmatrix} 0 \\ 0 \\ 0 \\ k \end{bmatrix}. \tag{7.11}$$

As explained in Chap. 14, the variable to be controlled is x; thus, the output equation is given as:

$$y = Cz = x = z_1, \quad C = [1\ 0\ 0\ 0].$$

7.2   The Error Equation

Consider a linear state equation of the form:

$$\dot{x} = Ax + Bu.$$

The control objective for this plant can be stated by defining a constant desired state vector x^* and ensuring that $\lim_{t \to \infty} x(t) = x^*$. This means that $x(t)$ must converge to the constant x^*, which implies that $\lim_{t \to \infty} \dot{x}(t) = \dot{x}^* = 0$, i.e., the control problem has a solution if, and only if, there is a pair (x^*, u^*) such that:

$$\dot{x}^* = Ax^* + Bu^* = 0, \tag{7.12}$$

where $u^* = \lim_{t \to \infty} u(t)$ is also a constant. The pair (x^*, u^*) is known as the operation point. The error state vector is defined as $e = x - x^*$ and it is clear that the control objective can be restated as $\lim_{t \to \infty} e(t) = 0$. To analyze the evolution of $e(t)$, i.e., to find the conditions ensuring that the above limit is satisfied, it is important to have a state equation in terms of the state error. In the following examples, it is shown how to proceed to obtain this equation, which is known as the error equation.

Example 7.3 Consider the ball and beam system studied in Example 7.2. The system state equation is given in (7.11). However, to obtain the error equation, the expression in (7.10) is more convenient, i.e.,:

$$\frac{d}{dt} \begin{bmatrix} z_1 \\ z_2 \\ z_3 \\ z_4 \end{bmatrix} = \begin{bmatrix} z_2 \\ \rho z_3 \\ z_4 \\ -a z_4 + ku \end{bmatrix}.$$

Mimicking (7.12), to obtain the possible operation points, find all possible solutions of:

$$\begin{bmatrix} 0 \\ 0 \\ 0 \\ 0 \end{bmatrix} = \begin{bmatrix} z_2^* \\ \rho z_3^* \\ z_4^* \\ -a z_4^* + ku^* \end{bmatrix}.$$

It is clear that:

$$
z^* = \begin{bmatrix} z_1^* \\ z_2^* \\ z_3^* \\ z_4^* \end{bmatrix} = \begin{bmatrix} c \\ 0 \\ 0 \\ 0 \end{bmatrix}, \quad u^* = 0,
$$

where c is any real constant because no constraint is imposed on z_1^*. Once the operation points (z^*, u^*) are defined, the error state is given as:

$$
e = \begin{bmatrix} e_1 \\ e_2 \\ e_3 \\ e_4 \end{bmatrix} = z - z^* = \begin{bmatrix} z_1 - c \\ z_2 \\ z_3 \\ z_4 \end{bmatrix}. \tag{7.13}
$$

The error equation is obtained differentiating the error state once, i.e.,:

$$
\dot{e} = \begin{bmatrix} \dot{e}_1 \\ \dot{e}_2 \\ \dot{e}_3 \\ \dot{e}_4 \end{bmatrix} = \begin{bmatrix} \dot{z}_1 - \dot{z}_1^* \\ \dot{z}_2 - \dot{z}_2^* \\ \dot{z}_3 - \dot{z}_3^* \\ \dot{z}_4 - \dot{z}_4^* \end{bmatrix} = \begin{bmatrix} \dot{z}_1 \\ \dot{z}_2 \\ \dot{z}_3 \\ \dot{z}_4 \end{bmatrix} = \begin{bmatrix} z_2 \\ \rho z_3 \\ z_4 \\ -a z_4 + k u \end{bmatrix}.
$$

From the last expression, $u^* = 0$, and (7.13), we have:

$$
\dot{e} = \begin{bmatrix} e_2 \\ \rho e_3 \\ e_4 \\ -a e_4 + k(u - u^*) \end{bmatrix}.
$$

Thus, defining $v = u - u^*$, the error equation can be written as:

$$
\dot{e} = Ae + Bv,
$$

with matrix A and vector B defined in (7.11).

Example 7.4 Consider the mechanical system in Example 7.1. The corresponding state equation is given in (7.8). However, to obtain the error equation, it is preferable to consider (7.7), i.e.,:

$$
\frac{d}{dt}\begin{bmatrix} x_1 \\ x_2 \\ x_3 \\ x_4 \end{bmatrix} = \begin{bmatrix} x_3 \\ x_4 \\ -\frac{b}{m_1}(x_3 - x_4) - \frac{K_1}{m_1}x_1 - \frac{K_2}{m_1}(x_1 - x_2) + \frac{1}{m_1}u \\ \frac{b}{m_2}(x_3 - x_4) - \frac{K_3}{m_2}x_2 + \frac{K_2}{m_2}(x_1 - x_2) \end{bmatrix}.
$$

Mimicking (7.12), to obtain the possible operation points find all the possible solutions of:

$$
\begin{bmatrix} 0 \\ 0 \\ 0 \\ 0 \end{bmatrix} = \begin{bmatrix} x_3^* \\ x_4^* \\ -\frac{b}{m_1}(x_3^* - x_4^*) - \frac{K_1}{m_1}x_1^* - \frac{K_2}{m_1}(x_1^* - x_2^*) + \frac{1}{m_1}u^* \\ \frac{b}{m_2}(x_3^* - x_4^*) - \frac{K_3}{m_2}x_2^* + \frac{K_2}{m_2}(x_1^* - x_2^*) \end{bmatrix},
$$

i.e., $x_3^* = 0$, $x_4^* = 0$, and the solutions of:

$$
-\frac{K_1}{m_1}x_1^* - \frac{K_2}{m_1}(x_1^* - x_2^*) + \frac{1}{m_1}u^* = 0,
$$

$$
-\frac{K_3}{m_2}x_2^* + \frac{K_2}{m_2}(x_1^* - x_2^*) = 0.
$$

Solving these expressions, it is found that the operation points are given as:

$$
x^* = \begin{bmatrix} x_1^* \\ x_2^* \\ x_3^* \\ x_4^* \end{bmatrix} = \begin{bmatrix} \frac{K_2 + K_3}{K_2}x_2^* \\ c \\ 0 \\ 0 \end{bmatrix}, \quad u^* = \left(\frac{(K_1 + K_2)(K_2 + K_3)}{K_2} - K_2 \right) x_2^*,
$$

(7.14)

where c is any real constant. Once the operation points (x^*, u^*) are defined, the error state is given as:

$$
e = \begin{bmatrix} e_1 \\ e_2 \\ e_3 \\ e_4 \end{bmatrix} = x - x^* = \begin{bmatrix} x_1 - x_1^* \\ x_2 - x_2^* \\ x_3 \\ x_4 \end{bmatrix}.
$$

(7.15)

The error equation is obtained differentiating the error state once, i.e.,:

$$
\dot{e} = \begin{bmatrix} \dot{e}_1 \\ \dot{e}_2 \\ \dot{e}_3 \\ \dot{e}_4 \end{bmatrix} = \begin{bmatrix} \dot{x}_1 - \dot{x}_1^* \\ \dot{x}_2 - \dot{x}_2^* \\ \dot{x}_3 \\ \dot{x}_4 \end{bmatrix} = \begin{bmatrix} \dot{x}_1 \\ \dot{x}_2 \\ \dot{x}_3 \\ \dot{x}_4 \end{bmatrix}
$$

$$
= \begin{bmatrix} x_3 \\ x_4 \\ -\frac{b}{m_1}(x_3 - x_4) - \frac{K_1}{m_1}x_1 - \frac{K_2}{m_1}(x_1 - x_2) + \frac{1}{m_1}u \\ \frac{b}{m_2}(x_3 - x_4) - \frac{K_3}{m_2}x_2 + \frac{K_2}{m_2}(x_1 - x_2) \end{bmatrix}.
$$

Adding and subtracting terms $\frac{K_1}{m_1}x_1^*$, $\frac{K_2}{m_1}x_2^*$, $\frac{K_2}{m_1}x_1^*$ and $\frac{K_3}{m_2}x_2^*$, $\frac{K_2}{m_2}x_2^*$, $\frac{K_2}{m_2}x_1^*$ in the third row and fourth row respectively of the last expression and using (7.14), (7.15), yields:

$$
\dot{e} = \begin{bmatrix} e_3 \\ e_4 \\ -\frac{b}{m_1}(e_3 - e_4) - \frac{K_1}{m_1}e_1 - \frac{K_2}{m_1}(e_1 - e_2) + \frac{1}{m_1}(u - u^*) \\ \frac{b}{m_2}(e_3 - e_4) - \frac{K_3}{m_2}e_2 + \frac{K_2}{m_2}(e_1 - e_2) \end{bmatrix}.
$$

Thus, defining $v = u - u^*$, the error equation can be written as:

$$
\dot{e} = Ae + Bv,
$$

with A and B defined in (7.8).

7.3 Approximate Linearization of Nonlinear State Equations

Nonlinear state equations are frequently found. These are state equations given as:

$$
\dot{x} = f(x, u), \tag{7.16}
$$

$$
f = \begin{bmatrix} f_1(x, u) \\ f_2(x, u) \\ \vdots \\ f_n(x, u) \end{bmatrix}, \quad x = \begin{bmatrix} x_1 \\ x_2 \\ \vdots \\ x_n \end{bmatrix}, \quad u = \begin{bmatrix} u_1 \\ u_2 \\ \vdots \\ u_p \end{bmatrix},
$$

where at least one the of functions $f_i(x, u)$, $i = 1, \ldots, n$, is a scalar nonlinear function of x or u, i.e., this function cannot be written as a linear combination of the components of x or u. For instance, in a state equation with three state variables and two inputs:

$$
f_i(x, u) = 3\sin(x_3) + x_1 u_2,
$$

$$
f_i(x, u) = \frac{e^{x_1}}{5x_2},
$$

$$
f_i(x, u) = u_1^2.
$$

It is important to stress that the notation $f_i(x, u)$ means that the i-th component of vector f is in general a function of the components of the state vector x and the input vector u. However, as shown in the previous examples, it is not necessary for $f_i(x, u)$ to be a function of all the state variables and all the inputs. Moreover, it is also possible that $f_i(x, u)$ is not a function of any state variable, but it is a function of some of the inputs and vice versa. A state equation of the form (7.16) possessing the above features is known as a nonlinear time invariant state equation, i.e., it has no explicit dependence on time.

The study of nonlinear state equations is much more complex than the study of linear state equations. However, most physical systems are inherently nonlinear, i.e., their models have the form given in (7.16). Hence, suitable analysis and design methodologies must be developed for this class of state equations. One of the simplest methods is to find a linear state equation, such as that in (7.3), which is approximately equivalent to (7.16). The advantage of this method is that the linear state equations such as that in (7.3) can be studied much more easily. Hence, the analysis and design of controllers for (7.16) are performed on the basis of the much simpler model in (7.3). Although this method is very useful, it must be stressed that its main drawback is that the results obtained are only valid in a small region around the point where the linear approximation is performed. These ideas are clarified in the following sections.

7.3.1 Procedure for First-order State Equations Without Input

Consider a nonlinear first-order differential equation without input:

$$\dot{x} = f(x), \tag{7.17}$$

where x is a scalar and $f(x)$ is a scalar nonlinear function of x. It is desired to obtain a linear differential equation representing (7.17), at least approximately. Define an *equilibrium point* x^* of (7.17) as that value of x satisfying $f(x^*) = 0$. According to (7.17), this means that at an equilibrium point $\dot{x}^* = f(x^*) = 0$, i.e., the system can remain at *rest* at that point forever.

Suppose that $f(x)$ is represented by the curve shown in Fig. 7.2. The dashed straight line $h(x)$ is tangent to $f(x)$ at the point x^*. Because of this feature, $h(x)$ and $f(x)$ are approximately equal for values of x that are close to x^*, i.e., if $x - x^* \approx 0$. However, the difference between $h(x)$ and $f(x)$ increases as x and x^* become far apart, i.e., for large values of $x - x^*$. Hence, $h(x)$ can be employed instead of $f(x)$ if x is constrained to only take values that are close to x^*. In Fig. 7.2, it is observed that:

$$m = \frac{h(x) - f(x^*)}{x - x^*}, \quad m = \left.\frac{df(x)}{dx}\right|_{x=x^*},$$

where the constant m represents the slope of $h(x)$, i.e., the derivative of $f(x)$ evaluated at the equilibrium point x^*. Then, from the first expression:

$$h(x) = m(x - x^*) + f(x^*), \tag{7.18}$$

Fig. 7.2 Approximation of a nonlinear function $f(x)$ using a straight line $h(x)$

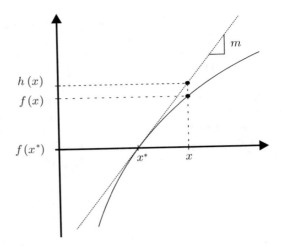

and, as $h(x) \approx f(x)$, it is possible to write:

$$f(x) \approx m(x - x^*) + f(x^*).$$

As x^* is an equilibrium point, then $f(x^*) = 0$ and:

$$f(x) \approx m(x - x^*).$$

Hence, (7.17) can be approximated by:

$$\dot{x} = m(x - x^*).$$

Defining the new variable $z = x - x^*$, then $\dot{z} = \dot{x}$, because $\dot{x}^* = 0$. Hence, the linear approximation of (7.17) is given by:

$$\dot{z} = m\,z, \tag{7.19}$$

, which is valid only if $x - x^* \approx 0$. Once (7.19) is employed to compute z, the definition $z = x - x^*$ can be used to obtain x as the equilibrium point x^* is known. Finally, note that the only requirement for all the above to stand is that $\frac{df(x)}{dx}$ exists and is continuous, i.e., that $f(x)$ is, at least once, continuously differentiable.

7.3.2 General Procedure for Arbitrary Order State Equations with an Arbitrary Number of Inputs

Now consider the state equation:

$$\dot{x} = f(x, u), \tag{7.20}$$

where $x \in R^n$ is a vector with n components and $u \in R^p$ is a vector with p components standing for the state equation input. Note that $f(x, u)$ must be a vectorial function with n components. This means that (7.20) has the form defined in (7.16). In this case, it is preferable to employ the concept of *operation point* instead of *equilibrium point* because the latter concept is only defined for state equations without inputs. An *operation point* is defined as the pair (x^*, u^*) satisfying $f(x^*, u^*) = 0$. This means that the solution of the state equation can remain at *rest* at x^*, because $\dot{x}^* = f(x^*, u^*) = 0$, if the suitable inputs u^* (p inputs) are applied, which are also constant.

Note that $f(x, u)$ depends on $n + p$ variables. Define the following vector:

$$y = \begin{bmatrix} x \\ u \end{bmatrix},$$

with $n + p$ components. Then (7.20) can be written as:

$$\dot{x} = f(y). \tag{7.21}$$

Following the ideas in the previous section, it is desired to approximate the nonlinear function $f(y)$ by a function $h(y)$ that is tangent to $f(y)$ at the operation point y^* defined as:

$$y^* = \begin{bmatrix} x^* \\ u^* \end{bmatrix}.$$

Extending the expression in (7.18) to the case of $n + p$ variables, it is possible to write:

$$h(y) = M(y - y^*) + f(y^*),$$

where $M = \left. \frac{\partial f(y)}{\partial y} \right|_{y=y^*}$ is a constant matrix defined as:

$$M = \begin{bmatrix} \frac{\partial f_1(x,u)}{\partial x_1} & \frac{\partial f_1(x,u)}{\partial x_2} & \cdots & \frac{\partial f_1(x,u)}{\partial x_n} & \frac{\partial f_1(x,u)}{\partial u_1} & \frac{\partial f_1(x,u)}{\partial u_2} & \cdots & \frac{\partial f_1(x,u)}{\partial u_p} \\ \frac{\partial f_2(x,u)}{\partial x_1} & \frac{\partial f_2(x,u)}{\partial x_2} & \cdots & \frac{\partial f_2(x,u)}{\partial x_n} & \frac{\partial f_2(x,u)}{\partial u_1} & \frac{\partial f_2(x,u)}{\partial u_2} & \cdots & \frac{\partial f_2(x,u)}{\partial u_p} \\ \vdots & \vdots & \ddots & \vdots & \vdots & \vdots & \ddots & \vdots \\ \frac{\partial f_n(x,u)}{\partial x_1} & \frac{\partial f_n(x,u)}{\partial x_2} & \cdots & \frac{\partial f_n(x,u)}{\partial x_n} & \frac{\partial f_n(x,u)}{\partial u_1} & \frac{\partial f_n(x,u)}{\partial u_2} & \cdots & \frac{\partial f_n(x,u)}{\partial u_p} \end{bmatrix}_{x=x^*, u=u^*}.$$

Taking advantage of $f(y) \approx h(y)$, (7.21) can be approximated as:

$$\dot{x} = M(y - y^*) + f(y^*).$$

As $f(y^*) = f(x^*, u^*) = 0$ and $\dot{x} - \dot{x}^* = \dot{x}$, because $\dot{x}^* = 0$, the following is found:

$$\dot{z} = Az + Bv, \tag{7.22}$$

$$A = \begin{bmatrix} \frac{\partial f_1(x,u)}{\partial x_1} & \frac{\partial f_1(x,u)}{\partial x_2} & \cdots & \frac{\partial f_1(x,u)}{\partial x_n} \\ \frac{\partial f_2(x,u)}{\partial x_1} & \frac{\partial f_2(x,u)}{\partial x_2} & \cdots & \frac{\partial f_2(x,u)}{\partial x_n} \\ \vdots & \vdots & \ddots & \vdots \\ \frac{\partial f_n(x,u)}{\partial x_1} & \frac{\partial f_n(x,u)}{\partial x_2} & \cdots & \frac{\partial f_n(x,u)}{\partial x_n} \end{bmatrix}_{x=x^*,u=u^*}, \tag{7.23}$$

$$B = \begin{bmatrix} \frac{\partial f_1(x,u)}{\partial u_1} & \frac{\partial f_1(x,u)}{\partial u_2} & \cdots & \frac{\partial f_1(x,u)}{\partial u_p} \\ \frac{\partial f_2(x,u)}{\partial u_1} & \frac{\partial f_2(x,u)}{\partial u_2} & \cdots & \frac{\partial f_2(x,u)}{\partial u_p} \\ \vdots & \vdots & \ddots & \vdots \\ \frac{\partial f_n(x,u)}{\partial u_1} & \frac{\partial f_n(x,u)}{\partial u_2} & \cdots & \frac{\partial f_n(x,u)}{\partial u_p} \end{bmatrix}_{x=x^*,u=u^*}, \tag{7.24}$$

where $z = x - x^*$ and $v = u - u^*$ have been defined. The expressions in (7.22), (7.23), (7.24) constitute the *linear approximate model* employed in the analysis and design of controllers for the nonlinear system in (7.20). Note that this linear approximate model is valid only if the state x and the input u remain close to the operation point (x^*, u^*), i.e., only if $z = x - x^* \approx 0$ and $v = u - u^* \approx 0$. From a practical point of view, it is commonly difficult to determine the values of x and u satisfying these conditions; hence, they are employed heuristically. Finally, note that the only requirement for all the above to stand is that all the partial derivatives defining the matrices A and B exist and are continuous, i.e., that $f(x, u)$ is, at least once, continuously differentiable.

Example 7.5 In Fig. 7.3, a simple pendulum is shown. In Example 2.6, Chap. 2, it was found that the corresponding mathematical model is given by the following nonlinear differential equation:

$$ml^2\ddot{\theta} + b\dot{\theta} + mgl\sin(\theta) = T(t). \tag{7.25}$$

Fig. 7.3 Simple pendulum

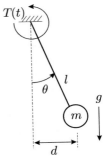

The feature that renders this differential equation nonlinear is the function $\sin(\theta)$ representing the effect of gravity. As it is a second-order differential equation, then only two state variables exist: the unknown θ and its first-time derivative $\dot{\theta}$:

$$x = \begin{bmatrix} x_1 \\ x_2 \end{bmatrix} = \begin{bmatrix} \theta \\ \dot{\theta} \end{bmatrix}, \quad u = T(t).$$

Using this nomenclature, the differential equation in (7.25) can be written as:

$$\ddot{\theta} = \dot{x}_2 = -\frac{b}{ml^2} x_2 - \frac{g}{l} \sin(x_1) + \frac{1}{ml^2} u,$$

i.e.,

$$\frac{d}{dt} \begin{bmatrix} x_1 \\ x_2 \end{bmatrix} = \begin{bmatrix} x_2 \\ -\frac{b}{ml^2} x_2 - \frac{g}{l} \sin(x_1) + \frac{1}{ml^2} u \end{bmatrix}.$$

Hence, the following state equation is found:

$$\dot{x} = f(x, u)$$

$$f(x, u) = \begin{bmatrix} f_1(x, u) \\ f_2(x, u) \end{bmatrix} = \begin{bmatrix} x_2 \\ -\frac{b}{ml^2} x_2 - \frac{g}{l} \sin(x_1) + \frac{1}{ml^2} u \end{bmatrix}.$$

Note that in this case, it is not possible to obtain the linear form $\dot{x} = Ax + Bu$ because of the function $\sin(\theta)$. To obtain a linear approximation for this nonlinear state equation, the operation points are first obtained, i.e., those pairs (x^*, u^*) satisfying $f(x^*, u^*) = [0\ 0]^T$:

$$f(x^*, u^*) = \begin{bmatrix} x_2^* \\ -\frac{b}{ml^2} x_2^* - \frac{g}{l} \sin(x_1^*) + \frac{1}{ml^2} u^* \end{bmatrix} = \begin{bmatrix} 0 \\ 0 \end{bmatrix}.$$

From this, the following is found:

$$x_2^* = 0, \quad u^* = -mgl \sin(x_1^*).$$

This means that the input at the operation point depends on the pendulum position at the operation point. This is because, to maintain the pendulum at rest, it is necessary to have an input that exactly compensates for the effect of gravity. To continue, a particular value for x_1^* must be selected. Two cases are considered here:

- When $x_1^* = 0$, $x_2^* = 0$, $u^* = 0$. Using (7.22), (7.23), (7.24), the following is found:

$$\dot{z} = Az + Bv,$$

$$A = \begin{bmatrix} \frac{\partial f_1(x,u)}{\partial x_1} & \frac{\partial f_1(x,u)}{\partial x_2} \\ \frac{\partial f_2(x,u)}{\partial x_1} & \frac{\partial f_2(x,u)}{\partial x_2} \end{bmatrix}_{x=x^*,u=u^*} = \begin{bmatrix} 0 & 1 \\ -\frac{g}{l}\cos(x_1) & -\frac{b}{ml^2} \end{bmatrix}_{x_1=x_2=u=0},$$

$$= \begin{bmatrix} 0 & 1 \\ -\frac{g}{l} & -\frac{b}{ml^2} \end{bmatrix},$$

$$B = \begin{bmatrix} \frac{\partial f_1(x,u)}{\partial u} \\ \frac{\partial f_2(x,u)}{\partial u} \end{bmatrix}_{x=x^*,u=u^*} = \begin{bmatrix} 0 \\ \frac{1}{ml^2} \end{bmatrix},$$

where $z = x - x^* = x$ and $v = u - u^* = u$ have been defined. Note that, according to this state equation:

$$\dot{z}_1 = z_2, \quad \dot{z}_2 = -\frac{g}{l}z_1 - \frac{b}{ml^2}z_2 + \frac{1}{ml^2}v.$$

Combining these first-order differential equations the following linear, second-order differential equation with constant coefficients is obtained:

$$\ddot{z}_1 + \frac{b}{ml^2}\dot{z}_1 + \frac{g}{l}z_1 = \frac{1}{ml^2}v. \tag{7.26}$$

This differential equation can be analyzed using the concepts studied in Sects. 3.3 (complex conjugate roots with a negative or zero real part), 3.4.2 (real and repeated negative roots) and 3.4.1 (real, different and negative roots), Chap. 3, as it corresponds to a differential equation with the form:

$$\ddot{y} + 2\zeta\omega_n\dot{y} + \omega_n^2 y = k\omega_n^2 v,$$

where:

$$y = z_1, \quad 2\zeta\omega_n = \frac{b}{ml^2}, \quad \omega_n^2 = \frac{g}{l}, \quad k = \frac{1}{mgl},$$

with $\zeta \geq 0$, $\omega_n > 0$ and $k > 0$ because $m > 0$, $g > 0$, $l > 0$ and $b \geq 0$.

- When $x_1^* = \pm\pi$, $x_2^* = 0$, $u^* = 0$. Using (7.22), (7.23), (7.24), the following is found:

$$\dot{z} = Az + Bv,$$

$$A = \begin{bmatrix} \frac{\partial f_1(x,u)}{\partial x_1} & \frac{\partial f_1(x,u)}{\partial x_2} \\ \frac{\partial f_2(x,u)}{\partial x_1} & \frac{\partial f_2(x,u)}{\partial x_2} \end{bmatrix}_{x=x^*,u=u^*} = \begin{bmatrix} 0 & 1 \\ -\frac{g}{l}\cos(x_1) & -\frac{b}{ml^2} \end{bmatrix}_{x_1=\pm\pi,x_2=u=0},$$

$$= \begin{bmatrix} 0 & 1 \\ \frac{g}{l} & -\frac{b}{ml^2} \end{bmatrix},$$

$$B = \begin{bmatrix} \frac{\partial f_1(x,u)}{\partial u} \\ \frac{\partial f_2(x,u)}{\partial u} \end{bmatrix}_{x=x^*,u=u^*} = \begin{bmatrix} 0 \\ \frac{1}{ml^2} \end{bmatrix},$$

where $z = x - x^* = [x_1 - x_1^*, x_2]^T$ and $v = u - u^* = u$ have been defined. Note that this state equation implies:

$$\dot{z}_1 = z_2, \quad \dot{z}_2 = \frac{g}{l}z_1 - \frac{b}{ml^2}z_2 + \frac{1}{ml^2}v.$$

Combining these first-order differential equations the following linear, second-order differential equation with constant coefficients is obtained:

$$\ddot{z}_1 + \frac{b}{ml^2}\dot{z}_1 - \frac{g}{l}z_1 = \frac{1}{ml^2}v. \tag{7.27}$$

This differential equation can be analyzed using the concepts studied in Sect. 3.4.1, Chap. 3 (real, different roots, one positive and the other negative, see Example 3.11) as it corresponds to a differential equation of the form:

$$\ddot{y} + c\dot{y} + dy = ev,$$

where:

$$y = z_1, \quad c = \frac{b}{ml^2} \geq 0, \quad d = -\frac{g}{l} < 0, \quad e = \frac{1}{ml^2}.$$

Note that, according to (7.26) and Sect. 3.3, Chap. 3, the pendulum oscillates with a natural frequency given as $\omega_n = \sqrt{\frac{g}{l}}$ when operating around $x_1^* = 0$, $x_2^* = 0$, $u^* = 0$. This means that a longer pendulum oscillates slower whereas a shorter pendulum oscillates faster. These observations are useful, for instance, when it is desired to adjust the rate of a clock that leads or lags in time. Moreover, rearranging (7.26) the following is found:

$$ml^2\ddot{z}_1 + b\dot{z}_1 + mglz_1 = v. \tag{7.28}$$

According to Example 2.5, Chap. 2, which analyzes the rotative spring-mass-damper system shown in Fig. 7.4, (7.28) means that factor $K = mgl > 0$ is equivalent to the stiffness constant of a spring because of the effect of gravity.

Fig. 7.4 A rotative spring-mass-damper system

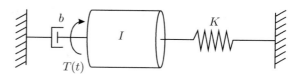

On the other hand, using similar arguments, it is observed that, rearranging (7.27):

$$ml^2\ddot{z}_1 + b\dot{z}_1 - mglz_1 = v.$$

When the pendulum operates around $x_1^* = \pm\pi$, $x_2^* = 0$, $u^* = 0$, it behaves as a mass-spring-damper system possessing a negative stiffness constant $K = -mgl < 0$. This can be seen as follows. A positive stiffness constant $K > 0$ indicates that the force exerted by the spring is always in the opposite sense to its deformation and, because of that, the spring forces the body to move back to the point where the spring is not deformed. This is the reason why the pendulum oscillates around $x_1^* = 0$, $x_2^* = 0$, $u^* = 0$. On the other hand, because of the sign change, in a spring with a negative stiffness constant $K < 0$, the opposite must occur, i.e., the force exerted by the spring must now be directed in the same sense as the deformation. This means that the deformation of a spring with $K < 0$ forces the deformation to increase further, i.e., the body moves away from the point where the spring deformation is zero (i.e., where $z_1 = 0$). This is exactly what happens to a pendulum around $x_1^* = \pm\pi$, $x_2^* = 0$, $u^* = 0$: if $z_1 \neq 0$ then the gravity forces the pendulum to fall; hence, to move away from $z_1 = 0$, i.e., where $x_1 = x_1^* = \pm\pi$. This behavior is the experimental corroboration of what is analytically demonstrated by a characteristic polynomial with a positive root: the operation points $x_1^* = \pm\pi$, $x_2^* = 0$, $u^* = 0$ are unstable.

Note that using the above methodology, a linear approximate model of a nonlinear system is already given in terms of the equation error introduced in Sect. 7.2. The reader is referred to Sect. 13.2, Chaps. 13, Sect. 15.3, Chap. 15, and Sect. 16.4, Chap. 16, to see some other examples of how to employ (7.22), (7.23), (7.24), to find linear approximate state equations for nonlinear state equations.

7.4   Some Results from Linear Algebra

Some results that are useful for studying the state variables approach are presented in this section.

Fact 7.1 ([1], chapter 2) *Let* $w_1, w_2, \ldots, w_n$, *be n vectors with n components. These vectors are linearly dependent if, and only if, there exist n scalars* $\alpha_1, \alpha_2, \ldots, \alpha_n$, *not all zero such that:*

$$\alpha_1 w_1 + \alpha_2 w_2 + \cdots + \alpha_n w_n = 0, \tag{7.29}$$

where zero at the right represents a vector with n zero components. If the only way of satisfying the latter expression is that $\alpha_1 = \alpha_2 = \cdots = \alpha_n = 0$, *then it is said that* $w_1, w_2, \ldots, w_n$, *are linearly independent vectors.*

In the simplest version of the linear dependence of vectors, two vectors with two components are linearly dependent if they are parallel. This can be shown as follows. Let w_1 and w_2 be two vectors, each one having $n = 2$ components. These vectors are linearly dependent if there are two constants $\alpha_1 \neq 0$ and $\alpha_2 \neq 0$ such that:

$$\alpha_1 w_1 + \alpha_2 w_2 = 0,$$

which means that:

$$w_1 = -\frac{\alpha_2}{\alpha_1} w_2.$$

Note that α_1 cannot be zero for obvious reasons. On the other hand, if α_2 is zero, then $w_1 = 0$ is the only solution in this case that is not of interest. Hence, it is assumed that both $\alpha_1 \neq 0$ and $\alpha_2 \neq 0$. Thus, as the factor $-\frac{\alpha_2}{\alpha_1}$ is a scalar (positive or negative), the above means that w_1 and w_2 are vectors with the same direction, i.e., they are parallel.[1]

In the case of three or more vectors, linear dependence means that one of vectors can be computed as the addition of the other vectors when suitably affected by factors α_i (this is what is known as a linear combination of vectors), i.e.,:

$$w_1 = -\frac{\alpha_2}{\alpha_1} w_2 - \frac{\alpha_3}{\alpha_1} w_3,$$

is directly obtained from (7.29) when $n = 3$ and $\alpha_1 \neq 0$. As $n = 3$, the above means that w_1, w_2 and w_3 are in the same plane, which is embedded in a volume (in R^3), i.e., that the linear combination of w_1, w_2 and w_3 does not allow another vector to be obtained that is not contained in the same plane. According to this, three vectors with three components are linearly independent if their linear combination spans three dimensions. Thus, the linear combination of n linearly independent vectors (with n components) spans an n−dimensional space. The linear combination of n linearly dependent vectors (with n components) spans a space with fewer than n dimensions.

Fact 7.2 ([1], chapter 2, [2], chapter 10) *Consider the following $n \times n$ matrix:*

$$E = \begin{bmatrix} w_1 & w_2 & \dots & w_n \end{bmatrix},$$

where $w_1, w_2, \dots, w_n$, are n column with n components. The vectors $w_1, w_2, \dots, w_n$ are linearly independent if, and only if, matrix E is nonsingular, i.e., if, and only if, the determinant of E is different from zero ($\det(E) \neq 0$).

To explain this result, the following is employed.

---

[1]Some authors use the term *antiparallel* to indicate that two vectors have the same direction but the opposite sense. In this book, the term *parallel* is employed to designate two vectors with the same direction no matter what their senses are.

Fact 7.3 ([1], chapter 2) *Consider an $n \times n$ matrix E. The only solution of $Ew = 0$, where w represents an n−dimensional vector, is $w = 0$ if, and only if, matrix E is nonsingular.*

Note that (7.29) can be written as:

$$\alpha_1 w_1 + \alpha_2 w_2 + \cdots + \alpha_n w_n = \begin{bmatrix} w_1 & w_2 & \ldots & w_n \end{bmatrix} \begin{bmatrix} \alpha_1 \\ \alpha_2 \\ \vdots \\ \alpha_n \end{bmatrix} = \begin{bmatrix} 0 \\ 0 \\ \vdots \\ 0 \end{bmatrix}.$$

According to Fact 7.3, the vector $[\alpha_1 \ \alpha_2 \ \ldots \ \alpha_n] = [0 \ 0 \ \ldots \ 0]$ is the only solution of this homogeneous system if, and only if, the matrix $E = [w_1 \ w_2 \ \ldots \ w_n]$ is nonsingular. This implies that the vectors $w_1, w_2, \ldots, w_n$ are linearly independent if, and only if, $\det(E) \neq 0$.

Fact 7.4 ([1], chapter 2) *The rank of an $n \times n$ matrix E is the order of the largest nonzero determinant obtained as submatrices of E. The rank of E is n if, and only if, E is nonsingular. The inverse matrix E^{-1} exists if, and only if, the matrix E is nonsingular [2], chapter 10.*

Fact 7.5 ([1], chapter 2) *The eigenvalues of an $n \times n$ matrix E are those scalars λ satisfying:*

$$\det(\lambda I - E) = 0,$$

where I stands for the $n \times n$ identity matrix. The expression $\det(\lambda I - E)$ is a n−degree polynomial in the variable λ and it is known as the characteristic polynomial of the matrix E. The eigenvalues of a matrix E can be either real or complex numbers. Matrix E has exactly n eigenvalues, which can be repeated.

Fact 7.6 ([3], chapter 7) *Suppose that two $n \times n$ matrices, E and $\bar{E}$, are related as:*

$$\bar{E} = PEP^{-1}, \tag{7.30}$$

where P is an $n \times n$ constant and nonsingular matrix, then both matrices, E and $\bar{E}$, possess identical eigenvalues. This means that:

$$\det(\lambda I - E) = \det(\lambda I - \bar{E}) \tag{7.31}$$

The following is useful for explaining the above.

Fact 7.7 ([4], pp. 303) *Let D and G be two $n \times n$ matrices. Then:*

$$\det(DG) = \det(D)\det(G). \tag{7.32}$$

As $PP^{-1} = I$ and $\det(I) = 1$, then (7.32) can be used to find that $\det(P)\det(P^{-1}) = 1$, i.e., $\det(P^{-1}) = \frac{1}{\det(P)}$. On the other hand, according to (7.30), (7.32), the *characteristic polynomial* of $\bar{E}$ is:

$$\det(\lambda I - \bar{E}) = \det(\lambda I - PEP^{-1}) = \det(P[\lambda I - E]P^{-1}),$$
$$= \det(P)\det(\lambda I - E)\det(P^{-1}) = \det(\lambda I - E).$$

Note that (7.31) has been retrieved in the last expression, which verifies that result.

Fact 7.8 ([4], pp. 334)) *Let E and F be two $n \times n$ matrices with $F = E^T$. The eigenvalues of F are identical to eigenvalues of E.*

The following is useful for explaining the above.

Fact 7.9 ([2], pp. 504) *The determinant of a matrix E is equal to the addition of the products of the elements of any row or column of E and their corresponding cofactors.*

Fact 7.10 ([2], pp. 507) *If $\det(D)$ is any determinant and $\det(G)$ is the determinant whose rows are the columns of $\det(D)$, then $\det(D) = \det(G)$.*

According to the above, $\det(\lambda I - E)$ can be computed using, for instance, the first column of matrix $\lambda I - E$, whereas $\det(\lambda I - F)$ can be computed using, for instance, the first row of matrix $\lambda I - F$. Note that the first column of $\lambda I - E$ is equal to the first row of $\lambda I - F$ as $(\lambda I - E)^T = \lambda I - F$ because $F = E^T$. Similarly, the rows of the cofactor of entry at row i and column 1 in matrix $\lambda I - E$ are equal to the columns of the cofactor of entry at row 1 and column i in matrix $\lambda I - F$. This shows that E and F have the same characteristic polynomial, i.e.,:

$$\det(\lambda I - E) = \det(\lambda I - F)$$

; thus, the eigenvalues of F are identical to those of E.

7.5 Solution of a Linear Time Invariant Dynamical Equation

Finding the solution of the dynamical equation:

$$\dot{x} = Ax + Bu, \quad x \in R^n, \quad u \in R, \tag{7.33}$$
$$y = Cx, \quad y \in R, \tag{7.34}$$

is a complex problem requiring the knowledge of advanced linear algebra and calculus concepts that are outside the scope of this book. However, the fundamental ideas that are involved in the corresponding procedure can be understood if resorting to the solution procedure of the following first-order differential equation:

$$\dot{x} = ax + bu, \quad x, u \in R.$$

Use of the Laplace transform yields:

$$sX(s) - x(0) = aX(s) + bU(s),$$

where $X(s)$, $U(s)$, stand for Laplace transforms of x and u respectively. This can be written as follows:

$$X(s) = \frac{b}{s-a}U(s) + \frac{x(0)}{s-a}. \tag{7.35}$$

Using the transformed pairs [5]:

$$\mathcal{L}\left\{e^{at}\right\} = \frac{1}{s-a} = G(s), \tag{7.36}$$

$$\mathcal{L}\left\{\int_0^t g(t-\tau)bu(\tau)d\tau\right\} = \frac{b}{s-a}U(s), \quad \text{(convolution)},$$

$$g(t) = \mathcal{L}^{-1}\{G(s)\},$$

it is found that, applying the inverse Laplace transform to (7.35), yields:

$$x(t) = \int_0^t g(t-\tau)bu(\tau)d\tau + e^{at}x(0).$$

Finally, using (7.36) again:

$$x(t) = e^{at}x(0) + \int_0^t e^{a(t-\tau)}bu(\tau)d\tau, \quad x, u, a, b \in R.$$

The solution of the state equation in (7.33) is similar to this last equation and we just need to use matrix notation, i.e., the solution of (7.33) is [1], pp. 142:

$$x(t) = e^{At}x(0) + \int_0^t e^{A(t-\tau)}Bu(\tau)d\tau, \quad x \in R^n, \ u \in R, \tag{7.37}$$

where e^{At} is an $n \times n$ matrix[2] and it is explained in the following how it is given.

- If λ is a (positive, negative or zero) real nonrepeated eigenvalue of matrix A, then at least one of the entries of matrix e^{At} includes the following function of time:

$$c\,e^{\lambda t},$$

where c is a real constant.

---

[2]Note that $e^{A(t-\tau)} = e^{Ar}|_{r=t-\tau}$.

- If λ is a (positive, negative or zero) real and r times repeated eigenvalue of matrix A, then each one of the following functions of time:

$$c_0 e^{\lambda t}, \quad c_1 t e^{\lambda t}, \quad c_2 t^2 e^{\lambda t}, \quad \ldots, \quad c_{r-1} t^{r-1} e^{\lambda t},$$

where $c_k, k = 1, 2, \ldots, r - 1$, are real constants, are included in at least one of the entries of matrix e^{At}.
- If $\lambda = a \pm jb$ is a nonrepeated complex eigenvalue of matrix A, where $j = \sqrt{-1}$, a is a (positive, negative or zero) real number and b is a strictly positive real number, then each one of the following functions of time:

$$c\, e^{at} \sin(bt), \quad d\, e^{at} \cos(bt),$$

where c and d are real constants, are included in at least one of the entries of matrix e^{At}.
- If $\lambda = a \pm jb$ is a complex eigenvalue of matrix A that repeats r times, where $j = \sqrt{-1}$, a is a (positive, negative or zero) real number and b is a strictly positive real number, then each one of the following functions of time:

$$
\begin{aligned}
&c_0 e^{at} \sin(bt), \quad , c_1 t e^{at} \sin(bt), \quad c_2 t^2 e^{at} \sin(bt), \quad \ldots, c_{r-1} t^{r-1} e^{at} \sin(bt), \\
&d_0 e^{at} \cos(bt), \quad , d_1 t e^{at} \cos(bt), \quad d_2 t^2 e^{at} \cos(bt), \quad \ldots, d_{r-1} t^{r-1} e^{at} \cos(bt),
\end{aligned}
$$

$$\tag{7.38}$$

where c_k and d_k, $k = 1, 2, \ldots, r - 1$, are real constants, and included in at least one of the entries of matrix e^{At}.

Finally, the output of the dynamical equation in (7.33), (7.34), is computed using (7.37), i.e.,:

$$y(t) = C e^{At} x(0) + C \int_0^t e^{A(t-\tau)} Bu(\tau) d\tau. \tag{7.39}$$

7.6   Stability of a Dynamical Equation

It is important to study the stability of the following dynamical equation without input:

$$\dot{x} = Ex, \tag{7.40}$$

$$y = Cx,$$

where $x \in R^n$, $y \in R$, E is an $n \times n$ constant matrix and C is a constant row vector with n components. Although a dynamical equation without input may seem unrealistic, it is shown in Sect. 7.11 that a closed-loop dynamical equation can be written as in (7.40), because in a feedback system the input is chosen as a function of the state, i.e., $u = -Kx$ where K is a row vector with n components. Then, replacing this input in (7.33) and defining $E = A - BK$, (7.40) is retrieved. This is the main motivation for studying this dynamical equation without input.

Although a formal definition of stability of (7.40) is elaborated, it can be simplified, stating that the dynamical equation (7.40) is stable if $\lim_{t \to \infty} x(t) = 0$ (hence, $\lim_{t \to \infty} y(t) = 0$) for any initial state $x(0) \in R^n$. This means that, if:

$$x(t) = \begin{bmatrix} x_1(t) \\ x_2(t) \\ \vdots \\ x_n(t) \end{bmatrix},$$

then $\lim_{t \to \infty} x_i(t) = 0$ for all $i = 1, 2, \ldots, n$, for any initial state. According to the solution in (7.37), the solution of (7.40) is obtained by defining $u = 0$ and $A = E$ in (7.37), i.e.,:

$$x(t) = e^{Et} x(0).$$

As $x(0)$ is a constant vector, the latter expression implies that, to satisfy $\lim_{t \to \infty} x(t) = 0$, then $\lim_{t \to \infty} e^{Et} = 0$ where "0" stands for an $n \times n$ matrix all of whose entries are equal to zero. Recalling how the entries of matrix e^{At}, $A = E$, are given (see the the previous section) the following is proven:

Theorem 7.1 ([1], pp. 409) *The solution of (7.40) satisfies $\lim_{t \to \infty} x(t) = 0$, no matter what the initial state $x(0)$ is, if, and only if, all eigenvalues of matrix E have a strictly negative real part. Under these conditions, it is said that the origin $x = 0$ of (7.40) is globally asymptotically stable.*

Recall that in Sect. 3.4.2, Chap. 3, it has been proven that:

$$\lim_{t \to \infty} t^j e^{pt} = 0,$$

for any integer $j > 0$ and any real number $p < 0$.

The reader may wonder why it is of interest to ensure that $\lim_{t \to \infty} x(t) = 0$ in a practical control problem. To answer this question is important to stress that the above result is very important if the dynamical equation without input given in (7.40) represents the error equation (see Sect. 7.2) of the closed-loop system. In such a case, the state x represents the error state and, as explained in Sect. 7.2, ensuring that the state error converges to zero implies that the plant state converges to its desired value.

7.7   Controllability and Observability

Two important properties of the dynamical equation:

$$\dot{x} = Ax + Bu, \quad x \in R^n, u \in R, \tag{7.41}$$

$$y = Cx, \quad y \in R, \tag{7.42}$$

that are employed throughout this chapter are controllability and observability.

7.7.1   Controllability

Definition 7.2 ([1], pp. 176)   A state equation is *controllable* at t_0 if there exists a finite time $t_1 > t_0$ such that given any states x_0 and x_1 exists and input u that is applied from $t = t_0$ to $t = t_1$ transfers the state from x_0 at $t = t_0$ to x_1 at $t = t_1$. Otherwise, the state equation is not controllable.

Note that this definition does not specify the trajectory to be tracked to transfer the state from x_0 to x_1. Furthermore, it is not necessary for the state to stay at x_1 for all $t > t_1$. Also note that controllability is a property that is only concerned with the input, i.e., with the state equation, and the output equation is not considered. Finally, this definition is very general as it also considers the possibility that the state equation is time variant, i.e., that the entries of A and B are functions of time.

In the case of a time-invariant state equation, such as that shown in (7.41), i.e., when all entries of A and B are constant, if the state equation is controllable then it is controllable for any $t_0 \geq 0$ and t_1 is any value such that $t_1 > t_0$. This means that the transference from x_0 to x_1 can be performed in any nonzero time interval. This allows us to formulate a simple way of checking whether a state equation is controllable:

Theorem 7.2 ([6], pp. 145)   *The state equation in (7.41) is controllable if, and only if, any of the following equivalent conditions are satisfied (see Exercises 10 and 11 at the end of this Chapter):*

1. The following $n \times n$ matrix:

$$W_c(t) = \int_0^t e^{A\tau} B B^T \left(e^{A\tau}\right)^T d\tau = \int_0^t e^{A(t-\tau)} B B^T \left(e^{A(t-\tau)}\right)^T d\tau,$$

$$\tag{7.43}$$

is nonsingular for any $t > 0$.

2. The $n \times n$ controllability matrix:

$$\left[B \ AB \ A^2B \ \cdots \ A^{n-1}B \right], \tag{7.44}$$

has rank n, i.e., its determinant is different from zero.

The reason for this result can be explained as follows. The following input:

$$u(t) = -B^T \left(e^{A(t_1-t)}\right)^T W_c^{-1}(t_1)[e^{At_1}x_0 - x_1] \qquad (7.45)$$

transfers the system state from $x_0 = x(0)$ at $t_0 = 0$ to x_1 at $t_1 > 0$. This can be verified by replacing in the solution shown in (7.37) but evaluated at $t = t_1$, i.e.,:

$$x(t_1) = e^{At_1}x(0) + \int_0^{t_1} e^{A(t_1-\tau)}Bu(\tau)d\tau,$$

the input in (7.45) but evaluated at $t = \tau$, i.e.,:

$$u(\tau) = -B^T \left(e^{A(t_1-\tau)}\right)^T W_c^{-1}(t_1)[e^{At_1}x_0 - x_1],$$

to obtain:

$$x(t_1) = e^{At_1}x(0)$$
$$+ \int_0^{t_1} e^{A(t_1-\tau)}B\left\{-B^T \left(e^{A(t_1-\tau)}\right)^T W_c^{-1}(t_1)[e^{At_1}x_0 - x_1]\right\}d\tau,$$
$$= e^{At_1}x(0)$$
$$- \underbrace{\left\{\int_0^{t_1} e^{A(t_1-\tau)}BB^T \left(e^{A(t_1-\tau)}\right)^T d\tau\right\}}_{W_c(t_1)} W_c^{-1}(t_1)[e^{At_1}x_0 - x_1],$$
$$= x_1.$$

Note that this result needs the $n \times n$ matrix $W_c(t)$, defined in (7.43), to be nonsingular, which is true for any $t_1 = t > 0$ if the state equation is controllable. The fact that t_1 is any positive time means that the state may go from x_0 to x_1 in any nonzero time interval.

On the other hand, the equation in (7.43) can be corroborated as follows. Defining $v = t - \tau$:

$$\int_0^t e^{A(t-\tau)}BB^T \left(e^{A(t-\tau)}\right)^T d\tau = \int_{\tau=0}^{\tau=t} e^{A(t-\tau)}BB^T \left(e^{A(t-\tau)}\right)^T d\tau,$$
$$= \int_{v=t}^{v=0} e^{Av}BB^T \left(e^{Av}\right)^T (-dv),$$
$$= \int_{v=0}^{v=t} e^{Av}BB^T \left(e^{Av}\right)^T dv,$$

which corroborates equation in (7.43) if τ is used instead of v in last integral, which is valid because the use of v or τ as the integration variable in the last integral does not affect the result.

Finally, the fact that the rank of matrix in (7.44) is n and the nonsingularity of $W_c(t)$ are equivalent conditions is very useful because it is easier to verify that (7.44) has rank n than the nonsingularity of $W_c(t)$.

7.7.2 Observability

Definition 7.3 ([1], pp. 193) A dynamical equation is *observable* at t_0 if there is a finite $t_1 > t_0$ such that for any unknown state x_0 at $t = t_0$, the knowledge of the input u and the output y on the time interval $[t_0, t_1]$ suffices to uniquely determine the state x_0. Otherwise the dynamical equation is not observable.

Note that observability is a property that is concerned with the possibility of knowing the system state (a variable that is internal to the system) only from measurements of the variables that are external to the system, i.e., the input and the output. Also note that this definition is very general as it also considers the possibility that the dynamical system is time variant, i.e., that the entries of A, B, and C are functions of time. In the case of a time invariant dynamical equation, such as that shown in (7.41), (7.42), with constant entries of A, B and C, if the dynamical equation is observable, then it is observable for all $t_0 \geq 0$ and t_1 is any time such that $t_1 > t_0$. This means that the determination of the initial state can be performed in any nonzero time interval. This allows us to obtain a very simple way of checking whether a dynamical equation is observable:

Theorem 7.3 ([6], pp. 155, 156) *The dynamical equation in (7.41), (7.42), is observable if, and only if, any of the following equivalent conditions are satisfied:*

1. *The following $n \times n$ matrix:*

$$W_o(t) = \int_0^t \left(e^{A\tau}\right)^T C^T C e^{A\tau} d\tau, \tag{7.46}$$

is nonsingular for any $t > 0$.
2. *The $n \times n$ observability matrix:*

$$\begin{bmatrix} C \\ CA \\ CA^2 \\ \vdots \\ CA^{n-1} \end{bmatrix}, \tag{7.47}$$

has a rank n, i.e., its determinant is different from zero.

The reason for this result is explained as follows. Consider the solution given in (7.39) and define:

$$\overline{y}(t) = Ce^{At}x(0), \tag{7.48}$$

$$= y(t) - C \int_0^t e^{A(t-\tau)} Bu(\tau)d\tau. \tag{7.49}$$

Note that, according to (7.49), the function $\overline{y}(t)$ can be computed from the exclusive knowledge of the input and the output. Multiplying both sides of (7.48) by $\left(e^{At}\right)^T C^T$ and integrating on $[0, t_1]$ it is found that:

$$\underbrace{\left\{ \int_0^{t_1} \left(e^{At}\right)^T C^T Ce^{At} dt \right\}}_{W_o(t_1)} x(0) = \int_0^{t_1} \left(e^{At}\right)^T C^T \overline{y}(t)dt.$$

If the matrix $W_o(t_1)$ is nonsingular, which is true if the system is observable, then the initial state $x(0) = x_0$ can be uniquely computed as:

$$x_0 = W_o^{-1}(t_1) \int_0^{t_1} \left(e^{At}\right)^T C^T \overline{y}(t)dt.$$

Note that this result needs the $n \times n$ matrix $W_o(t)$ defined in (7.46) to be nonsingular, which is true for any $t_1 = t > 0$ if the dynamical equation is observable. The fact that t_1 is any positive time means that the state x_0 can be computed with information obtained from the input and the output in any nonzero time interval.

Finally, the fact that the rank of matrix in (7.47) is n and the nonsingularity of $W_o(t)$ are equivalent conditions is very useful, as it is easier to check (7.47) to have rank n than the nonsingularity of $W_o(t)$.

Example 7.6 One way of estimating the actual state $x(t)$ (not the initial state $x(0)$) is using input and output measurements in addition to some of their time derivatives. It is shown in the following that a necessary condition for this is, again, that the matrix in (7.47) has the rank n. Consider the dynamical equation in (7.41), (7.42). The first $n - 1$ time derivatives of the output are computed as:

$$y = Cx,$$
$$\dot{y} = C\dot{x} = CAx + CBu,$$
$$\ddot{y} = CA\dot{x} + CB\dot{u} = CA^2x + CABu + CB\dot{u},$$
$$y^{(3)} = CA^2\dot{x} + CAB\dot{u} + CB\ddot{u} = CA^3x + CA^2Bu + CAB\dot{u} + CB\ddot{u},$$

$$\vdots$$

$$y^{(i)} = CA^ix + CA^{i-1}Bu + CA^{i-2}B\dot{u} + \cdots + CBu^{(i-1)},$$

$$\vdots$$

$$y^{(n-1)} = CA^{n-1}x + CA^{n-2}Bu + CA^{n-3}B\dot{u} + \cdots + CBu^{(n-2)},$$

where the exponent between brackets stands for the order of a time derivative. Using matrix notation:

$$\dot{Y} = Dx(t) + \dot{U},$$

where:

$$\dot{Y} = \begin{bmatrix} y \\ \dot{y} \\ \ddot{y} \\ y^{(3)} \\ \vdots \\ y^{(n-1)} \end{bmatrix}, \quad D = \begin{bmatrix} C \\ CA \\ CA^2 \\ CA^3 \\ \vdots \\ CA^{n-1} \end{bmatrix},$$

$$\dot{U} = \begin{bmatrix} 0 \\ CBu \\ CABu + CB\dot{u} \\ CA^2Bu + CAB\dot{u} + CB\ddot{u} \\ \vdots \\ CA^{n-2}Bu + CA^{n-3}B\dot{u} + \cdots + CBu^{(n-2)} \end{bmatrix}.$$

Thus, if the matrix in (7.47) has the rank n, then the $n \times n$ matrix D is invertible and the actual state can be computed as:

$$x(t) = D^{-1}(\dot{Y} - \dot{U}), \qquad (7.50)$$

i.e., by only employing input and output measurements in addition to some of their time derivatives. A drawback of (7.50) is that measurements of any variable have significant noise content in practice and this problem becomes worse as higher order time derivatives are computed from these measurements. This is the reason why (7.50) is not actually employed in practice to compute $x(t)$ and asymptotic methods such as that presented in Sect. 7.12 are preferred.

7.8   Transfer Function of a Dynamical Equation

Consider the following single-input single-output dynamical equation:

$$\dot{x} = Ax + Bu, \qquad (7.51)$$

$$y = Cx,$$

where $x \in R^n$, $u \in R$, $y \in R$, are functions of time, A is an $n \times n$ constant matrix, B is a constant column vector with n components and C is a constant row vector with n components. Applying the Laplace transform to (7.51), the following is obtained:

$$sX(s) - x(0) = AX(s) + BU(s), \tag{7.52}$$

$$Y(s) = CX(s), \tag{7.53}$$

where $X(s)$, $Y(s)$, $U(s)$ represent the Laplace transform of the state vector x, the scalar output y, and the scalar input u respectively, whereas $x(0)$ is the initial value of the state x. Recall that, given a vectorial function of time $x = [x_1, x_2, \ldots, x_n]^T$, its Laplace transform is obtained by applying this operation to each entry of the vector, i.e., [3], chapter 4:

$$X(s) = \begin{bmatrix} \mathcal{L}\{x_1\} \\ \mathcal{L}\{x_2\} \\ \vdots \\ \mathcal{L}\{x_n\} \end{bmatrix}.$$

A transfer function is always defined assuming zero initial conditions. Replacing $x(0) = 0$ in (7.52) and solving for $X(s)$:

$$X(s) = (sI - A)^{-1} BU(s), \tag{7.54}$$

where I represents the $n \times n$ identity matrix. Replacing (7.54) in (7.53) it is found that:

$$\frac{Y(s)}{U(s)} = G(s) = C(sI - A)^{-1} B. \tag{7.55}$$

Note that the transfer function $G(s)$ given in (7.55) is a scalar. Now, proceed to analyze this transfer function. In the following, it is assumed that $n = 3$ for ease of exposition of ideas, i.e., that A is an 3×3 matrix, I is the 3×3 identity matrix, and B and C are column and row vectors respectively, with 3 components:

$$A = \begin{bmatrix} a_{11} & a_{12} & a_{13} \\ a_{21} & a_{22} & a_{23} \\ a_{31} & a_{32} & a_{33} \end{bmatrix}, \quad B = \begin{bmatrix} b_1 \\ b_2 \\ b_3 \end{bmatrix}, \quad C = \begin{bmatrix} c_1 & c_2 & c_3 \end{bmatrix}.$$

The reader can verify that the same procedure is valid from any arbitrary n. First, compute the matrix:

$$sI - A = \begin{bmatrix} s - a_{11} & -a_{12} & -a_{13} \\ -a_{21} & s - a_{22} & -a_{23} \\ -a_{31} & -a_{32} & s - a_{33} \end{bmatrix},$$

whose inverse matrix is given as $(sI - A)^{-1} = \frac{Adj(sI-A)}{\det(sI-A)}$ where [2], chapter 10:

$$Adj(sI - A) = Cof^T(sI - A) = \begin{bmatrix} Cof_{11} & Cof_{12} & Cof_{13} \\ Cof_{21} & Cof_{22} & Cof_{23} \\ Cof_{31} & Cof_{32} & Cof_{33} \end{bmatrix}^T, \qquad (7.56)$$

with $Cof(sI-A)$ the cofactors matrix of $sI-A$, and $Cof_{ij} = (-1)^{i+j} M_{ij}$ with M_{ij} the $(n-1) \times (n-1)$ determinant (2×2 in this case) resulting when eliminating the i-th row and the j-th column from the determinant of matrix $sI - A$ [2], chapter 10. Explicitly computing these entries, it is realized that Cof_{ij} is a polynomial in s whose degree is strictly less than $n = 3$. On the other hand, solving through the first row it is found that:

$$\det(sI - A) = (s - a_{11}) \begin{vmatrix} s - a_{22} & -a_{23} \\ -a_{32} & s - a_{33} \end{vmatrix} + a_{12} \begin{vmatrix} -a_{21} & -a_{23} \\ -a_{31} & s - a_{33} \end{vmatrix}$$

$$- a_{13} \begin{vmatrix} -a_{21} & s - a_{22} \\ -a_{31} & -a_{32} \end{vmatrix},$$

i.e., $\det(sI - A)$ is a polynomial in s whose degree equals $n = 3$. From these observations, it is concluded that all entries of the matrix:

$$(sI - A)^{-1} = \frac{Adj(sI - A)}{\det(sI - A)} = \begin{bmatrix} Inv_{11} & Inv_{12} & Inv_{13} \\ Inv_{21} & Inv_{22} & Inv_{23} \\ Inv_{31} & Inv_{32} & Inv_{33} \end{bmatrix}, \qquad (7.57)$$

are given as the division of two polynomials in s such that the degree of the polynomial at the numerator is strictly less than $n = 3$ and the polynomial at the denominator is $\det(sI - A)$, which has the degree $n = 3$.

On the other hand, the product $(sI - A)^{-1}B$ is a column vector with $n = 3$ components. Each one of these components is obtained as:

$$(sI - A)^{-1}B = \begin{bmatrix} d_1 \\ d_2 \\ d_3 \end{bmatrix} = \begin{bmatrix} Inv_{11} b_1 + Inv_{12} b_2 + Inv_{13} b_3 \\ Inv_{21} b_1 + Inv_{22} b_2 + Inv_{23} b_3 \\ Inv_{31} b_1 + Inv_{32} b_2 + Inv_{33} b_3 \end{bmatrix}.$$

According to the previous discussion on the way in which the entries Inv_{ij} of matrix $(sI - A)^{-1}$ are given, and resorting to the addition of fractions with the same denominator, it is concluded that each one of the entries d_i of the column vector $(sI - A)^{-1}B$ is also given as a division of two polynomials in s such that the polynomial at the numerator has a degree strictly less than $n = 3$ whereas the polynomial at the denominator is $\det(sI - A)$, whose degree equals $n = 3$. Finally, the following is found:

$$C(sI - A)^{-1}B = c_1 d_1 + c_2 d_2 + c_3 d_3.$$

Reasoning as before, the following is concluded, which is valid for any arbitrary n:

Fact 7.11 ([1], chapters 6, 7) *The transfer function in (7.55) is a scalar, which is given as the division of two polynomials in s such that the degree of the polynomial at the numerator is strictly less than n and the polynomial at the denominator, given as* $\det(sI - A)$, *has the degree n, i.e.,*

$$\frac{Y(s)}{U(s)} = G(s) = C(sI - A)^{-1}B = \frac{b_m s^m + b_{m-1}s^{m-1} + \cdots + b_1 s + b_0}{s^n + a_{n-1}s^{n-1} + \cdots + a_1 s + a_0},$$

(7.58)

$$\det(sI - A) = s^n + a_{n-1}s^{n-1} + \cdots + a_1 s + a_0,$$

for some real constants b_k, a_l, $k = 0, 1, \ldots, m$, $l = 0, 1, \ldots, n - 1$, *with* $n > m$.

However, this statement is completely true if the following is satisfied [1], chapter 7:

Condition 7.1 *The dynamical equation in (7.51) must be controllable, i.e., the determinant of the matrix in (7.44) must be different from zero.*

Moreover, if the following is also true, then the polynomials $b_m s^m + b_{m-1}s^{m-1} + \cdots + b_1 s + b_0$ and $s^n + a_{n-1}s^{n-1} + \cdots + a_1 s + a_0$ have no common roots [1], chapter 6:

Condition 7.2 *The dynamical equation in (7.51) is observable, i.e., the determinant of the matrix in (7.47) is different from zero.*

Finally, the reader can revise the discussion presented above to verify that:

$$\frac{Y(s)}{U(s)} = G(s) + D,$$

is the transfer function of the dynamical equation:

$$\dot{x} = Ax + Bu,$$
$$y = Cx + Du,$$

where $x \in R^n$, $u \in R$, $y \in R$, are functions of time, A is an $n \times n$ constant matrix, B is a constant column vector with n components, C is a constant row vector with n components, and D is a constant scalar. This means that the polynomials at the numerator and the denominator of $\frac{Y(s)}{U(s)}$ have the same degree in this case.

Example 7.7 According to Chap. 10, the mathematical model of a permanent magnet brushed DC motor is given as (see (10.8)):

$$J\ddot{\theta} + b\dot{\theta} = n\,k_m\,i^*,$$

when no external disturbance is present. Furthermore, the electric current i^* can be considered to be the input signal if an internal high-gain electric current loop is employed. Defining the state variables as the position and the velocity, it is not difficult to realize that the following dynamical equation is obtained:

$$\dot{x} = Ax + Bu, \quad y = Cx, \tag{7.59}$$

$$x = \begin{bmatrix} x_1 \\ x_2 \end{bmatrix} = \begin{bmatrix} \theta \\ \dot{\theta} \end{bmatrix}, \quad A = \begin{bmatrix} 0 & 1 \\ 0 & -\frac{b}{J} \end{bmatrix}, \quad B = \begin{bmatrix} 0 \\ \frac{nk_m}{J} \end{bmatrix}, \quad u = i^*,$$

where C is a row vector to be defined later depending on the output to be considered. Note that the following matrix:

$$[B \ AB] = \begin{bmatrix} 0 & \frac{nk_m}{J} \\ \frac{nk_m}{J} & -\frac{nk_m}{J}\frac{b}{J} \end{bmatrix},$$

has the rank $n = 2$ because its determinant is equal to $-\left(\frac{nk_m}{J}\right)^2 \neq 0$. From the definitions in (7.59) the following results are obtained:

$$sI - A = \begin{bmatrix} s & -1 \\ 0 & s + \frac{b}{J} \end{bmatrix}, \quad \det(sI - A) = s\left(s + \frac{b}{J}\right),$$

$$Cof(sI - A) = \begin{bmatrix} s + \frac{b}{J} & 0 \\ 1 & s \end{bmatrix}, \quad Adj(sI - A) = Cof^T(sI - A) = \begin{bmatrix} s + \frac{b}{J} & 1 \\ 0 & s \end{bmatrix},$$

$$(sI - A)^{-1} = \frac{Adj(sI - A)}{\det(sI - A)} = \frac{1}{s(s + \frac{b}{J})}\begin{bmatrix} s + \frac{b}{J} & 1 \\ 0 & s \end{bmatrix}.$$

To obtain the corresponding transfer function, the following cases are considered:

- The output is the position, i.e., $y = \theta = x_1 = Cx$, where $C = [1 \ 0]$. Then:

$$G(s) = C(sI - A)^{-1}B = [1 \ 0]\frac{1}{s(s + \frac{b}{J})}\begin{bmatrix} s + \frac{b}{J} & 1 \\ 0 & s \end{bmatrix}\begin{bmatrix} 0 \\ \frac{nk_m}{J} \end{bmatrix},$$

$$G(s) = \frac{\frac{nk_m}{J}}{s(s + \frac{b}{J})}.$$

In this case, the matrix:

$$\begin{bmatrix} C \\ CA \end{bmatrix} = \begin{bmatrix} 1 & 0 \\ 0 & 1 \end{bmatrix},$$

has the rank $n = 2$ because its determinant is unity, i.e., it is different from zero.

- The output is velocity, i.e., $y = \dot{\theta} = x_2 = Cx$, where $C = [0\ 1]$. Then:

$$G(s) = C(sI - A)^{-1}B = [0\ 1]\frac{1}{s(s+\frac{b}{J})}\begin{bmatrix} s+\frac{b}{J} & 1 \\ 0 & s \end{bmatrix}\begin{bmatrix} 0 \\ \frac{nk_m}{J} \end{bmatrix},$$

$$G(s) = \frac{\frac{nk_m}{J}s}{s(s+\frac{b}{J})}.$$

The following matrix:

$$\begin{bmatrix} C \\ CA \end{bmatrix} = \begin{bmatrix} 0 & 1 \\ 0 & -\frac{b}{J} \end{bmatrix},$$

has a rank less than $n = 2$ because its determinant is zero.

Note that in both cases the transfer function obtained satisfies what was established in (7.58): the polynomial at the denominator is equal to $\det(sI - A)$. Also note that this is true because matrix $[B\ AB]$ has the rank $n = 2$ in both cases. Finally, the effect due to the fact that the system is not observable is also present: in the second case the system is not observable; hence, the corresponding transfer function has one pole and one zero, which cancel each other out and this does not occur in the first case because the system is observable. Such pole-zero cancellation in the second case results in a first-order transfer function, which means that only one of the states (velocity) is described by this transfer function. This implies that the motor position has no effect when velocity is the variable to control.

The above results are useful for giving an interpretation of observability in the following. When the output is the position, the system is observable because, once the position is known, the velocity can be computed (the other state variable) by position differentiation, for instance. But, if the output is velocity, the knowledge of this variable does not suffice to compute the position (the other state variable): although it may be argued that the position $\theta(t)$ can be computed by simple velocity integration, note that:

$$\theta(t) - \theta(0) = \int_0^t \dot{\theta}(r)dr,$$

means that besides the knowledge of velocity, it is necessary to know the initial position $\theta(0)$, which is also unknown. Thus, the system is not observable when the output is velocity. However, as previously stated, if controlling the velocity is the only thing that matters, then this can be performed without any knowledge of the position. Observability is important in other classes of problems that are studied later in Sects. 7.12 and 7.13.

Finally, the reader can verify the correctness of the above results by comparing the transfer functions obtained with those in Chaps. 10 and 11.

7.9   A Realization of a Transfer Function

The realization of a transfer function consists in finding a dynamical equation that describes the transfer function in an identical way. The importance of this problem arises, for instance, when a controller is given that is expressed as a transfer function. Then, expressing a transfer function in terms of a dynamical equation is important for practical implementation purposes because the latter can be solved numerically, for instance.

An important property of the state variables representation is that it is not unique. This means that several dynamical equations exist correctly describing the same physical system. This also means that given a transfer function, several dynamical equations exist correctly describing this transfer function. In the following, a way of obtaining one of these dynamical equations is presented. Consider the following transfer function:

$$\frac{Y(s)}{U(s)} = G(s) = \frac{b_m s^m + b_{m-1} s^{m-1} + \cdots + b_1 s + b_0}{s^n + a_{n-1} s^{n-1} + \cdots + a_1 s + a_0}. \tag{7.60}$$

Solving for the output:

$$Y(s) = \frac{b_m s^m + b_{m-1} s^{m-1} + \cdots + b_1 s + b_0}{s^n + a_{n-1} s^{n-1} + \cdots + a_1 s + a_0} U(s).$$

Define the following variable:

$$V(s) = \frac{1}{s^n + a_{n-1} s^{n-1} + \cdots + a_1 s + a_0} U(s), \tag{7.61}$$

to write:

$$Y(s) = (b_m s^m + b_{m-1} s^{m-1} + \cdots + b_1 s + b_0) V(s). \tag{7.62}$$

Applying the inverse Laplace transform to (7.61) and solving for the highest order time derivative:

$$v^{(n)} = -a_{n-1} v^{(n-1)} - \cdots - a_1 \dot{v} - a_0 v + u, \tag{7.63}$$

where $V(s) = \mathcal{L}\{v\}$. According to Sect. 7.1, define the state variables as the unknown variable in (7.63), v, and its first $n - 1$ time derivatives:

$$\bar{x}_1 = v, \quad \bar{x}_2 = \dot{v}, \quad \bar{x}_3 = \ddot{v}, \quad \ldots, \quad \bar{x}_n = v^{(n-1)}. \tag{7.64}$$

Then, (7.63) can be written as:

$$\dot{\bar{x}}_n = -a_{n-1} \bar{x}_n - \cdots - a_1 \bar{x}_2 - a_0 \bar{x}_1 + u. \tag{7.65}$$

On the other hand, using the inverse Laplace transform in (7.62), the following is obtained:

$$y = b_m v^{(m)} + b_{m-1} v^{(m-1)} + \cdots + b_1 \dot{v} + b_0 v.$$

If it is assumed that m takes its maximal value, i.e., $m = n - 1$ (recall that $n > m$) then (7.64) can be used to write:

$$y = b_{n-1} \bar{x}_n + b_{n-2} \bar{x}_{n-1} + \cdots + b_1 \bar{x}_2 + b_0 \bar{x}_1. \tag{7.66}$$

Using (7.64), (7.65) and (7.66), the following can be written:

$$\dot{\bar{x}} = \bar{A} \bar{x} + \bar{B} u, \tag{7.67}$$

$$y = \bar{C} \bar{x},$$

$$\bar{A} = \begin{bmatrix} 0 & 1 & 0 & 0 & \cdots & 0 & 0 \\ 0 & 0 & 1 & 0 & \cdots & 0 & 0 \\ 0 & 0 & 0 & 1 & \cdots & 0 & 0 \\ 0 & 0 & 0 & 0 & \cdots & 0 & 0 \\ \vdots & \vdots & \vdots & \vdots & \ddots & \vdots & \vdots \\ 0 & 0 & 0 & 0 & \cdots & 0 & 1 \\ -a_0 & -a_1 & -a_2 & -a_3 & \cdots & -a_{n-2} & -a_{n-1} \end{bmatrix}, \quad \bar{B} = \begin{bmatrix} 0 \\ 0 \\ 0 \\ 0 \\ \vdots \\ 0 \\ 1 \end{bmatrix},$$

$$\bar{C} = \begin{bmatrix} b_0 & b_1 & b_2 & b_3 & \cdots & b_{n-2} & b_{n-1} \end{bmatrix},$$

$$\bar{x} = \begin{bmatrix} \bar{x}_1 & \bar{x}_2 & \bar{x}_3 & \bar{x}_4 & \cdots & \bar{x}_{n-1} & \bar{x}_n \end{bmatrix}^T.$$

The dynamical equation in (7.67) represents identically the transfer function in (7.60) and it is said that it is written in the *controllability canonical form*. As is shown later, the controllability canonical form is very useful for finding a procedure for designing a state feedback controller.

7.10 Equivalent Dynamical Equations

It has been shown in Sect. 7.8 that any controllable single-input single-output dynamical equation (7.51) can be written as the transfer function in (7.58). Note that requiring observability is merely to ensure that no pole-zero cancellation is present in (7.58). On the other hand, it has been shown in Sect. 7.9 that the transfer function in (7.58) can be written as the dynamical equation in (7.67). This means that any of the dynamical equations in (7.51) or (7.67) can be obtained from the other, i.e., these dynamical equations are equivalent.

In the following, it is shown how to obtain (7.51) from (7.67) and vice versa, without the necessity of the intermediate step of a transfer function. Consider the single-input single-output dynamical equation in (7.51) and define the following *linear transformation*:

$$\bar{x} = Px, \tag{7.68}$$

where P is an $n \times n$ constant matrix defined from its inverse matrix as [1], chapter 7:

$$P^{-1} = \begin{bmatrix} q_1 & \cdots & q_{n-2} & q_{n-1} & q_n \end{bmatrix}, \tag{7.69}$$

$$q_n = B,$$

$$q_{n-1} = AB + a_{n-1}B,$$

$$q_{n-2} = A^2B + a_{n-1}AB + a_{n-2}B,$$

$$\vdots$$

$$q_1 = A^{n-1}B + a_{n-1}A^{n-2}B + \cdots + a_1B,$$

$$\det(sI - A) = s^n + a_{n-1}s^{n-1} + \cdots + a_1s + a_0.$$

It is important to stress that the matrix P^{-1} is invertible, i.e., its inverse matrix P always exists if the matrix defined in (7.44) has the rank n [1], chapter 7, or equivalently if (7.51) is controllable. This can be explained as follows. If the matrix defined in (7.44) has the rank n, i.e., all its columns are linearly independent, then when adding its columns as in the previous expressions defining vectors $q_1, \ldots, q_{n-2}, q_{n-1}, q_n$, the columns $q_1, \ldots, q_{n-2}, q_{n-1}, q_n$ are linearly independent (see fact 7.1). This means that the determinant of P^{-1} is different from zero; hence, its inverse matrix P exists.

Using (7.68) in (7.51), i.e., $\dot{\bar{x}} = P\dot{x}$, it is found that:

$$\dot{\bar{x}} = \bar{A}\bar{x} + \bar{B}u,$$

$$y = \bar{C}\bar{x},$$

$$\bar{A} = PAP^{-1}, \quad \bar{B} = PB, \quad \bar{C} = CP^{-1}. \tag{7.70}$$

The matrix $\bar{A}$ and the vectors $\bar{B}$, $\bar{C}$ in (7.70) are given as in the controllability canonical form (7.67) no matter what the particular form of A, B and C is, as long as the matrix defined in (7.44) has the rank n. A formal proof for this latter statement requires mathematical tools that are outside of the scope of this book; hence, it is not presented. The reader is advised to see [1] for a complete solution to this problem. The reader can also verify these ideas by proposing numerical values for the matrices A, B, and C, and performing the corresponding computations. The above means that each one of the dynamical equations (7.51) and (7.67) can be obtained from the other and the relationship among the matrices involved is given in (7.70), i.e., these dynamical equations are equivalent. Note that the fundamental condition for the existence of this equivalence is that the matrix defined in (7.44) has the rank n [1], chapter 5. This is summarized as follows:

Theorem 7.4 ([1], chapter 7) *If (7.51) is controllable, then this dynamical equation is equivalent to (7.67) through the linear transformation (7.68), (7.69), i.e., through (7.70).*

Example 7.8 The linear approximate model of a mechanism known as the Furuta pendulum is obtained in Chap. 15. This model is presented in (15.17), (15.18), and it is rewritten here for ease of reference:

$$\dot{z} = Az + Bv, \tag{7.71}$$

$$A = \begin{bmatrix} 0 & 1 & 0 & 0 \\ 0 & 0 & \dfrac{-gm_1^2 l_1^2 L_0}{I_0(J_1+m_1 l_1^2)+J_1 m_1 L_0^2} & 0 \\ 0 & 0 & 0 & 1 \\ 0 & 0 & \dfrac{(I_0+m_1 L_0^2)m_1 l_1 g}{I_0(J_1+m_1 l_1^2)+J_1 m_1 L_0^2} & 0 \end{bmatrix}, \quad B = \begin{bmatrix} 0 \\ \dfrac{J_1+m_1 l_1^2}{I_0(J_1+m_1 l_1^2)+J_1 m_1 L_0^2} \dfrac{k_m}{r_a} \\ 0 \\ \dfrac{-m_1 l_1 L_0}{I_0(J_1+m_1 l_1^2)+J_1 m_1 L_0^2} \end{bmatrix}.$$

To render the algebraic manipulation easier, the following constants are defined:

$$a = \frac{-gm_1^2 l_1^2 L_0}{I_0(J_1+m_1 l_1^2)+J_1 m_1 L_0^2}, \qquad b = \frac{(I_0+m_1 L_0^2)m_1 l_1 g}{I_0(J_1+m_1 l_1^2)+J_1 m_1 L_0^2},$$

$$c = \frac{J_1+m_1 l_1^2}{I_0(J_1+m_1 l_1^2)+J_1 m_1 L_0^2}\frac{k_m}{r_a}, \qquad d = \frac{-m_1 l_1 L_0}{I_0(J_1+m_1 l_1^2)+J_1 m_1 L_0^2}\frac{k_m}{r_a}.$$

The following matrix has the form:

$$C_o = \begin{bmatrix} B & AB & A^2 B & A^3 B \end{bmatrix} = \begin{bmatrix} 0 & c & 0 & ad \\ c & 0 & ad & 0 \\ 0 & d & 0 & bd \\ d & 0 & bd & 0 \end{bmatrix}.$$

After a straightforward but long algebraic procedure, it is found that:

$$\det(C_o) = \frac{m_1^4 l_1^4 L_0^2 g^2}{[I_0(J_1+m_1 l_1^2)+J_1 m_1 L_0^2]^4}\frac{k_m^4}{r_a^4} \neq 0.$$

Thus, the dynamical equation in (7.71) is controllable for any set of the Furuta pendulum parameters. This result does not change if the mechanism is large or small, heavy or light. This also means that the matrix P introduced in (7.68) is nonsingular and it is computed in the following using (7.69). To simplify the computations, the following numerical values are employed:

$$I_0 = 1.137 \times 10^{-3}[\text{kgm}^2], \quad J_1 = 0.38672 \times 10^{-3}[\text{kgm}^2], \quad g = 9.81[\text{m/s}^2]$$

$$l_1 = 0.1875[\text{m}], \quad m_1 = 0.033[\text{kg}], \quad L_0 = 0.235[\text{m}], \quad \frac{k_m}{r_a} = 0.0215[\text{Nm/V}].$$

Hence:

$$A = \begin{bmatrix} 0 & 1 & 0 & 0 \\ 0 & 0 & -35.81 & 0 \\ 0 & 0 & 0 & 1 \\ 0 & 0 & 72.90 & 0 \end{bmatrix}, \quad B = \begin{bmatrix} 0 \\ 13.4684 \\ 0 \\ -12.6603 \end{bmatrix}.$$

With these data, the roots of the characteristic polynomial $\det(\lambda I - A)$ are found to be:

$$\lambda_1 = 0, \quad \lambda_2 = 0, \quad \lambda_3 = 8.5381, \quad \lambda_4 = -8.5381.$$

Then the following product is computed:

$$(\lambda - \lambda_1)(\lambda - \lambda_2)(\lambda - \lambda_3)(\lambda - \lambda_4) = \lambda^4 + a_3\lambda^3 + a_2\lambda^2 + a_1\lambda + a_0,$$

to find, by equating coefficients on both sides:

$$a_3 = 0, \quad a_2 = -72.9, \quad a_1 = 0, \quad a_0 = 0. \tag{7.72}$$

Using these values in (7.69) the following is found:

$$P^{-1} = \begin{bmatrix} -528.481 & 0 & 13.4684 & 0 \\ 0 & -528.481 & 0 & 13.4684 \\ 0 & 0 & -12.6603 & 0 \\ 0 & 0 & 0 & -12.6603 \end{bmatrix},$$

hence:

$$P = \begin{bmatrix} -0.0019 & 0 & -0.0020 & 0 \\ 0 & -0.0019 & 0 & -0.0020 \\ 0 & 0 & -0.0790 & 0 \\ 0 & 0 & 0 & -0.0790 \end{bmatrix}. \tag{7.73}$$

Finally, to verify (7.70), the following products are performed:

$$PAP^{-1} = \begin{bmatrix} 0 & 1.0000 & 0 & 0 \\ 0 & 0 & 1.0000 & 0 \\ 0 & 0 & 0 & 1.0000 \\ 0 & 0 & 72.9 & 0 \end{bmatrix}, \quad PB = \begin{bmatrix} 0 \\ 0 \\ 0 \\ 1 \end{bmatrix}.$$

Note that these matrices have exactly the same form as that presented in (7.67) for $\bar{A} = PAP^{-1}$ and $\bar{B} = PB$. The above computations have been performed by executing the following MATLAB code in an m-file:

```
clc
A=[0 1 0 0;
0 0 -35.81 0;
0 0 0 1;
0 0 72.90 0];
B=[0;
13.4684;
0;
-12.6603];
U=[B A*B A^2*B A^3*B];
det(U)
v=eig(A)
f=conv([1 -v(1)],[1 -v(2)]);
g=conv(f,[1 -v(3)]);
h=conv(g,[1 -v(4)]);
a3=h(2);
a2=h(3);
a1=h(4);
a0=h(5);
q4=B;
q3=A*B+a3*B;
q2=A^2*B+a3*A*B+a2*B;
q1=A^3*B+a3*A^2*B+a2*A*B+a1*B;
invP=[q1 q2 q3 q4]
P=inv(invP)
Ab=P*A*invP
Bb=P*B
```

7.11 State Feedback Control

In this section, it is assumed that the plant to be controlled is given in terms of the error equation (see Sect. 7.2), i.e., given a plant as in (7.51), it is assumed that the state x stands for the state error. Hence, the control objective is defined as ensuring that the state error tends toward zero as time increases, i.e., $\lim_{t \to \infty} x(t) = 0$.

Consider the single-input state equation in (7.51), in closed-loop with the following controller:

$$u = -Kx = -(k_1 x_1 + k_2 x_2 + \cdots + k_n x_n), \tag{7.74}$$

$$K = \begin{bmatrix} k_1 & k_2 & \dots & k_n \end{bmatrix},$$

where k_i, $i = 1, \dots, n$, are n constant scalars. Substituting (7.74) in (7.51), it is found that:

$$\dot{x} = (A - BK)x. \tag{7.75}$$

This state equation has no input like that in (7.40). Hence, the stability criterion for (7.40) can be applied to (7.75). This means that the vector $x(t)$, the solution of (7.75), satisfies $\lim_{t \to \infty} x(t) = 0$ if, and only if, all eigenvalues of the matrix $A - BK$ have a strictly negative real part. However, from the point of view of a practical application, this is not enough as a good closed-loop performance is also required. It is important to stress that the waveform of the solution $x(t)$ of (7.75) depends on the exact location of the eigenvalues of the matrix $A - BK$. Hence, it must be possible to arbitrarily assign the eigenvalues of the matrix $A - BK$. As the row vector K can be chosen as desired, it is of interest to know how to choose K such that the eigenvalues of $A - BK$ are assigned as desired. The solution to this problem is presented in the following.

Consider the linear transformation (7.68), (7.69), and substitute it in (7.75) to obtain:

$$\dot{\bar{x}} = \bar{A}\bar{x} - \bar{B}\bar{K}\bar{x}, \tag{7.76}$$
$$\bar{K} = KP^{-1},$$

where $\bar{A}$ and $\bar{B}$ are given as in (7.70) and (7.67). Note that $\bar{K}\bar{x}$ is a scalar given as:

$$\bar{K}\bar{x} = \bar{k}_1\bar{x}_1 + \bar{k}_2\bar{x}_2 + \cdots + \bar{k}_n\bar{x}_n,$$
$$\bar{K} = \begin{bmatrix} \bar{k}_1 & \bar{k}_2 & \dots & \bar{k}_n \end{bmatrix},$$

, which, according to (7.67), only affects the last row of (7.76); hence, the following can be written:

$$\dot{\bar{x}} = (\bar{A} - \bar{B}\bar{K})\bar{x},$$

$$\bar{A} - \bar{B}\bar{K} =$$

$$\begin{bmatrix} 0 & 1 & 0 & 0 & \cdots & 0 & 0 \\ 0 & 0 & 1 & 0 & \cdots & 0 & 0 \\ 0 & 0 & 0 & 1 & \cdots & 0 & 0 \\ 0 & 0 & 0 & 0 & \cdots & 0 & 0 \\ \vdots & \vdots & \vdots & \vdots & \ddots & \vdots & \vdots \\ 0 & 0 & 0 & 0 & \cdots & 0 & 1 \\ -\bar{k}_1 - a_0 & -\bar{k}_2 - a_1 & -\bar{k}_3 - a_2 & -\bar{k}_4 - a_3 & \cdots & -\bar{k}_{n-1} - a_{n-2} & -\bar{k}_n - a_{n-1} \end{bmatrix}.$$

If the following is chosen:

$$\bar{K} = \begin{bmatrix} \bar{a}_0 - a_0 & \bar{a}_1 - a_1 & \dots & \bar{a}_{n-1} - a_{n-1} \end{bmatrix},$$ (7.77)

the following is obtained:

$$\bar{A} - \bar{B}\bar{K} = \begin{bmatrix} 0 & 1 & 0 & 0 & \cdots & 0 & 0 \\ 0 & 0 & 1 & 0 & \cdots & 0 & 0 \\ 0 & 0 & 0 & 1 & \cdots & 0 & 0 \\ 0 & 0 & 0 & 0 & \cdots & 0 & 0 \\ \vdots & \vdots & \vdots & \vdots & \ddots & \vdots & \vdots \\ 0 & 0 & 0 & 0 & \cdots & 0 & 1 \\ -\bar{a}_0 & -\bar{a}_1 & -\bar{a}_2 & -\bar{a}_3 & \cdots & -\bar{a}_{n-2} & -\bar{a}_{n-1} \end{bmatrix}.$$

Note that, according to (7.70) and (7.76), the following can be written:

$$\bar{A} - \bar{B}\bar{K} = PAP^{-1} - PBKP^{-1} = P(A - BK)P^{-1},$$

i.e., the matrices $\bar{A} - \bar{B}\bar{K}$ and $A - BK$ satisfy (7.30); hence, they possess identical eigenvalues. According to previous sections, the eigenvalues λ of $\bar{A} - \bar{B}\bar{K}$ satisfy:

$$\det(\lambda I - [\bar{A} - \bar{B}\bar{K}]) = \lambda^n + \bar{a}_{n-1}\lambda^{n-1} + \cdots + \bar{a}_1\lambda + \bar{a}_0,$$

$$= \prod_{i=1}^{n}(\lambda - \bar{\lambda}_i),$$

where $\bar{\lambda}_i$, $i = 1, \dots, n$ stand for the desired eigenvalues and these are proposed by the designer. It is stressed that when a complex conjugate desired eigenvalue is proposed, then its corresponding complex conjugate pair must also be proposed. This is to ensure that all the coefficients $\bar{a}_i$, $i = 0, 1, \dots, n - 1$ are real, which also ensures that all the gains $\bar{K}$ and K are real. The following is the procedure [1], chapter 7, suggested to compute the vector of the controller gains K, which assigns the desired eigenvalues for the closed-loop matrix $A - BK$.

- Check that the state equation in (7.51) is controllable. If this is the case, proceed as indicated in the remaining steps. Otherwise, it is not possible to proceed.
- Find the polynomial:

$$\det(\lambda I - A) = \lambda^n + a_{n-1}\lambda^{n-1} + \cdots + a_1\lambda + a_0.$$

- Propose the desired closed-loop eigenvalues $\bar{\lambda}_1, \bar{\lambda}_2, \dots, \bar{\lambda}_n$. If some of them are complex, then also propose its corresponding complex conjugate pair as one of the desired eigenvalues.

- Compute:

$$\prod_{i=1}^{n}(\lambda - \bar{\lambda}_i) = \lambda^n + \bar{a}_{n-1}\lambda^{n-1} + \cdots + \bar{a}_1\lambda + \bar{a}_0.$$

- Compute:

$$\bar{K} = \begin{bmatrix} \bar{a}_0 - a_0 & \bar{a}_1 - a_1 & \ldots & \bar{a}_{n-1} - a_{n-1} \end{bmatrix}.$$

- Compute the vectors $q_1, \ldots, q_{n-2}, q_{n-1}, q_n$ according to (7.69), obtain the matrix P^{-1} defined in that expression and obtain its inverse matrix P.
- Compute the vector of controller gains in (7.74) as $K = \bar{K}P$.

Example 7.9 Let us continue with the Example 7.8, in which the Furuta pendulum is studied. It is desired, now, to find the vector of controller gains K, which, when used in the controller (7.74) (with $x = z$), assigns the eigenvalues of matrix $A - BK$ at:

$$\bar{\lambda}_1 = -94, \quad \bar{\lambda}_2 = -18, \quad \bar{\lambda}_3 = -0.5, \quad \bar{\lambda}_4 = -1.$$

To this aim, the coefficients of the following polynomial:

$$(\lambda - \bar{\lambda}_1)(\lambda - \bar{\lambda}_2)(\lambda - \bar{\lambda}_3)(\lambda - \bar{\lambda}_4) = \lambda^4 + \bar{a}_3\lambda^3 + \bar{a}_2\lambda^2 + \bar{a}_1\lambda + \bar{a}_0,$$

are found to be:

$$\bar{a}_3 = 113.5, \quad \bar{a}_2 = 1860.5, \quad \bar{a}_1 = 2594, \quad \bar{a}_0 = 846.$$

Using these values and those obtained in (7.72), $\bar{K}$ is computed according to (7.77), i.e.,:

$$\bar{K} = \begin{bmatrix} 846 & 2594 & 1933.3 & 113.5 \end{bmatrix}.$$

Finally, using this and matrix P shown in (7.73) the vector of the controller gains K is computed using $K = \bar{K}P$:

$$K = \begin{bmatrix} -1.5997 & -4.9138 & -154.4179 & -14.1895 \end{bmatrix}.$$

The above computations are performed by executing the following MATLAB code after executing the MATLAB code at the end of Example 7.8:

```
lambda1d=-94;
lambda2d=-18;
lambda3d=-0.5;
lambda4d=-1;
```

```
fd=conv([1 -lambda1d],[1 -lambda2d]);
gd=conv(fd,[1 -lambda3d]);
hd=conv(gd,[1 -lambda4d]);
a3b=hd(2);
a2b=hd(3);
a1b=hd(4);
a0b=hd(5);
Kb=[a0b-a0 a1b-a1 a2b-a2 a3b-a3];
K=Kb*P
eig(A-B*K)
```

The command "eig(A-B*K)" is employed to verify that matrix A-B*K has the desired eigenvalues. Some simulation results are presented in Figs. 7.5 and 7.6 when controlling the Furuta pendulum using the vector gain $K = [-1.5997\ -4.9138\ -$

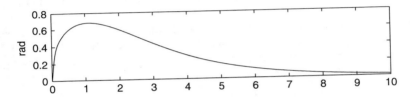

Fig. 7.5 Simulation result for the state z_1 of the Furuta pendulum

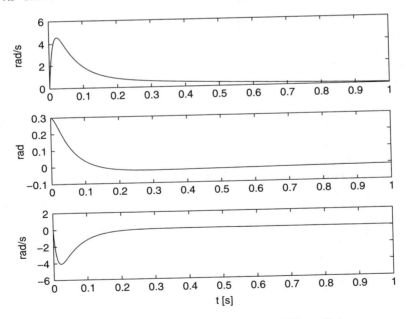

Fig. 7.6 Simulation results for the Furuta pendulum. Top: z_2. Middle: z_3. Bottom: z_4

Fig. 7.7 MATLAB/Simulink
diagram for the Furuta
pendulum

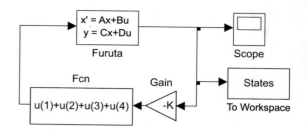

154.4179 $-$ 14.1895] and $z_1(0) = z_2(0) = z_4(0) = 0$, $z_3(0) = 0.3$ as initial
conditions. Note that all the state variables converge to zero, as expected. These
simulations were performed using the MATLAB/Simulink diagram in Fig. 7.7.
Block Furuta has the following parameters:

$$A \Rightarrow A, \quad B \Rightarrow B, \quad [1\ 0\ 0\ 0; 0\ 1\ 0\ 0; 0\ 0\ 1\ 0; 0\ 0\ 0\ 1] \Rightarrow C, \quad [0; 0; 0; 0] \Rightarrow D,$$

and $[0\ 0\ 0.3\ 0]$ as initial conditions. To obtain these results, one has to proceed
as follows: (1) Execute the above MATLAB code to compute all the required
parameters. (2) Run the simulation in Fig. 7.7. (3) Execute the following MATLAB
code in an m-file:

```
nn=length(States(:,1));
n=nn-1;
Ts=1/n;
t=0:Ts:1;
figure(1)
subplot(3,1,1)
plot(t,States(:,1),'b-');
ylabel('rad')
figure(2)
subplot(3,1,1)
plot(t,States(:,2),'b-');
ylabel('rad/s')
subplot(3,1,2)
plot(t,States(:,3),'b-');
ylabel('rad')
subplot(3,1,3)
plot(t,States(:,4),'b-');
ylabel('rad/s')
xlabel('t [s]')
```

At this point, Figs. 7.5 and 7.6 are drawn.

7.12   State Observers

In this section, it is assumed again that the plant to be controlled is given in terms of the error equation (see Sect. 7.2), i.e., given a plant as in (7.51), it is assumed that the state x stands for the error state. Hence, the control objective is defined as ensuring that the state error tends toward zero as time increases, i.e., $\lim_{t \to \infty} x(t) = 0$.

As pointed out in Sect. 7.1, the state x is composed of variables that are internal to the system; hence, they are not known in general. On the other hand, the output y represents a variable that is always known. This means that the practical implementation of the controller (7.74) is only possible if the complete state is measured. Because of this, it is important to know an estimate of the state $x(t)$, which is represented by $\widehat{x}(t)$, allowing a controller to be constructed of the form:

$$u = -K\widehat{x} = -(k_1\widehat{x}_1 + k_2\widehat{x}_2 + \cdots + k_n\widehat{x}_n), \qquad (7.78)$$

$$K = \begin{bmatrix} k_1 \ k_2 \ \dots \ k_n \end{bmatrix}, \quad \widehat{x} = \begin{bmatrix} \widehat{x}_1 \ \widehat{x}_2 \ \dots \ \widehat{x}_n \end{bmatrix}^T.$$

The estimate $\widehat{x}$ is computed using a *state observer* or, simply, an *observer*. An observer must compute $\widehat{x}$ exclusively employing information provided by the system input and output. On the other hand, if the controller (7.78) has to replace the controller in (7.74), then a fundamental property that an observer must satisfy is that the estimate $\widehat{x}$ converges to x as fast as possible or, at least, asymptotically, i.e., such that:

$$\lim_{t \to \infty} \widehat{x}(t) = x(t).$$

An observer satisfying this property is the following:

$$\dot{\widehat{x}} = (A - LC)\widehat{x} + Ly + Bu, \qquad (7.79)$$

where $L = [L_1, L_2, \ldots, L_n]^T$ is a constant column vector. This can be explained as follows. Define the estimation error as $\tilde{x} = x - \widehat{x}$. Then, subtracting (7.79) from (7.51) the following is obtained:

$$\dot{\tilde{x}} = Ax - (A - LC)\widehat{x} - Ly.$$

Using $y = Cx$:

$$\dot{\tilde{x}} = A\tilde{x} - LC\tilde{x},$$
$$= (A - LC)\tilde{x}.$$

Using the results in Sect. 7.6, it is concluded that $\lim_{t \to \infty} \tilde{x}(t) = 0$; hence, $\lim_{t \to \infty} \widehat{x}(t) = x(t)$ if, and only if, all the eigenvalues of matrix $A - LC$ have strictly negative real parts. Hence, the only problem that remains is how to select the gain column vector L such that all the eigenvalues of matrix $A - LC$ have strictly negative real parts and that they can be arbitrarily assigned. This problem is solved as follows:

Theorem 7.5 ([1], pp. 358) *The state of the dynamical equation in (7.51) can be estimated using the observer in (7.79) and all the eigenvalues of matrix $A - LC$ can be arbitrarily assigned, if, and only if, (7.51) is observable.*

Again, the complex eigenvalues must appear as complex conjugate pairs. To explain the above result, the following theorem is useful.

Theorem 7.6 ([1], pp. 195) *Consider the following dynamical equations:*

$$\dot{x} = Ax + Bu, \qquad\qquad (7.80)$$

$$y = Cx,$$

$$\dot{z} = -A^T z + C^T u, \qquad\qquad (7.81)$$

$$\gamma = B^T z,$$

where $u, y, \gamma \in R$, whereas $x, z \in R^n$. The dynamical equation in (7.80) is controllable (observable) if, and only if, the dynamical equation in (7.81) is observable (controllable).

Hence, going back to the observer problem, as the pair (A, C) is observable, then the pair $(-A^T, C^T)$ and the pair (A^T, C^T) are controllable (because sign "$-$" of matrix A^T does not affect the linear independence of the columns of the matrix in (7.44), see Sect. 7.4). From this, it is concluded that, following the procedure introduced in Sect. 7.11, it is always possible to find a constant row vector K such that the matrix $A^T - C^T K$ has any desired set of eigenvalues (arbitrary eigenvalues). As the eigenvalues of any matrix are identical to those of its transposed matrix (see Sect. 7.4), then the matrix $A - K^T C$ has the same eigenvalues as matrix $A^T - C^T K$ (the desired eigenvalues). Defining $L = K^T$, the gain column vector required to design the observer in (7.79) is computed.

It is stressed that assigning the eigenvalues to the matrix $A - LC$ arbitrarily means that they can be located where desired by the designer. It is obvious that all eigenvalues must have strictly negative real parts; however, they must be located on regions of the complex plane ensuring a fast convergence of $\widehat{x}(t)$ to $x(t)$. To determine where the eigenvalues of matrix $A - LC$ must be located, the concepts of the transient response studied in Chap. 3, where a relationship has been established with a location of poles of the transfer function in (7.58), are very important.

Finally, note that given the plant $\dot{x} = Ax + Bu$ we can try to construct an observer as $\dot{\widehat{x}} = A\widehat{x} + Bu$, which results in the following estimation error dynamics:

$$\dot{\tilde{x}} = A\tilde{x}, \quad \tilde{x} = x - \widehat{x}.$$

If all the eigenvalues of the matrix A have negative real parts, then $\lim_{t\to\infty} \tilde{x}(t) = 0$ is ensured and the observer works. However, this result is very limited because it is constrained to be used only with plants that are open-loop asymptotically stable. Moreover, even in such a case, the convergence speed of $\tilde{x} \to 0$ depends on the eigenvalue locations of the open-loop plant, i.e., it is not possible to render such a convergence arbitrarily fast. Note that, according to the previous arguments in this section, these drawbacks are eliminated when using the observer in (7.79). Also see Example 7.6.

7.13 The Separation Principle

When a dynamical equation (plant):

$$\dot{x} = Ax + Bu, \tag{7.82}$$

$$y = Cx,$$

and an observer:

$$\dot{\widehat{x}} = (A - LC)\widehat{x} + Ly + Bu, \tag{7.83}$$

are connected using the following controller (see Fig. 7.8):

$$u = -K\widehat{x} = -(k_1\widehat{x}_1 + k_2\widehat{x}_2 + \cdots + k_n\widehat{x}_n), \tag{7.84}$$

the question arises regarding how the stability of the closed-loop system (7.82), (7.83), (7.84), is affected, i.e., it is important to verify whether ensuring that $\lim_{t\to\infty}\widehat{x}(t) = x(t)$ and $\lim_{t\to\infty} x(t) = 0$ are enough to ensure stability

Fig. 7.8 A feedback system employing an observer to estimate the state

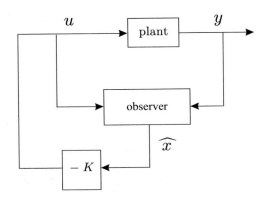

of the complete closed-loop system. The answer to this question is known as the *separation principle*, which claims the following:

Fact 7.12 ([1], pp. 367) *The eigenvalues of the closed-loop system (7.82), (7.83), (7.84), are the union of the eigenvalues of matrices $A - BK$ and $A - LC$.*

This means that the eigenvalues of the observer are not affected by the feedback in (7.84) and, at least in what concerns the eigenvalues, there is no difference between feeding back the estimate $\hat{x}(t)$ and feeding back the actual state $x(t)$. However, the reader must be aware that the system's transient response is often different if $x(t)$ is employed or $\hat{x}(t)$ is employed to compute the input. The important fact is that both $\lim_{t \to \infty} \hat{x}(t) = x(t)$ and $\lim_{t \to \infty} x(t) = 0$ are still true and the complete closed-loop system is stable. Thus, the design of the vector of controller gains K and the vector of gains L for the observer can be performed independently of each other.

Example 7.10 Consider the DC motor in Example 7.7, i.e., consider the dynamical equation:

$$\dot{x} = Ax + Bu, \quad y = Cx, \tag{7.85}$$

$$x = \begin{bmatrix} x_1 \\ x_2 \end{bmatrix} = \begin{bmatrix} \theta \\ \dot{\theta} \end{bmatrix}, \quad A = \begin{bmatrix} 0 & 1 \\ 0 & -\frac{b}{J} \end{bmatrix}, \quad B = \begin{bmatrix} 0 \\ \frac{nk_m}{J} \end{bmatrix}, \quad C = [1 \ 0], \quad u = i^*,$$

where it is assumed that the output is the position whereas velocity cannot be measured; hence, it shall be estimated using an observer. Suppose that it is desired to control the motor to reach a constant position, i.e., the desired values of the position and velocity are:

$$x_{1d} = \theta_d, \quad x_{2d} = 0,$$

where θ_d is a constant representing the desired position and the desired velocity is zero because the desired position is constant. Defining the state variables as:

$$\tilde{x}_1 = x_1 - x_{1d}, \quad \tilde{x}_2 = x_2,$$

and computing $\dot{\tilde{x}}_1 = \dot{x}_1 - \dot{x}_{1d} = x_2 = \tilde{x}_2$, $\dot{\tilde{x}}_2 = \dot{x}_2 = \ddot{\theta}$, the following dynamical equation is obtained:

$$\dot{\tilde{x}} = A\tilde{x} + Bu, \tag{7.86}$$

$$y = C\tilde{x},$$

where y is the new output and the matrix A and vectors B, C, are defined as in (7.85). Note that the measured output is now the position error $y = \tilde{x}_1$. From this point on, the numerical values of the motor controlled in Chap. 11 are considered, i.e.,:

$$k = \frac{nk_m}{J} = 675.4471, \quad a = \frac{b}{J} = 2.8681. \tag{7.87}$$

The corresponding observer is designed in the following. In the Example 7.7, it was shown that the dynamical equation in (7.85) is observable; hence, the dynamical equation in (7.86) is also observable. According to theorem 7.6, this implies that pairs $(-A^T, C^T)$ and (A^T, C^T) are controllable, i.e., that the following matrices have the rank $n = 2$:

$$[\, C^T \; -A^T C^T \,], \quad [\, C^T \; A^T C^T \,],$$

because the sign "$-$" of the matrix A does not affect the linear independence of these columns (see Sect. 7.4). Using the numerical values in (7.87), the roots of the polynomial $\det(\lambda I - A^T)$ are computed:

$$\lambda_1 = 0, \quad \lambda_2 = -2.8681.$$

Then, the following product is performed:

$$(\lambda - \lambda_1)(\lambda - \lambda_2) = \lambda^2 + a_1\lambda + a_0,$$

and equating coefficients:

$$a_1 = 2.8681, \quad a_0 = 0. \tag{7.88}$$

Using $B = C^T$ and A^T, instead of A, (7.69) becomes:

$$P^{-1} = [\, q_1 \; q_2 \,],$$
$$q_2 = C^T,$$
$$q_1 = A^T C^T + a_1 C^T.$$

Using the numerical values in (7.87), (7.88), in addition to the matrix A and vector C defined in (7.85) the following is obtained:

$$P^{-1} = \begin{bmatrix} 2.8681 & 1.0000 \\ 1.0000 & 0 \end{bmatrix},$$

hence:

$$P = \begin{bmatrix} 0 & 1.0000 \\ 1.0000 & -2.8681 \end{bmatrix}. \tag{7.89}$$

Suppose that it is desired to assign the following eigenvalues to matrix $A - LC$:

$$\bar{\lambda}_1 = -150, \quad \bar{\lambda}_2 = -100. \tag{7.90}$$

Then, the following polynomial is computed:

$$(\lambda - \bar{\lambda}_1)(\lambda - \bar{\lambda}_2) = \lambda^2 + \bar{a}_1\lambda + \bar{a}_0,$$

to find, equating coefficients, that:

$$\bar{a}_1 = 250, \quad \bar{a}_0 = 15000.$$

Using these data and values in (7.88), the vector of gains $\bar{K}$ is computed according to (7.77):

$$\bar{K} = \begin{bmatrix} 15000 & 247 \end{bmatrix}.$$

Using this and the matrix P shown in (7.89), the vector of the controller gains K is computed as $K = \bar{K}P$ and it is assigned $L = K^T$:

$$L = \begin{bmatrix} 247 \\ 14291 \end{bmatrix}. \tag{7.91}$$

On the other hand, using a similar procedure to those in examples 7.14.1 and 7.14.2, it is found that the vector of the controller gains:

$$K = \begin{bmatrix} 1.6889 & 0.0414 \end{bmatrix}, \tag{7.92}$$

assigns at:

$$-15.4 + 30.06j, \quad -15.4 - 30.06j, \tag{7.93}$$

the eigenvalues of the matrix $A - BK$. Note that the gains in (7.92) are identical to the proportional and velocity feedback gains for the controller designed and experimentally tested in Sect. 11.2.1 Chap. 11, where the closed-loop poles are assigned at the values in (7.93). The above computations have been performed by executing the following MATLAB code in an m-file:

```
clc
k=675.4471;
a=2.8681;
A=[0 1;
 0 -a];
B=[0;
k];
```

```
C=[1  0];
v=eig(A');
h=conv([1  -v(1)],[1  -v(2)]);
a0=h(3);
a1=h(2);
q2=C';
q1=A'*C'+a1*C';
invP=[q1 q2]
P=inv(invP)
lambda1d=-150;
lambda2d=-100;
hd=conv([1 -lambda1d],[1 -lambda2d]);
a0b=hd(3);
a1b=hd(2);
Kb=[a0b-a0 a1b-a1];
K=Kb*P;
L=K'

v=eig(A);
h=conv([1  -v(1)],[1  -v(2)]);
a0=h(3);
a1=h(2);
q2=B;
q1=A*B+a1*B;
invP=[q1 q2]
P=inv(invP)
lambda1d=-15.4+30.06*j;
lambda2d=-15.4-30.06*j;
hd=conv([1 -lambda1d],[1 -lambda2d]);
a0b=hd(3);
a1b=hd(2);
Kb=[a0b-a0 a1b-a1];
K=Kb*P
F=[L B];
```

Finally, it is stressed that the observer to construct is given as:

$$\dot{z} = (A - LC)z + L\gamma + Bu, \tag{7.94}$$

where $z = [z_1 \ z_2]^T$ is the estimate of vector $\tilde{x} = [\tilde{x}_1 \ x_2]^T$ y $u = i^*$. The controller is given as:

$$u = -K \begin{bmatrix} z_1 \\ z_2 \end{bmatrix}, \tag{7.95}$$

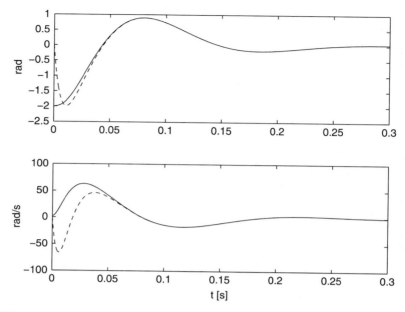

Fig. 7.9 Simulation results. Use of the observer in (7.94) to control a DC motor Top figure: continuous $\tilde{x}_1$, dashed z_1 Bottom figure: continuous $\tilde{x}_2 = x_2$, dashed z_2

where L and K take the values indicated in (7.91) and (7.92). Note that the controller must employ z_1, i.e., the estimate of $\tilde{x}_1$, despite $\tilde{x}_1$ being a known measured variable. This is because the theory above assumes that the whole state is to be estimated and fed back. In this respect, it is worth saying that the so-called reduced-order observers also exist that estimate only the unknown part of the estate if the other states are known. The reader is referred to [1] for further details.

In Fig. 7.9, some simulation results are shown when using the observer in (7.94) and the feedback in (7.95) together with the gains in (7.91) and (7.92) to control a DC motor whose parameters are shown in (7.87). The desired position is $\theta_d = 2$, the observer initial conditions are $z(0) = [0\ 0]^T$, whereas $\theta(0) = 0$ (i.e., $\tilde{x}_1(0) = -2$) and $\dot{\theta}(0) = 2$. It is observed that the estimates z_1 and z_2 asymptotically converge to the real values $\tilde{x}_1$ and x_2 and that both $\tilde{x}_1$ and x_2 converge to zero, i.e., θ converges to θ_d. It is interesting to realize that this convergence is achieved before the motor finishes responding. This has been accomplished by selecting the eigenvalues of the matrix $A - LC$ (shown in (7.90)) to be much faster than the eigenvalues assigned to the matrix $A - BK$ (shown in (7.93)). This is a criterion commonly used to assign the eigenvalues for an observer.

In Fig. 7.10 the MATLAB/Simulink diagram used to perform the above simulations is presented. Block step represents $\theta_d = 2$. Block DC motor has the following parameters:

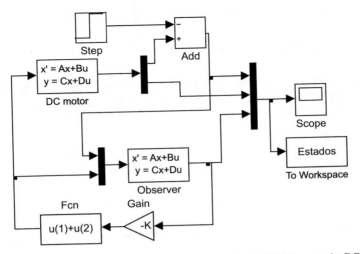

Fig. 7.10 MATLAB/Simulink diagram for use of the observer in (7.94) to control a DC motor

$$A \Rightarrow A, \quad B \Rightarrow B, \quad [1\ 0; 0\ 1] \Rightarrow C, \quad [0; 0] \Rightarrow D,$$

and [0 2] as initial conditions. The observer block has the following parameters:

$$A - L * [1\ 0] \Rightarrow A, \quad F \Rightarrow B, \quad [1\ 0; 0\ 1] \Rightarrow C, \quad [0\ 0; 0\ 0] \Rightarrow D,$$

and [0 0] as initial conditions, where the matrix F is computed at the end of the last MATLAB code above in an m-file. The simulation must be performed following these steps: (1) Execute last MATLAB code above in an m-file. (2) Run the simulation in Fig. 7.10. (3) Execute the following MATLAB code in an m-file:

```
nn=length(Estados(:,1));
n=nn-1;
Ts=0.5/n;
t=0:Ts:0.5;
figure(1)
subplot(2,1,1)
plot(t,Estados(:,1),'b-',t,Estados(:,3),'r--');
axis([0 0.3 -2.5 1])
ylabel('rad')
subplot(2,1,2)
plot(t,Estados(:,2),'b-',t,Estados(:,4),'r--');
axis([0 0.3 -100 100])
xlabel('t [s]')
ylabel('rad/s')
```

At this point, Fig. 7.9 is drawn.

7.14 Case Study: The Inertia Wheel Pendulum

7.14.1 Obtaining Forms in (7.67)

In Chap. 16 the linear approximate model of a mechanism known as the inertia wheel pendulum is obtained. This model is shown in (16.25), (16.28), and it is rewritten here for ease of reference:

$$\dot{z} = Az + Bw, \tag{7.96}$$

$$A = \begin{bmatrix} 0 & 1 & 0 \\ \overline{d}_{11}\overline{m}g & 0 & 0 \\ \overline{d}_{21}\overline{m}g & 0 & 0 \end{bmatrix}, \quad B = \begin{bmatrix} 0 \\ \overline{d}_{12} \\ \overline{d}_{22} \end{bmatrix} \frac{k_m}{R}.$$

To simplify the algebraic manipulation, the following constants are defined:

$$a = \overline{d}_{11}\overline{m}g, \quad b = \overline{d}_{21}\overline{m}g,$$

$$c = \overline{d}_{12}\frac{k_m}{R}, \quad d = \overline{d}_{22}\frac{k_m}{R}.$$

The following matrix has the form:

$$C_o = \begin{bmatrix} B & AB & A^2B \end{bmatrix} = \begin{bmatrix} 0 & c & 0 \\ c & 0 & ac \\ d & 0 & bc \end{bmatrix}.$$

After a straightforward procedure, the following is found:

$$\det(C_o) = (\overline{d}_{12})^2 \left(\frac{k_m}{R}\right)^3 \overline{m}g(\overline{d}_{11}\overline{d}_{22} - \overline{d}_{12}\overline{d}_{21}) \neq 0,$$

because $\overline{d}_{11}\overline{d}_{22} - \overline{d}_{12}\overline{d}_{21} \neq 0$ is a property of the mechanism, as explained in Chap. 16. Hence, the dynamical equation in (7.96) is controllable for any set of the inertia wheel parameters. This means that the result does not change whether the mechanism is large or small, heavy or light. The matrix P introduced in (7.68) is nonsingular and it is obtained in the following using (7.69). To simplify the computations, the following numerical values are considered:

$$d_{11} = 0.0014636, \quad d_{12} = 0.0000076,$$

$$d_{21} = 0.0000076, \quad d_{22} = 0.0000076,$$

$$\overline{m}g = 0.12597,$$

where:

$$D^{-1} = \begin{bmatrix} \bar{d}_{11} & \bar{d}_{12} \\ \bar{d}_{21} & \bar{d}_{22} \end{bmatrix} = \frac{1}{d_{11}d_{22} - d_{12}d_{21}} \begin{bmatrix} d_{22} & -d_{12} \\ -d_{21} & d_{11} \end{bmatrix},$$

, which correspond to parameters of the inertia wheel pendulum that is built and experimentally controlled in Chap. 16. Thus:

$$A = \begin{bmatrix} 0 & 1.0000 & 0 \\ 86.5179 & 0 & 0 \\ -86.5179 & 0 & 0 \end{bmatrix}, \quad B = \begin{bmatrix} 0 \\ -1.2758 \\ 245.6998 \end{bmatrix}.$$

Using these data, the roots of the polynomial $\det(\lambda I - A)$ are found to be:

$$\lambda_1 = 0, \quad \lambda_2 = 9.3015, \quad \lambda_3 = -9.3015.$$

Then, the following product is performed:

$$(\lambda - \lambda_1)(\lambda - \lambda_2)(\lambda - \lambda_3) = \lambda^3 + a_2\lambda^2 + a_1\lambda + a_0,$$

and, equating coefficients, the following is found:

$$a_2 = 0, \quad a_1 = -86.5179, \quad a_0 = 0. \tag{7.97}$$

Use of this in (7.69) yields:

$$P^{-1} = \begin{bmatrix} 0 & -1.275 & 0 \\ 5.467 \times 10^{-5} & 0 & -1.275 \\ -21147.049 & 0 & 245.6998 \end{bmatrix},$$

i.e.,

$$P = 10^{-5} \times \begin{bmatrix} 0 & -910.6662 & -4.7287 \\ -78379.7677 & 0 & 0 \\ 0 & -78379.8067 & -0.0002 \end{bmatrix}. \tag{7.98}$$

Finally, to verify the expressions in (7.70), the following products are performed:

$$PAP^{-1} = \begin{bmatrix} 0 & 1.0000 & 0 \\ -0.0000 & 0 & 1.0000 \\ 0 & 86.5179 & 0 \end{bmatrix}, \quad PB = \begin{bmatrix} 0 \\ 0 \\ 1 \end{bmatrix}.$$

Note that these matrices have exactly the forms shown in (7.67) for $\bar{A} = PAP^{-1}$ and $\bar{B} = PB$.

7.14.2 State Feedback Control

Now, the vector of controller gains K is computed. It is desired that using the controller (7.74) (with $x = z$ and $u = w$), the eigenvalues of matrix $A - BK$ are assigned at:

$$\bar{\lambda}_1 = -5.8535 + 17.7192j, \quad \bar{\lambda}_2 = -5.8535 - 17.7192j, \quad \bar{\lambda}_3 = -0.5268.$$

Hence, these values are used to compute the polynomial:

$$(\lambda - \bar{\lambda}_1)(\lambda - \bar{\lambda}_2)(\lambda - \bar{\lambda}_3) = \lambda^3 + \bar{a}_2\lambda^2 + \bar{a}_1\lambda + \bar{a}_0,$$

and, equating coefficients:

$$\bar{a}_2 = 12.2338, \quad \bar{a}_1 = 354.4008, \quad \bar{a}_0 = 183.4494.$$

These data and the values obtained in (7.97) are employed to compute the vector of the gains $\bar{K}$ according to (7.77):

$$\bar{K} = \begin{bmatrix} 183.44 & 440.91 & 12.23 \end{bmatrix}.$$

Finally, using this and the matrix P shown in (7.98) the vector of the controller gains K is computed using $K = \bar{K}P$ as:

$$K = \begin{bmatrix} -345.5910 & -11.2594 & -0.0086 \end{bmatrix},$$

which, except for some rounding errors, is the vector of the controller gains used to experimentally control the inertia wheel pendulum in Chap. 16. Some simulation results are presented in Fig. 7.11 where $z_1(0) = 0.3$ and $z_2(0) = z_3(0) = 0$ were set as initial conditions. It is observed that the state variables converge to zero, as desired. These simulations were performed using the MATLAB/Simulink diagram shown in Fig. 7.12. The inertia wheel pendulum block has the following parameters:

$$A \Rightarrow A, \quad B \Rightarrow B, \quad [1\ 0\ 0; 0\ 1\ 0; 0\ 0\ 1] \Rightarrow C, \quad [0; 0; 0] \Rightarrow D,$$

and $[0.3\ 0\ 0]$ as initial conditions. The results in Fig. 7.11 were obtained following these steps: 1) Execute the following MATLAB code in an m-file to obtain all the above computations:

```
clc
A=[0 1 0;
86.5179 0 0;
-86.5179 0 0];
B=[0;
-1.2758;
```

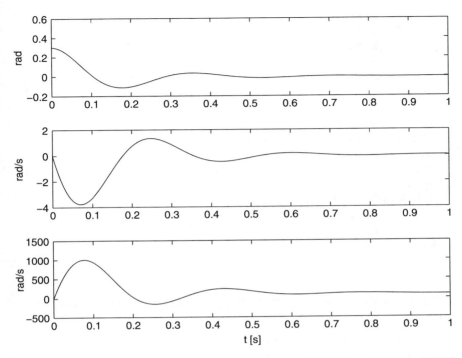

Fig. 7.11 Simulation results when using the gain vector $K = [-345.5910 \ -11.2594 \ -0.0086]$ to control the inertia wheel pendulum. Top: z_1. Middle: z_2. Bottom: z_3

Fig. 7.12
MATLAB/Simulink diagram used to simulate the control of the inertia wheel pendulum

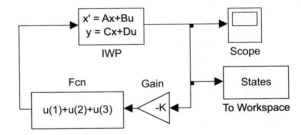

```
245.6998];
U=[B A*B A^2*B];
det(U)
v=eig(A);
f=conv([1 -v(1)],[1 -v(2)]);
h=conv(f,[1 -v(3)]);
a2=h(2);
a1=h(3);
a0=h(4);
q3=B;
```

```
q2=A*B+a2*B;
q1=A^2*B+a2*A*B+a1*B;
invP=[q1 q2 q3]
P=inv(invP)
Ab=P*A*invP
Bb=P*B

lambda1d=-5.8535+17.7192*j;
lambda2d=-5.8535-17.7192*j;
lambda3d=-0.5268;
fd=conv([1 -lambda1d],[1 -lambda2d]);
hd=conv(fd,[1 -lambda3d]);
a2b=hd(2);
a1b=hd(3);
a0b=hd(4);
Kb=[a0b-a0 a1b-a1 a2b-a2];
K=Kb*P
eig(A-B*K)
```

2) Run the simulation in Fig. 7.12. 3) Execute the following MATLAB code in an m-file:

```
nn=length(States(:,1));
n=nn-1;
Ts=1/n;
t=0:Ts:1;
figure(1)
subplot(3,1,1)
plot(t,States(:,1),'b-');
ylabel('rad')
subplot(3,1,2)
plot(t,States(:,2),'b-');
ylabel('rad/s')
subplot(3,1,3)
plot(t,States(:,3),'b-');
ylabel('rad/s')
xlabel('t [s]')
```

At this point, Fig. 7.11 is drawn.

7.15 Summary

The mathematical models used in the state variable approach are sets of first-order differential equations that must be solved simultaneously. Hence, the analysis and design are performed in the time domain and the Laplace transform is no longer used. This allows the study of nonlinear control systems, i.e., those systems represented by nonlinear differential equations (see Chap. 16 for an example of such applications). Recall that the Laplace transform cannot be employed when the differential equations are nonlinear.

Although an equivalence exists between the state space representation and the transfer function, the former is more general. This can be seen in the fact that the transfer function only represents the controllable and observable part of a state space representation. This means that there are some parts of a system that cannot be described by the transfer function. However, if a system is controllable and observable, the analysis and the design of the control system are simplified. In fact, there are powerful results for this case, which are presented in Sects. 7.11 and 7.12.

An advantage of the state space approach is that it gives simple solutions to problems that have a more complicated result when using the transfer function approach. Two examples of this situation are the experimental prototypes that are controlled in Chaps. 15 and 16, where two variables have to be controlled simultaneously: the arm and the pendulum positions (in Chap. 15) and the pendulum position and the wheel velocity (in Chap. 16). Another advantage of the state space approach is that it allows us to develop a general methodology to obtain a linear approximate model of nonlinear systems (see Sect. 7.3 and Chaps. 13, 15 and 16).

7.16 Review Questions

1. What is a controllable state equation?
2. What is an observable dynamical equation?
3. How can you check controllability and observability?
4. How useful is a controllable dynamical equation?
5. How useful is an observable dynamical equation?
6. What is the difference between the system state and the system output?
7. Suppose that a dynamical equation is controllable and observable. What is the relationship between the poles of the corresponding transfer function and the eigenvalues of matrix A?
8. What does global asymptotic stability of the origin mean?
9. What are the conditions required for the origin to be globally asymptotically stable?
10. Why is it important for the origin of a dynamical equation without input to be globally asymptotically stable?
11. What does state feedback control mean?
12. What is a state observer and what is it employed for?

7.17   Exercises

1. Say what you understand by state and propose a physical system to indicate its state.
2. Verify that the following dynamical equations are controllable:

$$A = \begin{bmatrix} 2 & 1 & 3 \\ 5 & 9 & 7 \\ 0 & 2 & 8 \end{bmatrix}, \quad B = \begin{bmatrix} 0 \\ 0 \\ 2 \end{bmatrix}, \quad C = \begin{bmatrix} 4 & 7 & 2 \end{bmatrix},$$

$$A = \begin{bmatrix} 0 & 1 & 0 \\ 0 & -20 & 50 \\ 0 & -5 & -250 \end{bmatrix}, \quad B = \begin{bmatrix} 0 \\ 0 \\ 100 \end{bmatrix}, \quad C = \begin{bmatrix} 1 & 0 & 0 \end{bmatrix},$$

$$A = \begin{bmatrix} 0 & 1 & 0 & 0 \\ 0 & 0 & -0.5 & 0 \\ 0 & 0 & 0 & 1 \\ 0 & 0 & 50 & 0 \end{bmatrix}, \quad B = \begin{bmatrix} 0 \\ 1 \\ 0 \\ -5 \end{bmatrix}, \quad C = \begin{bmatrix} 1 & 0 & 1 & 0 \end{bmatrix},$$

- Compute the matrix P defined in (7.69).
- Compute the matrices and vectors $\bar{A} = PAP^{-1}$, $\bar{B} = PB$, $\bar{C} = CP^{-1}$ defined in (7.70).
- Corroborate that these matrices and vectors have the forms defined in (7.67).
- Using these results, find the transfer function corresponding to each dynamical equation.
- Employ the MATLAB "tf(·)" command to compute the transfer function corresponding to each dynamical equation. Verify that this result and that in the previous item are identical.
- Use MATLAB/Simulink to simulate the response of the dynamical equation and the corresponding transfer function when the input is a unit step. Plot both outputs and compare them. What can be concluded?

3. Elaborate a MATLAB program to execute the procedure at the end of Sect. 7.11 to compute the vector of controller gains K assigning the desired eigenvalues to the closed-loop matrix $A - BK$. Employ this program to compute the gains of the state feedback controllers in Chaps. 15 and 16. Note that this program is intended to perform the same task as MATLAB's "acker(·)" command.
4. Modify the MATLAB program in the previous item to compute the observer vector gain L, assigning the desired eigenvalues to the observer matrix $A - LC$. Employ this program to design an observer for the system in Chap. 15. Define the system output as the addition of the pendulum and the arm positions.

5. The following expression constitutes a filter where $y(t)$ is intended to replace the time derivative of $u(t)$.

$$Y(s) = \frac{bs}{s+a}U(s), \quad b > 0, \ a > 0.$$

In fact, $y(t)$ is known as the *dirty* derivative of $u(t)$ and it is often employed to replace velocity measurements in mechanical systems. To implement this filter in practice, the corresponding dynamical equation is obtained. Find such a dynamical equation. Recall that $u(t)$ must be the input. Can you use frequency response arguments to select a value for a?

6. Find a dynamical equation useful for implementing the lead controller:

$$\frac{U(s)}{E(s)} = \frac{s+b}{s+c}, \quad 0 < b < c,$$

where $U(s)$ is the plant input and $E(s)$ is the system error. Compare with the result in Sect. F.3.

7. Consider the following system without input:

$$\dot{x} = Ax, \ x \in R^n,$$

Applying the Laplace transform to both sides, it is possible to find that:

$$x(t) = \mathcal{L}^{-1}\left\{(sI - A)^{-1}\right\}x(0), \text{ i.e., } e^{At} = \mathcal{L}^{-1}\left\{(sI - A)^{-1}\right\}.$$

Recall (7.57) and (7.56) to explain that, in general, the following stands for the solution of the closed-loop state equation:

$$\dot{x} = (A - BK)x, \ x \in R^n, \ K = [k_1, \ldots, k_n],$$

- $x_i(t)$, for any $i = 1, \ldots, n$, depends on all the initial conditions $x_i(0)$, $i = 1, \ldots, n$, i.e., the system is coupled.
- $x_i(t)$, for any $i = 1, \ldots, n$, depends on all the eigenvalues of the matrix $A - BK$.
- A particular k_i does not affect only a particular $x_i(t)$, $i = 1, \ldots, n$.
- A particular k_i does not affect only a particular eigenvalue of the matrix $A - BK$.

8. Given an $n \times n$ matrix A, an eigenvector w is an n–dimensional vector such that $Aw = \lambda w$ where the scalar λ is known as the eigenvalue associated with w. From this definition $(\lambda I - A)w = 0$ follows, where I is the $n \times n$ identity matrix. Hence, a nonzero w is ensured to exist if, and only if, $\det(\lambda I - A) = 0$. As $\det(\lambda I - A)$ is an n degree polynomial in λ, the matrix A has n eigenvalues that

can be real, complex, different or repeated. In the case of n real and different eigenvalues, λ_i, $i = 1, \ldots, n$, n linearly independent eigenvectors w_i, $i = 1, \ldots, n$ exist. Thus, an invertible $n \times n$ matrix:

$$Q^{-1} = [w_1 \ w_2 \ldots w_n],$$

and a linear coordinate transformation:

$$z = Qx,$$

can be defined such that:

$$\dot{x} = Ax, \quad \dot{z} = Ez,$$

where $x, z \in R^n$, are related by $E = QAQ^{-1}$. Show that the matrix E is given as:

$$E = \begin{bmatrix} \lambda_1 & 0 & \cdots & 0 \\ 0 & \lambda_2 & \cdots & 0 \\ \vdots & \vdots & \ddots & \vdots \\ 0 & 0 & \cdots & \lambda_n \end{bmatrix},$$

i.e., that:

$$z(t) = \begin{bmatrix} e^{\lambda_1 t} z_1(0) \\ e^{\lambda_2 t} z_2(0) \\ \vdots \\ e^{\lambda_n t} z_n(0) \end{bmatrix},$$

and:

$$x(t) = Q^{-1} \begin{bmatrix} e^{\lambda_1 t} & 0 & \cdots & 0 \\ 0 & e^{\lambda_2 t} & \cdots & 0 \\ \vdots & \vdots & \ddots & \vdots \\ 0 & 0 & \cdots & e^{\lambda_n t} \end{bmatrix} Qx(0).$$

9. Consider the matrix $\bar{A}$ and the vector $\bar{B}$, defined in (7.67), when $n = 5$.

   • Suppose that $\bar{a}_i = 0$ for $i = 0, 1, \ldots, 4$. Show that:

$$
\bar{B} = \begin{bmatrix} 0 \\ 0 \\ 0 \\ 0 \\ 1 \end{bmatrix}, \quad
\bar{A}\bar{B} = \begin{bmatrix} 0 \\ 0 \\ 0 \\ 1 \\ 0 \end{bmatrix}, \quad
\bar{A}^2\bar{B} = \begin{bmatrix} 0 \\ 0 \\ 1 \\ 0 \\ 0 \end{bmatrix}, \quad
\bar{A}^3\bar{B} = \begin{bmatrix} 0 \\ 1 \\ 0 \\ 0 \\ 0 \end{bmatrix}, \quad
\bar{A}^4\bar{B} = \begin{bmatrix} 1 \\ 0 \\ 0 \\ 0 \\ 0 \end{bmatrix},
$$

i.e., these vectors are linearly independent, the matrix $[\bar{B},\ \bar{A}\bar{B},\ \bar{A}^2\bar{B},\ \bar{A}^3\bar{B},\ \bar{A}^4\bar{B}]$ has a determinant that is different from zero, and, thus, the system in (7.67) is controllable.

- Suppose that $\bar{a}_i \neq 0$ for $i = 0, 1, \ldots, 4$. Show that:

$$
\bar{B} = \begin{bmatrix} 0 \\ 0 \\ 0 \\ 0 \\ 1 \end{bmatrix}, \quad
\bar{A}\bar{B} = \begin{bmatrix} 0 \\ 0 \\ 0 \\ 1 \\ * \end{bmatrix}, \quad
\bar{A}^2\bar{B} = \begin{bmatrix} 0 \\ 0 \\ 1 \\ * \\ * \end{bmatrix}, \quad
\bar{A}^3\bar{B} = \begin{bmatrix} 0 \\ 1 \\ * \\ * \\ * \end{bmatrix}, \quad
\bar{A}^4\bar{B} = \begin{bmatrix} 1 \\ * \\ * \\ * \\ * \end{bmatrix},
$$

where the symbol "$*$" stands for some numerical values depending on $\bar{a}_i$ for $i = 0, 1, \ldots, 4$. Show that these vectors are still linearly independent, i.e., that the matrix $[\bar{B},\ \bar{A}\bar{B},\ \bar{A}^2\bar{B},\ \bar{A}^3\bar{B},\ \bar{A}^4\bar{B}]$ has a determinant that is different from zero; thus, the system in (7.67) is controllable.

10. (Taken from [1]) A set of n functions of time $f_i(t)$, $i = 1, \ldots, n$, is said to be linearly dependent on the interval $[t_1, t_2]$ if there are numbers $\alpha_1, \alpha_2, \ldots, \alpha_n$, not all zero such that:

$$
\alpha_1 f_1(t) + \alpha_2 f_2(t) + \cdots + \alpha_n f_n(t) = 0,
$$

for all $t \in [t_1, t_2]$. Otherwise, the set of functions is said to be linearly independent of $[t_1, t_2]$. Use these arguments to explain why the following functions:

$$
f_1(t) = t,\ t \in [-1, 1], \quad \text{and } f_2(t) = \begin{cases} t, & t \in [0, 1] \\ -t, & t \in [-1, 0] \end{cases},
$$

are linearly dependent on $[0, 1]$ and on $[-1, 0]$. However, they are linearly independent of $[-1, 1]$.

11. Consider the results in Example 7.7.

- Recalling that $e^{At} = \mathcal{L}^{-1}\left\{(sI - A)^{-1}\right\}$ show that:

$$
e^{At} = \begin{bmatrix} 1 & \frac{J}{b}\left(1 - e^{-\frac{b}{J}t}\right) \\ 0 & e^{-\frac{b}{J}t} \end{bmatrix}.
$$

- Show that $\det\left(e^{At}\right) \neq 0$ for all $t \in R$, i.e., that e^{At} is nonsingular for all $t \in R$.
- Define:

$$F(t) = \begin{bmatrix} f_1(t) \\ f_2(t) \end{bmatrix} = e^{At} B. \qquad (7.99)$$

  Use the definition in the previous exercise to prove that $f_1(t)$ and $f_2(t)$ are linearly independent for all $t \in R$.
- The following theorem is taken from [1]. Let $f_i(t), i = 1, 2, \ldots, n$, be $1 \times p$ continuous functions defined on $[t_1, t_2]$. Let F(t) be the $n \times p$ matrix with $f_i(t)$ as its i−th row. Define:

$$W(t_1, t_2) = \int_{t_1}^{t_2} F(t) F^T(t) dt.$$

  Then, $f_1(t), f_2(t), \ldots, f_n((t),$ are linearly independent of $[t_1, t_2]$ if, and only if, the $n \times n$ constant matrix $W(t_1, t_2)$ is nonsingular.

  Use this theorem to prove that the two scalar functions defining the two rows of matrix $e^{At} B$ defined in the previous item are linearly independent of $t \in R$.
- The following theorem is taken from [1]. Assume that for each i, f_i is analytic on $[t_1, t_2]$. Let $F(t)$ be the $n \times p$ matrix with f_i as its i−th row, and let $F^{(k)}(t)$ be the k−th derivative of $F(t)$. Let t_0 be any fixed point in $[t_1, t_2]$. Then, the f_is are linearly independent of $[t_1, t_2]$ if, and only if, the rank of the following matrix, with an infinite number of columns, is n:

$$\begin{bmatrix} F(t_0) \;\vdots\; F^{(1)}(t_0) \;\vdots\; \cdots \;\vdots\; F^{(n-1)}(t_0) \;\vdots\; \cdots \end{bmatrix}.$$

  Using the definition in (7.99), show that $\begin{bmatrix} F(t_0) \;\vdots\; F^{(1)}(t_0) \end{bmatrix}$, $n = 2, t_0 = 0$, has rank 2.
- Using the facts that all the entries of $e^{At} B$ are analytic functions, $\frac{d}{dt} e^{At} = e^{At} A$, A^m, for $m \geq n$, can be written as a linear combination of $I, A, \ldots, A^{n-1}$ (Cayley–Hamilton theorem), and $e^{At}|_{t=0} = I$ (see [1] for an explanation of all these properties), show that the theorems in the two previous items establish the equivalence between the two conditions in theorem 7.2 to conclude controllability.

12. Consider the ball and beam system studied in Example 7.2, in this chapter, together with the following numerical values:

$$k = 16.6035, \quad a = 3.3132, \quad \rho = 5.$$

- Obtain the corresponding dynamical equation when the state is defined as $\tilde{z} = [x - x_d, \dot{x}, \theta, \dot{\theta}]^T$ and the output is $y = x - x_d$, where x_d is a constant standing for the desired value for x.
- Design a state feedback controller to stabilize the system at $x = x_d$, $\dot{x} = \theta = \dot{\theta} = 0$. Choose the desired eigenvalues as follows. Propose two pairs of desired rise time and overshoot. Using the expression in (3.71), Chap. 3, determine two pairs of complex conjugate eigenvalues such that the two pairs of desired rise time and overshoot are achieved.
- Design an observer to estimate the complete system state. Recall that the eigenvalues of the matrix $A - LC$ must be several times faster than those for the matrix $A - BK$.
- Fix all initial conditions to zero, except for $x(0) - x_d \neq 0$. Test the closed-loop system employing the above observer and the above state feedback controller to stabilize the system at $x = x_d$, $\dot{x} = \theta = \dot{\theta} = 0$. Verify through simulations whether the desired transient response characteristics have been accomplished.

References

1. C.-T. Chen, *Linear system theory and design*, Holt, Rinehart, and Winston, New York, 1984.
2. C. Ray Wylie, *Advanced engineering mathematics*, McGraw-Hill, 1975.
3. W. L. Brogan, *Modern control theory*, 3rd. edition, Prentice Hall, Upper Saddle River, 1991.
4. E. Kreyszig, *Advanced engineering mathematics*, Vol. 1, 3rd. edition, John Wiley and Sons, New Jersey, 1972.
5. M. R. Spiegel, *Mathematical handbook of formulas and tables*, McGraw-Hill, Schaum series, New York, 1974.
6. C.-T. Chen, *Linear system theory and design*, Oxford University Press, New York, 1999.

Chapter 8
Advanced Topics in Control

8.1 Structural Limitations in Classical Control

Consider the closed-loop control system shown in Fig. 8.1. The open-loop transfer function $G(s)$ is assumed to be given as the cascade connection of the controller $G_c(s)$ and the plant $G_p(s)$, i.e., $G(s) = G_c(s)G_p(s)$. The system error is defined as $e(t) = r(t) - c(t)$, where $e(t) = \mathcal{L}^{-1}\{E(s)\}$, $r(t) = \mathcal{L}^{-1}\{R(s)\}$ and $c(t) = \mathcal{L}^{-1}\{C(s)\}$.

Some results are presented in the following explaining some intrinsic limitations of the transient response of closed-loop classical control systems. These limitations are due to the structure of both the plant and the controller when connected, as in Fig. 8.1. This means that such limitations may be avoided if a different structure for the closed-loop system is chosen.

8.1.1 Open-Loop Poles at the Origin

The following result is taken from [7].

Lemma 8.1 *Consider the closed-loop system in Fig. 8.1. Suppose that $G(s)$ has i poles at the origin, i.e., plant and controller have together i integrators. If $r(t)$ is a step reference, then:*

$$\lim_{t \to \infty} e(t) = 0, \quad \forall i \geq 1, \tag{8.1}$$

$$\int_0^\infty e(t)dt = 0, \quad \forall i \geq 2. \tag{8.2}$$

© Springer International Publishing AG, part of Springer Nature 2019
V. M. Hernández-Guzmán, R. Silva-Ortigoza, *Automatic Control with Experiments*,
Advanced Textbooks in Control and Signal Processing,
https://doi.org/10.1007/978-3-319-75804-6_8

Fig. 8.1 Closed-loop system with unit feedback

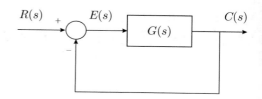

The proof for (8.1) is immediate from Table 4.15, in Sect. 4.4, concerning the steady-state error when the reference is a step signal. On the other hand, it was also shown in Sect. 4.4 that the system error is given as:

$$E(s) = \frac{1}{1 + G(s)} R(s). \tag{8.3}$$

Dividing by s yields:

$$\frac{E(s)}{s} = \frac{1}{s(1 + G(s))} R(s).$$

Using the inverse Laplace transform (3.3), it is found that:

$$\int_0^t e(v)\,dv = \mathcal{L}^{-1} \left\{ \frac{1}{s(1 + G(s))} R(s) \right\}.$$

Hence, applying the final value theorem (3.4):

$$\int_0^\infty e(t)\,dt = \lim_{t \to \infty} \int_0^t e(v)\,dv = \lim_{s \to 0} \frac{s}{s(1 + G(s))} R(s),$$

$$= \lim_{s \to 0} \frac{1}{1 + G(s)} R(s), \quad R(s) = \frac{A}{s},$$

$$= \frac{A}{\lim_{s \to 0} s G(s)}, \tag{8.4}$$

with A a constant. As the number of open-loop integrators is identical to the system type, then we can write:

$$G(s) = \frac{k(s - z_1)(s - z_2) \cdots (s - z_m)}{s^i(s - p_1)(s - p_2) \cdots (s - p_{n-i})}, \tag{8.5}$$

as in Sect. 4.4. Replacing (8.5) in (8.4), the expression in (8.2) is retrieved. Note that (8.2) implies that the system response presents overshoot. Overshoot is necessary to render $e(t)$ negative for some $t > 0$. This, together with the fact that $e(t)$ is initially positive, makes it possible for the integral in (8.2) to be zero.

Example 8.1 Consider the following transfer functions used in two different closed-loop systems whose structure is that in Fig. 8.1:

1) $G_{p1}(s) = \dfrac{8}{s(s+0.5)}$, $G_{c1}(s)=k_d s+k_p$, $\dfrac{C(s)}{R(s)} = \dfrac{8(k_d s+k_p)}{s^2+(0.5+8k_d)s+8k_p}$,

2) $G_{p2}(s) = \dfrac{8}{s^2}$, $G_{c2}(s) = k_d s + k_p$, $\dfrac{C(s)}{R(s)} = \dfrac{8(k_d s + k_p)}{s^2 + 8k_d s + 8k_p}$.

Note that $i = 1$ for system (1) and $i = 2$ for system (2). Hence, according to Lemma 8.1, the response of system (2) always presents overshoot when a step reference is applied, but this is not ensured for system (1). As both closed-loop transfer functions are very similar, it is natural to wonder about these different conclusions. To answer this question, in Figs. 8.2 and 8.3 the root loci and the closed-loop time responses corresponding to systems (1) and (2) are presented. There, $\frac{k_p}{k_d} = 1$ and three different values for k_d are employed to determine the corresponding closed-loop poles, which are also indicated in Figs. 8.2 and 8.3. The closed-loop time responses are shown for these values of k_d.

It is observed that for small values of k_d, system (1) has two real closed-loop poles; hence, the closed-loop time response does not exhibit overshoot. In the case of system (2), it has two complex conjugate closed-loop poles for small values of k_d; hence, the closed-loop time response exhibits overshoot. As observed in Fig. 8.2, the explanation for these different behaviors relies on the fact that the root locus for system (1) belongs to the real axis for small values of k_d, which results from the real open-loop poles at $s = 0$ and $s = -0.5$, whereas the root locus for system (2) does not belong to the real axis for small values of k_d, because of the two open-loop poles at $s = 0$.

For medium values of k_d, both systems have complex conjugate closed-loop poles; hence, the time response of both systems exhibits overshoot. Finally, for large values of k_d both systems have two real closed-loop poles, but, despite this, both of them present a time response with overshoot.[1] Thus, it is corroborated that response of system (2) always presents overshoot for a step reference, as predicted in Lemma 8.1.

Figures 8.2 and 8.3 were drawn by executing the following MATLAB code in an m-file several times:

```
clc
gp1=tf(8,[1 0.5 0]);
gc1=tf([1 1],1);
figure(1)
rlocus(gc1*gp1);
hold on
kd=0.01;% 0.01, 0.1,0.4
```

---

[1] The reason for this overshoot is explained in Sect. 8.1.2.

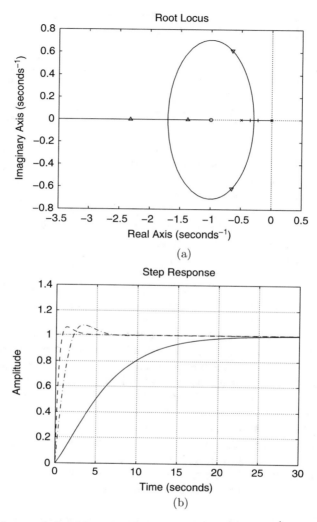

Fig. 8.2 Root locus and closed-loop time response for system (1), when $\frac{k_p}{k_d} = 1$ and k_d takes different values. (**a**) Closed-loop poles $+$: $k_d = 0.01$; triangle down: $k_d = 0.1$; triangle up: $k_d = 0.4$. (**b**) Continuous: $k_d = 0.01$; Dash–dot: $k_d = 0.1$; Dashed: $k_d = 0.4$

```
plot( -0.3540,0,'b+',-0.2260,0,'b+');
%plot(-0.6500,0.6144,'bv',-0.6500,-0.6144,'bv');
%plot(-2.3217,0,'b^',-1.3783,0,'b^');
gc1=tf(kd*[1 1],1);
M1=feedback(gc1*gp1,1,-1);
pole(M1)
figure(2)
```

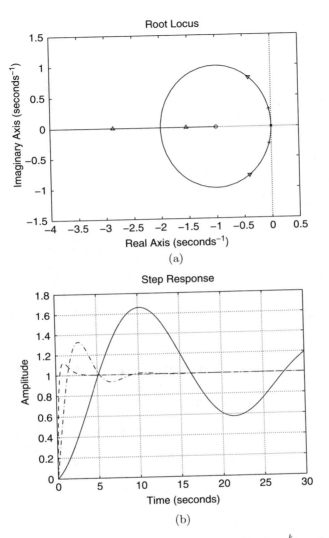

Fig. 8.3 Root locus and closed-loop time response for system 2), when $\frac{k_p}{k_d} = 1$ and k_d takes different values. (a) Closed-loop poles. +: $k_d = 0.01$; triangle down: $k_d = 0.1$; triangle up: $k_d = 0.55$. (b) Continuous: $k_d = 0.01$; Dash-dot: $k_d = 0.1$; Dashed: $k_d = 0.55$

```
step(M1,30)
grid on
hold on

gp2=tf(8,[1 0 0]);
gc2=tf([1 1],1);
```

```
figure(3)
rlocus(gc2*gp2);
hold on
kd=0.01;% 0.01, 0.1, 0.55
plot(-0.0400,0.2800,'b+',-0.0400,-0.2800,'b+')
%plot(-0.4000, 0.8,'bv',-0.4000,-0.8,'bv')
%plot( -2.8633,0,'b^',-1.5367,0,'b^');
gc2=tf(kd*[1 1],1);
M2=feedback(gc2*gp2,1,-1);
pole(M2)
figure(4)
step(M2,30)
grid on
hold on
```

Example 8.2 Consider the ball and beam system studied and experimentally controlled in Chap. 14. In particular, see the block diagrams in Fig. 14.8. Note that the open-loop transfer function has two poles at the origin. Thus, overshoot is unavoidable when a step reference x_d is commanded. This is the reason why the design performed in Sect. 5.2.9, Chap. 5, and experimentally tested in Chap. 14, propose a desired response with overshoot, i.e., $M_p = 25\%$.

8.1.2 Open-Loop Poles and Zeros Located Out of the Origin

The following result is also taken from [7].

Lemma 8.2 *Consider the closed-loop system in Fig. 8.1. Suppose that all the closed-loop poles are located at the left of $-\alpha$ for some real number $\alpha > 0$. Also assume that the open-loop transfer function $G(s)$ has at least one pole at the origin. Then, for an uncancelled plant zero z_0 or an uncancelled plant pole p_0 located at the right of the closed-loop poles, i.e., such that $Re(z_0) > -\alpha$ or $Re(p_0) > -\alpha$ respectively:*

1. For a positive step reference $R(s) = A/s$, $A > 0$, the following holds:

$$\int_0^\infty e(t)e^{-z_0 t}dt = \frac{A}{z_0},\tag{8.6}$$

$$\int_0^\infty e(t)e^{-p_0 t}dt = 0.\tag{8.7}$$

2. For z_0 in the right half-plane and a positive step reference $R(s) = A/s$, $A > 0$, the following holds:

$$\int_0^\infty c(t)e^{-z_0 t}dt = 0.\tag{8.8}$$

Note that, according to the definition of the Laplace transform in (3.1):

$$\int_0^\infty e(t)e^{-z_0 t}\,dt = E(s)|_{s=z_0} = \mathcal{L}\{e(t)\}|_{s=z_0}, \tag{8.9}$$

which is ensured to exist according to the arguments in Appendix A, i.e., see (A.18), because $\mathrm{Re}(z_0) > \mathrm{Re}(-\alpha) = -\alpha$. Thus, using (8.3) and (8.9):

$$\int_0^\infty e(t)e^{-z_0 t}\,dt = E(s)|_{s=z_0} = \frac{1}{1+G(s)}\frac{A}{s}\Big|_{s=z_0},$$

$$= \frac{1}{1+G(z_0)}\frac{A}{z_0} = \frac{A}{z_0},$$

because $G(z_0) = 0$, i.e., z_0 is a zero of $G(s)$. Thus, (8.6) is retrieved. A similar procedure shows that:

$$\int_0^\infty e(t)e^{-p_0 t}\,dt = E(s)|_{s=p_0} = \frac{1}{1+G(s)}\frac{A}{s}\Big|_{s=p_0},$$

$$= \frac{1}{1+G(p_0)}\frac{A}{p_0} = 0,$$

because $G(p_0) = \infty$, i.e., p_0 is a pole of $G(s)$. Thus, (8.7) is retrieved. Finally, the use of:

$$C(s) = \frac{G(s)}{1+G(s)}R(s),$$

yields:

$$\int_0^\infty c(t)e^{-z_0 t}\,dt = C(s)|_{s=z_0} = \frac{G(s)}{1+G(s)}\frac{A}{s}\Big|_{s=z_0},$$

$$= \frac{G(z_0)}{1+G(z_0)}\frac{A}{z_0} = 0,$$

because $G(z_0) = 0$, i.e., z_0 is a zero of $G(s)$. Thus, (8.8) is retrieved. Note that the assumption on the open-loop transfer function $G(s)$ with at least one pole at the origin ensures a zero steady-state error.

Carefully analyzing (8.6), (8.7), (8.8), the authors in [7] arrive at the following conclusions:

1. If z_0 is a negative real number, i.e., it represents a stable zero, then the error must change sign in order to render the integral in (8.6) negative. Recall that the error is initially positive. This implies that the transient response must overshoot. In [7], it is demonstrated that this overshoot is larger as $|z_0|$ is smaller.

2. If z_0 is a positive real number, i.e., it is an unstable zero, the integral in (8.6) is positive; hence, the error does not need to change sign. A small $z_0 > 0$ results in a large integral in (8.6). Furthermore, it is clear from the integral in (8.8) that for z_0 on the right half-plane the closed-loop response must undershoot, i.e., it must exhibit an initial response in the opposite direction of the reference, to render the integral in (8.8) zero. It is also demonstrated in [7] that the peak undershoot is larger as $|z_0|$ is smaller.

3. It is observed from (8.7) that any open-loop pole at the right of the closed-loop poles produces a change of sign in the error; hence, overshoot. Moreover, for a large positive p_0 compared with the location of the closed-loop poles, the function $e^{-p_0 t}$ vanishes much faster than the error $e(t)$. Hence, to compensate for a large positive initial value of $e(t)$, either a large overshoot or a slowly varying overshoot is expected.

Example 8.3 Consider the closed-loop systems studied in Example 8.1 for the case of large values of k_d. It is observed in Figs. 8.2 and 8.3 that the closed-loop time response overshoots for systems (1) and (2) despite both of them having only two real closed-loop poles. This behavior is explained by item 1 above in the present section: for a large k_d the open-loop zero at $s = z_0 = -\frac{k_p}{k_d} = -1$ is located at the right of both closed-loop poles for systems (1) and (2).

Example 8.4 Consider the system studied in Sect. 6.8. This system has an unstable open-loop pole at $s = 35.9 = p_0 > 0$. Thus, according to item 3 above in the present section, a large overshoot is expected. This is corroborated in the time response shown in Fig. 6.106 where $M_p = 89\%$ is observed. To improve this performance, it is recommended in [7] to increase the system bandwidth to be larger than $p_0 = 35.9$. This is also performed in Sect. 6.8, in a second design, and the obtained time response is presented in Fig. 6.109 where overshoot has been reduced to $M_p = 44\%$.

Example 8.5 Another plant with an unstable open-loop pole is studied in Chap. 13. In Fig. 13.11, it is observed that a considerable overshoot exists in the closed-loop system response for the designed controller gains. However, according to (13.33), (13.35), (13.37), all the closed loop poles are real; hence, they should not produce any overshoot. Similar results are found experimentally in Figs. 13.14, 13.16 and 13.17. The explanation for these phenomena is given, again, by item 3 above in the present section.

8.2 Differential Flatness

Differential flatness is a property that is useful for control design purposes. A practical example of such applications is presented in Chap. 15. In the present section, the main ideas of this approach are described. Differential flatness is defined for the general case of nonlinear systems as follows. A nonlinear system:

$$\dot{x} = f(x, u), \quad x \in R^n, \quad u \in R^m,$$

is *differentially flat* if all its states x and the input u can be written as functions of a variable y_f and a finite number of its derivatives [1], i.e.,:

$$x = g(y_f, \dot{y}_f, \ldots, y_f^{(\alpha)}), \quad u = h(y_f, \dot{y}_f, \ldots, y_f^{(\beta)}), \tag{8.10}$$

for some functions $g(\cdot)$ and $h(\cdot)$. Symbols α and β represent some finite positive integers and y_f is known as the *flat output*. However, there is no well-defined procedure to find the flat output in the case of nonlinear systems. As usual, the differential flatness approach is simplified in the case of linear time invariant systems. A linear time invariant system:

$$\dot{x} = Ax + Bu, \quad x \in R^n, \quad u \in R, \tag{8.11}$$

is differentially flat if it is controllable [2] and the flat output can be found as follows.

Proposition 8.1 ([2]) *The flat output of the linear controllable system (8.11) is proportional to the linear combination of the states obtained from the last row of the inverse of the controllability matrix $C_o = [B, AB, A^2 B, \ldots, A^{n-1} B]$, i.e.,:*

$$y_f = \bar{\lambda} [0 \, 0 \ldots 0 \, 1] [B \ AB \ A^2 B \ \ldots \ A^{n-1} B]^{-1} x, \tag{8.12}$$

where $\bar{\lambda}$ is any nonzero constant.

This statement can be understood as follows. Recall, from Chap. 7, that system (8.11) can be written in the form:

$$\dot{\bar{x}} = \bar{A}\bar{x} + \bar{B}u, \tag{8.13}$$

$$\bar{A} = \begin{bmatrix} 0 & 1 & 0 & 0 & \cdots & 0 & 0 \\ 0 & 0 & 1 & 0 & \cdots & 0 & 0 \\ 0 & 0 & 0 & 1 & \cdots & 0 & 0 \\ 0 & 0 & 0 & 0 & \cdots & 0 & 0 \\ \vdots & \vdots & \vdots & \vdots & \ddots & \vdots & \vdots \\ 0 & 0 & 0 & 0 & \cdots & 0 & 1 \\ -a_0 & -a_1 & -a_2 & -a_3 & \cdots & -a_{n-2} & -a_{n-1} \end{bmatrix}, \quad \bar{B} = \begin{bmatrix} 0 \\ 0 \\ 0 \\ 0 \\ \vdots \\ 0 \\ 1 \end{bmatrix},$$

$$\bar{x} = \begin{bmatrix} \bar{x}_1 & \bar{x}_2 & \bar{x}_3 & \bar{x}_4 & \cdots & \bar{x}_{n-1} & \bar{x}_n \end{bmatrix}^T,$$

if, and only if, it is controllable by using the coordinate change:

$$\bar{x} = Px, \tag{8.14}$$

where P is the $n \times n$ constant nonsingular matrix defined as:

$$P^{-1} = \begin{bmatrix} q_1 & \cdots & q_{n-2} & q_{n-1} & q_n \end{bmatrix}, \tag{8.15}$$

$$q_n = B,$$

$$q_{n-1} = AB + a_{n-1}B,$$

$$q_{n-2} = A^2 B + a_{n-1}AB + a_{n-2}B,$$

$$\vdots$$

$$q_1 = A^{n-1}B + a_{n-1}A^{n-2}B + \cdots + a_1 B$$

$$\det(sI - A) = s^n + a_{n-1}s^{n-1} + \cdots + a_1 s + a_0.$$

According to (8.13), if we define the flat output as $y_f = \lambda \bar{x}_1$, for a nonzero constant scalar λ, we can write:

$$\bar{x} = \frac{1}{\lambda} \begin{bmatrix} y_f \\ \dot{y}_f \\ \ddot{y}_f \\ \vdots \\ y_f^{(n-1)} \end{bmatrix}.$$

This and (8.14) imply that we can write

$$x = P^{-1}\bar{x} = \frac{1}{\lambda} P^{-1} \begin{bmatrix} y_f \\ \dot{y}_f \\ \ddot{y}_f \\ \vdots \\ y_f^{(n-1)} \end{bmatrix} = g(y_f, \dot{y}_f, \ldots, y_f^{(n-1)}), \tag{8.16}$$

where $g(\cdot)$ is an n-dimensional linear function in this case. Moreover, from the last row in (8.13) we have:

$$u = \frac{1}{\lambda} \left(a_0 y_f + a_1 \dot{y}_f + a_2 \ddot{y}_f + \cdots + a_{n-2} y_f^{(n-2)} + a_{n-1} y_f^{(n-1)} + y_f^{(n)} \right).$$

$$= h(y_f, \dot{y}_f, \ldots, y_f^{(n)}). \tag{8.17}$$

The following remarks are in order. The latter expressions and (8.10) imply that the controllable system in (8.11) is differentially flat. It is clear from (8.16) that the flat output, y_f, and its first $n-1$ time derivatives can be written in terms of the state x in this case. Furthermore, according to (8.17), the controllable linear system in (8.11) can be represented by the following transfer function without zeros:

$$\frac{Y_f(s)}{U(s)} = \frac{\lambda}{s^n + a_{n-1}s^{n-1} + a_{n-2}s^{n-2} + \cdots + a_2 s^2 + a_1 s + a_0}.$$

As shown in Chap. 15, this is a useful representation for designing a state feedback controller $u = -Kx$ for (8.11) using classical control design tools to select the gain vector K. On the other hand, the following can be written:

$$P = \begin{bmatrix} q_1 & \cdots & q_{n-2} & q_{n-1} & q_n \end{bmatrix}^{-1} = (C_o F)^{-1} = F^{-1} C_o^{-1},$$

$$F = \begin{bmatrix} a_1 & a_2 & \cdots & a_{n-2} & a_{n-1} & 1 \\ a_2 & a_3 & \cdots & a_{n-1} & 1 & 0 \\ a_3 & a_4 & \cdots & 1 & 0 & 0 \\ \vdots & \vdots & \ddots & \vdots & \vdots & \vdots \\ a_{n-1} & 1 & \cdots & 0 & 0 & 0 \\ 1 & 0 & \cdots & 0 & 0 & 0 \end{bmatrix},$$

where:

$$F^{-1} = \frac{Cof^T(F)}{\det(F)}, \quad \det(F) = \mu, \quad Cof(F) = \begin{bmatrix} 0 & \star & \cdots & \star & \star & \star \\ 0 & \star & \cdots & \star & \star & \star \\ 0 & \star & \cdots & \star & \star & \star \\ \vdots & \vdots & \ddots & \vdots & \vdots & \vdots \\ 0 & \star & \cdots & \star & \star & \star \\ \bar{\mu} & \star & \cdots & \star & \star & \star \end{bmatrix},$$

with $Cof(F)$ the cofactors matrix of F, T represents matrix transposition, μ and $\bar{\mu}$ are equal to either -1 or $+1$ (depending on n), "$\star$" indicates the entries of the matrix that we do not care for and C_o is defined in the Proposition 8.1. Hence:

$$P = \frac{1}{\det(F)} \begin{bmatrix} 0 & 0 & \cdots & 0 & 0 & \bar{\mu} \\ \star & \star & \cdots & \star & \star & \star \\ \star & \star & \cdots & \star & \star & \star \\ \vdots & \vdots & \ddots & \vdots & \vdots & \vdots \\ \star & \star & \cdots & \star & \star & \star \\ \star & \star & \cdots & \star & \star & \star \end{bmatrix} \begin{bmatrix} B & AB & A^2 B & \cdots & A^{n-1} B \end{bmatrix}^{-1}.$$

Using the above expressions, $y_f = \lambda \bar{x}_1$ and (8.14), it is concluded that:

$$y_f = \lambda \begin{bmatrix} 1 & 0 & \ldots & 0 & 0 \end{bmatrix} Px,$$

$$= \lambda \frac{\bar{\mu}}{\mu} \begin{bmatrix} 0 & 0 & \ldots & 0 & 1 \end{bmatrix} \begin{bmatrix} B & AB & A^2 B & \cdots & A^{n-1} B \end{bmatrix}^{-1} x.$$

Thus, (8.12) is retrieved by defining $\bar{\lambda} = \lambda \frac{\bar{\mu}}{\mu}$.

8.3 Describing Function Analysis

One important application of the describing function method is the study of limit
cycles in "almost" linear systems. Roughly speaking, a *limit cycle* can be understood
as a system sustained oscillation with respect to time. For our purposes an "almost"
linear system is a closed loop system with two components (see Fig. 8.4): one
of the components, $G(s)$, is a linear, time-invariant system, whereas the second
component, $c = f(e)$, is a "static" hard nonlinearity. A *"static" hard nonlinearity*
is a nonlinear, time-invariant function that is not continuously differentiable, whose
description does not involve any differential equation.

8.3.1 The Dead Zone Nonlinearity [3, 4]

An important example of a "static" hard nonlinearity is that represented in Fig. 8.5,
which is known as the dead zone. This means that there is no output of the
nonlinearity, $c = 0$, when the nonlinearity input e is "small enough," i.e., when
$|e| \leq \delta$. For larger values of e, the nonlinearity output represents a shifted, amplified
version of the nonlinearity input $|c| = k|(e - \delta)|, k > 0$. This behavior is common
in servomechanisms where no movement is produced when the applied voltage (or
the generated torque) is below a certain threshold determined by static friction.

The describing function method is based on the frequency response approach
of control systems. Although an exact frequency response description of nonlinear
functions is not known, an approximate frequency response description can be
obtained for "static" hard nonlinearities. In the following, it is explained how this is
possible. First, let us assume that a sustained oscillation exists. This means that the
input of the nonlinearity is given as:

Fig. 8.4 Standard closed
loop system for a limit cycle
study using the describing
function method

Fig. 8.5 The *dead zone*
"static" hard nonlinearity

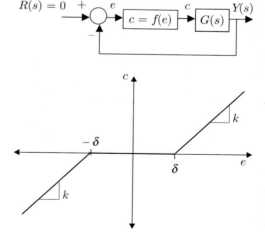

Fig. 8.6 Output of the dead zone nonlinearity when the input is a sinusoidal function of time

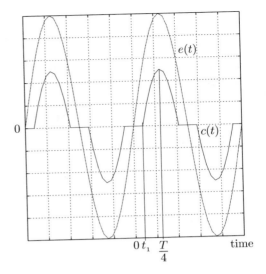

$$e(t) = A \sin(\omega t). \tag{8.18}$$

According to Fig. 8.5, the input and the output of the nonlinearity have the waveforms shown in Fig. 8.6. It is clear that the output of the nonlinearity is periodic; hence, it admits the Fourier series expansion:

$$c(t) = \frac{a_0}{2} + \sum_{n=1}^{\infty} (a_n \cos(n\omega t) + b_n \sin(n\omega t)), \tag{8.19}$$

$$a_0 = \frac{2}{T} \int_{-T/2}^{T/2} c(t)dt, \quad \omega = \frac{2\pi}{T}.$$

Note that, according to Fig. 8.6, $c(t)$ is an odd function, which implies that $a_0 = 0$.[2] On the other hand, if it is assumed that the magnitude of the linear system $G(j\omega)$ has low-pass filter properties, then the high order harmonics in (8.19) can be neglected to approximate $c(t)$ only by the fundamental frequency component, i.e.,:

$$c(t) \approx a_1 \cos(\omega t) + b_1 \sin(\omega t),$$

$$a_1 = \frac{2}{T} \int_{-T/2}^{T/2} c(t) \cos(\omega t)dt,$$

[2]This is because the dead zone characteristic presented in Fig. 8.5 is odd. Furthermore, $a_0 = 0$ for any odd "static" hard nonlinearity.

$$b_1 = \frac{2}{T} \int_{-T/2}^{T/2} c(t) \sin(\omega t) dt.$$

From these expressions, it is clear that $a_1 = 0$ because $c(t)$ is an odd function. Thus, the following can be written:

$$c(t) \approx b_1 \sin(\omega t).$$

According to (8.18) and Figs. 8.5 and 8.6:

$$c(t) = \begin{cases} 0, & 0 \le t \le t_1 \\ k(A \sin(\omega t) - \delta), & t_1 \le t \le \frac{T}{4} \end{cases}.$$

Taking advantage of the odd symmetry of $c(t)$ and by direct integration:

$$b_1 = \frac{2}{T} \int_{-T/2}^{T/2} c(t) \sin(\omega t) dt = \frac{4 \times 2}{T} \int_0^{T/4} c(t) \sin(\omega t) dt,$$

$$= \frac{2kA}{\pi} \left[\frac{\pi}{2} - \arcsin\left(\frac{\delta}{A}\right) - \frac{\delta}{A} \sqrt{1 - \left(\frac{\delta}{A}\right)^2} \right],$$

where $\omega t_1 = \arcsin\left(\frac{\delta}{A}\right)$ and $\cos\theta = \sqrt{1 - \sin^2\theta}$ have been used. According to the frequency response approach, any system component can be represented as the ratio between the output amplitude, i.e., b_1, and the input amplitude, i.e., A. This means that the dead zone nonlinearity can be represented by the frequency response "transfer function":

$$N(A) = \frac{b_1}{A} = \frac{2k}{\pi} \left[\frac{\pi}{2} - \arcsin\left(\frac{\delta}{A}\right) - \frac{\delta}{A} \sqrt{1 - \left(\frac{\delta}{A}\right)^2} \right].$$

$N(A)$ is called the *describing function* of the dead zone nonlinearity. We stress that this "transfer function" is real, positive, and frequency independent, but dependent on the input amplitude A. Moreover, the maximal value is $N(A) = k > 0$, which is reached as $A \to \infty$. The minimal value tends toward zero and it is approached as $A \to \delta$.

On the other hand, given a linear time invariant, negative feedback, closed-loop system with open-loop transfer function $G(s)H(s)$, the closed loop poles satisfy $1 + G(s)H(s) = 0$, which can be rearranged to write $G(s)H(s) = -1$. According to the Nyquist stability criterion, if we replace $s = j\omega$ such that

$G(j\omega)H(j\omega) = -1$ has a solution[3] for a nonzero ω, this means that that there are imaginary closed-loop poles. Thus, a sustained oscillation exists and the oscillation frequency ω is represented by the imaginary part of these closed-loop poles. Based on these arguments, going back to Fig. 8.4 and replacing the nonlinearity with its frequency response transfer function, $N(A)$, it is concluded that the condition $N(A)G(j\omega) = -1$ or, equivalently:

$$G(j\omega) = -\frac{1}{N(A)},\qquad (8.20)$$

indicates that a sustained oscillation may exist.[4] The condition (8.20) is commonly tested graphically by verifying that the polar plot of $G(j\omega)$ and the plot of $-1/N(A)$ intersect at some point. Recall that in the case of a dead zone nonlinearity $N(A)$ is real and positive. Hence, $-1/N(A)$ is real and negative. Thus, according to the above discussion, a limit cycle may exist if the polar plot of $G(j\omega)$ intersects the negative real axis in the open interval $(-\infty, -1/k)$. The oscillation frequency and the amplitude of the oscillation are found as the values of ω, in $G(j\omega)$, and A, in $-1/N(A)$, at the point where these plots intersect.

8.3.2   An Application Example

As an application example consider the closed loop system in Fig. 8.7 where:

$$d(a + bh) = -1.2186 \times 10^5, \qquad b = 93.9951.$$

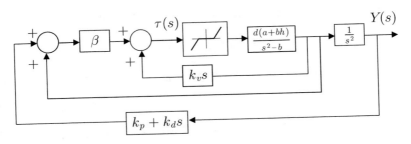

Fig. 8.7   A closed-loop system including a dead zone nonlinearity

---

[3]This implies that point $(-1, j0)$ belongs to the polar plot of $G(j\omega)H(j\omega)$, i.e., the closed loop system is marginally stable.
[4]Note that because of the approximate nature of the method the predicted results are valid only to a certain extent.

Fig. 8.8 An equivalent
representation for the block
diagram in Fig. 8.7

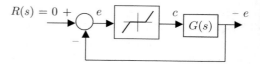

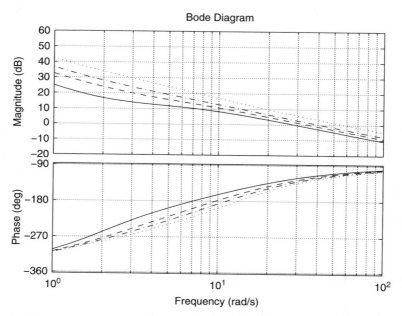

Fig. 8.9 Bode diagrams of $G(s)$ in (8.21) when using the numerical values in Table 8.1. (**a**) Continuous. (**b**) Dashed. (**c**) Dash–dot. (**d**) Dotted

Note that a dead zone "static" nonlinearity exists where it is assumed that:

$$\delta = 0.0115, \quad k = 1.$$

Using some block algebra, the block diagram in Fig. 8.8 is obtained where $e = \tau(s)$ and:

$$G(s) = \frac{-d(a + bh)(k_v s^3 + \beta s^2 + \beta k_d s + \beta k_p)}{s^2(s^2 - b)}, \quad -d(a+bh) > 0. \quad (8.21)$$

Hence, the closed loop system at hand has been written in the form shown in Fig. 8.4, an important step in applying the describing function method. Also note that $G(s)$ has four poles and only three zeros. Thus, its magnitude behaves as a low-pass filter. Recall that this is an important condition for the application of the describing function method. In Fig. 8.9, the Bode diagrams of the transfer function in (8.21) are presented when using the numerical values in Table 8.1. Note that, according to the above discussion, a limit cycle exists in all the cases in Table 8.1

Table 8.1 Numerical values used for $G(s)$ defined in (8.21)

	Controller gains		Controller gains
a)	$k_v = 2.3755 \times 10^{-4},$ $\beta = 0.0041,$ $k_d = 2.88,$ $k_p = 2.88$	b)	$k_v = 1.2 \times 2.3755 \times 10^{-4},$ $\beta = 1.4 \times 0.0041,$ $k_d = 4.59,$ $k_p = 4.59$
c)	$k_v = 1.4 \times 2.3755 \times 10^{-4},$ $\beta = 1.8 \times 0.0041,$ $k_d = 5.62,$ $k_p = 5.62$	d)	$k_v = 2 \times 2.3755 \times 10^{-4},$ $\beta = 2.8 \times 0.0041,$ $k_d = 7,$ $k_p = 7$

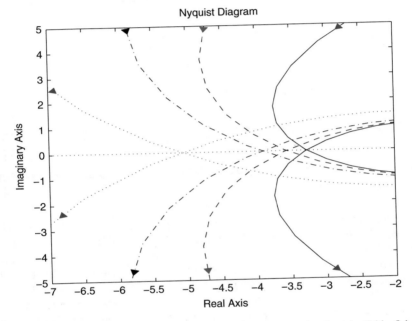

Fig. 8.10 Polar plot of $G(s)$ in (8.21) when using the numerical values in Table 8.1. (**a**) Continuous. (**b**) Dashed. (**c**) Dash–dot. (**d**) Dotted

because all the phase plots in Fig. 8.9 intersect the $-180°$ phase line. Furthermore, this happens when the magnitude is larger than 0[dB]. Recall that the open interval $(-\infty, -1/k)$ corresponds to the open interval $(0, \infty)$[dB] (because $k = 1$) when the phase is $-180°$. Note that the $-180°$ phase line is intersected at a larger frequency as we go from a) to d) in Table 8.1. Also note that, according to Fig. 8.10 the negative real axis is intersected by the polar plot of $G(j\omega)$ at some point that moves to the left as we go from a) to d) in Table 8.1. Recall that $-1/N(A)$ moves to the left as A approaches δ. Thus, it is concluded that the amplitude of the oscillations decreases and the frequency of the oscillations increases as we go from a) to d) in Table 8.1.

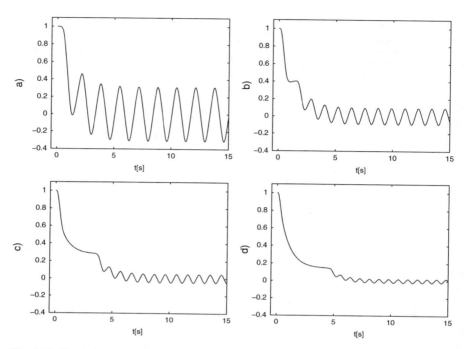

Fig. 8.11 Simulation response of the output y in Fig. 8.7 when using the numerical values in Table 8.1

In Fig. 8.11, the evolution obtained in the simulation of the output $y(t)$ in Fig. 8.7 is presented when using the numerical values in Table 8.1. All the initial conditions are set to zero except for $y(0) = 1$. These results verify all the above predictions: a sustained oscillation exists in all cases, the amplitude decreases and the frequency increases as we go from a) to d) in Table 8.1. It is important to note that the above theoretical discussion on limit cycles was intended to explain the amplitude and the frequency of the oscillation at the input of the "static" nonlinearity $e = \tau(s)$. However, in Fig. 8.11, we present the oscillation of the different variable y, which relates to $e = \tau(s)$ through $e = -(k_v s^3 + \beta s^2 + \beta k_d s + \beta k_p)Y(s)$. Hence, it is clear that, because of the linear relation between e and $y(t)$, these results are also valid for $y(t)$. Some of the results in this section were reported for the first time in [5].

Figures 8.9 and 8.10 were drawn by executing the following MATLAB code in an m-file several times:

```
clc
g=9.81;
km=0.0368;
l1= 0.129;
m1= 0.02218;
L0=0.1550;
```

```
J1=0.0001845;
I0=0.00023849;
den=I0*(J1+m1*l1^2)+J1*m1*L0^2;
h=(J1+m1*l1^2)/(m1*l1*L0);
a=-m1^2*l1^2*L0*g/den;
b=(I0+m1*L0^2)*m1*l1*g/den;
c=(J1+m1*l1^2)/den;
d=-m1*l1*L0/den;
%{
kv=2.3755e-4;
beta=0.0041;
kd=2.88;
kp=kd;
%}
%{
kv=2.3755e-4*1.2;
beta=0.0041*1.4;
kd=4.59;
kp=kd;
%}
%{
kv=2.3755e-4*1.4;
beta=0.0041*1.8;
kd=5.62;
kp=kd;
%}
%%{
kv=2.3755e-4*2;
beta=0.0041*2.8;
kd=7;
kp=kd;
%}
g=tf(-(a+b*h)*d*[kv beta beta*kd beta*kp],[1 0 -b
  0 0]);
figure(1)
bode(g,'b:',{1,100});
grid on
hold on
figure(2)
nyquist(g,'b:');
axis([-7 -2 -5 5])
hold on
```

Simulations in Fig. 8.11 were obtained using the MATLAB/Simulink diagram in Fig. 8.12. The dead zone block has as a parameter $\delta = 0.0115$, whereas the To work space block is programmed to store data in an array and 0.001 as sampling time.

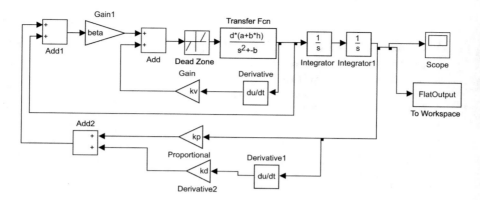

Fig. 8.12 MATLAB/Simulink diagram used for the simulations in Fig. 8.11

To perform these simulations, proceed as follows: (1) Execute the above MATLAB code for the first set of parameters k_v, β, k_p, k_d. (2) Run the simulation in Fig. 8.12. (3) Assign Y2_88_data=FlatOutput. (4) Repeat the steps (1), (2), (3), for the remaining sets of parameters k_v, β, k_p, k_d, and assign $Y4\_59\_data = FlatOutput$, $Y5\_62\_data = FlatOutput$ and $Y7\_data = FlatOutput$ respectively. (5) Execute the following MATLAB code in an m-file:

```
nn=length(Y2_88_data);
n=nn-1;
t=0:1:n;
t=t*0.001;
figure(3)
subplot(2,2,1)
plot(t,Y2_88_data)
axis([-0.5 15 -0.4 1.1])
ylabel('a)')
xlabel('t[s]')
subplot(2,2,2)
plot(t,Y4_59_data)
axis([-0.5 15 -0.4 1.1])
ylabel('b)')
xlabel('t[s]')
subplot(2,2,3)
plot(t,Y5_62_data)
axis([-0.5 15 -0.4 1.1])
ylabel('c)')
xlabel('t[s]')
subplot(2,2,4)
plot(t,Y7_data)
axis([-0.5 15 -0.4 1.1])
```

Fig. 8.13 The *saturation* "*static*" *hard nonlinearity*

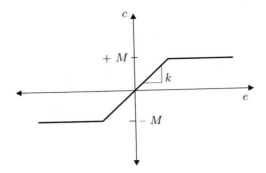

```
ylabel('d)')
xlabel('t[s]')
```

At this point, Fig. 8.11 is drawn.

8.3.3   The Saturation Nonlinearity [3, 4]

Another important example of a "static" hard nonlinearity is that represented in Fig. 8.13, which is known as saturation. This means that the output is equal to the input amplified k times when the absolute value of the output is less than a positive constant M and the output equals $\pm M$ if the absolute value of k times the input is larger than M. The sign of the output equals the sign of the input. This behavior is common in control practice because actuators are always capable of delivering power only within a restricted range of values.

As before, first assume that a sustained oscillation exists. This means that the input of the nonlinearity is given as in (8.18). According to Fig. 8.13, the input and the output of the nonlinearity have the waveforms shown in Fig. 8.14. It is clear that the output of the nonlinearity is periodic; hence, it admits a Fourier series expansion such as that in (8.19).

Note that, according to Fig. 8.14, $c(t)$ is an odd function, which implies that $a_0 = 0$.[5] On the other hand, if it is assumed that the magnitude of the linear system $G(j\omega)$[6] has low-pass filter properties, then the high order harmonics in (8.19) can be neglected to approximate $c(t)$ only by the fundamental frequency component, i.e.,

$$c(t) \approx a_1 \cos(\omega t) + b_1 \sin(\omega t),$$

---

[5]This is because the saturation characteristic presented in Fig. 8.13 is odd. Furthermore, $a_0 = 0$ for any odd "static" hard nonlinearity.

[6]See Fig. 8.4.

Fig. 8.14 Output of the
saturation nonlinearity when
the input is a sinusoidal
function of time

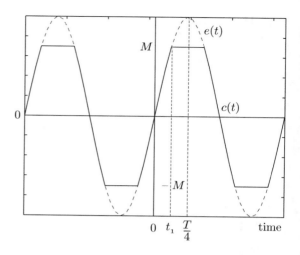

$$a_1 = \frac{2}{T} \int_{-T/2}^{T/2} c(t) \cos(\omega t) dt,$$

$$b_1 = \frac{2}{T} \int_{-T/2}^{T/2} c(t) \sin(\omega t) dt.$$

From these expressions, it is clear that $a_1 = 0$ because $c(t)$ is an odd function. Thus, the following can be written:

$$c(t) \approx b_1 \sin(\omega t).$$

According to (8.18) and Figs. 8.13 and 8.14:

$$c(t) = \begin{cases} kA \sin(\omega t), & 0 \le t \le t_1 \\ M, & t_1 \le t \le \frac{T}{4} \end{cases}.$$

Taking advantage of the odd symmetry of $c(t)$ and by direct integration:

$$b_1 = \frac{2}{T} \int_{-T/2}^{T/2} c(t) \sin(\omega t) dt = \frac{4 \times 2}{T} \int_0^{T/4} c(t) \sin(\omega t) dt,$$

$$= \frac{8}{T} \int_0^{t_1} kA \sin^2(\omega t) dt + \frac{8}{T} \int_{t_1}^{T/4} M \sin(\omega t) dt,$$

$$= \frac{2kA}{\pi} \left[\arcsin\left(\frac{M}{kA}\right) + \frac{M}{kA} \sqrt{1 - \left(\frac{M}{kA}\right)^2} \right],$$

where $\omega t_1 = \arcsin\left(\frac{M}{kA}\right)$ and $\cos\theta = \sqrt{1 - \sin^2\theta}$ have been used. According to the frequency response approach, any system component can be represented as the ratio between the output amplitude, i.e., b_1, and the input amplitude, i.e., A. This means that the saturation nonlinearity can be represented by the frequency response "transfer function":

$$N(A) = \frac{b_1}{A} = \frac{2k}{\pi}\left[\arcsin\left(\frac{M}{kA}\right) + \frac{M}{kA}\sqrt{1 - \left(\frac{M}{kA}\right)^2}\right], \text{ if } A > \frac{M}{k},$$

or:

$$N(A) = \frac{b_1}{A} = k, \text{ if } A \leq \frac{M}{k}.$$

$N(A)$ is the *describing function* of the saturation nonlinearity. This "transfer function" is real, positive, and frequency independent, but dependent on the input amplitude A. Moreover, the maximal value is $N(A) = k > 0$, which is reached for $A \leq \frac{M}{k}$. The minimal value tends toward zero and it is approached as $A \to \infty$. As explained in Sect. 8.3.1, a sustained oscillation may exist if (8.20) is satisfied, which is a condition to be tested graphically.

Recall that in the case of a saturation nonlinearity, $N(A)$ is real and positive. Hence, $-1/N(A)$ is real and negative. Thus, according to the above discussion, a limit cycle may exist if the polar plot of $G(j\omega)$ intersects the negative real axis in the open interval $(-\infty, -1/k]$. The oscillation frequency and the amplitude of the oscillation are found as the values of ω, in $G(j\omega)$, and A, in $-1/N(A)$, at the point where these plots intersect.

8.3.4 An Application Example

As an application example consider the closed-loop system studied in Sect. 8.3.2. But now consider that the torque signal, $\tau(s)$ in Fig. 8.7, is produced by the electric dynamics of a DC motor, which is provided with an electric current loop driven by a proportional–integral controller, i.e.,

$$L\frac{di}{dt} + Ri + k_b z_2 = u, \tag{8.22}$$

$$u = k_{pi}(i^* - i) + k_{ii}\int_0^t (i^* - i)dt, \tag{8.23}$$

where L, R, k_b, are positive constants standing for the armature inductance and resistance, in addition to the motor counter-electromotive force constant. i and u represent the armature electric current and voltage applied at the motor terminals, whereas $z_2 = \dot{y} - \frac{h}{a+bh} y^{(3)}$, where y is the inverse Laplace transform of $Y(s)$ in Fig. 8.7. Finally $I^*(s) = \frac{\tau^*(s)}{k_m}$ is the Laplace transform of the commanded electric current i^* and $\tau^*(s) = (k_v s^3 + \beta s^2 + \beta k_d s + \beta k_p)Y(s)$ is the commanded torque. Replacing (8.23) in (8.22), using:

$$s^2 Y(s) = \frac{d(a+bh)}{s^2 - b}\tau(s),$$

from Fig. 8.7, $\tau(s) = k_m I(s)$, and after an algebraic procedure, it is possible to find:

$$I(s) = G_I(s)I^*(s), \quad G_I(s) = \frac{(s^2 - b)(k_{pi}s + k_{ii})}{N(s)},$$

$$N(s) = Ls^4 + (R + k_{pi})s^3 + \left[k_{ii} - bL - \frac{k_b hd(a+bh)k_m}{a+bh} \right] s^2$$
$$- (R + k_{pi})bs - k_{ii}b + k_b d(a+bh)k_m.$$

Using $\tau(s) = k_m I(s)$ and $\tau^*(s) = k_m I^*(s)$ again, the following is finally found:

$$\tau(s) = G_I(s)\tau^*(s).$$

Thus, the block diagram in Fig. 8.7 now becomes the block diagram in Fig. 8.15. The dead zone nonlinearity is no longer considered because it is observed that in some experiments in Chap. 15, the dead zone does not have the dominant effect. Using some block algebra, the block diagram in Fig. 8.16 is obtained where $e = \tau(s)$ and:

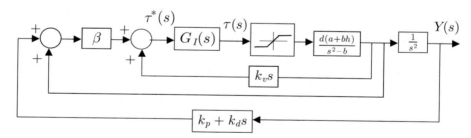

Fig. 8.15 A closed-loop system including a saturation nonlinearity and the electric dynamics of a direct current motor used to generate torque

Fig. 8.16 An equivalent representation for the block diagram in Fig. 8.15

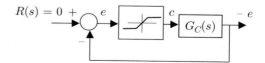

$$G_C(s) = G_I(s) \frac{-d(a+bh)(k_v s^3 + \beta s^2 + \beta k_d s + \beta k_p)}{s^2(s^2 - b)}, \quad -d(a+bh) > 0,$$

$$= \frac{-d(a+bh)(k_v s^3 + \beta s^2 + \beta k_d s + \beta k_p)(k_{pi} s + k_{ii})}{s^2 N(s)}. \tag{8.24}$$

Hence, the closed loop system at hand has been written in the form shown in Fig. 8.4, an important step to applying the describing function method. Also note that $G_C(s)$ has six poles and four zeros. Thus, its magnitude behaves as a low-pass filter. Recall that this is an important condition for the application of the describing function method. Using the following numerical values:

$$d(a+bh) = -1.2186 \times 10^5, \quad b = 93.9951, \quad h = 1.2483, \tag{8.25}$$

$$a = -54.0416, \quad k_m = k_b = 0.0368[\text{V/(rad/s)}], \quad R = 2.4[\text{Ohm}],$$

$$k_{ii} = 130 \quad L = 0.03[\text{H}],$$

$$k_v = 2 \times 2.3755 \times 10^{-4}, \quad \beta = 2.8 \times 0.0041, \quad k_d = 7, \quad k_p = 7,$$

the Bode diagrams in Fig. 8.17 and the polar plots in Fig. 8.18 are drawn. The continuous line is obtained using $k_{pi} = 0.03$ and the dashed line is obtained using $k_{pi} = 0.8$.

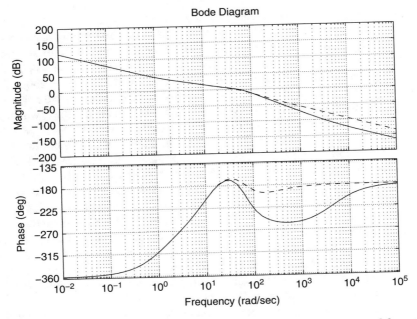

Fig. 8.17 Bode diagrams of $G_C(s)$ in (8.24). Continuous $k_{pi} = 0.03$. Dashed $k_{pi} = 0.8$

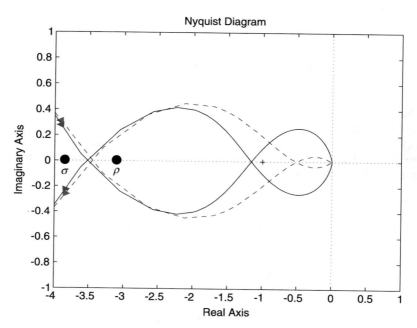

Fig. 8.18 Polar plot of $G_C(s)$ in (8.24). Continuous $k_{pi} = 0.03$. Dashed $k_{pi} = 0.8$

According to the discussion above and Fig. 8.18, two limit cycles exist for the case $k_{pi} = 0.03$ because the negative real axis is intersected twice by $G_C(j\omega)$ on the interval $(-\infty, -1/k]$. Recall that $k = 1$ in this case. Proceeding analogously, it is concluded that in the case of $k_{pi} = 0.8$ only one limit cycle exists. Limit cycles may be stable or unstable. To check limit cycle stability, the application of the Nyquist stability criterion is useful. Using the numerical values above, it is found that $N(s)$ has one real positive root and the remaining roots have a negative real part. Furthermore, $N(s)$ has two roots located at $+9.65$ and -9.62, which are approximately $\pm\sqrt{b} = \pm 9.69$. All roots of $(k_v s^3 + \beta s^2 + \beta k_d s + \beta k_p)(k_{pi} s + k_{ii})$ have a negative real part. Hence, $G_C(s)$ is nonminimum phase and it has two poles at $s = 0$. Therefore, the number of open-loop poles with a positive real part is $P = 1$ and the modified Nyquist path in Fig. 8.19 must be used. Following a similar procedure to that according to (6.65), in Chap. 6, the plots in Fig. 8.20 are obtained.

To study the stability of the limit cycles, the following modified version of the Nyquist stability criterion is useful [3]. Given a linear time-invariant, negative feedback, closed-loop system with an open-loop transfer function $kG(s)H(s)$, where k is any constant, real or complex, the closed-loop poles satisfy $1 + kG(s)H(s) = 0$, which can be rearranged to write $G(s)H(s) = -\frac{1}{k}$. Thus, the Nyquist stability criterion can be applied merely by taking into account the number of times that the polar plot $G(j\omega)H(j\omega)$ encircles the point $-\frac{1}{k}$, instead of the point -1.

In the case of limit cycle stability, proceed as follows [3]. Consider Fig. 8.4, replacing the nonlinearity with its frequency response transfer function $N(A)$. The condition:

Fig. 8.19 Modified Nyquist
path used to draw plots in
Fig. 8.20

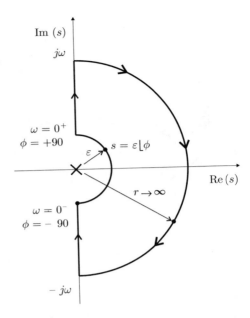

$$G(j\omega) = -\frac{1}{N(A)},$$

establishes that the point $-\frac{1}{N(A)}$, for a particular amplitude A, belongs to the polar plot $G(j\omega)$; hence, marginal stability is concluded, i.e., the possible existence of a limit cycle. Note that the point $-\frac{1}{N(A)}$ shifts if the oscillation amplitude A changes, i.e., in a similar manner to the point -1 shifting to the point $-\frac{1}{k}$ when a constant k is included in the loop, as explained above.[7] Thus, closed-loop stability can be analyzed by taking into account the number of times that the polar plot of $G(j\omega)$ encircles the shifted point $-\frac{1}{N(A)}$. These ideas are applied in the following to the problem at hand.

Consider the limit cycle at the left of Fig. 8.18 for both $k_{pi} = 0.03$ and $k_{pi} = 0.8$. Suppose that the oscillation is performed with an amplitude A_σ that is larger than the oscillation amplitude predicted by the point where $G_C(j\omega)$ and $-1/N(A)$ intersect. This means that $-1/N(A_\sigma)$ shifts to, say, the point σ in Fig. 8.18. Recall that, according to the describing function analysis for a saturation nonlinearity, a larger oscillation amplitude is obtained by moving to the left on the negative real axis in Fig. 8.18. Locate the point σ in any of the plots in Fig. 8.20 and determine the number N of times that these plots encircle the point σ. It is not difficult to verify

---

[7]This does not mean that another limit cycle with a different amplitude A exists. Recall that $N(A)$ is a linear approximation of a nonlinear function; hence, $N(A)$ changes if A changes and this does not require A to represent an oscillation amplitude in a steady state.

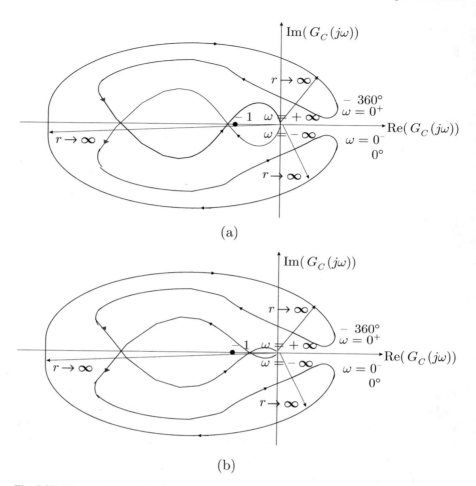

Fig. 8.20 Mapping the modified Nyquist path in Fig. 8.19 to the plane $G_C(j\omega)$. (a) $k_{pi} = 0.03$. (b) $k_{pi} = 0.8$

that $N = 1$ (clockwise). Applying the Nyquist stability criterion, the number of closed-loop unstable poles is $Z = N + P = 2$. Hence, the closed-loop system is unstable. This means that the oscillation amplitude increases; thus, the point σ further moves to the left, i.e., the system evolution abandons the limit cycle. If the oscillation starts at an amplitude A_ρ, which is less than the oscillation amplitude predicted by the point where $G_C(j\omega)$ and $-1/N(A)$ intersect, then the point ρ in Fig. 8.18 must be considered. Proceeding analogously as before, it is not difficult to verify that stability is concluded in this case. Thus, the oscillation amplitude decreases; hence, the point ρ moves further to the right, i.e., the system evolution abandons the limit cycle again. Thus, the limit cycle at the left of Fig. 8.18 is unstable for both $k_{pi} = 0.03$ and $k_{pi} = 0.8$.

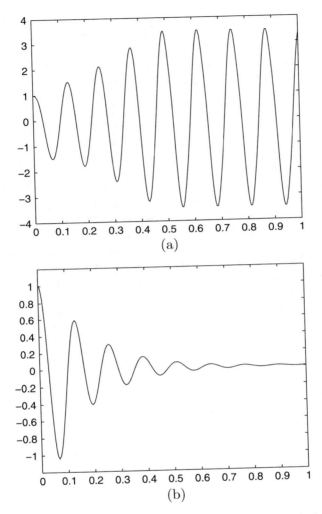

Fig. 8.21 Simulation response of signal at the input of the saturation nonlinearity in Fig. 8.15. The horizontal axis shows time in seconds. (**a**) $k_{pi} = 0.03$. (**b**) $k_{pi} = 0.8$

In the case of $k_{pi} = 0.03$, a second limit cycle exists at the right of Fig. 8.18. It is left as an exercise for the reader to proceed as above to verify that this limit cycle is stable.

In Fig. 8.21, the evolution obtained in simulation of the signal at the input of the saturation nonlinearity in Fig. 8.15 is presented when using the numerical values in (8.25). All the initial conditions are set to zero except for the signal at the input of the nonlinearity, which is set to 1. Note that, in the case of $k_{pi} = 0.8$ no limit cycle appears. This is a consequence that the limit cycle due to the saturation is unstable, but the closed-loop system is stable. This can be verified by applying the Nyquist stability criterion in Fig. 8.20b, i.e., considering the number of times that the point -1 on the real axis is encircled. Furthermore, limit cycles are not observed either in the experiments performed in Chap. 15, where it is explained that the controller gains used in this case are chosen such that any limit cycle due to a dead zone nonlinearity does not appear.

In the case of $k_{pi} = 0.03$, a limit cycle appears and the oscillation increases at the beginning to reach this limit cycle. Apply the Nyquist stability criterion to Fig. 8.20a, i.e., considering the number of times that the point -1 on the real axis is encircled, to verify that the closed-loop system is unstable. This explains the initial growth in the oscillation. Moreover, from Fig. 8.21a, it is possible to measure the oscillation frequency: $f \approx 8[\text{cycles/s}] = 8[\text{Hz}]$, then $\omega = 2\pi f \approx 50.26[\text{rad/s}]$. This limit cycle corresponds to the point in Fig. 8.17 where the $-180°$ phase line is crossed at $\omega \approx 48[\text{rad/s}]$ because, according to Fig. 8.18, this is the limit cycle appearing at a higher frequency. Hence, this corroborates the existence and stability of such a limit cycle. Note that this also shows that a saturation nonlinearity may convert closed-loop instability into a limit cycle, i.e., into sustained oscillations. This also explains the operation of the oscillator electronic circuit based on a transistor presented in Sect. 9.3.9.

Figures 8.17 and 8.18 were drawn using the following MATLAB code in an m-file:

```
clear all
clc
g=9.81;
l1= 0.129;
m1= 0.02218;
L0=0.1550;
J1=0.0001845;
I0=0.00023849;
kpi=0.8;%0.03;%
kii=130;
km=0.0368;%0.05
R=2.4;
L=0.030; %0.084; %0.163; %
kb=km;
den=I0*(J1+m1*l1^2)+J1*m1*L0^2;
h=(J1+m1*l1^2)/(m1*l1*L0);
a=-m1^2*l1^2*L0*g/den;
b=(I0+m1*L0^2)*m1*l1*g/den;
c=(J1+m1*l1^2)/den;
d=-m1*l1*L0/den;
kv=2*2.3755e-4;
beta=2.8*0.0041;
kd=7;
```

```
kp=7;
alla=-kii*b+kb*d*(a+b*h)*km;
deni=[L R+kpi kii-b*L-kb*h*d*(a+b*h)*km/(a+b*h)
      -(R+kpi)*b alla];
num1=conv([1 0 -b],[kpi kii])
numi1=conv(num1,[kv beta beta*kd beta*kp])
Gi=tf(numi1,deni)
pole(Gi)
zero(Gi)
G=tf(-d*(a+b*h),[1 0 -b 0 0]);
GH=Gi*G;
figure(1)
bode(GH,'r--')
grid on
hold on
figure(2)
nyquist(GH,'r--')
hold on
axis([-4 1 -1 1])
den2=conv([1 0 -b 0],deni)
```

The simulations in Fig. 8.21 were performed using the Simulink diagram in Fig. 8.22. The saturation block has ±2.5 as saturation limits, the Datos block saves data in an array, whereas the Datos1 block saves data as a structure with time. To obtain these graphical results, proceed as follows: (1) Execute the above MATLAB code for kpi=0.03. (2) Run the simulation in Fig. 8.22. (3) Execute the following MATLAB code in an m-file:

```
clc
tiempo=Datos1.time;
figure(3)
plot(tiempo,Datos(:,1))
axis([0 1 -1.2 1.2])
```

(4) repeat (1), (2), (3), for kpi=0.8.

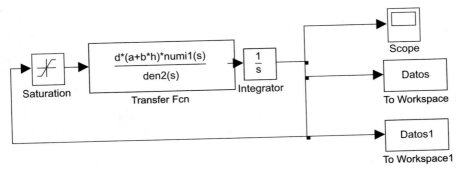

Fig. 8.22 MATLAB/Simulink diagram for the simulations in Fig. 8.21

8.4   The Sensitivity Function and Limitations When Controlling Unstable Plants

Given the closed loop system in Fig. 8.23, the *sensitivity function* is defined as:

$$S(s) = \frac{C(s)}{D(s)} = \frac{1}{1 + G(s)H(s)},$$

where $D(s)$ is a disturbance. The transfer functions $H(s)$ and $G(s)$ represent the plant and the controller respectively. From the above definition we can write:

$$|S(j\omega)| = \frac{1}{|G(j\omega)H(j\omega) - (-1)|}.$$

This means that the magnitude of the sensitivity function represents the inverse of the distance between point $(-1, j0)$ and the polar plot of $G(j\omega)H(j\omega)$. Hence, according to the Nyquist stability criterion, a bad performance is obtained if $|S(j\omega)|$ is large for some $\omega > 0$. Furthermore, large sensitivity values put the polar plot of $G(j\omega)H(j\omega)$ too close to the critical point $(-1, j0)$. This may result in closed-loop instability because of imperfections in control system implementation. Thus, a necessary condition for closed-loop stability is that $|S(j\omega)|$ is "small" for all $\omega > 0$.

A fundamental property of the feedback connection in Fig. 8.23 is represented by the so-called *Bode integrals*, which are given as follows [6]:

$$\int_0^\infty \ln|S(j\omega)|d\omega = 0, \tag{8.26}$$

$$\int_0^\infty \ln|S(j\omega)|d\omega = \pi \sum_{p \in P} R_e(p). \tag{8.27}$$

The properties in (8.26) and (8.27) stand for the cases when the plant transfer function $H(s)$ is stable and unstable respectively. In this respect, P represents the set of all the unstable poles of $H(s)$. Because of the low-pass filtering property of most plants, it has been suggested in [6] to replace (8.26) and (8.27) by:

$$\int_0^{\Omega_a} \ln|S(j\omega)|d\omega = \delta, \tag{8.28}$$

Fig. 8.23   Closed-loop system

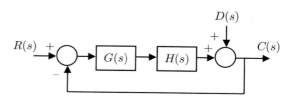

$$\int_0^{\Omega_a} \ln |S(j\omega)| d\omega = \pi \sum_{p \in P} \text{Re}(p) + \delta. \tag{8.29}$$

δ is a small constant (positive or negative) and $\Omega_a > 0$ is a constant known as the *available bandwidth*. The introduction of the available bandwidth concept is justified in [6] by the fact that modeling simplifications and hardware limitations render the system frequency response above some constant Ω_a uncertain. This means that it is only within the finite frequency range $\omega \in [0, \Omega_a]$ that the designer is able to suitably modify the system frequency response. In this respect, the small constant δ quantifies the system's uncertainties and the hardware imperfections.

To simplify the computation of the integrals in (8.28) and (8.29) suppose that the magnitude of the sensitivity function is kept constant at a minimal value for $\omega \in [0, \Omega_a]$, i.e., $|S(j\omega)| = S_{min}$, and $\delta \approx 0$ is negligible. Then, the following is obtained:

$$S_{min} = 1, \tag{8.30}$$

$$S_{min} = e^{\frac{\pi \xi}{\Omega_a}} > 1, \quad \xi = \sum_{p \in P} \text{Re}(p) > 0, \tag{8.31}$$

for (8.28) and (8.29) respectively.

Going back to (8.26) and (8.27), it is concluded that, although the sensitivity can be rendered small at certain specific values of ω, the price of achieving this is that the sensitivity becomes large at some other values of ω. This is because (8.26) and (8.27) represent some kind of "conservation laws." Moreover, this phenomenon becomes worse when controlling unstable plants whose unstable poles are located farther to the right. It is very important to stress that these properties are independent of the particular controller that is employed [6]. Note that, according to (8.31), in the case of unstable plants too small an available bandwidth Ω_a may also render the sensitivity function large enough to avoid good performances or to produce closed-loop instability, as explained above. Thus, attention must be paid to designing a suitable available bandwidth Ω_a.

The above ideas are employed in Sect. 15.9, Chap. 15, to experimentally control a Furuta pendulum.

8.5   Summary

This chapter is included to study the theory required to understand some interesting phenomena appearing during experiments when controlling some of the mechanisms in the last eight chapters of the book. This is the case of large overshoots observed in a magnetic levitation system, even when all the closed-loop poles are real. The explanation of this phenomenon is given in Sect. 8.1 "Structural limitations in classical control," where it is concluded that the plant open-loop unstable poles are responsible for this. It is also explained in that section that an open-loop double pole at the origin may contribute to such behavior and this is the case of experiments with the ball and beam system. On the other hand, limit cycle is a phenomenon induced by hard nonlinearities in some mechanical systems such as the Furuta pendulum. This is explained in Sect. 8.3 "Describing function analysis," where two different nonlinearities are considered: the dead zone and the saturation. Moreover, the saturation nonlinearity is responsible for sustained oscillations in sine wave oscillator circuits studied in Chap. 9. Furthermore, a methodology useful for selecting the controller gains such that limit cycles are avoided is proposed using the ideas presented in Sect. 8.2 "Differential flatness" and Sect. 8.3.

8.6   Review Questions

1. What is a limit cycle?
2. What is a dead zone nonlinearity and why does it appear in mechanical systems?
3. Consider the model of the ball and beam system presented in Chap. 14. Can a dead zone nonlinearity induce a limit cycle in a closed-loop stable ball and beam system? Explain.
4. Can a saturation nonlinearity induce limit cycles in a ball and beam system? Explain.
5. Try to tune a linear state feedback controller for the Furuta pendulum using a root locus, but without resorting to differential flatness. Is it possible? Why?
6. What are the fundamental ideas behind the methods proposed in this chapter to avoid limit cycles induced by the dead zone and saturation?

8.7   Exercises

1. The following mathematical model is introduced in (12.1), (12.2):

$$I_1\ddot{\theta}_1 + b_1\dot{\theta}_1 + K(\theta_1 - \theta_2) = T, \quad T = k_m(u - k_b\dot{\theta}_1)/R,$$
$$I_2\ddot{\theta}_2 + b_2\dot{\theta}_2 - K(\theta_1 - \theta_2) = 0,$$

where u is the input.

   • Write this model in the form $\dot{x} = Ax + Bu$, with $x = [\theta_1, \dot{\theta}_1, \theta_2, \dot{\theta}_2]^T$, and use (8.12) to show that θ_2 is the flat output. Do you find the expression in (12.7) useful for confirming this result?
   • Consider a controller of the form:

$$u = k_1(\theta_{2d} - \theta_2) - k_2\dot{\theta}_2 - k_3\ddot{\theta}_2 - k_4\theta_2^{(3)}, \qquad (8.32)$$

and select the controller gains such that all the poles of the closed-loop system (12.7), (8.32) are assigned arbitrarily.

- Using the fact that θ_2 is the flat output, find a way of expressing:

$$\ddot{\theta}_2 = L_2(\theta_2, \dot{\theta}_2, \theta_1, \dot{\theta}_1), \quad \theta_2^{(3)} = L_3(\theta_2, \dot{\theta}_2, \theta_1, \dot{\theta}_1),$$

where $L_2(\cdot)$ and $L_3(\cdot)$ are linear functions of their arguments, and show that (8.32) can be written as:

$$u = \gamma_1(\theta_{2d} - \theta_2) - \gamma_2\dot{\theta}_2 - \gamma_3(\theta_1 - \theta_{1d}) - \gamma_4\dot{\theta}_1, \tag{8.33}$$

where $\gamma_1, \gamma_2, \gamma_3, \gamma_4$, are some constants and $\theta_{1d} = \theta_{2d}$. Use the procedure before the Example 7.9 to find the controller gains $\gamma_1, \gamma_2, \gamma_3, \gamma_4$, arbitrarily assigning the eigenvalues to the closed-loop system $\dot{\tilde{x}} = A\tilde{x} + Bu$, $\tilde{x} = [\theta_1 - \theta_{1d}, \dot{\theta}_1, \theta_2 - \theta_{2d}, \dot{\theta}_2]^T$, for u given in (8.33).

- Verify through numerical examples that the procedures in the previous two items can assign the same eigenvalues and that the sets of gains in both approaches can be obtained from each other.
- Study the possible existence of limit cycles, induced by a dead zone nonlinearity, when using the control scheme in (8.32) in the closed-loop system with (12.7). Explain why these results are still valid when using the closed-loop system $\dot{\tilde{x}} = A\tilde{x} + Bu$, $\tilde{x} = [\theta_1 - \theta_{1d}, \dot{\theta}_1, \theta_2 - \theta_{2d}, \dot{\theta}_2]^T$, for u given in (8.33).

2. Compare Fig. 8.7, when the dead zone nonlinearity is not present, and Fig. 6.83. Show that a control scheme similar to that in Fig. 6.83, i.e., with a lead compensator in the outer loop instead of a proportional–derivative controller, can be designed for the plant in Fig. 8.7. Use the frequency response method presented after (6.53) to perform the design.

3. Consider a dead zone nonlinearity at the input of the ball and beam plant in Fig. 6.83 and study the possible existence of limit cycles.

References

1. M. Fliess, J. Levine, P. Martin, and P. Rouchon, Flatness and defect of non-linear systems: introductory theory and examples, *International Journal of Control*, Vol. 61, No. 6, pp. 1327–1361, 1995.
2. H. Sira-Ramírez and S.K. Agrawal, *Differentially flat systems*, Marcel Dekker, New York, 2004.
3. J. J. Slotine and W. Li, *Applied nonlinear control*, Prentice Hall, 1989.
4. E. Usai, Describing function analysis of nonlinear systems. University of Leicester. Department of Engineering and University of Cagliari Department of Electrical and Electronic Engineering, 2008.

5. V. M. Hernández-Guzmán, M. Antonio-Cruz, and R. Silva-Ortigoza, Linear state feedback regulation of a Furuta pendulum: design based on differential flatness and root locus, *IEEE Access*, Vol. 4, pp. 8721–8736, December, 2016

6. G. Stein, Respect the unstable, *IEEE Control Systems Magazine*, August, pp. 12–25, 2003.

7. G. C. Goodwin, S. F. Graebe, and M. E. Salgado, *Control system design*, Prentice Hall, Upper-Saddle River, 2001.

Chapter 9
Feedback Electronic Circuits

Several important applications of feedback systems to designing electronic circuits are presented in this chapter. Some of these applications are employed in designing some of the control systems to be presented in subsequent chapters in this book. Some other applications, such as those of oscillator circuits, constitute by themselves complete control systems requiring the employment of the control theory concepts introduced in the previous chapters.

9.1 Reducing the Effects of Nonlinearities in Electronic Circuits

Consider the closed-loop system depicted in Fig. 9.1. The corresponding closed-loop transfer function is:

$$\frac{C(s)}{R(s)} = \frac{A}{1 + \beta A},$$

where A and β are positive constants. Suppose that $\beta A \gg 1$, then:

$$\frac{C(s)}{R(s)} = \frac{A}{1 + \beta A} \approx \frac{1}{\beta}. \tag{9.1}$$

The applications of this fact are important in electronic circuits, as explained in the following.

Suppose that A is the gain of a power amplifier. These devices are capable of handling large amounts of power; hence, their components must be capable of working under large temperature variations. This implies that the components of a power amplifier are not high-precision and large changes in their parameters

© Springer International Publishing AG, part of Springer Nature 2019
V. M. Hernández-Guzmán, R. Silva-Ortigoza, *Automatic Control with Experiments*,
Advanced Textbooks in Control and Signal Processing,
https://doi.org/10.1007/978-3-319-75804-6_9

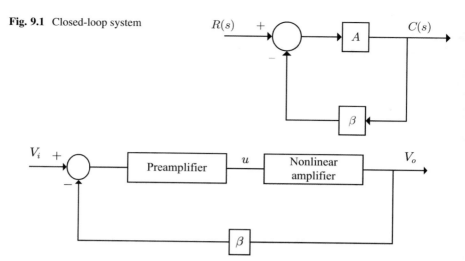

Fig. 9.1 Closed-loop system

Fig. 9.2 Use of feedback in a nonlinear amplifier to reduce distortion

are expected. Thus, it is natural to assume that A changes under normal operation conditions. According to (9.1), the use of feedback around an amplifier with gain A has the effect of rendering the circuit gain changes small, despite A exhibiting large variations. Hence, feedback can solve the problem of a power amplifier with large gain changes. An important feature is that the gain β must not exhibit significant changes and this is achieved if such a gain depends only on low-power components, i.e., that they can be composed only of precision devices.

In the following, some applications are presented where feedback is used to reduce the gain variations in some electronic circuits, which can also be understood as the reduction of the effects of the circuit nonlinearities.

9.1.1 Reducing Distortion in Amplifiers

Consider the block diagram in Fig. 9.2. The nonlinear amplifier is an amplifier with gain $A_1 = 2$ when the input voltage u is negative and a gain $A_2 = 0.5$ when the input voltage is positive. This property can be represented as in Fig. 9.3. Hence, this amplifier delivers at its output a distorted version of the signal applied at its input. In Fig. 9.4, the signal at the nonlinear amplifier output is shown, when a sinusoidal signal is applied at its input. This is a good example of distortion. Combining the effect of all amplifiers in the direct path, it is concluded that the block diagram in Fig. 9.2 can be represented by a block diagram such as that in Fig. 9.1 where:

$$A = A_1 A_0, \quad u < 0,$$

Fig. 9.3 Characteristics of
the nonlinear amplifier in
Fig. 9.2

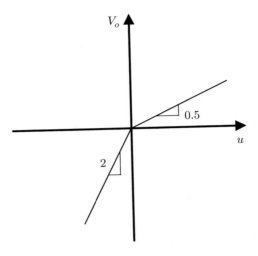

Fig. 9.4 Voltages at the input
u and the output V_o of the
nonlinear amplifier in Fig. 9.3

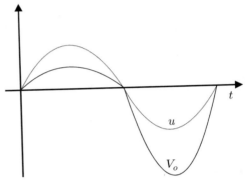

$$A = A_2 A_0, \quad u > 0,$$
$$\beta = 1, \quad \text{for instance,}$$

with A_0 the gain of the preamplifier, which is very large. Note that the following closed-loop transfer function is obtained:

$$\frac{C(s)}{R(s)} = \frac{A}{1 + \beta A} \approx \frac{1}{\beta} = 1,$$

if gains A_0 and β are such that $\beta A \gg 1$ for both values A_1, A_2. Hence, the closed-loop circuit in Fig. 9.2 works as an amplifier with constant gain β for both signs of the signal to be amplified. This means that the distortion due to the different amplifier gains A_1 and A_2 has been eliminated. This is the main advantage of feedback shown in Fig. 9.2 despite the total gain of the amplifier now being changed to $1/\beta = 1$. In Fig. 9.5, the waveform obtained for V_o at the output of the feedback

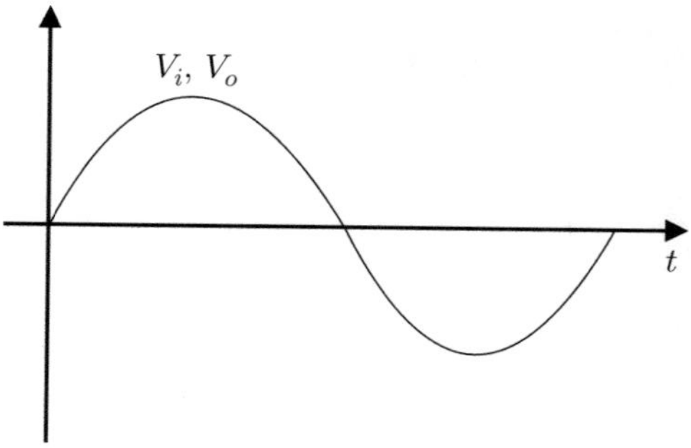

Fig. 9.5 Voltages at the input V_i and the output V_o in Fig. 9.2

V_i, V_o

t

Fig. 9.6 A complementary symmetry nonlinear power amplifier

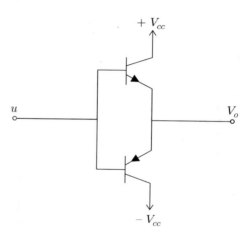

system depicted in Fig. 9.2 is shown when V_i is a sinusoidal signal. Note that V_o has the same amplitude for both semicircles when a nonlinear amplifier (with gain A) is embedded in the feedback system.

9.1.2 Dead Zone Reduction in Amplifiers

A power amplifier implemented using two transistors connected in complementary symmetry is shown in Fig. 9.6. This circuit has a dead zone between -0.6[V] and $+0.6$ [V] due to the polarization voltage level required by the base-emitter transistor junctions. This means that the output signal V_o remains at zero as long as the input signal u remains within the range $[-0.6, +0.6]$[V]. In Fig. 9.7, the characteristic of this nonlinear amplifier is shown. In Fig. 9.8, the signal obtained at the amplifier output V_o is shown, when a sinusoidal signal is applied at the input u. Note the effect of the dead zone for values of u that are close to zero. The circuit in Fig. 9.9 is employed to reduce the effect of such a dead zone. The objective is to render the waveform at the output V_o identical or very similar to the waveform of the applied

Fig. 9.7 Characteristics of
the nonlinear amplifier in
Fig. 9.6

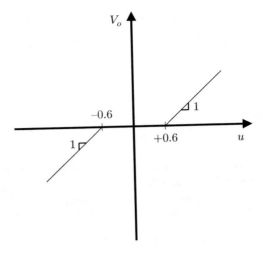

Fig. 9.8 Voltages at the input
u and the output V_o of the
amplifier in Fig. 9.6

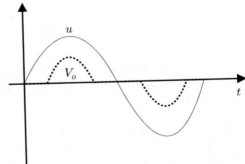

Fig. 9.9 Feedback circuit to
reduce the effect of the
nonlinearity present in the
amplifier in Fig. 9.6

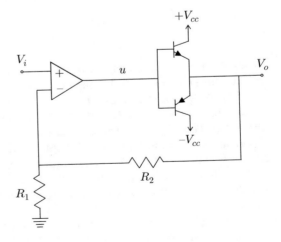

Fig. 9.10 Voltages at the
input V_i and the output V_o of
the circuit in Fig. 9.9

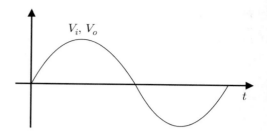

signal V_i, as shown in Fig. 9.10; hence, to eliminate the effect of the dead zone by
employing feedback. This is explained as follows.

The circuit in Fig. 9.9 can be represented by the block diagram in Fig. 9.1. This
feedback system has a closed-loop transfer function given as:

$$\frac{C(s)}{R(s)} = \frac{A}{1 + \beta A}, \quad C(s) = V_o(s), \quad R(s) = V_i(s),$$

$$\beta = \frac{R_1}{R_1 + R_2} \le 1,$$

where $A = \alpha(u)A_0$, with A_0 the open-loop transfer function of the operational
amplifier used to implement the subtraction point and $\alpha(u)$ is the u dependent gain
of the characteristic shown in Fig. 9.7, i.e., the dead zone can be represented as
$V_o = \alpha(u)u$. It is well known that A_0 is a large real scalar, in the order of 100,000,
whereas $\alpha(u)$ has a value in the open interval $(0, 1]$. Recall that the dead zone is
an idealized *hard* behavior of the real smooth behavior of a transistor. Thus, $\beta A = \alpha(u)A_0\beta \gg 1$ and the following can be approximated, with a small error:

$$\frac{V_o(s)}{V_i(s)} = \frac{A}{1 + \beta A} \approx \frac{1}{\beta}. \tag{9.2}$$

The expression in (9.2) indicates that the whole circuit behaves as an amplifier with
the constant gain $1/\beta$ such that the effect of the dead zone is eliminated and the
waveform of V_o is identical to that of V_i.

Note that R_1 and R_2 can be chosen to be large to ensure that only a small electric
current flows through them; hence, precision resistances can be employed. It is also
important to stress the following. According to Fig. 9.7 $\alpha(u)$ is zero for u in the
interval $[-0.6, +0.6][V]$. This suggests that the condition $\beta A = \alpha(u)A_0\beta \gg 1$ is
not true. However, it is important to understand that the characteristic in Fig. 9.7 is
an idealization of what really happens in practice where V_o is zero only when u is
also zero.

In Fig. 9.11 the experimental voltages u and V_o corresponding to Fig. 9.6 are
shown. The Darlington transistors TIP 141 (NPN) and TIP 145 (PNP) are employed.
This means that the voltage at the base-emitter junction has a nominal direct
polarization voltage of about 1.2[V]; hence, the dead zone is in this case in the

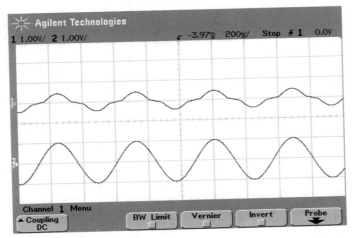

Fig. 9.11 Experimental results. Voltages at the input and the output of the circuit in Fig. 9.6. Upper line: V_o, bottom line: u

interval $[-1.2, +1.2][V]$ of voltage at the input u. Voltage u is a sinusoidal voltage at 2.083[KHz] and a peak amplitude of 1[V]. Thus, voltage at the base-emitter junction is clearly below its direct polarization nominal value. It is important to state that a 1[KOhm] resistance is used as the load between the connection of both emitters and the ground. Note that the voltage V_o has a waveform that is very different from the waveform of a sinusoidal signal (compare with the dotted line in Fig. 9.8). This experimentally verifies that the complementary symmetry connection of two transistors strongly distorts the input signal.

In Fig. 9.12, the waveforms of the signals V_i and V_o corresponding to Fig. 9.9 are shown. These measurements were obtained in an experiment where the Darlington transistors TIP 141 (NPN) and TIP 145 (PNP) are employed and voltage V_i is a sinusoidal signal at 2.083[KHz] with a peak amplitude of 1[V]. The resistances $R_1 = R_2 = 10$[KOhm] are also employed, in addition to a 1[KOhm] load resistance between the connection of both emitters and the ground. A TL081 operational amplifier is used. Note that the waveform of V_o is similar to that of V_i, i.e., both signals are sinusoidal signals of the same frequency, although the amplitude of V_o is twice the amplitude of V_i. This is because, according to (9.2):

$$\frac{V_o(s)}{V_i(s)} = \frac{1}{\beta} = 2, \quad \beta = \frac{R_1}{R_1 + R_2} = 0.5.$$

Some other applications of feedback in electronic circuits are presented in [1].

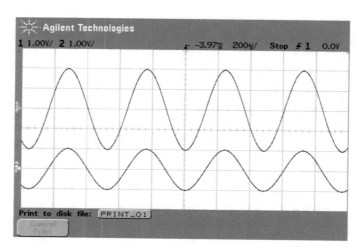

Fig. 9.12 Experimental results. Voltages at the input and the output of the circuit in Fig. 9.9. Upper line: V_o, bottom line: V_i

Fig. 9.13 Implementation of an analog controller using an operational amplifier

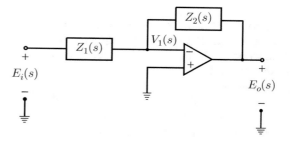

9.2   Analog Controllers with Operational Amplifiers

Consider the circuit in Fig. 9.13 where $Z_1(s)$ and $Z_2(s)$ stand for the impedances of two passive networks located at those places. In an operational amplifier, it is true that:

$$E_o(s) = (V_+(s) - V_-(s))A_0,$$

where $V_+(s)$ and $V_-(s)$ stand respectively for voltages at the terminals "+" and "−" of the operational amplifier, whereas A_0 is the open-loop gain of the operational amplifier, which is very large (about 100,000 [2], pp. 500). As $V_+(s) = 0$ and $V_-(s) = V_1(s)$, then:

$$E_o(s) = (0 - V_1(s))A_0 = -V_1(s)A_0. \tag{9.3}$$

Fig. 9.14 Equivalent circuit
to that shown in Fig. 9.13

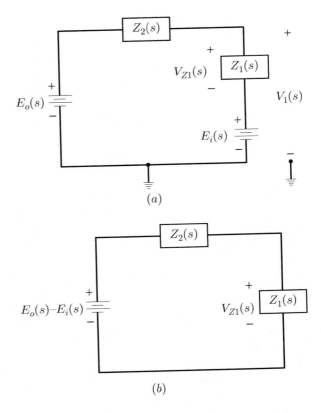

(a)

(b)

To compute V_1, the auxiliary circuits in Figs. 9.14a and b are employed. There, it is found that:

$$V_1(s) = \frac{Z_1(s)}{Z_1(s) + Z_2(s)}(E_o(s) - E_i(s)) + E_i(s),$$

$$= \frac{Z_1(s)}{Z_1(s) + Z_2(s)} E_o(s) + E_i(s)\left(1 - \frac{Z_1(s)}{Z_1(s) + Z_2(s)}\right).$$

Using (9.3), the following is found:

$$E_o(s) = -\left[\frac{Z_1(s)}{Z_1(s) + Z_2(s)} E_o(s) + E_i(s)\left(1 - \frac{Z_1(s)}{Z_1(s) + Z_2(s)}\right)\right] A_0,$$

hence:

$$\frac{E_o(s)}{E_i(s)} = \frac{-A_0\left(1 - \frac{Z_1(s)}{Z_1(s) + Z_2(s)}\right)}{1 + A_0\frac{Z_1(s)}{Z_1(s) + Z_2(s)}}.$$

Rearranging the terms:

$$\frac{E_o(s)}{E_i(s)} = \frac{-A_0 \left(\frac{Z_2(s)}{Z_1(s)+Z_2(s)}\right)}{1 + A_0 \frac{Z_1(s)}{Z_1(s)+Z_2(s)}}.$$

It is possible to assume that $A_0 \frac{Z_1(s)}{Z_1(s)+Z_2(s)} \gg 1$, because A_0 is very large; thus:

$$\frac{E_o(s)}{E_i(s)} = -\frac{Z_2(s)}{Z_1(s)}. \tag{9.4}$$

Note, however, that the condition $A_0 \frac{Z_1(s)}{Z_1(s)+Z_2(s)} \gg 1$ is true depending on A_0. In this respect, it is important to point out that, in practical operational amplifiers, the gain A_0 is not a constant, but changes with the frequency of the signals the operational amplifier is processing. It is common that A_0 decreases when such a frequency increases in a similar manner to the magnitude of a low-pass filter [2], pp. 500. Hence, the conditions $A_0 \frac{Z_1(s)}{Z_1(s)+Z_2(s)} \gg 1$ and (9.4) are not satisfied for high frequencies. On the other hand, the manner in which A_0 changes with the frequency depends on the particular operational amplifier that is employed. This means that attention must be paid when selecting an operational amplifier for a given application, i.e., taking into account how fast the plant to be controlled is.

In Table 9.1, some examples are shown on the possible networks employed to implement $Z_1(s)$ and $Z_2(s)$, in addition to the analog controller that results when using them. It is left as an exercise for the reader to compute $Z_1(s)$ and $Z_2(s)$ for each one of these networks to verify, using (9.4), that the transfer function of the controller in the column at the right of this table is obtained. See [3] for a more complete table including additional controllers. It is stressed, however, that these controllers can also be implemented using either a digital computer or a microcontroller.

Table 9.1 Implementation of analog controllers

Controller	$Z_1(s)$	$Z_2(s)$	$-\frac{Z_2(s)}{Z_1(s)}$
PI	R_1	R_2, C series	$-\frac{R_2}{R_1}\frac{\left(s+\frac{1}{R_2C}\right)}{s}$
PD	R_1, C parallel	R_2	$-R_2C\left(s+\frac{1}{R_1C}\right)$
PID	R_1, C_1 parallel	R_2, C_2 series	$-\left[\left(\frac{R_2}{R_1}+\frac{C_1}{C_2}\right)+R_2C_1s+\frac{\frac{1}{R_1C_2}}{s}\right]$
Lead compensator	R_1, C_1 parallel	R_2, C_2 parallel	$-\frac{C_1}{C_2}\frac{\left(s+\frac{1}{R_1C_1}\right)}{\left(s+\frac{1}{R_2C_2}\right)}$, $R_1C_1 > R_2C_2$

9.3 Design of Sinusoidal Waveform Oscillators

The purpose of this section is to show how control theory can be employed to design feedback electronic circuits generating sinusoidal signals. As sustained oscillations are present only in marginally stable systems, a necessary requirement for these feedback electronic circuits, from the linear control theory point of view, is that their characteristic polynomial must possess a pair of imaginary conjugate roots, i.e., the closed-loop system has to possess imaginary conjugate poles. This is known as the oscillation condition. Once this is achieved, the oscillation frequency, in radians/second, is equal to the imaginary part of such roots or poles. This part of the oscillator electronic circuit analysis and the design is performed using both classical control theory approaches: the time response and the frequency response. In the present section, some designs are presented based on either operational amplifiers or transistors.

An oscillator based on operational amplifiers simplifies the analysis and design from the control point of view and it is useful when operating at low frequencies as it avoids the necessity for inductance. It is important to stress that inductances required to generate low-frequency oscillations are large, resulting in bulky designs. The main drawback of this design is that it cannot work at high frequencies as in the radiofrequency bands because of the limitations of operational amplifiers.

On the other hand, the use of transistors results in a more complex analysis and design from the control point of view. However, the main advantage of the use of transistors is that it makes it possible to design oscillators for high frequencies, i.e., for the radiofrequency bands. The use of transistors is also interesting because of the following feature. As shown in the remainder of this section, that when operational amplifiers are employed, the resulting circuit is linear, requiring the introduction, in an artificial manner, of a nonlinear circuit component to render the oscillation possible. On the contrary, as the transistor is a nonlinear device it renders the oscillation possible in a rather natural manner.

9.3.1 Design Based on an Operational Amplifier: The Wien Bridge Oscillator

Consider the circuit in Fig. 9.15. In Example 2.14, Chap. 2, it was found that the voltages $E_o(s)$ and $E_i(s)$ are related as in (2.73). This expression is rewritten here for ease of reference:

$$\frac{E_o(s)}{E_i(s)} = G_T(s) = \frac{Ts}{T^2s^2 + 3Ts + 1}, \quad T = RC. \quad (9.5)$$

Consider now the circuit in Fig. 9.16 where an operational amplifier-based non-inverter amplifier is included, see Fig. 9.17. In the operational amplifier:

$$V_0(s) = (V_+(s) - V_-(s))A_0,$$

Fig. 9.15 *RC* Series parallel circuit

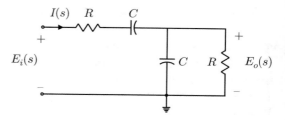

Fig. 9.16 An operational amplifier-based oscillator circuit

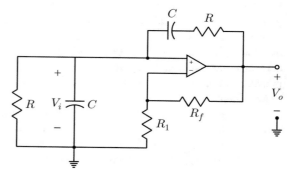

with $V_+(s) = V_i(s)$ and $V_-(s) = \frac{R_1}{R_1+R_f} V_0(s)$. Then:

$$V_0(s) = \left(V_i(s) - \frac{R_1}{R_1 + R_f} V_0(s) \right) A_0,$$

and rearranging:

$$V_0(s) \left(\frac{R_1 + R_f + R_1 A_0}{R_1 + R_f} \right) = A_0 V_i(s).$$

as A_0 is large, then:

$$V_0(s) \left(\frac{R_1 A_0}{R_1 + R_f} \right) = A_0 V_i(s),$$

and, finally:

$$V_0(s) = A V_i(s), \quad A = \frac{R_1 + R_f}{R_1}.$$

Note that the block diagrams in Figs. 9.18a and b can be obtained from this circuit. It is important to observe that, according to these block diagrams, the circuit in Fig. 9.16 is a positive feedback circuit. As the theory presented in the previous chapters assumes that negative feedback control systems are designed, such as that

Fig. 9.17 Non-inverter
amplifier

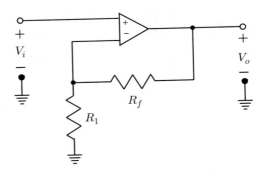

in Fig. 9.18d, it is necessary to transform the block diagram in Fig. 9.18b into a more convenient form. This is achieved in Fig. 9.18c where a negative sign is included in the feedback path, which is compensated for by a sign change in the transfer function on the direct path. Thus, a negative feedback system has been obtained that is equivalent to the systems in Figs. 9.18b and d. Thus, the open-loop transfer function is given as:

$$G(s)H(s) = -G_T(s)A. \tag{9.6}$$

9.3.2   Time Response-Based Analysis

This method studies the circuit at hand through the location of the closed-loop poles. From the study of differential equations in Chap. 3, it is known that sustained oscillations are possible only if the closed-loop poles are imaginary, i.e., with a zero real part. Under these conditions, the oscillation frequency, in radians/second, is equal to the imaginary part of these poles.

It is also known that the closed-loop poles satisfy:

$$1 + G(s)H(s) = 0.$$

Replacing (9.6) and (9.5):

$$1 + G(s)H(s) = 1 - G_T(s)A = 1 - \frac{TAs}{T^2s^2 + 3Ts + 1},$$

$$= \frac{T^2s^2 + 3Ts + 1 - TAs}{T^2s^2 + 3Ts + 1} = 0,$$

; hence, closed-loop poles satisfy:

$$T^2s^2 + (3 - A)Ts + 1 = 0.$$

Fig. 9.18 Block diagrams equivalent to the circuit in Fig. 9.16

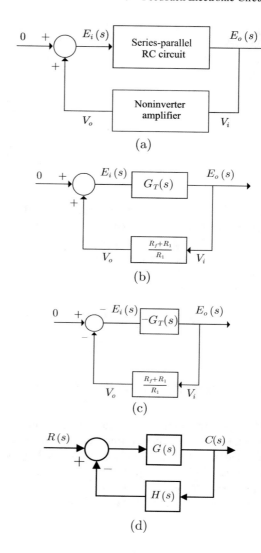

The closed-loop poles are the roots of this polynomial, i.e.,

$$s_1 = \frac{-(3-A)T + \sqrt{(3-A)^2 T^2 - 4T^2}}{2T^2},$$

$$s_2 = \frac{-(3-A)T - \sqrt{(3-A)^2 T^2 - 4T^2}}{2T^2}.$$

Note the following:

- Both poles have a nonzero imaginary part; hence, the circuit oscillates if:

$$(3-A)^2 T^2 - 4T^2 < 0,$$

i.e., if $5 > A > 1$.

- If $A > 3$, then both poles have a positive real part, i.e., the circuit is unstable, which is not desirable.
- If $A < 3$, then both poles have a negative real part, i.e., the circuit is stable. This, however, is not desirable either because circuit oscillation disappears as time increases.
- If $A = 3$, then both poles have a zero real part; thus, they are imaginary as desired. These poles are located at $s_{1,2} = \pm j\frac{1}{T}$. Hence, the oscillation frequency is $\omega = \frac{1}{T} = \frac{1}{RC}$.

9.3.3 Frequency Response-Based Analysis

Replace the variable change $s = j\omega$ in (9.5):

$$G_T(j\omega) = \frac{jT\omega}{T^2(j\omega)^2 + 3jT\omega + 1}.$$

Evaluating at the frequency $\omega = 1/T = 1/(RC)$:

$$G_T(j/T) = \frac{1}{3}.$$

To plot the Bode diagrams of $G_T(s)$ rearrange:

$$G_T(s) = Ts\frac{\frac{1}{T^2}}{s^2 + \frac{3}{T}s + \frac{1}{T^2}}. \tag{9.7}$$

The Bode diagrams of $G_T(s)$ are shown in Fig. 9.19. The magnitude of $G_T(j\omega)$ is maximal (equal to $\frac{1}{3}$) at the frequency $\omega = 1/T = 1/(RC)$, exactly when the phase of $G_T(j\omega)$ is zero.

The polar plot of $G(s)H(s)$, given in (9.6), is depicted in Fig. 9.20. From the Bode diagrams of $G_T(s)$, shown in Fig. 9.19, it is concluded that the polar plot of $G(s)H(s)$, shown in Fig. 9.20, consists of two clockwise contours. Note that the negative sign in (9.6) changes 180° the phase at every point of the Bode diagram of $G_T(s)$. Also note that the polar plot in Fig. 9.20 crosses the negative real axis at the point $-\frac{R_1+R_f}{3R_1}$ when the frequency is $\omega = 1/T = 1/(RC)$. On the other hand, because the polar plot includes positive and negative frequencies and it is symmetrical with respect to the real axis for negative and positive frequencies, then the negative real axis is also crossed at the point $-\frac{R_1+R_f}{3R_1}$ when the frequency is $\omega = -1/T = -1/(RC)$. Also note that the number of unstable open-loop poles of the function given in (9.6) is zero, i.e., $P = 0$, because all the coefficients of the polynomial at the denominator of $G_T(s)$ are positive (see Sect. 4.2.1).

Fig. 9.19 Bode diagrams of
$G_T(s)$ in (9.7). (*i*) T, (*ii*) s,
(*iii*) $\dfrac{\frac{1}{T^2}}{s^2+\frac{3}{T}s+\frac{1}{T^2}}$, (*iv*) $G_T(s)$

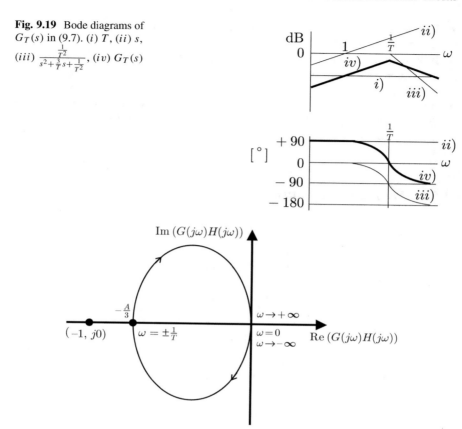

Fig. 9.20 Polar plot of $G(s)H(s)$ in (9.6)

Finally, before applying the Nyquist stability criterion, the following must be noted. As $G(s)H(s)$ has one zero at $s = 0$, i.e., on the imaginary axis, a contour must be included such as that in Fig. 6.79. The only difference is that in this case it is a zero (instead of a pole) that must be contoured. However, along this contour $s = \varepsilon\angle\phi$ where $\varepsilon \to 0$; hence, $G(s)H(s) \to 0$ also on this contour. This means that $G(s)H(s)$ represents a unique point on the origin of the complete contour. Thus, when applying the Nyquist stability criterion, three cases exist:

1. If $\frac{R_1+R_f}{3R_1} > 1$, then the number of contours around the point $(-1, j0)$ is $N = 2$, i.e., the number of closed-loop unstable poles is $Z = N + P = 2$. This implies circuit instability, which is undesirable.

2. If $\frac{R_1+R_f}{3R_1} < 1$, then the number of contours around the point $(-1, j0)$ is $N = 0$ and the number of unstable closed-loop poles is $Z = N + P = 0$. Although this implies circuit stability, it is also undesirable in this application because the circuit oscillation disappears as time increases.

3. If $\frac{R_1+R_f}{3R_1} = 1$, then the circuit is marginally stable, i.e., there are closed-loop poles on the imaginary axis. This can also be understood as follows. The condition to be satisfied by the closed-loop poles is $1 + G(s)H(s) = 0$, or $G(s)H(s) = -1$. Then, if the closed-loop poles are imaginary, it suffices to perform the variable change $s = j\omega$. Note that this is what happens when the polar plot in Fig. 9.20 crosses the negative real axis at the point $(-1, j0)$ when the frequency is $\omega = 1/T = 1/(RC)$ and $\omega = -1/T = -1/(RC)$. This means that $G(j\omega)H(j\omega)|_{\omega=1/T} = -1$ and $G(j\omega)H(j\omega)|_{\omega=-1/T} = -1$. Hence, it is concluded that there are two closed-loop poles that are imaginary conjugate and placed at $s = j/T$ and $s = -j/T$. Thus, the circuit exhibits permanent oscillations at the frequency $\omega = 1/T$.

9.3.4 A Practical Oscillator

From the above discussion, it is concluded that the gain of the operational amplifier-based non-inverter amplifier must be chosen as:

$$A = \frac{R_1 + R_f}{R_1} = 3.$$

However, the circuit in Fig. 9.16 does not oscillate correctly. The reason for this is that commercial values for R_f and R_1 cannot be found to exactly satisfy $\frac{R_1+R_f}{R_1} = 3$. Moreover, any small change in their nominal values would render the circuit either unstable or stable (oscillation vanishes as time increases). This is a well-known fact in electronics: a linear circuit cannot correctly oscillate in practice, i.e., all practical oscillators are nonlinear.

The way to build a practical oscillator using the above ideas is by employing the circuit in Fig. 9.21, which includes a nonlinearity introduced by two Zener diodes. Suppose that $A > 3$. Hence, the circuit is unstable and oscillates with an ever increasing amplitude. The job of the Zener diodes is to put a short circuit at the 10[KOhm] resistance when they reach the avalanche voltage. This decreases the gain A of the amplifier when the output voltage reaches a certain threshold. This renders $A < 3$, and the oscillation amplitude decreases. Hence, the voltage at the Zener diodes decreases and the short circuit at the 10[KOhm] resistance disappears. Then, $A > 3$ again. Finally, this process ends at a steady state where a sinusoidal voltage with constant amplitude is produced. Note that, using the values of resistances in Fig. 9.21, $A = 45/10 = 4.5$ if the Zener diodes are open and $A = 35/10 = 3.5$ if the Zener diodes are on. Hence, the 25.6[KOhm] potentiometer must be adjusted to obtain a value slightly larger than 3 for A to obtain a sustained oscillation. Note that the expected oscillation frequency is:

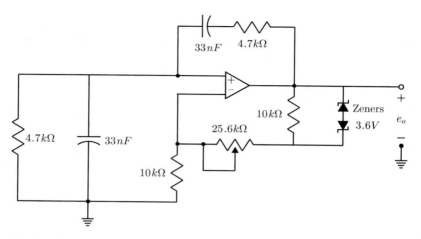

Fig. 9.21 Practical oscillator circuit

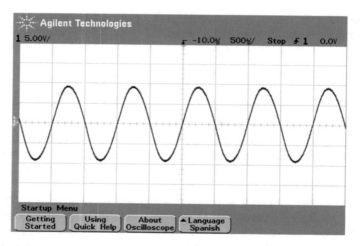

Fig. 9.22 Waveform of voltage at the output of the operational amplifier in Fig. 9.21. The potentiometer is at 20[KOhm]

$$\omega = \frac{1}{T} = \frac{1}{RC} = \frac{1}{(4.7 \times 10^3)(0.033 \times 10^{-6})} = 6447[\text{rad/s}],$$

$$f = \frac{\omega}{2\pi} = \frac{6447}{2\pi} = 1.026[\text{KHz}].$$

Some experimental results obtained with circuit in Fig. 9.21 are shown in Figs. 9.22 and 9.23. The UA741 operational amplifier has been employed. The frequency measured in experiments is 1[KHz]. The oscillation amplitude can be modified if the resistance of the 25.6[KOhm] potentiometer is increased, which also increases

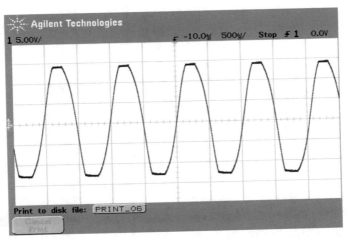

Fig. 9.23 Waveform of the voltage at the output of the operational amplifier in Fig. 9.21. The potentiometer is at 25.6[KOhm]

Fig. 9.24 Phase shift network

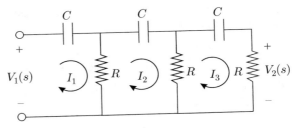

the amplifier gain A. However, if this gain is too large, the sinusoidal waveform is distorted. This is what happens in Fig. 9.23 where the 25.6[KOhm] poten-tiometer is adjusted to its maximal resistance value. On the contrary, in Fig. 9.22 the 25.6[KOhm] potentiometer is adjusted to approximately 20[KOhm]. This is evidence that the behavior of Zener diodes is smooth instead of abrupt, as the amplitude of voltage at the operational amplifier output increases. Thus, a larger loop gain in (9.6) reduces to 1 if the amplitude of oscillations increases, i.e., if the Zener diodes work deeper in the avalanche region.

Finally, note that, according to the block diagrams in Fig. 9.18, this oscillator circuit is a closed-loop system without an input, i.e., the circuit oscillates when a zero input is applied. This must not be a surprise as it is clearly explained in Sect. 3.3, Chap. 3, that a second-order (or a larger order) circuit may oscillate, despite the input being zero if the initial conditions are different from zero. In this respect, note that despite the fact that the circuit has no stored energy initially, some small initial conditions that are different from zero are produced as a consequence of a perturbation in the circuit when this is turned on. These initial conditions that are different from zero, although small, suffice to make the circuit oscillate given the initial instability of the circuit, because $A > 3$ when the circuit is turned on.

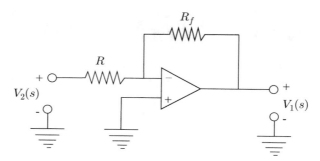

Fig. 9.25 Inverter amplifier

9.3.5 Design Based on an Operational Amplifier: The Phase Shift Oscillator

Consider the circuit shown in Fig. 9.24. The relationship between the voltages $V_1(s)$ and $V_2(s)$ is shown in (2.76) in Example 2.15, Chap. 2. This expression is rewritten here for ease of reference:

$$\frac{V_2(s)}{V_1(s)} = \frac{R^3 C^3 s^3}{R^3 C^3 s^3 + 6R^2 C^2 s^2 + 5RCs + 1} = F(s). \tag{9.8}$$

On the other hand, according to Sect. 9.2, the inverter amplifier shown in Fig. 9.25 performs the operation:

$$V_1(s) = -\frac{R_f}{R} V_2(s). \tag{9.9}$$

Now, consider the circuit shown in Fig. 9.26. Note that this circuit can be represented using the block diagram in Fig. 9.27a. Moreover, according to (9.8) and (9.9), the block diagram in Fig. 9.27b is obtained, which can be represented as in Fig. 9.27c. This is convenient for seeing that it is a negative feedback closed-loop system. Hence, the open-loop transfer function is given as:

$$G(s)H(s) = \frac{R_f}{R} F(s). \tag{9.10}$$

It is well known that the closed-loop poles satisfy:

$$1 + G(s)H(s) = 0, \tag{9.11}$$

i.e.,

$$1 + \frac{R_f}{R} F(s) = 0. \tag{9.12}$$

Fig. 9.26 Phase shift
oscillator

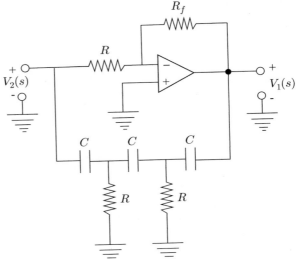

Table 9.2 Applying Routh's
criterion to the characteristic
polynomial in (9.13)

s^3	$R^4C^3 + R_f R^3 C^3$	$5R^2C$
s^2	$6R^3C^2$	R
s^1	$\frac{(6R^3C^2)(5R^2C)-R(R^4C^3+R_f R^3 C^3)}{6R^3C^2}$	0
s^0	R	

9.3.6   Time Domain-Based Analysis

In the following, we analyze the closed-loop circuit to find the conditions for the
existence of a pair of imaginary conjugate closed-loop poles. From (9.12), we obtain
the following characteristic polynomial:

$$(R^4C^3 + R_f R^3 C^3)s^3 + 6R^3C^2 s^2 + 5R^2 Cs + R = 0. \qquad (9.13)$$

According to Chap. 3, the behavior of this circuit depends on the roots of the
characteristic polynomial in (9.13). As this polynomial is third-degree, it is useful
to employ Routh's stability criterion (Sect. 4.3) to study its roots; hence, Table 9.2
is obtained. Three different behaviors can be predicted from this table:

- $(6R^3C^2)(5R^2C) - R(R^4C^3 + R_f R^3 C^3) > 0$, i.e., $R_f < 29R$. In this case,
  all the entries in the first column of Table 9.2 are positive, ensuring that all the
  roots of the polynomial in (9.13) have a negative real part, i.e., circuit stability is
  concluded. This means that, although the circuit could oscillate, this oscillation
  disappears as time increases.

Fig. 9.27 Equivalent block diagrams for the circuit in Fig. 9.26

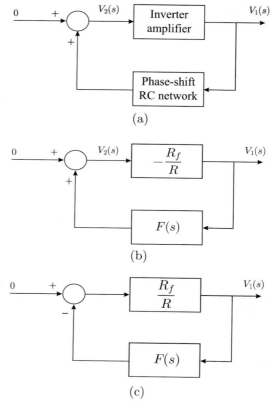

(a)

(b)

(c)

Table 9.3 Applying Routh's criterion to the characteristic polynomial in (9.13) (cont.)

s^3	$R^4C^3 + R_f R^3 C^3$	$5R^2C$
s^2	$6R^3C^2$	R
s^1	$12R^3C^2$	0
s^0	R	

- $(6R^3C^2)(5R^2C) - R(R^4C^3 + R_f R^3 C^3) < 0$, i.e., $R_f > 29R$. In this case, there are two changes of sign in the first column of Table 9.2. This means that the polynomial in (9.13) has two roots with a positive real part, i.e., circuit instability is concluded: although the circuit could oscillate, the amplitude of this oscillation would increase without a limit as time increases.
- $(6R^3C^2)(5R^2C) - R(R^4C^3 + R_f R^3 C^3) = 0$, i.e., $R_f = 29R$. In this case, there is a row in Table 9.2 that is composed only of zeros (the row corresponding to s^1). In this case, Routh's criterion establishes (see Sect. 4.3, Example 4.14) that the derivative of the polynomial obtained from the row s^2 must be computed, i.e., $\frac{dP(s)}{ds} = 12R^3C^2s$, where $P(s) = 6R^3C^2s^2 + R$, to replace these coefficients in the row s^1 and to continue constructing the table as shown in Table 9.3. As there is no change of sign in the first column of Table 9.3, it is concluded that no roots

exist with positive real parts; hence, imaginary conjugate roots exist. Thus:

$$R_f = 29R, \tag{9.14}$$

is a necessary condition for circuit in Fig. 9.26 to exhibit sustained oscillations. To determine the oscillation frequency, it is useful to recall another property of data shown in Table 9.2 (see Sect. 4.3, Example 4.14): "if one row is composed only of zeros, then the roots of the polynomial obtained with data in the row immediately above the row composed of zeros are also roots of the characteristic polynomial in (9.13)". Hence, the roots of:

$$6R^3C^2s^2 + R = 0,$$

are also roots of the characteristic polynomial shown in (9.13). Solving the latter expression, it is found that the corresponding roots are imaginary, as expected:

$$s_1 = j\frac{1}{\sqrt{6}RC}, \quad s_2 = -j\frac{1}{\sqrt{6}RC}.$$

This means that $\omega = \frac{1}{\sqrt{6}RC} = 2\pi f$, i.e., that the oscillation frequency in Hertz is given as:

$$f = \frac{1}{2\pi\sqrt{6}RC}. \tag{9.15}$$

Thus, the oscillator circuit in Fig. 9.26 must be designed by choosing R and C such that the oscillation frequency is computed as in (9.15), and then oscillation is ensured by choosing R_f according to (9.14).

9.3.7 Frequency Domain Analysis

In what follows, the closed-loop circuit is studied by applying the Nyquist stability criterion to find the conditions for closed-loop marginal stability. Let us first apply Routh's criterion to the characteristic polynomial in (9.8) (see Table 9.4). As there are no changes of sign in the first column, we conclude that the characteristic

Table 9.4 Routh's criterion applied to the characteristic polynomial in (9.8)

s^3	R^3C^3	$5RC$
s^2	$6R^2C^2$	1
s^1	$\frac{29R^3C^3}{6R^2C^2}$	0
s^0	1	

Fig. 9.28 Bode diagrams of $F(s)$ in (9.8)

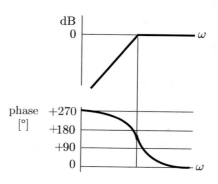

polynomial in (9.8) has no pole with a positive real part. As $G(s)H(s) = \frac{R_f}{R}F(s)$, this means that $P = 0$. In Fig. 9.28, the Bode diagrams for $F(s)$ defined in (9.8) are depicted. From Fig. 9.28 the polar plot of $F(s)$ is obtained, which is depicted in Fig. 9.29. Note that, according to $G(s)H(s) = \frac{R_f}{R}F(s)$, the polar plot of the open-loop transfer function is identical to the polar plot of $F(s)$ in Fig. 9.29, but it is only required to increase the magnitude by a factor $\frac{R_f}{R}$. Hence, depending on the particular value of $\frac{R_f}{R}$, there are three possibilities, which are depicted in Fig. 9.30. Thus, closed-loop marginal stability is obtained for a value $\frac{R_f}{R}$ that satisfies $G(j\omega)H(j\omega) = -1$ for some ω. This means that the closed-loop poles are imaginary. Moreover, as $1 + G(s)H(s) = 0$ is the condition that defines the closed-loop poles s, then $1 + G(j\omega)H(j\omega) = 0$, or $G(j\omega)H(j\omega) = -1$, implies that the closed-loop poles can be written as $s = \pm j\omega$ where ω is the frequency at which the polar plot of $G(j\omega)H(j\omega)$ crosses the point $(-1, j0)$. The value of $\frac{R_f}{R}$ that produces marginal stability is known as the oscillation condition and can be obtained from $G(j\omega)H(j\omega) = \frac{R_f}{R}F(j\omega) = -1$, i.e.,

$$\frac{R_f}{R} = -\frac{1}{F(j\omega_2)}.$$

where ω_2 is the frequency when $\angle F(j\omega_2) = 180°$.

In Fig. 9.31, the Bode diagrams of $F(s)$ when $RC = 0.001$ are presented. These plots have been drawn using the following MATLAB code:

```
RC=0.001;
F=tf([RC^3 0 0 0],[RC^3 6*RC^2 5*RC 1]);
bode(F)
grid on
```

Note that $\angle F(j\omega_2) = 180°$ is obtained when $\omega_2 = 411[\text{rad/s}]$ (the oscillation frequency) and $|F(j\omega_2)|_{[dB]} = -29.2[\text{dB}]$. Thus:

$$\frac{R_f}{R} = \frac{1}{10^{-29.2/20}} = 28.8403 \approx 29,$$

Fig. 9.29 Polar plot of $F(s)$ in (9.8)

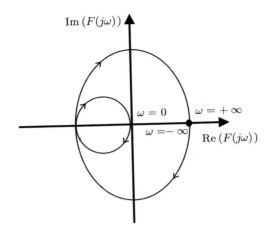

a value that is very close to the oscillation condition established in (9.14). Also note that, according to the previous section, the oscillation frequency is given as $\omega = \frac{1}{\sqrt{6}RC} = 408.2483[\text{rad/s}]$, where $RC = 0.001$ has been used, which is very close to $\omega_2 = 411[\text{rad/s}]$ found in Fig. 9.31. Note that the frequency domain analysis does not provide closed expressions to compute the exact value of $\frac{R_f}{R}$ or the oscillation frequency. However, very precise numerical values are given for both parameters.

9.3.8   A Practical Oscillator Circuit

From the above discussion, it is concluded that R_f and R must satisfy (9.14). However, similar to the oscillator circuit in Sect. 9.3.1, the oscillator in Fig. 9.26 must be modified, as shown in Fig. 9.32 to accomplish a satisfactory oscillation. Note that, using the values of resistance in Fig. 9.32, $\frac{R_f}{R} = \frac{80000}{2200} = 36.36$ if the Zener diodes are open and $\frac{R_f}{R} = \frac{47000}{2200} = 21.36$ if the Zener diodes conduct. Hence, the 47[KOhm] potentiometer must be adjusted to obtain a value for $\frac{R_f}{R}$ slightly larger than 29 to obtain sustained oscillations. Note that the expected oscillation frequency is:

$$f = \frac{1}{2\pi\sqrt{6}RC} = \frac{1}{2\pi\sqrt{6}(2200)(0.01 \times 10^{-6})} = 2.9534[\text{KHz}].$$

Some experimental results are shown in Fig. 9.33 that were obtained with the circuit in Fig. 9.32. The UA741 operational amplifier is employed. Frequency measured in experiments is 2.7027[KHz], which is very close to the design value: 2.9534[KHz].

Fig. 9.30 Polar plots of
$G(s)H(s)$ for different values
of $\frac{R_f}{R}$. Recall that $P = 0$. (**a**)
$N = 0$, $Z = N + P = 0$,
closed-loop stability. (**b**)
$N = 2$, $Z = N + P = 2$,
closed-loop instability. (**c**)
Closed-loop marginal
stability

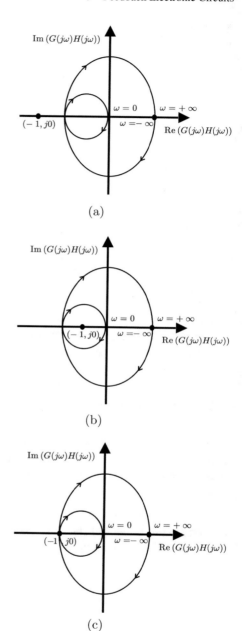

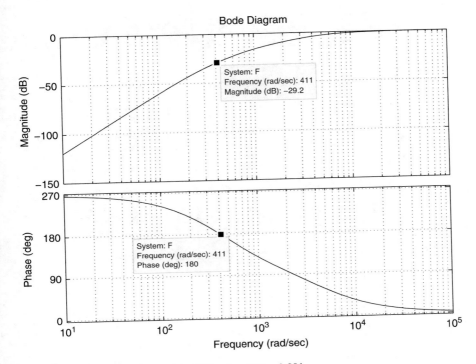

Fig. 9.31   Bode diagrams of $F(s)$ in (9.8), when $RC = 0.001$

Fig. 9.32   A practical
oscillator circuit

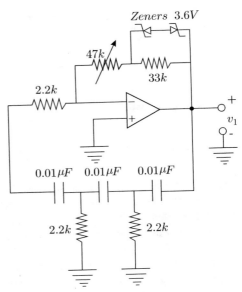

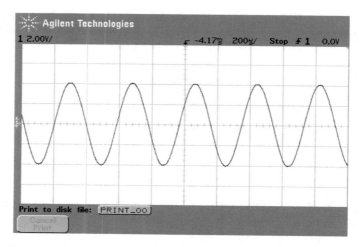

Fig. 9.33 Waveform of the voltage at the output of the operational amplifier in Fig. 9.32

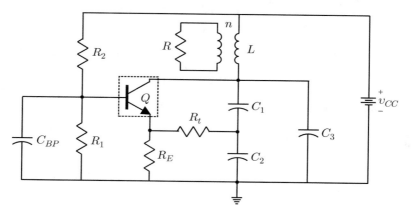

Fig. 9.34 Colpitts oscillator

9.3.9   A Transistor-Based Design

The oscillator circuit shown in Fig. 9.34 is studied in this section. This circuit is known as the Colpitts oscillator. The transistor is a nonlinear device. Because of that, the problem is studied by proceeding as in the case of nonlinear differential equations. It is assumed that the circuit in Fig. 9.34 works around an operation point, i.e., only small variations of signals are allowed around it. The operation point is determined by the direct current operation of the circuit in Fig. 9.34, whereas the variations around the operation point are analyzed using an equivalent small-signal circuit for the circuit in Fig. 9.34. The *small-signal circuit* is similar to approximate linear models obtained for nonlinear systems (nonlinear differential equations) in

Sect. 7.3, which is valid if only small variations around the selected operation point are allowed. The circuit is studied under direct current conditions in the next section. The equivalent small-signal circuit is analyzed in subsequent sections.

Given the large number of variables involved (voltages and currents at each circuit element), the following nomenclature is employed. v_{R1Q} stands for the (constant) voltage at the resistance R_1 at the operation point, v_{r1} represents the voltage variations at the resistance R_1 around the operation point and v_{R1} is the total voltage at the resistance R_1, i.e., $v_{R1} = v_{R1Q} + v_{r1}$. Currents and voltages at the other circuit elements are defined analogously. On the other hand, upper case letters are employed to represent the Laplace transform of a time function represented with a lower case letter, e.g., $I(s) = \mathcal{L}\{i(t)\}$.

9.3.10 Direct Current Analysis

The operation point in a transistor is determined by the constant values of the collector current i_{CQ} and the collector to emitter voltage v_{CEQ}. First, note that $v_{LQ} = L\frac{di_{LQ}}{dt} = 0, i_{C1Q} = C\frac{dv_{C1Q}}{dt} = 0, i_{C2Q} = C\frac{dv_{C2Q}}{dt} = 0$ e $i_{C3Q} = C\frac{dv_{C3Q}}{dt} = 0$ because i_{LQ}, v_{C1Q}, v_{C2Q} and v_{C3Q} are constants. Hence, the circuit in Fig. 9.35 is considered for the direct current analysis.

The fundamental direct current laws for a transistor establish that [1], Chap. 5:

$$i_{CQ} = \beta i_{BQ}, \ i_{EQ} = (1+\beta)i_{BQ}, \ \beta \gg 1, \ i_{EQ} \approx i_{CQ}, \ v_{BEQ} = 0.7[V], \text{ silicon.}$$

The Kirchhoff voltage law applied to mesh defined by J_c in Fig. 9.35 yields:

$$v_{cc} = v_{CEQ} + v_{REQ},$$
$$v_{cc} = v_{CEQ} + R_E i_{EQ}. \tag{9.16}$$

Fig. 9.35 Direct current equivalent circuit

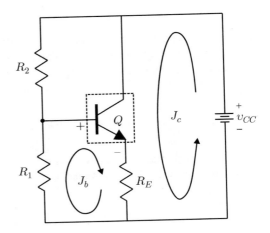

The Kirchhoff voltage law applied to mesh defined by J_b in Fig. 9.35 yields:

$$v_{R1Q} = v_{BEQ} + v_{REQ}.$$

$$v_{R1Q} = 0.7 + R_E i_{EQ}. \tag{9.17}$$

Suppose that i_{CQ}, v_{CEQ}, β and v_{cc} are given as known values. R_E is computed using (9.16) and $i_{EQ} \approx I_{CQ}$. Then, i_{R1Q} is proposed from $i_{CQ} = \beta i_{BQ}$ and the following important assumption:

$$i_{R1Q} \gg i_{BQ}. \tag{9.18}$$

Hence, (9.17) can be employed to compute:

$$R_1 = \frac{v_{R1Q}}{i_{R1Q}}, \tag{9.19}$$

$$R_2 = \frac{v_{cc} - v_{R1Q}}{i_{R1Q}}. \tag{9.20}$$

The values for L, C_1, C_2, C_3, C_{BP} and R_t are computed from the analysis of the equivalent small-signal circuit, as shown in the following.

9.3.11 Equivalent Small-Signal Circuit

The equivalent small-signal model for the transistor is found first. Although several small-signal models for the transistor exist, in this book the following is considered. The emitter current, i.e., the current through the diode at the base-emitter junction, is given by the Shockley equation [1], chapter 5:

$$i_E = I_{ES} \left[e^{\frac{v_{BE}}{V_T}} - 1 \right],$$

where I_{ES} is a constant with a value between 10^{-12}[A] and 10^{-16}[A] whereas $V_T = 0.026$[V] for a temperature of 300 Kelvin degrees. An important equation in the transistor establishes that $i_B = (1 - \alpha)i_E$, where α is a positive constant slightly less than 1. Hence:

$$i_B = (1 - \alpha)I_{ES} \left[e^{\frac{v_{BE}}{V_T}} - 1 \right]. \tag{9.21}$$

As the transistor is assumed to work in its active region, then the number 1 subtraction in the latter expression can be neglected to write:

$$i_B = (1 - \alpha)I_{ES} \left[e^{\frac{v_{BE}}{V_T}} \right].$$

Recalling that $i_B = i_{BQ} + i_b$ and $v_{BE} = v_{BEQ} + v_{be}$, the following can be written:

$$i_{BQ} + i_b = (1 - \alpha)I_{ES}\left[e^{\frac{v_{BEQ}+v_{be}}{V_T}}\right],$$

$$= (1 - \alpha)I_{ES}e^{\frac{v_{BEQ}}{V_T}}e^{\frac{v_{be}}{V_T}}. \tag{9.22}$$

Also recall that small changes in i_B are due to small changes in v_{BE}, i.e., $i_b = 0$ if $v_{be} = 0$. Then, according to the last expression:

$$i_{BQ} = (1 - \alpha)I_{ES}\left[e^{\frac{v_{BEQ}}{V_T}}\right].$$

Hence, (9.22) can be written as:

$$i_{BQ} + i_b = I_{BQ}e^{\frac{v_{be}}{V_T}}. \tag{9.23}$$

If only small values of v_{be} are allowed, the following approximation is possible [4], pp. 942:

$$e^{\frac{v_{be}}{V_T}} \approx 1 + \frac{v_{be}}{V_T},$$

which, together with (9.23), implies that:

$$i_b = \frac{v_{be}}{r_\pi}, \qquad r_\pi = \frac{V_T}{I_{BQ}}. \tag{9.24}$$

On the other hand, the following expression is also valid for small variations:

$$i_c = \beta i_b. \tag{9.25}$$

Using (9.24) and (9.25) the small-signal model of the transistor shown in Fig. 9.36a is obtained. It is possible to find an equivalence between this model and that shown in Fig. 9.36b. Note that the only difference is that the base-emitter voltage is now given as a voltage drop due to the emitter current, i.e., it must depend on the new resistance h_{ib}:

$$v_{be} = h_{ib}i_e. \tag{9.26}$$

Comparing (9.24), (9.26), and using $i_e = (1 + \beta)i_b$, the following is found:

$$h_{ib} = \frac{v_{be}}{i_e} = \frac{v_{be}}{(1+\beta)i_b} = \frac{r_\pi}{1+\beta} = \frac{V_T}{(1+\beta)I_{BQ}} = \frac{V_T}{I_{EQ}}. \tag{9.27}$$

Fig. 9.36 Equivalent
small-signal transistor models

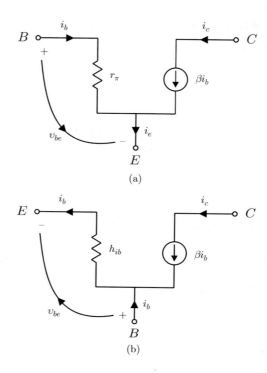

(a)

(b)

Note that value of h_{ib} depends on I_{EQ}, i.e., on the operating point. Once the small-signal model of the transistor has been found, we proceed to finding the equivalent small-signal circuit for the whole oscillator circuit.

Consider the circuit in Fig. 9.37a. The Kirchhoff voltage law applied to mesh J_3 establishes that:

$$v_{cc} = L\frac{di_L}{dt} + v_{CE} + v_{RE},$$

$$i_L = i_{LQ} + i_l,$$

$$v_{CE} = v_{CEQ} + v_{ce},$$

$$v_{RE} = v_{REQ} + v_{re}.$$

Suitably arranged, the latter expression can be written as:

$$v_{cc} = L\frac{d}{dt}(i_{LQ} + i_l) + v_{CEQ} + v_{ce} + v_{REQ} + v_{re}.$$

Using (9.16) and $v_{LQ} = L\frac{di_{LQ}}{dt} = 0$ in the previous expression, the following is found:

$$0 = L\frac{di_l}{dt} + v_{ce} + v_{re}.$$

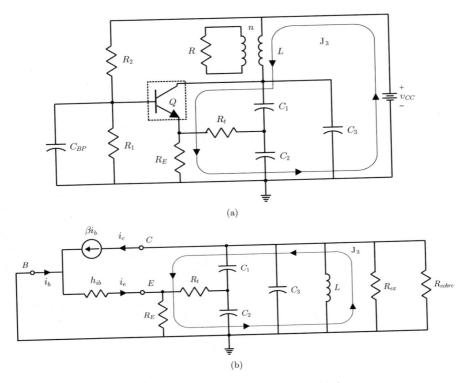

Fig. 9.37 Equivalent small-signal circuit for the whole oscillator circuit

Hence, the equivalent small-signal circuit must represent a mesh such as that indicated by J_3 in Fig. 9.37b. Proceeding in a similar manner for all possible meshes and nodes in the circuit depicted in Fig. 9.37a, it is found that the equivalent small-signal circuit is given as in Fig. 9.37b. It is important to state that R_{copper} is the equivalent parallel resistance of the inductance internal resistance (due to copper), whereas $R_{ex} = n^2 R$, where R is the external load for the oscillator circuit and n is the ratio of the turns number of L and the turns number of the inductance connected to R.

Finally, the resistances R_1 and R_2 in Fig. 9.37a disappear in Fig. 9.37b because it is assumed that the impedance of C_{BP} (parallel to R_1 and R_2) is very small compared with the equivalent parallel resistance of R_1 and R_2 at the oscillation frequency ω_1, i.e., $\frac{1}{\omega_1 C_{BP}} \ll \frac{R_1 R_2}{R_1 + R_2}$. Moreover, if $\frac{1}{\omega_1 C_{BP}} \ll h_{ib}$, the capacitor impedance can be neglected.

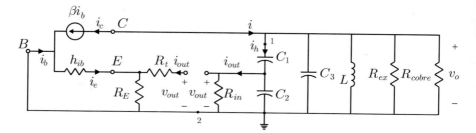

Fig. 9.38 Circuit employed for analysis purposes

9.3.12 Closed-Loop System Equations

The circuit in Fig. 9.34 is analyzed in this and the subsequent sections on the basis of the equivalent small-signal circuit in Fig. 9.37b. The reader must remember that all that is predicted by this analysis only stands around the operation point, which is determined in the direct current analysis.

For analysis purposes, the feedback path through the emitter resistance is opened as shown in Fig. 9.38 [5], pp. 265, [6], pp. 64. R_{in} in Fig. 9.38 is given as:

$$R_{in} = R_t + \frac{R_E h_{ib}}{R_E + h_{ib}}. \tag{9.28}$$

Now, we proceed to analyze the circuit in Fig. 9.38. The impedance between points 1 and 2 is given as:

$$Z_f(s) = \frac{1}{Y_f(s)} = \frac{1}{sC_1} + \frac{R_{in}\frac{1}{sC_2}}{R_{in} + \frac{1}{sC_2}} = \frac{sR_{in}(C_1 + C_2) + 1}{sC_1(sC_2 R_{in} + 1)}, \tag{9.29}$$

where $Y_f(s)$ is the admittance between points 1 and 2. On the other hand, $V_o(s)$ and $I(s)$ are related as:

$$V_o(s) = Z_c(s)I(s), \tag{9.30}$$

where:

$$\frac{1}{Z_c(s)} = Y_f(s) + \frac{1}{sL} + sC_3 + \frac{1}{R_{ex}} + \frac{1}{R_{copper}}. \tag{9.31}$$

After an algebraic procedure, it is found that:

$$Z_c(s) = \frac{sL(sR_{in}(C_1 + C_2) + 1)}{a_3 s^3 + a_2 s^2 + a_1 s + 1}, \tag{9.32}$$

$$a_3 = L(C_1C_2 + C_3(C_1 + C_2))R_{in},$$

$$a_2 = LC_1 + LC_3 + \frac{R_{copper} + R_{ex}}{R_{copper}R_{ex}}LR_{in}(C_1 + C_2),$$

$$a_1 = R_{in}(C_1 + C_2) + \frac{R_{copper} + R_{ex}}{R_{copper}R_{ex}}L.$$

Voltage $V_{out}(s)$ can be computed as:

$$V_{out}(s) = M(s)V_o(s), \tag{9.33}$$

where $V_{out}(s)$ and $V_o(s)$ are related to current $I_h(s)$ through:

$$V_o(s) = \frac{1}{Y_f(s)}I_h(s), \quad V_{out}(s) = \frac{1}{\frac{1}{R_{in}} + sC_2}I_h(s). \tag{9.34}$$

Combining both expressions in (9.34), the following is found:

$$V_o(s) = \frac{1}{Y_f(s)}\frac{1 + R_{in}sC_2}{R_{in}}V_{out}(s).$$

Use of (9.29) yields:

$$M(s) = \frac{V_{out}(s)}{V_o(s)} = \frac{s(R_{in}^2C_1C_2s + R_{in}C_1)}{R_{in}^2C_2(C_1 + C_2)s^2 + R_{in}(2C_2 + C_1)s + 1}. \tag{9.35}$$

According to (9.33) and (9.30), the following is found:

$$I_{out}(s) = \frac{V_{out}(s)}{R_{in}} = \frac{1}{R_{in}}M(s)V_o(s) = \frac{1}{R_{in}}M(s)Z_c(s)I(s). \tag{9.36}$$

On the other hand, applying the current divisor to the emitter circuit in Fig. 9.38, the following is found:

$$I_e(s) = \frac{-R_E}{R_E + h_{ib}}I_{out}(s).$$

Thus, if:

$$R_E \gg h_{ib}, \quad \beta \gg 1, \tag{9.37}$$

then it can be approximated:

$$I_{out}(s) \approx -I_e(s) = -(\beta + 1)I_b(s) \approx -\beta I_b(s) = I(s), \tag{9.38}$$

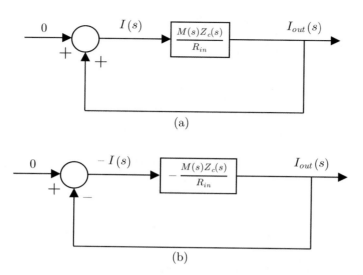

Fig. 9.39 Equivalent block diagrams for the circuit in Fig. 9.38

i.e., the transistor current gain in the common base configuration is unitary. According to the expressions in (9.36) and (9.38), the block diagram in Fig. 9.39a is obtained. This block diagram has positive feedback, which is an important feature for producing sustained oscillations. In Fig. 9.39b an equivalent block diagram is shown, which, however, is now expressed as a negative feedback system. This allows the application of classical control analysis and design tools when applied to a closed-loop system such as that in Fig. 9.18d. Hence, the open-loop transfer function is:

$$G(s)H(s) = -\frac{1}{R_{in}}M(s)Z_c(s). \tag{9.39}$$

9.3.13 Conditions for Ensuring Sustained Oscillations

Let us first study the transfer functions $M(s)$ and $Z_c(s)$. Note that all the coefficients of these transfer functions are positive. According to the criteria studied in Sects. 4.2.1 and 4.2.2 on the coefficient signs of first- and second-order polynomials and their corresponding roots, it is concluded that $M(s)$ has two poles with a negative real part, one zero with a negative real part and one zero at $s = 0$.

However, according to Sect. 4.2.3, the coefficient signs criterion cannot be applied to $Z_c(s)$ because its denominator is a third-degree polynomial. Nevertheless, a well-known result of the linear electric circuit theory[1] can be employed. This result

---

[1] Routh's criterion can also be used, if preferred.

states [7], Chapt. 19, Sects. 5 and 6, that the impedance of any network composed only of passive circuit elements (resistances, capacitance, and inductances) is a transfer function that only has poles with real parts, which are less than or equal to zero. Note that this is the case of $Z_C(s)$. Furthermore, as $Z_C(s)$ has a denominator with a term that is independent of s, then it is ensured that $Z_C(s)$ has three poles with a negative real part, one zero with a negative real part, and one zero at $s = 0$.

According to the above discussion and what was presented in Chap. 6, it is concluded that the Bode and polar plots of $Z_C(s)$ and $M(s)$ have the shapes shown in Figs. 9.40 and 9.41 respectively, whereas the polar plot of $G(s)H(s)$ given in (9.39) is shown in Fig. 9.42. It is important to stress that the polar plot of $G(s)H(s)$ is composed of two clockwise closed contours obtained as the frequency goes from $-\infty$ to $+\infty$. The real axis is crossed twice at the frequencies $\omega = \pm w_1$. Note that, at these frequencies, the transfer function $G(j\omega)H(j\omega)$ has a zero imaginary part. Finally, as $G(s)H(s)$ has two zeros at $s = 0$, to apply the Nyquist criterion the contour shown in Fig. 6.79 must be performed around such zeros at origin. However, as $s = \varepsilon \angle \phi$, with $\varepsilon \to 0$, then $G(s)H(s) \to 0$ along the complete contour; hence, it is represented by a single point at the origin. Applying the Nyquist criterion, the following conclusions are in order:

- If $|G(j\omega)H(j\omega)|_{\omega=w_1} < 1$, then closed-loop stability is ensured and sustained oscillations are not possible. This is because $Z = P + N$, where $P = 0$ represents the number of poles with positive real parts in $G(s)H(s)$, $N = 0$ is the number of clockwise contours around the point $(-1, 0)$ in Fig. 9.42 and $Z = 0$ is the number of closed-loop unstable poles.
- If $|G(j\omega)H(j\omega)|_{\omega=w_1} > 1$, then the closed-loop system is unstable because $Z = N + P = 2$, with $P = 0$ and $N = 2$. This situation is clearly undesirable.
- If $|G(j\omega)H(j\omega)|_{\omega=w_1} = 1$, then there are two imaginary closed-loop poles that allows the existence of sustained oscillations. The imaginary part of these poles is $\pm\omega_1$ representing the circuit oscillation frequency.

From these observations, a way of computing the values for the circuit elements is presented next. However, given the complexity of the expressions defining $G(s)H(s)$, some assumptions about the circuit components must be considered.

9.3.14 Computing the Circuit Components

First, compute the following expression:

$$\frac{1}{Z_c(j\omega)} = \frac{\omega^2 C_1^2 R_{in}}{(\omega R_{in}(C_1 + C_2))^2 + 1} + \frac{1}{R_{ex}} + \frac{1}{R_{copper}}$$
$$+ j\left[\omega C_1 \frac{\omega^2 C_2 R_{in}^2(C_1 + C_2) + 1}{(\omega R_{in}(C_1 + C_2))^2 + 1} + \omega C_3 - \frac{1}{\omega L}\right], \qquad (9.40)$$

Fig. 9.40 Frequency response plots for $Z_c(s)$ in (9.32). (**a**) Bode diagrams of $Z_c(s)$. (**b**) Polar plot of $Z_c(s)$ (two turns)

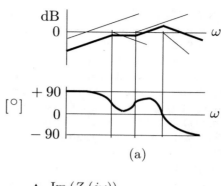

(a)

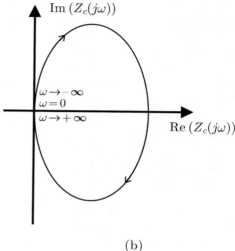

(b)

by performing the variable change $s = j\omega$ in (9.31), and the following expression:

$$M(j\omega) = \frac{R_{in}^2 \omega^2 (C_1^2 + C_1 C_2) + j\omega R_{in} C_1}{1 + R_{in}^2 \omega^2 (C_1 + C_2)^2}.$$ (9.41)

by performing the variable change $s = j\omega$ in (9.35). To render the procedure for obtaining this expression easier, it is important to state that, during such a procedure, the factor $1 + R_{in}^2 \omega^2 C_2^2$ appears at both the numerator and the denominator of $M(j\omega)$; hence, they cancel each other out to finally obtain (9.41).

As previously observed, the polar plot of $G(j\omega)H(j\omega)$ crosses the negative real axis at the frequency ω_1. This means that the imaginary part of (9.40) and (9.41) must be zero when evaluated at $\omega = \omega_1$. Applying this condition in (9.40), the following is found:

Fig. 9.41 Frequency
response plots for $M(s)$ in
(9.35). (**a**) Bode diagrams of
$M(s)$. (**b**) Polar plot of $M(s)$

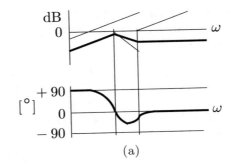

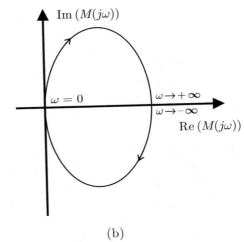

(b)

Fig. 9.42 Polar plot of
$G(s)H(s)$ in (9.39)

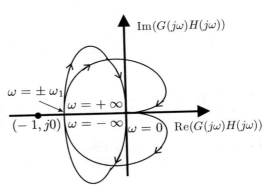

$$\omega_1 = \frac{1}{\sqrt{L\left[\frac{C_1 C_2}{C_1 + C_2} + C_3\right]}}, \tag{9.42}$$

, which represents the circuit oscillation frequency. It is very important to state that, to simplify these expressions to obtain (9.42), the following assumptions have to be considered:

$$R_{in}^2 \gg \frac{1}{\omega_1^2 C_2 (C_1 + C_2)}, \quad R_{in}^2 \gg \frac{1}{\omega_1^2 (C_1 + C_2)^2}. \tag{9.43}$$

On the other hand, the phase of (9.41) is given as:

$$\angle M(j\omega) = \arctan\left(\frac{\frac{1}{\omega(C_1 + C_2)}}{R_{in}}\right). \tag{9.44}$$

Note that this phase is different from zero for any frequency. The above-cited condition requiring the imaginary part of $M(j\omega)$ to be zero is equivalent to asking for the phase in (9.44) to be zero. Although this is not possible, a phase that is close to zero can be obtained for $\omega = \omega_1$ if:

$$R_{in} \gg \frac{1}{\omega_1 (C_1 + C_2)}. \tag{9.45}$$

This and the other approximations that have been considered result in only small differences between the computed values and those obtained experimentally on the oscillation frequency and the gain of the open-loop transfer function.

Finally, $G(j\omega)H(j\omega)|_{\omega=w_1}$ is obtained as:

$$|G(j\omega)H(j\omega)|_{\omega=w_1} = \frac{1}{R_{in}} \text{Re}(M(j\omega))|_{\omega=\omega_1} \text{Re}(Z_c(j\omega))|_{\omega=\omega_1}, \tag{9.46}$$

where $\text{Re}(x)$ stands for the real part of x. From (9.46) and taking into account the assumption in (9.45), the following is found:

$$|G(j\omega)H(j\omega)|_{\omega=w_1} = \frac{1}{R_{in} \frac{C_1+C_2}{C_1}} \left[\frac{1}{\frac{C_1^2}{R_{in}(C_1+C_2)^2} + \frac{1}{R_{ex}} + \frac{1}{R_{copper}}} \right].$$

Hence, according to the condition $|G(j\omega)H(j\omega)|_{\omega=w_1} = 1$, it is concluded that the circuit presents sustained oscillations at the frequency given in (9.42) if:

$$\frac{1}{R_{in} \frac{C_1+C_2}{C_1}} \left[\frac{1}{\frac{C_1^2}{R_{in}(C_1+C_2)^2} + \frac{1}{R_{ex}} + \frac{1}{R_{copper}}} \right] = 1. \tag{9.47}$$

Similar to the case of an oscillator based on an operation amplifier, it is important to stress the following. It is not possible to satisfy the condition (9.47) in practice

Fig. 9.43 Real characteristic between the emitter current and the collector current in a transistor

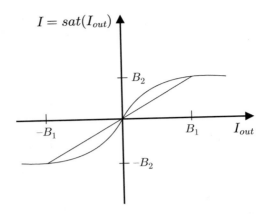

because of the uncertainties present in the commercial values of the components involved in such a condition. Furthermore, these parameters may present changes during normal circuit operation. The way of solving this problem in practice is to use the following condition instead of (9.47):

$$\frac{1}{R_{in}\frac{C_1+C_2}{C_1}} \left[\frac{1}{\frac{C_1^2}{R_{in}(C_1+C_2)^2} + \frac{1}{R_{ex}} + \frac{1}{R_{copper}}} \right] > 1. \tag{9.48}$$

This renders the circuit unstable when turned on and, because of that, the amplitude of the oscillation increases. However, this amplitude growth does not stand forever because the transistor reaches saturation (zero collector to emitter voltage and maximal collector current) and cutoff (zero collector current and maximal collector to emitter voltage) regions. Hence, I_{out} and I in Fig. 9.39b are related through a saturation function as $I = sat(I_{out})$ shown in Fig. 9.43. The transistor current gain corresponds to the slope $\frac{B_2}{B_1}$, where B_1 is the amplitude of I_{out} and B_2 is the amplitude of I. Then, the block diagram in Fig. 9.39b must be changed by that in Fig. 9.44 and we now have:

$$G(j\omega)H(j\omega)|_{\omega=w_1} = \frac{B_2}{B_1}\frac{1}{R_{in}\frac{C_1+C_2}{C_1}} \left[\frac{1}{\frac{C_1^2}{R_{in}(C_1+C_2)^2} + \frac{1}{R_{ex}} + \frac{1}{R_{copper}}} \right].$$

$$\tag{9.49}$$

Note that the slope $\frac{B_2}{B_1}$ is equal to 1 for the amplitudes of I_{out}, i.e., B_1, which are close to zero, but $\frac{B_2}{B_1}$ decreases to zero as B_1 grows. Hence, if (9.48) is satisfied, there always exists an oscillation amplitude such that (9.49) becomes 1 and the desired sustained oscillations are obtained. Although this suggests that the left hand

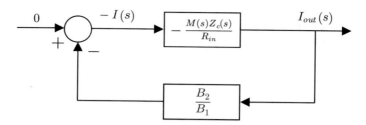

Fig. 9.44 Block diagram considering saturation in the collector current

in (9.48) can be designed to be as large as desired, this is not recommended and a value close to 1 must be designed. The reason for this is that large values for the left-hand side of (9.48) require a strong reduction in the current gain, which implies that the transistor works deep in its nonlinear region, i.e., the sinusoidal waveform is strongly distorted.

Another more formal way of explaining the sustained oscillations in this oscillator circuit is based on the limit cycle analysis presented in Sects. 8.3.3 and 8.3.4. According to Fig. 9.42, the polar plot of $G(s)H((s)$ crosses the negative real axis in the open interval $(-\infty, -1]$, if closed-loop instability is designed, i.e., as desired in an oscillator circuit. On the other hand, according to Sect. 8.3.3, the plot of $-1/N(A)$, where $N(A)$ is the describing function of transistor saturation, is represented by the open real interval $(-\infty, -1]$ in Fig. 9.42. According to Sects. 8.3.3 and 8.3.4, these conditions suffice for a stable limit cycle to appear. This means that sustained oscillations appear instead of closed-loop instability. Moreover, if the left-hand side of (9.48) is larger, then the point where $G(j\omega)H(j\omega)$ crosses the real open interval $(-\infty, -1]$ moves to the left, which, according to Sects. 8.3.3 and 8.3.4, implies that a larger oscillation amplitude of the limit cycle is produced. This, however, results in a more distorted waveform because of the saturation nonlinearity.

On the other hand, note that the conditions (9.43), (9.45) are immediately satisfied if:

$$R_{in} \gg \frac{1}{\omega_1 C_2}. \tag{9.50}$$

The conditions (9.37) and (9.50) are important for design purposes as explained in the following. The first of these conditions ensures that the current gain of the common base configuration is unitary, whereas the second one ensures that the capacitor C_2 is not put into the short circuit by the resistance R_{in}. If this were the case, the feedback through the capacitive network composed of C_1 and C_2 would not work correctly. Also note that the expression between brackets in (9.48) stands for the parallel equivalent resistance of R_{ex}, R_{copper} and R_{in} multiplied by the constant factor $[(C_1 + C_2)/C_1]^2$. R_{in} is commonly small compared with R_{ex} and R_{copper}. If the factor $(C_1 + C_2)/C_1$ is chosen to be large through:

$$C_2 \geq 10C_1, \tag{9.51}$$

then it is ensured that the value of the factor between brackets in (9.48) is approximately the equivalent parallel resistance of R_{ex} and R_{copper}. Hence, the left hand in (9.48) is given by this equivalent parallel resistance (which is large compared with R_{in}) divided by the factor $R_{in}(C_1 + C_2)/C_1$. This ratio can be rendered slightly larger than 1 if (9.51) is used. Thus, the design rules for the oscillator are summarized by (9.37), (9.50), (9.51), (9.42), and (9.48). It is interesting to state that these conditions appear in the books concerned with the electronic design of this class of oscillator [6], pp. 65, [8], pp. 9–13.

Finally, note the advantages of the frequency response techniques with respect to the time response techniques (based on the location of the closed-loop system poles) in this problem. When trying to apply the latter of these methods, a fifth-order characteristic polynomial is found. When trying to use Routh's criterion to establish the conditions to obtain imaginary closed-loop poles, a complex problem is found: the resulting conditions are expressed in a very complex manner and it is difficult to find clear design rules such as those in (9.37), (9.42), (9.48), (9.50), (9.51).

9.3.15   Experimental Results

In this section, an oscillator circuit is designed and experimentally tested to generate a sinusoidal waveform at 1.8[MHz]. The design data are the following. $v_{CC} = 12[V]$, $i_{EQ} \approx i_{CQ} = 1.3[mA]$, $v_{CEQ} = 10.7[V]$.

A PN2222A transistor is selected that has $\beta = 200$. Using (9.16), the following is found:

$$R_E = 1[KOhm].$$

From $i_{CQ} = \beta i_{BQ}$ it is found that $i_{BQ} = 6.5 \times 10^{-3}[mA]$. According to (9.18), $i_{R1Q} = 1[mA]$ is proposed. With this and (9.17), (9.19), (9.20), the following is computed:

$$R_1 = 2[KOhm], \quad R_2 = 10[KOhm].$$

The inductance is $L = 0.328 \times 10^{-3}[Hy]$, with an equivalent parallel internal resistance $R_{copper} = 101677[Ohm]$. Capacitances $C_1 = 27 \times 10^{-12}[F]$, $C_2 = 200 \times 10^{-12}[F]$ are employed and it is assumed that the capacitor C_3 is not present, i.e., $C_3 = 0$. Note that this selection of capacitors approximately satisfies (9.51). Hence, using (9.42), it is found that the circuit oscillation frequency is $f_1 = 1.8018[MHz]$, i.e., $\omega_1 = 2\pi f_1 = 1.132 \times 10^7[rad/s]$.

On the other hand, it is also assumed that the oscillator has no external load i.e., $R_{ex} \to \infty$ or $1/R_{ex} = 0$. Using $R_t = 10000[Ohm]$, $V_T = 0.026[V]$, $i_{EQ} = 1.3[mA]$, (9.27), and (9.28), $R_{in} = 10020[Ohm]$ is found. With these data, the following is found:

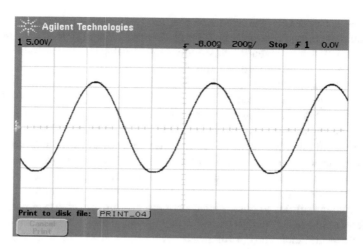

Fig. 9.45 Voltage waveform at the inductance terminals when $R_t = 10000$[Ohm]

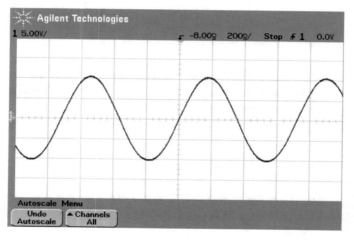

Fig. 9.46 Voltage waveform at the inductance terminals when $R_t = 11000$[Ohm]

$$\frac{1}{R_{in}\frac{C_1+C_2}{C_1}}\left[\frac{1}{\frac{C_1^2}{R_{in}(C_1+C_2)^2}+\frac{1}{R_{ex}}+\frac{1}{R_{copper}}}\right] = 1.1363,$$

i.e., (9.48), (9.50), (9.37), are satisfied. On the other hand, $C_{BP} = 0.1 \times 10^{-6}$[F] is used and $\frac{1}{\omega_1 C_{BP}} = 0.883$[Ohm]$\ll \frac{R_1 R_2}{R_1+R_2} = 1666$[Ohm]. Furthermore, 0.883[Ohm]$\ll h_{ib} = 20$[Ohm]. Thus, all the design conditions established above are satisfied. The voltage v_o measured at the inductance terminals is shown in Fig. 9.45. The frequency measured in this experiment is 1.388[MHz].

The voltage v_o is shown in Fig. 9.46 when $R_t = 11000$[Ohm], which yields:

$$\frac{1}{R_{in}\frac{C_1+C_2}{C_1}}\left[\frac{1}{\frac{C_1^2}{R_{in}(C_1+C_2)^2}+\frac{1}{R_{ex}}+\frac{1}{R_{copper}}}\right]=1.0460. \qquad (9.52)$$

This means that the loop gain is smaller. Note that having a smaller loop gain requires narrower oscillations; hence, a smaller incursion of the signal within the transistor nonlinear region to achieve a unit loop gain, i.e., to ensure sustained oscillations. Obtaining a larger oscillation amplitude in Fig. 9.45, when $R_t = 10000$[Ohm], is also explained in this manner. In fact, a slight distortion can be observed at the bottom part of the sinusoidal waveform when $R_t = 10000$[Ohm].

Finally, in Figs. 9.47a, b and 9.48a, the polar plots are shown for $M(s)$, $Z_c(s)$ and $G(s)H(s)$ respectively, obtained with the numerical values referred to above and $R_t = 11000$[Ohm]. Note the similarities between Figs. 9.47a and 9.41b, and between Figs. 9.47b and 9.40b. Figures 9.42 and 9.48a are also very similar. The apparent differences for the frequencies close to zero are because such a part cannot be appreciated in Fig. 9.48a. This is verified in Fig. 9.48b, where a zoom-in is presented in Fig. 9.48a, and it is observed that the polar plot is tangent to the positive real axis in Fig. 9.42. It is observed in Fig. 9.48a that (9.52) is satisfied.

Figures 9.47a, b, 9.48a, and b were drawn using the following MATLAB code in an m-file:

```
beta=200;
RE=1e3;
R1=2e3;
R2=10e3;
L=0.328e-3;
Rcobre=101.677e3;
C1=27e-12;
C2=200e-12;
C3=0;
Rex=100*Rcobre;
Rt=10e3;
VT=0.026;
IEQ=1.3e-3;
hib=VT/IEQ
Rin=Rt+RE*hib/(RE+hib);
num=[Rin^2*C1*C2 Rin*C1 0];
den=[Rin^2*C2*(C1+C2) Rin*(2*C2+C1) 1];
M=tf(num,den)

numzc=[L*Rin*(C1+C2) L 0];
a3=(L*C1*C2+L*C3*(C1+C2))*Rin;
a2=L*C1+L*C3+(Rcobre+Rex)/(Rcobre*Rex)*L*Rin*(C1+C2);
```

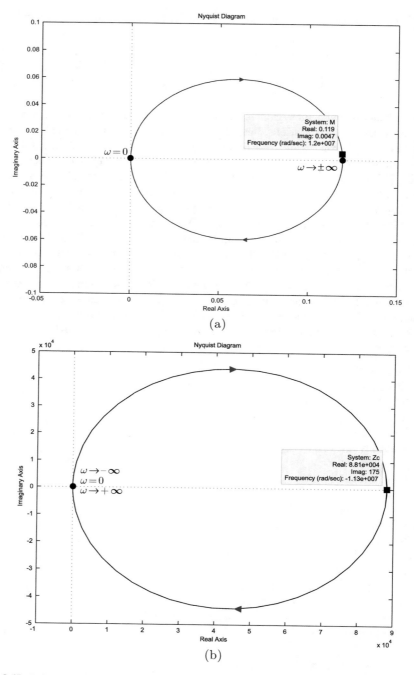

Fig. 9.47 Polar plots using the numerical values for the circuit experimentally tested. (**a**) $M(s)$.
(**b**) $Z_c(s)$

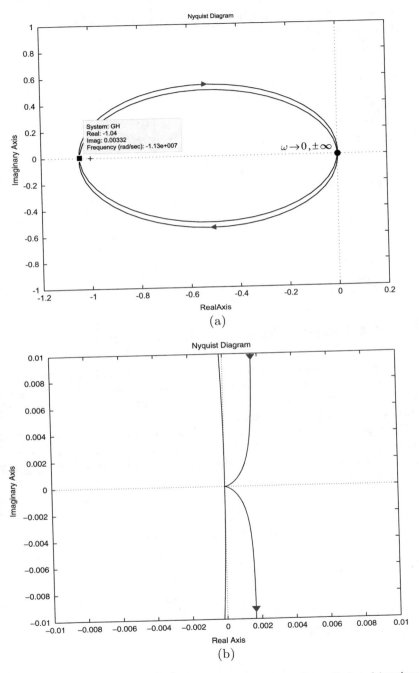

Fig. 9.48 Polar plots using the numerical values of the circuit experimentally tested (continued).
(a) $G(s)H(s)$. (b) $G(s)H(s)$ (zoom in)

```
a1=Rin*(C1+C2)+(Rcobre+Rex)/(Rcobre*Rex)*L;
denzc=[a3 a2 a1 1];
Zc=tf(numzc,denzc)

GH=-M*Zc/Rin

figure(1)
nyquist(M)
axis([-0.05 0.15 -0.1 0.1])

figure(2)
nyquist(Zc)

figure(3)
nyquist(GH)
axis([-.01 .01 -.01 .01])
w=1/sqrt(L*( C1*C2/(C1+C2)+C3 ))
f=w/(2*pi)
```

9.4   A Regenerative Radiofrequency Receiver

Regenerative radiofrequency (RF) receivers are recognized to possess very large gain. This is useful because radio signals collected in an antenna are very weak. This large amplification is achieved thanks to the use of positive feedback in the tuning circuit, as explained in this section.

The regenerative RF receiver depicted in Fig. 9.49 was introduced in [9]. The equivalent small-signal circuit is depicted in Fig. 9.50. This circuit is obtained by using the equivalent small-signal circuit for the transistor presented in Fig. 9.36a. From the circuit in Fig. 9.51a, the following is found:

$$v_r = r_\pi i_b + v_1, \tag{9.53}$$

$$v_1 = R_3 i_e = R_3(1 + \beta)i_b,$$

i.e.,

$$i_b = \frac{v_r}{r_\pi + R_3(1 + \beta)}, \qquad R = \frac{v_r}{i_b} = r_\pi + R_3(1 + \beta).$$

Replacing this in (9.53), we obtain:

$$v_r = \frac{r_\pi}{r_\pi + R_3(1 + \beta)} v_r + v_1,$$

Fig. 9.49 A regenerative radio receiver

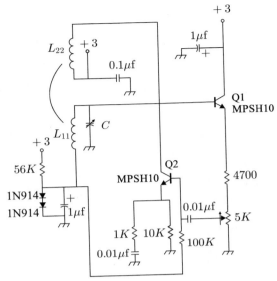

Fig. 9.50 Small-signal equivalent circuit for the receiver in Fig. 9.49

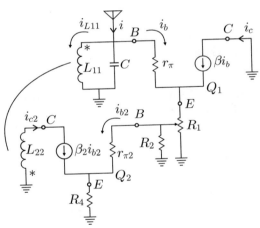

$$v_r \left(1 - \frac{r_\pi}{r_\pi + R_3(1 + \beta)}\right) = v_1,$$

$$v_1 = \frac{R_3(1 + \beta)}{r_\pi + R_3(1 + \beta)} v_r. \tag{9.54}$$

On the other hand, from Fig. 9.51b, it is clear that:

$$R_3 = R_1(1 - \alpha) + (\alpha R_1 || R_2 || z_{in}),$$

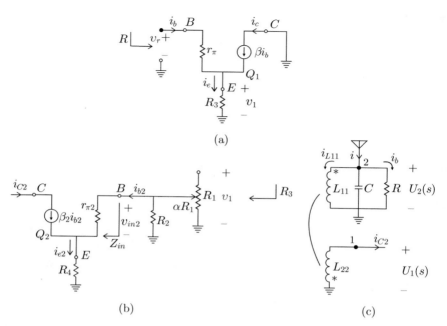

Fig. 9.51 Different sections of the circuit in Fig. 9.50. **(a)** Transistor Q_1. **(b)** Transistor Q_2. **(c)** Tuning and feedback circuit

where $(\alpha R_1 || R_2 || z_{in})$ stands for the parallel equivalent resistance of αR_1, R_2, and z_{in}. Furthermore:

$$v_{in2} = r_{\pi 2} i_{b2} + R_4 i_{e2},$$

$$= r_{\pi 2} i_{b2} + R_4 (1 + \beta_2) i_{b2},$$

$$= (r_{\pi 2} + R_4 (1 + \beta_2)) i_{b2},$$

$$z_{in} = \frac{v_{in2}}{i_{b2}} = r_{\pi 2} + R_4 (1 + \beta_2).$$

Moreover:

$$v_{in2} = \frac{(\alpha R_1 || R_2 || z_{in})}{R_3} v_1 = z_{in} i_{b2},$$

$$i_{b2} = \frac{(\alpha R_1 || R_2 || z_{in})}{z_{in} R_3} v_1 = \frac{i_{c2}}{\beta_2},$$

$$i_{c2} = \beta_2 \frac{(\alpha R_1 || R_2 || z_{in})}{z_{in} R_3} v_1. \tag{9.55}$$

Use of (9.54) and (9.55) yields:

$$i_{c2} = \gamma v_r, \quad \gamma = \beta_2 \frac{(\alpha R_1 || R_2 || z_{in})}{z_{in} R_3} \frac{R_3(1 + \beta)}{r_\pi + R_3(1 + \beta)}. \tag{9.56}$$

On the other hand, from Fig. 9.51c, the inductance matrix is found to be:

$$M = \begin{bmatrix} L_{11} & L_{12} \\ L_{21} & L_{22} \end{bmatrix},$$

because the arc depicted in Fig. 9.49 between L_{11} and L_{22} indicates that these coils are magnetically coupled. Note that according to Sect. D.2, in Appendix D:

$$L_{12} = L_{21} > 0,$$

because i_{L11} and i_{c2} enter into the inductances L_{11} and L_{22} respectively at the terminals with the polarity mark "$*$". This also means that the magnetic fluxes due to i_{L11} and i_{c2} have the same sense. This is a necessary condition for a successful operation of a regenerative RF receiver as it establishes positive feedback of the electric current from the collector of transistor Q_2 to the tuning circuit. Using (D.4), in Appendix D, i.e.,

$$\Gamma_{kj} = \frac{\text{cof}(M_{jk})}{\det(M)},$$

the invertances are computed to be:

$$\Gamma_{11} = \frac{L_{22}}{L_{11}L_{22} - L_{12}L_{21}} > 0, \tag{9.57}$$

$$\Gamma_{22} = \frac{L_{11}}{L_{11}L_{22} - L_{12}L_{21}} > 0,$$

$$\Gamma_{12} = \frac{-L_{12}}{L_{11}L_{22} - L_{12}L_{21}} < 0,$$

$$\Gamma_{21} = \frac{-L_{21}}{L_{11}L_{22} - L_{12}L_{21}} < 0,$$

$$\det(M) = L_{11}L_{22} - L_{12}L_{21} > 0.$$

Solving by nodes the circuit in Fig. 9.51c, the following is found:

$$I_{c2}(s) = \frac{\Gamma_{12}}{s} U_2(s) + \frac{\Gamma_{22}}{s} U_1(s), \tag{9.58}$$

$$I(s) = \left(sC + \frac{\Gamma_{11}}{s} + \frac{1}{R} \right) U_2(s) + \frac{\Gamma_{21}}{s} U_1(s), \tag{9.59}$$

where $U_1(s), U_2(s)$ are the Laplace transforms of voltages at nodes 1 and 2 respectively, as indicated in Fig. 9.51c, whereas $I_{c2}(s)$ and $I(s)$ are the Laplace transforms of the collector current in the transistor Q_2, i.e., i_{c2}, and the electric current entering from the antenna, i.e., i respectively.

According to Fig. 9.51c and to the convention adopted to define the polarity marks in the coil terminals (see Sect. D.2, in Appendix D), if $U_2(s) > 0$ increases then the terminal with the polarity mark in the coil L_{22} is positive, producing a negative $I_{c2}(s)$, i.e., in the opposite direction defined for this current in Fig. 9.51c. This fact is correctly predicted by the term $\frac{\Gamma_{12}}{s}$ in (9.58), as $\Gamma_{12} < 0$. A similar argument justifies the term $\frac{\Gamma_{21}}{s} U_1(s)$ in (9.59), as $\Gamma_{21} < 0$ too.

Solving (9.58) for $U_1(s)$, replacing in (9.59), using definitions in (9.57) and rearranging, the following is found:

$$U_2(s) = \frac{\frac{1}{C}s}{s^2 + \frac{1}{RC}s + \frac{1}{L_{11}C}} \left(I(s) + \frac{L_{12}}{L_{11}} I_{c2}(s) \right).$$

As $U_2(s) = V_r(s)$, where $V_r(s) = \mathcal{L}\{v_r\}$, we can use (9.56) to obtain the block diagram in Fig. 9.52 and:

$$\frac{I_{c2}(s)}{I(s)} = \frac{\gamma \frac{1}{C}s}{s^2 + \left(\frac{1}{RC} - \gamma \frac{L_{12}}{L_{11}C} \right)s + \frac{1}{L_{11}C}}.$$

It is clear from Fig. 9.52 that positive feedback exists from the collector circuit of transistor Q_2 to the tuning circuit. Furthermore, it is interesting to observe that feedback is performed through the mutual inductance $L_{12} > 0$. This demonstrates that positive feedback is present thanks to the magnetic coupling between the inductances in the collector circuit of the transistor Q_2 and the tuning circuit.

Note that the transfer function:

$$G(s) = \frac{\gamma \frac{1}{C}s}{s^2 + \left(\frac{1}{RC} - \gamma \frac{L_{12}}{L_{11}C} \right)s + \frac{1}{L_{11}C}}, \tag{9.60}$$

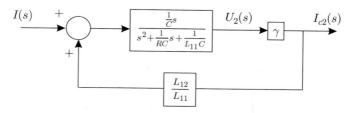

Fig. 9.52 Equivalent closed-loop block diagram of the circuit in Fig. 9.50

represents a *band-pass filter*, which has a very large gain when:

$$\omega \approx \sqrt{\frac{1}{L_{11}C}}, \quad \frac{1}{RC} - \gamma \frac{L_{12}}{L_{11}C} \approx 0,$$

i.e.,

$$|G(j\omega)|_{\omega=\sqrt{\frac{1}{L_{11}C}}} = \frac{\gamma \frac{1}{C}}{\frac{1}{RC} - \gamma \frac{L_{12}}{L_{11}C}} \to \infty, \quad \text{as} \quad \left(\frac{1}{RC} - \gamma \frac{L_{12}}{L_{11}C}\right) \to 0.$$

Note that the last condition can be accomplished by suitably selecting γ, i.e., according to (9.56) by adjusting the potentiometer R_1, which implies modification of α. This explains why the regenerative receivers are recognized to have very large amplification gains. Moreover, it is recommended in the literature [10] to select the feedback gain of a regenerative receiver such as that obtained just before sustained oscillations appear. Note that $G(s)$ in (9.60) is a second-order system with zero damping if $\frac{1}{RC} - \gamma \frac{L_{12}}{L_{11}C} = 0$, which implies that sustained oscillations appear, even when $i = 0$. This, however, is not good for a regenerative receiver as these oscillations would dominate the circuit response; hence, the received radiofrequency signal could not be processed. On the other hand, if the feedback gain is chosen such that $\frac{1}{RC} - \gamma \frac{L_{12}}{L_{11}C} > 0$ is small, then sustained oscillations are avoided, i.e., the circuit is stable, but $|G(j\omega)|_{\omega=\sqrt{\frac{1}{L_{11}C}}}$ is still very large. Note that $\omega = \sqrt{\frac{1}{L_{11}C}}$ represents the frequency of the received signal that is selected to be processed, i.e., the frequency of the desired radio broadcast station.

For simulation purposes, the circuit in Fig. 9.49, introduced in [9], has the following numerical values: $L_{11} = 300 \times 10^{-6}$[H], $C = 100 \times 10^{-12}$[F], $L_{12} = 2 \times 10^{-6}$[H], $R_1 = 9.7$[KOhm], $R_2 = 100$[KOhm], $R_4 = 1000 \times 10 \times 10^3/(11 \times 10^3)$[OHm], $\beta = \beta_2 = 60$, $I_{BQ} = 10 \times 10^{-6}$[A]. This allows a radio signal of about 900[KHz] to be received. Using these values, we obtain the Bode diagrams presented in Figs. 9.53, 9.54, 9.55, for different values of α. Note that the magnitude of $G(j\omega)$ is greater than 30[dB] for all three values of α that have been tested. Moreover, for $\alpha = 0.2619$, the magnitude of $G(j\omega)$ is about 100[dB], which represents an amplification of about 10^5[A]/[A]. Thus, a very large gain is obtained in this case.

Figures 9.53, 9.54, and 9.55 were drawn by executing the following several times: MATLAB code in an m-file:

```
clc
clear all
L11=300e-6;
C=100e-12;
L12=2e-6;
R1=9.7e3;
```

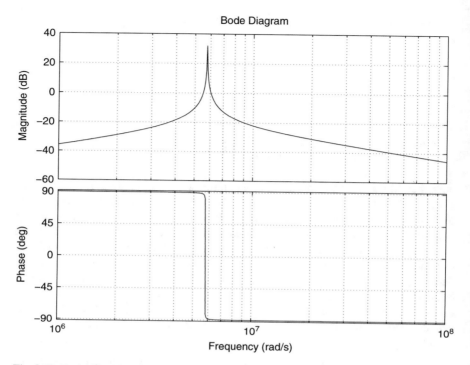

Fig. 9.53 Bode diagrams of $G(s)$ in (9.60) for $\alpha = 0.0515$

```
R2=100e3;
R4=1000*10e3/(11e3);
beta=60;
beta2=beta;
VT=0.026;
IBQ=10e-6;
rpi=VT/IBQ;
rpi2=rpi;
alfa=0.5/9.7%0.5/9.7%1.5/9.7%;%2.54/9.7;
zin=rpi2+R4*(1+beta2);
paralelo=1/( 1/(alfa*R1)+1/R2+1/zin );
R3=R1*(1-alfa)+paralelo;
gamma=beta2*paralelo/(zin*R3)*R3*(1+beta)/(rpi+R3*
    (1+beta));
R=rpi+R3*(1+beta);
G=tf([gamma/C 0],[1 1/(R*C)-gamma*L12/(L11*C)
1/(L11*C)]);
w=2.4e8:0.000001e8:3.4e8;
bode(G)
```

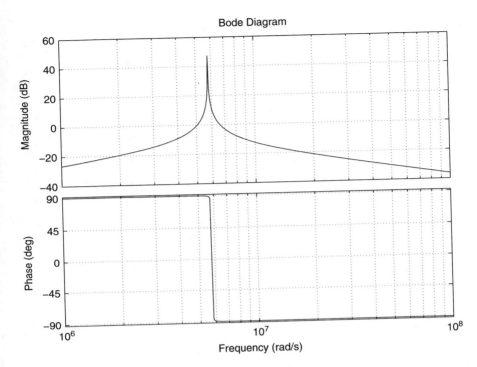

Fig. 9.54 Bode diagrams of $G(s)$ in (9.60) for $\alpha = 0.1546$

```
grid on
1/(R*C)-gamma*L12/(L11*C)
```

9.5 Summary

In this chapter, we have shown how to reduce the distortion produced by nonlinear electronic circuits, how to implement analog controllers in practice using feedback electronic circuits, and how to design and build electronic oscillator circuits producing a sinusoidal waveform. The fundamental tool for solving these problems is classical control, which includes the time response and the frequency response approaches.

In the case of reducing distortion in nonlinear electronic circuits, the basic idea is the use of feedback and a high loop gain to render the closed-loop system robust to open-loop system uncertainties, i.e., nonlinearities. No stability analysis is required for this solution because the circuits involved are assumed to have no dynamics, i.e., they can be modeled as static gains instead of differential equations.

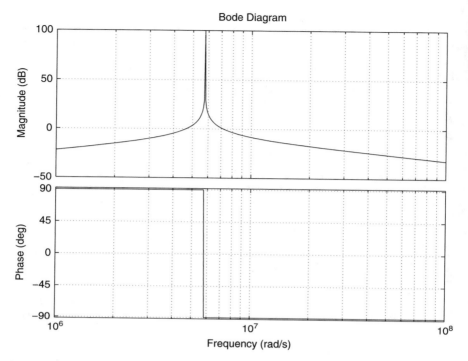

Fig. 9.55 Bode diagrams of $G(s)$ in (9.60) for $\alpha = 0.2619$

In the case of implementing analog controllers, it is important to choose an operational amplifier suitable for the particular application. Recall that the operational amplifier open-loop gain, designated by A_0 in the present chapter, is not really a constant, but it changes with frequency in a similar manner to the magnitude of a low-pass filter changing with frequency. This means that A_0 is reduced at high frequencies. Hence, an operational amplifier with a suitable bandwidth must be selected.

Finally, in the case of oscillator circuits, marginal stability is a fundamental concept. Hence, the circuit components must be chosen such that marginal stability is accomplished. The fundamental tools for analyzing and designing these circuits are the time response approach (location of closed-loop poles) in the case of operational amplifier-based circuits, and the frequency response approach (Nyquist stability criterion) in the case of bipolar transistor-based circuits.

9.6 Review Questions

1. When is a system marginally stable? Why is this concept important in the design of oscillator circuits?
2. Why is it said that it is not possible to design a completely linear oscillator circuit? Why does a nonlinear electronic circuit solve this problem?
3. A bipolar junction transistor is a nonlinear electronic device, then why are linear control techniques employed to design bipolar junction transistor-based oscillator circuits?
4. Why are Zener diodes required in operational amplifier-based oscillator circuits? Why are Zener diodes not employed in bipolar junction transistor-based oscillator circuits?
5. What is the so-called transistor small-signal model?
6. Why do you think oscillator circuits are positive feedback systems?
7. Why must an oscillator circuit be designed to be slightly unstable?

References

1. A. R. Hambley, *Electronics. A top-down approach to computer-aided circuit design*, Macmillan Publishing Company, New York, 1994.
2. R. F. Coughlin and F. F. Driscoll, *Operational amplifiers and linear integrated circuits*, 4th Edition, Prentice Hall, Englewood Cliffs, 1991.
3. N. S. Nise, *Control systems engineering*, 5th edition, John Wiley and Sons, New Jersey, 2008.
4. M. Alonso and E. J. Finn, *Physics*, Addison-Wesley, Reading, MA, 1992.
5. W. Hayward, *Introduction to radio frequency design*, The American Radio Relay League Inc., 1994.
6. P. H. Young, *Electronic communication techniques*, Merrill Publishing Company, 1990.
7. C. A. Desoer and E. S. Kuh, *Basic circuit theory*, International Student Edition, McGraw-Hill, Tokyo, 13th printing 1983.
8. M. Kaufman and A. H. Seidman, *Handbook of electronics calculations for engineers and technicians*, McGraw-Hill, USA, 1979.
9. R. Quan, *Build your own transistor radios*, McGraw-Hill, New York, 2013.
10. W. Hayward, R. Campbell, and B. Larkin, *Experimental methods is RF design*, The American Radio Relay League Inc., 2009.

Chapter 10
Velocity Control of a Permanent Magnet Brushed Direct Current Motor

An electric motor is a device that is often employed to provide motion to mechanical systems. There are several classes of electric motors: induction motors [1], Chaps. 10 and 11, PM stepper motors [2], PM synchronous motors [3], switched reluctance motors [4, 5], brushless DC motors [6], chapter 10, etc. However, the PM brushed DC motor is the simplest motor to control. Moreover, as its mathematical model is linear and single–input–single–output, it is the ideal motor for using as an experimental prototype in the first stages of education in automatic control. In this chapter, velocity control of a PM brushed DC motor is studied. Position control of this class of motors is studied in the next chapter.

10.1 Mathematical Model

A PM brushed DC motor actuating on a load through a gear box is depicted in Fig. 10.1. The nomenclature employed is the following.

- u is the voltage applied at the motor armature terminals.
- i is the electric current through the armature circuit.
- Θ is the motor angular position.
- θ is the load angular position.
- $e_a = k_e \dot{\Theta}$ is the counter electromotive force where k_e is the counter electromotive force constant.
- $T = k_m i$ is the generated torque, or electromagnetic torque, where k_m is the motor torque constant.
- T_c is the load equivalent torque seen at the motor shaft. This torque opposes the motor movement (it is applied in the opposite sense to the definition of Θ).
- T_L is the torque applied by the motor on the load.

© Springer International Publishing AG, part of Springer Nature 2019
V. M. Hernández-Guzmán, R. Silva-Ortigoza, *Automatic Control with Experiments*,
Advanced Textbooks in Control and Signal Processing,
https://doi.org/10.1007/978-3-319-75804-6_10

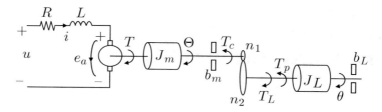

Fig. 10.1 A loaded permanent magnet (PM) brushed direct current (DC) motor

- T_p is the torque due to an external disturbance. It is assumed that this torque opposes the load movement (it is applied in the opposite sense to the definition of θ). If this is not the case, then T_p is negative.
- L is the armature inductance.
- R is the armature resistance.
- J_m is the inertia of the motor's rotor.
- b_m is the motor viscous friction coefficient.
- J_L is the load inertia.
- b_L is the load viscous friction coefficient.
- n_1 and n_2 stand for the teeth number at the motor axis and at the load axis respectively.

Note that a PM brushed DC motor is an electromechanical system with an electrical subsystem and a mechanical subsystem. The corresponding mathematical model is found by analyzing these individual subsystems to join them afterward. The model of each of these parts is obtained in the following.

10.1.1 The Motor Electrical Subsystem Model

Applying Kirchhoff's voltage law (see Fig. 10.1), the following is found:

applied voltage $= \sum$ voltage drops in the mesh,

$u = $ voltage drop at inductance $+$ voltage drop at resistance $+$

$+$counter electromotive force,

$$u = L \frac{di}{dt} + R\,i + e_a, \tag{10.1}$$

where, according to (2.92) in Chap. 2, the counter electromotive force is given as:

$$e_a = k_e \dot{\Theta},$$

where dot "`·`" stands for the first time derivative.

10.1.2 *The Motor Mechanical Subsystem Model*

Newton's Second Law must be applied separately to each one of the two bodies: the motor's rotor and its load.

10.1.2.1 Motor's Rotor Model

$$\text{inertia} \times \text{angular acceleration} = \sum \text{torques applied to inertia } J_m,$$

$$J_m \ddot{\Theta} = \text{generated torque} - \text{friction torque} - \text{load torque},$$

$$J_m \ddot{\Theta} = T - b_m \dot{\Theta} - T_c, \tag{10.2}$$

where, according to (2.93), Chap. 2, the generated torque is given as:

$$T = k_m \, i.$$

The addition on the right-hand side of (10.2) is algebraic, i.e., the signs indicate the sense of the corresponding applied torque. A negative sign indicates that the torque opposes the motion of the inertia J_m, whereas a positive sign indicates that the torque favors the motion of the inertia J_m.

10.1.2.2 Load Model

Because of the presence of a gear box, it is necessary to use (2.86) and (2.87), Chap. 2, to conclude that:

$$\Theta = n\,\theta, \qquad n = \frac{n_2}{n_1}, \tag{10.3}$$

$$T_L = n\,T_c, \tag{10.4}$$

Applying Newton's Second Law to the load:

$$\text{inertia} \times \text{angular acceleration} = \sum \text{torques on inertia } J_L,$$

$$J_L \ddot{\theta} = T_L - b_L \dot{\theta} - T_p. \tag{10.5}$$

The motor's mechanical subsystem model is obtained using (10.2), (10.5), (10.3), and (10.4) as:

$$J_m \ddot{\Theta} = k_m i - b_m \dot{\Theta} - \frac{1}{n} T_L,$$

$$J_m \ddot{\Theta} = k_m i - b_m \dot{\Theta} - \frac{1}{n} \left(J_L \ddot{\theta} + b_L \dot{\theta} + T_p \right),$$

$$n J_m \ddot{\theta} = k_m i - b_m n\dot{\theta} - \frac{1}{n} \left(J_L \ddot{\theta} + b_L \dot{\theta} + T_p \right),$$

$$\left(n^2 J_m + J_L \right) \ddot{\theta} + \left(n^2 b_m + b_L \right) \dot{\theta} = n k_m i - T_p. \tag{10.6}$$

Defining:

$$J = n^2 J_m + J_L, \qquad b = n^2 b_m + b_L, \tag{10.7}$$

(10.6) can be written as:

$$J\ddot{\theta} + b\dot{\theta} = n k_m i - T_p. \tag{10.8}$$

Finally, the complete motor mathematical model is given by (10.1) and (10.8), i.e.,

$$L \frac{di}{dt} = u - R i - n k_e \dot{\theta}, \tag{10.9}$$

$$J\ddot{\theta} = -b \dot{\theta} + n k_m i - T_p. \tag{10.10}$$

Note that the differential equation in (10.10) is identical to that presented in (2.118) corresponding to the system shown in Fig. 2.29a, Chap. 2. The only differences are because (2.118) is expressed in terms of an applied external torque (which is equal to $k_m i$ in a DC motor) and (2.118) does not take into account any external torque applied on body 2 ($T_p = 0$), i.e., on the load.

The differential equations in (10.9) and (10.10) represent the mathematical model of the DC motor-load system because the load position θ can be found as a function of time by solving these differential equations if all the constants involved are known in addition to the voltage applied at the armature terminals. Using the Laplace transform in (10.9) and (10.10) by assuming zero initial conditions, the following is found:

$$I(s) = \frac{1}{sL + R}[U(s) - n k_e s\theta(s)], \tag{10.11}$$

$$\theta(s) = \frac{1}{s^2 J + sb}[n k_m I(s) - T_p(s)], \tag{10.12}$$

where $I(s)$, $U(s)$, $\theta(s)$ and $T_p(s)$, stand for Laplace transforms of i, u, θ and T_p respectively.

10.2 Power Amplifier

It is important to stress the necessity of employing a power amplifier. A control system can be divided into two parts: (*i*) the control part, and (*ii*) the power part. The control part includes all the hardware devoted to processing information (sensors and control algorithms) to be used to determine the command to be sent to the controlled process. This task requires precise electronic circuits that handle weak electrical signals, i.e., they merely process information but they do not handle high power levels. However, these commands must be capable of producing significant variations of the signal applied at the plant (motor) input. Furthermore, the plant input handles important power levels to be capable of producing the expected changes in the plant. This is the reason for including a power stage in a control system.

The power stage is commonly composed of a power amplifier that receives at its input the weak signal delivered by the control stage (control signal) and it must deliver at its output a strong high-power signal that is directly applied to the plant. For instance, in electromechanical systems a power amplifier must deliver the voltage to be applied directly to electric motors, which convert such an electric signal into torque to produce the mechanism motion. Hence, a power amplifier is placed between the weak signal delivered by the controller and the strong signal applied to the motor. According to this, a power amplifier can be modeled as a constant A_p relating to voltage at its input u_i (control signal) and voltage delivered at its output u (at the motor's armature terminals):

$$u = A_p \, u_i. \tag{10.13}$$

10.3 Electric Current Control

A common technique for motor control in industry [6], pp. 76, [7, 8], is to employ an internal electric current loop. In the simplest version, this internal loop employs a proportional electric current controller. This means that the control signal is computed as:

$$u_i = K(i^* - i), \tag{10.14}$$

where K is a positive constant standing for the proportional electric current controller gain, whereas i^* represents the commanded current that is given as the output of another controller (the external loop) intended to control a mechanical variable, i.e., velocity or position. The design of this external loop is explained later. For now, just consider that i^* is the desired value for the current through the motor, which is specified below. Using (10.13) and (10.14) it is found that voltage applied at the motor armature terminals is given as:

$$u = A_p \, K(i^* - i). \tag{10.15}$$

Applying the Laplace transform to (10.15), the following is found:

$$U(s) = A_p K(I^*(s) - I(s)), \tag{10.16}$$

where $I^*(s)$ is the Laplace transform of i^*. Combining (10.11), (10.12), and (10.16), the block diagram in Fig. 10.2a is found. As demonstrated in the Example 4.4, Sect. 4.1, Fig. 10.2a can be simplified as shown in Figs. 10.2b and c. It is also revealed in Example 4.4, Sect. 4.1, that from Fig. 10.2c, the following can be written:

$$\theta(s) = G_1(s)I^*(s) + G_2(s)T_p(s), \tag{10.17}$$

where:

$$G_1(s) = \frac{nk_m}{\left[\left(\frac{sL+R}{KA_p} + 1\right)(sJ+b) + \frac{n^2 k_m k_e}{KA_p}\right]s}, \tag{10.18}$$

$$G_2(s) = \frac{-\left(\frac{sL+R}{KA_p} + 1\right)}{\left[\left(\frac{sL+R}{KA_p} + 1\right)(sJ+b) + \frac{n^2 k_m k_e}{KA_p}\right]s}. \tag{10.19}$$

If the proportional gain K is chosen to be large enough, then it can be assumed that [6], pp. 76:

$$\frac{sL+R}{KA_p} \approx 0, \qquad \frac{n^2 k_m k_e}{KA_p} \approx 0,$$

to approximate (10.18) and (10.19) as:

$$G_1(s) = \frac{nk_m}{s^2 J + sb}, \tag{10.20}$$

$$G_2(s) = \frac{-1}{s^2 J + bs}. \tag{10.21}$$

Using (10.20) and (10.21) in (10.17), the following is found:

$$\theta(s) = \frac{1}{s^2 J + sb}[nk_m I^*(s) - T_p(s)], \tag{10.22}$$

which is represented by the block diagram in Fig. 10.2d. Note that the objective of the electric current control shown in (10.15) is to render the motor's electrical dynamics negligible such that the motor model is now represented only by the motor mechanical subsystem. This can be clearly seen if (10.22) and (10.12) are compared. These expressions indicate that electric current can now be used as the control variable instead of voltage. It is clear that voltage is still the variable that is directly manipulated, as indicated in (10.15); however, for analysis and

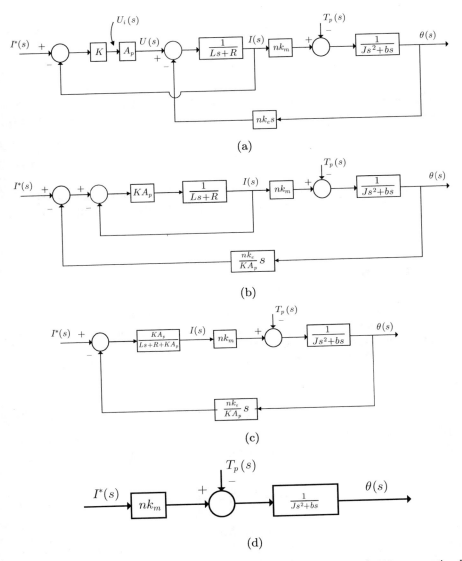

Fig. 10.2 Equivalent block diagrams for a PM brushed DC motor provided with a proportional current loop and a power amplifier

design purposes it can be assumed that electric current is the variable that is directly manipulated through the commanded current i^*. The purpose of the electric current control (10.15) is to accomplish that the actual electric current i tracks the commanded current i^*. A large value of K is very useful for this. Note that the expression in (10.22) can be rewritten as:

$$\theta(s) = \frac{1}{s(s+a)}[kI^*(s) - \frac{1}{J}T_p(s)] \tag{10.23}$$

$$a = \frac{b}{J}, \quad k = \frac{nk_m}{J}.$$

Finally, as velocity ω and position θ are related through $\omega = \frac{d\theta}{dt}$, then the use of the Laplace transform under zero initial conditions yields $\omega(s) = s\theta(s)$. Thus, (10.23) can be rewritten as:

$$\omega(s) = s\theta(s) = \frac{1}{s+a}[kI^*(s) - \frac{1}{J}T_p(s)]. \tag{10.24}$$

In the remainder of this chapter, the model in (10.24) is used to design velocity controllers. Some experimental results obtained with these controllers are also presented.

10.4   Identification

In this section we are interested in performing an experiment that allows us to compute the values of the constants k and a present in the model (10.24). This experiment is described in the following.

Suppose that there is no external disturbance, i.e., $T_p(s) = 0$; hence, the model in (10.24) becomes:

$$\omega(s) = \frac{k}{s+a}I^*(s). \tag{10.25}$$

This is a first-order system such as that studied in Sect. 3.1. Hence, if $i^* = A$ is applied, i.e., $I^*(s) = \frac{A}{s}$, then velocity ω evolves over time as shown in Fig. 10.3, i.e.,

$$\omega(t) = \frac{kA}{a}\left(1 - e^{-at}\right). \tag{10.26}$$

The time constant τ can be measured as shown in Fig. 10.3 and, using $\tau = \frac{1}{a}$, the numerical value of a can be computed. On the other hand, defining the final value of the velocity as $\omega_f = \lim_{t\to\infty} \omega(t) = \frac{kA}{a}$ and, using the value of a just computed, $k = \frac{a\omega_f}{A}$ can be found.

Some results obtained when performing the experiment described above are shown in Fig. 10.4. The noise content in velocity measurements is evidence for a well-known fact in control systems: velocity is a noisy signal. See Sect. 10.6 for an explanation of why ω is measured in volts. In Fig. 10.4, $i^* = A = 0.3[A]$ is used and by direct measurement in Fig 10.4, the following is found:

$$\tau = 2.7[s], \quad \omega_f = 2[V]. \tag{10.27}$$

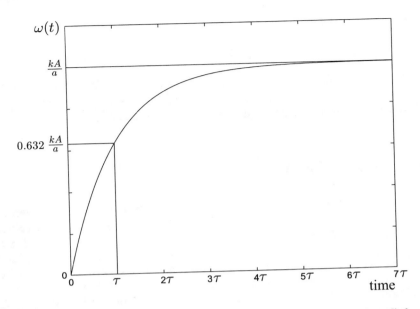

Fig. 10.3 Velocity evolution in a PM brushed DC motor when a constant $i^* = A$ is applied

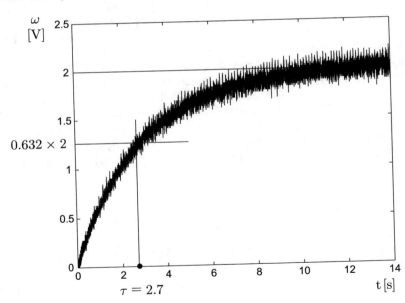

Fig. 10.4 Experimental identification of a PM brushed DC motor

These measurements are simplified and rendered more exact if a computer program is employed to perform them. Using these values and the procedure described above, the following is found:

$$a = 0.3704, \qquad k = 2.4691. \tag{10.28}$$

Finally, it is not important to compute the numerical value of $1/J$, appearing as the coefficient of the disturbance $T_p(s)$ in (10.24). The reason for this is that T_p is commonly unknown and, despite this, its effect can be compensated for.

10.5   Velocity Control

The main objective of control design is to ensure that the desired transient and steady-state response specifications are achieved. The transient response specifications refer to how fast the closed-loop system is desired to respond. This is achieved by suitably assigning the closed-loop poles. On the other hand, the steady-state specifications refer to the difference that is desired to exist between the measured velocity ω and the desired velocity ω_d when time increases. Moreover, this must stand despite the presence of the external disturbance $T_p(s)$. The controller is the device responsible for ensuring that all these specifications are satisfied.

Suppose that the desired velocity is a constant, i.e., a step function. Then, ω_d is a constant that may be positive, negative or zero. As the plant in (10.24) has no poles at $s = 0$, then the system type is 0. This means that the use of a proportional velocity controller with the form $i^* = k_p(\omega_d - \omega)$, with $k_p > 0$, cannot achieve $\omega = \omega_d$ in a steady state even if no external torque disturbance is present, i.e., even when $T_p = 0$. The solution to this problem is a controller with an integral action to render the system type 1, i.e., a proportional–integral (PI) controller. However, as explained in Sect. 5.2.4, a PI controller cannot simultaneously achieve the desired transient response specifications and a satisfactory rejection of external disturbances $T_p(s)$. It is for this reason that the design of a modified PI velocity controller is presented in the following, which is capable of simultaneously achieving these requirements.

10.5.1   A Modified PI Controller

Applying the inverse Laplace transform to (10.24), the following is found:

$$\dot{\omega} + a\omega = ki^* - \frac{1}{J}T_p. \tag{10.29}$$

In the following, it is shown that the use of the controller:

$$i^* = k_p(\omega_d - \omega) + k_i \int_0^t (\omega_d - \omega(r))dr + \frac{a}{k}\omega_d + k_1(\omega(0) - \omega_d), \tag{10.30}$$

where $\omega(0)$ is the initial velocity, ensures that: 1) The motor velocity ω reaches the desired constant velocity ω_d in a steady state. 2) This is achieved with a time constant that is arbitrarily assigned. 3) The effects of a constant disturbance T_p are compensated for as fast as desired.

Replacing (10.30) in (10.29) and rearranging:

$$\dot{\omega} + (a + k_p k)(\omega - \omega_d) + k_i k \int_0^t (\omega(r) - \omega_d) dr = k_1 k(\omega(0) - \omega_d) - \frac{1}{J} T_p.$$

Defining:

$$k_p = k_p' + k_1,$$

the following can be written:

$$\dot{\omega} + (a + k_p' k)(\omega - \omega_d) + k_i k \int_0^t (\omega(r) - \omega_d) dr$$

$$+ k_1 k[\omega - \omega_d - (\omega(0) - \omega_d)] = -\frac{1}{J} T_p.$$

Note that:

$$\omega - \omega_d - (\omega(0) - \omega_d) = \int_0^t (\dot{\omega}(r) - \dot{\omega}_d) dr,$$

and defining:

$$k_i = k_i' k_1,$$

then:

$$\dot{\omega} + (a + k_p' k)(\omega - \omega_d) + k_1 k \int_0^t \left[k_i'(\omega(r) - \omega_d) + (\dot{\omega}(r) - \dot{\omega}_d) \right] dr$$

$$= -\frac{1}{J} T_p. \tag{10.31}$$

As ω_d is a constant, then $\dot{\omega}_d = 0$. Hence, if:

$$k_i' = a + k_p' k,$$

$$\xi = \int_0^t \left[\dot{\omega}(r) + k_i'(\omega(r) - \omega_d) \right] dr,$$

are defined, then (10.31) can be written as:

$$\dot{\xi} + k_1 k \xi = -\frac{1}{J} T_p.$$

Applying the Laplace transform with zero initial conditions:

$$\xi(s) = \frac{-\frac{1}{J}}{s + k_1 k} T_p(s),$$

or:

$$s\xi(s) = \frac{-\frac{1}{J}s}{s + k_1 k} T_p(s). \tag{10.32}$$

If $T_p = t_d$ is a constant, i.e., $T_p(s) = \frac{t_d}{s}$, then:

$$s\xi(s) = \frac{-\frac{1}{J}s}{s + k_1 k} \frac{t_d}{s}.$$

Applying the final value theorem:

$$\lim_{t \to \infty} \dot{\xi}(t) = \lim_{s \to 0} s \left[s\xi(s) \right],$$

$$= \lim_{s \to 0} s \left[\frac{-\frac{1}{J}s}{s + k_1 k} \frac{t_d}{s} \right] = 0.$$

Hence, if $k_1 k > 0$, the filter in (10.32) is stable and $\lim_{t \to \infty} \dot{\xi}(t) = 0$ is ensured. Furthermore, $\dot{\xi}(t)$ approaches zero faster as $k_1 k > 0$ is larger. Note that $\lim_{t \to \infty} \dot{\xi}(t) = 0$ also implies that:

$$\frac{d}{dt} \int_0^t \left[\dot{\omega}(r) + k_i'(\omega(r) - \omega_d) \right] dr = \dot{\omega} + k_i'(\omega - \omega_d) \to 0.$$

This means that the effect of the constant disturbance disappears as time increases and the remaining differential equation $\dot{\omega} + k_i'\omega = k_i'\omega_d$, which is stable if $k_i' > 0$, possesses the time constant $\tau = \frac{1}{k_i'}$ and has a unit gain in a steady state. This means that:

- If no perturbation exists ($T_p = 0$ and $\dot{\xi}(t) = 0$ for all $t \geq 0$), the velocity response is as that of a first-order system with time constant $\tau = \frac{1}{k_i'}$. The velocity ω reaches the desired velocity ω_d (which is constant) in a steady state.
- If a constant disturbance appears ($T_p = t_d \neq 0$), the deviation produced by such a disturbance vanishes as time increases. Moreover, if a larger k_1 is chosen (such that the product $k_1 k > 0$ is larger), then the deviation due to the disturbance vanishes faster. If once the deviation due to the disturbance is taken to zero (when $\omega = \omega_d$), a step change is commanded in the desired velocity ω_d (a change in ω_d is commanded when the disturbance T_p is present), then the velocity responds as if the disturbance T_p is not present, i.e., as in the previous item: with $\tau = \frac{1}{k_i'}$ as the time constant and velocity ω reaches its desired value ω_d in a steady state.

As the initial conditions are always assumed to be zero in classical control, then $\omega(0) = 0$ can be assumed and the controller in (10.30) becomes:

$$i^* = k_p(\omega_d - \omega) + k_i \int_0^t (\omega_d - \omega(r))dr + \left(\frac{a}{k} - k_1\right)\omega_d, \qquad (10.33)$$

which constitutes a simple PI controller with a constant feedforward term. Finally, the tuning rule is summarized by:

$$k_p = k'_p + k_1, \qquad (10.34)$$

$$k_i = k'_i k_1,$$

$$k'_i = a + k'_p k,$$

where:

- $k'_p > 0$ is chosen to define the desired time constant $\tau = \frac{1}{k'_i} = \frac{1}{a + k'_p k}$, which is always less than the time constant of the plant to be controlled $\frac{1}{a}$, i.e., a very convenient feature.
- $k_1 > 0$ is chosen large for a fast rejection of the disturbance effects. This can be done by trial and error or it can be computed recalling that $\frac{1}{k_1 k}$ is the time constant of the filter in (10.32), which is responsible for the elimination of the deviation due to the disturbance.

Another, better known, way of designing a PI controller with the same advantages found above is by using two-degrees-of-freedom controllers. These are presented in the following section.

10.5.2   A Two-Degrees-of-Freedom Controller

A closed-loop system with a two-degrees-of-freedom controller is shown in Fig. 10.5. The controller has two components with transfer functions $G_{c1}(s)$ and $G_{c2}(s)$, whereas $G_p(s)$ is the plant to be controlled. An external disturbance $D(s)$ applied at the plant input is also considered. In Example 4.5, Sect. 4.1, it is shown that the output of the closed-loop system is given as:

$$\omega(s) = G_1(s)\omega_d(s) + G_2(s)D(s),$$

where:

$$G_1(s) = \frac{G_{c1}(s)G_p(s)}{1 + (G_{c1}(s) + G_{c2}(s))G_p(s)},$$

$$G_2(s) = \frac{G_p(s)}{1 + (G_{c1}(s) + G_{c2}(s))G_p(s)}, \qquad (10.35)$$

are the transfer functions defined as:

$$G_1(s) = \frac{\omega(s)}{\omega_d(s)}, \quad \text{when } D(s) = 0,$$

$$G_2(s) = \frac{\omega(s)}{D(s)}, \quad \text{when } \omega_d(s) = 0.$$

The reason why this control scheme is called *two-degrees-of-freedom control* is that the transfer functions defined in (10.35) can be tuned by maintaining them independent from each other using the two controller components $G_{c1}(s)$ and $G_{c2}(s)$ as independent variables, i.e., as degrees-of-freedom. As a PM brushed DC motor is to be controlled in this chapter, assume that:

$$G_p(s) = \frac{k}{s+a}.$$

Also assume that $G_{c1}(s)$ and $G_{c2}(s)$ are PI controllers, i.e.,

$$G_{c1}(s) = \frac{k_{p1}s + k_{i1}}{s}, \quad G_{c2}(s) = \frac{k_{p2}s + k_{i2}}{s},$$

$$G_c(s) = G_{c1}(s) + G_{c2}(s) = \frac{k_p s + k_i}{s}, \quad k_p = k_{p1} + k_{p2}, \quad k_i = k_{i1} + k_{i2}.$$

Hence, the first transfer function in (10.35) can be written as:

$$\frac{\omega(s)}{\omega_d(s)} = \frac{G_{c1}(s)\frac{k}{s+a}}{1 + \frac{k_p s + k_i}{s}\frac{k}{s+a}},$$

$$= \frac{skG_{c1}(s)}{s^2 + (a + k_p k)s + k_i k}. \tag{10.36}$$

Suppose that the closed-loop response is desired to be as that of a first- order with a pole at $s = -p_1$, with $p_1 > 0$. Then, the characteristic polynomial in the last expression must satisfy:

$$s^2 + (a + k_p k)s + k_i k = (s + p_1)(s + f),$$

$$= s^2 + (p_1 + f)s + p_1 f,$$

where the root at $s = -f$, $f > 0$, is introduced merely to obtain a second-degree polynomial on the right-hand side of this expression. Equating the coefficients on both sides, the following is found:

$$a + k_p k = p_1 + f, \quad k_i k = p_1 f,$$

i.e.,

$$k_p = \frac{p_1 + f - a}{k}, \quad k_i = \frac{p_1 f}{k}.$$

On the other hand, to obtain a response of a first order with a pole at $s = -p_1$, the transfer function in (10.36) must satisfy:

$$\frac{\omega(s)}{\omega_d(s)} = \frac{p_1(s + f)}{s^2 + (a + k_p k)s + k_i k} = \frac{p_1(s + f)}{(s + p_1)(s + f)} = \frac{p_1}{s + p_1}. \quad (10.37)$$

Thus, equating (10.36) and (10.37), the following is found:

$$skG_{c1}(s) = p_1(s + f).$$

It is concluded that $G_{c1}(s)$ is a PI controller:

$$G_{c1}(s) = \frac{k_{p1}s + k_{i1}}{s}, \quad k_{p1} = \frac{p_1}{k}, \quad k_{i1} = \frac{p_1 f}{k}.$$

Based on this result, $G_{c2}(s)$ can be computed as:

$$G_{c2}(s) = G_c(s) - G_{c1}(s) = \frac{k_{p2}s + k_{i2}}{s}, \quad k_{p2} = \frac{f - a}{k}, \quad k_{i2} = 0.$$

This means that $G_{c2}(s) = k_{p2}$ is a proportional controller. On the other hand, the second transfer function in (10.35) becomes:

$$\frac{\omega(s)}{D(s)} = \frac{\frac{k}{s+a}}{1 + \frac{k_p s + k_i}{s} \frac{k}{s+a}},$$

$$= \frac{ks}{s^2 + (a + k_p k)s + k_i k},$$

$$= \frac{ks}{(s + p_1)(s + f)}. \quad (10.38)$$

Using this expression and the final value theorem, it can be verified that the effect of a constant external disturbance $D(s) = -\frac{t_d}{Jks}$ vanishes as time increases. Furthermore, this is achieved faster as f and p_1 are larger. As p_1 is fixed by the desired response to a constant velocity command, then f is the free parameter. Thus, f must be chosen to be large. According to Fig. 10.5, the controller is given as:

$$I^*(s) = G_{c1}(s)(\omega_d(s) - \omega(s)) - G_{c2}(s)\omega(s),$$

$$= k_{p1}(\omega_d(s) - \omega(s)) + k_{i1}\frac{(\omega_d(s) - \omega(s))}{s} - k_{p2}\omega(s),$$

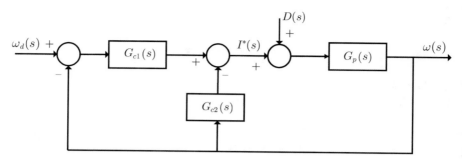

Fig. 10.5 Closed-loop system with a two-degrees-of-freedom controller

$$= (k_{p1} + k_{p2})(\omega_d(s) - \omega(s)) + k_{i1}\frac{(\omega_d(s) - \omega(s))}{s} - k_{p2}\omega_d(s),$$

$$= k_p(\omega_d(s) - \omega(s)) + k_{i1}\frac{(\omega_d(s) - \omega(s))}{s} - k_{p2}\omega_d(s). \qquad (10.39)$$

Using the inverse Laplace transform, the following is finally obtained:

$$i^* = k_p(\omega_d - \omega) + k_{i1}\int_0^t (\omega_d - \omega(r))dr - k_{p2}\omega_d. \qquad (10.40)$$

Note that controllers in (10.40) and (10.33) are identical if:

$$k_1 = \frac{f}{k}, \quad p_1 = k_i',$$

are defined. Thus, it is up to the reader to decide which of the approaches presented in Sect. 10.5.1 or 10.5.2 is preferred.

10.6   Experimental Prototype

The following are the main components of the control system.

- Microcontroller PIC16F877A [9].
- Digital/analog converter, DAC0800LCN, 8 bits.
- 2 PM brushed DC motors. Nominal voltage 24[V], nominal current 2.3[A].

Both motor shafts are coupled such that one motor is employed as a generator. The generated voltage is used as measurement of the other motor's velocity which is controlled according to the control law in (10.33).

The electric diagram of the complete velocity control system is shown in Fig. 10.6. The control algorithm in (10.33) is computed by a microcontroller

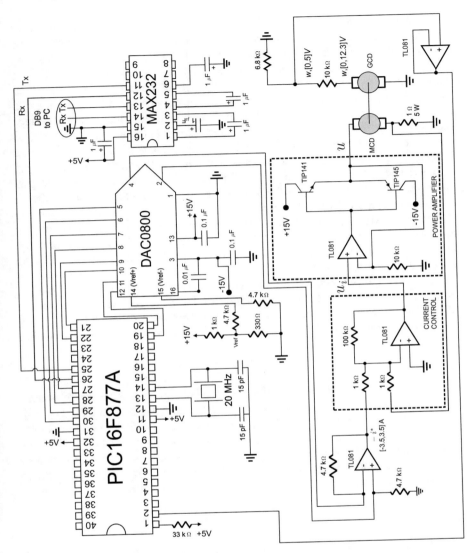

Fig. 10.6 Electric diagram of the complete velocity control system for a PM brushed DC motor

PIC16F877A. Once the desired velocity ω_d and the velocity measurement ω are known, the microcontroller computes the control signal i^* in (10.33). The microcontroller program is listed in Sect. 10.9.

In the following, the way in which the microcontroller PIC16F877A can be used as a controller is described. It is stressed that it is not the purpose to present detailed information on the microcontroller operation. A simple description is presented on how the components of this microcontroller are employed to implement the control law being tested.

The motor velocity is measured through canal 0 of the microcontroller analog/digital converter. Only the eight highest bits of the converter are employed. The input analog range of the converter is $[0, +5][V]$. Hence, the datum delivered by the converter t_0 is used to compute w=0.0196*t_0 (where $0.0196 = 5/255$) to retrieve, as the variable w, voltage at the analog input of the analog/digital converter. This voltage is delivered by the motor used as a generator and it is proportional to the velocity of the motor to be controlled. Hence, it is assumed that this voltage represents the measured velocity. Note that before entering the analog/digital converter (through microcontroller pin 2), the generator voltage passes through a tension divisor used to adapt the maximal generated voltage (12.3[V] at maximal velocity) to 5[V]. This is the upper limit of the analog input of the analog/digital converter. The operational amplifier TL081 is used to protect the microcontroller.

The controller in (10.33) is computed with the instructions "iast=kp*error+ki*ie +cte*wd; ie=ie+0.002*error;," where the second expression represents the integral of the velocity error. See Appendix F for an explanation of why this expression represents a discrete iterative version of the integral on the velocity error. As indicated in Fig. 10.6, $-i^*$ must appear at the input of the operational amplifier TL081 inside the part entitled "current control." This is accomplished as follows. First, iast=-iast is computed in the program listed in Sect. 10.9, to change the sign of i^*. The microcontroller must deliver an 8-bit digital code (iastd) to the 8-bit digital/analog converter DAC0800LCN. Hence, the microcontroller must compute:

$$iastd = 36.4286 * iast + 127,$$

which results from the graphical relation presented in Fig. 10.7. Once "iastd" is computed, it is delivered to the digital/analog converter through port D (PORTD=iastd). The digital/analog converter works together with an operational amplifier TL081. These devices are connected following the manufacturer's suggestions [10]. This ensures that, at the operational amplifier output, a voltage is obtained whose numerical value corresponds to $-i^*$. This voltage is received by another operational amplifier TL081 devoted to implementing the current controller, i.e., to compute:

$$u_i = 100(i^* - i).$$

The reason why $-i^*$ must appear at the indicated place in Fig. 10.6 has to do with the sign operations performed by the operational amplifiers in this figure. On the other hand, a unit gain power amplifier is composed of a third operational amplifier TL081 together with two power transistors TIP141 and TIP145 in a complementary symmetry connection.

The circuit MAX232 is devoted to sending the measured velocity ω and the control signal i^* to a portable computer (through a USB to a series adapter) whose unique purpose is the graphical representation of those variables. This is performed through the variables "cuentaH=0x00" and "cuentaL=t_0" (to send velocity) or "cuentaH=0x00" and "cuentaL=(iastd) &(0xFF)" (to send the control

Fig. 10.7 Conditioning of the signal i^*, which must be sent to the digital/analog converter

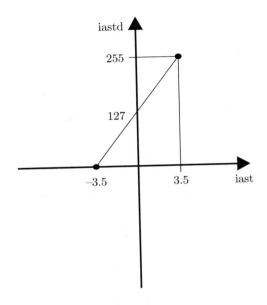

signal). These data are sent to the portable computer through the instructions: putc(0xAA); putc(cuentaH); putc(cuentaL);. Finally, the timer TMR0 is employed to fix the sample time $T = 0.002$[seg] (40 counts of timer 0).

10.6.1 Electric Current Control

The electrical loop shown in (10.14) is implemented as follows. A 1[Ohm] power resistance is series connected to the motor armature. Hence, voltage between terminals of this resistance is numerically equal to the electric current i flowing through the motor. Then, the operational amplifier at the block "current control" in Fig. 10.6 performs the operation indicated in (10.14) with $K = 100$.

10.6.2 Power Amplifier

According to Sect. 9.1.2, the block labeled "power amplifier," in Fig. 10.6, allows us to reduce the dead zone introduced by the power transistors in a complementary symmetry connection. Moreover, with the indicated values $A_p = 1$ is accomplished in (10.13) and (10.15). This results in $A_p K = 100$. This value has been chosen because a good performance was obtained. It is important to stress that large values such as $A_p K = 700$ are employed in commercial drivers for industrial applications [8].

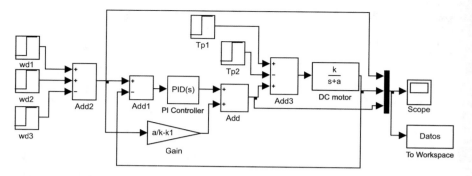

Fig. 10.8 MATLAB/Simulink diagram used to obtain the results in Fig. 10.9

10.7   Simulation Results

Some simulation results obtained with the controller in (10.33) (or, equivalently, with the controller in (10.40)), employing the MATLAB/Simulink diagram in Fig. 10.8 are presented in Fig. 10.9. Starting from a zero initial velocity, the following desired velocity is commanded:

$$\omega_d = \begin{cases} 1.5, & 0 \le t < 4 \\ 2.5, & 4 \le t < 12 \\ 1.5, & t \ge 12 \end{cases}. \tag{10.41}$$

Also, the constant disturbance $T_p = 2.5$ is applied at $t = 8[s]$, via software, which disappears at $t = 17[s]$. This means that T_p appears as an electric current signal, which adds to the electric current commanded by controller. The DC motor parameters are the same as those in (10.28). The gains of the controller in (10.33) are computed according to (10.34) together with numerical values in (10.28) and:

$$k_p' = 0.5, \quad k_1 = 4.0,$$

, which fixes the desired time constant as $\tau = 0.6231[s]$. The reader can verify that the labels in Fig. 10.9 prove that the desired time constant is accomplished when applying the desired velocity steps. Also note that the effects of the disturbance T_p are kept small and disappear very fast.

The wd and Tp blocks in Fig. 10.8 are suitably programmed to produce the desired velocity defined in (10.41) and the disturbance T_p defined above respectively. The PI controller block has kp and ki as the proportional and integral gains respectively, whereas the To workspace block is programmed to save data as an array with 0.001 as sample time. The graphical results in Fig. 10.9 were obtained by executing the following MATLAB code in an m-file:

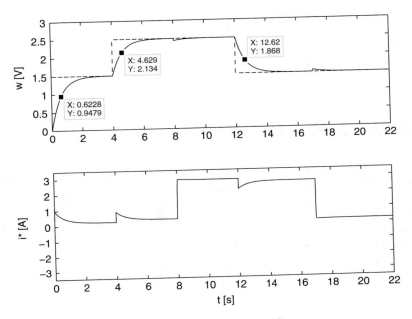

Fig. 10.9 Simulation results when using the controller in (10.33)

```
a=0.3704;
k=2.4691;
T=0.6231; %time constant, sec
kpp=(1/T-a)/k
k1=40;
kp=kpp+k1;
kip=a+kpp*k;
ki=kip*k1;
Tp=2.5;
```

then simulation in Fig. 10.8 is run, and, finally, the following code is executed in an m-file:

```
Ts=0.001;
t=0:Ts:22;
figure(1)
subplot(2,1,1)
plot(t,Datos(:,1),'b--');
hold on
plot(t,Datos(:,2),'b-');
hold off
axis([0 22 0 3])
ylabel('w [V]')
```

```
subplot(2,1,2)
plot(t,Datos(:,3),'b-');
axis([0 22 -3.5 3.5])
xlabel('t [s]')
ylabel('i* [A]')
```

10.8   Experimental Results

Some experimental results obtained with the controller in (10.33) (or, equivalently, with the controller in (10.40)), employing the experimental prototype described in Sect. 10.6, are presented in Fig. 10.10. Starting from a zero initial velocity, the desired velocity in (10.41) is commanded. Also, disturbance T_p and the controller gains are the same as those described in Sect. 10.7. The velocity measured in the experiment is shown in Fig. 10.10. The white line represents the response obtained in simulation for the system without disturbances:

$$\dot{\omega} + k_i'\omega = k_i'\omega_d, \tag{10.42}$$

with $k_i' = \frac{1}{0.6231}$, i.e., the value computed according to (10.34), and ω_d given in (10.41). Note that the velocity experimental response exactly tracks the theoretical response defined by (10.42) and this is still true during the transition from $\omega_d = 2.5$ to $\omega_d = 1.5$ at $t = 12[s]$ despite the constant disturbance $T_p = 2.5$ being present during this transition. This corroborates the conclusions stated just before (10.33). Moreover, it is important to stress that the desired transient and steady-state response specifications have been exactly achieved and the deviations observed when an external disturbance appears and disappears vanish very fast. Note that the simulation results in Fig. 10.9 and the experimental results in Fig. 10.10 are very similar.

Next, the results obtained with a classical PI controller are presented to compare with results obtained in Fig. 10.10. The results obtained with a classical PI controller are presented in Fig. 10.11. The controller gains are computed as:

$$\frac{k_i}{k_p} = a, \quad \tau = \frac{1}{k_p k} = 0.6231[s], \tag{10.43}$$

according to the discussion in Sect. 5.2.4, with τ the desired closed-loop system time constant. The desired velocity ω_d is identical to that shown in (10.41) and a disturbance is applied that is identical to that used in Fig. 10.10, i.e., it is applied via software with a constant value $T_p = 2.5$ at $t = 8[s]$ and disappears at $t = 17[s]$. The following is observed:

- Although the transient response is close to the desired one, it is not as good as that in Fig. 10.10 (the thin line represents the desired transient response).

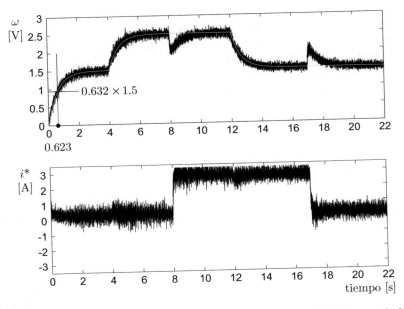

Fig. 10.10 Velocity control using the modified PI controller shown in (10.33) (or, equivalently, with the controller in (10.40))

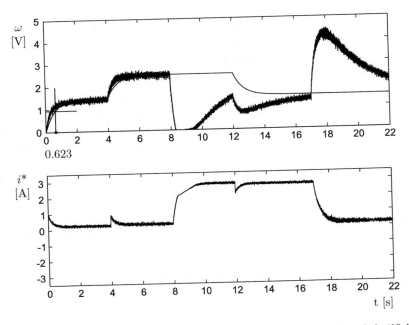

Fig. 10.11 Classical proportional–integral (PI) velocity control using the tuning rule in (10.43)

- The measured velocity reaches the desired velocity in a steady state when no external disturbance is applied, i.e., before $t = 8[s]$.
- The deviation produced by the external disturbance is much larger than in Fig. 10.10 and it vanishes much more slowly. In fact, this deviation is not compensated for before a new change in the desired velocity appears at $t = 12[s]$. The same is observed when the disturbance disappears at $t = 17[s]$. According to Sect. 5.2.4, this is because the dynamics of the disturbance rejection has a dominant time constant that is equal to the motor open-loop time constant, $\frac{1}{a}$, which is 2.7[s] as indicated in (10.27).

The experimental results shown in Fig. 10.12 are obtained using a classical PI controller whose gains are chosen according to:

$$\frac{k_i}{k_p} = d, \; d = 1.4 < p_1, \; k_p = \frac{l_2 l_3}{l_1 k}, \tag{10.44}$$

$$l_1 = abs(-p_1 + d), \; l_2 = abs(-p_1), \; l_3 = abs(-p_1 + a), \; p_1 = \frac{1}{0.6231} = 1.6.$$

This tuning criterion is introduced in (5.25), Sect. 5.2.4, where $\tau = 0.6231[s]$ is the slowest time constant present in the disturbance rejection dynamics. This time constant is chosen to be equal to the desired time constant when the system responds to the desired velocity because, according to (10.38), this is the slowest time constant that is present in the disturbance rejection shown in Fig. 10.10 ($f > p_1$ for the numerical values employed). The desired velocity ω_d is identical to that shown in (10.41) and a disturbance is applied that is identical to that used in Fig. 10.10, i.e., it is constant with value $T_p = 2.5$, it is applied via software at $t = 8[s]$, and disappears at $t = 17[s]$. The following is observed:

- The deviation due to the disturbance vanishes almost as fast as in Fig. 10.10.
- Although the transient response to the reference of velocity is very fast, it does not correspond to the desired transient response represented by the thin line in Fig. 10.12. As explained in Sect. 5.2.4, this is because the closed-loop system has two poles, one assigned at $s = -p_1$, but the other is far to the left of this point. Moreover, the pole located at $s = -\frac{1}{0.6231} = -1.6 = -p_1$ is very close to the zero at $s = -d = -1.4$, i.e., its effects are significantly compensated for (cancelled) and only the dynamics of the fast pole located far to the left is observed.

The experimental results in Figs. 10.11 and 10.12 corroborate the observations presented in Sect. 5.2.4 on classical PI velocity control: there is no tuning rule allowing us to compute the proportional and integral gains to simultaneously achieve the desired specifications for the transient response to a velocity reference and for a satisfactory disturbance rejection. This motivates the design of the modified PI controller presented in (10.33) whose experimental results are shown in Fig. 10.10.

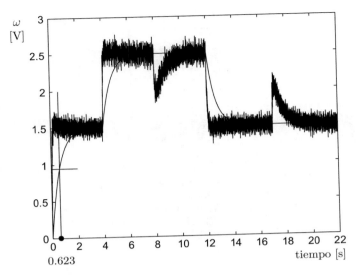

Fig. 10.12 Classical PI velocity control using the tuning rule in (10.44)

10.9 Microcontroller PIC16F877A Programming

The flow diagram for the Microcontroller PIC16F877A programming when controlling velocity in a PM brushed DC motor is presented in Fig. 10.13. The complete
code is listed in the following. The reader is referred to [11] for a precise explanation
of each one of the instructions in the program.

```
// PM brushed DC motor velocity control microcontroller program
#include<16f877A.h>
#include<stdlib.h>
#include<math.h>

#fuses HS,NOWDT,PUT,NOBROWNOUT,NOLVP,NOWRT,NOPROTECT,NOCPD

#use delay(clock=20000000) //time basis
// Config. series port
#use rs232(baud=115200,XMIT=PIN_C6,RCV=PIN_C7,BITS=8,PARITY=N)
// Ports and registers addresses
#byte OPTION= 0x81
#byte TMR0 = 0x01
#byte PORTA = 0x05

#byte PORTB = 0x06

#byte PORTC = 0x07
#byte PORTD = 0x08
#byte PORTE = 0x09
```

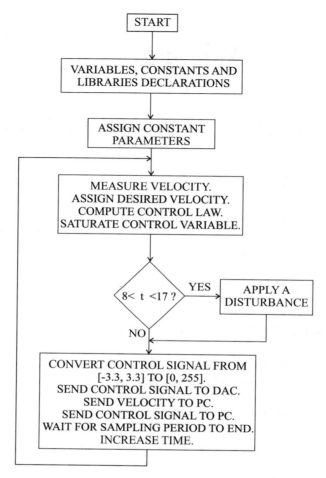

Fig. 10.13 Flow diagram for microcontroller PIC16F877A programming when controlling velocity in a PM brushed DC motor

```
#byte ADCON0= 0x1F
#byte ADCON1= 0x9F
#bit PC0 = 0x07.0

#bit PC1 = 0x07.1
//-----------Variables declaration-----------//
int16 inter,cuenta;

int8 cuentaH,cuentaL,puerto,AB,AB_1,aux;

int8 cont,iastd,cont2,cont3,cont4,t_0;
```

```
float w,error,iast,wm1,kp,ki,ie,k,a,wd,tiempo,kpp,k3,kib,cte;
unsigned int u,i;
//-----------Interrupt subroutine------------//
#int_rb void rb_isr() {
puerto=PORTB;
AB=((puerto)&(0x30))>>4;
aux=AB^AB_1;
if(aux!=0)
if(aux!=3)
if(((AB_1<<1)^AB)&(0x02))
cuenta--;
else
cuenta++;
AB_1=AB;
}
//-----------Main program------------//
void main(void) {
setup_adc(ADC_CLOCK_INTERNAL ); //ADC (internal clock)
set_tris_a(0b11111111);
set_tris_b(0b11111111);
set_tris_c(0b10000000); //Serial communication
set_tris_d(0b00000000);
set_tris_e(0b11111111);
OPTION=0x07; //pre_scaler timer0, 1:256
PORTC=0;
TMR0=0;
cuenta=0;
AB=0;
AB_1=0;
enable_interrupts(global);
enable_interrupts(int_rb);
cont=4;
cont2=0;
cont3=0;
cont4=0;
PORTD=127;
k=2.4691;
a=1/2.7;
kpp=0.5;
k3=4.0;
kp=kpp+k3;
kib=a+k*kpp;
ki=kib*k3;
cte=a/k-k3;
tiempo=0.0;
wm1=0.0;
ie=0.0;
ADCON1=0x04;
ADCON0=0x81;
while(cont2<3)
{
while(cont3<255)
{
TMR0=0;
```

```
while(TMR0<255)
{
}
cont3++;
}
cont2++;
}
TMR0=0;
while(TRUE)
{
cont++;
ADCON0=0x81; //Changing to CH0
for(i=0;i<37;i++); //Stabilizing internal capacitor
t_0=read_adc(); //Read ADC
w=0.0196*t_0;
if(tiempo<4.0)
wd=1.5;
else
if(tiempo<12.0)
wd=2.5;
else
wd=1.5;
// wd=2.5;
error=wd-w;
/* Controller */
iast=kp*error+ki*ie+cte*wd;
ie=ie+0.002*error;
/* _____ */
if(iast<-3.3)
iast=-3.3;
if(iast>3.3)
iast=3.3;
// iast=0.3;
iast=-iast;
if(tiempo>8.0)
{
if(tiempo<17.0)
iast=iast+2.5;
}
iastd=(unsigned int)(36.4286*iast+127);
PORTD=iastd;
wm1=w;
/*
// inter=(cuenta)&(0xFF00);
cuentaH=0x00; //inter>>8;
cuentaL=t_0;//(cuenta)&(0x00FF);
putc(0xAA); //Recognition serial port
putc(cuentaH); //Sending to serial port
putc(cuentaL);
if(cont==5)
{
cont=0;
cont4++;
if(cont4==255)
```

```
cont4=0;
}
*/
if(tiempo>8.0)
{
if(tiempo<17.0)
iast=iast-2.5;
}
iast=-iast;
iastd=(unsigned int)(36.4286*iast+127);
// inter=(cuenta)&(0xFF00);
cuentaH=0x00; //inter>>8;
cuentaL=(iastd)&(0xFF);
putc(0xAA); //Recognition serial port
putc(cuentaH); //sending to serial port
putc(cuentaL);
if(cont==5)
{
cont=0;
cont4++;
if(cont4==255)
cont4=0;
}
PC1=1; //waiting for sampling time
while(TMR0<40) // 1 count=(4/FXtal)*256 seg, 40=2 ms
{
}
PC1=0; //sampling time has arrived
TMR0=0;
tiempo=tiempo+0.002;
} //closing the infinite while
} //closing mian
```

10.10 Frequency Response-Based Design

10.10.1 Model Identification

Another experimental prototype for velocity control in a PM brushed DC motor has been built. The identification of the corresponding model has been performed using a frequency response experiment. The details on the steps that have been followed to perform this task are given in the following.

1. The following voltage signal has been applied at the motor armature terminals $u = 2 + A \sin(\bar{\omega}t)$[V] and the measured velocity of the motor has been observed to have the following form $\omega = \omega_0 + B \sin(\bar{\omega}t + \phi)$[rad/s], where ω_0 is a constant. The variable ω is used to designate the motor angular velocity and $\bar{\omega}$ to designate the angular frequency. ω is used to designate any of these variables when it is evident from the context which of these variables is referred to.

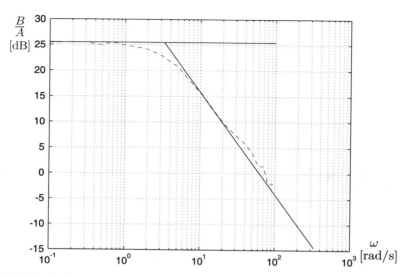

Fig. 10.14 Bode diagrams from the experimental data used for model identification. Continuous: asymptotes. Dashed: experimental data (Table 10.1)

2. This experiment has been performed for several values of frequency ω, and the results are presented in Table 10.1. Note that only the amplitude of the sinusoidal functions is taken into account. Moreover, the phase ϕ is not considered.
3. The Bode diagrams obtained from Table 10.1 are presented in Fig. 10.14. One 0[dB/dec] asymptote and one -20[dB/dec] asymptote fit these experimental data well. This means that the velocity motor model must have the following structure:

$$\frac{\omega(s)}{U(s)} = \frac{k}{s + a}. \tag{10.45}$$

4. From the intersection of these asymptotes,
   it is concluded that:

$$a = 3.3 \tag{10.46}$$

5. From the zero frequency response magnitude: $20 \log(k/a) = 25.5$[dB], i.e.,

$$k = 3.3 \times 10^{\frac{25.5}{20}} = 62.1604. \tag{10.47}$$

Table 10.1 Frequency response experimental data for model identification of a PM brushed DC motor

$\bar{\omega}$[rad/s]	$2A$ [V]	$2B$ [rad/s]
0.1	1	18.68
0.2	1	18.85
0.3	1	18.85
0.4	1	18.07
0.5	1	17.68
0.6	1	18.46
0.7	1	18.46
0.8	1	18.46
0.9	1	17.67
1	1	17.67
2	1	16.1
3	1	14.14
4	1	12.17
5	1	11
6	1	9.42
7	1	8.25
8	1	7.46
9	1	6.68
10	1	6.29
20	1	3.15
30	1	2.35
40	1	1.96
50	1	1.57
60	1	1.18
70	1	1.18
80	1	0.79
90	1	0.79
100	1	0.79

Fig. 10.15 Proportional–integral control of velocity

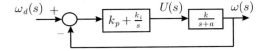

10.10.2   *Proportional–Integral Control Design*

As the motor model in 10.45 has no pole at $s = 0$, then a controller with integral action is required to render the system type 1. This ensures a zero steady-state error if the desired velocity is constant. Hence, given the first-order structure of the motor model, the design of a proportional–integral (PI) controller is well suited (see Fig. 10.15).

This means that the open loop transfer function is given as:

$$G(s)H(s) = \frac{k_p s + k_i}{s}\frac{k}{s + a}.$$

Fig. 10.16 Integral control
of velocity $k_i = 1$, $k_p = 0$

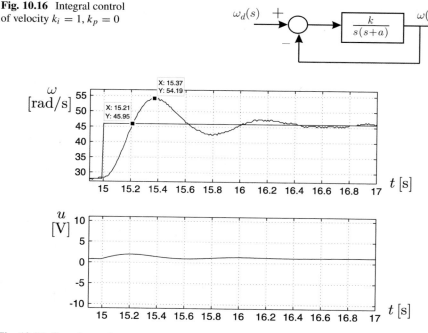

Fig. 10.17 Experimental response of the control system in Fig. 10.16, i.e., an integral controller
of velocity with $k_i = 1$, $k_p = 0$

Proceeding as in Sect. 6.7.1, this transfer function can be rewritten as:

$$G(s)H(s) = k_p b(z + 1)\frac{k}{s(s+a)}, \quad z = \frac{s}{b}, \quad b = \frac{k_i}{k_p}.$$

Hence, the factor $\frac{k}{s(s+a)}$ is to be considered as the "plant to be controlled" and
the factor $k_p b(z + 1)$ as the "controller." Thus, it is of interest to know how a
closed system such as that shown in Fig. 10.16 would respond. Note that this can
be achieved in practice by implementing an integral controller with a unit gain, i.e.,
$k_i = 1$, $k_p = 0$. In Fig. 10.17, we present the experimental results obtained when
testing the response of such a control system. We observe that 45.6% overshoot and
a rise time of $t_r = 0.21$[s] are produced.

It is chosen to maintain the same rise time and to reduce overshoot to 15%.
Hence, the crossover frequency of the factor $\frac{k}{s(s+a)}$ alone must be kept, i.e., the
magnitude of the factor $k_p b(z + 1)$ must be 0[dB] at the crossover frequency of
the factor $\frac{k}{s(s+a)}$. On the other hand, to achieve a 15% overshoot, we can use this
value and:

$$\zeta = \sqrt{\frac{\ln^2(M_p/100)}{\ln^2(M_p/100) + \pi^2}},$$

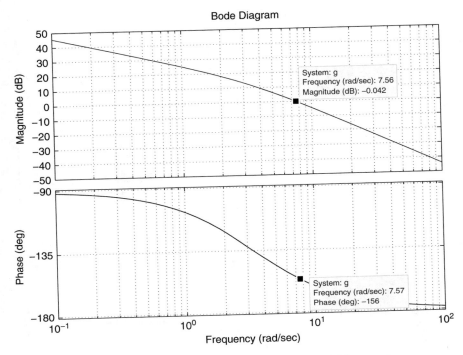

Fig. 10.18 Bode diagrams of the factor $\frac{k}{s(s+a)}$, where (10.46) and (10.47) are used

to find $\zeta = 0.5169$. According to Table 6.5, this requires a $52°$ phase margin. In Fig. 10.18 the Bode diagrams of the factor $\frac{k}{s(s+a)}$ are presented where (10.46) and (10.47) are used. It is observed that the crossover frequency is $\omega_1 = 7.55[\text{rad/s}]$ (to be kept) and the phase is $-156°$. This means that a $28°$ phase lead has to be contributed by the controller, i.e., by the factor $z + 1$, when $\omega = \omega_1 = 7.55[\text{rad/s}]$. In Fig. 10.19, it is found that a $28°$ phase lead is obtained when $\frac{\omega}{b} = 0.53$. Hence, using $\omega = \omega_1 = 7.55[\text{rad/s}]$ $b = 14.2453$ is obtained. On the other hand, from the magnitude condition:

$$|k_p b(z + 1)|_{dB} = 20 \log(k_p b) + |z + 1|_{dB} = 0[\text{dB}],$$

where $|z + 1|_{dB} = 1.09[\text{dB}]$ according to Fig. 10.19. Thus:

$$k_p = \frac{1}{b} 10^{\frac{-1.09}{20}} = 0.0619.$$

Finally, from $k_i = bk_p$, the following is found:

$$k_i = 0.8821.$$

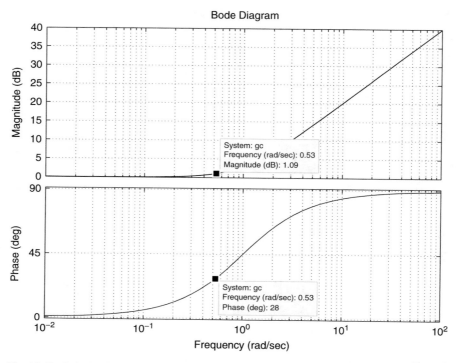

Fig. 10.19 Bode diagrams of the factor $z + 1$. The horizontal axis represents $\frac{\omega}{b}$

In Fig. 10.20, a 7.55[rad/s] crossover frequency and a 52° phase margin is observed. This means that the frequency response design is correct. On the other hand, the simulations presented in Fig. 10.21 show a 21% overshoot, instead of the desired 15%, and 0.235[s] as rise time, instead of the desired value $t_r = 0.21$[s]. In Fig. 10.22, some experimental results are presented where 15.22% is obtained as overshoot and 0.21[s] is the value of the rise time. Note that these values are very close to the desired values. Finally, according to Sect. 6.5.2, an additional open-loop phase lag of $-\omega T$[rad] must be considered to take into account the time required for controller computation. For experiments in Fig. 10.22 a sampling period of $T = 0.01$[s] was employed, which represents the worst case time delay. Thus, the additional phase lag at the crossover frequency $\omega = 7.57$[rad/s] is $-\omega T \times 180°/\pi = -4.33°$. As this phase lag is small, this explains why the response in Fig. 10.22 is not deteriorated by the time delay induced by the controller digital implementation.

In Fig. 10.23 this PI controller is compared with the PI control scheme presented (10.33). The controller gains were computed using:

$$k_1 = 0.1, \quad k_i' = 1/0.1,$$

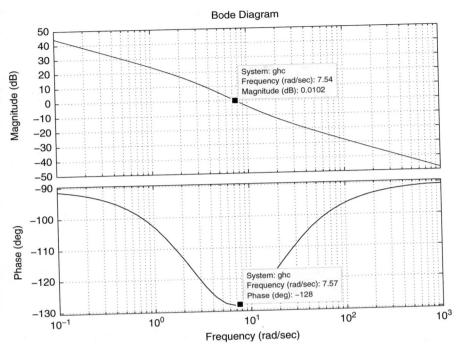

Fig. 10.20 Bode diagrams of the transfer function $\frac{k_p s + k_i}{s} \frac{k}{s+a}$. $k_p = 0.0619$, $k_i = 0.8821$, and (10.46), (10.47) are used

and the relations in (10.34). These gains were selected such that the time responses are approximately the same. Note that the controller in (10.33) forces a first-order response when step changes in the desired velocity are commanded. A constant disturbance is applied for $t \geq 17[s]$ via software. This is done by adding the constant 2.5[V] to the voltage that is delivered by the controller, which has to be applied to the motor. We realize that the effects of the disturbance are a little smaller for the controller in (10.33), although the rejection time is almost the same.

Note that the disturbance rejection with both controllers is not so different than what happens with the responses obtained in Sect. 10.8 for a classical PI controller and controller in (10.33). The key to understanding this is to recall that the PI controller in Sect. 10.8 is designed by imposing $\frac{k_i}{k_p} = a$, which results in a slow rejection of external disturbances, as explained in Sect. 5.2.4. On the other hand, the classical PI controller in the present section is designed without imposing such a constraint. Thus, a faster disturbance rejection is accomplished. Note, however, that this design has been performed without taking into consideration the desired specifications for the disturbance rejection response.

Figures 10.14, and 10.18 to 10.21, in addition to all computations involved in controller design, were performed using the following MATLAB code in an m-file:

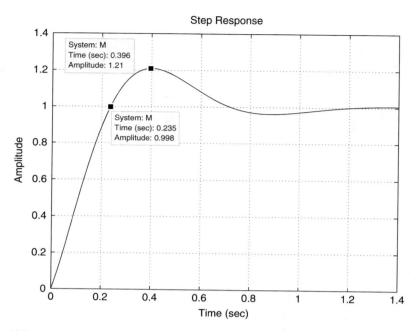

Fig. 10.21 Simulation response of the control system in Fig. 10.15 when a unit step in the desired velocity is applied. $k_p = 0.0619$, $k_i = 0.8821$, (10.46), and (10.47) are used

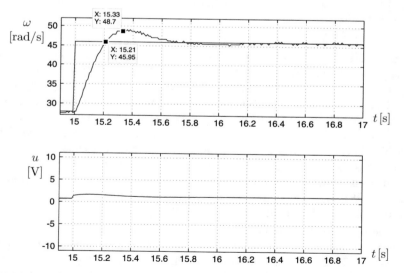

Fig. 10.22 Experimental response when using the control scheme in Fig. 10.15, with gains $k_p = 0.0619$ and $k_i = 0.8821$

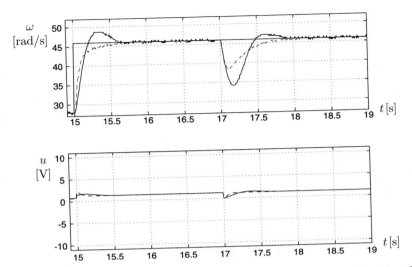

Fig. 10.23 Experimental comparison of the control scheme in Fig. 10.15 (continuous) and the controller in (10.33) (dashed)

```
clc
W=[0.1 0.2 0.3 0.4 0.5 0.6 0.7 0.8 0.9 1 2 3 4 5 6 7 8 9 10
20 30 40 50 60 70 80 90 100];
B=[18.68 18.85 18.85 18.07 17.68 18.46 18.46 18.46 17.67
17.67 16.1 14.14 12.17 11 9.42 8.25 7.46 6.68 6.29 3.15
2.35 1.96 1.57 1.18 1.18 0.79 0.79 0.79];
A=[1 1 1 1 1 1 1 1 1 1 1 1 1 1 1 1 1 1 1 1 1 1 1 1 1 1 1];
magnitude=20*log10(B./A);
figure(1)
semilogx(W,magnitude,'r-')
grid on
line([0.1 100],[25.5 25.5])
line([3.3 330],[25.5 -14.5])
a=3.3;
k=a*(10^(25.5/20))
trd=0.21; % tr desired
g=tf(k,[1 a 0]); % estimate DC motor model + integral ctrl
figure(2)
bode(g)
grid on
w1=7.55;
phase=-156;
Mpd=15;
zd=sqrt(log(Mpd/100)^2/(log(Mpd/100)^2+pi^2))
Kfd=52;
phased=-(180-52);
phase_lead=(156+phased) % 28 degrees
gc=tf([1 1],1)
figure(3)
```

```
bode(gc)
grid on
b=w1/0.53;
kp=10^(-1.09/20)/b
ki=kp*b
ctrl=tf([kp ki],[1 0]);
motor=tf(k,[1 a]);
ghc=ctrl*motor;
figure(4)
bode(ghc)
grid on
M=feedback(ghc,1,-1)
figure(5)
step(M)
grid on
```

10.10.3 Prototype Construction

The electric diagram, in addition to the computer program used to evaluate control algorithms, is identical to the electric diagram and the Builder 6 C++ code listed in Chap. 12. The only difference is that in Chap. 12, two encoders have to be read, whereas in the present chapter only one encoder is to be read. The sampling period used in the present section was 0.01[s]. Also, a microcontroller PI16F877A is used in Chap. 12 and in the present chapter. The code for programming this microcontroller is identical in both chapters. Again, in the present chapter, only one encoder has to be read. Thus, the code of the program is simpler in the present case.

In particular, the frequency response experiments performed for model identification in Sect. 10.10.1 require the following voltage $u = 2 + A \sin(\bar{\omega}t)[V]$ to be applied at the armature terminals of the motor. This is accomplished by writing the program line "iast=2.0+A*sin(w t);," provided that A=0.5 and w are defined as constants in the non-executable part of the program. This is all we need, because, according to Chap. 12, the control program and hardware are so contrived that the numerical value of the variable "iast" appears as a voltage signal at the motor armature terminals.

This prototype has been built because it presents some advantages when performing the frequency response experiments described in Sect. 10.10.1. When using the prototype depicted in Fig. 10.6 for frequency response identification, the velocity of the motor has a large drift, which poses difficulties for the measurement of the velocity amplitude. Such a problem is eliminated with the new prototype.

10.11   Summary

Two velocity controllers for a PM brushed DC motor have been experimentally tested: a classical PI controller and a modified PI controller. The latter controller has been motivated by some of the drawbacks of the former and the experimental results have corroborated the good performance of the new controller. Clear explanations have also been provided on how to build the interfaces required by the complete control system and how a microcontroller can be programmed to implement the designed controller.

It is important to stress that it is common in the motor control literature to assume that either torque or electric current are the motor input signal, despite it the voltage being the actual motor input signal and torque being a variable that results from the evolution of the motor's electrical dynamics. It has been explained in this chapter that use of an internal electric current loop makes it possible to use electric current, or the generated torque, as the input variable.

A significant problem that must be solved before designing a controller is to determine the numerical values of the plant parameters. Two ways of solving this problem in a PM brushed DC motor are presented in this chapter. The fundamental idea in the first approach is to apply a step input to the motor and to observe its response. This response is found to be similar to that of a first-order system. As this response is well known, then the parameters of a first-order system are identified from such a time response. In the second approach, a frequency response experiment is performed: (1) Several sinusoidal voltages are applied at the motor terminals, each one with a different frequency. (2) The ratio between the output and the input amplitudes are plotted with respect to frequency. (3) From this magnitude Bode diagram, it is possible to estimate the numerical parameters of the DC motor model.

10.12   Review Questions

1. What is an electric current loop and what is it used for?
2. Why does a PI controller suffice to regulate velocity in a PM brushed DC motor? Why not use a proportional–integral–derivative (PID) controller?
3. Is a PI velocity controller useful for taking the steady-state error to zero when the desired velocity is not constant? What happens if the disturbance is not constant? Explain.
4. How would you use the ideas in the present chapter to regulate voltage produced by an electric generator that is actuated by a PM brushed DC motor?
5. How would you use the ideas in the present chapter to control a temperature control system and a level control system?
6. What controllers would you use for the control systems in the previous question? Would you use a PID controller? Why?

7. If a classical PI controller slowly compensates for the effect of a disturbance, what changes would you make to the controller gains?
8. What are the effects of the proportional and the integral gains in a classical PI controller?

References

1. R. Ortega, A. Loría, P. J. Nicklasson, and H. Sira-Ramírez, *Passivity-based control of Euler-Lagrange Systems*, Springer, London, 1998.
2. V. M Hernández Guzmán and R. V. Carrillo Serrano, Global PID position control of PM stepper motors and PM synchronous motors, *International Journal of Control*, vol. 84, no. 11, pp. 1807–1816, 2011.
3. V. M Hernández Guzmán and R. Silva-Ortigoza, PI control plus electric current loops for PM synchronous motors, *IEEE Transactions on Control Systems Technology*, vol. 19, no. 4, pp. 868–873, 2011.
4. G. Espinosa-Pérez, P. Maya-Ortiz, M. Velasco-Villa, and H. Sira-Ramírez, Passivity-based control of switched reluctance motors with nonlinear magnetic circuits, *IEEE Transactions on Control Systems Technology*, Vol. 12, pp. 439–448, 2004.
5. V. M. Hernández-Guzmán, J. Orrante-Sakanassi, and F. Mendoza-Mondragón, Velocity regulation in switched reluctance motors under magnetic flux saturation conditions, *Mathematical Problems in Engineering*, Hindawi, Vol. 2018, 13 pp., Article ID 3280468.
6. J. Chiasson, *Modeling and high-performance control of electric machines*, IEEE Press-Wiley Interscience, New Jersey, 2005.
7. Parker Automation, Position systems and Controls, Training and Product Catalog DC-ROM, *Compumotor's Virtual Classroom*, 1998.
8. R. Campa, E. Torres, V. Santibáñez, and R. Vargas, Electromechanical dynamics characterization of a brushless direct-drive servomotor. *Proc. VII Mexican Congress on Robotics, COMRob 2005*, México, D.F., October 27–28, 2005.
9. PIC16F877A Enhanced Flash Microcontroller, Data sheet, Microchip Technology Inc., 2003.
10. DAC0800 8-bit Digital-to-Analog Converter, Data sheet, National Semiconductor Corporation, 1995.
11. Custom Computer Services Incorporated, CCS C Compiler Reference Manual, 2003.

Chapter 11
Position Control of a PM Brushed DC Motor

In this chapter, we continue with the control of a permanent magnet (PM) brushed DC motor, but now the position control problem is considered. To this aim, the motor model in (10.23) is considered after a power amplifier and a current loop have been included (see Sects. 10.2 and 10.3). This model is rewritten here for ease of reference:

$$\theta(s) = \frac{1}{s(s+a)}[kI^*(s) - \frac{1}{J}T_p(s)], \tag{11.1}$$

$$a = \frac{b}{J}, \quad k = \frac{nk_m}{J},$$

where $\theta(s)$ is the angular position to be controlled, $I^*(s)$ is the input or the control variable, and $T_p(s)$ is an external torque disturbance.

11.1 Identification

In this section, an experiment is designed that allows us to know the values of the constants k and a in the model (11.1). In the following, the way of performing this experiment is explained. Suppose that no external disturbance is applied, i.e., $T_p(s) = 0$; hence, the model (11.1) becomes:

$$\theta(s) = \frac{k}{s(s+a)}I^*(s). \tag{11.2}$$

Suppose that i^* is designed as a proportional position controller, i.e.,

$$i^* = k_p(\theta_d - \theta), \tag{11.3}$$

© Springer International Publishing AG, part of Springer Nature 2019
V. M. Hernández-Guzmán, R. Silva-Ortigoza, *Automatic Control with Experiments*,
Advanced Textbooks in Control and Signal Processing,
https://doi.org/10.1007/978-3-319-75804-6_11

Fig. 11.1 Proportional
position control

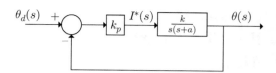

or, using the Laplace transform:

$$I^*(s) = k_p(\theta_d(s) - \theta(s)),\tag{11.4}$$

where k_p is a positive constant. A block diagram combining the expressions (11.2) and (11.4) is shown in Fig. 11.1. The closed-loop transfer function is:

$$\frac{\theta(s)}{\theta_d(s)} = \frac{k_p k}{s^2 + as + k_p k} = \frac{\omega_n^2}{s^2 + 2\zeta\omega_n s + \omega_n^2},\tag{11.5}$$

$$\omega_n = \sqrt{k_p k}, \quad \zeta = \frac{a}{2\sqrt{k_p k}}.\tag{11.6}$$

According to (11.5) and Sect. 3.3.1, if the desired position θ_d is a step the obtained time response $\theta(t)$ presents an overshoot and a rise time given as:

$$M_p(\%) = 100 \times e^{-\frac{\zeta}{\sqrt{1-\zeta^2}}\pi},\tag{11.7}$$

$$t_r = \frac{1}{\omega_d}\left[\pi - \arctan\left(\frac{\sqrt{1-\zeta^2}}{\zeta}\right)\right],\tag{11.8}$$

if $0 \le \zeta < 1$. This can always be achieved using a large enough value for k_p. From (11.7) and (11.8), the following is found:

$$\zeta = \sqrt{\frac{\ln^2\left(\frac{M_p(\%)}{100}\right)}{\ln^2\left(\frac{M_p(\%)}{100}\right) + \pi^2}},\tag{11.9}$$

$$\omega_d = \frac{1}{t_r}\left[\pi - \arctan\left(\frac{\sqrt{1-\zeta^2}}{\zeta}\right)\right],\tag{11.10}$$

$$\omega_n = \frac{\omega_d}{\sqrt{1-\zeta^2}}.\tag{11.11}$$

This allows us to compute k and a using the following procedure:

- Implement the proportional controller shown in (11.4) or (11.3). Ensure that no external disturbance is applied to the motor. Propose some known positive value for k_p. This value must be large enough to produce a clear overshoot on the

position response θ when a step change is commanded in the desired position θ_d.
Plot the position θ versus time when a step change is commanded in the desired
position θ_d.

- Measure $M_p(\%)$ and t_r.
- Use (11.9)–(11.11) to compute ζ and ω_n.
- Using the expressions in (11.6) and k_p employed in the experiment, compute k
  and a.

The results obtained when performing the experiment described in the above
procedure are shown in Fig. 11.2. $k_p = 0.5$ was employed. By direct measurement
in Fig. 11.2, the following is found:

$$M_p(\%) = 78.2\%, \quad t_r = 0.09[\text{s}].$$

These measurements are simplified and rendered more exact if a computer program
is employed. Following the above procedure, the following is found:

$$k = 675.4471, \quad a = 2.8681. \tag{11.12}$$

Finally, it is important to state that, as in the case of the velocity model, it is not
necessary to compute the value of the factor $1/J$ appearing as coefficient of the
disturbance $T_p(s)$ in (11.1). This is because T_p is also unknown and, despite this,
its effect can be compensated for.

11.2   Position Control When External Disturbances Are Not Present ($T_p = 0$)

The controller design is intended to ensure that the closed-loop transient and steady-
state response specifications is achieved. The transient response specifications refer
to the desired rise time and overshoot. These specifications are satisfied by suitably
assigning the closed-loop poles. On the other hand, the steady-state specifications
are achieved by ensuring that the measured position θ reaches its desired value θ_d
as time increases. The controller is a device designed in such a way that the desired
transient and steady-state response specifications are satisfied.

Suppose that the desired position is a constant or a step, i.e., $\theta_d = A$ is a constant.
As the plant in (11.1) has one pole at $s = 0$, then the system type is 1, which
ensures that $\theta = \theta_d = A$ in a steady state, if a proportional position controller is
used and no external disturbance is present, i.e., if $T_p = 0$. This means that the
requirement to reach the desired position is naturally satisfied by the plant, i.e., the
DC motor. Hence, the controller design is to be performed merely to satisfy the
transient response specifications, i.e., to suitably assign the closed-loop poles. In the
case of a DC motor, there are two simple controllers allowing the closed-loop poles
to be suitably located: (i) Proportional position control with velocity feedback. (ii)
A lead compensator.

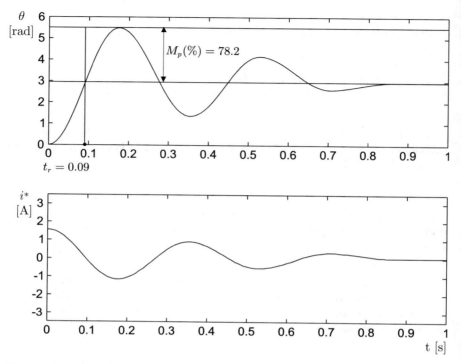

Fig. 11.2 Experimental identification

11.2.1   *Proportional Position Control with Velocity Feedback*

This controller is given as:

$$i^*(t) = k_p(\theta_d - \theta) - k_v\dot\theta, \tag{11.13}$$

where k_p and k_v are positive constants, or using the Laplace transform:

$$I^*(s) = k_p(\theta_d(s) - \theta(s)) - k_v s\theta(s). \tag{11.14}$$

The block diagram resulting from the combination of (11.14) and (11.2) is depicted in Fig. 11.3. The closed-loop transfer function is:

$$\frac{\theta(s)}{\theta_d(s)} = \frac{k_p k}{s^2 + (a + k_v k)s + k_p k} = \frac{\omega_n^2}{s^2 + 2\zeta\omega_n s + \omega_n^2},$$

$$\omega_n = \sqrt{k_p k}, \quad \zeta = \frac{a + k_v k}{2\sqrt{k_p k}}. \tag{11.15}$$

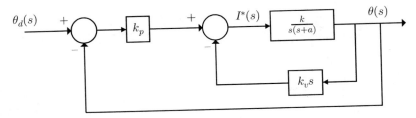

Fig. 11.3 Proportional position control with velocity feedback

Thus, it is always possible to find positive values for k_v and k_p such that the desired ζ and ω_n are obtained. This means that it is always possible to assign the closed-loop poles $(s + \zeta\omega_n + j\omega_d)(s + \zeta\omega_n - j\omega_d) = s^2 + 2\zeta\omega_n s + \omega_n^2$, at any point on the left half-plane, by a suitable selection of k_p and k_v. For instance, consider the numerical values in (11.12). Suppose that overshoot and rise time are desired to be:

$$M_p(\%) = 20\%, \quad t_r = 0.068[s]. \tag{11.16}$$

Using these data and (11.9)–(11.11), it is found that the closed-loop poles must be located at:

$$s = -\zeta\omega_n \pm j\omega_d = -15.4009 \pm j30.0623,$$

with $\zeta = 0.4559$ and $\omega_n = 33.7776$. Finally, using these values, (11.12), (11.15), and (11.16), it is found that the following controller gains are required to achieve the desired rise time and overshoot:

$$k_p = 1.6891, \quad k_v = 0.0414. \tag{11.17}$$

Some simulation results obtained when using these controller gains, the motor parameters in (11.12), and the control scheme in Fig. 11.3 are shown in Fig. 11.4. The desired position is $\theta_d = 1.5[rad]$. From these results, it is possible to verify that overshoot is 20% and rise time is 0.068[s], as desired. This simulation is performed using the following MATLAB code in an m-file:

```
k=675.4471;
a=2.8681;
kp=1.6891;
kv=0.0414;
gm=tf(k,[1 a 0]);
velfeedck=tf(kv*[1 0],1);
gc1=feedback(gm,velfeedck,-1);
M=feedback(kp*gc1,1,-1);
figure(1)
```

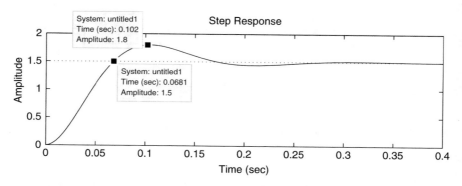

Fig. 11.4 Proportional position control with velocity feedback. Simulation results

```
subplot(2,1,1)
step(M*1.5)
```

The experimental results obtained with controller (11.13) and gains in (11.17), when θ_d is a step, are shown in Fig. 11.5. Overshoot and rise time measured in this experiment are:

$$M_p(\%) = 16\%, \quad t_r = 0.068[\text{s}],$$

which are close to the design specifications in (11.16).

It is worth stating that the use of a dedicated sensor to measure velocity implies several mechanical modifications that complicate the prototype construction and increase cost. Because of this, it is common to estimate velocity from the measured position in practice. In the experiments shown in Fig. 11.5, the velocity $\dot{\theta}$ has been estimated by numeric differentiation of the measured position (see Appendix F). However, this process has the effect of a *high-pass filter*: high-frequency signals, i.e., noise, are strongly amplified by the velocity feedback of the controller. The noise content may result in control system performance deterioration because motor vibration may occur, even at rest. Thus, the gain k_v must always be kept small. In the following section, a controller is proposed that needs neither velocity measurements nor position differentiation to suitably assign the closed-loop poles.

11.2.2 A Lead Compensator

This controller is defined by the transfer function:

$$\frac{I^*(s)}{E(s)} = \gamma \frac{s + d}{s + c}, \tag{11.18}$$

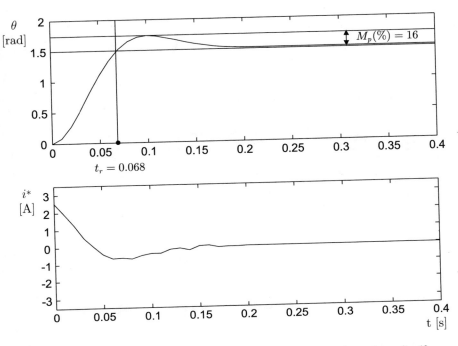

Fig. 11.5 Proportional position control with velocity feedback. Experimental results ($\theta_d =$ 1.5[rad])

where $E(s)$ is the Laplace transform of the difference $e = \theta_d - \theta$, i.e., the system error, whereas γ, d and c are positive constants to be computed and they are constrained to satisfy $d < c$.

The block diagram resulting from the connection of (11.18) and (11.2) is shown in Fig. 11.6. Note that the open-loop transfer function:

$$G(s)H(s) = \frac{\gamma k(s + d)}{s(s + c)(s + a)},$$ (11.19)

has three poles, i.e., the closed-loop transfer function also has three poles and one zero at $s = -d$. It is left as an exercise for the reader to find the closed-loop transfer function to verify that it has the form:

$$\frac{\theta(s)}{\theta_d(s)} = \frac{\gamma k(s + d)}{(s - p_1)(s - p_2)(s - p_3)}.$$ (11.20)

In this case, besides assigning the complex conjugate poles p_1 and p_2 at the desired locations, it must be ensured that the third pole, at p_3, and the zero at $s = -d$ do not significantly modify the transient response. As discussed in Sect. 5.2.3, this

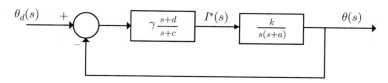

Fig. 11.6 Position control with a lead compensator

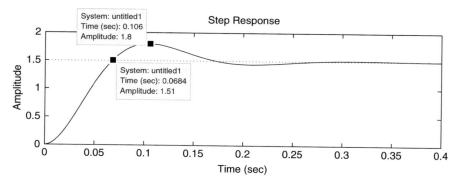

Fig. 11.7 Position control with a lead compensator. Simulation results ($\theta_d = 1.5$[rad])

is achieved by setting:

$$d = a, \quad c = 2\zeta\omega_n, \quad \gamma = \frac{\omega_n^2}{k}.$$

Using this tuning rule and the parameters in (11.12), it is found that use of the gains:

$$c = 30.8018, \qquad \gamma = 1.6891, \qquad d = 2.8681,$$

in the controller in (11.18) assigns two closed-loop poles at:

$$s = -\zeta\omega_n \pm j\omega_d = -15.4009 \pm j30.0623.$$

This specifies by design a 0.068[s] rise time and a 20% overshoot.

The simulation results with c, d, γ, as above, the motor parameters in (11.12), and the control scheme in 11.6 are presented in Fig. 11.7. The reader can verify that overshoot is 20% and rise time is 0.068[s], as desired. These simulations were performed by executing the following MATLAB code in an m-file:

```
k=675.4471;
a=2.8681;
gm=tf(k,[1 a 0]);
d=a;
z=0.4559;
```

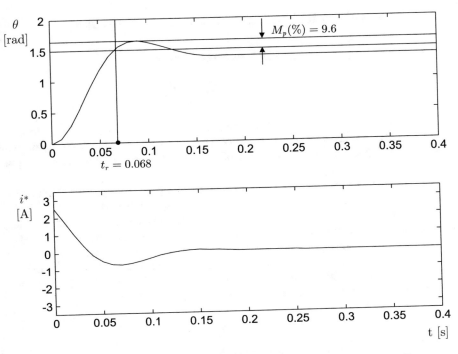

Fig. 11.8 Position control with a lead compensator. Experimental results ($\theta_d = 1.5$[rad])

```
wn=33.7776;
c=2*z*wn
gamma=wn^2/k
gc=tf(gamma*[1 d],[1 c]);
M=feedback(gc*gm,1,-1)
figure(1)
subplot(2,1,1)
step(M*1.5,0.4)
```

Some experimental results obtained when $\theta_d = 1.5$[rad] is a step are shown in Fig. 11.8. There, a $M_p(\%) = 9.6\%$ overshoot and a $t_r = 0.068$[s] rise time are measured. Note that these values are close to the design values: $t_r = 0.068$[s] and $M_p(\%) = 20\%$. Finally, it is important to stress that the use of the lead compensator (Fig. 11.8) also results in a small steady-state error. This steady-state error is due to the presence of nonlinear friction: static and Coulomb (see (2.28)).

11.3   Control Under the Effect of External Disturbances

In this section, a constant external disturbance is considered, i.e.,

$$T_p = t_d, \quad T_p(s) = \frac{t_d}{s},$$

where t_d is a constant different from zero. This means that the possibility that an external agent applies a torque at the motor shaft is considered. The controller design must be performed such that the effect of such a disturbance on the system output must vanish as fast as possible.

In this class of applications, it is common to employ a proportional–integral–derivative (PID) controller because its integral part is intended to cope with the deviations due to the constant disturbance. However, as stated in Sect. 5.2.5, a tuning rule is not at the disposal of the designer, allowing us to compute the PID controller gains to simultaneously achieve the desired transient response specifications (rise time and overshoot) and a satisfactory external disturbance rejection. Motivated by this situation, two new controllers are introduced in the following sections that are specially designed to simultaneously achieve these specifications.

11.3.1   A Modified PID Controller

The use of the inverse Laplace transform in (11.1) yields the following differential equation:

$$\ddot{\theta} + a\dot{\theta} = ki^* - \frac{1}{J}T_p. \tag{11.21}$$

It is shown next that the following controller allows us to find a tuning rule, solving the problem described above:

$$i^* = \kappa_p(\theta_d - \theta) - \kappa_d\dot{\theta} + k_1\kappa_p k \int_0^t (\theta_d - \theta(r))dr - k_1(\dot{\theta} - \dot{\theta}(0)),$$

$$-k_1(a + \kappa_d k)(\theta - \theta(0)), \tag{11.22}$$

where θ_d is a constant standing for the desired position whereas $\theta(0)$ and $\dot{\theta}(0)$ represent the position and the velocity initial values respectively. Using the fact that:

$$\dot{\theta} - \dot{\theta}(0) = \int_0^t \ddot{\theta}(r)dr,$$

$$\theta - \theta(0) = \int_0^t \dot{\theta}(r)dr,$$

the above-mentioned controller can be written as:

$$i^* = \kappa_p(\theta_d - \theta) - \kappa_d\dot{\theta} + k_1\kappa_p k \int_0^t (\theta_d - \theta(r))dr - k_1 \int_0^t \dot{\theta}(r)dr$$

$$-k_1(a + \kappa_d k) \int_0^t \dot{\theta}(r)dr,$$

$$= \kappa_p(\theta_d - \theta) - \kappa_d\dot{\theta} - k_1 \int_0^t \left[\ddot{\theta}(r) + (a + \kappa_d k)\dot{\theta}(r) + \kappa_p k(\theta(r) - \theta_d)\right]dr.$$

Replacing this in (11.21), the following is found:

$$\ddot{\theta} + (a + \kappa_d k)\dot{\theta} + \kappa_p k(\theta - \theta_d)$$

$$+k_1 k \int_0^t \left[\ddot{\theta}(r) + (a + \kappa_d k)\dot{\theta}(r) + \kappa_p k(\theta(r) - \theta_d)\right]dr = -\frac{1}{J}T_p.$$

This can be written as:

$$\dot{\xi} + k_1 k\xi = -\frac{1}{J}T_p, \tag{11.23}$$

if the following is defined:

$$\xi = \int_0^t \left[\ddot{\theta}(r) + (a + \kappa_d k)\dot{\theta}(r) + \kappa_p k(\theta(r) - \theta_d)\right]dr.$$

Applying the Laplace transform to (11.23) and assuming all initial conditions to be zero, the following is obtained:

$$\xi(s) = \frac{-\frac{1}{J}}{s + k_1 k}T_p(s),$$

or:

$$s\xi(s) = \frac{-\frac{1}{J}s}{s + k_1 k}T_p(s). \tag{11.24}$$

If the disturbance is constant, i.e., $T_p(s) = \frac{t_d}{s}$, then the final value theorem can be used to find:

$$\lim_{t\to\infty} \dot{\xi}(t) = \lim_{s\to 0} s[s\xi(s)],$$

$$= \lim_{s\to 0} s\frac{-\frac{1}{J}s}{s + k_1 k}\frac{t_d}{s} = 0.$$

Hence, if $k_1 k > 0$, the filter in (11.24) is stable and it is ensured that $\lim_{t \to \infty} \dot{\xi}(t) = 0$. Moreover, the larger is $k_1 k > 0$ the faster $\dot{\xi}(t)$ approaches zero. Note that $\lim_{t \to \infty} \dot{\xi}(t) = 0$ implies that:

$$\frac{d}{dt} \int_0^t \left[\ddot{\theta}(r) + (a + \kappa_d k)\dot{\theta}(r) + \kappa_p k(\theta(r) - \theta_d) \right] dr,$$

$$= \dddot{\theta} + (a + \kappa_d k)\ddot{\theta} + \kappa_p k(\dot{\theta} - \theta_d) \to 0.$$

This means that the effect of a constant disturbance vanishes as time increases and the only thing that remains is the differential equation $\ddot{\theta} + (a + \kappa_d k)\dot{\theta} + \kappa_p k\theta = \kappa_p k\theta_d$. This differential equation is stable if $\kappa_p > 0$ and $\kappa_d > 0$, has a unit gain in a steady state, and the gains κ_p, κ_d, can be used to arbitrarily assign the poles of the system. Thus:

- If no disturbance is present ($T_p = 0$ and $\dot{\xi}(t) = 0$ for all $t \geq 0$), the response in the position is as the response of a second-order system with rise time and overshoot that can be arbitrarily assigned by suitably selecting κ_p and κ_d. The position θ reaches the desired position θ_d (constant) in a steady state.
- If a constant disturbance appears ($T_p = t_d \neq 0$), the deviation that it produces vanishes as time increases. Moreover, if a larger k_1 is chosen (such that the product $k_1 k > 0$ is larger), then the deviation due to the disturbance vanishes faster. If, once the deviation due to a disturbance vanishes (once $\theta = \theta_d$), a step change in the desired position θ_d is commanded (i.e., when the disturbance T_p is still present), then the position response is identical to that obtained when $T_p = 0$, i.e., the rise time and overshoot are identical as before and the position θ reaches its desired value θ_d in a steady state.

As all initial conditions are assumed to be zero in classical control, then $\theta(0) = 0$, $\dot{\theta}(0) = 0$, can be assumed and the controller in (11.22) becomes:

$$i^* = \kappa_p(\theta_d - \theta) - \kappa_d \dot{\theta} + k_1 \kappa_p k \int_0^t (\theta_d - \theta(r)) dr - k_1 \dot{\theta} - k_1(a + \kappa_d k)\theta,$$

$$= [\kappa_p + k_1(a + \kappa_d k)](\theta_d - \theta) - (\kappa_d + k_1)\dot{\theta} + k_1 \kappa_p k \int_0^t (\theta_d - \theta(r)) dr$$

$$\quad - k_1(a + \kappa_d k)\theta_d,$$

$$= K_P(\theta_d - \theta) - K_D \dot{\theta} + K_I \int_0^t (\theta_d - \theta(r)) dr - k_1(a + \kappa_d k)\theta_d, \quad (11.25)$$

$$K_P = \kappa_p + k_1(a + \kappa_d k), \quad K_D = \kappa_d + k_1, \quad K_I = k_1 \kappa_p k,$$

which constitutes a simple PID controller with a constant feedforward term. Note that the derivative term does not contain a time derivative of the desired position. Finally, the tuning rule of the controller in (11.25) is summarized as follows:

- $\kappa_p > 0$ and $\kappa_d > 0$ are chosen to fix the desired rise time and overshoot by suitably assigning the roots of the characteristic polynomial $s^2 + (a + \kappa_d k)s + \kappa_p k$, i.e.,

$$a + \kappa_d k = 2\zeta\omega_n, \quad \kappa_p k = \omega_n^2. \tag{11.26}$$

- $k_1 > 0$ is chosen large for a fast disturbance rejection. This can be done by trial and error or it can be computed recalling that $\frac{1}{k_1 k}$ is the time constant of the filter in (11.24), which is responsible for the disturbance rejection.

Another, better known, way of designing a PID controller with the same advantages is by using two-degrees-of-freedom controllers. These is presented in the next section.

11.3.2   A Two-Degrees-of-Freedom Controller

The block diagram of a two-degrees-of-freedom controller is shown in Fig. 11.9. $G_p(s)$ stands for the plant whereas the controller is composed of two parts: $G_{c1}(s)$ and $G_{c2}(s)$. In Example 4.5, Sect. 4.1, it is shown that the closed-loop system output is given as:

$$\theta(s) = G_1(s)\theta_d(s) + G_2(s)D(s),$$

where:

$$G_1(s) = \frac{G_{c1}(s)G_p(s)}{1 + (G_{c1}(s) + G_{c2}(s))G_p(s)},$$

$$G_2(s) = \frac{G_p(s)}{1 + (G_{c1}(s) + G_{c2}(s))G_p(s)}, \tag{11.27}$$

are the transfer functions defined as:

$$G_1(s) = \frac{\theta(s)}{\theta_d(s)}, \quad \text{when } D(s) = 0,$$

$$G_2(s) = \frac{\theta(s)}{D(s)}, \quad \text{when } \theta_d(s) = 0.$$

The reason why this control scheme is called *two-degrees-of-freedom-controller* is because the transfer functions defined in (11.27) can be tuned independently from each other using for that, as independent variables (or degrees-of-freedom), the two controller components $G_{c1}(s)$ and $G_{c2}(s)$. It is desired to design $G_{c1}(s)$ and $G_{c2}(s)$ such that: (*i*) The transient and steady-state response specifications to a position reference $\theta_d(s)$ are satisfied. (*ii*) The effect of an external disturbance, $D(s)$, is taken to zero as fast as desired.

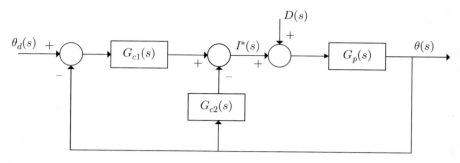

Fig. 11.9 Closed-loop system with a two-degrees-of-freedom controller

As the plant under study is a PM brushed DC motor, assume that:

$$G_p(s) = \frac{k}{s(s+a)}. \tag{11.28}$$

The controller design presented in this section assumes that $G_{c1}(s)$ and $G_{c2}(s)$ are PID controllers [5], chapter 10; hence:

$$G_{c1}(s) = \frac{k_{d1}s^2 + k_{p1}s + k_{i1}}{s},$$

$$G_{c2}(s) = \frac{k_{d2}s^2 + k_{p2}s + k_{i2}}{s},$$

$$G_c(s) = G_{c1}(s) + G_{c2}(s) = \frac{k_d s^2 + k_p s + k_i}{s},$$

$$= \frac{k_d(s+\alpha)(s+\beta)}{s} = \frac{k_d(s^2 + (\alpha+\beta)s + \alpha\beta)}{s}, \tag{11.29}$$

$$k_d = k_{d1} + k_{d2}, \quad k_p = k_{p1} + k_{p2}, \quad k_i = k_{i1} + k_{i2},$$

for some constants α and β. Thus:

$$\frac{\theta(s)}{\theta_d(s)} = \frac{G_{c1}(s)G_p(s)}{1 + G_c(s)G_p(s)}.$$

Replacing (11.28) and (11.29) in the latter expression:

$$\frac{\theta(s)}{\theta_d(s)} = \frac{G_{c1}(s)\frac{k}{s(s+a)}}{1 + \frac{k_d(s^2+(\alpha+\beta)s+\alpha\beta)}{s}\frac{k}{s(s+a)}},$$

$$= \frac{skG_{c1}(s)}{s^2(s+a) + k_d k(s^2 + (\alpha+\beta)s + \alpha\beta)},$$

$$= \frac{skG_{c1}(s)}{s^3 + (a + k_d k)s^2 + k_d k(\alpha + \beta)s + k_d k\alpha\beta}. \tag{11.30}$$

Suppose that the desired transient response is described by two dominant complex conjugate poles given as $s = -\sigma + j\omega_d$ and $s = -\sigma - j\omega_d$, $\sigma > 0$, $\omega_d > 0$. Then, the characteristic polynomial must satisfy:

$$s^3 + (a + k_d k)s^2 + k_d k(\alpha + \beta)s + k_d k\alpha\beta$$
$$= (s + \sigma + j\omega_d)(s + \sigma - j\omega_d)(s + f),$$
$$= s^3 + (2\sigma + f)s^2 + (\sigma^2 + \omega_d^2 + 2\sigma f)s + (\sigma^2 + \omega_d^2)f,$$

where the following must be satisfied:

$$a + k_d k = 2\sigma + f, \tag{11.31}$$
$$k_d k(\alpha + \beta) = \sigma^2 + \omega_d^2 + 2\sigma f, \tag{11.32}$$
$$k_d k\alpha\beta = (\sigma^2 + \omega_d^2)f. \tag{11.33}$$

To avoid the pole in $s = -f$, $f > 0$, to have an effect on the transient response of (11.30), a transfer function such as the following is proposed:

$$\frac{\theta(s)}{\theta_d(s)} = \frac{\rho(s + f)}{s^3 + (2\sigma + f)s^2 + (\sigma^2 + \omega_d^2 + 2\sigma f)s + (\sigma^2 + \omega_d^2)f}, \tag{11.34}$$

where, to obtain a transfer function with a unit gain in a steady state, the following must be satisfied:

$$\rho f = (\sigma^2 + \omega_d^2)f,$$

i.e.,

$$\rho = \sigma^2 + \omega_d^2. \tag{11.35}$$

On the other hand, to render the transfer functions in (11.30) and (11.34) equal, the following must be imposed:

$$skG_{c1}(s) = \rho(s + f),$$

where:

$$G_{c1}(s) = \frac{\rho s + \rho f}{sk} = k_{p1} + k_{i1}\frac{1}{s}, \tag{11.36}$$

$$k_{p1} = \frac{(\sigma^2 + \omega_d^2)}{k}, \quad k_{i1} = \frac{(\sigma^2 + \omega_d^2)f}{k}.$$

This means that $G_{c1}(s)$ must be a proportional–integral (PI) controller. The controller $G_{c2}(s)$ is obtained solving (11.29):

$$G_{c2}(s) = G_c(s) - G_{c1}(s),$$

$$= k_d s + k_d(\alpha + \beta) + k_d \alpha \beta \frac{1}{s} - \frac{(\sigma^2 + \omega_d^2)}{k} - \frac{(\sigma^2 + \omega_d^2)f}{k}\frac{1}{s},$$

$$= k_{d2}s + k_{p2}, \quad k_{p2} = \frac{2\sigma f}{k}, \quad k_{d2} = k_d = \frac{2\sigma + f - a}{k}, \quad (11.37)$$

where (11.31), (11.32), (11.33), have been used. This means that $G_{c2}(s)$ is a PD controller. Thus, according to Fig. 11.9, the controller is given as:

$$I^*(s) = G_{c1}(s)(\theta_d(s) - \theta(s)) - G_{c2}(s)\theta(s),$$

$$= (k_{p1} + k_{i1}\frac{1}{s})(\theta_d(s) - \theta(s)) - (k_{d2}s + k_{p2})\theta(s),$$

$$= (k_{p1} + k_{p2})(\theta_d(s) - \theta(s)) - k_{d2}s\theta(s) + k_{i1}\frac{1}{s}(\theta_d(s) - \theta(s)) - k_{p2}\theta_d(s).$$

Applying the inverse Laplace transform:

$$i^* = (k_{p1} + k_{p2})(\theta_d - \theta) - k_{d2}\dot{\theta} + k_{i1}\int_0^t (\theta_d - \theta(r))dr - k_{p2}\theta_d. \quad (11.38)$$

As in the previous sections, it is desired that the response to a position step command has a rise time of 0.068[s] and a 20% overshoot, which corresponds to closed-loop poles at $s = -\sigma \pm j\omega_d$, where:

$$\sigma = 15.4009, \quad \omega_d = 30.0623. \quad (11.39)$$

On the other hand, the value of f does not affect the transient response to a desired reference because its effect is canceled (see (11.34)). What is then the effect of f and how to compute it? The key to an answer is the following:

- According to (11.31), once σ, a, k, are fixed, the derivative gain k_d is larger if f is larger. According to the previous sections, large derivative gains are not desirable because of noise amplification. Hence, a criterion is to select f as small as possible.
- The transfer function in (11.27) is computed as:

$$\frac{\theta(s)}{D(s)} = \frac{ks}{(s + \sigma + j\omega_d)(s + \sigma - j\omega_d)(s + f)}.$$

Note that the pole at $s = -f$ cannot be canceled in this transfer function, i.e., its effect is very important in the transient response to an external disturbance.

Hence, the smaller f is, the slower the disturbance effect vanishes. In this respect, note that the final value theorem can be used to verify that the zero that this transfer function has at $s = 0$ ensures that the effect of a constant disturbance vanishes in a steady state.

Thus, f must be selected such that a trade-off between a fast disturbance rejection and noise attenuation is satisfied. Propose:

$$f = 40. \tag{11.40}$$

Using the plant parameters:

$$k = 675.4471, \quad a = 2.8681,$$

and (11.36), (11.37), (11.39), (11.40), the following is found:

$$G_{c1}(s) = 1.6891 + 67.5659\frac{1}{s}, \tag{11.41}$$

$$G_{c2}(s) = 0.1006s + 1.8241. \tag{11.42}$$

According to (11.36), (11.37) and (11.38), the controller is given as:

$$i^* = 3.5132(\theta_d - \theta) - 0.1006\dot{\theta} + 67.5659 \int_0^t (\theta_d - \theta(r))dr - 1.8241\theta_d. \tag{11.43}$$

One last observation. The reader can verify that the controller in (11.38) is identical to the controller in (11.25) if (11.26) is satisfied and the following is established:

$$k_1 = \frac{f}{k}, \quad \sigma = \zeta\omega_n, \quad \sigma^2 + \omega_d^2 = \omega_n^2.$$

It is up to the reader to decide which of the approaches presented in Sects. 11.3.1 or 11.3.2 is preferred.

Some simulation results are presented in Fig. 11.10 when using the controller in (11.43) and the motor parameters $k = 675.4471$, $a = 2.8681$. There, a step change in the desired position is commanded and $t_r = 0.068[s]$, $M_p(\%) = 19.6\%$, are measured, which are almost identical to the values specified above, i.e., $t_r = 0.068[s]$ and $M_p(\%) = 20\%$. A constant disturbance is also considered $D(s) = -T_p(s)/(Jk) = -0.5/s$. This disturbance is introduced via software as a signal that adds to i^*, given in (11.43), from $t = 0.7[s]$ on. Note that the steady-state error is zero. On the other hand, a very good response is observed when the external disturbance appears.

Some other simulation results are shown in Fig. 11.11 when using the controller (11.43). In this case, the following desired position is commanded:

$$\theta_d = \begin{cases} 1.5, & 0 \leq t < 1.5 \\ 3, & 1.5 \leq t < 4 \\ 1.5, & t \geq 4 \end{cases} . \tag{11.44}$$

The following external disturbance is also applied (via software):

$$d(t) = \begin{cases} 0, & 0 \leq t < 0.7 \\ 0.5, & 0.7 \leq t < 2.55 \\ 0, & 2.55 \leq t < 3.25 \\ 0.5, & 3.25 \leq t < 5.1 \\ 0, & 5.1 \leq t < 5.8 \\ 0.5, & t \geq 5.8 \end{cases} . \tag{11.45}$$

Comparing Figs. 11.10 and 11.11, it is clear that the desired transient and steady-state response specifications are satisfied. Moreover, note that the step changes in the reference commanded at $t = 1.5$[s] and $t = 4$[s] occur when the disturbance $d(t)$ is present and, despite this, the transient response specifications are still satisfied. Also note that the deviations due to the disturbances vanish very fast.

Some experimental results are presented in Fig. 11.12 under the same conditions as in Fig. 11.10. The following values are measured $t_r = 0.068$[s] and $M_p(\%) = 19.3\%$. Note that these experimental results are very similar to the simulation results in Fig. 11.10. It is stressed that the very good response observed when an external disturbance appears is thanks to $f = 40$, and that such a large value is possible because of the low noise content in the measured position θ.

Some other experimental results are shown in Fig. 11.13 that were performed under the same conditions as the simulations in Fig. 11.11. Aside from the measured position (continuous line) the response obtained in simulation (dashed line) is also shown in Fig. 11.13. The simulation response is obtained with the second-order system without disturbances $\ddot{\theta} + 30.8018\dot{\theta} + 1140.9\theta = 1140.9\theta_d$, which satisfies the design specifications $t_r = 0.068$[s] and $M_p(\%) = 20\%$, with poles located at $s = -15.4009 \pm j30.0623$. It is observed that both responses are almost identical and they cannot be distinguished from each other in Fig. 11.13. This shows that the desired transient and steady-state response specifications are satisfied in the experiment. These experimental results corroborate the observations stated just before (11.25).

The simulations in Figs. 11.10 and 11.11 were performed using the MAT-LAB/Simulink diagram shown in Fig. 11.14 and the following steps: (1) Execute the following MATLAB code in an m-file:

```
a=2.8681;
k=675.4471;
f=40;
Mp=20; %
tr=0.068; % sec
```

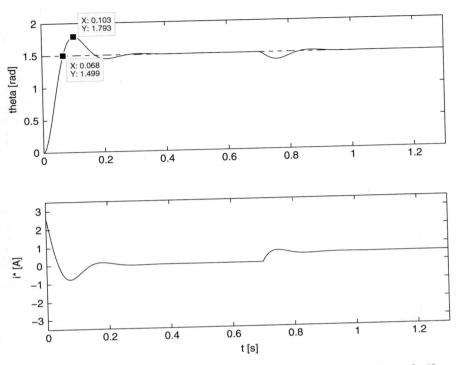

Fig. 11.10 Position control using a two-degrees-of-freedom controller. Simulation results ($\theta_d = 1.5$[rad])

```
z=sqrt( log(Mp/100)^2/( log(Mp/100)^2 + pi^2) );
wn=1/(tr*sqrt(1-z^2))*(pi-atan(sqrt(1-z^2)/z));
sigma=z*wn;
wd=wn*sqrt(1-z^2);
kp1=(sigma^2+wd^2)/k;
ki1=(sigma^2+wd^2)*f/k;
kp2=2*sigma*f/k;
kd2=(2*sigma+f-a)/k;
kp=kp1+kp2
kd=kd2
ki=ki1
D=0.5;
```

(2) Run the simulation in Fig. 11.14. (3) Run the following MATLAB code in an m-file:

```
Ts=0.001;
t=0:Ts:6.5;
figure(1)
```

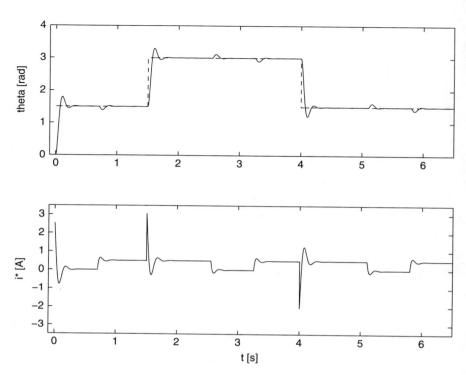

Fig. 11.11 Position control with a two-degrees-of-freedom controller. Simulation results

```
subplot(2,1,1)
plot(t,Datos(:,1),'b--');
hold on
plot(t,Datos(:,2),'b-');
hold off
axis([-0.1 6.5 0 4])
ylabel('theta [rad]')
subplot(2,1,2)
plot(t,Datos(:,3),'b-');
axis([-0.1 6.5 -3.5 3.5])
xlabel('t [s]')
ylabel('i* [A]')

figure(2)
subplot(2,1,1)
plot(t,Datos(:,1),'b--');
hold on
plot(t,Datos(:,2),'b-');
hold off
```

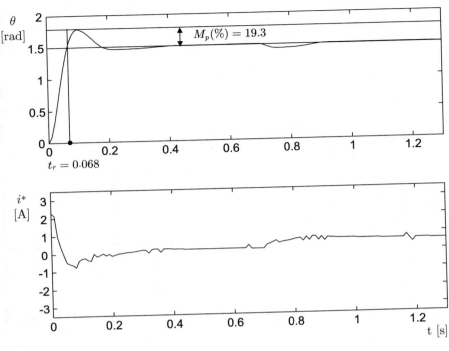

Fig. 11.12 Position control using a two-degrees-of-freedom controller. Experimental results ($\theta_d = 1.5[\text{rad}]$)

```
axis([0 1.3 0 2])
ylabel('theta [rad]')
subplot(2,1,2)
plot(t,Datos(:,3),'b-');
axis([0 1.3 -3.5 3.5])
xlabel('t [s]')
ylabel('i* [A]')
```

The blocks in the MATLAB/Simulink diagram shown in Fig. 11.14 were programmed as follows. Three theta step blocks and five step D blocks were suitably programmed to generate the desired position in (11.44) and the disturbance in (11.45). The To workspace block saves data as an array with 0.001 as the sample time.

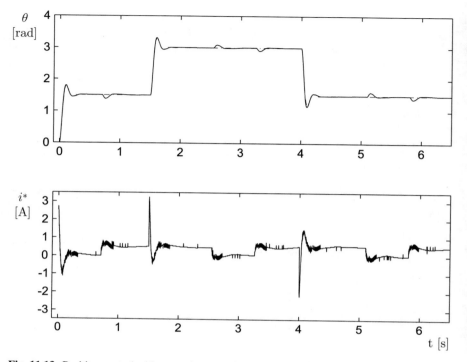

Fig. 11.13 Position control with a two-degrees-of-freedom controller. Experimental results

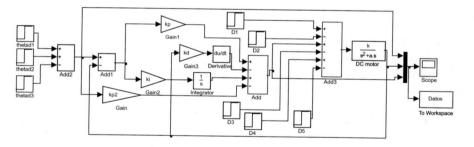

Fig. 11.14 MATLAB/Simulink diagram used for simulations in Figs. 11.10 and 11.11

11.3.3   A Classical PID Controller

To appreciate the advantages of the controller designed in Sects. 11.3.1 and 11.3.2, some experimental results are presented in Figs. 11.15 and 11.16, when using the following classical PID controller:

$$i^* = k_p(\theta_d - \theta) + k_d \frac{d}{dt}(\theta_d - \theta) + k_i \int_0^t (\theta_d - \theta(r))dr.$$

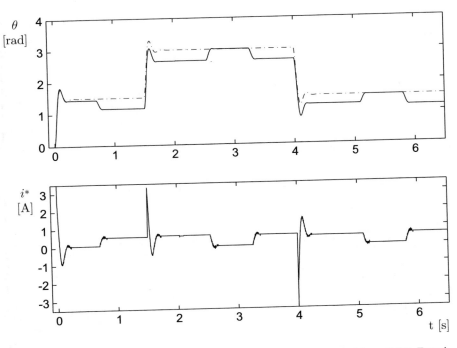

Fig. 11.15 Classical proportional–integral–derivative (PID) position control ($\alpha = 0.01$). Experimental results

The desired position and the external disturbance are the same used in Fig. 11.13, i.e., those defined in (11.44) and (11.45). The measured position (continuous line) is presented in addition to the response obtained in the simulation (dashed line). The simulation response is, again, obtained from the system without disturbances $\ddot{\theta} + 30.8018\dot{\theta} + 1140.9\theta = 1140.9\theta_d$, which achieves $t_r = 0.068[s]$ and $M_p(\%) = 20\%$, with poles located at $s = -15.4009 \pm j30.0623$.

The classical PID controller design is performed according to Sect. 5.2.5. Using the plant parameters:

$$k = 675.4471, \quad a = 2.8681,$$

the desired closed-loop poles located at:

$$s = -15.4009 \pm j30.0623,$$

to achieve 0.068[s] as the rise time and 20% overshoot, the design parameter $\alpha = 0.01$, and (5.30), (5.32), the following controller gains are obtained:

$$k_d = 0.0414, \quad k_p = 1.6896, \quad k_i = 0.0169.$$

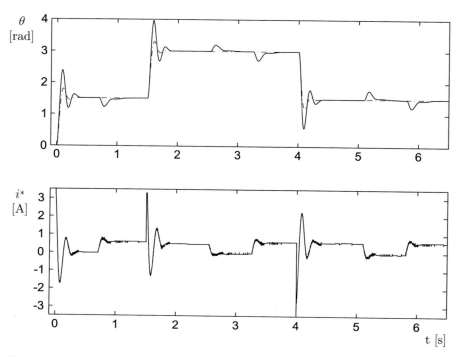

Fig. 11.16 Classical PID position control ($\alpha = 10$). Experimental results

It is important to stress that, according to the design criterion presented in Sect. 5.2.5, $\alpha = 0.01$ is chosen very close to zero to ensure that the transient response satisfies the desired specifications. As observed in Fig. 11.15, this is achieved for the first step change in the reference, i.e., before a disturbance appears. However, problems arise when an external disturbance is applied:

The use of a small α results in a very small integral gain, which produces such slow integral adjustments that the disturbance effect is not compensated for in Fig. 11.15. Moreover, although the measured position reaches the desired position at $t \approx 3[s]$ and $t \approx 5.5[s]$, this is because the external disturbance is not present at those points in time.

As explained in Sect. 5.2.5, larger integral gains can be obtained if α is chosen to be large. Thus, $\alpha = 10$ is proposed, which results in the following controller gains:

$$k_d = 0.0414, \quad k_p = 2.1027, \quad k_i = 16.8915.$$

The corresponding experimental results are shown in Fig. 11.16. Note that the deviation due to the disturbance varnishes very fast. However, the price to pay for this improvement is that the response to a step change in the desired position does not satisfy the desired specifications. This is the main drawback of classical PID control: there is no tuning rule at the disposal of the designer to compute controller gains that simultaneously achieve the desired transient response specifications to a desired reference change and a satisfactory disturbance rejection. On the contrary, this problem is solved by the controllers designed in Sects. 11.3.1 and 11.3.2 whose experimental results are shown in Fig. 11.13.

11.4 Trajectory Tracking

In this section, a controller is designed that allows the motor position to reach a time-varying desired position, i.e., $\theta_d(t)$ is no longer constant, but changes over time. It is assumed that no external disturbance exists, i.e., $T_p(s) = 0$, and the first and second time derivatives of $\theta_d(t)$ can be computed, i.e., $\dot{\theta}_d(t)$ and $\ddot{\theta}_d(t)$ are known and they are continuous. Consider the motor model in (11.1) once the inverse Laplace transform is applied with $T_p(s) = 0$, i.e.,

$$\ddot{\theta} + a\dot{\theta} = ki^*. \tag{11.46}$$

Also consider the following controller, where θ_d, $\dot{\theta}_d$ and $\ddot{\theta}_d$ are known:

$$i^*(t) = \frac{1}{k}\left[\frac{d^2\theta_d(t)}{dt^2} - k_d\left(\frac{d\theta(t)}{dt} - \frac{d\theta_d(t)}{dt}\right) - k_p\left(\theta(t) - \theta_d(t)\right) + a\frac{d\theta(t)}{dt}\right].$$
$$\tag{11.47}$$

Note that $\frac{d\theta(t)}{dt}$ stands for the measured motor velocity. Replacing (11.47) in (11.46), the following is found:

$$\frac{d^2\theta(t)}{dt^2} - \frac{d^2\theta_d(t)}{dt^2} + k_d\left(\frac{d\theta(t)}{dt} - \frac{d\theta_d(t)}{dt}\right) + k_p\left(\theta(t) - \theta_d(t)\right) = 0.$$
$$\tag{11.48}$$

Define the tracking error and its Laplace transform as $\varepsilon(t) = \theta(t) - \theta_d(t)$ and $E(s)$ respectively. Then, (11.48) can be written as:

$$s^2 E(s) - s\varepsilon(0) - \frac{d\varepsilon(0)}{dt} + k_d s E(s) - k_d \varepsilon(0) + k_p E(s) = 0, \tag{11.49}$$

and finally:

$$E(s) = \frac{(k_d + s)\varepsilon(0) + \frac{d\varepsilon(0)}{dt}}{s^2 + k_d s + k_p}. \tag{11.50}$$

According to Chap. 3, the solution of (11.50) has the form:

$$\varepsilon(t) = \varepsilon_n(t) + \varepsilon_f(t)$$

where the forced response is zero $\varepsilon_f(t) = 0$ because the differential equation in (11.49) is not excited. Hence, $\varepsilon(t) = \varepsilon_n(t)$, which tends toward zero as time grows if the roots of the characteristic polynomial $s^2 + k_d s + k_p$ have a negative real part. This situation is desirable because if $\varepsilon(t)$ tends toward zero, then $\theta(t)$ tends toward $\theta_d(t)$ as time grows. On the other hand, as the characteristic polynomial $s^2 + k_d s + k_p$ is second-degree, according to Sect. 4.2, its two roots have a negative real part if, and only if, all the coefficients of this polynomial have the same sign, i.e., if, and only if, $k_p > 0$ and $k_d > 0$. This is the criterion for selecting the gains for this controller. However, it is important to stress that the controller gains achieving a good performance in practice must be selected within this set of gains by trial and error. Some experimental results are shown in Fig. 11.17 when using a motor with $k = 25.26$, $a = 3.5201$, together with the controller in (11.47), the controller gains:

$$k_p = 363.88, \quad k_d = 22,$$

and the desired trajectory:

$$\theta_d(t) = 0.8 \sin(5t) + 0.5 \cos(3t) + 3.0. \tag{11.51}$$

Note that θ reaches the desired position $\theta_d(t)$ despite these variables having different initial values. It is important to state that, as a mathematical expression is known for θ_d, which is given in (11.51), the corresponding expressions for $\dot{\theta}_d$ and $\ddot{\theta}_d$ are computed analytically by differentiation. This avoids problems due to noise amplification.

11.5 Prototype Construction

The electrical diagram of the complete control system is presented in Fig. 11.18. The main components are the following:

- PM brushed DC motor. Nominal voltage 20[V], nominal current 3[A]. Provided with an optical encoder with 400 ppr, which requires +5 DC voltage.
- A portable computer with USB port.
- USB-series adapter.
- Microchip PIC16F877A Microcontroller.
- DAC0800LCN Digital/analog converter.
- TL084 quadruple operational amplifier.
- TL081 operational amplifier.
- MAX232 driver.
- TIP141 and TIP145 transistors.

Microcontroller PIC16F877A computes the control algorithm. Once the desired position θ_d and the measured position θ are known, the microcontroller computes the control signal i^*. This value is assigned to the variable "iast" according to program listed in Sect. 11.6.

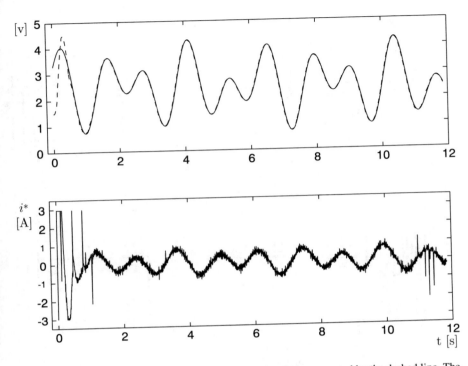

Fig. 11.17 Trajectory tracking. Motor measured position θ is represented by the dashed line. The continuous line represents $\theta_d(t)$

As indicated in Fig. 11.18, $-i^*$ must appear at the input of the operational amplifier TL081 inside the block labeled "current control." This is accomplished as follows. First, the program listed in Sect. 11.6 assigns iast=-iast to change the sign of i^*. The microcontroller must deliver an 8-bit digital code to the digital/analog converter DAC0800LCN. With this aim, the microcontroller computes:

$$\text{iastd} = 36.4286 * \text{iast} + 127,$$

which corresponds to the relation shown in Fig. 11.19. The digital/analog converter works together with an operational amplifier TL081. These devices are connected according to the manufacturer's specifications [1]. This ensures that, at the operational amplifier output, a voltage appears whose numerical value corresponds to $-i^*$. This value is received by an operational amplifier TL081, which evaluates the electric current controller [6, 7]:

$$u_i = 100(i^* - i).$$

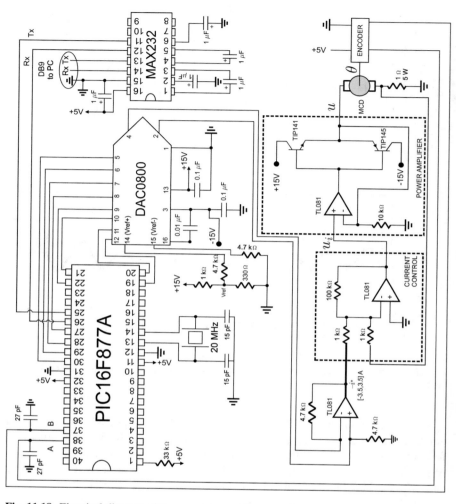

Fig. 11.18 Electrical diagram of the complete position control system based on a PIC16F877A microcontroller

Finally, a power amplifier with a unit gain is implemented with a third operational amplifier TL081 and two power transistors TIP141 and TIP145 in a complementary symmetry connection.

The driver MAX232 sends the measured position θ and the control signal i^* to a portable computer (through an USB series adapter) whose unique task is to plot these variables. In the following, it is described how the microcontroller PIC16F877A performs the controller task. It is stressed that the following exposition is not intended to present deep explanations of how the microcontroller works. We are merely presenting a general description of the microcontroller components involved in the specific task to be performed. The program employed by the microcontroller is listed in the next section.

Fig. 11.19 Conditioning of
signal i^* that must be sent to
the digital/analog converter

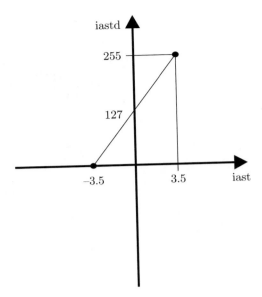

The microcontroller PIC16F877A [2] has a maximal clock frequency of 20
MHz, it has five input–output configurable ports and three timers. Timer TMR0 is
employed in this application to fix the sampling period: $T = 0.001[s]$ (20 counts of
timer 0) for experiments in Sects. 11.1, 11.2.1, and 11.2.2, whereas $T = 0.002[seg]$
(40 counts of timer 0) for the experiment in Sect. 11.3.2.

The motor position θ is measured from the encoder using a subroutine and it
is stored in the 16-bit variable "cuenta2." The position in radians is obtained by
computing pos=cuenta2*0.0039 where 0.0039=3.1416 rad/(2*400 cuentas). This
is because the encoder has 400 ppr, but 4×400 counts per revolution are really
obtained (each 2π radians), i.e., 2×400 counts each π radians. Once the control
signal "iast" is computed, it is delivered to the digital/analog converter through port
D (PORTD=iastd). Finally, either the motor position (through the variable "cuenta")
or the control signal (through the variable "iastd") is sent to the portable computer
through the instructions: putc(0xAA); putc(cuentaH); putc(cuentaL). The electric
current control and the power amplifier are implemented identically as described in
Sects. 10.6.1 and 10.6.2 respectively.

11.6   Microcontroller PIC16F877A Programming

The flow diagram for the programming of microcontroller PIC16F877A when
controlling the position in a PM brushed DC motor is shown in Fig. 11.20.
The complete code is listed below. The reader is referred to [3] for a complete
explanation of each of the instructions appearing in the following program.

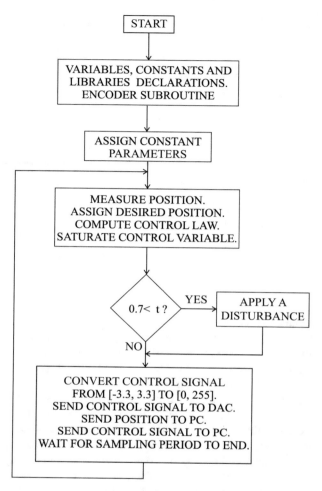

Fig. 11.20 Flow diagram for the programming of microcontroller PIC16F877A when controlling the position in a permanent magnet (PM) brushed DC motor

```
// PM brushed DC motor position control program
#include<16f877A.h>
#device adc=10 //ADC, 10 bits
#include<stdlib.h>
#include<math.h>

#fuses HS,NOWDT,PUT,NOBROWNOUT,NOLVP,NOWRT,NOPROTECT,NOCPD
#usedelay(clock=20000000) //time basis

#use
rs232(baud=115200,XMIT=PIN_C6,RCV=PIN_C7,BITS=8,PARITY=N)
//Config.
```

```
//serial port
//ports and registers addresses
#byte OPTION= 0x81
#byte TMR0 = 0x01
#byte PORTA = 0x05
#byte PORTB = 0x06
#byte PORTC = 0x07
#byte PORTD = 0x08
#byte PORTE = 0x09
#bit PC0 = 0x07.0
#bit PC1 = 0x07.1
//-----------Variables declaration------------//
int16 inter,cuenta;
int8 cuentaH,cuentaL,puerto,AB,AB_1,aux,cont,iastd,cont2,
cont3,cont4;
float pos,error,iast,cuenta2,posm1,kp,kv,c,delta,v,kp1,ki1,
kp2,kd2,ie;
unsigned int u;
//int1 ban;
//-----------Encoder interrupt subroutine------------//
#int_rb void rb_isr() {
puerto=PORTB;
AB=((puerto)&(0x30))>>4;
aux=AB^AB_1;
if(aux!=0)
if(aux!=3)
if(((AB_1<<1)^AB)&(0x02))
cuenta--;
else
cuenta++;
AB_1=AB;
}
//-----------Main program------------//
void main(void) {
setup_adc(ADC_CLOCK_INTERNAL ); //ADC (internal clock)
set_tris_a(0b11111111);
set_tris_b(0b11111111);
set_tris_c(0b10000000); //serial communication
set_tris_d(0b00000000);
set_tris_e(0b11111111);
OPTION=0x07; //pre_scaler timer0, 1:256
PORTC=0;
TMR0=0;
cuenta=0;
AB=0;
AB_1=0;
enable_interrupts(global);
enable_interrupts(int_rb);
cont=4;
cont2=0;
cont3=0;
cont4=0;
PORTD=127; // iast=0 is set
kp=1.6891;
```

```
kv=0.0414;
c=30.8018;
delta=1.6891;
v=0.0;
kp1=1.6891;
ki1=67.5659;
kp2=1.8241;
kd2=0.1006;
posm1=0.0;
ie=0.0;
while(cont2<3) // a 10-s waiting period is introduced
{
while(cont3<255)
{
TMR0=0;
while(TMR0<255)
{
}
cont3++;
}
cont2++;
}
TMR0=0;
while(TRUE) // control loop
{
cont++;
cuenta2=(signed int16)cuenta;
pos=cuenta2*0.0039; // 3.1416 rad/(2*400 cuentas)
error=1.5-pos;
/* Identification */
// iast=0.5*error;
/* _____ */

/* Proportional control plus velocity feedback */
// iast=kp*error-kv*(pos-posm1)/0.001;
/* _____ */

/* Lead compensator */
// iast=delta*(error+v);
// v=(-c*v+(2.8-c)*error)*0.001+v;
/* _____ */

/* 2-degrees-of-freedom controller */
iast=kp1*error+ki1*ie-kp2*pos-kd2*(pos-posm1)/0.002;
ie=ie+0.002*error;
/* _____ */

if(iast<-3.3) // iast is constrained to [-3.3,+3.3] amperes
iast=-3.3; // (DAC delivers within the range [-3.5,+3.5])
if(iast>3.3)
iast=3.3;
iast=-iast; //-i*, TL081 at ''current control''
if(cont4>70) //time>0.7 s, disturbance (-0.5)
iast=iast+0.5;
```

```
iastd=(unsigned int)(36.4286*iast+127);
PORTD=iastd;
posm1=pos;
/* // uncomment if it is desired to send position to computer
// every 10 ms
if(cont==5)
{
cont=0;
inter=(cuenta)&(0xFF00);
cuentaH=inter>>8;
cuentaL=(cuenta)&(0x00FF);
putc(0xAA); //recognition serial port
putc(cuentaH); //sending to serial port
putc(cuentaL);
cont4++;
if(cont4==255)
cont4=0;
}
*/
/* // uncomment if it is desired to send control signal to
// computer every 10 ms
iast=-iast; // plot +i*
if(cont4>70) // plot control signal without disturbance
iast=iast+0.5;
iastd=(unsigned int)(36.4286*iast+127);
if(cont==5)
{
cont=0;
inter=(cuenta)&(0xFF00);
cuentaH=0x00; //inter>>8;
cuentaL=(iastd)&(0xFF);
putc(0xAA); //recognition serial port
putc(cuentaH); //sending to serial port
putc(cuentaL);
cont4++;
if(cont4==255)
cont4=0;
}
*/
PC1=1; //waiting for sampling time
while(TMR0<40) //each cuenta=(4/FXtal)*256 s,
// (40=2 ms, sampling period)
{
}
PC1=0; //sampling time has arrived
TMR0=0;
} //closing the infinite while
} //closing main
```

11.7   Personal Computer-Based Controller Implementation

The controller presented in Sect. 11.4, for trajectory tracking, was experimentally implemented using a personal computer and a data acquisition board. The reason for this is that the sine and cosine trigonometric functions involved in the controller are time-consuming for the microcontroller PIC16F877A. This problem disappears when a personal computer is employed. The following are the main control system components.

- Personal computer Pentium II, 233 MHz
- Data acquisition board Advantech PCL-812PG [4]. Fifteen analog/digital input channels with a unique 12-bit successive approximations analog/digital converter, with an analog range $[-5, +5][V]$. Two digital/analog output channels, each based on a 12-bit digital/analog converter with analog range $[0, +5][V]$. It is also provided with two programmable timers employed to fix the sampling period. In Fig. 11.17, the sampling period was set to de 0.005[s].
- PM brushed DC motor. Nominal voltage 24[V], nominal current 2.3[A].
- Position sensor. 1[KOhm] Precision potentiometer, 10 turns.

The block diagram of the complete control system is shown in Fig. 11.21.

11.7.1   Input Interface

The potentiometer used as a position sensor, Fig. 11.21, delivers a voltage θ in the range $[0, +5][V]$. This is very convenient as the analog/digital converter has an analog range $[-5, +5][V]$. From the voltage at the potentiometer, the analog/digital converter delivers a code, "$code_i$," within the range 0 to $2^n = 4096$ because it is a 12-bit converter. From this code, the variable u_a is retrieved, which must be numerically identical to the voltage at the potentiometer θ. According to Fig. 11.22, this can be performed by computing:

$$u_a = \frac{10.0}{4096.0} code_i - 5.0.$$

This must be performed by the computer.

11.7.2   Output Interface

According to (11.1) the control signal is the current command $-i^*$, which, depending on the motor specifications, may take values in the range -3[A] to +3[A]. $-i^*$ is sent by the computer to the data acquisition board, which must deliver $-i^*$ as a voltage signal.

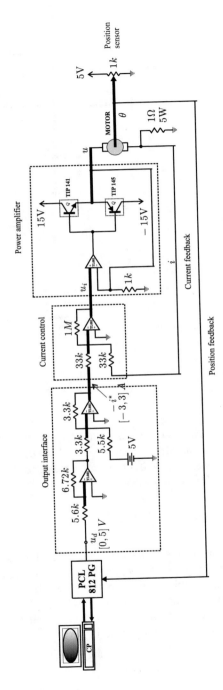

Fig. 11.21 Block diagram and interfaces of the personal computer-based position control system

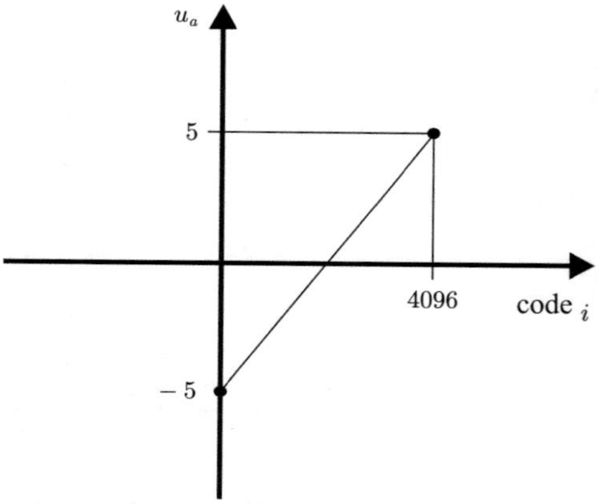

Fig. 11.22 Conditioning of
the position signal delivered
by the analog/digital
converter

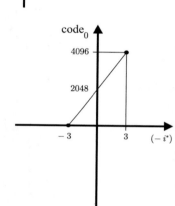

Fig. 11.23 Conditioning of
the signal $-i^*$, which is to be
sent to the digital/analog
converter

Once the computer obtains $-i^*$, as a result of the corresponding control
algorithm evaluation, a set of instructions constrains $-i^*$ to the range $[-3, +3][V]$.
After that, taking into account that the digital/analog converter is 12-bit, Fig. 11.23
is drawn to realize that the following datum:

$$\text{code}_o = \frac{4096}{6}(-i^*) + 2048,$$

has to be sent to the digital/analog converter. Hence, the digital/analog converter
delivers a voltage u_d in the range $0[V]$ to $+5[V]$. Thus, an analog electronic circuit
must be employed to convert this voltage in the range $0[V]$ to $+5[V]$ into the
range $-3[V]$ to $+3[V]$, which corresponds to the values taken by $-i^*$. According
to Fig. 11.24 this is accomplished by an operational amplifier-based analog circuit
performing the following mathematical operation:

$$-i^* = \frac{6}{5}u_d - 3.$$

Fig. 11.24 Design of the analog interface that must be placed at the output of the digital/analog converter

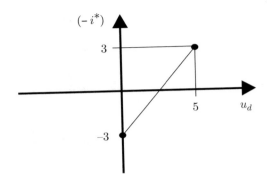

This function is performed by the block labeled "output interface" in Fig. 11.21.[1]

The operational amplifier shown in the block labeled "current control" in Fig. 11.21 performs the mathematical operation indicated in (10.14) with $K = 30$. The power amplifier is identical to that presented in Sect. 10.6.2.

11.7.3 Computer Program

The flow diagram for the PC programming when controlling the position in a PM brushed DC motor is presented in Fig. 11.25. All the control algorithms were programmed in Borland C++. The following is the code that was employed.

```
#include <stdio.h>
#include <conio.h>
#include <math.h>

#include <stdlib.h>
#include <string.h>
#include <iostream.h>

int alto,bajo,dato,salida,i;

volatile float Ts=0.005;//sampling period

/*------------------encoder variables----------*/
//volatile int alto_dig=0,bajo_dig=0,dato_dig=0;
volatile float y,ir,ym1,v,dv,nv,e,ninte;

volatile float inte,gc1,gc2,tiempo,perturbacion;
/*-------------------------------------------------*/
```

---

[1] All operational amplifiers used in Fig. 11.21 are TL081. The first three amplifiers have their terminals "+" to ground whereas the operational amplifier connected to the transistor bases has its terminal "+" connected to u_i.

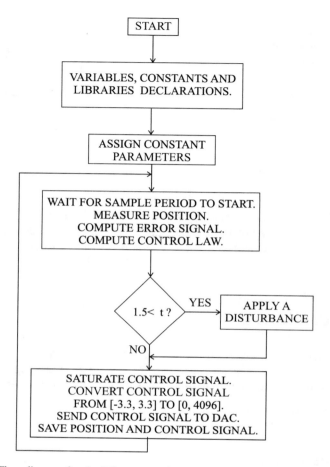

Fig. 11.25 Flow diagram for the PC programming when the controlling the position in a PM brushed DC motor

```
volatile float r_des=1.0,Kp=1.9434,Kv=0.1777;
//volatile float gama=2.3853,c=12.2734,b=4.8935;
volatile float gama=2.0465,c=10.3672,b=3.8935;

volatile float kp1=1.9434,ki1=9.7170;

volatile float kp2=1.3878,kd2=0.3195;
volatile float k=25.26,a=3.5201;
volatile float kp=363.88,kd=22;
volatile float ddthetad,dthetad,thetad,dy;
volatile long double y0;
/*-Base address and sampling period for acquisition board--*/
```

```c
int base=0x210;//Acquisition board base address

int lsb1=0x00;//sampling period

int msb1=0x01;//sampling period

int lsb2=0x00;//sampling period

int msb2=0x01;//sampling period

volatile float yy[3000];
volatile float uu[3000];

volatile float ref[3000];
FILE *fp;
char Direccion[30];

void Inicializa(void);

void EscribeArchivo(float datos1,floatdatos2,float datos3);

void Cierra(void); void main()
{
clrscr();

/*---------Acquisition board programming--------*/

/*---See PCL 812-PG Advantech manual------*/

outportb(base+11,6);
//CAD starts by clock and datum software transfer
outportb(base+9,0);// gain 1

outportb(base+10,15);// Channel AD No. 15
outportb(base+3,0x77);//Programming sampling period
outportb(base+1,lsb1);//Programming sampling period
outportb(base+1,msb1);//Programming sampling period
outportb(base+3,0xb7);//Programming sampling period
outportb(base+2,lsb2);//Programming sampling period
outportb(base+2,msb2);//Programming sampling period
i=1;
do
// control cycle begins
{
/*------------Conversion AD---------------------------*/
do
{
} while (inportb(base+5)>15);
alto=inportb(base+5);
bajo=inportb(base+4);
dato=256*alto+bajo;
y=(10.0/4096.0)*dato-5.0;//Position theta
tiempo=(i-1)*Ts;
/*--------encoder reading begins -------------------*/
```

```
// for the digital inputs configuration
/*
alto_dig=inport(base+7);
alto_dig+=0xff+1;
bajo_dig=inport(base+6);
bajo_dig=(bajo_dig)&(0x00ff);
dato_dig=256*alto_dig+bajo_dig;
po_rad=((6.283185307*dato_dig)/1600);
// 1600 is no. of pulses per revolution.
// gotoxy(10,10); printf("Posicion en radianes: %f",po_rad);
y=po_rad;
*/
/*-------encoder reading ends----------------------*/
if(i==1)
{ //initial conditions
y0=y;
ym1=0.0;
v=0.0;
inte=0.0;
}
y=y-y0;//output is fixed to zero
e=r_des-y;
/*------ Proportional control plus velocity feedback------*/
/*
ir=Kp*(r_des-y)-Kv*(y-ym1)/Ts;
*/
/*---------- Lead compensator --------------*/
/*
dv=(-c*v+(b-c)*e )*Ts;
nv=v+dv;
ir=gama*(e+v);
*/
/*---------- 2-degrees-of-freedom controller----------*/
/* gc1=kp1*e+inte;
gc2=kp2*y+kd2*(y-ym1)/Ts;
ir=gc1-gc2;
ninte=inte+ki1*e*Ts; */
/*---------- Trajectory tracking controller----------*/
dy=(y-ym1)/Ts;
thetad=0.8*sin(5.0*tiempo)+0.5*cos(3.0*tiempo)+3.0;
dthetad=4.0*cos(5.0*tiempo)-1.5*sin(3.0*tiempo);
ddthetad=(-1)*20.0*sin(5.0*tiempo)-4.5*cos(3.0*tiempo);
ir=1/k*(ddthetad-kd*(dy-dthetad)-kp*(y-thetad)+a*dy);
perturbacion=0.0;
if(tiempo>1.5)
{perturbacion=0.5;}
ir=-ir+perturbacion;
// Saturation of ir [-3,3][A]
if(ir>2.9 )
{ ir=2.9; }
if(ir<-2.9)
{ir=-2.9;}
salida=floor(682.66*ir+2048.0);
bajo=salida & 0xff;
```

```
alto=salida & 0x0f00;
alto=floor(alto/256.0);
outportb(base+4,bajo); //send code_o to CDA
outportb(base+5,alto);
yy[i]=y;
uu[i]=ir-perturbacion;
ref[i]=r_des;
ym1=y;
v=nv;
inte=ninte;
i++;
}while (!kbhit());// control cycle ends
// while (i<=3000);
ir=0.0;// motor is stopped
salida=floor(682.66*ir+2048.0);
bajo=salida & 0xff;

alto=salida & 0x0f00;
alto=floor(alto/256.0);
outportb(base+4,bajo);
outportb(base+5,alto);
// Save data to hard disk
cout<<i<<endl;
Inicializa();
i-=1;
for(int jj=1;jj<i;jj++)
{
EscribeArchivo(yy[jj],uu[jj],ref[jj]);
}
Cierra();
}

void Inicializa(void) {
strcpy(Direccion,"c:\\motorcd\\sigue.txt");
if( (fp=fopen(Direccion,"w+")) == NULL)
{
cout<<"File cannot be created";
exit(1);
}
}

void EscribeArchivo(float datos1,float datos2,float datos3) {
fprintf(fp,"%f %f %f \n",datos1,datos2,datos3);
}

void Cierra(void) {
fclose(fp);
}
```

11.8   Frequency Response-Based Design

11.8.1   Model Identification

Another experimental prototype for position control in a PM brushed DC motor has been built. This prototype is the same used in Sect. 10.10, Chap. 10. The identification of the corresponding model has been performed using a frequency response experiment. The details on the steps that have been followed to perform this task are given in the following.

1. The following voltage signal has been applied at the motor armature terminals $u = A \sin(\omega t)$[V] and it has been observed that the measured position of the motor has the following form $\theta = B \sin(\omega t + \phi)$[rad].
2. This experiment has been performed for several values of the frequency ω, and the results are presented in Table 11.1. Note that the phase ϕ is not considered.
3. In Fig. 11.26, the Bode diagram obtained from Table 11.1 is presented. One asymptote with a slope of -20[dB/dec] and one asymptote with a slope of -40[dB/dec] fit these experimental data well. This means that the motor model must have the following structure:

$$\frac{\theta(s)}{U(s)} = \frac{k}{s(s+a)}. \tag{11.52}$$

4. From the intersection of these asymptotes we conclude that:

$$a = 3.653 \tag{11.53}$$

5. The parameter k is found by trial and error until the data obtained from the model in (11.52) fit the experimental data well:

$$k = 62. \tag{11.54}$$

11.8.2   Proportional–Integral–Derivative Control Design

Although the motor model in 11.52 has one pole at $s = 0$, a controller with integral action is desirable to compensate for constant torque disturbances, e.g., the friction effects. Hence, given the second-order structure of the motor model, the design of a PID controller is well suited (see Fig. 11.27).

This means that the open loop transfer function is given as:

$$G(s)H(s) = \frac{k_d s^2 + k_p s + k_i}{s} \frac{k}{s(s+a)}.$$

Table 11.1 Experimental data for model identification of a PM brushed DC motor

ω[rad/s]	$2A$ [V]	$2B$ [rad]
1.5	6	64.97
2	6	46.535
3	6	27.335
4	6	17.305
5	6	12.045
6	6	8.585
7	6	6.29
8	6	5.065
9	6	4.2
10	6	3.355
20	6	0.86
30	6	0.415
40	6	0.235
50	6	0.16
60	6	0.115
70	6	0.08
80	6	0.06
90	6	0.045
100	6	0.04

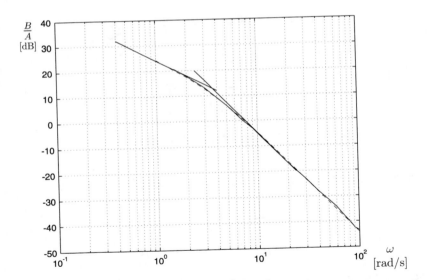

Fig. 11.26 Bode diagram from experimental data used for model identification. Continuous: experimental data (Table 11.1). Dashed: data from the model in (11.52)

Fig. 11.27 Proportional–integral–derivative control of the position

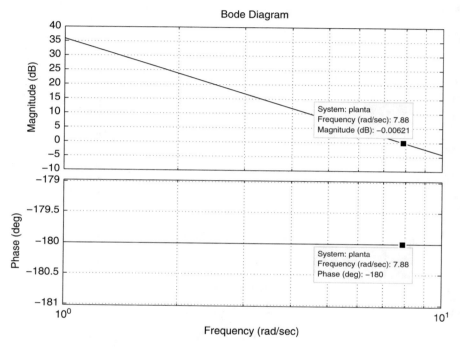

Fig. 11.28 Bode diagrams of the factor $\frac{k}{s^2}$, where (11.54) is used

Proceeding as in Sect. 6.7.3, we can rewrite this transfer function as:

$$G(s)H(s) = k_d z_2(z+1)\frac{k}{s^2}, \quad z = \frac{s}{z_2}, \quad z_1 = a,$$

$$k_p = k_d(z_1 + z_2), \quad k_i = k_d z_1 z_2. \quad (11.55)$$

Hence, the factor $\frac{k}{s^2}$ is considered the "plant to be controlled" and the factor $k_d z_2(z + 1)$ the "controller."

The Bode diagrams of the factor $\frac{k}{s^2}$ are presented in Fig. 11.28 where (11.54) is used. The crossover frequency is $\omega_1 = 7.88[\text{rad/s}]$ and the phase is $-180°$. This means that the phase margin is $0°$. It is chosen to maintain the same crossover frequency and to design a 20% overshoot. Hence, using:

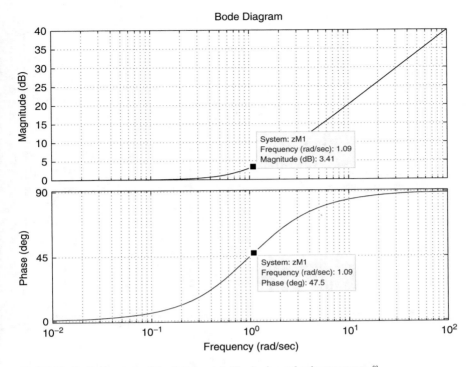

Fig. 11.29 Bode diagrams of the factor $z + 1$. The horizontal axis represents $\frac{\omega}{z_2}$

$$\zeta = \sqrt{\frac{\ln^2(M_p/100)}{\ln^2(M_p/100) + \pi^2}},$$

it is found that $\zeta = 0.4559$ is required. According to Table 6.5, this means that a 47.5° phase margin is needed. Hence, a 47.5° phase lead has to be contributed by the controller, i.e., by the factor $z + 1$, when $\omega = \omega_1 = 7.88[\text{rad/s}]$. It is observed in Fig. 11.29 that a 47.5° phase lead is obtained when $\frac{\omega}{z_2} = 1.09$. Hence, using $\omega = \omega_1 = 7.88[\text{rad/s}]$, the following is obtained: $z_2 = 7.2294$. On the other hand, from the magnitude condition:

$$|k_d z_2(z + 1)|_{dB} = 20\log(k_d z_2) + |z + 1|_{dB} = 0[\text{dB}],$$

where $|z + 1|_{dB} = 3.41[\text{dB}]$, according to Fig. 11.29. Thus, we have:

$$k_d = \frac{1}{z_2}10^{\frac{-3.41}{20}} = 0.0935.$$

Finally, from (11.55) and $z_1 = a$, the following is found:

$$k_p = 1.0177, \quad k_i = 2.4697.$$

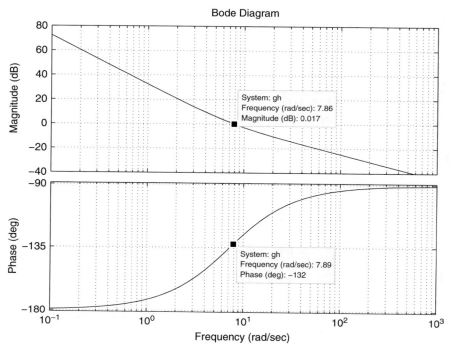

Fig. 11.30 Bode diagrams of the transfer function $\frac{k_d s^2 + k_p s + k_i}{s} \cdot \frac{k}{s(s+a)}$. $k_d = 0.0935$, $k_p = 1.0177$, $k_i = 2.4697$, and (11.53), (11.54) are used

In Fig. 11.30, a 7.88[rad/s] crossover frequency is observed and a 48° phase margin. This means that the frequency response design is correct. On the other hand, the simulations presented in Fig. 11.31 show a 33% overshoot, instead of the desired 20%, and 0.192[s] as rise time. See Sect. 6.7.1 for an explanation for this difference between the desired and the obtained overshoots. Some experimental results are presented in Fig. 11.32 where 29% is obtained as overshoot and 0.21[s] as rise time. Note that these values are very close to the simulation values.

According to Sect. 6.5.2 an additional phase lag of $-\omega T$ [rad] must be considered in Fig. 11.30 to take into account the effects of time delay induced by the numerical computation of the control law in experiments. As the sampling period used in experiments is 0.01[s], then a time delay of $T = 0.01$[s] may be assumed to take into account the worst case and $\omega = 7.88$[rad/s] is the crossover frequency. Then $-\omega T \times 180°/\pi = 4.51°$ is the maximal additional phase lag contributed by the digital implementation of the controller. This shifts the phase margin to 43.49°; hence, no performance deterioration is expected in the experiments, which is corroborated in Fig. 11.32.

Finally, in Fig. 11.33, we compare the control scheme in Fig. 11.27, $k_d = 0.0935$, $k_p = 1.0177$, $k_i = 2.4697$, with that presented in (11.38). The controller

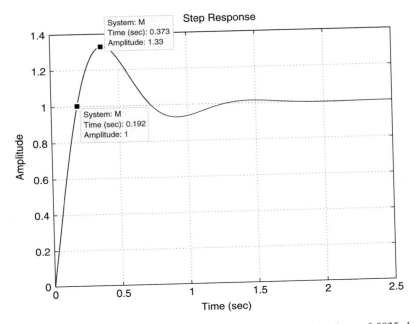

Fig. 11.31 Simulation response of the closed loop system in Fig. 11.27. $k_d = 0.0935$, $k_p = 1.0177$, $k_i = 2.4697$

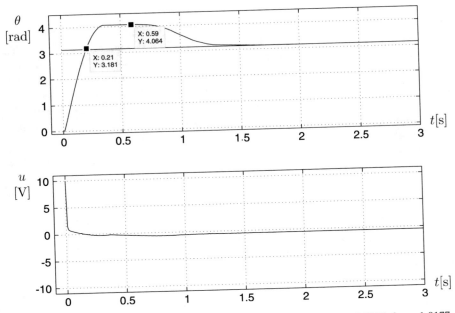

Fig. 11.32 Experimental results for the control scheme in Fig. 11.27. $k_d = 0.0935$, $k_p = 1.0177$, $k_i = 2.4697$

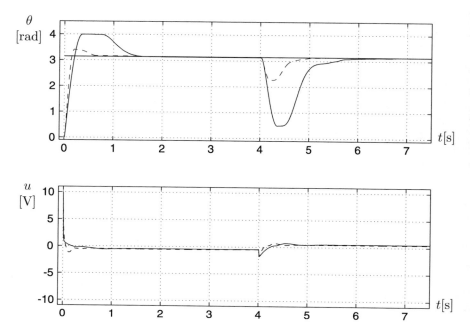

Fig. 11.33 Experimental comparison of the control scheme in Fig. 11.27 (continuous line) and controller in (11.38) (dashed line)

gains of the controller in (11.38) were computed using $t_r = 0.19$[s], $M_p = 20\%$, (11.9), (11.10), (11.11), $\sigma = \zeta\omega_n$, (11.36), (11.37), and $f = 5$. This yields $k_{p1} = 2.3571$, $k_{p2} = 0.889$, $k_{i1} = 11.7855$ and $k_{d2} = 0.1995$. A constant disturbance is applied for $t \geq 4$[s] via software. This has been done by adding a constant 2.5[V] to the voltage delivered by the controllers, which has to be sent to the motor. We realize that the effects of the disturbance are much smaller when the controller in (11.38) is used. Also, a faster response is achieved with the controller in (11.38).

Figures 11.26, and 11.28 to 11.31, have been plotted using the following MATLAB code in an m-file:

```
clc
clear
W=[1.5 2 3 4 5 6 7 8 9 10 20 30 40 50 60 70 80 90 100];
B=[64.97 46.535 27.335 17.305 12.045 8.585 6.29 5.065 4.2
3.355 0.86 0.415 0.235 0.16 0.115 0.08 0.06 0.045 0.04];
A=[6 6 6 6 6 6 6 6 6 6 6 6 6 6 6 6 6 6 6];
magnitude=20*log10(B./A);
figure(1)
semilogx(W,magnitude,'r-')
grid on
hold on
line([0.4 4],[32 12])
```

```
line([2.4 24],[20 -20])
k=62;
a=3.653;
g=tf(k,[1 a 0]);
ww=1:1:100;
[mg,fase]=bode(g,ww);
for i=1:100,
mag(i)=20*log10(mg(1,1,i));
end
semilogx(ww,mag,'k--')
hold off
plant=tf(k,[1 0 0]);
figure(2)
bode(planta)
grid on
z1=a;
w1=7.88;
Mpd=20;
PhaseLead_d=47.5;
zM1=tf([1 1],1)
figure(3)
bode(zM1)
grid on
z2=w1/1.09;
kd=10^(-3.4/20)/z2
kp=kd*(z1+z2)
ki=kd*z1*z2
PID=tf([kd kp ki],[1 0]);
plant2=tf(k,[1 a 0]);
gh=PID*plant2;
figure(4)
bode(gh)
grid on
M=feedback(gh,1,-1);
figure(5)
step(M)
grid on
%%%%%%%%%%%%%%% 2-DOF controller design
tr=0.19;
Mp=20;
z=sqrt(log(0.2)^2/(log(0.2)^2+pi^2))
wd=(pi-atan(sqrt(1-z^2)/z))/tr
wn=wd/sqrt(1-z^2)
sigma=z*wn;
f=5;
kp1=(sigma^2+wd^2)/k
ki1=(sigma^2+wd^2)*f/k
kp2=2*sigma*f/k
kd2=(2*sigma+f-a)/k
```

11.8.3 Prototype Construction

The electric diagram in addition to the computer program used to evaluate the control algorithms is identical to the electric diagram and the Builder 6 C++ code presented in Chap. 12. The only difference is that in Chap. 12 two encoders have to be read, whereas in the present chapter only one encoder is to be read. Also, a microcontroller PIC16F877A is used in Chap. 12 and in the present chapter. The code for programming this microcontroller is identical in both chapters. Again, in the present chapter only one encoder has to be read. Thus, the code for programming is simpler in the present chapter.

In particular, the identification experiments described in Sect. 11.8.1 require to the voltage signal $u = A \sin(\omega t)$[V] to be applied at the motor armature terminals. This is performed by writing the code line "iast=A*sin(w t)," provided that A=3 and w are defined as constants at the beginning of the program. This is all that we have to do as software and hardware are so contrived that the numerical value of "iast" appears as the voltage at the motor armature terminals.

This prototype has been built because it presents some advantages when performing the frequency response experiments described in Sect. 11.8.1. When using the prototype depicted in Fig. 11.18 for frequency response identification, the motor position has a large drift, which poses difficulties for the measurement of the position amplitude. Such a problem is eliminated with the new prototype. See Chap. 12.

11.9 Summary

The position of a PM brushed DC motor has been controlled under the effect of an external torque disturbance. Two controllers have been tested: a classical PID controller and a modified PID controller. Although both controllers solve the problem, the modified PID controller has a better performance. As this controller requires numerical values of the motor parameters, an experimental identification of the motor model is also presented. In this chapter, it has also been shown how to tune a classical PID controller using the numerical values of the plant parameters. However, the main advantage of this controller is that it can be tuned by a trial and error procedure, i.e., without any information about the plant parameters (see Sect. 5.3 in Chap. 5). On the other hand, a PD position controller has also been designed to solve the trajectory tracking problem. This is a complex task requiring the exact values of the motor parameters. Detailed explanations are included on how to implement these controllers using a microcontroller and a personal computer. The reason for using a personal computer is that the implementation of the PD controller for trajectory tracking requires computation of sine and cosine functions, which consumes too much time when using a microcontroller.

When a control system is implemented in practice, it is very important that the signal processing performed by the software and hardware precisely corresponds to the mathematical operations indicated by the controller. Control systems that are built in this book satisfy this requirement and this can be corroborated by following their careful descriptions.

11.10   Review Questions

1. Consider the PID position control of a DC motor. What would happen if either the torque disturbance or the desired position were not constant?
2. Industrial robots employ PID controllers, even for trajectory tracking. How do you think that this task can be accomplished (see Sect. 6.7.4)?
3. How does each one of the classical PID controller gains affect the response of a position control system? If a disturbance were rejected very slowly, how would you modify the controller gains to improve performance?
4. In Chap. 10, it was shown that a PI controller is suitable for compensating for the effects of an external torque disturbance in a velocity control system, Why do you think that a PID controller has to be used instead of a PI controller in the case of position control?
5. List some advantages and disadvantages when using a microcontroller or a personal computer to implement digital controllers.
6. A PM brushed DC motor can also be controlled without using an internal electric current loop if it is assumed that the armature inductance is negligible, especially in small motors. What changes must be performed in the circuits of Figs. 11.18 and 11.21 to work without an electric current loop?
7. Describe the time response and the frequency response methods presented in this chapter to experimentally identify the parameters of a DC motor.
8. What is a lead compensator?
9. What are the advantages of using lead compensators as position controllers?
10. From the theoretical point of view, the response of a position control system can be rendered as fast and as well damped as desired using a PID controller. List some problems arising in practice when the PID controller gains are too large.

References

1. DAC0800 8-bit Digital-to-Analog Converter, Data sheet, National Semiconductor Corporation, 1995.
2. PIC16F877A Enhanced Flash Microcontroller, Data sheet, Microchip Technology Inc., 2003.
3. Custom Computer Services Incorporated, CCS C Compiler Reference Manual, 2003.
4. PCL-812PG User's Manual, *Advantech*, Taiwan, 1996.
5. K. Ogata, *Modern control engineering*, 4th edition, Prentice-Hall, Upper Saddle River, 2002.
6. J. Chiasson, *Modeling and high-performance control of electric machines*, IEEE Press-Wiley Interscience, New Jersey, 2005.
7. Parker Automation, Position Systems and Controls, Training and Product Catalog DC-ROM, *Compumotor's Virtual Classroom*, 1998.

Chapter 12
Control of a Servomechanism with Flexibility

In mechanical systems, the mechanical power is often provided by an electric motor. This power is supplied to a load by means of a shaft or a mechanical transmission that is made of steel. Despite the mechanical strength of steel, flexibility becomes a significant phenomenon when large loads are present. This situation is common in large radio telescopes, for instance.

Flexibility complicates the control design task because a second-order plant becomes a fourth-order plant. Hence, as shown in this chapter, the closed-loop system becomes unstable when simple controllers such as proportional–derivative (PD) or proportional–integral–derivative (PID) are used to directly control the load position. Thus, a more elaborate controller design is required. In this chapter, some approaches are presented to accomplish this task. However, to carry out practical experimentation with our designs, a radio telescope cannot be employed but instead a laboratory mechanical system that has been built. The mathematical model of this prototype is presented in the following.

12.1 Mathematical Model

In Fig. 12.1 a PM brushed DC motor is presented, with rotor inertia I_1, which actuates on a load, with inertia I_2, through a rotative spring, with a stiffness constant K. Both the DC motor and the load receive friction torques characterized by the viscous friction coefficients b_1 and b_2. Finally, the torque generated by the DC motor is designated by $T(t)$, whereas θ_1 and θ_2 are positions of the DC motor shaft and the load respectively.

The mathematical model is obtained using Newton's Second Law:

$$I_1 \ddot{\theta}_1 = \sum \text{applied torques on the DC motor rotor,}$$

© Springer International Publishing AG, part of Springer Nature 2019
V. M. Hernández-Guzmán, R. Silva-Ortigoza, *Automatic Control with Experiments*,
Advanced Textbooks in Control and Signal Processing,
https://doi.org/10.1007/978-3-319-75804-6_12

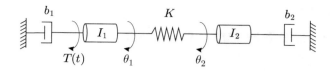

Fig. 12.1   A mechanical system with two bodies and one spring

$$I_2\ddot{\theta}_2 = \sum \text{applied torques on the load,}$$

which yields:

$$I_1\ddot{\theta}_1 = -b_1\dot{\theta}_1 - K(\theta_1 - \theta_2) + T,$$
$$I_2\ddot{\theta}_2 = -b_2\dot{\theta}_2 + K(\theta_1 - \theta_2).$$

Finally:

$$I_1\ddot{\theta}_1 + b_1\dot{\theta}_1 + K(\theta_1 - \theta_2) = T, \tag{12.1}$$
$$I_2\ddot{\theta}_2 + b_2\dot{\theta}_2 - K(\theta_1 - \theta_2) = 0. \tag{12.2}$$

Using the Laplace transform in (12.1), assuming that all initial conditions are zero and solving for $\theta_1(s)$, the following is found:

$$\theta_1(s) = \frac{\frac{1}{I_1}}{s^2 + \frac{b_1}{I_1}s + \frac{K}{I_1}} T(s) + \frac{\frac{K}{I_1}}{s^2 + \frac{b_1}{I_1}s + \frac{K}{I_1}} \theta_2(s). \tag{12.3}$$

Using the Laplace transform in (12.2), assuming that all initial conditions are zero and solving for $\theta_2(s)$, yields:

$$\theta_2(s) = \frac{\frac{K}{I_2}}{s^2 + \frac{b_2}{I_2}s + \frac{K}{I_2}} \theta_1(s). \tag{12.4}$$

Replacing (12.3) in (12.4) and solving for $\theta_2(s)$:

$$\theta_2(s) = G_2(s)T(s), \tag{12.5}$$

$$G_2(s) = \frac{\frac{K}{I_1 I_2}}{s^4 + \left(\frac{b_1}{I_1} + \frac{b_2}{I_2}\right)s^3 + \left(\frac{b_1}{I_1}\frac{b_2}{I_2} + \frac{K}{I_1} + \frac{K}{I_2}\right)s^2 + \left(\frac{b_2}{I_2}\frac{K}{I_1} + \frac{b_1}{I_1}\frac{K}{I_2}\right)s}.$$

Replacing (12.4) in (12.3) and solving for $\theta_1(s)$:

$$\theta_1(s) = G_1(s)T(s), \tag{12.6}$$

$$G_1(s) = \frac{\left(s^2 + \frac{b_2}{I_2}s + \frac{K}{I_2}\right)\frac{1}{I_1}}{s^4 + \left(\frac{b_1}{I_1} + \frac{b_2}{I_2}\right)s^3 + \left(\frac{b_1}{I_1}\frac{b_2}{I_2} + \frac{K}{I_1} + \frac{K}{I_2}\right)s^2 + \left(\frac{b_2}{I_2}\frac{K}{I_1} + \frac{b_1}{I_1}\frac{K}{I_2}\right)s}.$$

On the other hand, the dynamic model of the electrical subsystem of a PM brushed DC motor is:

$$L\frac{di}{dt} = u - Ri - k_b\dot{\theta}_1, \quad T = k_m i,$$

where L, R, k_b, k_m are the armature inductance and resistance, the counter electromotive force constant, and the torque constant respectively. Finally, i, u, T, are the electric current through the armature circuit, the applied voltage at the armature terminals and the generated torque respectively. If the motor to be employed is small, then it is reasonable to assume that the inductance can be neglected, i.e., $L \approx 0$. Thus, the above expression becomes:

$$i = \frac{1}{R}(u - k_b\dot{\theta}_1).$$

Recalling $T = k_m i$, it is found that $T = k_m(u - k_b\dot{\theta}_1)/R$. Applying the Laplace transform to this expression, replacing in (12.5) and using (12.4) yields:

$$\theta_2(s) = G_{2u}(s)U(s), \tag{12.7}$$

$$G_{2u}(s) = \frac{\frac{K k_m}{I_1 I_2 R}}{s^4 + \alpha_3 s^3 + \alpha_2 s^2 + \alpha_1 s},$$

$$\alpha_3 = \frac{b_1}{I_1} + \frac{b_2}{I_2} + \frac{k_m k_b}{R I_1}, \quad \alpha_2 = \frac{b_1}{I_1}\frac{b_2}{I_2} + \frac{K}{I_1} + \frac{K}{I_2} + \frac{k_b k_m b_2}{R I_2 I_1},$$

$$\alpha_1 = \frac{b_2}{I_2}\frac{K}{I_1} + \frac{b_1}{I_1}\frac{K}{I_2} + \frac{k_m k_b K}{R I_1 I_2},$$

whereas substitution in (12.6) yields:

$$\theta_1(s) = G_{1u}(s)U(s), \tag{12.8}$$

$$G_{1u}(s) = \frac{\left(s^2 + \frac{b_2}{I_2}s + \frac{K}{I_2}\right)\frac{k_m}{R I_1}}{s^4 + \alpha_3 s^3 + \alpha_2 s^2 + \alpha_1 s}.$$

These last expressions are the dynamical model of the servomechanism and their parameters are experimentally identified in the next section.

12.2   Experimental Identification

In this section, some experiments are performed to obtain the numerical parameters of the plant mathematical models presented in Sect. 12.1. It is important to state that the measurement of the individual parameters such as inertias, stiffness constants or viscous friction coefficients is not a practical solution in this plant. This is because such parameter identification procedures would imply either plant disassembly or complicated and inaccurate measurement procedures, i.e., for viscous friction coefficient measurement. Thus, in this section a more convenient model identification procedure for this plant is employed: a frequency response-based model identification procedure. This is described in what follows.

The following identification procedure has been performed on the experimental prototype described in Sect. 12.4 whose dynamical model has been presented in Sect. 12.1.

1. A sinusoidal voltage signal is applied at the armature terminals of the PM brushed DC motor used as an actuator: $u(t) = A\sin(\omega t)$, where t is time in seconds, ω is in radians/second, and $A = 3[V]$.
2. Under the action of this applied voltage, it is found that the motor rotor position θ_1 and the load position θ_2 are given, in a steady state, as $\theta_1(t) = B_1\sin(\omega t + \phi_1)$ and $\theta_2(t) = B_2\sin(\omega t + \phi_2)$.
3. This experiment is repeated for the angular frequencies ω shown in Table 12.1.
4. The corresponding values of B_1 and B_2 are measured and they are shown in Table 12.1. The phase angles ϕ_1 and ϕ_2 are not presented because the minimum phase property of the model renders the measurement of these variables unnecessary.
5. Define the magnitude functions $|G_{1u}(j\omega)|$ and $|G_{2u}(j\omega)|$ as the ratios B_1/A and B_2/A respectively, for each ω in Table 12.1. Using these data compute and plot, in Figs. 12.2 and 12.3, the decibel valued magnitudes $|G_{1u}(j\omega)|_{dB} = 20\log(|G_{1u}(j\omega)|)$ and $|G_{2u}(j\omega)|_{dB} = 20\log(|G_{2u}(j\omega)|)$.
6. Fit to these data some straight lines whose slopes are either 0[dB/dec] or an integer multiple of ±20[dB/dec], as shown in Figs. 12.2 and 12.3.
7. Obtain the zero and pole locations of the corresponding transfer functions from the corner frequencies defined by these straight lines and the corresponding transfer function gain as the constant that has to be introduced to fit the data obtained from these transfer functions to the experimental data.

According to the identification procedure listed above, we realize that, in Fig. 12.2, two lines with slopes of -20[dB/dec] and -40[dB/dec] are enough to fit the experimental data. Hence, from Fig. 12.2 we conclude that real poles exist for $G_{1u}(s)$, which are located at $s_1 = 0$ and $s_2 = -5$. Moreover, the resonance peak implies that a corner frequency exists at $\omega_{n1} = 62.83$[rad/s], which must be due to a pair of complex conjugate poles. This means that $G_{1u}(s)$ includes the factor:

Table 12.1 Frequency response experimental data

f [Hz]	$\omega = 2\pi f$ [rad/s]	A [V]	B_1 [rad]	B_2 [rad]
0.4	2.5133	3	30	27.5
0.5	3.1416	3	24	22.5
0.6	3.7699	3	16	16
0.7	4.3982	3	13.25	13.25
0.8	5.0265	3	11.5	11.5
0.9	5.6549	3	8.5	8.5
1	6.2832	3	7.5	7.5
2	12.5664	3	2.25	2.25
3	18.8496	3	0.95	1.25
4	25.1327	3	0.525	0.675
5	31.4159	3	0.315	0.525
6	37.6991	3	0.19	0.44
7	43.9823	3	0.1	0.45
8	50.2655	3	0.045	0.465
8.5	53.4071	3	0.025	0.55
9	56.5487	3	0.039	0.56
10	62.8319	3	0.31	0.95
12	75.3982	3	0.095	0.055
15	94.2478	3	0.05	0.015
20	125.6637	3	0.025	0

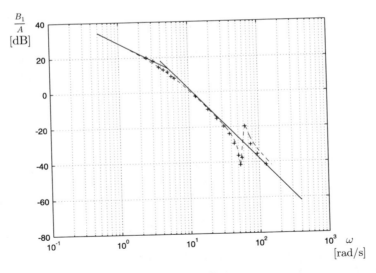

Fig. 12.2 Experimental identification of $G_{1u}(s) = \frac{\theta_1(s)}{U(s)}$. Points marked with "+" represent the experimental data. The dashed line represents data obtained from the identified model. The straight lines represent the estimated asymptotes

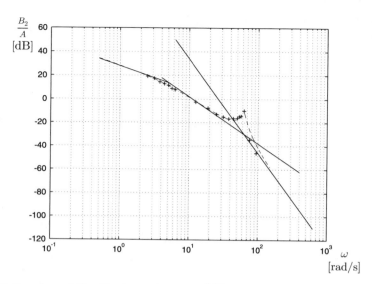

Fig. 12.3 Experimental identification of $G_{2u}(s) = \frac{\theta_2(s)}{U(s)}$. Points marked with "+" represent the experimental data. The dashed line represents data obtained from the identified model. The straight lines represent the estimated asymptotes

$$\frac{1}{s^2 + 2\zeta_1\omega_{n1}s + \omega_{n1}^2},$$

where ζ_1 is unknown. Hence, the coefficient $2\zeta_1\omega_{n1}$ has to be found by trial and error until the data obtained from the above factor fit the resonance peak. Thus, the following is found:

$$\frac{1}{s^2 + 2\zeta_1\omega_{n1}s + \omega_{n1}^2} = \frac{1}{s^2 + 8s + 62.83^2}.$$

Finally, note that the $-40[\text{dB/dec}]$ asymptote remains unchanged, despite the above-mentioned second-order factor appearing. Moreover, a notch exists just before $\omega = \omega_{n1}$. These features are clear evidence that the following factor is present in $G_{1u}(s)$:

$$s^2 + 2\zeta_2\omega_{n2}s + \omega_{n2}^2 = s^2 + 3s + 53.68^2.$$

The numerical coefficients in this factor have been found by trial and error until this model fits the notch. Finally, to match the gain of the experimental data, it was necessary to include 175 as the transfer function gain. Thus, we have found that:

$$G_{1u}(s) = \frac{175(s^2 + 3s + 53.68^2)}{s(s+5)(s^2 + 8s + 62.83^2)}. \tag{12.9}$$

In Fig. 12.2, we can see that the data obtained from this mathematical model fit the experimental data very well. Moreover, the structure of $G_{1u}(s)$ in (12.9) perfectly fits the structure of $G_{1u}(s)$ in (12.8), i.e., both of them have four poles, one of them at the origin, and two zeros.

On the other hand, in Fig. 12.3 the experimental data used to identify $G_{2u}(s)$ are presented. We realize that these data fit the asymptotes with slopes $-20[\text{dB/dec}]$, $-40[\text{dB/dec}]$ and $-80[\text{dB/dec}]$ very well. This implies that $G_{2u}(s)$ includes poles at $s_1 = 0$, $s_2 = -5$ and a corner frequency exists at $\omega_{n1} = 62.83[\text{rad/s}]$, which is due to a pair of complex conjugate poles. This means that $G_{2u}(s)$ includes the factor:

$$\frac{1}{s^2 + 2\zeta_1\omega_{n1}s + \omega_{n1}^2} = \frac{1}{s^2 + 8s + 62.83^2},$$

where the coefficient $2\zeta_1\omega_{n1} = 8$ has been estimated by trial and error, until data obtained from this model fit the experimental resonance peak. Finally, the gain of the transfer function is found to be 520000, as this factor renders it possible that data obtained from the following transfer function:

$$G_{2u}(s) = \frac{520000}{s(s+5)(s^2 + 8s + 62.83^2)}, \tag{12.10}$$

fit the experimental data, as shown in Fig. 12.3. Note that the structure of $G_{2u}(s)$ in (12.10) fits the structure of $G_{2u}(s)$ in (12.7) very well, i.e., both of them have four poles, one at the origin, and no zeros are present.

Figures 12.2 and 12.3 were drawn using the following MATLAB code in an m-file:

```
clc
f=[0.4 0.5 0.6 0.7 0.8 0.9 1 2 3 4 5 6 7 8 8.5 9 10 12 15 20];
thetam=[60 48 32 26.5 23 17 15 4.5 1.9 2.4-1.35 1.03-0.4
0.68-0.3 0.2 0.09 0.05 0.2-0.122 0.8-0.18 0.57-0.38
0.1 0.05]/2;
theta2=[55 45 32 26.5 23 17 15 4.5 2.5 2.6-1.25 1.2-0.15
0.9-0.02 0.9 0.93 1.1 0.65+0.47 1.9 0.11 0.03 0]/2
u=[6 6 6 6 6 6 6 6 6 6 6 6 6 6 6 6 6 6 6 6]/2;
w=2*pi*f;
magm=thetam./u;
mag2=theta2./u;
magmdB=20*log10(magm);
mag2dB=20*log10(mag2);
c=conv([1 0],[1 5]);
d=conv(c,[1 8 62.83^2]);
gm=tf(175*[1 3 53.68^2],d)
g2=tf(520000,d)
w2=0.5:0.1:150;
[m2,f2]=bode(g2,w2);
[m,f]=bode(gm,w2);
for i=1:length(w2),
```

Fig. 12.4 Proportional control of the system in (12.10)

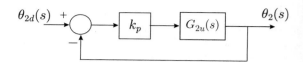

```
mm(i)=20*log10(m(1,1,i));
mm2(i)=20*log10(m2(1,1,i));
fm(i)=f(1,1,i);
fm2(i)=f2(1,1,i);
end
length(w2)
length(magmdB)
length(mm)
figure(1)
semilogx(w,magmdB,'k+',w2,mm,'r--')
grid on
line([4 400],[18 -62])
line([0.5 5],[34 14])
figure(2)
semilogx(w,mag2dB,'k+',w2,mm2,'r--')
grid on
line([4 400],[18 -62])
line([0.5 5],[34 14])
line([6.3 630],[50 -110])
```

12.3   Controller Design

From (12.10), it is easy to find that the poles of $G_{2u}(s)$ are located at:

$$s_1 = 0, \quad s_2 = -5, \quad s_3 = -4 + 62.7025j, \quad s_4 = -4 - 62.7025j.$$

The simplest controller that we should test is a proportional one, i.e., to obtain a closed-loop system such as that shown in Fig. 12.4. From the root locus diagram presented in Fig. 12.5, it is clear that very oscillatory responses are expected or even instability.

This may suggest the use of a PD controller with transfer function:

$$PD(s) = k_p + k_d s = k_d \left(s + \frac{k_p}{k_d} \right) = k_d(s + b), \quad b = \frac{k_p}{k_d}.$$

This means that a PD controller introduces one open-loop zero at $s = -b$. The corresponding closed-loop block diagram is shown in Fig. 12.6. From the root locus diagram depicted in Fig. 12.7, we realize that this is not a good solution either as instability problems or a very slow and oscillatory response are expected to appear no matter what the value of b. To corroborate these observations, in Fig. 12.8,

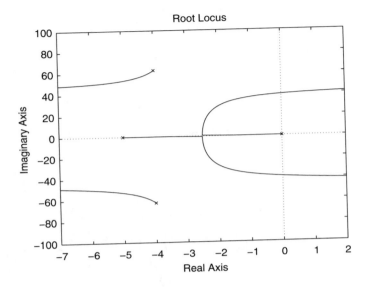

Root Locus

Fig. 12.5 Root locus for $G(s)H(s) = k_p G_{2u}(s)$

Fig. 12.6
Proportional–derivative (PD)
control of system in (12.10)

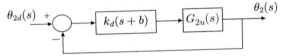

$\theta_{2d}(s)$ $+$ ... $k_d(s+b)$ ... $G_{2u}(s)$ ... $\theta_2(s)$

we present the experimental responses when using a PD controller for the load position. Note that instability is observed even when a very small proportional gain is used. These findings motivate us to design more complex controllers for θ_2. Some different approaches are tested in the following sections.

12.3.1 Multi-Loop Control

To motivate the design based on this approach, let us first use intuitive ideas to explain the main steps of the procedure. From (12.9), it is easy to find that $G_{1u}(s) = \frac{\theta_1(s)}{U(s)}$ has poles located at:

$$s_1 = 0, \quad s_2 = -5, \quad s_3 = -4 + 62.7025j, \quad s_4 = -4 - 62.7025j,$$

and zeros located at:

$$s_5 = -1.5 + 53.659j, \quad s_6 = -1.5 - 53.659j.$$

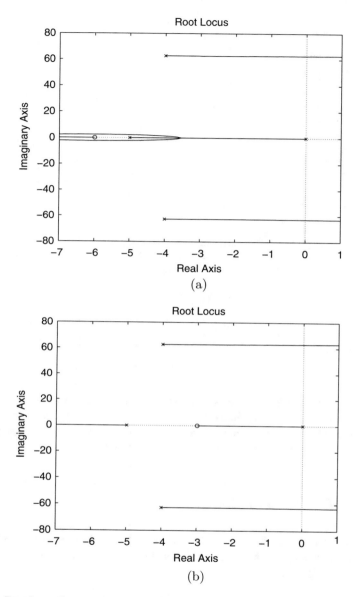

Fig. 12.7 Root locus diagrams for the closed-loop system in Fig. 12.6

Suppose that the following PD controller is employed to control θ_1:

$$PD_1(s) = k_{p1} + k_{d1}s = k_{d1}(s + b_1), \quad b_1 = \frac{k_{p1}}{k_{d1}}.$$

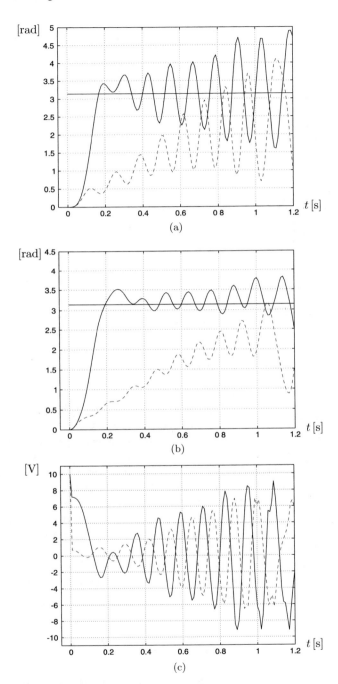

Fig. 12.8 PD control of θ_2. Experimental results. Continuous: $k_{p2} = 2.3$, $k_{d2} = 0.1$. Dashed: $k_{p2} = 0.23$, $k_{d2} = 0.1$. (**a**) θ_2. (**b**) θ_1. (**c**) Applied voltage, u

Fig. 12.9 PD control of the motor position

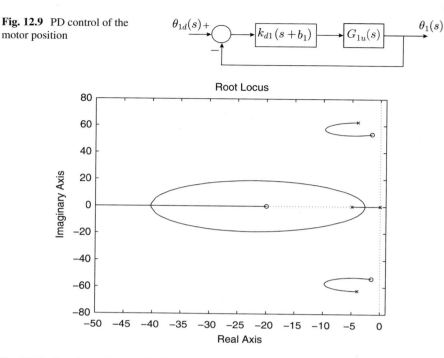

Fig. 12.10 Root locus diagram for $G(s)H(s) = k_{d1}(s + b_1)G_{1u}(s)$

The equivalent closed-loop block diagram and the corresponding root locus diagram are presented in Figs. 12.9 and 12.10 respectively. Note that two complex conjugate closed-loop poles exist, which are close to the complex conjugate zeros at $s_5 = -1.5 + 53.659j$ and $s_6 = -1.5 - 53.659j$. It is not difficult to verify that these open-loop zeros are also closed-loop zeros. Hence, any oscillation produced by these closed-loop poles on $\theta_1(t)$ would be kept small if a large value for k_{d1} is used. Furthermore, the closed-loop poles on the other two branches of the root locus diagram in Fig. 12.10 move away from the origin and approach the negative real axis as k_{d1} is chosen to be larger. Therefore, a fast and well-damped response for θ_1 can be ensured by using a large value for k_{d1}.

Assume that the control of the motor position θ_1 has been accomplished such that a desired value θ_{1d} is tracked very fast and without oscillations, i.e., assume that $\theta_1 \approx \theta_{1d}$ all the time. Hence, using (12.4) we can write:

$$\theta_2(s) = \frac{\frac{K}{I_2}}{s^2 + \frac{b_2}{I_2}s + \frac{K}{I_2}}\theta_{1d}(s). \qquad (12.11)$$

From this expression, $\theta_{1d}(s)$ can be designed as the output of the following PID controller, which is driven by the position error $e = \theta_{2d} - \theta_2$:

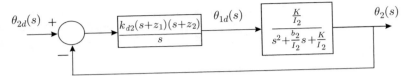

Fig. 12.11 Proportional–integral–derivative (PID) control of the load position (12.11)

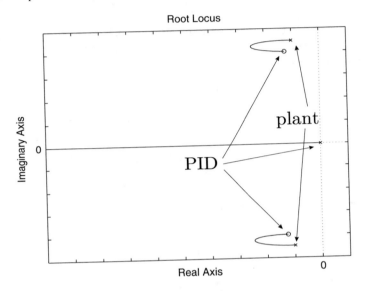

Fig. 12.12   Root locus diagram for the closed-loop system in Fig. 12.11

$$\theta_{1d}(s) = PID_2(s)E(s), \quad E(s) = \theta_{2d}(s) - \theta_2(s), \tag{12.12}$$

$$PID_2(s) = k_{p2} + k_{d2}s + \frac{k_{i2}}{s} = \frac{k_{d2}\left(s^2 + \frac{k_{p2}}{k_{d2}}s + \frac{k_{i2}}{k_{d2}}\right)}{s}$$

$$= \frac{k_{d2}(s+z_1)(s+z_2)}{s}, \quad \frac{k_{p2}}{k_{d2}} = z_1 + z_2, \quad \frac{k_{i2}}{k_{d2}} = z_1 z_2.$$

The corresponding closed-loop block diagram and the root locus diagram are depicted in Figs. 12.11 and 12.12 respectively. The main idea is to force the branches starting at the roots of $s^2 + \frac{b_2}{I_2}s + \frac{K}{I_2}$ to be attracted by the open-loop zeros contributed by the PID controller in (12.12). This would reduce any oscillation due to these closed-loop poles [2], section 10–8. Thus, the remaining closed-loop pole can be chosen to be real, negative, and far away from the origin, by selecting a large k_{d2}.

Now, the above ideas are applied to control the load position θ_2. In the following, the design is performed without neglecting any open-loop pole or zero. A control system with two loops is to be designed. The internal loop is designed as a PD controller for the motor position θ_1. The main objective is to strongly reduce the oscillations in this variable. The external loop must be designed as a PID controller for the load position θ_2. The output of this PID controller is to be used as the desired value for the motor position θ_1.

The internal loop is defined by the following PD controller:

$$U(s) = (k_{p1} + k_{d1}s)(V_1(s) - \theta_1(s)), \tag{12.13}$$

where $V_1(s)$ represents the desired value for θ_1, which is defined when designing the external loop. Replacing $T(s) = k_m U(s)/R - k_m k_b s \theta_1(s)/R$ and (12.13) in (12.3), and solving for $\theta_1(s)$:

$$\theta_1(s) = \frac{(k_{d1}s + k_{p1})\frac{k_m}{RI_1}}{s^2 + \left(\frac{b_1}{I_1} + \frac{(k_{d1}+k_b)k_m}{RI_1}\right)s + \left(\frac{K}{I_1} + \frac{k_{p1}k_m}{RI_1}\right)} V_1(s)$$

$$+ \frac{\frac{K}{I_1}}{s^2 + \left(\frac{b_1}{I_1} + \frac{(k_{d1}+k_b)k_m}{RI_1}\right)s + \left(\frac{K}{I_1} + \frac{k_{p1}k_m}{RI_1}\right)} \theta_2(s).$$

Replacing this in (12.4) and solving for $\theta_2(s)$:

$$\theta_2(s) = G_3(s)V_1(s), \quad G_3(s) = \frac{\frac{Kk_m}{RI_1 I_2}(k_{d1}s + k_{p1})}{N(s)}, \tag{12.14}$$

$$N(s) = s^4 + a_3 s^3 + a_2 s^2 + a_1 s + a_0,$$

$$a_3 = \frac{b_1}{I_1} + \frac{k_{d1}k_m}{RI_1} + \frac{k_b k_m}{RI_1} + \frac{b_2}{I_2},$$

$$a_2 = \frac{K}{I_1} + \frac{k_{p1}k_m}{RI_1} + \frac{b_2}{I_2}\left(\frac{b_1}{I_1} + \frac{k_{d1}k_m}{RI_1} + \frac{k_b k_m}{RI_1}\right) + \frac{K}{I_2},$$

$$a_1 = \frac{b_2}{I_2}\left(\frac{K}{I_1} + \frac{k_{p1}k_m}{RI_1}\right) + \frac{K}{I_2}\left(\frac{b_1}{I_1} + \frac{k_{d1}k_m}{RI_1} + \frac{k_b k_m}{RI_1}\right),$$

$$a_0 = \frac{K}{I_2}\frac{k_{p1}k_m}{RI_1}.$$

The external loop is defined by the following PID controller:

$$V_1(s) = \left(k_{p2} + k_{d2}s + \frac{k_{i2}}{s}\right)(\theta_{2d}(s) - \theta_2(s)), \tag{12.15}$$

where θ_{2d} is a constant representing the desired value for θ_2. The closed-loop block diagram is depicted in Fig. 12.13.

The PD controller shown in (12.13) is designed experimentally by trial and error using step changes for $v_1(t)$ ($\mathcal{L}\{v_1(t)\} = V_1(s)$). However, we stress that the load is

12.3 Controller Design

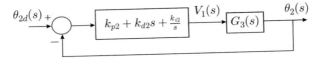

Fig. 12.13 External loop for the load position control

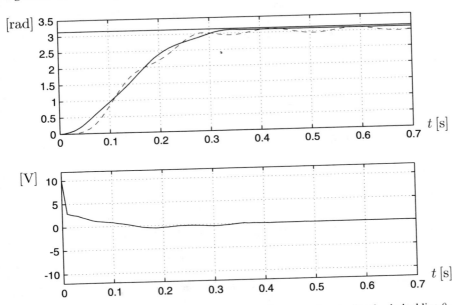

Fig. 12.14 PD control of θ_1. Experimental results. Top figure: continuous line θ_1, dashed line θ_2. Bottom figure: applied voltage u. The desired position is set to $\theta_{1d} = \pi$ [rad]

present in the mechanism when performing these experiments. The values for k_{p1} and k_{d1} are selected as those achieving a fast and well-damped response for $\theta_1(t)$. The limit for the achievable fastest response is keeping, for most of the time, the applied voltage below 3[V], which has been used as the maximal applied voltage during the identification experiments. Hence, we have found the fastest and well-damped response in θ_1 (see Fig. 12.14), using the following controller gains:

$$k_{p1} = 1.0, \quad k_{d1} = 0.1. \tag{12.16}$$

Note that $G_{1u}(s)$ in (12.9) can be rewritten as:

$$G_{1u}(s) = \frac{175s^2 + 525s + 5.043 \times 10^5}{s^4 + 13s^3 + 3988s^2 + 1.974 \times 10^4 s}.$$

Comparing this expression, (12.8), (12.9) and $G_3(s)$ in (12.14), we conclude that:

$$\frac{K k_m}{R I_1 I_2} = 5.043 \times 10^5,$$

$$a_3 = 13 + 175 k_{d1},$$

$$a_2 = 3988 + 175 k_{p1} + 3 \times 175 k_{d1},$$

$$a_1 = 1.974 \times 10^4 + 3 \times 175 k_{p1} + 175 \times 53.68^2 k_{d1},$$

$$a_0 = 5.043 \times 10^5 k_{p1}.$$

Using (12.16), we find that:

$$G_3(s) = \frac{5.043 \times 10^4 s + 5.043 \times 10^5}{s^4 + 30.5 s^3 + 4215 s^2 + 7.069 \times 10^4 s + 5.043 \times 10^5},$$

which has one zero at:

$$s_7 = -10,$$

and poles at:

$$s_8 = -6.304 + 61.7989 j, \quad s_9 = -6.304 - 61.7989 j,$$
$$s_{10} = -8.946 + 7.1167 j, \quad s_{11} = -8.946 - 7.1167 j.$$

As explained before, $V_1(s)$ has to be selected as the output of the PID controller.

$$V_1(s) = PID_2(s) E(s), \quad E(s) = \theta_{2d}(s) - \theta_2(s),$$

$$PID_2(s) = k_{p2} + k_{d2} s + \frac{k_{i2}}{s} = \frac{k_{d2}\left(s^2 + \frac{k_{p2}}{k_{d2}} s + \frac{k_{i2}}{k_{d2}}\right)}{s},$$

$$= \frac{k_{d2}(s + z_1)(s + z_2)}{s}, \quad \frac{k_{p2}}{k_{d2}} = z_1 + z_2, \quad \frac{k_{i2}}{k_{d2}} = z_1 z_2, (12.17)$$

such that the open-loop transfer function is given as:

$$PID_2(s) G_3(s) = \frac{k_{d2}(s + z_1)(s + z_2)}{s}$$

$$\times \frac{5.043 \times 10^4 s + 5.043 \times 10^5}{s^4 + 30.5 s^3 + 4215 s^2 + 7.069 \times 10^4 s + 5.043 \times 10^5}.$$

$$(12.18)$$

Thus:

$$z_1 = 6.304 + 61.7989 j, \quad z_2 = 6.304 - 61.7989 j, \quad (12.19)$$

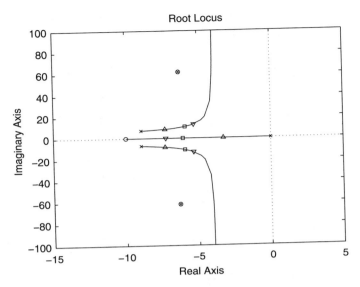

Fig. 12.15 Root locus diagram for the open-loop transfer function in (12.18), (12.19). Some closed-loop poles are shown corresponding to three particular values of k_{d2}: triangle up $k_{d2} = 0.0008$, square $k_{d2} = 0.0018$, triangle down $k_{d2} = 0.0028$

are chosen to render the cancellation of these badly damped plant poles with the zeros contributed by the PID controller possible. The derivative gain is computed as follows. In Fig. 12.15 we present the root locus diagram corresponding to the open-loop transfer function in (12.18), (12.19), as k_{d2} goes from 0 to $+\infty$. Note that, although the open-loop poles at $s_8 = -6.304 + 61.7989j$ and $s_9 = -6.304 - 61.7989j$, are successfully cancelled out, branches starting at the poles $s_{10} = -8.946 + 7.1167j$ and $s_{11} = -8.946 - 7.1167j$ converge to an asymptote close to the imaginary axis as k_{d2} increases. Hence, the deteriorating performance. On the other hand, a large k_{d2} moves a real closed-loop pole toward the left, thus rendering the system response faster. The following values $k_{d2} = 0.0008$, $k_{d2} = 0.0018$, and $k_{d2} = 0.0028$ have been chosen. Using these data, (12.19) and (12.17), the following PID controller gains are obtained for each one of these cases:

$$k_{p2} = 0.01009, \quad k_{d2} = 0.0008, \quad k_{i2} = 3.087, \tag{12.20}$$

$$k_{p2} = 0.02269, \quad k_{d2} = 0.0018, \quad k_{i2} = 6.946, \tag{12.21}$$

$$k_{p2} = 0.03530, \quad k_{d2} = 0.0028, \quad k_{i2} = 10.8. \tag{12.22}$$

These three sets of PID controller gains must be employed together with the PD gains in (12.16). The system response is presented in Fig. 12.16 when using these three sets of PID controller gains. From these experimental results, it is

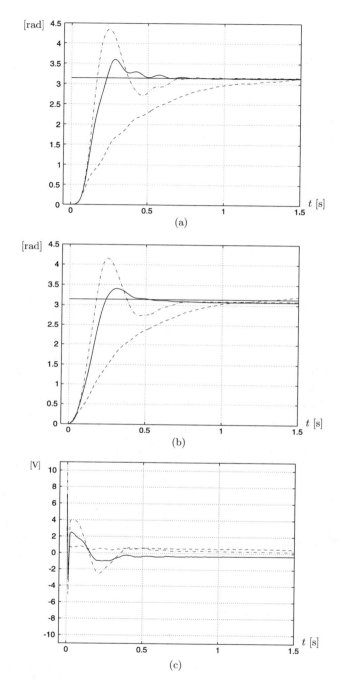

Fig. 12.16 PID control of θ_2. Experimental results. Continuous: PID gains in (12.21). Dashed: PID gains in (12.20). Dash–dot: PID gains in (12.22). Desired load position $\theta_{2d} = \pi$. (a) θ_2. (b) θ_1. (c) Applied voltage, u

concluded that the PID controller gains in (12.21) achieve the best performance. In Fig. 12.17, the simulation response obtained when a unit step change is commanded on the desired load position is presented. The Bode diagrams obtained from the corresponding closed-loop transfer function are also presented. The reason for this is to show the reader that the same information can be obtained from the time response and from the frequency response: i) The PID gains in (12.20) produce the slowest response (the smallest bandwidth), without overshoot (without a resonance peak). ii) The PID gains in (12.22) produce the fastest response (the largest bandwidth) with a clear overshoot (a clear resonance peak). iii) The PID gains in (12.21) produce an intermediate response time (an intermediate bandwidth) with a very small overshoot (a negligible resonance peak).

In Fig. 12.16, it is observed that $t_r = 0.23[s]$ and $M_p = 14.53\%$ have been obtained experimentally as rise time and overshoot respectively for the PID gains in (12.21). On the other hand, in Fig. 12.17, it is observed that $t_r = 0.294[s]$ and $M_p = 3\%$ have been obtained in simulation as the rise time and overshoot respectively for the PID gains in (12.21). The differences between these values are reasonable; hence, it is concluded that the identified model is accurate enough for our purposes. Furthermore, the experimental responses in Fig. 12.16 and the closed-loop poles shown in Fig. 12.15 for each set of the above PID gains, corroborate the predictions in the previous paragraphs on the system performance. Also note that a zero steady-state error is accomplished. This is possible, despite the presence of nonlinear static friction, because of the integral action of the PID controller used in the external loop.

On the other hand, recall that the proposed design procedure is intended to cancel out the plant's badly damped open-loop poles with the PID controller zeros. Hence, it is natural to wonder what would happen if such poles are not successfully cancelled out. This problem may appear because of inaccuracies in model identification that are always present to some extent. In Figs. 12.18 and 12.19 some experimental results are presented when using the internal loop PD controller gains in (12.16), but the external loop PID controller gains are computed such that the PID zeros do not cancel out the open-loop poles at $-6.304 \pm 61.7989j$. In Figs. 12.20 and 12.21 some root locus diagrams are presented corresponding to the closed-loop diagram in Fig. 12.13 when the PID zeros are selected as in Figs. 12.18 and 12.19. The location of the corresponding closed-loop poles are indicated with triangles.

In Figs. 12.22 and 12.23, the simulation closed-loop response and Bode diagrams for the above cases are presented. Note that the experimental and simulation time responses in Figs. 12.18 and 12.22 respectively are similar for the purposes of this study. However, what we want to compare are the experimental time responses to information provided by the Bode diagrams. Consider, for instance, the case of the PID zeros at $-14 \pm 77j$. It is observed in Fig. 12.23 that the resonance peak exists at $\omega \approx 10[rad/s]$. By measuring the oscillation period in Fig. 12.18, it can be verified that the overshoot in time response is due to a frequency component of approximately $10[rad/s]$. On the other hand, it is also observed that the experimental time response contains small-amplitude, high-frequency oscillations.

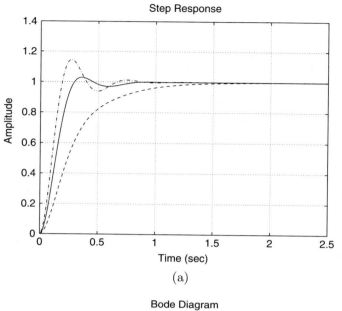

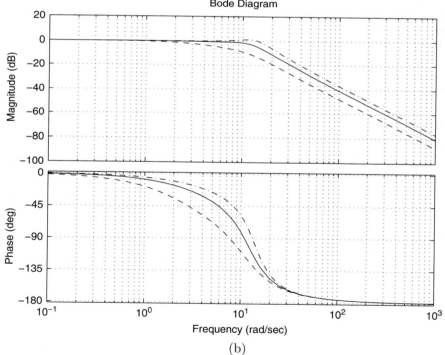

Fig. 12.17 PID control of θ_2. Simulation results. Continuous: PID gains in (12.21). Dashed: PID gains in (12.20). Dash–dot: PID gains in (12.22). (**a**) Time response of θ_2 when a unit step change is commanded. (**b**) Closed-loop Bode diagrams

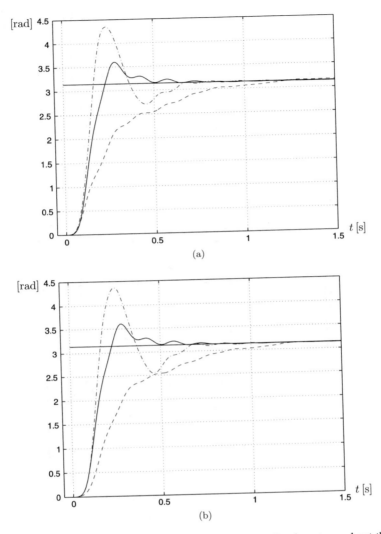

Fig. 12.18 PID control of θ_2 when the zeros of the PID controller do not cancel out the open-loop poles of the plant located at $-6.304 \pm 61.7989\,j$ exactly. Experimental results. Continuous: PID controller gains in (12.21). (**a**) Dashed: PID zeros at $-14 \pm 45\,j$. Dash–dot: PID zeros at $-14 \pm 77\,j$. (**b**) Dashed: PID zeros at $-3.2 \pm 45\,j$. Dash–dot: PID zeros at $-3.2 \pm 77\,j$. In all cases $k_{d2} = 0.0018$. (**a**) θ_2. (**b**) θ_2

These oscillations can be explained by the small resonance peak appearing at $\omega \approx 60$[rad/s], which is due to the inaccurate cancellation of the badly damped open-loop poles at $-6.304 \pm 61.7989\,j$. From Fig. 12.21a, note that there are two closed-loop complex conjugate poles, which are close to these open-loop complex conjugate poles. We stress that, although appreciable, the effects of these closed-

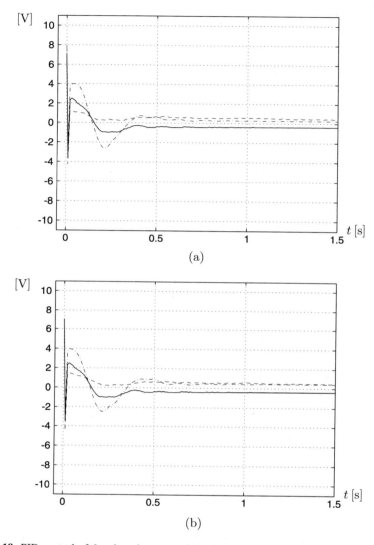

Fig. 12.19 PID control of θ_2 when the zeros of the PID controller do not cancel out the open-loop poles of the plant located at $-6.304 \pm 61.7989j$ exactly. Experimental results. Continuous: PID controller gains in (12.21). (**a**) Dashed: PID zeros at $-14 \pm 45j$. Dash–dot: PID zeros at $-14 \pm 77j$. (**b**) Dashed: PID zeros at $-3.2 \pm 45j$. Dash–dot: PID zeros at $-3.2 \pm 77j$. In all cases $k_{d2} = 0.0018$. (**a**) Applied voltage, u. (**b**) Applied voltage, u

loop complex conjugate poles are compensated for very fast by the presence of the PID zeros, which are also closed-loop zeros. The effects of these zeros are (see Fig. 12.23a): i) A phase lag contributed by a pair of complex conjugate closed-loop poles is only observed in a limited range of frequencies, i.e., $\omega \in [40, 200][\text{rad/s}]$.

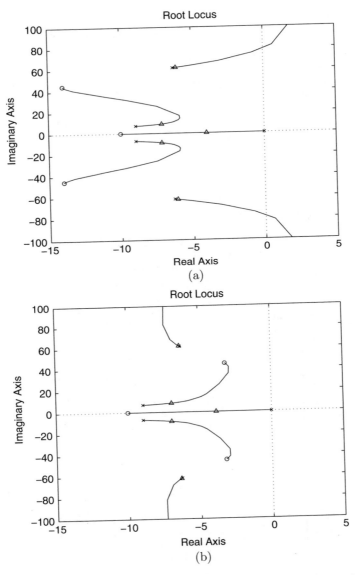

Fig. 12.20 Root locus diagrams when the PID zeros do not cancel out the undesired open-loop poles at $-6.304 \pm 61.7989j$ exactly. Triangles represent closed-loop poles in each case. In all cases $k_{d2} = 0.0018$. (a) PID zeros at $-14 \pm 45j$. (b) PID zeros at $-3.2 \pm 45j$

ii) The resonance peak at $\omega \approx 60[\text{rad/s}]$ is kept small by the pair of complex conjugate closed-loop zeros. This means that even an inexact cancellation of the badly damped poles is useful to attenuate the undesired oscillations due to the

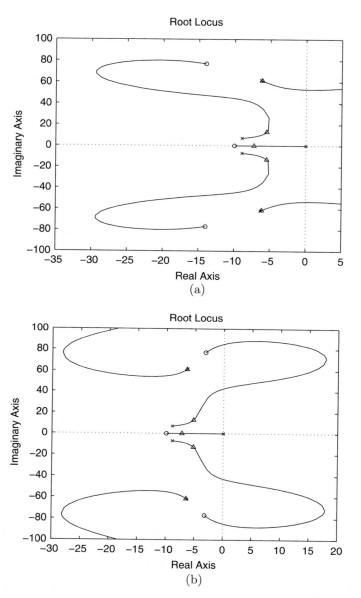

Fig. 12.21 Root locus diagrams when the PID zeros do not cancel out the undesired open-loop poles at $-6.304 \pm 61.7989j$ exactly. Triangles represent closed-loop poles in each case. In all cases $k_{d2} = 0.0018$. (**a**) PID zeros at $-14 \pm 77j$. (**b**) PID zeros at $-3.2 \pm 77j$

system flexibility. Hence, these results support the arguments that we have employed to propose our design procedure. Note that similar conclusions can be obtained by analyzing the system responses for the other cases in the figures referred to.

12.3.1.1   Design Based on the Root Locus

Once the problem is understood, the controller grains can be computed using the root locus. The internal PD controller gains and the external PID controller gains are computed. According to (12.17) and (12.18), the external PID controller gains can be computed using the root locus once the poles of $G_3(s)$ are known. Recall that $G_3(s)$ has four poles, a pair of them are the badly damped open-loop poles, which have to be cancelled out by zeros of the PID controller. Suppose that this is the case. Hence, it can be assumed that the cascade connection of the external PID controller and $G_3(s)$ only has two poles located at p_1 and p_2 on the left half-plane and one pole at the origin. Also note that $G_3(s)$ has one zero at $-k_{p1}/k_{d1}$. According to Fig. 12.24, given a desired closed-loop pole at s_d, we can use the root locus to compute the required location of the poles at p_1 and p_2. Recall that the root locus method performs a design by satisfying two conditions: the magnitude condition and the angle condition. Thus, these two conditions are enough to solve for two unknowns, i.e., p_1 and p_2. However, $G_3(s)$ possesses another unknown: the zero at $-k_{p1}/k_{d1}$. To eliminate this third unknown it is proposed to modify the internal loop: instead of using a PD controller, the following proportional controller with velocity feedback is designed to control the motor position θ_1:

$$U(s) = k_{p1}(V_1(s) - \theta_1(s)) - k_{d1}s\theta_1(s).$$

The reader can verify that the use of this controller modifies $G_3(s)$ to:

$$G_3(s) = \frac{\frac{Kk_m}{RI_1I_2}k_{p1}}{N(s)},$$

with $N(s)$ a polynomial that is identical to the polynomial given in (12.14). This means that the poles of $G_3(s)$ remain unchanged, but now $G_3(s)$ does not possess any zeros. Hence, it is now possible to apply the root locus method to only compute p_1 and p_2, as in Fig. 12.24.

As a result of the procedure in the previous paragraph, we obtain the PID controller gains k_{p2}, k_{d2}, k_{i2}, and the specific values p_1, p_2, for a pair of poles of $G_3(s)$. According to (12.14), the denominator coefficients of $G_3(s)$ depend on the design parameters k_{p1} and k_{d1}. Hence, two poles of $G_3(s)$ can be directly assigned at p_1 and p_2 by suitable selection of k_{p1} and k_{d1}. In the following, we explain how this procedure is to be performed.

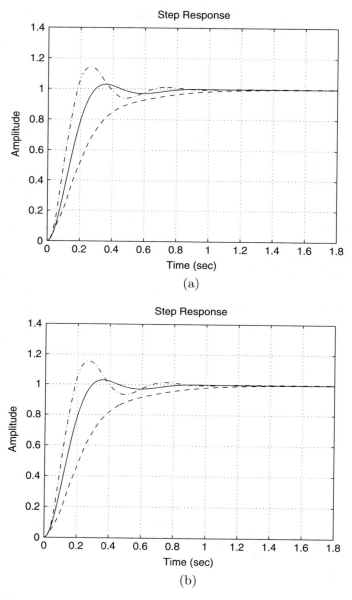

Fig. 12.22 Simulation results. Step response when the PID zeros do not cancel out the undesired open-loop poles at $-6.304 \pm 61.7989j$ exactly. Continuous: PID zeros at $-6.304 \pm 61.7989j$. (**a**) Dashed: PID zeros at $-14 \pm 45j$, dash–dot: PID zeros at $-14 \pm 77j$. (**b**) Dashed: PID zeros at $-3.2 \pm 45j$, dash–dot: PID zeros at $-3.2 \pm 77j$. In all cases $k_{d2} = 0.0018$. (**a**) θ_2. (**b**) θ_2

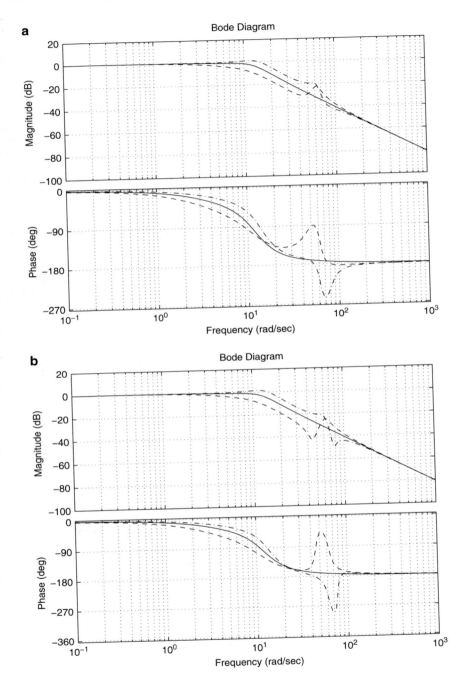

Fig. 12.23 Closed-loop Bode diagrams when the PID zeros do not cancel out badly damped open-loop poles at $-6.304 \pm 61.7989\,j$ exactly. Continuous: PID zeros at $-6.304 \pm 61.7989\,j$. (**a**) Dashed: PID zeros at $-14 \pm 45\,j$, dash–dot: PID zeros at $-14 \pm 77\,j$. (**b**) Dashed: PID zeros at $-3.2 \pm 45\,j$, dash–dot: PID zeros at $-3.2 \pm 77\,j$. In all cases $k_{d2} = 0.0018$

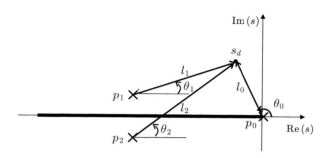

Fig. 12.24 Computation of the open-loop poles at p_1 and p_2, given a desired closed-loop pole at s_d

Given some values for the desired rise time t_r and overshoot M_p, the desired closed-loop pole, $s_d = \text{Re}(s_d) \pm \text{Im}(s_d)j$, can be computed as:

$$\zeta = \sqrt{\frac{\ln^2(M_p/100)}{\ln^2(M_p/100) + \pi^2}}, \quad \omega_n = \frac{1}{t_r\sqrt{1-\zeta^2}}\left(\pi - \arctan\left(\frac{\sqrt{1-\zeta^2}}{\zeta}\right)\right),$$

$$\text{Re}(s_d) = -\zeta\omega_n, \quad \text{Im}(s_d) = \omega_n\sqrt{1-\zeta^2}.$$

Assume that p_1 and p_2 are complex conjugate, i.e., $p_1 = \text{Re}(p_1) + \text{Im}(p_1)j$ and $p_2 = \text{Re}(p_2) + \text{Im}(p_2)j$, with $\text{Re}(p_1) = \text{Re}(p_2)$ and $\text{Im}(p_1) = -\text{Im}(p_2)$. From Fig. 12.24 we have:

$$\theta_0 = 180° - \arctan\left(\frac{\text{Im}(s_d)}{-\text{Re}(s_d)}\right),$$

$$\theta_2 = 180° - (\theta_0 + \theta_1),$$

$$\tan(\theta_2) = \frac{\text{Im}(s_d) - \text{Im}(p_2)}{-(\text{Re}(p_2) - \text{Re}(s_d))}, \quad \tan(\theta_1) = \frac{\text{Im}(s_d) - \text{Im}(p_1)}{-(\text{Re}(p_1) - \text{Re}(s_d))}.$$

Based on the several root loci depicted above and to simplify the problem, we propose θ_1 to be:

$$\theta_1 = 15°.$$

Solving the above equations we have:

$$\text{Re}(p_1) = \frac{\text{Re}(s_d)\left(1 + \frac{\tan(\theta_1)}{\tan(\theta_2)}\right) - 2\frac{\text{Im}(s_d)}{\tan(\theta_2)}}{1 + \frac{\tan(\theta_1)}{\tan(\theta_2)}},$$

$$\text{Im}(p_1) = \text{Im}(s_d) + [\text{Re}(p_1) - \text{Re}(s_d)]\tan(\theta_1).$$

Proposing $t_r = 0.23[s]$ and $M_p = 10\%$, the above formulas yield:

$$p_1 = -24.9126 + 4.7855j, \quad p_2 = -24.9126 - 4.7855j. \quad (12.23)$$

Finally:

$$l_0 = \sqrt{\text{Im}^2(s_d) + \text{Re}^2(s_d)},$$

$$l_1 = \sqrt{(\text{Im}(s_d) - \text{Im}(p_1))^2 + (\text{Re}(p_1) - \text{Re}(s_d))^2},$$

$$l_2 = \sqrt{(\text{Im}(s_d) + \text{Im}(p_1))^2 + (\text{Re}(p_1) - \text{Re}(s_d))^2},$$

and, from the magnitude condition:

$$k_{d2} = \frac{l_0 l_1 l_2}{\frac{K k_m}{R I_1 I_2} k_{p1}}. \quad (12.24)$$

The above procedure results in precise values for p_1 and p_2. Now, it is required to compute suitable values for k_{p1} and k_{d1}, which render p_1 and p_2 poles of $G_3(s)$, i.e., the residue must be zero when dividing $N(s)$ by a second-order polynomial with p_1 and p_2 as roots:

$$\frac{N(s)}{s^2 - (p_1 + p_2)s + p_1 p_2} =$$

$$s^2 + (a_3 + p_1 + p_2)s + c + \frac{[a_1 - p_1 p_2(a_3 + p_1 + p_2) + c(p_1 + p_2)]s + a_0 - p_1 p_2 c}{s^2 - (p_1 + p_2)s + p_1 p_2},$$

$$c = (a_2 - p_1 p_2) + (p_1 + p_2)(a_3 + p_1 + p_2),$$

$$a_1 - p_1 p_2(a_3 + p_1 + p_2) + c(p_1 + p_2) = 0, \quad a_0 - p_1 p_2 c = 0, \quad (12.25)$$

with the polynomial $N(s)$ and the coefficients a_0, a_1, a_2, a_3, defined in (12.14). Note that a_0, a_1, a_2, a_3 depend on k_{p1} and k_{d1}. Thus, both conditions in (12.25) can be solved for k_{p1} and k_{d1} to find:

$$n_3 = 5.043 \times 10^5, \quad d_4 = 1.974 \times 10^4, \quad d_3 = 3988, \quad d_2 = 13,$$

$$f_1 = p_1 p_2 d_3 + p_1 p_2(p_1 + p_2)d_2 - (p_1 p_2)^2 + p_1 p_2(p_1 + p_2)^2,$$

$$f_2 = n_3 - 175 p_1 p_2, \quad f = \frac{f_1}{f_2},$$

$$g = \frac{3 \times 175 p_1 p_2 + 175 p_1 p_2(p_1 + p_2)}{f_2},$$

$$d = (p_1 p_2 - (p_1 + p_2)^2)d_2 - d_4 - (p_1 + p_2)d_3 + 2p_1 p_2(p_1 + p_2)$$
$$-(p_1 + p_2)^3 - (3 \times 175 + 175(p_1 + p_2))f,$$
$$t = 175 \times 53.68^2 + 175((p_1 + p_2)^2 - p_1 p_2) + 3 \times 175(p_1 + p_2),$$
$$q = t + (3 \times 175 + 175(p_1 + p_2))g,$$
$$k_{d1} = \frac{d}{q},$$
$$u_a = p_1 p_2 d_3 + p_1 p_2(p_1 + p_2)d_2 - (p_1 p_2)^2 + p_1 p_2(p_1 + p_2)^2$$
$$+(3 \times 175 p_1 p_2 + 175 p_1 p_2(p_1 + p_2))k_{d1},$$
$$k_{p1} = \frac{u_a}{f_2}.$$

Using these expressions and (12.23), the following gains for the internal controller are found:

$$k_{p1} = 4.3038, \quad k_{d1} = 0.3124. \tag{12.26}$$

Using these values, the poles of $G_3(s)$ can be computed to be located at:

$$-8.9191 \pm 57.3838 j, \quad -24.9126 \pm 4.7855 j.$$

As the zeros of the external PID controller must be chosen to cancel out the badly damped poles of $G_3(s)$, choose:

$$z_1 = 8.9191 + 57.3838 j, \quad z_2 = 8.9191 - 57.3838 j$$

Using this, (12.24), (12.26), and:

$$\frac{k_{p2}}{k_{d2}} = z_1 + z_2, \quad \frac{k_{i2}}{k_{d2}} = z_1 z_2,$$

the following gains for the external PID controller are found:

$$k_{p2} = 0.04149, \quad k_{d2} = 0.002326, \quad k_{i2} = 7.844. \tag{12.27}$$

In Fig. 12.25, some experimental results obtained with the controller gains in (12.26) and (12.27) are presented. In this experiment, it is observed that the rise time is 0.26[s] and overshoot is 8.19%. Note that these values are close to the desired values $t_r = 0.23$[s] and $M_p = 10\%$. These results corroborate the correctness of our design procedure.

On the other hand, carefully observing Fig. 12.25 it is possible to note that the control signal obtained with the design based on the root locus, i.e., with the controller gains in (12.26), (12.27), does not remain constant once the mechanism

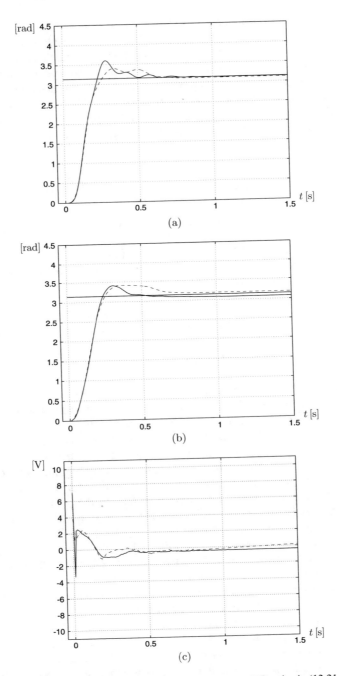

Fig. 12.25 PID control of θ_2. Experimental results. Continuous: PID gains in (12.21). Dashed: the design of the internal controller for θ_1 and the external PID controller for θ_2 is carried out using the root locus. The controller gains are given in (12.26) and (12.27) respectively. (**a**) θ_2. (**b**) θ_1. (**c**) Applied voltage, u

stops moving. Moreover, this response is shown on a larger time axis in Fig. 12.26 where some slow oscillations are observed in both θ_2 and the applied voltage, which eventually vanish. This behavior is reminiscent of a limit cycle that may be produced by a dead zone nonlinearity due to static friction at the motor shaft. To explain this phenomenon, in Sect. 8.3, it is suggested that the equivalent block diagrams shown in Fig. 12.27 might be obtained. The block diagram in Fig. 12.27a corresponds to the control scheme used to obtain the results represented by a continuous line in Fig. 12.25. This results in the equivalent block diagram in Fig. 12.27c, where:

$$H_1(s) = \left(k_{p2} + k_{d2}s + k_{i2}\frac{1}{s} \right) + \frac{s^2 + \frac{b_2}{I_2}s + \frac{K}{I_2}}{\frac{K}{I_2}},$$

$$G(s) = (k_{p1} + k_{d1}s)H_1(s)G_{2u}(s). \tag{12.28}$$

The block diagram in Fig. 12.27b corresponds to the control scheme used to obtain the results represented by a dashed line in 12.25. This results in the equivalent block diagram in Fig. 12.27c where:

$$H_2(s) = k_{p1}H_1(s) + k_{d1}s\frac{s^2 + \frac{b_2}{I_2}s + \frac{K}{I_2}}{\frac{K}{I_2}},$$

$$G(s) = H_2(s)G_{2u}(s). \tag{12.29}$$

The Bode diagrams and polar plots of $G(s)$ in (12.28) and (12.29), using the corresponding numerical values, are shown in Fig. 12.28. Note that the polar plot of $G(s)$ in (12.29) crosses the negative real axis at the left of -1. Hence, according to Sect. 8.3, a limit cycle exists. On the contrary, as this is not the case for the polar plot of $G(s)$ in (12.28) a limit cycle does not exist in that case. Moreover, the Bode diagrams for both cases are similar. The only significant difference is that the Bode diagrams for $G(s)$ in (12.29) show a small phase lag, which is responsible for crossing the $-180°$ phase line, forcing the polar plot to cross the negative real axis at the left of -1. Recall that, aside from the numerical values, the only significant difference between $G(s)$ in (12.28) and (12.29) is the use of a PD controller and a proportional controller with velocity feedback respectively in the internal loop. Moreover, the reader may plot the Bode diagrams and the polar plot for $G(s)$ in (12.28) using the numerical values in (12.26) and (12.27) to verify that a limit cycle does not exist in such a case. Hence, it is concluded that the zero of the internal PD controller is instrumental in avoiding the appearance of limit cycles. Note that, although the limit cycle eventually disappears, the induced oscillations worsen performance a little. Also note that using a PD controller instead of a proportional controller with velocity feedback together with the gains in (12.26) and (12.27) results in a transient response that does not satisfy the design requirements $t_r = 0.23$[s] and $M_p = 10\%$. Of course, additional fine tuning is always possible. Thus, if the observed performance deterioration is acceptable, the

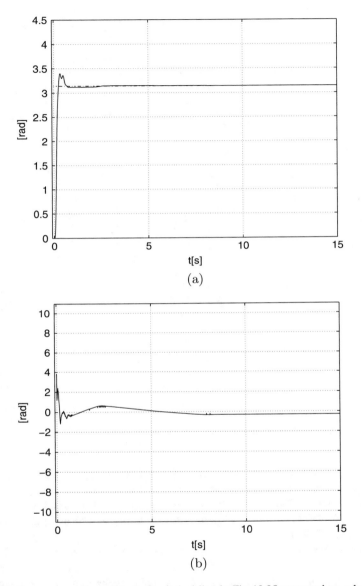

Fig. 12.26 Response corresponding to the dashed line in Fig. 12.25 presented on a larger time axis. (**a**) θ_2. (**b**) Applied voltage, u

control design based on the root locus is a good solution. If slow oscillation is not desired, the control scheme with an internal loop driven by a PD controller and an external loop driven by a PID controller, with controller gains in (12.16), (12.21), is preferred.

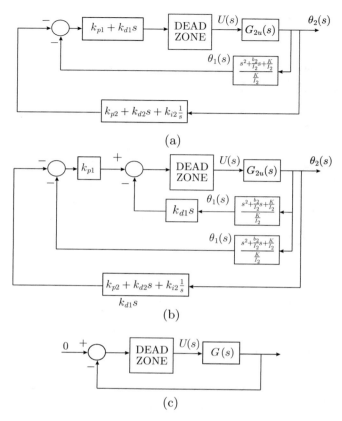

Fig. 12.27 Closed-loop diagrams for limit cycle analysis

Finally, according to Sect. 6.5.2 an additional phase lag of $-\omega T$ [rad] must be considered in Fig. 12.28a to take into account the effects of time delay induced by the digital implementation of controllers in experiments. The sampling period used in these experiments was 0.01[s]; hence, $T = 0.01$[s] may be assumed to consider the worst case time delay.

The crossover frequency in Fig. 12.28a for $G(s)$ in (12.28) is $\omega \approx 13$[rad/s] and $-\omega T \times 180°/\pi = -7.44°$ is the additional maximal phase lag contributed by the digital implementation of the controllers. This results in an approximately 43° phase margin. Thus, good performance is still expected in the experiments, which is corroborated in Fig. 12.25.

On the other hand, the crossover frequency in Fig. 12.28a for $G(s)$ in (12.29) is $\omega \approx 32$[rad/s] and $-\omega T \times 180°/\pi = -18.33°$ is the additional maximal phase lag contributed by the digital implementation of controllers. This results in an approximate 51° phase margin. Thus, good performance is still expected in experiments, which is corroborated again in Fig. 12.25.

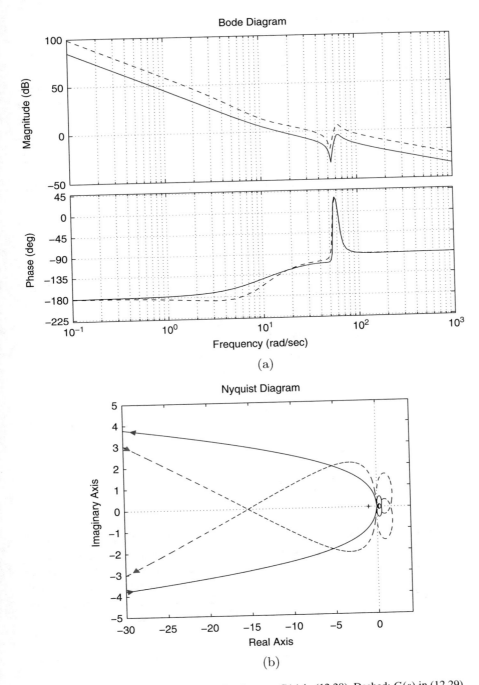

Fig. 12.28 (**a**) Bode and (**b**) polar plots. Continuous: $G(s)$ in (12.28). Dashed: $G(s)$ in (12.29)

12.3.1.2 Another Nested Control Scheme

In the previous study we designed an inner loop for the motor position θ_1, driven by a PD controller, and an external loop for the load position θ_2, driven by a PID controller. The reader may wonder why not design a PID controller for θ_1 and a PD controller for θ_2? Motivated by this question, such a control scheme is designed in the following. The proposed control scheme is described by:

$$U(s) = \left(k_{p1} + k_{d1}s + k_{i1}\frac{1}{s}\right)(V_2(s) - \theta_1(s)), \qquad (12.30)$$

$$V_2(s) = (k_{p2} + k_{d2}s)(\theta_{2d}(s) - \theta_2(s)), \qquad (12.31)$$

and is depicted in Fig. 12.29a. Replace $T(s) = k_m U(s)/R - k_m k_b s \theta_1(s)/R$ and (12.30) in (12.3), and solve for $\theta_1(s)$, then replace in (12.4) and solve for $\theta_2(s)$ to obtain:

$$\theta_2(s) = \frac{\frac{k_m K}{R I_1 I_2}(k_{d1}s^2 + k_{p1}s + k_{i1})}{M(s)} V_2(s), \qquad (12.32)$$

$$M(s) = s^5 + c_4 s^4 + c_3 s^3 + c_2 s^2 + c_1 s + c_0,$$

$$c_4 = \frac{b_2}{I_2} + \frac{b_1}{I_1} + \frac{k_m k_b}{R I_1} + \frac{k_m k_{d1}}{R I_1},$$

$$c_3 = \frac{K}{I_1} + \frac{k_{p1} k_m}{R I_1} + \frac{b_2}{I_2}\left(\frac{b_1}{I_1} + \frac{k_m k_b}{R I_1} + \frac{k_m k_{d1}}{R I_1}\right) + \frac{K}{I_2},$$

$$c_2 = \frac{k_m k_{i1}}{R I_1} + \frac{b_2}{I_2}\left(\frac{K}{I_1} + \frac{k_{p1} k_m}{R I_1}\right) + \frac{K}{I_2}\left(\frac{b_1}{I_1} + \frac{k_{d1} k_m}{R I_1} + \frac{k_b k_m}{R I_1}\right),$$

$$c_1 = \frac{b_2}{I_2}\frac{k_m k_{i1}}{R I_1} + \frac{K k_{p1} k_m}{R I_1 I_2},$$

$$c_0 = \frac{K k_m k_{i1}}{R I_1 I_2}.$$

This results in the closed-loop system shown in Fig. 12.29b. Note that the open-loop transfer function in Fig. 12.29b has no pole at the origin, as $c_0 \neq 0$. Hence, if the desired position θ_{2d} is a constant in time, then a nonzero steady-state error is obtained. Thus, to ensure a zero steady-state error at least one pole at the origin is required in the open-loop transfer function. This means that a PID controller is necessary for the load position θ_2, i.e., the following must be used:

$$V_2(s) = \left(k_{p2} + k_{d2}s + k_{i2}\frac{1}{s}\right)(\theta_{2d}(s) - \theta_2(s)), \qquad (12.33)$$

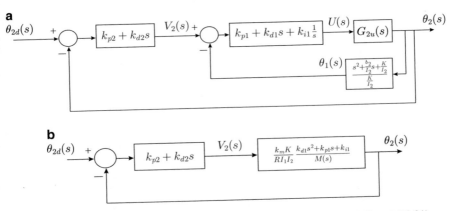

Fig. 12.29 Block diagram of the closed-loop control system when using (12.30) and (12.31)

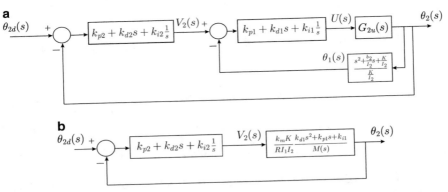

Fig. 12.30 Block diagrams of the closed-loop control system when using (12.30) and (12.33)

instead of (12.31). Then, the closed loop system in Fig. 12.30 is obtained. As proposed in the previous study, the idea is to use the zeros of the PID controller in (12.33) to cancel out two badly damped poles of (12.32). This defines two gains of the external PID controller, i.e., k_{p2} and k_{i2}. Hence, the closed-loop system has four poles and there are four remaining controller gains, i.e., $k_{p1}, k_{d1}, k_{i1}, k_{d2}$. It is possible to proceed to obtain the closed-loop characteristic polynomial to show that its four coefficients can be arbitrarily assigned using $k_{p1}, k_{d1}, k_{i1}, k_{d2}$. This means that the closed-loop poles can be assigned arbitrarily by suitably selecting these controller gains. Hence, it is possible to compute the controller gains exactly by equating the closed-loop characteristic polynomial as $s^4 + d_3 s^3 + d_2 s^2 + d_1 s + d_0 = (s - p_1)(s - p_2)(s - p_3)(s - p_4)$, where $p_i, i = 1, \ldots, 4$, are the desired closed-loop poles. However, this is not performed because it is more important for us to show that sustained oscillations exist in the closed-loop response. This is shown below.

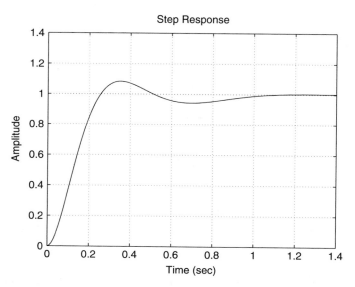

Fig. 12.31 Simulation results. Response of the load position θ_2 when using the controller gains in (12.34)

Using a trial and error procedure it is found that the following controller gains:

$$k_{p1} = 0.8, \quad k_{d1} = 0.12, \quad k_{i1} = 3, \tag{12.34}$$

$$k_{p2} = 0.02535, \quad k_{d2} = 0.002, \quad k_{i2} = 7.625,$$

produce the closed-loop response shown in Fig. 12.31, where $t_r = 0.254[\text{s}]$ and $M_p = 8\%$. In Fig. 12.32 the response obtained experimentally is presented when using the controller gains in (12.34). It is observed that the rise time and overshoot values are 0.22[s] and 16.25% respectively. Note that these experimental data are close to data obtained in simulations. However, it is observed in Fig. 12.32 that it seems that the system has not finished responding. Hence, the same response is presented over a larger interval of time in Fig. 12.33. Note that sustained oscillations exist. This is known as a *limit cycle*. According to Sect. 8.3, a limit cycle may appear when hard nonlinearities are present. In the present mechanism, a hard saturation called *dead zone* appears as an effect of static friction present at the motor shaft. To study this limit cycle, the closed loop system in Fig. 12.30 must be represented as shown in Fig. 12.34a. From this diagram Fig. 12.34b is obtained where:

$$G_4(s) = \left(k_{p2} + k_{d2}s + k_{i2}\frac{1}{s} \right) + \frac{s^2 + \frac{b_2}{I_2}s + \frac{K}{I_2}}{\frac{K}{I_2}},$$

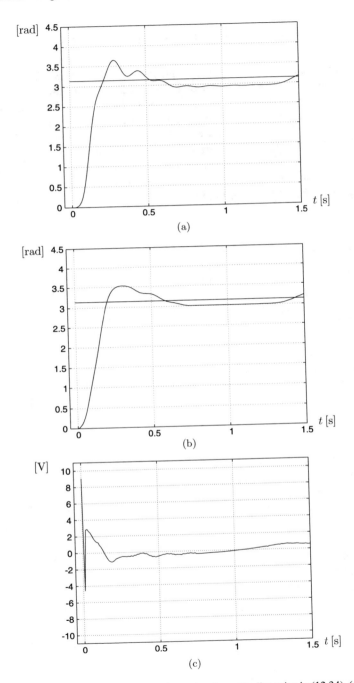

Fig. 12.32 Experimental results obtained when using the controller gains in (12.34). (**a**) θ_2. (**b**) θ_1. (**c**) Applied voltage, u

$$G_5(s) = \left(k_{p1} + k_{d1}s + k_{i1}\frac{1}{s} \right) G_4(s)G_{2u}(s).$$

The Bode and polar plots of the transfer function $G_5(s)$ are presented in Fig. 12.35. Note that the polar plot of $G_5(s)$ crosses the negative real axis at the left of $(-1, 0)$. Hence, according to Sect. 8.3, the existence of a limit cycle is theoretically justified. Moreover, as a $-180°$ phase is obtained at $\omega = 5.73$[rad/s], the limit cycle is theoretically predicted to appear at a frequency of 0.912[Hz]. A frequency of 0.5747[Hz] is measured in Fig. 12.33. Thus, taking into account that uncertainties are always present in theoretical models, it is concluded that these experimental results have been correctly predicted by theory.

The plots in Fig. 12.35 were drawn using the following MATLAB code in an m-file:

```
clc
c=conv([1  0],[1  5]);
d=conv(c,[1  8  62.83^2]);
num1=175*[1  3  53.68^2];
g1=tf(num1,d);
g2u=tf(520000,d);
%%% PID1
ki1=3;
kp1=0.8;
kd1=0.12;
PID1=tf([kd1 kp1 ki1],[1 0]);
%%% PID2
kd2=0.002;
kp2=0.02535;
ki2=7.625;
PID2=tf([kd2 kp2 ki2],[1 0]);
fg=tf([1  3  53.68^2],53.68^2);
g4=parallel(PID2,fg);
g5=PID1*g4*g2u;
figure(1)
bode(g5,{1,1000})
grid on
figure(2)
nyquist(g5,{5,1000})
t=0:0.001:15;
%%{
figure(3)
plot(t,Datos(:,1),'k-');
axis([-0.5 15 0 4])
ylabel('\theta_2 [rad]')
xlabel('t [s]')
```

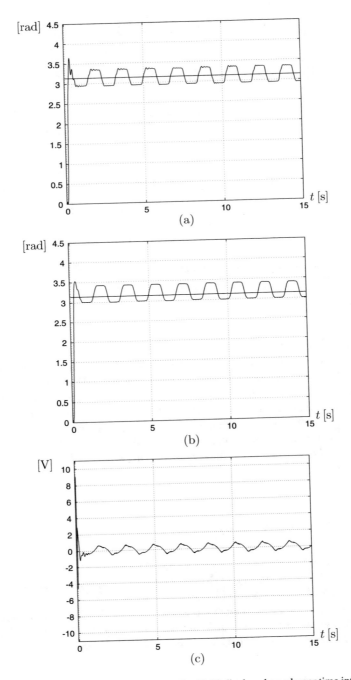

Fig. 12.33 Experimental results. Response in Fig. 12.32 displayed on a larger time interval. (**a**) θ_2.
(**b**) θ_1. (**c**) Applied voltage, u

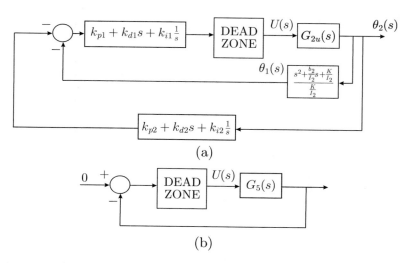

Fig. 12.34 Closed-loop diagrams for limit cycle analysis

```
grid on
%}
```

In Fig. 12.36 the MATLAB/Simulink diagram used to perform the simulation in Fig. 12.37 is shown. This simulation is executed under the same conditions as the experimental results in Fig. 12.33. This means that the PID controller parameters were chosen as in (12.34) and $\theta_{2d} = \pi$ [rad]. The To work space block saves data in an array with 0.001 as the sampling time. The dead zone block was programmed with $\delta = 0.3$ because this results in oscillations with an amplitude that is similar to that obtained in the experiments. After executing the MATLAB code above, run the simulation in Fig. 12.36 and, finally, execute the MATLAB code above again to obtain Fig. 12.37.

12.3.2   Direct Control of θ_2

According to the discussion at the beginning of this section, the main problem in controlling θ_2 directly is that $G_{2u}(s)$ has poles that are located close to the imaginary axis. Hence, a possible solution to this control problem may be to replace such open-loop poles for some others located at more convenient regions of the complex plane. In the following, we present two different ways of using this approach to design a closed-loop control system such as that depicted in Fig. 12.38.

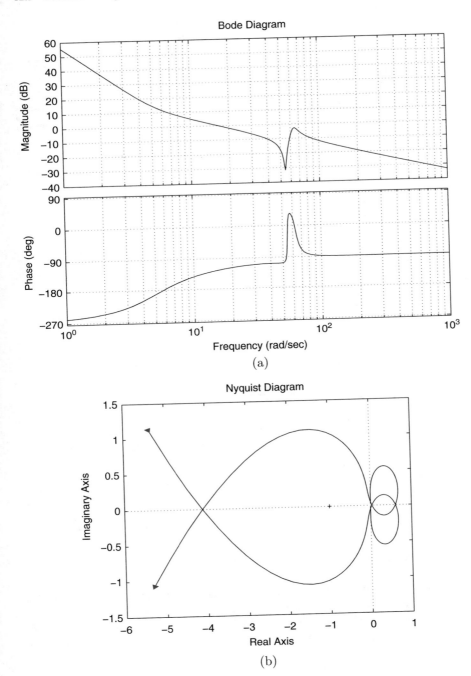

Fig. 12.35 (a) Bode and (b) polar plots of $G_5(s)$

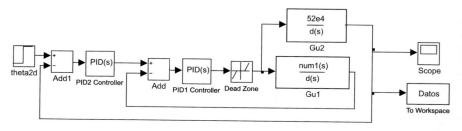

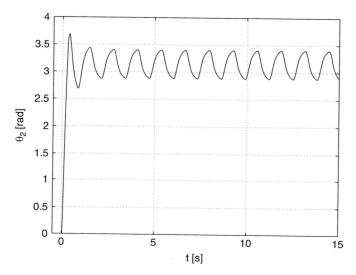

Fig. 12.36 MATLAB/Simulink diagram used to obtain results in Fig. 12.37

Fig. 12.37 Simulation results obtained under the same conditions as the experimental results in Fig. 12.33

Fig. 12.38 Direct control of the load position

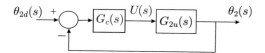

12.3.2.1   Design Based on the Root Locus

First note that it is desirable to maintain the open-loop pole at $s = 0$, as it is useful to satisfy the steady-state response requirements, i.e., the steady-state error. Hence, only three poles must be cancelled out, which has to be performed using three zeros of the controller. This means that, for realizability reasons, the controller must be represented by a transfer function of the form:

$$G_c(s) = \gamma \frac{(s + 5)(s^2 + 8s + 62.83^2)}{(s - p_1)(s - p_2)(s - p_3)}, \tag{12.35}$$

Fig. 12.39 Root locus diagram corresponding to the open-loop transfer function in (12.36)

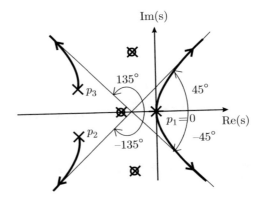

where the poles with nonpositive real parts p_1, p_2, p_3 and the constant $\gamma > 0$ must be determined by a design procedure, in this case, the root locus method.

First, suppose that $p_1 = 0$. The root locus diagram obtained from the open loop transfer function:

$$G_c(s)G_{2u}(s) = \gamma \frac{(s+5)(s^2+8s+62.83^2)}{s(s-p_2)(s-p_3)} \frac{520000}{s(s+5)(s^2+8s+62.83^2)},$$

$$= \gamma \frac{520000}{s^2(s-p_2)(s-p_3)}, \tag{12.36}$$

is shown in Fig. 12.39. Note that four asymptotes exist whose angles can be computed using:

$$\text{Asymptote angles} = \frac{\pm 180^\circ (2q+1)}{n-m}, \quad q = 0, 1, 2, \ldots, \quad n = 4, \ m = 0,$$

to find $\pm 45^\circ$ and $\pm 135^\circ$. As no open-loop zero exists, the branches starting at the pair of open-loop poles located at $s = 0$ are immediately pushed into the complex right half-plane. This means closed-loop instability for any $\gamma > 0$ and any p_2, p_3 in the complex left half-plane. Hence, to minimize the closed-loop oscillations, p_1, p_2, p_3 must be proposed to be real and negative. Moreover, the design based on the root locus method requires two conditions to be satisfied: the angle condition and the magnitude condition. This means that the design problem can be analytically solved if only two unknown quantities exist. In the control problem at hand, this can be accomplished if we propose $p_1 = p_2 = p_3 = p < 0$ as only $p < 0$ and $\gamma > 0$ are hence unknown.

Given some values for the desired rise time t_r and overshoot M_p, the desired closed loop pole $s_d = \text{Re}(s_d) \pm \text{Im}(s_d)j$ can be computed as:

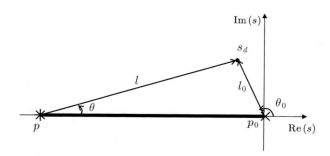

Fig. 12.40 Finding the location of the pole at $p < 0$

$$\zeta = \sqrt{\frac{\ln^2(M_p/100)}{\ln^2(M_p/100) + \pi^2}}, \quad \omega_n = \frac{1}{t_r\sqrt{1-\zeta^2}}\left(\pi - \arctan\left(\frac{\sqrt{1-\zeta^2}}{\zeta}\right)\right),$$

$$Re(s_d) = -\zeta\omega_n, \quad Im(s_d) = \omega_n\sqrt{1-\zeta^2}.$$

Under the condition $p_1 = p_2 = p_3 = p < 0$, the open-loop transfer function becomes:

$$G_c(s)G_{2u}(s) = \gamma\frac{520000}{s(s-p)^3}.$$

The solution of this problem using the root locus method requires us to consider the situation depicted in Fig. 12.40. Note that:

$$\theta_0 = 180° - \arctan\left(\frac{I_m(s_d)}{abs(R_e(s_d))}\right),$$

and, from the angle condition:

$$\theta = \frac{180° - \theta_0}{3}.$$

Moreover:

$$\tan(\theta) = \frac{I_m(s_d)}{-(p - R_e(s_d))} \Rightarrow p = -\left(\frac{I_m(s_d)}{\tan(\theta)} - R_e(s_d)\right),$$

$$l_0 = \sqrt{R_e^2(s_d) + I_m^2(s_d)}, \quad l = \sqrt{(p - R_e(s_d))^2 + I_m^2(s_d)},$$

and, from the magnitude condition:

$$\frac{\gamma 520000}{l^3 l_0} = 1 \Rightarrow \gamma = \frac{l^3 l_0}{520000}.$$

Proposing $t_r = 0.23[s]$ and $M_p = 10\%$ as desired values and using the above expressions the following is found:

$$p = -36.6439, \quad \gamma = 0.6892.$$

Thus, the controller is given as:

$$G_c(s) = 0.6892 \frac{(s+5)(s^2 + 8s + 62.83^2)}{(s + 36.6439)^3}. \tag{12.37}$$

The experimental results obtained with the controller in (12.37) are presented in Fig. 12.41. Note that a 0.28[s] rise time and a 15% overshoot are obtained. We note that a nonzero steady-state error is obtained that is due to nonlinear friction present in the real physical system but is not considered during the modeling stage. To render the steady- state error zero, closing an external loop around the closed-loop system shown in Fig. 12.38 was attempted. This external loop was designed to be driven by either a PI or a PID controller. However, in both cases, bad performances were found. In particular, when using the PI controller, a limit cycle was observed. It is left as an exercise for the reader to mathematically show that such a limit cycle is expected to appear.

12.3.2.2 Design Based on the Polynomial Approach

Another common way of designing the closed-loop control system in Fig. 12.38 is the so-called polynomial-based pole assignment approach, which is described below. First, assume that the controller and the plant are expressed as:

$$G_{2u}(s) = \frac{D(s)}{C(s)}, \quad G_c(s) = \frac{F(s)}{E(s)}, \tag{12.38}$$

$$D(s) = d_0,$$
$$C(s) = s^4 + c_3 s^3 + c_2 s^2 + c_1 s + c_0,$$
$$F(s) = f_3 s^3 + f_2 s^2 + f_1 s + f_0,$$
$$E(s) = e_3 s^3 + e_2 s^2 + e_1 s + e_0.$$

The closed-loop transfer function of the control system in Fig. 12.38 is given as:

$$M(s) = \frac{\theta_2(s)}{\theta_{2d}(s)} = \frac{F(s)D(s)}{E(s)C(s) + F(s)D(s)}.$$

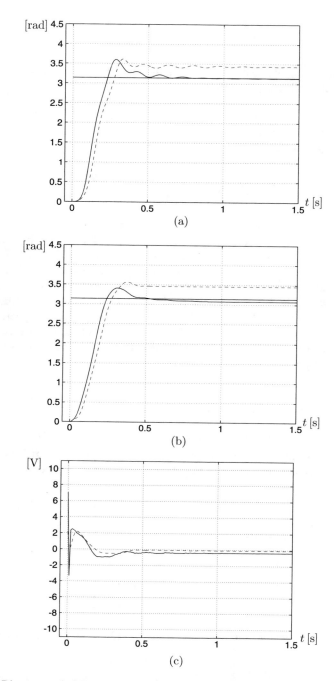

Fig. 12.41 Direct control of the load position θ_2. Experimental results. Continuous: PID controller with gains in (12.21). Dashed: use of controller in (12.37). (**a**) θ_2. (**b**) θ_1. (**c**) Applied voltage, u

The closed-loop poles are the roots of the polynomial at the denominator, i.e., $E(s)C(s) + F(s)D(s)$. This approach consists in assigning:

$$E(s)C(s) + F(s)D(s) = s^7 + a_6 s^6 + a_5 s^5 + a_4 s^4 + a_3 s^3 + a_2 s^2 + a_1 s + a_0,$$

$$= \prod_{i=1}^{7} (s - \lambda_i),$$

where λ_i, $i = 1, \ldots, 7$ are the desired closed-loop poles.

Replacing the polynomials defined in (12.38) in the latter expression and equating the coefficients of polynomials on both sides, the following system of linear algebraic equations is found:

$$Qx = y, \quad x = [e_3, e_2, e_1, e_0, f_3, f_2, f_1, f_0]^T, \quad y = [1, a_6, a_5, a_4, a_3, a_2, a_1, a_0]^T,$$

$$Q = \begin{bmatrix} 1 & 0 & 0 & 0 & 0 & 0 & 0 & 0 \\ c_3 & 1 & 0 & 0 & 0 & 0 & 0 & 0 \\ c_2 & c_3 & 1 & 0 & 0 & 0 & 0 & 0 \\ c_1 & c_2 & c_3 & 1 & 0 & 0 & 0 & 0 \\ c_0 & c_1 & c_2 & c_3 & d_0 & 0 & 0 & 0 \\ 0 & c_0 & c_1 & c_2 & 0 & d_0 & 0 & 0 \\ 0 & 0 & c_0 & c_1 & 0 & 0 & d_0 & 0 \\ 0 & 0 & 0 & c_0 & 0 & 0 & 0 & d_0 \end{bmatrix}.$$

According to theorem 7.1, pp. 181, and lemma 7.1, pp. 82, in [1], the matrix Q is nonsingular if, and only if, the polynomials $D(s)$ and $C(s)$ have no common root. In such a case, an unique solution exists for x, which is given as:

$$x = Q^{-1}y. \qquad (12.39)$$

It is not difficult to realize that in the problem at hand, $D(s)$ and $C(s)$ have no common root, as:

$$D(s) = d_0 = 520000,$$
$$C(s) = s^4 + c_3 s^3 + c_2 s^2 + c_1 s + c_0 = s(s + 5)(s^2 + 8s + 62.83^2).$$
$$= s^4 + 13s^3 + 3988s^2 + 1.974 \times 10^4 s.$$

Using these data to construct the matrix Q, (12.39) and proposing the following desired closed-loop poles:

$$\lambda_1 = -7.0212 + 9.5795j, \quad \lambda_2 = -7.0212 - 9.5795j,$$
$$\lambda_3 = -4 + 62.7025j, \quad \lambda_4 = -4 + 62.7025j,$$
$$\lambda_5 = -40 + 5j, \quad \lambda_6 = -40 - 5j, \quad \lambda_7 = -5,$$

the following is found:

$$e_3 = 1, \quad e_2 = 94.04, \quad e_1 = 2889, \quad e_0 = 3.41 \times 10^4,$$
$$f_3 = 0.4408, \quad f_2 = 5.734, \quad f_1 = 1758, \quad f_0 = 8701.$$

Thus, the controller is given as:

$$G_c(s) = \frac{0.4408s^3 + 5.734s^2 + 1758s + 8701}{s^3 + 94.04s^2 + 2889s + 3.41 \times 10^4}. \tag{12.40}$$

Is it important to stress that a stable solution for this problem is strongly dependent on proposing a suitable set of desired closed-loop poles. This mean that an unstable controller may result for some choices of the desired closed-loop poles. Although stability may be concluded for the closed-loop system, performance obtained in practice deteriorates severely if the controller is unstable or if some of its zeros are in the right half-plane. Hence, such solutions must be avoided.

Some experimental results obtained with the controller in (12.40) are presented in Fig. 12.42. Note that the rise time is 0.288[s] and overshoot is 10%. The reader can verify that the desired closed-loop complex conjugate poles located at $\lambda_1 = -7.0212 + 9.5795j$ and $\lambda_2 = -7.0212 - 9.5795j$, are intended to produce 0.23[s] and 10% as rise time and overshoot respectively. Thus, the experimental response is close to these desired values for the accuracy limits of these experiments. Finally, note that a nonzero steady-state error is obtained, which is due to nonlinear friction present in the real physical system but is not considered during the modeling stage. To render the steady- state error zero, closing an external loop around the closed-loop system shown in Fig. 12.38 was attempted. This external loop was designed to be driven by either a PI or a PID controller. However, in both cases, bad performances were found. In particular, when using the PI controller a limit cycle was observed to. It is left as an exercise for the reader to mathematically show that such a limit cycle is expected to appear.

12.4   Experimental Prototype Construction

The experimental prototype that has been built, consists of a PM brushed DC motor model Tohoko Ricoh 7K00011. The nominal voltage, electric current, and power of this motor are 24[V]CD, 2.3[A], and 55[W] respectively. This motor is provided with a 400 pulses per revolution optical encoder. The load is composed of a cylindrical body made in bronze, measuring 0.03[m] and 0.02[m] in radius and length respectively. The motor shaft and load are joined with a spring made of steel. This spring is normally used as a translational spring, but we use it as a torsional spring (see Fig. 12.43). The load position is measured using a 1000 pulses per revolution optical encoder model S1-1000-250-I-B-D from US Digital.

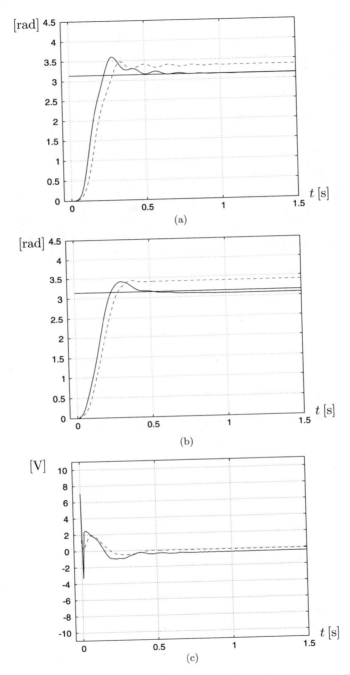

Fig. 12.42 Direct control of the load position θ_2. Experimental results. Continuous: PID controller with gains in (12.21). Dashed: use of controller in (12.40). (**a**) θ_2. (**b**) θ_1. (**c**) Applied voltage, u

Fig. 12.43 The experimental prototype that has been built

The electrical diagram of the complete control system is shown in Fig. 12.44. A PIC16F877A microcontroller is used as a data acquisition board. This microcontroller communicates with a personal computer (Pentium 4, Windows XP), via the serial port. The microcontroller receives the position of both the motor and the load and sends these data to the personal computer. The personal computer evaluates the control algorithm and sends the voltage signal to be applied to the motor back to the microcontroller . A MAX232IC renders communication of the microcontroller via the serial port possible. A L298N IC dual H bridge actuates as a power amplifier and applies the commanded voltage at the motor armature terminals. The L298N IC is configured such that both its H bridges are connected in parallel, to be capable of handling up to 4[A]. This integrated circuit is driven via one of the pulse-width modulation (PWM) modules of the microcontroller. A 74L08 IC, AND logic, is used to adapt the PWM signal to the logic of the L298N dual H bridge.

The C code used to program the microcontroller PIC16F877A is presented in Sect. 12.5. After defining some libraries and constants, an interruption is listed that is devoted to counting pulses from both encoders used to measure the motor and the load positions. In the executable part of the program, there is an infinite cycle represented by a "while(true)" instruction. Within this loop, a recognition code is first sent to the computer using "putc(0xAA);." Then, the motor position and, finally, the load position are sent to the computer via the instructions "putc(cuentaH1); putc(cuentaL1); putc(cuentaH2); putc(cuentaL2);." The instruction "do-while(TMR0<196)" has two purposes: i) To set the sample period to 10[ms]. ii) To wait for the voltage data to be sent back by the computer once the control algorithm has been evaluated. After that, the voltage to be applied to the motor is sent to the PWM module as a duty cycle using "set_pwm1_duty(pwm);."

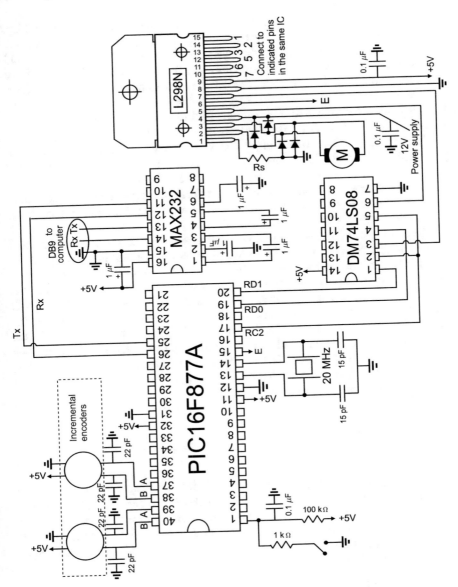

Fig. 12.44 Electric diagram of the complete control system

The conversion to duty cycle requires the voltage information from the lowest 7 bits of the variable "dato" to be extracted, as the highest bit contains the voltage sign information. Once the sample period has elapsed, the do while (TMR0<196) cycle repeats again.

The code of the program executed by the personal computer is listed in Sect. 12.6. After defining some libraries, constants, the controller gains among them, and variables, the executable part of the program starts with the instruction "void ProcessByte(BYTE byte)." The instructions that follow are devoted to waiting for the recognition code "0xAA" to be sent by the microcontroller. Once this code is recognized, the motor and the load positions are received. Then, the motor and the load velocities are computed by numerical differentiation. After that, four control algorithms are coded. The first one is a PD controller used to directly control the load position. This control algorithm was used to perform the experiments presented in Fig. 12.8. The second control algorithm is the multi-loop nested PID controllers used to perform all experiments in Sect. 12.3.1. The third and fourth control algorithms are controllers used to perform the experiments in Figs. 12.41 and 12.42 respectively. The reader is referred to Appendix F for an explanation on how these controllers are numerically implemented.

The control signal or voltage to be applied at the motor terminals is represented by the variable "iast." This variable is saturated, converted into the range [0, 255], and finally confined to the lowest 7 bits of an 8-bit word, as the highest bit is devoted to containing the voltage sign. A byte with this information is sent to the microcontroller by means of the instruction "MainForm->send_byte(pwm);." After that, several variables are saved in a text file named "MONIT.TXT." Finally, the current time is increased by 10[ms] using "t=t+Ts;," the program returns to the beginning of the instruction "void ProcessByte(BYTE byte)" and stays there waiting again for the recognition code "0xAA." We stress that the sample time is fixed by the microcontroller and the computer merely increases the time by 10[ms] when the executable part of program reaches its end.

12.5   Microcontroller PIC16F877A C Program

The flow diagram for the microcontroller PIC16F877A programming when used as an interface between the mechanism with flexibility and a PC, is shown in Fig. 12.45. We refer to [3, 4], for a complete explanation of instructions employed in the program listed below. We do not try to present a detailed explanation of this program. Our intention is merely to show the reader how a microcontroller can be used as part of a closed-loop control system.

```
#include<16f877A.h>
#include<stdlib.h>
#include<math.h>
#fuses HS,NOWDT,PUT,NOBROWNOUT,NOLVP,NOWRT,NOPROTECT,NOCPD
#use delay(clock=20000000) // Xtal freq
#use rs232(baud=115200,XMIT=PIN_C6,RCV=PIN_C7,BITS=8,PARITY=N)
//ports and registers addresses
#byte OPTION= 0x81
#byte TMR0 = 0x01
#byte PORTA = 0x05
```

Fig. 12.45 Flow diagram for microcontroller PIC16F877A programming when used as an interface between a mechanism with flexibility and a PC

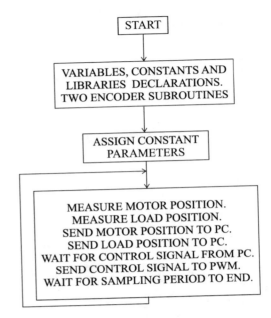

```
#byte PORTB = 0x06
#byte PORTC = 0x07
#byte PORTD = 0x08
#byte PORTE = 0x09

#bit PD0 = 0x08.0
#bit PD1 = 0x08.1
#bit PD2 = 0x08.2
#bit PD3 = 0x08.3
#bit PC0 = 0x07.0
//-----------Variables declaration------------//
int16 inter,cuenta1,cuenta2;
int8 pwm,dato,cuentaH1,cuentaL1,cuentaH2;
int8 cuentaL2,puerto,AB,AB_1,BC,BC_1,aux;
//-----------Encoder interrupt subroutine------------//
#int_rb
void rb_isr()
{
puerto=PORTB;
AB=((puerto)&(0x30))>>4;
aux=AB^AB_1;
if(aux!=0)
if(aux!=3)
if(((AB_1<<1)^AB)&(0x02))
{
cuenta1--;
}
else
{
```

```
cuenta1++;
}
AB_1=AB;

BC=((puerto)&(0xC0))>>6;

aux=BC^BC_1;
if(aux!=0)
if(aux!=3)
if(((BC_1<<1)^BC)&(0x02))
{
cuenta2--;
}
else
{
cuenta2++;
}
BC_1=BC;
}
//-----------main program------------//
void main(void)
{
set_tris_a(0b11111111);
set_tris_b(0b11111111);
set_tris_c(0b10000000); //pin config.
set_tris_d(0b00000000);
set_tris_e(0b00000000);
OPTION=0x07; //pre_scaler timer0, 1:256
setup_ccp1(CCP_PWM); //PWM configuration
setup_timer_2(T2_DIV_BY_16,255,1);//T=(1/clock)*4*t2div*(period+1)
PORTC=0;
TMR0=0;
cuenta1=0;
cuenta2=0;
set_pwm1_duty(0);
AB=0;
AB_1=0;
BC=0;
BC_1=0;

enable_interrupts(global);
enable_interrupts(int_rb);
PC0=1;

while(TRUE)
{
inter=(cuenta1)&(0xFF00);
cuentaH1=inter>>8;
cuentaL1=(cuenta1)&(0x00FF);

inter=(cuenta2)&(0xFF00);
cuentaH2=inter>>8;
cuentaL2=(cuenta2)&(0x00FF);
```

```
PD3=1;
putc(0xAA); //sending serial port
putc(cuentaH1);
putc(cuentaL1);
putc(cuentaH2);
putc(cuentaL2);

PD3=0;
PD2=1; //waits for sampling time
do
{
if(kbhit())
{
dato=getc();//getting datum
if(dato>=128)
{
PD0=0;
PD1=1;
dato=dato-128;
pwm=dato<<1;
}
else
{
PD0=1;
PD1=0;
pwm=dato<<1;
}
set_pwm1_duty(pwm); //sending pwm
}
}
while(TMR0<196); //each count = (4/FXtal)*256 sec
PD2=0; //sampling time has arrived
TMR0=0;
} //closing the infinite while
} //closing main
```

12.6 Personal Computer Builder C++ Program

The flow diagram for PC programming when used to control a mechanism with flexibility is presented in Fig. 12.46. We refer to [5] for a complete explanation of instructions employed in the program listed below. We do not try to present a detailed explanation of this program. Our intention is merely to show the reader how a computer program can be used to implement control algorithms.

```
#include <vcl.h>
#pragma hdrstop
#include "Main.h"
#include <math.h>
#include <stdio.h>
```

Fig. 12.46 Flow diagram for
PC programming when used
to control a mechanism with
flexibility

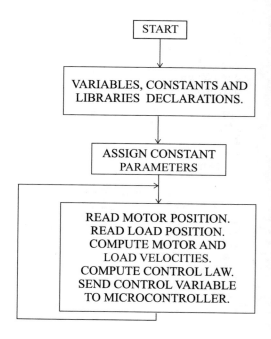

```
#include <share.h>
#include <conio.h>
#include <dos.h>
//---------------------------------------------------------------
#pragma package(smart_init)
#pragma link "CSPIN"
#pragma resource "*.dfm"
//---------------------------------------------------------------
#define Ts 0.01
#define pprp 1000.0
#define pprb 400.0
#define pi_ 3.14159265358
#define fte 11.5
#define cpH 1.5
//-------------controller gains-------------------
#define kp1 0.8
#define kd1 0.12
#define ki1 3.0

#define kp2 0.02535
#define kd2 0.002
#define ki2 7.625

FILE *ptrMonit;
TMainForm *MainForm;
unsigned char flagcom=0,flagfile=0,signo_sal,pwm,iastd;
unsigned short int pos,pos2; /
float VM=fte-cpH,ef=0.0,ef_1=0.0,e2_1=0.0,ef2,e2,iast;
```

```cpp
float pwmf,t=0.0,escp=pi_/(2*pprp);
float escb=pi_/(2*pprb),vo=fte-cpH,escs=127/VM,Voltaje;
float w,factor=6.0*pi_/255.0;
float fact2=2.0*pi_,alfa=1.0,a1,spvf=pi_,vv=0.0,ii,v;
float v2,pvff,pvff_1,ep_1=0.0;
float pvf_1=0.0,pvf2_1=0.0,p_1=0,p2_1=0,pvf=0,pvf2=0;
float s2t1,vc,v2c,V_q,theta1d,e,f,u,u1,ie=0.0,ie2=0.0;
float u2,A=0.3,iTs=1/Ts;
float x1,x2,x3,x1m=0.0,x2m=0.0,x3m=0.0,ief=0.0;
//-----------------------------------------------------
void ProcessByte(BYTE byte)
{
if(flagcom!=0)
flagcom++;
if((byte==0xAA)&&(flagcom==0))
{
pos=0;
pos2=0;
flagcom=1;
}
if(flagcom==2)
{
pos=byte;
pos=pos<<8;
}
if(flagcom==3)
{
pos=pos+byte;
}
if(flagcom==4)
{
pos2=byte;
pos2=pos2<<8;
}
if(flagcom==5)
{
pos2=pos2+byte;
pvf=(signed short int)pos;
pvf=escb*pvf; // motor
pvf2=(signed short int)pos2;
pvf2=-escp*pvf2; // load

// velocities computation
// *********************************
if(abs(pvf-pvf_1)<60.0)
{
v=(pvf-pvf_1)*iTs; // motor velocity
}
pvf_1=pvf;
// *****************************
if(abs(pvf2-pvf2_1)<60.0)
v2=(pvf2-pvf2_1)*iTs; // mass 2 velocity
ef2=spvf2-v2; // mass 2 velocity error
pvf2_1=pvf2;
```

```
// *************************************

/* ========= PD on load position ========================*/
// ef=spvf-pvf2;
// iast=kp1*ef+kd1*(ef-ef_1)/Ts;
// ef_1=ef;
/* ========= PD on load position ========================*/
/* ========= Multiple loop control PID1, PID2 =============*/
e2=spvf-pvf2;
u2=kp2*e2+ki2*ie2+kd2*(e2-e2_1)/Ts;
ie2=ie2+Ts*e2;

ef=u2-pvf;
iast=(kp1*ef+kd1*(ef-ef_1)/Ts+ki1*ief);
ief=ief+Ts*ef;
// iast=(kp1*ef-kd1*(pvf-ep_1)/Ts);

ef_1=ef;
e2_1=e2;
ep_1=pvf;

/* ========== Multiple loop control PID1, PID2=============*/
/* ========= control polynomials ========================*/
/*
ef=spvf-pvf2;
x1=(-109.9*x1m-62.94*x2m-24.03*x3m+8.0*ef)*Ts+x1m;
x2=64.0*x1m*Ts+x2m;
x3=32.0*x2m*Ts+x3m;

iast=-8.351*x1m-0.05483*x2m-1.24*x3m+0.6892*ef;
x1m=x1;
x2m=x2;
x3m=x3;
*/

/*
ef=spvf-pvf2;
x1=(-94.04*x1m-45.15*x2m-16.65*x3m+8.0*ef)*Ts+x1m;
x2=64.0*x1m*Ts+x2m;
x3=32.0*x2m*Ts+x3m;

iast=-4.465*x1m+0.9454*x2m-0.3866*x3m+0.4408*ef;
x1m=x1;
x2m=x2;
x3m=x3;
*/
/* ========= control polynomials ========================*/
/* =========== Frequency response ==================*/
// iast=3.0*sin(w*t);
/* =========== Frequency response==================*/
// *********************************************************
// * iast stands for voltage applied to motor *
// *********************************************************
if(iast>VM)
```

```
{iast=VM;}
if(iast<-VM)
{iast=-VM;}
pwmf=iast;

pwmf=escs*pwmf;
pwm=(unsigned char)abs(pwmf);
if(pwmf<0)
pwm=pwm+128;

//screen exhibition
MainForm->Edit1->Text = FloatToStr (v);
MainForm->Edit2->Text = FloatToStr (v2);
MainForm->Edit3->Text = FloatToStr (pvf);
MainForm->Edit4->Text = FloatToStr (pvf2);
MainForm->Edit5->Text = FloatToStr (iast);
MainForm->Edit6->Text = FloatToStr (t);
MainForm->Edit7->Text = FloatToStr (E);

//sending control signal byte
MainForm->Acknowledge();
MainForm->send_byte(pwm);
flagcom=0;

/*open/close a file*/
if(flagfile==0)
if((ptrMonit=fopen("MONIT.TXT", "w"))==NULL){}
flagfile=1;
/*write to file*/
fprintf(ptrMonit,"%3.3f\t%3.3f\t%3.3f\t%3.3f\t%3.3f\n",t,spvf,
%pvf,pvf2,iast);
t=t+Ts;
}//closing if flagcom 3 now flagcom 5
}
//-------------------------------------------------------------
__fastcall TMainForm::TMainForm(TComponent* Owner)
: TForm(Owner), SerialPort(1, ProcessByte), fAcknowledge(true)
{
if (SerialPort.IsReady() != TRUE)
MessageBox(NULL, "Problems with port", "Error", MB_OK);
}
//-------------------------------------------------------------
void TMainForm::send_byte(unsigned char byte_sal)
{
if (fAcknowledge == false)
return;
SerialPort.WriteByte(byte_sal);
fAcknowledge = false;
}
//-------------------------------------------------------------
void TMainForm::Acknowledge()
{
fAcknowledge = true;
}
```

```
void __fastcall TMainForm::Button1Click(TObject *Sender)
{
/* close the file */
fclose(ptrMonit);
Close();
}
//----------------------------------------------------------------
```

12.7 Summary

Position control of a rotative mechanism composed of two bodies that are joined at their shafts through a rotative spring is analyzed, designed, and experimentally tested in this chapter. Several approaches to controller design are used and all these designs are experimentally tested to appreciate their advantages and disadvantages. The reason for including this control problem is that this plant cannot be controlled using a single-loop PID controller; hence, a more elaborate control scheme is required. In this chapter, therefore, it is shown how to perform this task.

One approach to control design is to use a two-loop control scheme, with the external loop driven by a classical PID controller. The idea is to use the PID pair of zeros to approximately cancel out the badly damped plant poles, i.e., to reduce the oscillations. This approach successfully stabilizes the closed-loop system and renders the steady-state error zero. Another approach is to use a single-loop controller that is intended to replace the undesired badly damped dynamics of the plant by other more suitable dynamics. The drawback of this approach is the lack of an integral action in the controller that results in a steady-state error that is different from zero. However, this approach is useful for showing how to implement controllers that are given in terms of a transfer function: the realization of transfer functions using the steady-state representation is very useful in this respect.

12.8 Review Questions

1. The polynomial approach has been presented in this chapter as a way of designing controllers with many poles and zeros. Can the root locus method also be used to this end?
2. Try to use the frequency response approach to design a controller for this control problem.
3. Give a practical example of a mechanism with flexibility where the ideas in this chapter may be used.
4. What are the plant limitations of achieving a fast and well-damped response?
5. What is the physical phenomenon responsible for limit cycles in this mechanism?

6. The plant identification has been performed using the frequency response approach. What are the advantages of this approach in this example with respect to the identification method used in Chaps. 15 and 16?

References

1. G. C. Goodwin, S. F. Graebe, and M. E. Salgado, *Control system design*, Prentice-Hall, Upper Saddle River, 2001.
2. B. C. Kuo, *Automatic control systems*, Prentice-Hall, 1995.
3. Custom Computer Services Incorporated, CCS C Compiler Reference Manual, 2003.
4. PIC16F877A Enhanced Flash Microcontroller, Data sheet, Microchip Technology Inc., 2003.
5. Borland C++Builder 6 Language guide, 2002.

Chapter 13
Control of a Magnetic Levitation System

Magnetic levitation systems are very important at present because of their numerous applications. This equipment is also known as MagLev. One of the main reasons for the use of magnetic levitation systems is their capacity to eliminate friction between mechanical pieces. Some application examples are high-velocity trains, traveling by floating upon a rail [1], magnetic bearings [2] that avoid friction, and, hence, wear of the mechanical pieces and high-precision positioning control systems [3] thanks to friction elimination. Because of their practical importance, this equipment is often employed as a workbench to test the performance of classical control and some advanced control techniques [4–8, 10].

In this chapter, a simple magnetic levitation system is designed that may be seen as the basic principle of the more complex applications cited above. This magnetic levitation system is depicted in Fig. 13.1. It consists of a ferromagnetic ball that receives the upward effect of a magnetic force produced by an electromagnet. This ball also receives the downward effect of gravity. The ball must levitate at a desired distance from the electromagnet. The main problem in accomplishing this task is the open-loop instability of the system, which can be seen as follows.

Suppose that the electromagnet exerts a constant upward force on the ball that is equal to the downward force of gravity. Then, these forces exactly cancel out and the ball remains levitating in space. Everyday experience tells us that the electromagnet force on the ball increases as the ball approaches the electromagnet, whereas the gravity force is always the same. Suppose now that, because of some disturbance, the ball moves upward from the position where it was levitating. This renders the force of the electromagnet greater than the gravity force and, hence, the ball continues moving upward until the electromagnet is reached. On the other hand, if the ball moves slightly downward from the position where it was levitating, then the gravity force is greater than the electromagnet force and the ball continues to fall. This situation is a description of system instability: when an *unstable system* is slightly disturbed, the system variables diverge from the values that they possessed at the point where the system was at rest.

© Springer International Publishing AG, part of Springer Nature 2019
V. M. Hernández-Guzmán, R. Silva-Ortigoza, *Automatic Control with Experiments*,
Advanced Textbooks in Control and Signal Processing,
https://doi.org/10.1007/978-3-319-75804-6_13

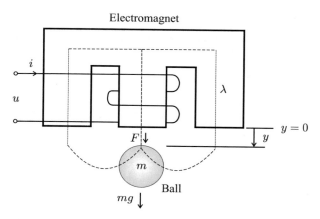

Fig. 13.1 Magnetic levitation system

The variables and parameters involved are as follows:

- u is voltage applied at the electromagnet terminals.
- i is electric current through the winding of the electromagnet.
- y is the ball position measured from the electromagnet's bottom surface to the ball's top surface. This distance is zero when the ball is in contact with the electromagnet and it increases (positive) as the ball moves downward.
- F is the magnetic force exerted by the electromagnet on the ball. Its sense is defined in the same sense where y increases. It is shown later that F is negative, i.e., it correctly represents an attractive force.
- m is the ball's mass.
- L stands for the N turns of electromagnet winding inductance. Inductance is dependent on the ball position and it is usual to write $L(y)$ to stress this fact.
- R is the electromagnet winding copper resistance.
- λ is the magnetic flux in the magnetic circuit, whereas the flux linkage in the electromagnet winding is given as $\psi = 2N\lambda = L(y)\, i$.

13.1 The Complete Nonlinear Mathematical Model

13.1.1 Electrical Subsystem Model

Note that the electromagnet electric circuit is merely a series RL circuits involving the applied voltage u, and the electromagnet winding inductance and resistance (see Fig. 13.1). Applying the *Kirchhoff's voltage law*, the following is found:

$$\text{applied voltage} = \sum \text{voltage drops in the mesh},$$

$$u = \text{voltage drop at inductance} + \text{voltage drop at resistance},$$

$$u = \frac{d\psi}{dt} + R\,i, \tag{13.1}$$

where *Faraday's Law* has been employed to express the voltage drop at the inductance given as the flux linkage time derivative $\frac{d\psi}{dt}$.

The flux linkage is related to the electric current through the inductance as $\psi = L(y)\,i$, where the inductance $L(y)$ is not constant but it is a function of the ball position y. It is important to express how the inductance depends on the ball position. In [7, 8], pp. 245, [11], pp. 31, it is suggested that the inductance is given as:

$$L(y) = k_0 + \frac{k}{1 + \frac{y}{a}}, \tag{13.2}$$

where a, k_0 and k are positive constants. It is interesting to realize that k_0 and $k_0 + k$ stand for the inductance values when the ball is not present ($y \rightarrow \infty$) and when the ball is in contact with the electromagnet ($y = 0$) respectively. Hence, as the ball moves away from the electromagnet, the inductance $L(y)$ changes, as depicted in Fig. 13.2. The parameter a quantifies how fast the inductance changes between $k_0 + k$ and k_0 as y increases. It is important to state that (13.2) is proposed empirically to describe the behavior shown in Fig. 13.2; hence, it is not the only way of representing $L(y)$. Another function that is often employed is [12, 13]:

$$L(y) = k_0 + ke^{-\frac{y}{a}}. \tag{13.3}$$

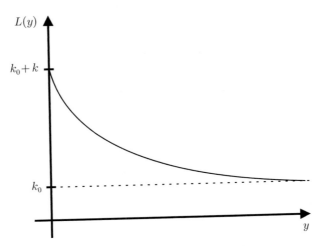

Fig. 13.2 Electromagnet inductance as a function of the ball position

Either of the functions (13.2) or (13.3) can be chosen to represent $L(y)$, as long as they fit the experimental data. It is explained in Sect. 13.4.2 how to perform an experiment to identify the inductance parameters. In this chapter, the general symbol $L(y)$ is still used to represent the inductance to allow the reader to employ the results that are presented next for any of the cases in (13.2) and (13.3). Only at the final design stage, when the numerical values need to be specific, will we proceed to use (13.2) because this expression fits our experimental data better (see Sect. 13.4.2).

13.1.2 Mechanical Subsystem Model

An important step in modeling the mechanical subsystem is to obtain an expression for the *magnetic force*, F, exerted by the electromagnet on the ball. This force, in the sense where y increases, is given by (2.98). Recall that $F = -f$, according to Sect. 2.4. This expression is rewritten in the following for ease of reference and it can also be found in previous works [7, 11] pp. 31, [12, 13]:

$$F = \frac{\partial E(i, y)}{\partial y},$$ (13.4)

$$E(i, y) = \frac{1}{2} L(y)\, i^2,$$ (13.5)

where $E(i, y)$ is the *magnetic energy* stored in the complete magnetic system (electromagnet and ball).

Applying *Newton's Second Law* to the ball, the mechanical subsystem model is obtained (see Fig. 13.1):

$$\text{mass} \times \text{acceleration} = \sum \text{forces on the ball},$$

$$m \frac{d^2 y}{dt^2} = \text{electromagnet force} + \text{ball weight},$$

$$m \frac{d^2 y}{dt^2} = F(i, y) + mg,$$

$$m \frac{d^2 y}{dt^2} = \frac{1}{2} \frac{\partial L(y)}{\partial y}\, i^2 + mg,$$ (13.6)

where (13.4) and (13.5) have been used to write:

$$F(i, y) = \frac{1}{2} \frac{\partial L(y)}{\partial y}\, i^2.$$

The reader can use either (13.2) or (13.3) to verify the following property:

$$\frac{\partial L(y)}{\partial y} < 0, \tag{13.7}$$

for all positive y, i.e., for any realizable ball position. Note that the weight mg appears with a positive sign in (13.6) because it favors the increment of y. On the other hand, the force $F(i, y)$ also appears affected by a positive sign because it is defined in the sense where y increases. However, its value is negative because of (13.7), which implies that its effect opposes the increments in y, i.e., it really opposes the ball weight. Thus, the property in (13.7) is instrumental in the ball levitating.

Replacing $\psi = L(y)i$ in (13.1), the following is obtained:

$$L(y)\frac{di}{dt} = u - Ri - \frac{\partial L(y)}{\partial y}\dot{y}\, i.$$

Then, the complete magnetic levitation system model is given as:

$$L(y)\frac{di}{dt} = u - Ri - \frac{\partial L(y)}{\partial y}\dot{y}\, i, \tag{13.8}$$

$$m\frac{d^2 y}{dt^2} = \frac{1}{2}\frac{\partial L(y)}{\partial y}i^2 + mg. \tag{13.9}$$

The mathematical model in (13.8), (13.9), is nonlinear because it includes the inverse function of y in addition to the quadratic function of i. These features do not allow the use of the Laplace transform for controller design. One way of solving this problem is by finding a linear model that represents, at least approximately, the nonlinear model in (13.8), (13.9). However, this approximate model and the controller designed from it are only useful in a small operation region. This becomes clear in the next section.

13.2   Approximate Linear Model

13.2.1   A State Variables Representation Model

According to Sect. 7.1, given a set of differential equations such as those in (13.8), (13.9), a state variables representation can be found by defining the state vector x components as the unknown variables in those differential equations and their first $q - 1$ time derivatives, where q is the corresponding differential equation order. The unknown variable in (13.8) is i and the differential equation is first-order; hence, only i is a component of the state vector. On the other hand, the unknown variable in (13.9) is y and the differential equation is second-order; hence, y and $\dot{y}$ are components of the state vector. Thus, the following state vector is proposed:

$$x = \begin{bmatrix} x_1 \\ x_2 \\ x_3 \end{bmatrix} = \begin{bmatrix} i \\ y \\ \dot{y} \end{bmatrix}. \tag{13.10}$$

Replacing in (13.8), (13.9), the variables defined in (13.10) it is possible to write:

$$\dot{x}_1 = \frac{1}{L(x_2)} \left(u - R\, x_1 - \frac{\partial L(x_2)}{\partial x_2} x_3\, x_1 \right),$$

$$\dot{x}_2 = x_3,$$

$$\dot{x}_3 = g + \frac{1}{2m} \frac{\partial L(x_2)}{\partial x_2} x_1{}^2.$$

Using vectorial notation:

$$\dot{x} = f(x, u), \tag{13.11}$$

$$f(x, u) = \begin{bmatrix} f_1(x, u) \\ f_2(x, u) \\ f_3(x, u) \end{bmatrix} = \begin{bmatrix} \frac{1}{L(x_2)} \left(u - R\, x_1 - \frac{\partial L(x_2)}{\partial x_2} x_3\, x_1 \right) \\ x_3 \\ g + \frac{1}{2m} \frac{\partial L(x_2)}{\partial x_2} x_1{}^2 \end{bmatrix}. \tag{13.12}$$

According to Sect. 7.3.2, the operation points are the pairs (x^*, u^*) satisfying:

$$f(x^*, u^*) = \begin{bmatrix} \frac{1}{L(x_2^*)} \left(u^* - R\, x_1^* - \frac{\partial L(x_2^*)}{\partial x_2} x_3^*\, x_1^* \right) \\ x_3^* \\ g + \frac{1}{2m} \frac{\partial L(x_2^*)}{\partial x_2} x_1^{*2} \end{bmatrix} = \begin{bmatrix} 0 \\ 0 \\ 0 \end{bmatrix}, \tag{13.13}$$

where $L(x_2^*)$ and $\frac{\partial L(x_2^*)}{\partial x_2}$ indicate that those functions are evaluated at $x_2 = x_2^*$. Hence:

$$x_3^* = 0, \tag{13.14}$$

$$x_1^* = \sqrt{\frac{2mg}{-\frac{\partial L(x_2^*)}{\partial x_2}}}, \tag{13.15}$$

$$u^* = R x_1^*. \tag{13.16}$$

Recall that $-\frac{\partial L(x_2^*)}{\partial x_2} > 0$, because of (13.7). The value $x_2^* = y_d$ is proposed by the designer as it stands for the ball's desired position.

13.2.2 Linear Approximation

According to (7.22), (7.23), (7.24) in Sect. 7.3.2, the nonlinear model given in (13.11), (13.12), can be approximated by the following linear model:

$$\dot{z} = Az + Bv, \tag{13.17}$$

$$A = \begin{bmatrix} \dfrac{\partial f_1(x, u)}{\partial x_1} & \dfrac{\partial f_1(x, u)}{\partial x_2} & \dfrac{\partial f_1(x, u)}{\partial x_3} \\[2mm] \dfrac{\partial f_2(x, u)}{\partial x_1} & \dfrac{\partial f_2(x, u)}{\partial x_2} & \dfrac{\partial f_2(x, u)}{\partial x_3} \\[2mm] \dfrac{\partial f_3(x, u)}{\partial x_1} & \dfrac{\partial f_3(x, u)}{\partial x_2} & \dfrac{\partial f_3(x, u)}{\partial x_3} \end{bmatrix}_{x=x^*,u=u^*},$$

$$B = \begin{bmatrix} \dfrac{\partial f_1(x,u)}{\partial u} \\[2mm] \dfrac{\partial f_2(x,u)}{\partial u} \\[2mm] \dfrac{\partial f_3(x,u)}{\partial u} \end{bmatrix}_{x=x^*,u=u^*},$$

where $z = x - x^*$ and $v = u - u^*$. Using (13.12), the following is found:

$$A = \begin{bmatrix} a_{11} & 0 & a_{13} \\ 0 & 0 & 1 \\ a_{31} & a_{32} & 0 \end{bmatrix}, \qquad B = \begin{bmatrix} \dfrac{1}{L(x_2^*)} \\ 0 \\ 0 \end{bmatrix}, \tag{13.18}$$

$$a_{11} = -\frac{R}{L(x_2^*)}, \qquad a_{13} = -\frac{1}{L(x_2^*)}\frac{\partial L(x_2^*)}{\partial x_2}x_1^*,$$

$$a_{31} = \frac{1}{m}\frac{\partial L(x_2^*)}{\partial x_2}x_1^*, \qquad a_{32} = \frac{1}{2m}\frac{\partial^2 L(x_2^*)}{\partial x_2^2}x_1^{*2}.$$

Recall that the operation point (x^*, u^*) represents four constant scalar values. Applying the Laplace transform to each row in (13.17), together with the values given in (13.18), the following is obtained:

$$Z_1(s) = \frac{1}{s - a_{11}}[a_{13}s Z_2(s) + \frac{1}{L(x_2^*)}V(s)], \tag{13.19}$$

$$Z_2(s) = \frac{a_{31}}{s^2 - a_{32}}Z_1(s), \tag{13.20}$$

where $Z_1(s)$, $Z_2(s)$, $V(s)$, stand for the Laplace transforms of the functions of time $z_1 = x_1 - x_1^*$, $z_2 = x_2 - x_2^*$ and $v = u - u^*$ respectively. These expressions constitute the approximate linear model to be used for analysis and controller design. Recall that, according to Sect. 7.3, all information obtained from (13.19) is valid only if both x and u remain close to the operation point (x^*, u^*).

13.3   Experimental Prototype Construction

13.3.1   Ball

The ball must be made of ferromagnetic material and, also, it must be light enough. On the other hand, to make the measurement of its position easy, the ball must not be too small. These requirements were satisfied by selecting a Styrofoam ball of approximately 0.03[m] in diameter, suitable for the position sensor to be constructed, and inserting iron tacks on its surface with the help of glue. Although the resulting surface is not perfectly even, this idea was proven to work in experiments.

13.3.2   Electromagnet

An electric transformer was employed as the main component. Commonly, the original transformer core was composed of E-shaped iron sheets laying one opposite the other. This allows a magnetic circuit with two branches to form, with negligible flux leakage. This core was modified to lay all the E-shaped iron sheets in the same direction to obtain a new E-shaped core, as depicted in Fig. 13.1. This allows the ball, when levitating, to suitably complete the electromagnet magnetic circuit; thus, a strong attractive force is applied on the ball. The electromagnet winding was built using number 19 AWG magnet wire. This must be performed such that the wire loops lay one beside the other, employing as many layers as possible. The transformer dimensions were selected by trial and error: once the ball is built, employ the available power supplies to verify whether a strong enough magnetic force is generated to attract the ball against the gravity effect. It should also be verified that the electromagnet temperature does not increase significantly.

13.3.3   Position Sensor

The main component is a simple photoconductive cell series connected to a fix carbon resistance to form a tension divider (see Fig. 13.3). The photoconductive cell is placed inside a ping-pong (table tennis) ball to take advantage of the light dispersion produced by the ball's surface. Hence, although the photoconductive cell is small, an uniform lighting variation is obtained on the photoconductive cell in a wide range for y (the ball's position). In this respect, the diameter of the Styrofoam ball used to construct the levitating ball in Sect. 13.3.1 is equal to the diameter of the ping-pong ball. The assembly is put inside a metal pipe casing (0.10[m] length and the same diameter as the ping-pong ball) at one end. The pipe's internal walls are black. The part of the ping-pong ball that is projected outside the metal

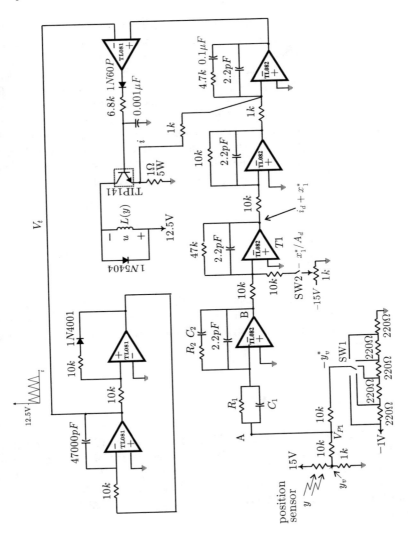

Fig. 13.3 Electric diagram of the complete control system

pipe casing is painted in black to avoid lighting disturbances. A visible light hand lamp is placed in front of the other end of the metal pipe casing. The assembly is placed under the electromagnet such that, when the levitating ball touches the electromagnet, the light is completely blocked, i.e., the photoconductive cell does not receive light and its electric resistance is very high. As the ball moves away from the electromagnet (downward), the light intensity on the photoconductive cell increases, i.e., its electric resistance decreases. This allows the voltage at the point y_v in Fig. 13.3 to increase as the ball position increases, i.e., when the ball moves downward. The voltage y_v represents the measurement, in volts, of the position y in meters (recall that $y = x_2$). The desired position is represented by the voltage y_v^* in Fig. 13.3, which is obtained by using a selector and an array of resistances. This allows step position commands y_v^*, i.e., $y^* = x_2^*$, to be produced. The resistances between the points y_v and y_v^* in Fig. 13.3 render the voltage V_{P1} in the same figure equal to $\frac{1}{2}z_{2v}$, where $z_{2v} = y_v - y_v^*$ is the position error in volts (see Exercise 5, Chap. 2). It is important for each of these two resistances (10[KOhm]) to be several times larger than the resistances series connected to the photoconductive cell and the potentiometer y_v^*. On the other hand, the 10[KOhm] resistances must be at least six times smaller than R_1. The ratio between z_{2v} and the position error in meters $z_2 = x_2 - x_2^*$ is represented by the optical sensor gain A_s, i.e.,

$$z_{2v} = A_s z_2. \tag{13.21}$$

13.3.4   Controller

According to Sect. 9.2, the operational amplifier placed between the points A and B in Fig. 13.3 implements a proportional–integral–derivative (PID) controller with a sign change. Later in this chapter, it is explained how to select R_1, R_2, C_1 and C_2 for correct controller gain selection.

13.3.5   Electric Current Loop

The two operational amplifiers after $T1$ in Fig. 13.3 are employed to implement a proportional–integral (PI) electric current controller that performs the mathematical operation $k_{pi}(i_d - i) + k_{ii} \int_0^t (i_d(r) - i(r))dr$ with $k_{pi} = 4.7$ and $k_{ii} = 10000$ (see Sect. 9.2). The result of this operation is delivered to the power amplifier at the input of the operational amplifier TL081 used as a comparator in a pulse-width modulator (PWM).

13.3.6   Power Amplifier

The power amplifier is PWM-based. This amplifier has a triangular wave generator whose value V_t changes between 0[V] and 12.5[V] (top of Fig. 13.3). The frequency of this signal determines the frequency of the PWM, i.e., 1.53[KHz]. As explained in Sect. 13.5, this frequency satisfies the design requirements. The triangular wave signal is compared (TL081, in Fig. 13.3) with the signal delivered by the electric current controller to generate a PWM signal. This signal is applied at the base of an NPN TIP141 transistor devoted to connecting and disconnecting the power voltage

$V_s = 12.5[V]$ (see Fig. 13.3). This power amplifier is modeled by the gain:

$$A_p = \frac{V_s}{\text{peak-to-peak voltage of } V_t} = \frac{12.5}{12.5} = 1. \tag{13.22}$$

The diode at the TL081 operational amplifier output, which is used as a comparator, is intended to avoid negative voltages. As the voltage delivered by the operational amplifier is alternating current at 1.53[kHz], a radiofrequency amplitude modulated detector 1N60P diode is employed. The 6.8[KOhm] resistance limits the electric current, whereas the capacitor at the transistor base is useful for noise filtering.

13.4 Experimental Identification of the Model Parameters

13.4.1 Electromagnet Internal Resistance, R

A set of constant voltages are applied at the electromagnet terminals and the resulting electric current is measured. These pairs of values are plotted in Fig. 13.4 with the symbol "+" and it is found that a straight line fits them. The slope of this straight line corresponds to the electromagnet internal resistance R that was found to be $R = 2.72[Ohm]$.

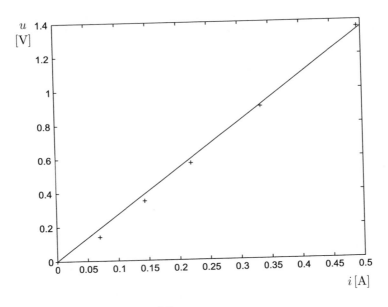

Fig. 13.4 Experimental measurement of R

13.4.2   Electromagnet Inductance, $L(y)$

The ball is placed at different positions, y, which are measured from the electromagnet's bottom surface to the ball's upper surface. Using an RLC meter, the inductance $L(y)$ is measured between the terminals of the electromagnet at each of these positions. As the electromagnet operates at 1.53[KHz], because this is the PWM power amplifier operation frequency, the inductance measurements were performed at 1[KHz]. This is the closest frequency to 1.53[KHz] allowed by the RLC meter used for the measurements. In Fig. 13.5, the symbol "+" represents the measured values. Then, several values are proposed for the parameters k_0, k, a and they are used to compute the function $L(y)$ given in (13.2) at each y used in the experiment. In Fig. 13.5, the continuous line represents the function $L(y)$ that fits the experimental data best. This was achieved using the following parameters to compute $L(y)$:

$$k_0 = 36.3 \times 10^{-3}[\text{H}], \quad k = 3.5 \times 10^{-3}[\text{H}], \quad a = 5.2 \times 10^{-3}[\text{m}]. \quad (13.23)$$

13.4.3   Position Sensor Gain, A_s

The ball is placed at different positions, y, from the electromagnet's bottom surface. The voltage y_v in Fig. 13.3 is measured for each of these positions. In Fig. 13.6, the symbol "+" represents the experimental data that were obtained. It is found that a

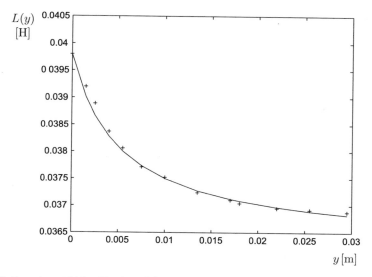

Fig. 13.5 Experimental identification of the parameters k, k_0, and a defining $L(y)$ in (13.2)

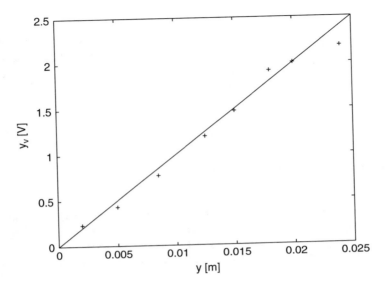

Fig. 13.6 Experimental identification of A_s

straight line with a 100[V/m] slope fits to these points well. This means that the sensor gain is:

$$A_s = \frac{y_v}{y} = \frac{z_{2v}}{z_2} = 100[\text{V/m}]. \tag{13.24}$$

Note that the sensor gain also relates z_{2v} and z_2 if $y_v^* = 100x_2^*$ is selected (recall that $y = x_2$ and $x_2^* = y_d$). Note that the sensor behavior is linear.

13.4.4 Ball Mass, m

It was found that $m = 0.018[\text{kg}]$, using a digital scale.

13.5 Control System Structure

The control system to be designed is depicted in Fig. 13.7. Using the expressions in (13.18), the section between $V(s)$ and $Z_2(s)$ in the block diagram in Fig. 13.8a is obtained. Note that the constant u^* is added and subtracted in Fig. 13.7. It is clear that adding and subtracting the same quantity is equivalent to zero, which correctly represents the situation shown in the electric diagram in Fig. 13.3 where

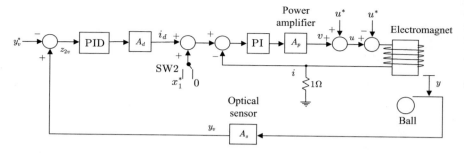

Fig. 13.7 Control system to be designed ($z_{2v} = y_v - y_v^*$)

the operation $u = v + u^*$ is not implemented (which results from the definition of v just after (13.17)). However, adding and subtracting u^* in Fig. 13.7 allows us to realize that $U^*(s)$ must be subtracted in Fig. 13.8a. Hence, not implementing the operation $u = v + u^*$ in Fig. 13.3 is equivalent to including the disturbance $-U^*(s)$ in Fig. 13.8a.

13.5.1 Internal Current Loop

A PI electric current controller is included in Fig. 13.7 just before the power amplifier. The reason for including this current loop is the closed-loop performance improvement achieved by rendering the electrical dynamics faster. Throughout Sect. 13.5.1, it is assumed that the switch SW2 selects "x_1^*".[1] Then, $i_d + x_1^*$ appears affected by a positive sign directly at the adder placed at the input of the PI electric current controller. In the following, the advantages of using a PI electric current controller instead of a simple proportional controller are explained.

13.5.2 Proportional Current Control

Suppose that the current loop is provided with a proportional electric current controller, i.e., suppose that the block PI in Fig. 13.7 is replaced by a simple gain k_{pi}. Then, replacing:

$$u = k_{pi} A_p (i_d + x_1^* - i),$$

in (13.8), the following is found:

---

[1]This means that the switch SW2 in Fig. 13.3 is closed.

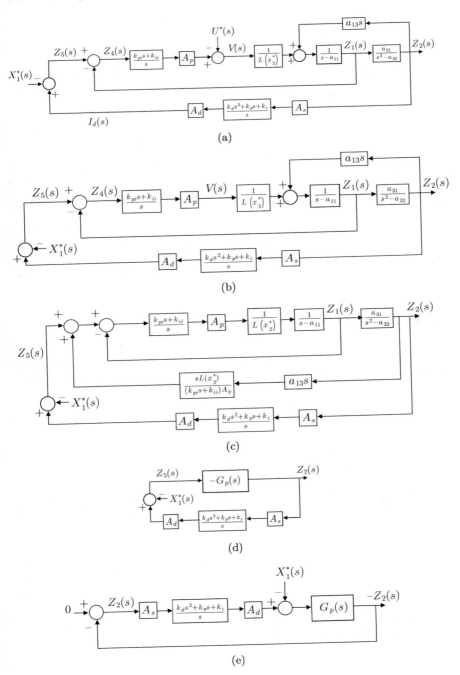

Fig. 13.8 Equivalent block diagrams of the control system. The disturbance $-X_1^*(s)$ is only taken into account when the switch SW2 is open in Fig. 13.3 or, equivalently, when the switch SW2 selects the signal "0" in Fig. 13.7

$$L(y)\frac{di}{dt} = k_{pi} A_p (i_d + x_1^* - i) - R\, i - \frac{\partial L(y)}{\partial y} \dot{y}\, i.$$

Once the steady state is reached (when all the variables remain constant in time), the latter expression implies that:

$$i = \frac{k_{pi} A_p}{k_{pi} A_p + R}(i_d + x_1^*).$$

We stress that $i_d + x_1^*$ represents the electric current desired to flow through the electromagnet to keep the ball levitating. This current is determined by a PID position controller and feedforward of the desired electric current at the operation point.

Note that $i \approx i_d + x_1^*$ if k_{pi} is chosen to be large. However, the drawback of using large values for k_{pi} is the amplification of the noise contained in i_d and i. As a consequence, the ball vibrates excessively and levitation is not possible. Hence, an alternative solution to a proportional electric current control must be employed.

13.5.3 PI Electric Current Control

Suppose that the inductance is not dependent on y in (13.8) and u is computed by a PI electric current controller and a power amplifier, i.e., the following can be written:

$$L\frac{di}{dt} = u - R\, i - \frac{\partial L(y)}{\partial y}\dot{y}\, i,$$

$$u = A_p \left(k_{pi}(i_d + x_1^* - i) + k_{ii} \int_0^t (i_d(r) + x_1^* - i(r))dr \right),$$

where k_{pi}, k_{ii}, are positive constants standing for the proportional and integral gains respectively of the PI electric current controller. Replacing u in the first equation and differentiating once with respect to time, the following is found:

$$L\frac{d^2 i}{dt^2} = -(R + A_p k_{pi})\frac{di}{dy} + A_p k_{ii}(i_d + x_1^* - i) + A_p k_{pi}\frac{di_d}{dt} - \frac{d}{dt}\left(\frac{\partial L(y)}{\partial y}\dot{y}\, i \right).$$

$$(13.25)$$

Assuming that the steady state has been reached, i.e., that all the variables have become constant, the latter expression implies:

$$i = i_d + x_1^*,$$

because A_p and k_{ii} are different from zero. This means that the use of a PI electric current controller allows the electric current through the electromagnet i to equal the desired current $i_d + x_1^*$, in a steady state. Note that this is achieved without requiring k_{pi} or k_{ii} to be large: the only requirements are $k_{ii} \neq 0$ and that the current loop is stable. Thus, the problems due to noise that are present in a proportional current controller are strongly attenuated by a PI electric current controller.

It is important to state that the use of a PID electric current controller is not recommended because of the additional noise amplification produced by the time derivative of the electric current error. Furthermore, in this application, the additional derivative term does not produce any improvement with respect to what is achieved with a PI electric current controller. See Sect. 3.8.5 for a discussion on this point. Thus, a PI electric current controller has been selected.

The electric current i_d is defined as the output of a PID position controller for the ball. A block with the gain A_d is included in Fig. 13.7 between the PID position controller and i_d. This is because in the electric diagram of Fig. 13.3 an adder is included, which introduces this gain, performing the operation $i_d + x_1^*$. This value represents the electric current that must flow through the electromagnet to levitate the ball at the desired position. This means that $i_d = 0$ is possible if $i = x_1^*$ in a steady state. However, as seen later, x_1^* is not exactly known and this motivates the use of a PID position controller, which achieves $i = x_1^*$ in a steady state, despite the fact that x_1^* is not exactly known. Finally, between the PID position controller and the ball position measurement the position sensor gain A_s is included (see Fig. 13.7). The block diagram of the complete system that has just been described is shown in Fig. 13.8a.

Note that $z_1 = x_1 - x_1^*$ is the signal that is fed back in the current loop of this block diagram. Hence, the input of the PI electric current controller is $z_4 = i_d + x_1^* - x_1$ in the block diagram of Fig. 13.8a if the switch SW2 is closed in Fig. 13.3 or, equivalently, the switch SW2 selects the signal x_1^* in Fig. 13.7. In this case, the signal $-X_1^*(s)$ on the left of the block diagrams in Fig. 13.8 must not be considered.

On the other hand, if the switch SW2 is open in Fig. 13.3 or, equivalently, the switch SW2 selects the signal "0" in Fig. 13.7, then the signal $-X_1^*(s)$ on the left of the block diagrams in Fig. 13.8 must be considered. This is to render $z_4 = z_5 - z_1 = (i_d - x_1^*) - (x_1 - x_1^*) = i_d - x_1$ the signal that appears at the input of the PI electric current controller.

13.5.4 External Position Loop

As described above, an external position loop is used in Fig. 13.7, which includes a PID controller. However, it is important to understand why a PID controller has been chosen.

As previously stated, the signal $U^*(s)$ represents a disturbance (constant in time as u^* is constant) in the block diagram of Fig. 13.8a. Hence, a PI electric current controller ensures that any deviation produced by such a disturbance vanishes in a steady state. This is concluded as follows. Suppose that in the internal-most loop in Fig. 13.8c the disturbance $-U^*(s)$ is considered between the blocks A_p and $\frac{1}{L(x_2^*)}$. Using superposition, i.e., assuming that the input to this internal-most loop in Fig. 13.8c is zero, the following is found:

$$Z_1(s) = -\frac{s}{L(x_2^*)s^2 + (-L(x_2^*)a_{11} + A_p k_{pi})s + A_p k_{ii}} U^*(s).$$

Using the final value theorem:

$$\lim_{t \to \infty} z_1(t) = \lim_{s \to 0} s Z_1(s),$$

$$= \lim_{s \to 0} s \left(-\frac{s}{L(x_2^*)s^2 + (-L(x_2^*)a_{11} + A_p k_{pi})s + A_p k_{ii}} \frac{d}{s} \right),$$

$$= 0,$$

where $U^*(s) = \frac{d}{s}$, with d a constant such that $u^*(t) = d$. This means that the deviation in $z_1(t)$, in a steady state, due to the constant disturbance $u^*(t) = d$, is zero for any $d \neq 0$. This proves that a PI electric current controller eliminates any deviation that such a disturbance may produce in a steady state. Thus, the presence of the disturbance $-U^*(s)$ in Fig. 13.8a is ignored in following. This allows the block diagram in Fig. 13.8b to be obtained.[2] The block diagram in Fig. 13.8c is obtained from Fig. 13.8b after some block algebra. From Figs. 13.8c, d is obtained, where:

$$G_p(s) = -\frac{a_{31} A_p (k_{pi}s + k_{ii})}{[sL(x_2^*)(s - a_{11}) + (k_{pi}s + k_{ii})A_p](s^2 - a_{32}) - a_{13}a_{31}L(x_2^*)s^2},$$

$$(13.26)$$

or:

$$G_p(s) = -\frac{a_{31} A_p (k_{pi}s + k_{ii})}{b_4 s^4 + b_3 s^3 + b_2 s^2 + b_1 s + b_0}, \qquad (13.27)$$

$$b_4 = L(x_2^*),$$

$$b_3 = -a_{11}L(x_2^*) + k_{pi} A_p,$$

$$b_2 = k_{ii} A_p - a_{32}L(x_2^*) - a_{13}a_{31}L(x_2^*),$$

$$b_1 = a_{11}a_{32}L(x_2^*) - k_{pi} A_p a_{32},$$

$$b_0 = -k_{ii} A_p a_{32}.$$

---

[2]If SW2 is closed in Fig. 13.3 or, equivalently, SW2 selects the signal x_1^* in Fig. 13.7, then the signal $-X_1^*(s)$ at the left of the block diagrams in Fig. 13.8 must not be considered.

Finally, an equivalent block diagram is shown in Fig. 13.8e such that the closed-loop system appears to be negative feedback.

It is not difficult to realize that the characteristic polynomial of $G_p(s)$ has positive and negative coefficients. This means that $G_p(s)$ is an unstable transfer function and this is a direct consequence of the fact that the magnetic levitation system is open-loop unstable. This behavior was intuitively explained in the introduction to this chapter. To render the closed-loop system stable, the position controller must include a derivative term. On the other hand, the position controller must also include an integral term to compose a PID controller with transfer function:

$$\frac{k_d s^2 + k_p s + k_i}{s}.$$

There are two main reasons for choosing a PID controller, which are explained as follows.

- Achieving a zero steady-state position error when a constant disturbance is present. As previously explained, when the switch SW2 is closed in Fig. 13.3 or, equivalently, when SW2 selects the signal x_1^* in Fig. 13.7, the signal $-X_1^*(s)$ on the left in block diagrams of Fig. 13.8 must not be considered. However, if the switch SW2 is open in Fig. 13.3 or, equivalently, if SW2 selects the signal "0" in Fig. 13.7, then the signal $-X_1^*(s)$ on the left in the block diagrams of Fig. 13.8 must be considered. In such a case, $-X_1^*(s)$ represents a constant disturbance in time, i.e., $x_1^* = f$ is constant. It is important to stress that a similar disturbance appears even when the switch SW2 is closed in Fig. 13.3, but x_1^* is not exactly known. This is the case when the ball weight or the inductance is not exactly known (see (13.15)). The steady-state position error is studied in the following when a PID controller is employed. From Fig. 13.8e the following is found:

$$Z_2(s) = \frac{sG_p(s)}{s + A_s A_d(k_d s^2 + k_p s + k_i)G_p(s)} X_1^*(s).$$

Applying the final value theorem:

$$\lim_{t \to \infty} z_2(t) = \lim_{s \to 0} s Z_2(s),$$

$$= \lim_{s \to 0} s \left(\frac{sG_p(s)}{s + A_s A_d(k_d s^2 + k_p s + k_i)G_p(s)} \frac{f}{s} \right),$$

$$= 0,$$

where $X_1^*(s) = \frac{f}{s}$. Note that $G_p(s)$ has no poles at $s = 0$; hence, the factor s at the numerator of $\frac{Z_2(s)}{X_1^*(s)}$ cannot be cancelled out. This means that the constant disturbance $x_1^*(t) = f$ does not produce any steady-state error thanks to the integral part of the PID position controller.

- The transfer function $G_p(s)$ has no poles at $s = 0$. Hence, the system type is 1 only because of the integral part of the controller. This ensures that the position error is zero in a steady state when constant position references are commanded.

The external position loop, driven by the PID controller, is positive feedback, as observed in Figs. 13.8a, b, c, and d. This is concluded from the fact that is $z_2 = x_2 - x_2^*$ (where $x_2 = y$), instead of $x_2^* - x_2$, the error used as input for the PID position controller once it passes through the position sensor (also see Fig. 13.7). As observed in Fig. 13.8e, this results in the following open-loop transfer function:

$$A_d A_s G_p(s) \frac{k_d s^2 + k_p s + k_i}{s},$$

which has a positive gain as a_{31}, a negative number because of $\frac{\partial L(y)}{\partial y} < 0$, appears multiplied by the sign "$-$" in the definition of $G_p(s)$. Hence, the positive feedback used in the external loop of Figs. 13.8a, b, c, and d is instrumental to canceling the negative gain that the magnetic levitation system naturally possesses (because of $a_{31} < 0$).

By the way, the reason why the magnetic levitation system has an open-loop negative gain can be explained using intuitive arguments:

The input of the magnetic levitation system is voltage u applied at the electromagnet terminals and the output is the ball position y, which increases if the ball moves (downward) away from the electromagnet. If u is increased, then the electric current through the electromagnet also increases, producing a stronger attractive force on the ball; hence, the ball approaches the electromagnet, i.e., y decreases. This is the effect of a negative gain: an input *increase* u produces an output *decrease* y.

13.6 Controller Design

According to the above discussion, the closed-loop control system has two controllers: a PI electric current controller and a PID position controller. To design the PID position controller using either the root locus or frequency response methods, the poles and zeros of $G_p(s)$ need to be known. However, these poles and zeros depend on k_{pi} and k_{ii}, the PI electric current controller gains. Hence, the design of the PID position controller requires knowledge of the PI electric current controller gains. However, this task can be simplified using a criterion that is commonly employed in electromechanical systems, and is described in the following.

In electromechanical systems, it is desirable to design the PI electric current controller such that the "electrical dynamics is fast enough". This means that the electric current through the electromagnet i must reach its desired value $i_d + x_1^*$ very fast and both must remain close to each other all the time. It was stated above that this is accomplished in a steady state using a PI electric current controller with

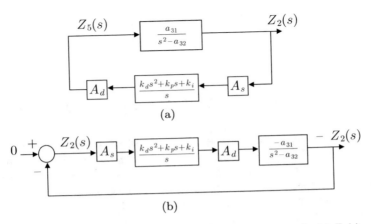

Fig. 13.9 Block diagrams when the electric current loop is fast enough. **(a)** $Z_5(s) = Z_1(s)$ because $I(s) = I_d(s) + X_1^*(s)$. **(b)** $-a_{31} > 0$

positive values for k_{pi} and k_{ii}. However, it is now required now that $i = i_d + x_1^*$, even during the system transient response. The way of achieving this is by using large positive values for both k_{pi} and k_{ii} because, in such a case, (13.26) is simplified to:

$$G_p(s) = -\frac{a_{31}}{s^2 - a_{32}}. \tag{13.28}$$

Thus, the block diagrams in Figs. 13.8d and e become those presented in Fig. 13.9. Note that the dynamics of the whole magnetic levitation system is reduced to the mechanical subsystem represented in (13.20) because $i = i_d + x_1^*$, i.e., $Z_5(s) = Z_1(s)$.

Then, if the electric current loop is fast enough, the PID controller can be designed by only taking into account the mechanical dynamics of the magnetic levitation system, i.e., using the block diagram in Fig. 13.9b. Recall that the PID controller transfer function can be written as:

$$\frac{k_d s^2 + k_p s + k_i}{s} = k_d \frac{s^2 + \frac{k_p}{k_d} s + \frac{k_i}{k_d}}{s}, \tag{13.29}$$

$$= k_d \frac{(s + q_1)(s + q_2)}{s}, \quad q_1 + q_2 = \frac{k_p}{k_d}, \quad q_1 q_2 = \frac{k_i}{k_d}.$$

where q_1, q_2 and k_d are some positive constants that have to be computed during the design stage. Once these values are known, k_p and k_i can be obtained using the expressions in (13.29). Note that, according to (13.29), the open-loop transfer function in Fig. 13.9b is:

$$G(s)H(s) = \frac{-a_{31}k_d A_d A_s (s + q_1)(s + q_2)}{s(s^2 - a_{32})}.$$ (13.30)

13.6.1   PID Position Controller Design Using the Root Locus

The root locus diagram corresponding to the block diagram in Fig. 13.9b is presented in Fig. 13.10. The location of $-q_1$ and $-q_2$ in Fig. 13.10 are proposed to pull the root locus diagram into the left half-plane, i.e., to render the closed-loop system stable. We stress that k_d is the gain to be adjusted from 0 to $+\infty$ to select the desired closed-loop poles belonging to the root locus (thick lines) in Fig. 13.10.

Consider the case when the desired position is $x_2^* = y_d = 0.006$[m]. Using this, (13.14), (13.15), (13.16), the expressions in (13.18), the numerical values in Sect. 13.4, and $A_d = 4.7$ (see the operation amplifier labeled $T1$ in Fig. 13.3), the following is found:

$$\frac{-a_{31}}{s^2 - a_{32}} A_d A_s = \frac{5911}{s^2 - 1752},$$

that has poles located at:

$$s = \sqrt{a_{32}} = 41.8569, \quad s = -\sqrt{a_{32}} = -41.8569.$$ (13.31)

Hence, according to Fig. 13.10, propose:

$$q_1 = 29.9355, \quad q_2 = 1.3045.$$ (13.32)

If $k_d = 0.0277$ is chosen, the closed-loop poles and zeros are located at:

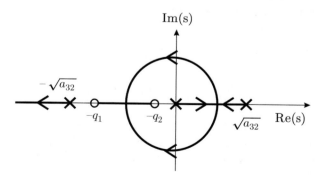

Fig. 13.10   Root locus diagram for the block diagram in Fig. 13.9b

$$\text{poles: } s = -140.1441, \quad s = -21.5812, \quad s = -2.1154, \qquad (13.33)$$
$$\text{zeros: } s = -29.9355, \quad s = -1.3045,$$

and, using (13.29), (13.32), the following controller gains are found:

$$k_d = 0.0277, \quad k_p = 0.866, \quad k_i = 1.082. \qquad (13.34)$$

If $k_d = 0.0277 \times 0.8 = 0.0222$ is chosen, the closed-loop poles and zeros are located at:

$$\text{poles: } s = -110.2411, \quad s = -18.2935, \quad s = -2.5381, \qquad (13.35)$$
$$\text{zeros: } s = -29.9355, \quad s = -1.3045,$$

and, using (13.29), (13.32), the following controller gains are found:

$$k_d = 0.0222, \quad k_p = 0.6928, \quad k_i = 0.866. \qquad (13.36)$$

Finally, if $k_d = 0.0277 \times 1.2 = 0.0333$ is chosen, the closed-loop poles and zeros are located at:

$$\text{poles: } s = -171.2313, \quad s = -23.4669, \quad s = -1.9107, \qquad (13.37)$$
$$\text{zeros: } s = -29.9355, \quad s = -1.3045,$$

and, using (13.29), (13.32), the following controller gains are found:

$$k_d = 0.0333, \quad k_p = 1.039, \quad k_i = 1.299. \qquad (13.38)$$

In Fig. 13.11 the time response when $z_1(0) = 0$, $z_2(0) = 1$, $z_3(0) = 0$ is presented and the controller gains in (13.34), (13.36), (13.38) are used. These simulations have been performed by executing the following MATLAB code several times in an m-file:

```
clc;
clear all;
factor=0.7*0.0396*1.2;%1, 0.8 ,1.2
q1=29.9355;
q2=1.3045;
nume=conv([1 q1],[1 q2]);
PID=tf(factor*nume,[1 0]);
g=tf(5911,[1 0 -1752]);
gh=g*PID;
M=feedback(gh,1,-1);
dn1=pole(M);
zero(M);
```

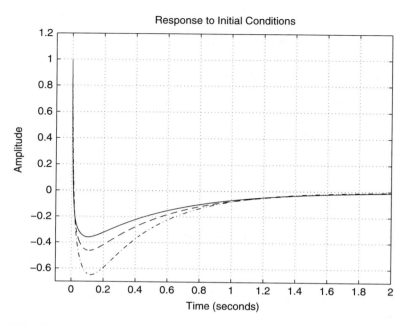

Fig. 13.11 Plot of $z_2(t)$ when $z_1(0) = 0$, $z_2(0) = 1$, $z_3(0) = 0$. Continuous: controller gains in (13.38). Dashed: controller gains in (13.34). Dotted line: controller gains in (13.36)

```
dn2=conv([1 -dn1(1)],[1 -dn1(2)]);
dn4=conv(dn2,[1 -dn1(3)]);
A=[0 1 0 ;
0 0 1 ;
-dn4(4) -dn4(3) -dn4(2)];
B=[ 0;
0;
1];
cc=factor*nume*5911;
C=[cc(3) cc(2) cc(1)];
estado=ss(A,B,C,0);
figure(1)
initial(estado,'b-',[1/cc(3) 0 0],2)
axis([-0.1 2 -0.7 1.2])
grid on
hold on
```

It is observed that the position error converges to zero as time increases in all the cases, i.e., $z_2(t) \to 0$ as $t \to \infty$, which is possible because the closed-loop system is stable and because we are using a position controller with an integral part, i.e., a PID position controller.

Also note, in Fig. 13.11, that a considerable overshoot exists in the position error $z_2(t)$ in all three cases. This overshoot is larger for the controller gains in (13.36) and smaller for the controller gains in (13.38). Note, however, that according to (13.33), (13.35), (13.37), all the closed-loop poles are real in all three cases; hence, they should not produce any overshoot. The explanation for this phenomenon is presented in Example 8.5, Chap. 8, in the present book and it follows the ideas exposed in section 8.6.5 of [16]:

From (13.31), it is observed that there is an open-loop unstable pole located at $s = \sqrt{a_{32}} = 41.8569$. The magnitude of this pole is much greater than that of the slowest (stable) closed-loop poles in (13.33), (13.35), (13.37). According to Sect. 8.1.2, in the present book, and section 8.6.5 in [16], the direct consequence of these facts is the overshoot observed in Fig. 13.11 and this does not require the existence of any complex conjugate closed-loop poles. Moreover, we also realize from (13.33), (13.35), (13.37), that the two slowest closed-loop poles approach the closed-loop zeros located at $s = -29.9355$, $s = -1.3045$ as k_d increases. This means that the effects of such slowest poles is less important in the response of $z_2(t)$ as k_d increases, i.e., this response tends to be dominated by the third closed-loop pole, which is located at the left of $s = -\sqrt{a_{32}} = -41.8569$ (see (13.33), (13.35), (13.37)). As the magnitude of this pole is larger than that of the unstable open-loop pole at $s = \sqrt{a_{32}} = 41.8569$, the overshoot decreases in Fig. 13.11 as k_d increases.

The above discussion suggests that choosing a large k_d might be a good design criterion. However, it is explained in Sect. 13.6.3 that a large value for k_d may result in instability because the closed-loop poles representing the electrical dynamics are pushed into the right half-plane as k_d increases. It is shown in Sect. 13.6.3 that a good performance is found in experiments when using the controller gains in (13.34).

13.6.2 Design of the PI Electric Current Controller

According to the previous sections, the gains for the PI electric current controller have to be chosen to be large. However, these controller gains cannot be chosen to be too large in practice because this would amplify noise, which would result in a poor closed-loop performance. Hence, the PI electric current controller gains must be selected as the largest gains that do not cause performance to deteriorate. Thus, the selection of this controller gains can only be achieved through experimental essays. The following criterion is suggested:

"Choose k_{pi} and k_{ii} as the largest positive values that do not introduce important vibrations as a consequence of noise amplification. For this reason, it is recommended to choose k_{ii} larger than k_{pi}".

The controller gains that have been tested experimentally are presented in Table 13.1. The PI electric current controller is implemented using the TL082 operational amplifier placed before the TL081 operational amplifier used as a

Table 13.1 Numerical values tested experimentally for the PI electric current controller

$y_d = x_2^* = 0.006[\text{m}]$	
PI gains	Circuit component values
$k_{pi} = 4.7$	$R = 4.7[\text{K}\Omega]$
$k_{ii} = 100$	$C = 10[\mu\text{F}]$
$k_{pi} = 4.7$	$R = 4.7[\text{K}\Omega]$
$k_{ii} = 1000$	$C = 1[\mu\text{F}]$
$k_{pi} = 4.7$	$R = 4.7[\text{K}\Omega]$
$k_{ii} = 10000$	$C = 0.1[\mu\text{F}]$

comparator (see Fig. 13.3). At the input of that operational amplifier there is a $1[\text{K}\Omega]$ resistance and there are a series resistance and capacitor at the feedback path. Thus, according to Table 9.1, a PI controller is correctly implemented with gains computed as:

$$k_{pi} = \frac{R}{1000}, \quad k_{ii} = \frac{1}{1000C}, \tag{13.39}$$

where R, C are the resistance (in Ohms) and the capacitance (in farads) at the feedback path. The corresponding circuit component values used to implement the controller gains for each case are also presented in Table 13.1.

On the other hand, recall that the PID position controller is implemented using the operational amplifier placed between points A and B in Fig. 13.3. According to Table 9.1, the gains of this controller are given as:

$$k_{pa} = \frac{R_2}{R_1} + \frac{C_1}{C_2}, \quad k_{da} = R_2 C_1, \quad k_{ia} = \frac{1}{R_1 C_2} \tag{13.40}$$

Moreover, in Sect. 13.3.3 it was explained that the $10[\text{KOhm}]$ resistances connected to the point V_{P1} introduce a $1/2$ constant gain. As this gain must be considered a part of the controller we conclude that:

$$k_p = \frac{k_{pa}}{2}, \quad k_d = \frac{k_{da}}{2}, \quad k_i = \frac{k_{ia}}{2} \tag{13.41}$$

In the experimental essays that are presented next, the numerical values shown in (13.34) were used for the PID position controller. In Table 13.2 we present the circuit component values used for resistances and capacitances located between points A and B in Fig. 13.3, which render the implementation of these controller gains possible. The values of R_1, R_2, C_1 and C_2 shown in Table 13.2 are computed using (13.40) and (13.41). These circuit component values are obtained as the series and the parallel combination of several resistances and capacitors: $330[\text{KOhm}]$ and $470[\text{kOhm}]$ in the series yield $800[\text{KOhm}]$, seven $0.01[\mu\text{F}]$ parallel capacitors yield $0.07[\mu\text{F}]$.

Table 13.2 Numerical values tested experimentally for the PID position controller

$y_d = x_2^* = 0.006[m]$	
PID gains	Circuit component values
$k_d = 0.0277$	$R_1 = 470[K\Omega]$
$k_p = 0.866$	$R_2 = 800[K\Omega]$
$k_i = 1.082$	$C_1 = 0.07[\mu F]$
	$C_2 = 1[\mu F]$

As indicated in both Tables 13.1 and 13.2, the experimental essays were performed setting $y_d = x_2^* = 0.006[m]$ as the desired position for the ball. This desired position was set by using the switch SW1 in Fig. 13.3 to select $y_v^* = 0.6[V]$. Also, in all experiments, the switch SW2 was kept open in Fig. 13.3.

It was observed experimentally that the best performance is obtained when $k_{ii} = 10000$ is used, i.e., the ball levitates at the desired position and any vibration due to noise is not present. Thus:

$$k_{pi} = 4.7, \quad k_{ii} = 10000, \tag{13.42}$$

were selected as the correct values for the PI electric current controller gains. This is why a 4.7[KΩ] resistance and a 0.1[μF] capacitance are included in Fig. 13.3 at the feedback path of the TL082 operational amplifier, which implements the PI electric current controller (that located before the TL081 operational amplifier used as a comparator).

Once the controller gains in (13.42) are fixed, it must be verified that the base frequency of the PWM power amplifier is correctly selected. Using $x_2^* = 0.006[m]$, (13.14), (13.15), (13.16), the expressions in (13.18), the numerical values given in (13.42), and Sect. 13.4, in addition to $A_d = 4.7$ (see the operational amplifier labeled $T1$ in Fig. 13.3), it is found that the internal-most loop in Fig. 13.8c, represented by the transfer function:

$$\frac{A_p(k_{pi}s + k_{ii})}{L(x_2^*)s^2 + (-a_{11}L(x_2^*) + A_pk_{pi})s + A_pk_{ii}}$$

has complex conjugate poles located at $s = -97.82 \pm 504.09j$.[3] This means that the electrical dynamics is a low-pass filter with $\omega_n = 513.4934[rad/s]$ as corner frequency. As stated earlier in Sect. 13.3.6, the PWM power amplifier employs 1.53[KHz] as the base frequency, which corresponds to 9613.3[rad/s]. As, 9613.3[rad/s] is more than ten times larger than the corner frequency $\omega_n = 513.4934[rad/s]$, we conclude that the PWM power amplifier base frequency is well selected.

The time response of $z_2(t)$ obtained in a simulation when $z_1(0) = 0$, $z_2(0) = 1$, $z_3(0) = 0$, and zero initial conditions for both integral parts of the controllers,

---

[3] These poles can also be obtained as the fastest poles in $G_p(s)$.

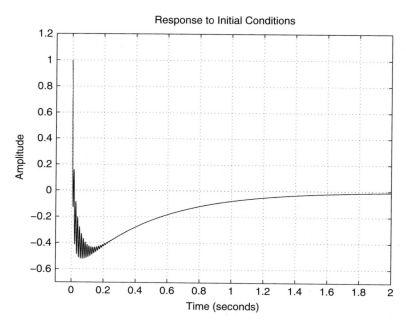

Fig. 13.12 Plot of $z_2(t)$ when $z_1(0) = 0$, $z_2(0) = 1$, $z_3(0) = 0$, and zero initial conditions for both integral parts of the controllers

is presented in Fig. 13.12. This was performed according to the block diagram in Fig. 13.8e, i.e., using the open-loop transfer function:

$$A_d A_s G_p(s) \frac{k_d s^2 + k_p s + k_i}{s}$$

with $G_p(s)$ given in (13.26), setting $x_2^* = 0.006\text{[m]}$ and the controller gains given in (13.42) and Table 13.2. The simulation in Fig. 13.12 was executed using the following MATLAB code in an m-file:

```
clc;
clear all;
%%%%% data %%%%%
k0=36.3e-3; % Hy
k=3.5e-3; % Hy
a=5.2e-3; % m
Ap=1; % Power amplifier
As=100; % V/m, Position sensor
Ad=4.7; %
m=0.018; % Kg, Ball mass
g=9.81; %m/s^2, gravity acceleration
Ra=2.72; % Winding resistance
```

```
yd=0.006; % m, desired position
factor=0.7*0.0396;
q1=29.9355;
q2=1.3045;
nume=conv([1 q1],[1 q2]);
gc=tf(factor*nume,[1 0])
C1=0.07e-6;
R1=470000;
C2=1e-6;
R2=800000;
kp=(R2/R1+C1/C2)/2;
kd=(R2*C1)/2;
ki=(1/(R1*C2))/2;
PID=tf([kd kp ki],[1 0]);
%PI current
kpi=4.7;
kii=10000;
Lyd=(k0+k/(1+yd/a));
dLyd=(-k*a/(a+yd)^2);
d2Lyd=(2*k*a/(a+yd)^3);
x1d=sqrt(2*m*g/abs(dLyd));
a11=-Ra/Lyd;
a13=-1/Lyd*dLyd*x1d;
a31=1/m*dLyd*x1d;
a32=1/(2*m)*d2Lyd*x1d^2;
numera=Ap*a31*[kpi kii];
denomina=[Lyd -a11*Lyd+kpi*Ap kii*Ap-a32*Lyd
    -a13*a31*Lyd
a11*a32*Lyd-kpi*Ap*a32 -kii*Ap*a32];
gp=tf(numera,denomina);
g=-Ad*As*gp;
gh=g*gc;
M=feedback(gh,1);
dn1=pole(M);
dn2=conv([1 -dn1(1)],[1 -dn1(2)]);
dn4=conv(dn2,[1 -dn1(3)]);
dn5=conv(dn4,[1 -dn1(4)]);
dn3=conv(dn5,[1 -dn1(5)]);
dn3(1);
A=[0 1 0 0 0;
0 0 1 0 0;
0 0 0 1 0;
0 0 0 0 1;
-dn3(6) -dn3(5) -dn3(4) -dn3(3) -dn3(2)];
B=[0;
```

```
0;
0;
0;
1];
nume1r=conv(nume,[kpi kii]);
cc=-factor*nume1r*Ad*As*Ap*a31;
C=[cc(4)  cc(3)  cc(2)  cc(1)  0];
estado=ss(A,B,C,0);
figure(1)
initial(estado,[1/cc(4)  0 0 0 0],2)
axis([-0.1 2 -0.7 1.2])
grid on
```

Note that the results in Fig. 13.12 are almost identical to those shown in Fig. 13.11. The only difference is constituted by some high-frequency oscillations present in Fig. 13.12, which must be produced by the fast electric dynamics that has been neglected in Fig. 13.11. As a matter of fact, in Fig. 13.13 we present the root locus diagram corresponding to the block diagram in Fig. 13.8e. In Fig. 13.13a a zoom-in is presented around the origin, which is very similar to the root locus diagram sketched in Fig. 13.10, i.e., when the electric dynamics is neglected. In Fig. 13.13b a zoom-out of the same root locus diagram is presented where it is possible to see two horizontal branches that are due to the fast electric dynamics. These branches place two closed-loop complex conjugate poles at about $s = -19 \pm 500j$. These poles are high-frequency and they are not very well damped. Thus, they are responsible for the high-frequency oscillations observed in Fig. 13.12.

Some experimental results are presented in Fig. 13.14 when step changes are applied from $y_d = z_2^* = 0.006$[m] to $y_d = z_2^* = 0.008$[m] and vice versa. The controller gains in (13.42) and Table 13.2 are employed. The following features, which have already been predicted by the simulations and explained using theoretical arguments, are observed. A zero steady-state error is obtained, a large overshoot of about 50% is present, high-frequency oscillations are present when a step change is applied from $y_d = z_2^* = 0.008$[m] to $y_d = z_2^* = 0.006$[m]. Also note that the peak and the settling times in Fig. 13.14 are very similar to those in Figs. 13.11 and 13.12. The reason why high-frequency oscillations are not present when a step change is applied from $y_d = z_2^* = 0.006$[m] to $y_d = z_2^* = 0.008$[m] is that the plant model is nonlinear; hence, it behaves differently when z_2^* is different. In the following section, we present further details on this topic.

13.6.3 Some Experimental Tests

In all the experiments presented in the following, the switch SW2 was kept open in Fig. 13.3. The experimental tests in the previous section were performed using a controller designed for $y_d = x_2^* = 0.006$[m]. Additional experimental tests were

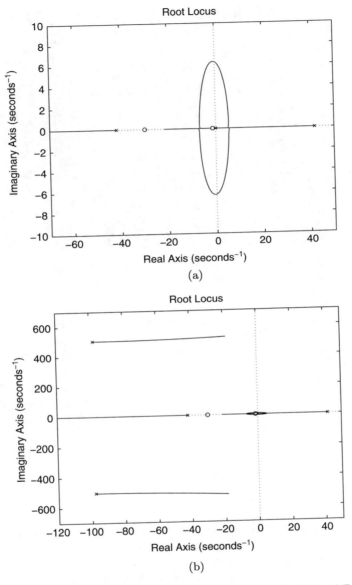

Fig. 13.13 Root locus diagram corresponding to the block diagram in Fig. 13.8e. (**a**) Zoom-in. (**b**) Zoom-out

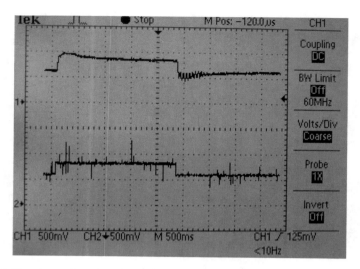

Fig. 13.14 Experimental results when step changes are applied from $y_d = z_2^* = 0.006$[m] to $y_d = z_2^* = 0.008$[m] and vice versa. Top line $x_{2v} = y_v$, bottom line $x_{2v}^* = y_v^*$

performed adjusting SW1, in Fig. 13.3, to select the desired position $y_d = x_2^* = 0.002$[m], i.e., $y_v^* = 0.2$[V]. However, it was observed in experiments that, under this condition, the ball cannot levitate when using the controller gains in (13.42) and Table 13.2.

Trying to understand what happens, we used $y_d = x_2^* = 0.002$[m], the controller gains in Table 13.2, (13.14), (13.15), (13.16), the expressions in (13.18), the numerical values in Sect. 13.4, $A_d = 4.7$ and the system equations when the electric dynamics is neglected, i.e., the block diagrams in Fig. 13.9, to find that the closed-loop poles and zeros are located at:

$$\text{poles: } s = -232.5255, \quad s = -20.2212, \quad s = -2.1167, \quad (13.43)$$

$$\text{zeros: } s = -29.9355, \quad s = -1.3045.$$

This means that the closed-loop system is still stable when neglecting the electrical dynamics. Thus, instability may be produced by the electrical dynamics. To verify this, in Figs. 13.15a, b, and c the root locus diagrams for the internal-most loop and the middle internal loop in Fig. 13.8c, and for the loop shown in Fig. 13.8e, are presented respectively. As is well known, each of these root locus diagrams is drawn based on the open-loop poles and zeros in the respective loop. We also stress that each one of these root locus diagrams takes into account, as open-loop poles and/or open-loop zeros, the closed-loop poles and closed-loop zeros that result from the root locus diagram corresponding to the previous internal loop. This is the reason why the poles at s_1 and s_2 appear in both Figs. 13.15a and b, for instance.

Fig. 13.15 Root locus diagrams. (**a**) Internal-most loop in Fig. 13.8c. (**b**) Middle internal loop in Fig. 13.8c. (**c**) Loop in Fig. 13.8e

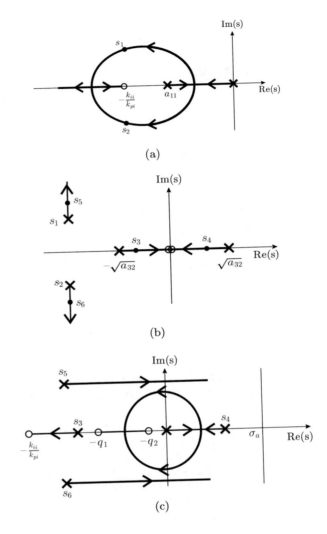

It is not difficult to realize that the open-loop poles at s_5 and s_6 in addition to the open-loop zero at $-\frac{k_{ii}}{k_{pi}}$, which are shown in Fig. 13.15c, are due to the electrical dynamics when taking into account the presence of the PI electric current controller whose gains are given in (13.42). Moreover, note that aside from these open-loop poles and zeros, the root locus diagram in Fig. 13.15c is identical to the root locus diagram in Fig. 13.10. Also note that the root locus branches starting at s_5 and s_6 in Fig. 13.15c end at two asymptotes on the right half-plane. This can be verified as follows.

Using (13.14), (13.15), (13.16), the expressions in (13.18), the numerical values given in (13.42) and Sect. 13.4 in addition to $A_d = 4.7$, $y_d = x_2^* = 0.002$[m], and q_1, q_2 given in (13.32), we find that the open-loop transfer function in Fig. 13.8e, i.e.,

$$A_d A_s G_p(s) \frac{k_d s^2 + k_p s + k_i}{s} \tag{13.44}$$

has poles located at:

$$s_9 = 0, \quad s_5 = -95.55 + 498.59 j, \quad s_6 = -95.55 - 498.59 j, \quad s_4 = 52.18,$$
$$s_3 = -52.18,$$

and zeros located at:

$$-\frac{k_{ii}}{k_{pi}} = -2127.7, \quad s_7 = -29.9, \quad s_8 = -1.3.$$

Hence, using (see Chap. 5):

$$\text{asymptote angle} = \frac{\pm 180° (2q + 1)}{n - m}, \quad q = 0, 1, 2, \ldots$$

and:

$$\sigma_a = \frac{s_9 + s_5 + s_6 + s_4 + s_3 - (-\frac{k_{ii}}{k_{pi}} + s_7 + s_8)}{2},$$

it is found that the root locus diagram in Fig. 13.15c has two asymptotes at $\pm 90°$ with respect to the positive real axis, that cross the real axis at:

$$\sigma_a = 983.9.$$

This proves that the root locus branches starting at s_5 and s_6 in Fig. 13.15c end at the asymptotes that we have just found on the right half-plane. Note that such branches do not exist when the electrical dynamics is not taken into account, i.e., in Fig. 13.10. Hence, this could be the reason for the instability observed in the experiment described above.

However, the reader can perform the corresponding computations to verify that the asymptotes on the right half-plane also exist when $y_d = x_2^* = 0.006$[m]. Hence, we may wonder why instability is not observed experimentally when using the controller gains in (13.42) and Table 13.2, together with $y_d = x_2^* = 0.006$[m]. The answer to this question arises from the gain of $G_p(s)$, which, according to (13.26), directly depends on $-a_{31}$. We can see from (13.18) that $-a_{31} > 0$ is proportional to $-\frac{\partial L(x_2^*)}{\partial x_2} > 0$, which, according to Fig. 13.2 increases as x_2^* decreases. As a matter of fact, using numerical values, we find that $-a_{31} = 9.2663$ for $x_2^* = 0.01$[m], $-a_{31} = 12.5757$ for $x_2^* = 0.006$[m] and $-a_{31} = 19.5621$ for $x_2^* = 0.002$[m]. Thus, a larger loop gain $-a_{31}$ pushes the branches starting at s_5 and s_6 in Fig. 13.15c into the right half-plane if the same PID position controller gains are used.

This explanation also provides a solution to the problem. According to (13.44) and (13.29), the gain of the loop also depends directly on the derivative gain k_d. Hence, if $-a_{31}$ increases, a smaller k_d can be used to keep the gain of the loop approximately constant. Thus, the reader can verify that using the same values for q_1 and q_2 presented in (13.32) and only proposing a different value for k_d, the following controller gains are found:

$$\text{for } y_d = x_2^* = 0.01[m] : \ k_d = 0.0396, \ k_p = 1.2371, \ k_i = 1.5464 \quad (13.45)$$

$$\text{for } y_d = x_2^* = 0.002[m] : \ k_d = 0.01584, \ k_p = 0.4948, \ k_i = 0.6186 \quad (13.46)$$

Moreover, we encourage the reader to verify that the use of these controller gains ensures the stability of the closed-loop system shown in Fig. 13.8e, with $G_p(s)$ given in (13.27), when considering the controller gains in (13.42), i.e., when the electrical dynamics is taken into account. Note that the ratio between the different values of $-a_{31}$ is $12.5757/9.2663 = 1.3571$, which is approximately equal to the ratio between the different values of k_d, $0.0396/0.02772 = 1.4286$, when $y_d = x_2^*$ passes from $0.01[m]$ to $0.006[m]$.

Similarly, the ratio between the different values of $-a_{31}$ is $12.5757/19.5621 = 0.6429$, which is approximately equal to the ratio between the different values of k_d, $0.01584/0.02772 = 0.5714$ when $y_d = x_2^*$ passes from $0.006[m]$ to $0.002[m]$.

A good performance has been observed in experiments when using the controller gains in (13.45), (13.46), the corresponding values for y_d and the PI electric current controller gains in (13.42). It is interesting to say that the ball cannot levitate in experiments when the controller gains in (13.45) are used and the desired position is set at $y_d = x_2^* = 0.002[m]$. However, the opposite situation does perform well, i.e., when the controller gains in (13.46) are used and the desired position is set at $y_d = x_2^* = 0.01[m]$ the closed-loop system has a good performance: the ball levitates without any problems.

Based on the above discussion, we recommend being careful when trying to neglect some part of the closed-loop dynamics, as instability may appear in some cases. This fact has been pointed out by some authors in the control theory literature [14, 15].

In Table 13.3 we present the PID position controller gains used in the above experiments in addition to the corresponding circuit component values used to implement such controller gains. These circuit component values were computed using (13.40) and (13.41).

In Fig. 13.16, the experimental results are presented when a step change in the desired position is commanded from $y_d = x_2^* = 0.002[m]$ to $y_d = x_2^* = 0.006[m]$ and the controller gains in (13.42) and (13.46) are used. It is observed that the ball position y_v reaches its desired value y_v^* in a steady state. Moreover, it is clear that this affirmation is true after and before the step change on the desired position is commanded.

The experimental results in Fig. 13.17 were obtained when a step change in the desired position is commanded from $y_d = x_2^* = 0.006[m]$ to $y_d = x_2^* = 0.01[m]$

Table 13.3 Numerical values tested experimentally for the PID position controller

	PID gains	Circuit component values
$y_d = x_2^* = 0.01[m]$	$k_d = 0.0396$ $k_p = 1.2371$ $k_i = 1.5464$	$R_1 = 330[K\Omega]$ $R_2 = 800[K\Omega]$ $C_1 = 0.1[\mu F]$ $C_2 = 1[\mu F]$
$y_d = x_2^* = 0.006[m]$	$k_d = 0.02772$ $k_p = 0.866$ $k_i = 1.082$	$R_1 = 470[K\Omega]$ $R_2 = 800[K\Omega]$ $C_1 = 0.07[\mu F]$ $C_2 = 1[\mu F]$
$y_d = x_2^* = 0.002[m]$	$k_d = 0.01584$ $k_p = 0.4948$ $k_i = 0.6186$	$R_1 = 800[K\Omega]$ $R_2 = 800[K\Omega]$ $C_1 = 0.04[\mu F]$ $C_2 = 1[\mu F]$

and the controller gains in (13.42) and Table 13.2 are used. It is observed again that the ball position y_v reaches its desired value y_v^* in a steady state and this affirmation is true after and before the step change on the desired position is commanded.

In both Figs. 13.16 and 13.17 a similar phenomenon is observed to that appearing in Fig. 13.11: a large overshoot of about 100% is present in Fig. 13.16 and about 50% in Fig. 13.17. This phenomenon is explained again using the arguments in Sect. 8.1.2 (see also the explanation for Fig. 13.11).

If the reader is trying to compare the time responses in Figs. 13.16 and 13.17, note that different time scales are used in these figures. In Fig. 13.18 a picture of the magnetic levitation system that has been built and used in the experiments is presented.

13.6.4 Pulse Width Modulation Power Amplifier

Recall that the closed-loop system performs well when the controller gains in (13.46), (13.42) are used at $y_d = x_2^* = 0.002[m]$. In Fig. 13.19, the electric current through the power transistor TIP 141 is shown, i.e., through the $1[\Omega]$, 5[W], resistance in Fig. 13.3, obtained when using these controller gains at this desired position. The PWM waveform of this electric current is clearly observed. In Fig. 13.20 is shown the electric current waveform obtained when $k_{ii} = 10^5$ is set and all the other controller gains remain the same as in (13.46), (13.42). Note that this electric current has no clear PWM waveform. Furthermore, this also happens when $k_{ii} = 10^6$ is used and in Fig. 13.21 the waveform of this electric current in this case is shown. The reason for this behavior is explained in the following.

The block diagram in Fig. 13.22a represents the section of Fig. 13.3 corresponding to the internal electric current loop and the PWM power amplifier. The symbol

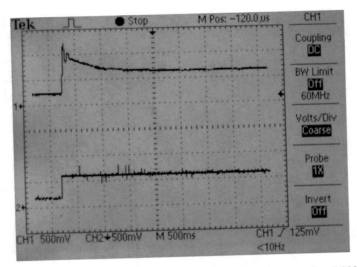

Fig. 13.16 A step change in the desired position is commanded from $y_d = x_2^* = 0.002[\text{m}]$ ($y_v^* = 0.2[\text{V}]$) to $y_d = x_2^* = 0.006[\text{m}]$ ($y_v^* = 0.6[\text{V}]$). Top line $x_{2v} = y_v$, bottom line $x_{2v}^* = y_v^*$

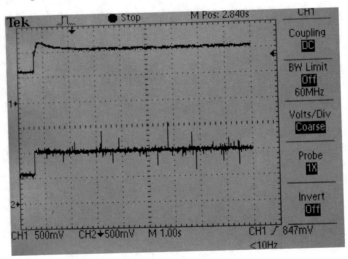

Fig. 13.17 A step change in the desired position is commanded from $y_d = x_2^* = 0.006[\text{m}]$ ($y_v^* = 0.6[\text{V}]$) to $y_d = x_2^* = 0.01[\text{m}]$ ($y_v^* = 1.0[\text{V}]$) Top line $x_{2v} = y_v$, bottom line $x_{2v}^* = y_v^*$

A_0 stands for the open-loop gain of the operational amplifier used as a comparator in the PWM power amplifier. From this block diagram we obtain the block diagram in Fig. 13.22b, which allows us to quantify the effect of the triangular wave $V_t(s)$ on the electric current through the power transistor $I_t(s)$. To obtain the block diagram in Fig. 13.22b it is assumed that $I_d(s) = 0$ in Fig. 13.22a (superposition). It is not

Fig. 13.18 The magnetic levitation system that has been built

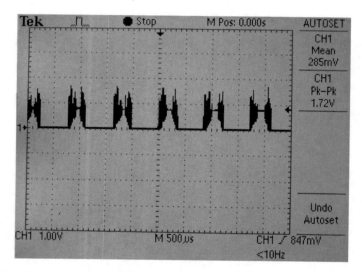

Fig. 13.19 Waveform of the electric current flowing through the power transistor TIP 141 when $k_{ii} = 10000$

difficult to see that:

$$I_t(s) = -\frac{A_0}{1 + A_0 \frac{k_{pi}s + k_{ii}}{s}} V_t(s).$$

As A_0 is very large, this expression is simplified to:

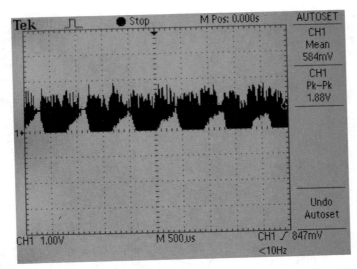

Fig. 13.20 Waveform of the electric current flowing through the power transistor TIP 141 when $k_{ii} = 10^5$

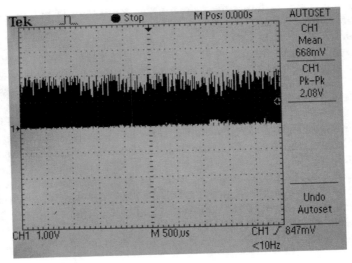

Fig. 13.21 Waveform of the electric current flowing through the power transistor TIP 141 when $k_{ii} = 10^6$

$$I_t(s) = -\frac{s}{k_{pi}s + k_{ii}} V_t(s) = -\frac{1}{k_{pi}} \frac{s}{\left(s + \frac{k_{ii}}{k_{pi}}\right)} V_t(s). \tag{13.47}$$

Thus, $I_t(s)$ is obtained from $V_t(s)$ through a first-order, high-pass filter with corner frequency at $\frac{k_{ii}}{k_{pi}}$ and $\frac{1}{k_{pi}}$ as the high-frequency gain. Using the controller gains in (13.42), it is found that the corner frequency is 2127.7[rad/s]. Hence, as the

Fig. 13.22 Effect of V_t in Fig. 13.3 on the electric current through the power transistor TIP 141

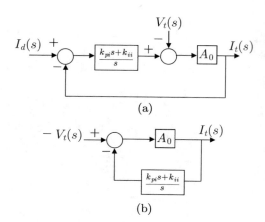

frequency of V_t is 9613.3[rad/s], i.e., 1.53[KHz], which is larger than 2127.7[rad/s], it is concluded that V_t is not significantly attenuated (by a gain of 0.2128, i.e., -13.7[dB], see Fig. 13.23) and this explains why the PWM signal is clearly observed in Fig. 13.19. We stress that the PWM waveform is not a triangle because the very large gain A_0 forces the output of the comparator to have a rectangular waveform with a frequency of 1.53[KHz].

When $k_{ii} = 10^5$ is used, the corner frequency is 21277[rad/s], which is much larger than the frequency of V_t, i.e., 9613.3[rad/s]. This means that the effect of V_t on $I_t(s)$ is strongly attenuated (by a gain of 0.087, i.e., -21.2[dB], see Fig. 13.24). This means that the PWM waveform in Fig. 13.20 should be $0.2128/0.087 = 2.4460$ times more attenuated than the PWM waveform in Fig. 13.19. This explains why the PWM signal is not clearly observed in Fig. 13.20. Moreover, from Fig. 13.24 it is concluded that what we see in Fig. 13.20 is noise. It is not difficult to realize that this situation is worse when $k_{ii} = 10^6$ (V_t is attenuated by a gain of 0.0096, i.e., -40.3[dB], see Fig. 13.25). This means that the PWM waveform in Fig. 13.21 should be $0.2128/0.0096 = 22.1667$ times more attenuated than the PWM waveform in Fig. 13.19. This also explains why the electric current has no clear PWM waveform, as observed in Fig. 13.21. Again, from Fig. 13.25 we conclude that what we see Fig. 13.21 is noise.

13.6.5 Design of the PID Position Controller Using the Frequency Response

The design using the frequency response is also performed based on the assumption that the electric subsystem dynamics is rendered fast enough using a suitable PI controller. Hence, the block diagrams in Fig. 13.9 and the expressions in (13.29) and (13.30) are also considered in the present section. Recall that $s^2 - a_{32} = (s -$

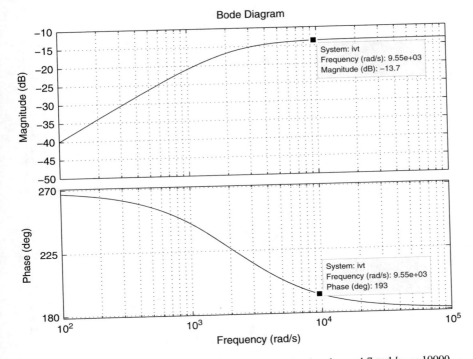

Fig. 13.23 Bode diagrams of the transfer function in (13.47) when $k_{pi} = 4.7$ and $k_{ii} = 10000$

$\sqrt{a_{32}})(s + \sqrt{a_{32}})$, as $a_{32} > 0$. Thus, one approach to controller design is to choose:

$$q_1 = \sqrt{a_{32}}, \qquad (13.48)$$

as this converts the open-loop transfer function in (13.30) into:

$$G(s)H(s) = \frac{-a_{31}k_d A_d A_s (s + q_2)}{s(s - \sqrt{a_{32}})}.$$

On the other hand, write:

$$s + q_2 = q_2(z + 1), \quad z = \frac{s}{q_2}.$$

Hence:

$$G(s)H(s) = k_d q_2 (z + 1) \frac{-a_{31} A_d A_s}{s(s - \sqrt{a_{32}})}. \qquad (13.49)$$

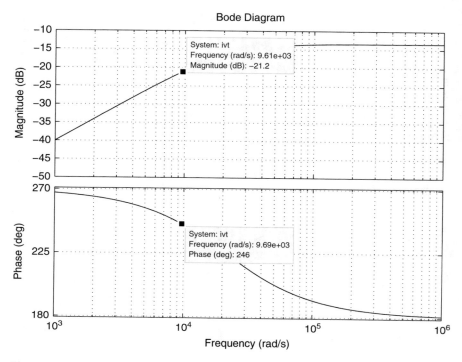

Fig. 13.24 Bode diagrams of the transfer function in (13.47) when $k_{pi} = 4.7$ and $k_{ii} = 10^5$

In the following, it is assumed that $k_d q_2(z+1)$ is the "controller" and $\frac{-a_{31} A_d A_s}{s(s-\sqrt{a_{32}})}$ is the "plant to be controlled." The polar plot for the transfer function in (13.49) is almost identical to the polar plot in Fig. 6.102. The only difference is that, in the present case, the polar plot reaches the origin parallel to the negative imaginary axis. This is because the electric dynamics is not taken into account in this section. In the case in which the polar plot has the form shown in Fig. 6.102, then closed-loop stability is ensured. Thus, in what follows, the crossover frequency and the phase margin are proposed such that the shape shown in Fig. 6.102 is ensured.

Using $x_2^* = 0.006$[m] to find the numerical values for a_{31} and a_{32} in addition to $A_s = 100$, $A_d = 4.7$, the Bode diagrams for $\frac{-a_{31} A_d A_s}{s(s-\sqrt{a_{32}})}$, which are presented in Fig. 13.26, can be plotted. Setting $\omega_1 = 71.5$[rad/s] as the desired crossover frequency and $40°$ as the desired phase margin, it is found that $70°$ is the phase lead that has to be contributed by the "controller" $k_d q_2(z+1)$ when $\omega = \omega_1$. In Fig. 13.27 the Bode diagrams for the factor $z + 1$ are presented, which is the only component of the "controller" that can contribute some phase lead. There, it is found that the desired $70°$ phase lead is produced when:

$$\frac{\omega}{q_2} = 2.76,$$

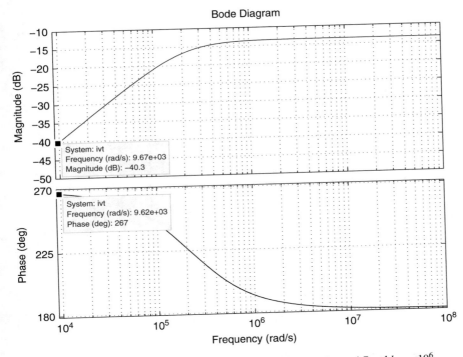

Fig. 13.25 Bode diagrams of the transfer function in (13.47) when $k_{pi} = 4.7$ and $k_{ii} = 10^6$

and that 9.35[dB] is the magnitude of $|z + 1|_{dB}$ at that frequency. Hence, replacing $\omega = \omega_1 = 71.5$[rad/s] in the previous expression, the following is found:

$$q_2 = \frac{\omega}{2.76} = \frac{71.5}{2.76} = 25.9058. \tag{13.50}$$

On the other hand, to fix the crossover frequency at $\omega_1 = 71.5$[rad/s], the magnitude of the transfer function in (13.49) is required to be equal to 0[dB] when $\omega = \omega_1$. As the magnitude of $\frac{-a_{31}A_d A_s}{s(s - \sqrt{a_{32}})}$ is 0[dB] when $\omega = \omega_1$ (see Fig. 13.26), this means that:

$$|k_d q_2(z + 1)|_{dB} = 20 \log(k_d q_2) + |z + 1|_{dB} = 0, \tag{13.51}$$

when $\omega = \omega_1$, i.e., when $|z + 1|_{dB} = 9.35$[dB]. Thus:

$$k_d = \frac{1}{q_2} 10^{\frac{-9.35}{20}} = 0.01309.$$

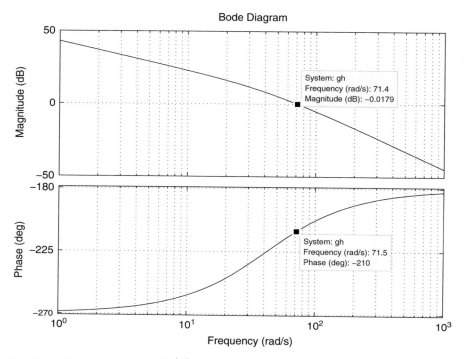

Fig. 13.26 Bode diagrams for $\frac{-a_{31}A_dA_s}{s(s-\sqrt{a_{32}})}$, using $x_2^* = 0.006[\text{m}]$, $A_s = 100$ and $A_d = 4.7$

Using this value, (13.29), (13.50) and (13.48), to find $q_1 = 41.8543$, the following is obtained:

$$k_p = 0.8873, \quad k_i = 14.2.$$

The Bode diagrams of the transfer function in (13.30) are presented in Fig. 13.28 when $q_1 = 41.8543$, $q_2 = 25.9058$, $k_d = 0.01309$ are used. Note that a $\omega_1 = 71.5[\text{rad/s}]$ crossover frequency and a $40°$ phase margin have been obtained. The corresponding time response is presented in Fig. 13.29 when $z_1(0) = 0$, $z_2(0) = 1$, $z_3(0) = 0$ and a zero initial condition is used for the integral part of the PID position controller. Note that a large overshoot of about 75% is present. This overshoot is related to the large resonance peak that can be observed in Fig. 13.30 where the corresponding Bode diagrams of the closed-loop system are presented. The obvious solution to reducing such an overshoot is significantly increasing the phase margin, i.e., let us propose a $70°$ phase margin. However, note, from Fig. 13.26, that keeping $\omega_1 = 71.5[\text{rad/s}]$ as the desired crossover frequency would require the "controller" to contribute a $100°$ phase lead, which is not possible, as is clear from Fig. 13.27. Carefully observing Fig. 13.26 it is concluded that a $70°$ phase margin

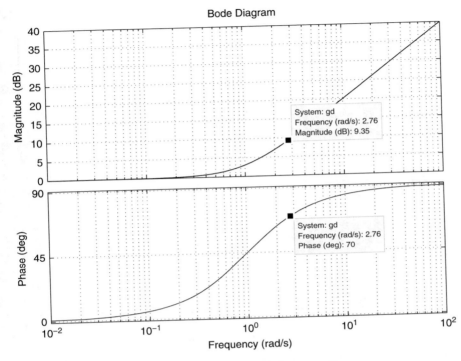

Fig. 13.27 Bode diagrams for $z + 1$. The horizontal axis represents $\frac{\omega}{q_2}$

can be obtained if the crossover frequency ω_1 is also significantly increased. In the following, it is shown how to perform such a design.

Bode diagrams in Fig. 13.31 are plotted using (13.48), (13.49), $x_2^* = 0.006[m]$, in addition to $A_s = 100$, $A_d = 4.7$. These are the same Bode diagrams as those shown in Fig. 13.26. The only difference is that the data are analyzed at a different frequency. Setting $\omega_1 = 130[rad/s]$ as the desired crossover frequency and 70° as the desired phase margin, it is found that 88° is the phase lead that has to be contributed by the "controller" $k_d q_2(z + 1)$ when $\omega = \omega_1$. The Bode diagrams for the factor $z + 1$ are presented in Fig. 13.32, which is the only component of the "controller" that can contribute some phase lead. There, it is found that the desired 88° phase lead is produced when:

$$\frac{\omega}{q_2} = 28.6,$$

and that 29.1[dB] is the magnitude of $|z + 1|_{dB}$ at that frequency. Hence, replacing $\omega = \omega_1 = 130[rad/s]$ in the previous expression:

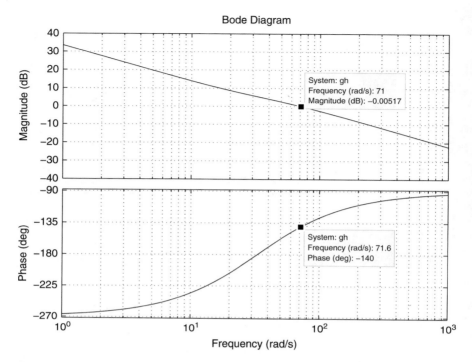

Fig. 13.28 Bode diagrams for $G(s)H(s)$ in (13.30) when $q_1 = 41.8543$, $q_2 = 25.9058$, $k_d = 0.01309$

$$q_2 = \frac{\omega}{28.6} = \frac{130}{28.6} = 4.4828. \tag{13.52}$$

On the other hand, to fix the crossover frequency at $\omega_1 = 130\,\text{[rad/s]}$, it is necessary for the magnitude of the transfer function in (13.49) to be equal to 0[dB] when $\omega = \omega_1$. As the magnitude of $\frac{-a_{31}A_dA_s}{s(s-\sqrt{a_{32}})}$ is -9.61[dB] when $\omega = \omega_1$ (see Fig. 13.31), this means that:

$$|k_dq_2(z+1)|_{dB} = 20\log(k_dq_2) + |z+1|_{dB} = 9.61, \tag{13.53}$$

when $\omega = \omega_1$, i.e., when $|z+1|_{dB} = 29.1\text{[dB]}$. Thus:

$$k_d = \frac{1}{q_2}10^{\frac{(9.61-29.1)}{20}} = 0.02293.$$

Using this value, (13.29), (13.52) and (13.48), to find $q_1 = 41.8543$, yields:

$$k_p = 1.063, \quad k_i = 4.303.$$

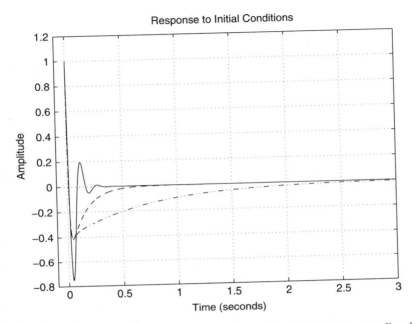

Fig. 13.29 Closed-loop time response of system in Fig. 13.9 (simulation). Continuous line: $k_d = 0.01309$, $k_p = 0.8873$, $k_i = 14.2$. Dashed: $k_d = 0.02293$, $k_p = 1.063$, $k_i = 4.303$. Line-dot: $k_d = 0.02259$, $k_p = 1.09$, $k_i = 1.068$

Bode diagrams of the transfer function in (13.30) are presented in Fig. 13.33 when $q_1 = 41.8543$, $q_2 = 4.4828$, $k_d = 0.02293$ are used. Note that a $\omega_1 = 130$[rad/s] crossover frequency and a 70° phase margin have been obtained. The corresponding time response is presented in Fig. 13.29 when $z_1(0) = 0$, $z_2(0) = 1$, $z_3(0) = 0$ and a zero initial condition is used for the integral part of the PID position controller. It is observed that the overshoot has decreased to about 40%. This overshoot is related to the smaller resonance peak observed in Fig. 13.30 where the corresponding Bode diagrams of the closed-loop system are presented.

Although the overshoot has been significantly improved, it was found in experiments (see Sect. 13.6.6) that the ball describes fast oscillations when commanding step changes in the desired position. Hence, the main objective is now to design a controller that can attenuate such oscillations. As the observed oscillations are fast, we try to slow them down by increasing the effect of low frequencies. In the previous designs q_1 has been computed using (13.48). This, however, has resulted in values for q_2 that are larger than 4.4828 and even larger values for q_1. The reader can verify that q_1 and q_2 fix the location of the low-frequency zeros of the closed-loop transfer function. Hence, as we want to increase the effect of low frequencies on the closed-loop transfer function we need a value for q_1 or q_2 that is smaller than 4.4828. With this aim q_1 is no longer computed as in (13.48), but, instead, the following is proposed:

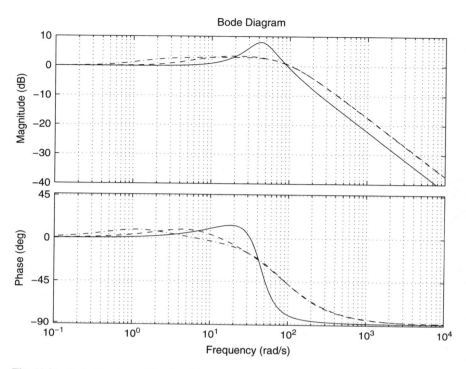

Fig. 13.30 Bode diagrams of the closed-loop system in Fig. 13.9. Continuous line: $k_d = 0.01309$, $k_p = 0.8873$, $k_i = 14.2$. Dashed: $k_d = 0.02293$, $k_p = 1.063$, $k_i = 4.303$. Line-dot: $k_d = 0.02259$, $k_p = 1.09$, $k_i = 1.068$

$$q_1 = 1$$

whereas q_2 is a result of the design method. In this case, the factor $(s + q_1)$ does not cancel with any of factors $(s - \sqrt{a_{32}})$ or $(s + \sqrt{a_{32}})$ and, because of that, the open-loop transfer function in (13.30) converts into:

$$G(s)H(s) = k_d q_2 (z + 1)\frac{-a_{31} A_d A_s (s + 1)}{s(s^2 - a_{32})}. \tag{13.54}$$

From now on the factor $\frac{-a_{31} A_d A_s (s+1)}{s(s^2 - a_{32})}$ is considered the "plant to be controlled" whereas the factor $k_d q_2 (z + 1)$ is considered the "controller". As the use of a $\omega_1 = 130$[rad/s] crossover frequency and a 70° phase margin have resulted in a good overshoot improvement, it is chosen to maintain these design parameters in the following.

Using $x_2^* = 0.006$[m], in addition to $A_s = 100$, $A_d = 4.7$, the Bode diagrams of the factor $\frac{-a_{31} A_d A_s (s+1)}{s(s^2 - a_{32})}$ can be plotted, which are shown in Fig. 13.34. To obtain a 70° phase margin and a $\omega_1 = 130$[rad/s] crossover frequency, the "controller"

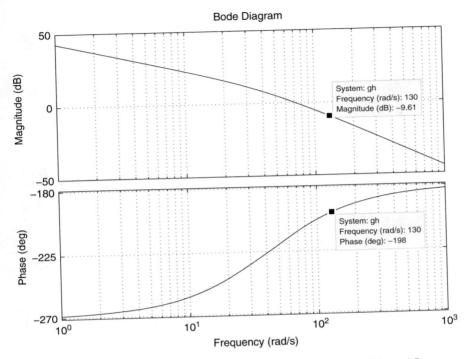

Fig. 13.31 Bode diagrams for $\frac{-a_{31}A_dA_s}{s(s-\sqrt{a_{32}})}$, using $x_2^* = 0.006[m]$, $A_s = 100$ and $A_d = 4.7$

$k_d q_2(z + 1)$ is required to contribute a 70° phase lead when $\omega = \omega_1$. The Bode diagrams for the factor $z+1$ are presented in Fig. 13.35, which is the only component of the "controller" that is capable of contributing some phase lead. There, it is found that the desired 70° phase lead is produced when:

$$\frac{\omega}{q_2} = 2.76,$$

and that 9.37[dB] is the magnitude of $|z + 1|_{dB}$ at that frequency. Hence, replacing $\omega = \omega_1 = 130[rad/s]$ in the previous expression:

$$q_2 = \frac{\omega}{2.76} = \frac{130}{2.76} = 47.2727. \tag{13.55}$$

On the other hand, to fix the crossover frequency at $\omega_1 = 130[rad/s]$, it is necessary for the magnitude of the transfer function in (13.54) to be equal to 0[dB] when $\omega = \omega_1$. As the magnitude of $\frac{-a_{31}A_dA_s(s+1)}{s(s^2-a_{32})}$ is $-9.94[dB]$ when $\omega = \omega_1$ (see Fig. 13.34), this means that:

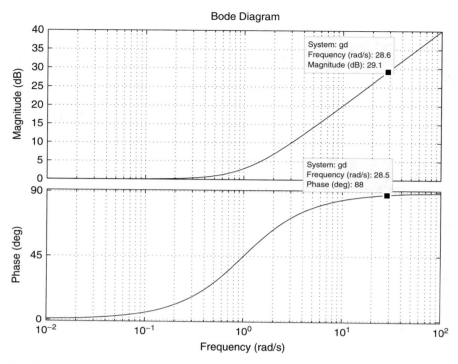

Fig. 13.32 Bode diagrams for $z + 1$. The horizontal axis represents $\frac{\omega}{q_2}$

$$|k_d q_2(z + 1)|_{dB} = 20 \log(k_d q_2) + |z + 1|_{dB} = 9.94, \qquad (13.56)$$

when $\omega = \omega_1$, i.e., when $|z + 1|_{dB} = 9.37[\text{dB}]$. Thus:

$$k_d = \frac{1}{q_2} 10^{\frac{(9.94-9.37)}{20}} = 0.02259.$$

Using this value, (13.29), (13.55) and $q_1 = 1$:

$$k_p = 1.09, \quad k_i = 1.068.$$

Bode diagrams of the transfer function in (13.54) are presented in Fig. 13.36, when $q_1 = 1$, $q_2 = 47.2727$, $k_d = 0.02259$ are used. Note that a $\omega_1 = 130[\text{rad/s}]$ crossover frequency and a $70°$ phase margin have been obtained. The corresponding time response is presented in Fig. 13.29, when $z_1(0) = 0$, $z_2(0) = 1$, $z_3(0) = 0$, and a zero initial condition is used for the integral part of the PID position controller. Note that the overshoot is kept at about 40%. Consider Fig. 13.30, where the corresponding Bode diagram of the closed-loop system is presented. Note that although the high-frequency part of the resonance peak remains the same as in the

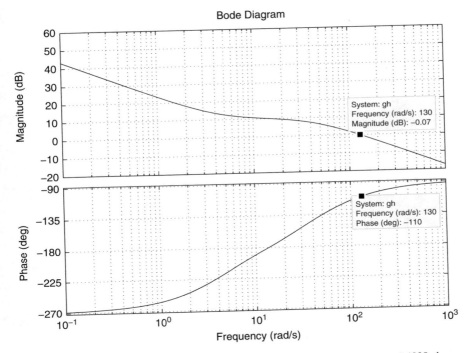

Fig. 13.33 Bode diagrams for $G(s)H(s)$ in (13.30) when $q_1 = 41.8543$, $q_2 = 4.4828$, $k_d = 0.02293$ are used

previous design, the low-frequency part of the resonance peak has grown. Hence, these Bode diagrams explain why the corresponding time response in Fig. 13.29 has the same rise time but a larger settling time.

13.6.6 Some Other Experimental Tests

In all the experiments presented in the following, the switch SW2 was kept open in Fig. 13.3. In this section, some experimental results are obtained when testing the PID position controller gains computed in the previous section, which are now shown in Table 13.4. The circuit component values used to implement the corresponding controller gains are also presented in this table. The resistances $R_1 = 300[K\Omega]$ and $R_1 = 350[K\Omega]$ were constructed as the series connection of three $100[K\Omega]$ resistances and the series connection of one $330[K\Omega]$ and one $22[K\Omega]$ resistances respectively. Eight $0.01[\mu F]$ and four $0.1[\mu F]$ capacitors were connected in parallel to obtain $C_1 = 0.08[\mu F]$ and $C_2 = 0.4[\mu F]$ respectively. Finally, $C_2 = 1.5[\mu F]$ was implemented as the parallel connection of one $1[\mu F]$ and five $0.1[\mu F]$ capacitors. The PI electric current controller gains were selected as

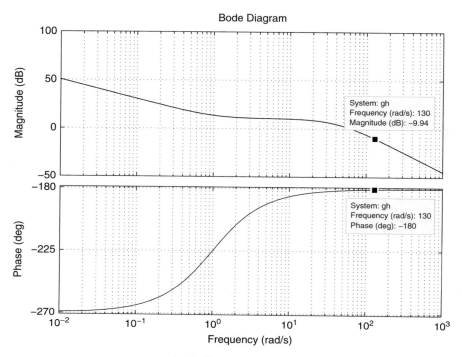

Fig. 13.34 Bode diagrams for $\frac{-a_{31}A_dA_s(s+1)}{s(s^2-a_{32})}$, using $x_2^* = 0.006$[m], $A_s = 100$ and $A_d = 4.7$

those presented in (13.42), which are rewritten here for ease of reference:

$$k_{pi} = 4.7, \quad k_{ii} = 10000.$$

When testing the controller gains in Table 13.4a) experimentally, it was found that the ball cannot levitate, i.e., the ball cannot stay at the desired position $y_d = z_2^* = 0.006$[m]. The reader can plot the corresponding polar diagram when taking into account the electric dynamics and the gains $k_{pi} = 4.7$, $k_{ii} = 10000$ to verify that the closed-loop system is stable from the theoretical point of view. Hence, the problem is that controller gains in Table 13.4a) do not achieve a good performance in practice. Recall that these controller gains were found to produce a very large overshoot in the previous section. During the theoretical design stage, this feature warned us that a bad performance may be obtained in practice and this has motivated us to look for better designs. Thus, the controller gains in Table 13.4b) and c) were computed and they are tested experimentally in the following.

In Fig. 13.37 the time response obtained experimentally is presented when step changes are applied from $y_d = z_2^* = 0.006$[m] to $y_d = z_2^* = 0.008$[m] and vice versa. The PID position controller gains are those in Table 13.4b). It is observed that the desired position is reached in a steady state. Also note the high-

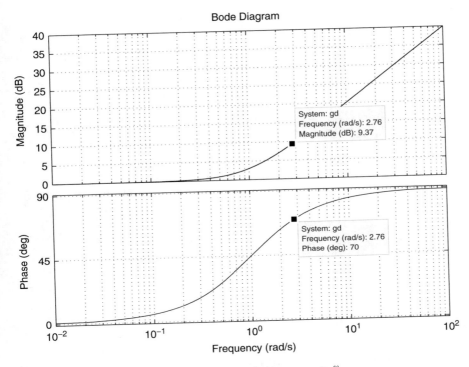

Fig. 13.35 Bode diagrams for $z + 1$. The horizontal axis represents $\frac{\omega}{q_2}$

frequency oscillations, which are especially important when $y_d = z_2^* = 0.006[m]$ is commanded. In Fig. 13.38 a simulation of the closed-loop system is presented when the electrical dynamics is taken into account. There, it is observed that high-frequency oscillations are present. Hence, such high-frequency oscillations are introduced by the electrical dynamics, which has been rendered fast by the PI electric controller gains that are employed. Moreover, in Fig. 13.39 the closed-loop Bode diagrams corresponding to both cases, i.e., when neglecting the electrical dynamics and when this dynamics is taken into account, are compared. Note that both Bode diagrams are identical for low frequencies but there is a difference at high frequencies, i.e., because of the effect of the fast electrical dynamics. Finally, in Fig. 13.40 the open-loop transfer function polar plot, when the electrical dynamics is taken into account, is presented. This is done to show that the closed-loop system is stable, despite the presence of the fast electrical dynamics, which was not taken into account during the design stage presented in the previous section.

In Fig. 13.41, the time response obtained experimentally is presented when step changes are applied from $y_d = z_2^* = 0.006[m]$ to $y_d = z_2^* = 0.008[m]$ and vice versa. The PID position controller gains are those in Table 13.4c). Note that the desired position is reached in a steady state. Also note that although some high-

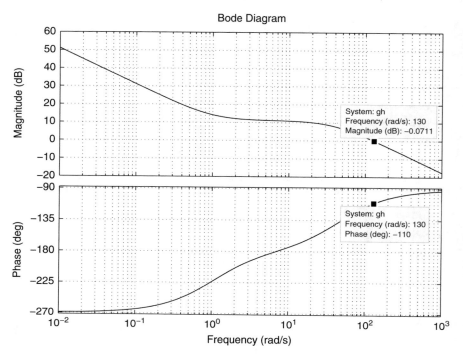

Fig. 13.36 Bode diagrams for $G(s)H(s)$ in (13.54) when $q_1 = 1$, $q_2 = 47.2727$, $k_d = 0.02259$ are used

Table 13.4 Numerical values tested experimentally for the PID position controller when $y_d = x_2^* = 0.006[m]$

	PID gains	Circuit component values
a)	$k_d = 0.01309$ $k_p = 0.8873$ $k_i = 14.2$	$R_1 = 350[K\Omega]$ $R_2 = 330[K\Omega]$ $C_1 = 0.08[\mu F]$ $C_2 = 0.1[\mu F]$
b)	$k_d = 0.02293$ $k_p = 1.063$ $k_i = 4.303$	$R_1 = 300[K\Omega]$ $R_2 = 560[K\Omega]$ $C_1 = 0.08[\mu F]$ $C_2 = 0.4[\mu F]$
c)	$k_d = 0.02259$ $k_p = 1.09$ $k_i = 1.068$	$R_1 = 300[K\Omega]$ $R_2 = 560[K\Omega]$ $C_1 = 0.08[\mu F]$ $C_2 = 1.5[\mu F]$

frequency oscillations are evident, they are much more attenuated than oscillations in Fig. 13.37. This corroborates the correctness of the ideas that have guided us in the previous section to perform this design. Finally, it is clear that the settling time

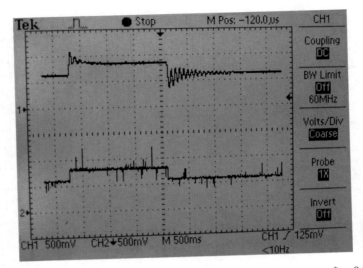

Fig. 13.37 Experimental results when step changes are commanded from $y_d = z_2^* = 0.006$[m] to $y_d = z_2^* = 0.008$[m] and vice versa. Top line $x_{2v} = y_v$, bottom line $x_{2v}^* = y_v^*$. Controller gains in Table 13.4b)

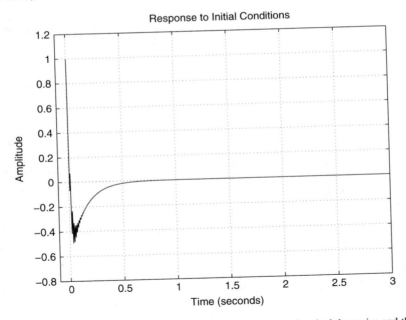

Fig. 13.38 Simulated time response when taking into account the electrical dynamics and the PI electric current controller. Controller gains in Table 13.4b) and $k_{pi} = 4.7$, $k_{ii} = 10000$

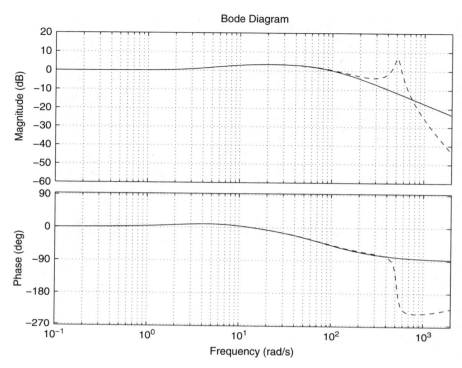

Fig. 13.39 Closed-loop system Bode diagrams. Continuous: the electrical dynamics is not considered. Dashed: the electrical dynamics is taken into account. Controller gains in Table 13.4b) and $k_{pi} = 4.7$, $k_{ii} = 10000$

is much larger in Fig. 13.41 than in Fig. 13.37, which was also predicted during the design stage in the previous section.

In Fig. 13.42 a simulation of the closed-loop system is presented when the electrical dynamics is taken into account. From this, it is concluded that such high-frequency oscillations are introduced by the electrical dynamics, which has been rendered fast by the PI electric controller gains that are employed. In Fig. 13.43 the closed-loop Bode diagrams corresponding to both cases, i.e., when neglecting the electrical dynamics and when this dynamics is taken into account, are compared. Note that both Bode diagrams are identical for low frequencies but there is a difference at high frequencies, i.e., because of the effect of the fast electrical dynamics. Finally, in Fig. 13.44 the open-loop transfer function polar plot, when the electrical dynamics is taken into account, is presented. This is done to show that the closed-loop system is stable despite the presence of the fast electrical dynamics, which was not taken into account during the design stage presented in the previous section.

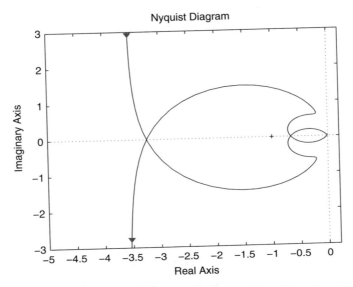

Fig. 13.40 Open-loop polar plot taking into account electrical dynamics. Controller gains in Table 13.4b) and $k_{pi} = 4.7$, $k_{ii} = 10000$

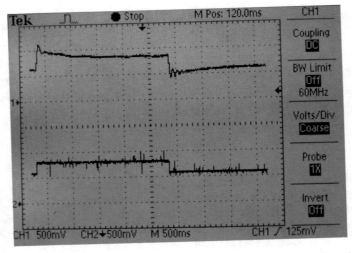

Fig. 13.41 Experimental results when step changes are applied from $y_d = z_2^* = 0.006$[m] to $y_d = z_2^* = 0.008$[m] and vice versa. Top line $x_{2v} = y_v$, bottom line $x_{2v}^* = y_v^*$. Controller gains in Table 13.4c)

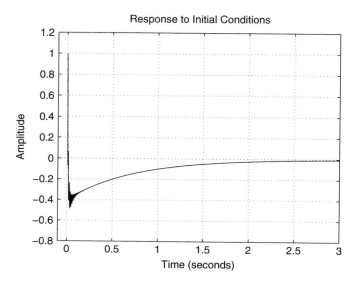

Fig. 13.42 Simulated time response when taking into account the electrical dynamic and the PI electric current controller. Controller gains in Table 13.4c) and $k_{pi} = 4.7$, $k_{ii} = 10000$

13.6.7   An Alternative Procedure for Designing the PI Electric Current Controller

In this section, an alternative procedure that is suggested in the literature [9] for designing the PI electric current controller is described. Consider the block diagram in Fig. 13.8c. It is suggested in [9] to choose:

$$\frac{k_{ii}}{k_{pi}} = -a_{11} > 0 \tag{13.57}$$

as this reduces the corresponding block diagram to that shown in Fig. 13.45.

In Figs. 13.46a, b, c the root locus diagrams are presented for the internal-most loop, the middle internal loop in addition to the external loop respectively in Fig. 13.45. As is well known, each one of these root locus diagrams is drawn based on the open-loop poles and zeros in the respective loop. We also stress that each of these root locus diagrams takes into account, as open-loop poles and/or open-loop zeros, the closed-loop poles and closed-loop zeros that result from the root locus diagram corresponding to the previous internal loop. This is the reason why the pole at s_1 appears in both Figs. 13.46a and b, for instance.

Note, from Figs. 13.45 and 13.46a, that k_{pi} can be used to move s_1 to the left on the left half-plane, i.e., to render the electric dynamics faster, which is one objective of the PI electric current controller, as explained in Sect. 13.6.2. Also note that the zero at $-\frac{k_{ii}}{k_{pi}}$ in Fig. 13.46c appears because of the pole at $-\frac{k_{ii}}{k_{pi}}$ contained in the

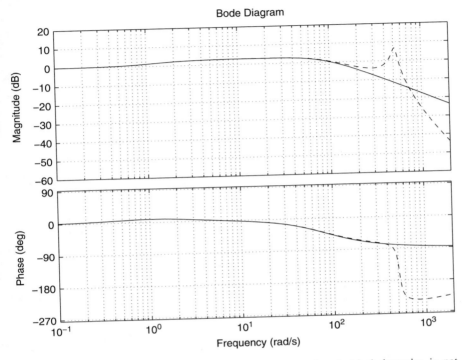

Fig. 13.43 Closed-loop system Bode diagrams. Continuous: the electrical dynamics is not considered. Dashed: the electrical dynamics is taken into account. Controller gains in Table 13.4c) and $k_{pi} = 4.7$, $k_{ii} = 10000$

feedback path of the middle internal loop in Fig. 13.45. Hence, according to:

$$\text{asymptote angle} = \frac{\pm 180°(2q + 1)}{n - m}, \quad q = 0, 1, 2, \ldots$$

and:

$$\sigma_a = \frac{s_5 + s_4 + s_3 + s_2 + 0 - (-q_1 - q_2 - \frac{k_{ii}}{k_{pi}})}{n - m}$$

with $n - m = 5 - 3 = 2$, the asymptotes crossing the real axis at σ_a are orthogonal to the real axis. Moreover, these asymptotes lay on the left half-plane, as σ_a can always be rendered negative because a large $s_1 < 0$ (using a large $k_{pi} > 0$) produces large negative real parts for s_4 and s_5. Furthermore, contrary to what happens in the design method described in Sect. 13.6.2, $\frac{k_{ii}}{k_{pi}} = -a_{11}$ is not very large in this case.

Recall that it has been noted in Sect. 13.6.3 that the design procedure introduced in Sect. 13.6.2 may result in instability problems because two branches of the root

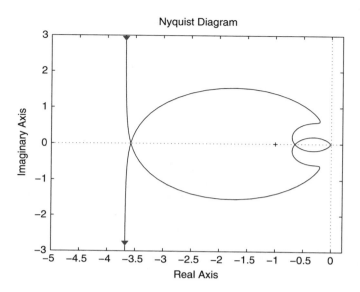

Fig. 13.44 Open-loop polar diagram taking into account electrical dynamics. Controller gains in Table 13.4c) and $k_{pi} = 4.7$, $k_{ii} = 10000$

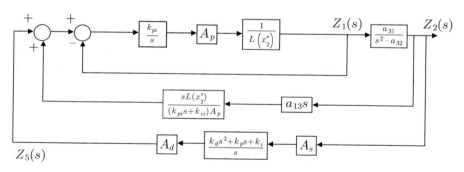

Fig. 13.45 Block diagram in Fig. 13.8c when (13.57) is considered

locus are pushed into the right half-plane. It is clear that such instability problems do not exist when using the design procedure in the present section, which represents an important advantage of this method.

However, the condition in (13.57) represents a restriction on the PI controller gains, which does not ensure that a good performance, is obtained in practice. Thus, to study the performance achievable when using (13.57), we tested this tuning procedure experimentally. Using $x_2^* = 0.01$[m], it is found that (13.57) becomes:

$$\frac{k_{ii}}{k_{pi}} = -a_{11} = 72.5384. \tag{13.58}$$

13.6 Controller Design

Fig. 13.46 Root locus diagrams for the block diagrams in Fig. 13.45. (**a**) Most internal loop. (**b**) Middle internal loop (recall that $a_{11} = -\frac{k_{ii}}{k_{pi}}$). (**c**) External loop

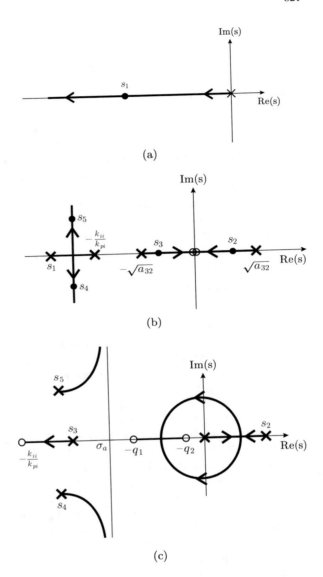

(a)

(b)

(c)

This yields $k_{ii} = 340.89$ when using $k_{pi} = 4.7$ as in (13.42). This value for k_{pi} is chosen to fairly compare with the results in Sects. 13.6.2 and 13.6.3. Hence, $R = 4.7[\text{K}\Omega]$ and $C = 3[\mu\text{F}]$ must be used in (13.39). Using the plant parameters corresponding to $x_2^* = 0.01[\text{m}]$ and the controller gains in (13.45), in addition to $k_{ii} = 340.89$, $k_{pi} = 4.7$, it is found that the closed-loop system has poles located at:

$$-48.25 \pm 123.26j, \quad -72.22, \quad -27.40, \quad -1.77,$$

and zeros located at:

$$-72.5298, \quad -29.9355, \quad -1.3045.$$

These numerical values verify the correctness of the root locus diagram shown in Fig. 13.46c.

When performing experiments setting $x_2^* = 0.01$[m], using the controller gains in (13.45) and $k_{ii} = 340.89$, $k_{pi} = 4.7$, a better performance was not found: the ball is prone to oscillate and most times it cannot stay levitating at the desired position. Moreover, the performance is worse when testing the system at $x_2^* = 0.006$[m] and $x_2^* = 0.002$[m]. Of course, in such cases, the controller gains in Table 13.2 and (13.46) respectively, and $k_{ii} = 340.89$, $k_{pi} = 4.7$ were used.

These practical results have motivated us to select the numerical values in (13.42) as the correct controller gains.

13.7   Summary

In this chapter, a magnetic levitation system has been controlled using exclusively analog electronics. The controller employed is a classical PID that has been implemented using operational amplifiers according to Sect. 9.2 in Chap. 9. As the plant is nonlinear, it is necessary to obtain a linear approximate model, which is useful only locally, i.e., the control system performs correctly only if the ball to levitate is initially located close to the position where it is desired to stay. To obtain this linear approximate model, the state variables approach in Chap. 7 is very useful. Once a linear approximate model is obtained, the PID controller is designed. This requires knowledge of the numerical values of the plant parameters. Hence, it is explained how these numerical values can be estimated experimentally.

As the plant is unstable and nonlinear, this is a control problem whose commissioning is somewhat complex. This means that attention is required to take into account all the gains introduced by each one of the control circuit components when modeling the system and when designing the controller. Any errors in this respect result in deterioration of the control system's performance.

It is very important to stress that the construction of the prototype that is presented in this chapter does not require expensive sophisticated sensors.

13.8   Review Questions

1. Why is an E-shaped core well suited for the electromagnet? Why not use a simple straight core?

2. In some recent works in the control literature, the electromagnet inductance is assumed to be constant to simplify the problem. Do you believe that this is correct?
3. In a magnetic levitation system, the use of an internal electric current loop has some significant advantages in the performance achieved. What are these advantages?
4. What is the advantage of using a PID controller instead of a proportional–derivative controller in this control problem?
5. What problems may appear if polarized capacitors are employed in the circuit used to implement the controller?
6. What must be done if it is desired to increase the gain of the pulse with a modulation-based power amplifier?
7. What must be done if it is desired to increase the frequency of the pulse width modulator?
8. What is the usefulness of the diode in parallel with the electromagnet?
9. What is the operation point in a magnetic levitation system and how is it determined?

References

1. J. D. Kraus, *Electromagnetics*, McGraw-Hill, Singapore, 1992.
2. Several authors, *Special issue on magnetic bearing control*, IEEE Transactions on Control Systems Technology, Vol.4, No. 5, 1996.
3. M. Y. Chen, C. F. Tsai, H. H. Huang, and L. C. Fu, Integrated design for a planar MagLev for micro positioning, in *Proc. American Control Conference*, pp. 3066–3071, Portland, 2005.
4. J. Lévine, J. Lottin, and J. C. Ponsart, A nonlinear approach to the control of magnetic bearings, *IEEE Transactions on Control Systems Technology*, vol. 4, no. 5, pp. 524–544, 1996.
5. M. G. Feemster, Y. Fang, and D. Dawson, Disturbance rejection for a magnetic levitation system, *IEEE Transactions on Mechatronics*, vol. 11, no. 6, pp. 709–717, 2006.
6. J.-C. Shen, H^{∞} control and sliding mode control of magnetic levitation system, *Asian Journal of Control*, vol. 4, no. 3, pp. 333–340, 2002.
7. R. Ortega, A. van der Schaft, I. Mareels, and B. Maschke, Putting energy back in control, *IEEE Control Systems Magazine*, pp. 18–33, April 2001.
8. R. Ortega, A. Loría, P. J. Nicklasson, and H. Sira-Ramírez, *Passivity-based control of Euler-Lagrange Systems*, Springer, London, 1998.
9. B. Lantos and L. Márton, *Nonlinear control of vehicles and robots*, Springer, London, 2011.
10. M. S. de Queiroz and D. Dawson, Nonlinear control of active magnetic bearings: a backstepping approach, *IEEE Transactions on Control Systems Technology*, vol. 4, no. 5, pp. 545–552, 1996.
11. H. Khalil, *Nonlinear Systems*, 3rd Edition, Prentice-Hall, Upper Saddle River, 2002.
12. W. Hurley, M. Hynes, and W. Wolfle, PWM Control of a magnetic suspension system, *IEEE Transactions on Education*, vol. 47, no. 2, pp. 165–173, 2004.
13. W. Hurley and W. Wolfle, Electromagnetic design of a magnetic suspension system, *IEEE Transactions on Education*, vol. 40, no. 2, pp. 124–130, 1997.
14. T.-J. Tarn, A. K. Bejczy, X. Yun, and Z. Li, Effect of motor dynamics on nonlinear feedback robot arm control. *IEEE Transactions on Robotics and Automation*, vol. 7, pp. 114–122, 1991.

15. S. Eppinger and W. Seering, Introduction to dynamic models for robot force control. *IEEE Control Systems Magazine*, vol. 7, pp. 48–52, 1987.
16. G. C. Goodwin, S. F. Graebe, and M. E. Salgado, *Control system design*, Prentice-Hall, Upper Saddle River, 2001.

Chapter 14
Control of a Ball and Beam System

Consider the situation represented in Fig. 14.1. It is a ball that rolls inside a channel on a beam. The beam angle is modified by a permanent magnet (PM) brushed direct current (DC) motor and this produces the ball movement by the effect of gravity. The objective of the mechanism control is to stabilize the ball at some desired position on the beam. This mechanism is a very common workbench for classical and advanced control techniques [1, 2]. As shown in the present chapter, this mechanism is unstable and the classical design methods commonly introduce an internal control loop. These are the reasons for presenting this control problem in the present chapter.

The nomenclature employed is the following:

- u is the applied voltage at the motor armature terminals.
- x is the ball position measured from the beam left end.
- θ is the beam angle measured with respect to the horizontal configuration.
- m is the ball mass.
- R and r stand respectively for the ball radius and the ball rotation radius on the channel edge.
- i is the electric current through the motor armature.
- L is the armature inductance.
- R_a is the armature resistance.
- k_e is the motor counter electromotive force constant.
- k_m is the motor torque constant.
- J_m is the motor rotor inertia.
- b_m is the motor viscous friction constant.
- J_L is the beam inertia.
- b_L is the beam viscous friction constant.
- n_1 and n_2 stand for the teeth number at the motor shaft and the beam shaft respectively, and $n = \frac{n_2}{n_1}$.

© Springer International Publishing AG, part of Springer Nature 2019
V. M. Hernández-Guzmán, R. Silva-Ortigoza, *Automatic Control with Experiments*,
Advanced Textbooks in Control and Signal Processing,
https://doi.org/10.1007/978-3-319-75804-6_14

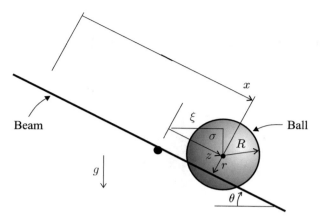

Fig. 14.1 Ball and beam system

14.1   Mathematical Model

14.1.1   Nonlinear Model

Consider Fig. 14.1. The beam length is L and suppose that the beam rotates around an horizontal axis at $x = L/2$. Note that:

$$z = x - \frac{L}{2}, \quad \xi = z\cos\theta, \quad \sigma = -z\sin\theta,$$

have been defined. From these expressions the following is obtained:

$$\dot{\sigma} = -\dot{z}\sin\theta - z\dot{\theta}\cos\theta,$$
$$\dot{\xi} = \dot{z}\cos\theta - z\dot{\theta}\sin\theta.$$

It is also important to stress that it is assumed that the ball rotates without slipping, which is established by constraining the ball angular velocity ω and the ball translational velocity $\dot{z}$ to satisfy:

$$r\omega = \dot{z}, \quad \text{or} \quad \dot{z} - r\omega = 0 \tag{14.1}$$

as r is the ball rotation radius.

   The computation of the *kinetic energy* for a body whose mass is distributed on a volume can be simplified if it is decomposed into two parts: one part is due to the translational movement of the body's center of mass (assumed to be a particle) and the other part is due to the body's rotative movement around its center of mass. For instance, the ball movement is easily described by its rotative movement around its center of mass (assumed to be located at its geometric center) and the translational movement of its center of mass. Thus, the kinetic energy of the ball is given as:

$$K_b = \frac{1}{2} J_b \omega^2 + \frac{1}{2} m(\dot{\xi}^2 + \dot{\sigma}^2).$$

where $J_b = \frac{2}{5} m R^2$ is the inertia of a solid sphere (the ball) with mass m and radius R rotating on its own axis [3, 4], pp. 271.

The total kinetic energy of the system is obtained just by adding the rotative kinetic energy of the beam, i.e.,

$$K = K_b + \frac{1}{2} J \dot{\theta}^2 = \frac{1}{2} J_b \omega^2 + \frac{1}{2} m(\dot{\xi}^2 + \dot{\sigma}^2) + \frac{1}{2} J \dot{\theta}^2,$$

$$J = n^2 J_m + J_L.$$

Note that the inertia J takes into account the beam inertia and the motor inertia coupled with a gear box. The total potential energy is due only to the ball, i.e.,

$$P = \sigma m g = -z m g \sin \theta.$$

Thus, the Lagrangian of the system $L = K - P$ is given as:

$$L = \frac{1}{2} J_b \omega^2 + \frac{1}{2} m(\dot{\xi}^2 + \dot{\sigma}^2) + \frac{1}{2} J \dot{\theta}^2 + z m g \sin \theta. \tag{14.2}$$

According to the above arguments, the system's generalized coordinates can be defined as the ball's translational position $z = \pm\sqrt{\xi^2 + \sigma^2}$, the beam position θ, and the ball's angular position γ, i.e. $\dot{\gamma} = \omega$. Hence, according to the nomenclature in Appendix E:

$$q = [z, \theta, \gamma]^T, \quad \dot{q} = [\dot{z}, \dot{\theta}, \omega]^T.$$

On the other hand, according to (E.2), the constraint on the velocities given in (14.1) can be written as:

$$A(q)\dot{q} = 0, \quad A(q) = [1, 0, -r].$$

Thus, according to (E.4), the Euler–Lagrange equations representing the system model can be written as:

$$\frac{d}{dt} \frac{\partial L}{\partial \dot{z}} - \frac{\partial L}{\partial z} = \lambda, \tag{14.3}$$

$$\frac{d}{dt} \frac{\partial L}{\partial \dot{\theta}} - \frac{\partial L}{\partial \theta} = \tau_\theta - b\dot{\theta},$$

$$\frac{d}{dt}\frac{\partial L}{\partial \omega} - \frac{\partial L}{\partial \gamma} = -r\lambda,$$

where $\tau_Q = [0, -b\dot{\theta}, 0]^T$, $b > 0$, has been assumed, i.e., no friction is considered to exist in the ball dynamics because it rotates without slipping, $\tau = [0, \tau_\theta, 0]^T$ with τ_θ the torque applied by the motor on the beam, and λ is an unknown scalar to be computed later. Replacing (14.2) in (14.3) it is found:

$$\frac{d}{dt}[mC\cos\theta + mB(-\sin\theta)]$$

$$- [mC(-\dot{\theta}\sin\theta) + mB(-\dot{\theta}\cos\theta) + mg\sin\theta] = \lambda,$$

$$\frac{d}{dt}[mC(-z\sin\theta) + mB(-z\cos\theta) + J\dot{\theta}]$$

$$- [mC(-\dot{z}\sin\theta - z\dot{\theta}\cos\theta) + mB(-\dot{z}\cos\theta + z\dot{\theta}\sin\theta) + zmg\cos\theta] = \tau_\theta - b\dot{\theta},$$

$$J_b\dot{\omega} = -r\lambda,$$

where:

$$C = \dot{z}\cos\theta - z\dot{\theta}\sin\theta, \quad B = -\dot{z}\sin\theta - z\dot{\theta}\cos\theta.$$

Performing the indicated time derivatives and using some trigonometric identities in addition to $\lambda = -\frac{1}{r}J_b\dot{\omega}$, from the last equation, the above equations simplify to:

$$m\ddot{z} - mz\dot{\theta}^2 - mg\sin\theta = -\frac{1}{r}J_b\dot{\omega},$$

$$J\ddot{\theta} + mz^2\ddot{\theta} + mz\dot{z}\dot{\theta} - mgz\cos\theta = \tau_\theta - b\dot{\theta}.$$

According to (14.1), $r\dot{\omega} = \ddot{z}$, and hence:

$$\left(m + \frac{1}{r^2}J_b\right)\ddot{z} - mz\dot{\theta}^2 - mg\sin\theta = 0, \tag{14.4}$$

$$J\ddot{\theta} + mz^2\ddot{\theta} + mz\dot{z}\dot{\theta} + b\dot{\theta} - mgz\cos\theta = \tau_\theta.$$

These equations represent the ball and beam system mathematical model. Notice that this model is nonlinear because of several terms such as $-mz\dot{\theta}^2$, $-mg\sin\theta$, $mz^2\ddot{\theta}$, $mz\dot{z}\dot{\theta}$, $-mgz\cos\theta$. As this book is concerned only with linear control techniques, a linear model for the ball and beam system is required. Such a model is obtained in the following section by considering some simplifications of the above model.

14.1.2 Linear Approximate Model

It is interesting to note that, according to the term $-mgz\cos\theta$ in the second equation in (14.4), the ball weight exerts a torque on the beam axis, which depends on the ball position on the beam. On the other hand, an interesting phenomenon appears because of the term $-mz\dot{\theta}^2$ in the first equation: if the ball position $z > 0$ is large, the beam angle θ must be rendered negative very fast, i.e., $\dot{\theta}^2$ is large, to avoid the ball escaping (see Fig. 14.1), then, according to $-mz\dot{\theta}^2$ a centripetal force is exerted on the ball that forces it to continue escaping instead of stopping. It is not difficult to verify that a similar situation occurs when $z < 0$.

To avoid these phenomena appearing, the following assumptions are considered:

- Torque on the beam, due to weight of the ball, is negligible, which is especially true if the motor actuates on the beam through a gear box with a high reduction ratio. Under these conditions: $mgz\cos\theta \approx 0$ and $mz^2\ddot{\theta} \approx 0$.
- The beam and the ball always move at low velocities $\dot{\theta}$, $\dot{z}$, and the beam angle θ only takes values around zero. Under these conditions $\sin\theta \approx \theta$, if θ is given in radians, and $\dot{\theta}^2 \approx 0$, $\dot{z}\dot{\theta} \approx 0$.

Thus, the model in (14.4) becomes:

$$\left(m + \frac{1}{r^2}J_b\right)\ddot{z} - mg\theta = 0,$$

$$J\ddot{\theta} + b\dot{\theta} = \tau_\theta.$$

Differentiating $z = x - \frac{L}{2}$ twice yields $\ddot{z} = \ddot{x}$. Thus, the above model becomes[1]:

$$\left(1 + \frac{2}{5}\frac{R^2}{r^2}\right)\ddot{x} = g\theta,$$

$$J\ddot{\theta} + b\dot{\theta} = \tau_\theta. \tag{14.5}$$

where $J_b = \frac{2}{5}mR^2$ has been used. As this model is linear, the Laplace transform can be used in the first expression to find:

$$\frac{X(s)}{\theta(s)} = \frac{\rho}{s^2}, \qquad \rho = \frac{g}{1 + \frac{2}{5}\frac{R^2}{r^2}}. \tag{14.6}$$

On the other hand, the electric motor can be modeled as in Sect. 10.1 taking into account that the load is represented by the beam, i.e.,

---

[1] Note that this model can also be expressed in terms of z: just replace $\ddot{z} = \ddot{x}$.

$$L \frac{di}{dt} = u - R_a\, i - n\, k_e\, \dot{\theta},$$

$$J\ddot{\theta} = -b\, \dot{\theta} + n\, k_m\, i, \quad n\, k_m\, i = \tau_\theta,$$

$$J = n^2\, J_m + J_L, \quad b = n^2\, b_m + b_L,$$

where the second expression corresponds to (14.5). Proceeding as in Sects. 10.1, 10.2 and 10.3, the following current loop:

$$u_i = K\,(i^* - i),$$

and power amplifier:

$$u = A_p\, u_i,$$

can be used to find that the model of the motor and the beam is given as:

$$\theta(s) = \frac{k}{s(s+a)} I^*(s), \tag{14.7}$$

$$a = \frac{b}{J}, \quad k = \frac{n k_m}{J},$$

where it is assumed that no external disturbance is present, i.e., $T_p = 0$. Finally, the linear approximate ball and beam system model is given by the combination of (14.7) and (14.6), i.e.,

$$\frac{X(s)}{\theta(s)} = \frac{\rho}{s^2}, \quad \theta(s) = \frac{k}{s(s+a)} I^*(s). \tag{14.8}$$

14.2   Prototype Construction

The main components are the following.

- Portable computer with USB port and Windows XP.
- USB to series adapter.
- PIC16F877A Microcontroller from Microchip.
- DAC0800 LCN digital/analog converter.
- TL081 operational amplifiers.
- MAX232 driver.
- PM brushed DC motor. Nominal voltage 24[V], nominal current 2.3[A]. Provided with an optical encoder with 400 ppr.
- Gear box with reduction ratio $n = 30$.

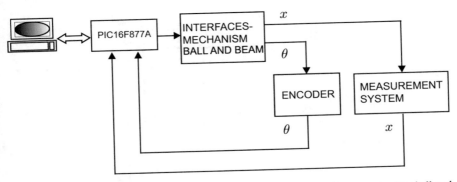

Fig. 14.2 Block diagram of the complete control system. Block interfaces–mechanism ball and beam is shown in Fig. 14.19. Block measurement system is shown in Fig. 14.3

- Beam of the system. Composed of: (*i*) A beam in wood with a circular section (rail number one), (*ii*) A beam in aluminum with a circular section (rail number two), (*iii*) A sheet of wood employed to hold both cylindrical beams.
- A ball position measurement system as described next.

A block diagram of the whole system is presented in Fig. 14.2.

14.2.1   Ball Position x Measurement System

One important and interesting component of the experimental prototype is the ball position measurement system. In fact, this is the most challenging component to construct. The most common way of performing the ball position measurements is employing a resistance distributed along one of the beam channel rails. The other rail is a simple electric conductor whose resistance can be considered to be zero. Hence, as the ball (made in an electric conductor material) rolls, putting both rails in electric contact, the second rail has a voltage that is equal to the voltage of the first rail at the point where the ball touches it. Thus, both rails work together as a potentiometer with the ball acting as the mobile contact.

The electric diagram of the ball position measurement system that has been built is shown in Fig. 14.3. Rail number one consists of a beam in wood with a circular section of about 0.01[m] in diameter and 0.7[m] in length. Magnetic wire is carefully wound onto this beam. All the wire loops must be tight to each other but without overlapping. Thus, a cylinder-shaped coil is obtained with a length of 0.7[m] and a diameter of 0.01[m] whose external surface is even enough to allow the ball to roll smoothly. The reason for using a beam in wood for this rail is to avoid the danger of putting the wire loops in a short circuit if the insulating paint fails.

Rail number two is a cylinder beam in aluminum with a length of 0.65[m] and a diameter of 0.01[m]. Both rails lay one in front of the other, in parallel, with

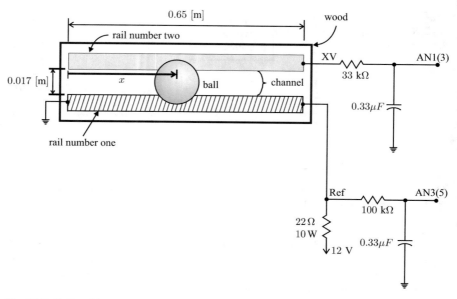

Fig. 14.3 Ball position measurement system

0.017[m] the minimal distance between them, on a sheet in wood. Hence, a channel measuring 0.65[m] in length and 0.017[m] in width is formed. This assembly constitutes the beam of the mechanism.

The ball dimensions are selected such that it is allowed to roll within the channel by putting both rails in electric contact but the channel width is enough to avoid the ball escaping. This feature is important, especially in the first experimental tests when it is usual for the ball to escape from the channel. Finally, the narrow strip where the ball touches rail number one is identified and the insulating paint is stripped with sandpaper along the entire length of rail number one. This allows electric contact, through the rolling ball, between the two rails. The ball that has been selected has the following dimensions (see Fig. 14.4):

$$2R = 0.0238[\text{m}], \quad 2z = 0.01895[\text{m}], \quad r = 0.0072[\text{m}]. \tag{14.9}$$

Finally, a gear box is employed to join the PM brushed DC motor and the beam to magnify the torque applied to the latter.

As stated above, both rails work together as a simple potentiometer delivering a voltage x_v that is proportional to the ball position x. The voltage x_v is measured using one of the analog/digital converters in the PIC16F877A microcontroller by connecting the terminal labeled AN1 to the microcontroller's pin 3. It is important to state that x_v has previously passed through a *low-pass filter* with 14.6148 [Hz] ($R = 33[\text{KOhm}]$, $C = 0.33[\mu\text{F}]$) as corner frequency. Furthermore, to render the measurement system robust with respect to temperature changes in rail number one,

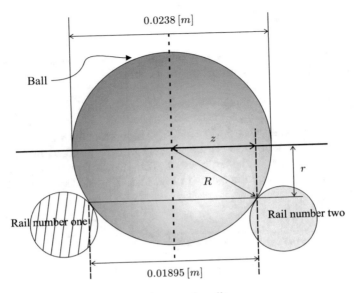

Fig. 14.4 Ball rolling in the channel, i.e., between the rails

voltage at the terminal labeled AN3 (connected to the microcontroller's pin 5) is employed as the reference voltage for the analog/digital converter used to measure x_v. It is also important to state that the signal at AN3 has previously passed through a low-pass filter with 4.8229 [Hz] ($R = 100$[KOhm], $C = 0.33[\mu F]$) as corner frequency.

14.2.2   Beam Angle θ Measurement System

The beam angle θ is measured using an optical encoder integrated into the PM brushed DC motor. The encoder is 400 ppr and is fixed to the motor shaft. Recall that a gear box with ratio 1 : 30, i.e., $n = 30$, is used between the motor shaft and the beam shaft. Then, the position measured by the encoder (the motor position) must be divided by n to obtain the corresponding beam position measurement θ.

14.3   Parameter Identification

The experimental identification of parameters in the model (14.8) can be performed from the separate study of the motor and the beam dynamics (14.7) and the ball dynamics (14.6). This is explained in the following.

14.3.1   Motor–Beam Subsystem

The procedure suggested in this part is similar to that introduced in Sect. 11.1. Consider the following proportional position control:

$$I^*(s) = k_p A_\theta (\theta_d(s) - \theta(s)), \tag{14.10}$$

where A_θ is included because $\theta_v(s) = A_\theta \theta(s)$ represents the variable delivered by the measurement system and it is the variable actually handled by the computer employed as controller in the experiment. This means that the closed-loop system (14.7), (14.10), is represented as in Fig. 14.5. The corresponding closed-loop transfer function is:

$$\frac{\theta_v(s)}{\theta_{vd}s} = \frac{A_\theta k_p k}{s^2 + as + A_\theta k_p k} = \frac{\omega_n^2}{s^2 + 2\zeta \omega_n s + \omega_n^2}.$$

$$\omega_n = \sqrt{A_\theta k_p k}, \quad a = 2\zeta \omega_n, \tag{14.11}$$

where $\theta_{vd}(s) = A_\theta \theta_d(s)$ is the desired beam position in terms of the variable delivered by the measurement system. Some experimental results are presented in Fig. 14.6 when $k_p = 4$ and $\theta_d = 0.2[\text{rad}]$ is the desired value for θ. This experiment is performed without the ball between the beam rails or channel. From Fig. 14.6, the following is measured $t_r = 0.252[\text{s}]$ and $M_p(\%) = 79$. These data are employed in:

$$\zeta = \sqrt{\frac{\ln^2 \left(\frac{M_p(\%)}{100}\right)}{\ln^2 \left(\frac{M_p(\%)}{100}\right) + \pi^2}},$$

$$\omega_d = \frac{1}{t_r}\left[\pi - \arctan\left(\frac{\sqrt{1 - \zeta^2}}{\zeta}\right)\right],$$

$$\omega_n = \frac{\omega_d}{\sqrt{1 - \zeta^2}},$$

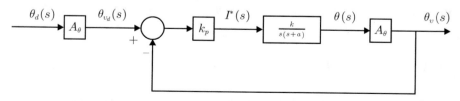

Fig. 14.5 Proportional control of the beam position

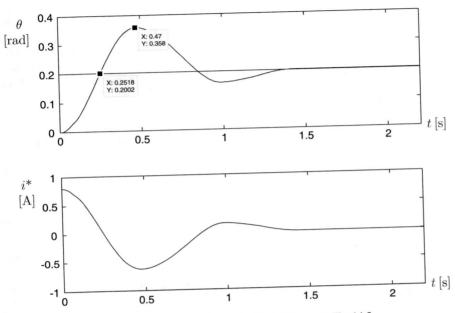

Fig. 14.6 Experimental results using $k_p = 4$ and the block diagram in Fig. 14.5

to find $\zeta = 0.0748$ and $\omega_n = 6.5489$[rad/s]. Finally, using the expressions in (14.11) with $k_p = 4$ and $A_\theta = 1$ (see Sect. 14.7), the following is obtained:

$$k = 10.7219, \quad a = 0.98. \tag{14.12}$$

14.3.2 Ball Dynamics

Applying the inverse Laplace transform to (14.6) and assuming all the initial conditions to be zero, the following is found:

$$\ddot{x} = \rho\,\theta(t). \tag{14.13}$$

Suppose that $\theta(t) = \theta_0$ where θ_0 is a constant. Integrating (14.13) twice:

$$x(t) = \frac{1}{2}\rho\,\theta_0\,t^2. \tag{14.14}$$

A procedure is presented next allowing us to experimentally estimate the parameter ρ.

- Fix the beam at a the constant angle $\theta = \theta_0$, where θ_0 is a small positive value. A good example is $\theta_0 = 0.2$[rad] because $\sin(0.2) = 0.1987$, i.e., $\sin(\theta_0) \approx \theta_0$.
- Place the ball at the beam left end and let it roll until reaching the beam end at the right (the beam must not move during this stage). Measure the ball position described in this experiment. This task is rendered easier if a computer is employed.
- Plot the ball position in this experiment versus time.
- Using θ_0 defined at the first step of this experiment and proposing some arbitrary positive value for ρ, plot the function $\frac{1}{2}\rho\,\theta_0\,t^2$ on the same axes employed at the previous step.
- Propose different values for ρ. The correct value of ρ is that allowing the plot of the function $\frac{1}{2}\rho\,\theta_0\,t^2$ to fit the plot of the experimental data.

The plots referred to in the previous procedure are shown in Fig. 14.7. when using $\theta_0 = 0.2$[rad] and:

$$\rho = 4.8. \tag{14.15}$$

It is important to verify that the experimental data x are plotted using meters as the measurement unit. This is rendered easier if $A_x = 1$, as explained in Sect. 14.7. Finally, it is interesting to state that the value for ρ shown in (14.15) has been corroborated using (14.6), (14.9) and $g = 9.81$[m/s^2], which yields $\rho = 4.687$.

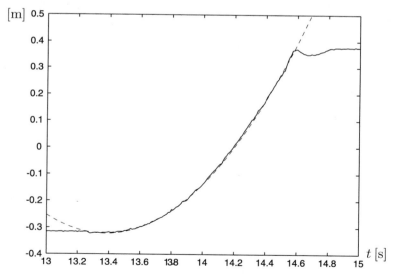

Fig. 14.7 Experimental identification of the parameter ρ. Continuous: $x(t)$ measured in the experiment. Dashed: $x(t) = \frac{1}{2}\rho\theta_0(t - 13.38)^2 - 0.322$, $\theta_0 = 0.2$[rad], $\rho = 4.8$

14.4 Controller Design

Note that the plant model (14.8) has three poles at $s = 0$, i.e., it is open-loop unstable. Hence, an important design objective is to achieve closed-loop stability. A block diagram of the control system to be designed is shown in Fig. 14.8a. This block diagram can be represented as in Fig. 14.8b. It is suggested that the reader might see Sect. 6.7.5 to understand the reason for this block diagram. In this design problem $\lim_{t \to \infty} x(t) = x_d$ must be accomplished, where x_d is a constant standing for the desired ball position on the beam.

Note that the open-loop transfer function of the system shown in Fig. 14.8b has two poles at $s = 0$, i.e., the system type is 2. Then, as x_d is constant, it is ensured that $\lim_{t \to \infty} x(t) = x_d$ if the closed-loop transfer function $X(s)/X_d(s)$ is stable (see Sects. 3.4 and 4.4). Hence, the only problem that remains is to find some constants α, k_v, γ, b and c such that the transfer function $X(s)/X_d(s)$ is stable. It is explained next how to achieve this.

The open-loop transfer function of the system shown in Fig. 14.8b is given as:

$$G(s)H(s) = \frac{A_x \, \rho \, \gamma b}{c A_\theta} \frac{s+b}{b} \frac{c}{s+c} \frac{\alpha k A_\theta}{s^2 + (a + k_v k A_\theta)s + \alpha k A_\theta} \frac{1}{s^2}, \quad (14.16)$$

where $c > b > 0$. In Fig. 14.9, the Bode diagrams of $G(s)H(s)$ and each one of the transfer functions that comprise it are shown. Notice that, if the constant $\alpha k A_\theta$ is large enough, i.e., if $\alpha k A_\theta \gg s^2 + (a + k_v k A_\theta)s$, then:

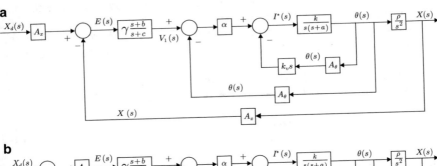

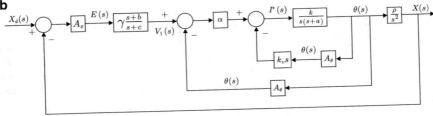

Fig. 14.8 Equivalent block diagrams of the control systems to be designed

Fig. 14.9 Bode diagrams of
$G(s)H(s)$ in (14.16)

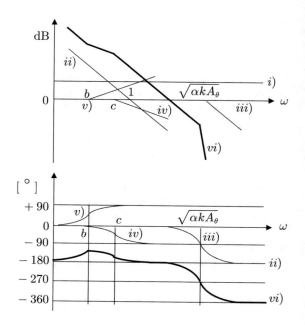

$$\frac{\alpha k A_\theta}{s^2 + (a + k_v k A_\theta)s + \alpha k A_\theta} \approx 1. \tag{14.17}$$

This can also be interpreted by thinking that the transfer function in (14.17) has 0[dB] as magnitude and 0 degrees as phase. It is interesting to observe that this situation can also be seen in Fig. 14.9: when $\alpha k A_\theta$ is large enough, i.e., when the corner frequency of the transfer function in (14.17) is located far to the right, this second-order factor has no effect on open-loop frequency response; thus, it can be neglected.

Hence, if $\alpha k A_\theta$ is large, the problem is simplified to achieve closed-loop stability for a system whose open-loop transfer function is given as:

$$\frac{A_x \, \rho \, \gamma b}{c A_\theta} \frac{s+b}{b} \frac{c}{s+c} \frac{1}{s^2}. \tag{14.18}$$

This is an important simplification because the phase lag of the transfer function in (14.17) is eliminated, which may take values between 0 and -180 degrees (when such a transfer function is not close to unity). This means that stabilizing (14.18) requires less phase lead than that required to stabilize (14.16). Recall that requiring more phase lead renders the controller design task more restrictive because the compensator $\gamma \frac{s+b}{s+c}$ would be more sensitive to noise in the position sensors.

Note that a large $\alpha k A_\theta$ can be achieved by enlarging α as k and A_θ are fixed once the prototype has been built. However, a large $\alpha k A_\theta$ causes the performance of the control system to deteriorate. In fact, large values for $\alpha k A_\theta$ were employed during the experimental tests and it was observed that this increases the deteriorating effect due to noise in sensors: the mechanism strongly vibrates and the control system does not work. Hence, α is limited by a maximal value imposed by the experimental prototype even when the condition (14.17) is not satisfied.

On the other hand, if $k_v = 0$ the system in (14.17) will be badly damped because $\zeta = a/(2\sqrt{\alpha k A_\theta})$ is small. This produces a resonance peak in the Bode diagrams of (14.16). The problem with a resonance peak is that it increases the effect of high frequencies producing, again, undesirable vibrations in the mechanism. Good results have been obtained using some $k_v > 0$ because this increases the damping factor $\zeta = (a + k_v k A_\theta)/(2\sqrt{\alpha k A_\theta})$ of the system in (14.17); hence, the resonance peak in the Bode diagrams of (14.16) is reduced.

According to the above arguments the following experiment was performed.

- Consider a closed-loop system to control only the beam position, i.e., only consider the block diagram between the signals $V_1(s)$ and $\theta(s)$ in any of the block diagrams in Fig. 14.8. This means that the ball must not be on the beam when performing the experiments.
- Set $V_1(s)$ as a step standing for the desired beam position and $k_v = 0$.
- Propose some values for $\alpha > 0$ such that the beam response is fast.
- When the oscillations are important, keep α constant and increase $k_v > 0$.
- Repeat the previous two items until the beam response is fast and well-damped but avoiding vibrations due to excessive noise amplification. Consider the resulting values of α and k_v as the correct values.

Once the above procedure had been performed, it was found that the use of the following controller gains achieves a satisfactory performance:

$$\alpha = 30, \quad k_v = 0.35 \tag{14.19}$$

Notice that the system in Fig. 14.8b, together with $A_x = A_\theta = 1$ (see Sect. 14.7), and (14.12), (14.15), (14.19) represent the control problem analyzed in Sect. 6.7.5.2, where the solution is given by the following compensator:

$$G_c(s) = \gamma \frac{s+b}{s+c}, \quad b = 0.6404, \quad c = 4.5746, \quad \gamma = 1.6406.$$

Finally, according to Sect. 6.5.2, to take into account the effects of time delay contributed by the digital implementation of the controller, the transfer function:

$$\frac{k}{s(s+a)},$$

in Fig. 14.8b may be replaced by:

$$\frac{k\,e^{-Ts}}{s(s+a)},$$

as the time delay T appears at the controller output, i.e., at the plant input. Using suitable block algebra it is possible to show that the corresponding open-loop transfer function is the same as $G(s)H(s)$ in (14.16) with e^{-Ts} as a new factor. This means that the phase lag $-\omega T$[rad] must be added to the phase in Fig. 6.87. There, the crossover frequency is $\omega = 1.71$[rad/s]. Also $T = 0.01$[s] can be set as this is the sampling period used in the experiments presented next, i.e., the worst case time delay. Hence, $-\omega T \times 180°/\pi = -0.97°$. This means that the phase margin remains almost the same as in Fig. 6.87, i.e., about $47.5°$, and the time delay induced by the digital implementation of the controller does not cause the performance to deteriorate.

14.5   Experimental Results

In Fig. 14.10 some experimental results are presented that are obtained when using the following desired ball position:

$$x_d = \begin{cases} 0.0, \ 0 \le t \le 10 \\ 0.1, \ 10 < t \le 20 \\ 0.2, \ 20 \le t \le 30 \\ 0.1, \ 30 \le t \le 40 \\ 0.0, \ 40 \le t \le 50 \\ 0.1, \ 50 \le t \le 60 \\ 0.2, \ 60 \le t \le 70 \\ 0.1, \ 70 \le t \le 80 \\ 0.0, \ 80 \le t \end{cases} .$$

where x_d is given in meters and t in seconds. It can be observed that, although the ball position tracks its desired value, a steady-state error that is different from zero is present. This steady-state error is due to nonlinear friction introduced by the gear box. Also notice that the overshoot is not the same when responding to different step commands. This effect is mainly due to beam inertia changes because of the ball staying in a different position on the beam. Note, however, that this effect has not been considered in the mechanism model.

To measure the rise time and overshoot to one of the commanded step references, a zoom-in in Fig. 14.10 is presented in Fig. 14.11. There, it is observed that $t_r = 0.79$[s] and $M_p(\%) = 33\%$. The steady-state value of x is 0.077[m] in this figure. It is interesting to recall that a 30% overshoot and a 0.927[s] rise time were obtained through simulations in Fig. 6.88. These values are very close to those obtained through the experiments in Fig. 14.11.

The nonzero steady-state error observed in the above experiments may be attributed to nonlinear friction but, also, recall that a torque on the beam due to the ball weight has been neglected in Sect. 14.1.2 to obtain a linear approximation

14.5 Experimental Results

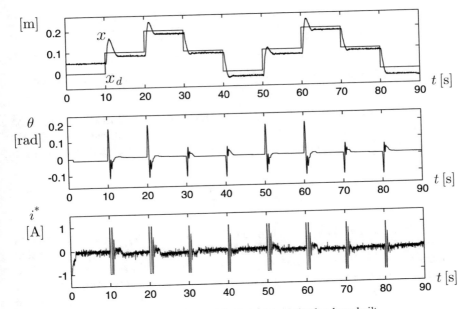

Fig. 14.10 Experimental results obtained with the prototype that has been built

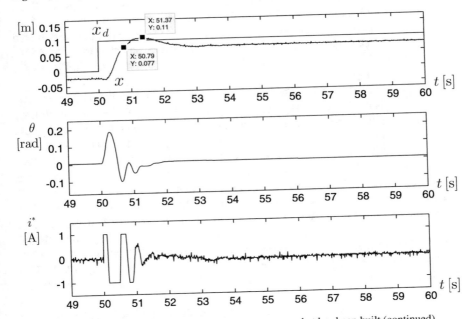

Fig. 14.11 Experimental results obtained with the prototype that has been built (continued)

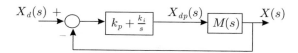

Fig. 14.12 Block diagram of a control scheme that includes a controller with integral action

model. Note that this torque disturbance becomes constant when the ball position is constant. Hence, a controller with an integral action is well suited to solving this problem and this is presented next. Designate by $M(s)$ the closed-loop function defined as:

$$M(s) = \frac{X(s)}{X_d(s)},$$

in the block diagram in Fig. 14.8b when all the parameters and gains take values according to the above design. In Fig. 14.12 a block diagram is presented that illustrates how an integral action is included in the control system. There, k_p and k_i are the gains of a PI controller that is included in the open-loop transfer function. Note that $X_{dp}(s)$ represents an intermediate variable that replaces $X_d(s)$ in all the previous block diagrams, whereas the desired ball position $X_d(s)$ is now shifted to the left to allow the PI controller to be driven by the ball position error $X_d(s) - X(s)$.

The controller gains $k_p = 0.5$ and $k_i = 1.5$ are selected to obtain a $\omega_1 = 1.98 \approx 1.71$[rad/s] crossover frequency and a $K_f = 45°$ phase margin, as depicted in Fig. 14.13. Recall that 1.71 is the crossover frequency selected in the previous design. In Fig. 14.14 a simulation response is presented when $X_d(s)$, in Fig. 14.12, is a unit step change.

Some experimental results are presented in Fig. 14.15 when the control scheme in Fig. 14.12 is used. Note that a zero steady- state error is reached now each time that a step change is commanded. Moreover, to further demonstrate the effectiveness of this approach, a constant disturbance is applied for approximately $t \geq 45$[s]. This disturbance is produced by adding a constant tilt angle to the mechanism base. This forces θ to be different from zero at a steady state to stabilize the ball at some constant position x. However, this increases the nonzero steady-state error in x if the controller does not possess an integral action. This is what happens in Fig. 14.16 where the ball position x stays far away from x_d once the disturbance appears, i.e., for $t \geq 45$[s]. In the case of the control scheme shown in Fig. 14.12 the integral action of the external loop controller achieves a zero steady-state error even when such a disturbance is applied, i.e., for $t \geq 45$[s], as can be seen in Fig. 14.15. This experimental test demonstrates the superiority of a control scheme with an integral action.

On the other hand, in Sect. 5.2.9, Chap. 5, the control scheme depicted in Fig. 5.53 was designed where:

$$\rho = 4.8, \quad a = 0.98, \quad k = 10.729, \quad k_v = 0.35, \quad \alpha = 30,$$

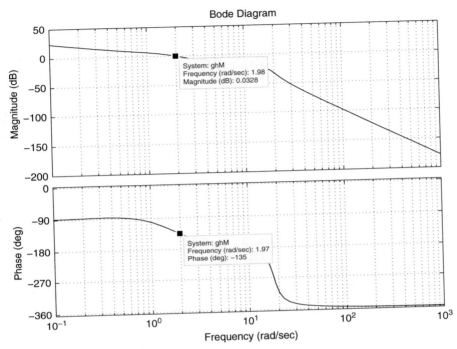

Fig. 14.13 Bode diagrams for the transfer function $\left(k_p + \frac{k_i}{s}\right) M(s)$, $k_p = 0.5$ and $k_i = 1.5$

$$A_\theta = A_x = 1, \quad b = 1.4, \quad c = 5.1138, \quad \gamma = 2.1868.$$

This design was performed to achieve a rise time $t_r = 1[s]$ and an overshoot $M_p = 25\%$. This control scheme is now tested experimentally using the ball and beam prototype described in this section. A step change in the desired position x_d is applied at $t = 0$, which changes from 0.05[m] to 0.15[m]. The corresponding experimental results are shown in Fig. 14.17, where $t_r = 1.169[s]$ and $M_p = 14\%$ has been obtained. These results are considered to be satisfactory given the accuracy of the prototype that has been built. Note that, contrary to the previous experiments shown in the present section, the steady-state error is zero without the necessity of an integral controller. Recall that friction is responsible for such a steady-state error and friction is also uncertain and changes during normal operation. Hence, a zero steady-state error is sometimes accomplished, despite the presence of friction. Moreover, the effect of the ball weight can be reduced if the experiment is performed when the ball is close to the center of the beam (where the beam axis is located) as in Fig. 14.17. Finally, it is shown through simulations in Sect. 5.2.9, Chap. 5, that the controller gains $b = 2$, $c = 11.7121$, $\gamma = 4.6336$, result in a better time response. However, it was observed in experiments that use of these controller gains results in

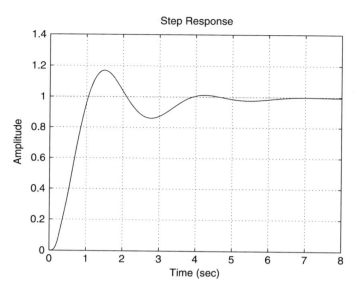

Fig. 14.14 Simulation response of the closed-loop system in Fig. 14.12, $k_p = 0.5$ and $k_i = 1.5$

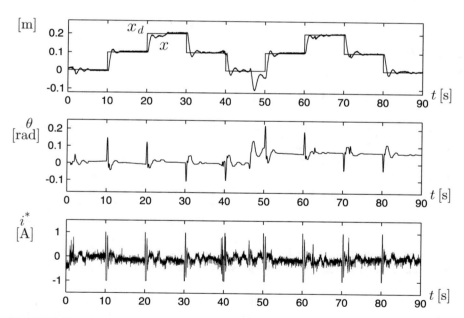

Fig. 14.15 Experimental response when the control scheme in Fig. 14.12 is used, $k_p = 0.5$ and $k_i = 1.5$. A constant disturbance exists for $t \geq 45[\mathrm{s}]$

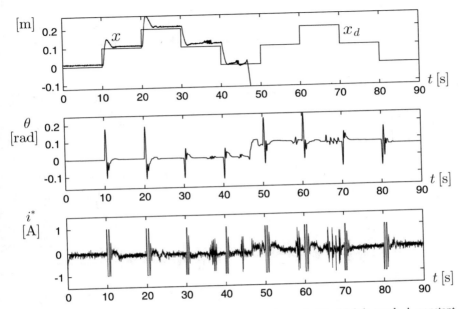

Fig. 14.16 Experimental response when the control scheme in Fig. 14.8 is used. A constant disturbance exists for $t \geq 45[s]$

noise amplification and the closed-loop control system has a bad performance. This is because these controller gains are large.

Finally, a picture of the ball and beam prototype that has been built is presented in Fig. 14.18.

14.6   Control System Electric Diagram

The control system electric diagram is shown in Fig. 14.19. This diagram is complemented with Fig. 14.3. A PIC16F877A Microchip microcontroller-based electronic board is employed as interface between the control program, implemented in a portable computer, and the ball and beam system. The communication is performed through the driver MAX232. This interface board receives pulses from the encoder and voltage delivered by the ball position x measurement system, and sends them to the portable computer. There, the control program handles these data numerically (which is explained later) to obtain the beam position θ in radians, and the ball position x in meters.

On the other hand, as the result of the control algorithm evaluation, the portable computer obtains the electric current that must flow though the DC motor and sends it back to interface board. There, the microcontroller sends such an electric current value to the power amplifier, which applies a suitable voltage at the motor

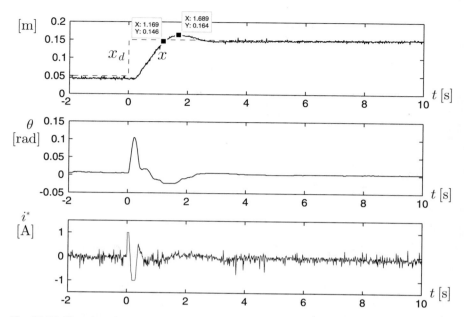

Fig. 14.17 Experimental response when the control scheme in Fig. 5.53 is used with $k_v = 0.35$, $\alpha = 30$, $b = 1.4$, $c = 5.1138$, $\gamma = 2.1868$

Fig. 14.18 Ball and beam system prototype that has been built

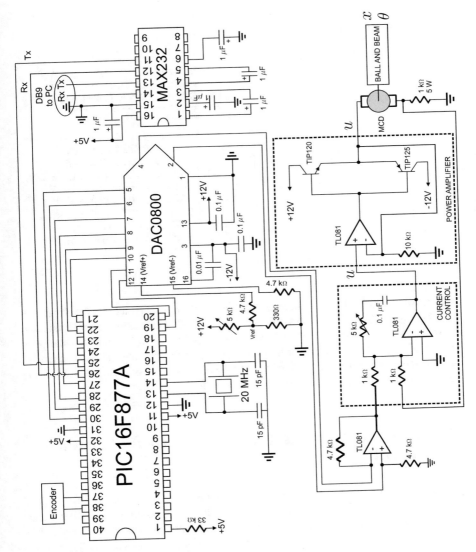

Fig. 14.19 Control system electric diagram

terminals, ensuring that such a desired current actually flows through the motor armature. To achieve this task, a DAC0800 digital/analog converter is employed which, working together with a TL081 operational amplifier, delivers $-i^*$ as an analog voltage signal. Then, an operational amplifier performs as a proportional–integral (PI) electric current controller with ≈ 5 as proportional gain and 10000 as integral gain. The voltage applied at the motor armature terminals, u, must be equal to the signal at the output of this PI controller, u_i. However, because of the dead zone

Fig. 14.20 Flow diagram for
PC programming when used
to control a ball and beam
system

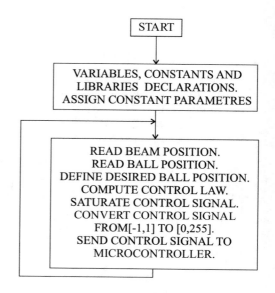

nonlinearity produced by a couple of bipolar junction transistors (BJTs), feedback
must be employed around such BJTs (see Sect. 9.1.2). The electric current through
the motor armature circuit is measured using a 1[Ohm], 5[W], power resistance in
series with the motor. The motor applies a torque to the ball and beam mechanism
and the measurement systems described in Sects. 14.2.1 and 14.2.2 deliver to the
microcontroller the positions of the beam and ball.

14.7   Builder 6 C++ Code Used to Implement the Control Algorithms

In this section we explain the main steps in implementing the control algorithms
using the Builder 6 C++ code listed below. The corresponding flow diagram is
shown in Fig. 14.20. Before defining some libraries and constants, the executable
part of the program begins with the instruction "void ProcessByte(BYTE byte)".
There, the program remains waiting for the recognition code "0xAA" to be sent by
the microcontroller PIC16F877A as an indicative that data from the mechanism is
to be sent in the subsequent instructions. The beam position is first received and
assigned to the variable "pos", then the ball position is received and assigned to the
variable "conv". After that, the desired ball position is generated and assigned to the
variable "xd". The beam position in radians is assigned in the variable "theta" where
"esctheta" is a factor converting pulses from encoder to radians using the relation:

$$2\pi\,[\text{rad}] \rightarrow 4 \times 400[\text{pulses/rev}]$$

theta[rad] $\rightarrow$ theta[number of pulses]$/n$.

$$\Rightarrow \quad \text{theta[rad]} = \frac{2\pi\,[\text{rad}]}{4n \times 400[\text{pulses/rev}]}\text{theta[number of pulses]}.$$

Note that the gear box ratio n is taken into account as the encoder is fixed to the motor shaft. This allows us to consider that $A_\theta = 1$. The ball position in meters is assigned to the variable "x" where "escx= 0.7/1022" is a factor converting from conv $\in [0, 1022]$ (the code from a 10-bit analog/digital converter) to $x \in [-0.7/2, 0.7/2]$ in meters. Note that the subtraction "conv-511" and the fact that 0.7[m] is the beam length are important to performing this task. This allows us to consider that $A_x = 1$. This also means that in all the experiments presented until now, the ball position x is measured from the middle point of the beam, i.e., instead of using the symbol x it would be more correct to use the variable z (which was not done, however) defined as $z = x - \frac{L}{2}$ at the beginning of Sect. 14.1.1. This has not affected the correctness of the designs presented so far, because, as noted in the footnote above (14.5), the dynamical model of the ball and beam is identical when written in terms of either x or z.

Two controllers are coded: the first one corresponds to the control scheme in Fig. 14.12 and the second one corresponds to the control scheme in Fig. 14.8. The electric current to be commanded by these controllers is represented by the variable "iast". This variable is saturated so as not to take values out of the admissible range and its sign is changed to deliver $-i^*$ at the output of the digital/analog converter stage described in Sect. 14.6. Then "iast" is converted into the range [0, 255] to send it to the microcontroller as an 8-bit word. This is performed via the instruction "MainForm− >send_byte(cuentas)", some variables are saved in the file MONIT.txt, and time is increased by 0.01[s]. Finally, the program returns and remains waiting for the recognition code "0xAA" again.

```
//Ball and beam control computer program
//---------------------------------
#include <vcl.h>
#pragma hdrstop
#include "Main.h"
#include <math.h>
#include <stdio.h>
#include <share.h>
#include <conio.h>
#include <dos.h>
//---------------------------------
#pragma package(smart_init)
#pragma link "CSPIN"
#pragma resource "*.dfm"
//---------------------------------
#define Ts 0.01 //sampling period in sec
#define pi_ 3.1416

#define gamma 1.6406 //0.5781
#define b 0.6404 //0.3745
```

```
#define c 4.5746 //2.6752
#define alfa 30.0
#define kv 0.35
#define kp2 0.5
#define ki2 1.5

#define maxI 2.0
#define IM 1.0
#define ppr 400.0
#define n 36.0
#define lriel 0.7
#define cADC 1022.0

FILE *ptrMonit;
TMainForm *MainForm;
unsigned char flagcom=0,flagfile=0,cuentas;
unsigned short int conv,pos; //de 16 bits
float t=0,inte=0,area=0,escs=255.0/(2.0*maxI);
float esctheta=pi_/(2*ppr*n),x,xd=0.0,theta,e,iast;
float v=0,thetad,theta_1=0,escx=lriel/cADC,iTs=1/Ts;
float ic=0.6,ee,k1,k2;
float integ=0.0,integ1=0.0,iast1,ese,ep,xd2;
//-----------------------------------------
void ProcessByte(BYTE byte)
{
if(flagcom!=0)
flagcom++;
if((byte==0xAA)&&(flagcom==0))
{
pos=0;
conv=0;
flagcom=1;
}
if(flagcom==2)
{
pos=byte;
pos=pos<<8;
}
if(flagcom==3)
pos=pos+byte;
if(flagcom==4)
{
conv=byte;
conv=conv<<8;
}
if(flagcom==5)
{
if(t>10){
xd=0.1;
}
if(t>20){
xd=0.2;
}
if(t>30){
```

```
xd=0.1;
}
if(t>40){
xd=0.0;
}
if(t>50){
xd=0.1;
}
if(t>60){
xd=0.2;
}
if(t>70){
xd=0.1;
}
if(t>80){
xd=0.0;
}
conv=conv+byte;
theta=(signed short int)pos;
theta=esctheta*theta;
x=(signed short int)(conv-511);
x=escx*x;
e=xd-x;
/*_____Controllers_____ */
/* ---------- Integral control ------*/
/* xd2=kp2*e+ki2*integ;
integ=integ+Ts*e;
ep=xd2-x;

thetad=gamma*(ep+v);//thetad=v1
v=(-c*v+(b-c)*ep)*Ts+v;
iast1=alfa*(thetad-theta)-kv*(theta-theta_1)*iTs;
iast=iast1;
theta_1=theta; */
/* ---------- Integral control ends ------*/
/* ---------- without integral control ------*/

thetad=gamma*(e+v);//thetad=v1
v=(-c*v+(b-c)*e)*Ts+v;
iast1=alfa*(thetad-theta)-kv*(theta-theta_1)*iTs;
iast=iast1;
theta_1=theta;
/* ---------- without integral control ends-----*/
/* _____Controllers ends _____ */
/* _____Output saturation_____ */
if(iast>IM)
iast=IM;
if(iast<-IM)
iast=-IM;
/* _____ */
MainForm->Edit3->Text = FloatToStr (iast);
iast=-iast;
cuentas=escs*(iast+maxI);
/* ___Physical constraint on the beam position___ */
```

```
if(theta>1.0)//0.5
cuentas=127;
if(theta<-1.0)// -0.5
cuentas=127;
/* _____ */
MainForm->Edit1->Text = FloatToStr (xd);
MainForm->Edit2->Text = FloatToStr (x);
MainForm->Edit4->Text = FloatToStr (theta);
MainForm->Edit5->Text = FloatToStr (t);
MainForm->Edit6->Text = IntToStr (conv);
//Sending output byte
MainForm->Acknowledge();
MainForm->send_byte(cuentas);
flagcom=0;
/* open/close a file */
if(flagfile==0)
if((ptrMonit=fopen("MONIT.TXT", "w"))==NULL){}
flagfile=1;
/*Write to file*/
fprintf(ptrMonit,"%3.3f\t%3.3f\t%3.3f\t%3.3f\t%3.3f\n",t,xd,
x,-iast,theta);
t=t+Ts;
}
}
//---------------------------------------------
__fastcall TMainForm::TMainForm(TComponent* Owner)
: TForm(Owner), SerialPort(1, ProcessByte), fAcknowledge(true)
{
if (SerialPort.IsReady() != TRUE)
MessageBox(NULL, "Problems with port", "Error", MB_OK);
}
//---------------------------------------------
void TMainForm::send_byte(unsigned char byte_sal)
{
if (fAcknowledge == false)
return;
SerialPort.WriteByte(byte_sal);
fAcknowledge = false;
}
//---------------------------------------------
void TMainForm::Acknowledge()
{
fAcknowledge = true;
}
void __fastcall TMainForm::Button1Click(TObject *Sender)
{
/* close the file */
fclose(ptrMonit);
Close();
}
//---------------------------------------------
```

14.8   PIC C Code Used to Program the Microcontroller PIC16F877A

In this section, the main steps are described to program the microcontroller PIC16F877A used to exchange data between the computer and the mechanism. This program is written in PIC C language and is listed below. The corresponding flow diagram is presented in Fig. 14.21. After defining some libraries and constants, the interruption "int_rb" is used to count pulses from the encoder. The number of pulses are represented by the variable "cuenta". The executable part of the program begins at the instruction "void main(void)". Then, "TMR0=0" is set to start the first sample period, "cuenta=0" is set to fix $\theta = 0$ as the initial position of the beam (the beam must be horizontal at this position) and "PORTD=127" is set to fix to zero the initial commanded electric current, i.e., to maintain the motor without movement as the starting point. The instruction "while(TRUE)" defines an infinite cycle where the control task is to be implemented. The ball position is measured via an analog/digital converter at "conv=read_adc()". The instructions "putc(0xAA); putc(cuentaH); putc(cuentaL); putc(convH); putc(convL);" send the recognition code "0xAA" first, and then the beam position and the ball position, in that order. The instruction "while(TMR0<197)" defines a cycle where the microcontroller remains waiting for two events: (i) For the computer to send back the computed desired current, which is taken and sent to the digital/analog converter via the instructions "if(kbhit()) PORTD=getc();", (ii) That the sample period elapses (0.01[s]), which is true when

Fig. 14.21 Flow diagram for microcontroller PIC16F877A programming when used as interface between a ball and beam system and a PC

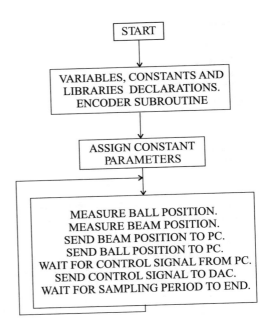

the variable TMR0 is equal to or larger than 197. Once this happens, TMR=0 is set again and the program returns to the beginning of the infinite cycle "while(TRUE)".

```c
// PIC16F877A Microcontroller program
#include<16f877A.h>
#device adc=10 //adc, 10 bits
#include<stdlib.h>
#include<math.h>
#fuses HS,NOWDT,PUT,NOBROWNOUT,NOLVP,NOWRT,NOPROTECT,NOCPD
#use delay(clock=20000000)// (frequency Xtal)
//Config. serial port
#use rs232(baud=115200,XMIT=PIN_C6,RCV=PIN_C7,BITS=8,PARITY=N)
//ports and registers addresses
#byte OPTION= 0x81
#byte TMR0 = 0x01
#byte PORTA = 0x05
#byte PORTB = 0x06
#byte PORTC = 0x07
#byte PORTD = 0x08
#byte PORTE = 0x09

#bit PC0 = 0x07.0
#bit PC1 = 0x07.1
//------------Variables declaration ------------//
int16 inter,cuenta,conv;
int8 cuentaH,cuentaL,convH,convL,puerto,AB,AB_1,aux;
//int1 ban;
//-----------Encoder interrupt subroutine --------//
#int_rb
void rb_isr()
{
puerto=PORTB;
AB=((puerto)&(0x30))>>4;
aux=AB^AB_1;
if(aux!=0)
if(aux!=3)
if(((AB_1<<1)^AB)&(0x02))
cuenta--;
else
cuenta++;
AB_1=AB;
}


//-----------main program------------//
void main(void)
{
set_tris_a(0b11111111);
set_tris_b(0b11111111);
set_tris_c(0b10000000); //pin config.
set_tris_d(0b00000000);
set_tris_e(0b11111111);
OPTION=0x07; //pre_scaler timer0, 1:256
PORTC=0;
TMR0=0;
```

```
cuenta=0;
AB=0;
AB_1=0;
PORTD=127;
setup_adc(ADC_CLOCK_INTERNAL); //Config. ADC
setup_adc_ports(ANALOG_RA3_REF);//Config. reference ADC
set_adc_channel(1);
delay_ms(3000);
enable_interrupts(global);
enable_interrupts(int_rb);

while(TRUE)
{
PC0=1; //sending starts
conv=read_adc();
inter=(conv)&(0xFF00);
convH=inter>>8;
convL=(conv)&(0x00FF);

inter=(cuenta)&(0xFF00);
cuentaH=inter>>8;
cuentaL=(cuenta)&(0x00FF);

putc(0xAA); //serial port acknowledgment
putc(cuentaH); //sending to serial port
putc(cuentaL);
putc(convH);
putc(convL);


PC0=0; //sending ends
PC1=1; //waiting for sampling time
while(TMR0<197) //each count= (4/FXtal)*256 s
{
if(kbhit())//asking for a datum at buffer
PORTD=getc();
}
PC1=0; //sampling time is arrived
TMR0=0;
} //closing infinite while
} //closing main
```

14.9 Control Based on a PIC16F877A Microcontroller

14.9.1 Prototype Construction

In this section, a PIC16F877A microcontroller [5] is used to evaluate the control algorithm for the ball and beam system. Although an computer is also employed, this is only to save the important data and to plot them. It is also remarked that the

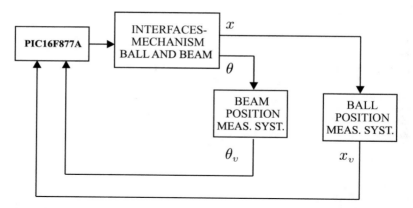

Fig. 14.22 Block interfaces and ball and beam mechanism are shown in Fig. 14.24. The beam position measurement system is composed of a potentiometer fixed to the beam shaft. The ball position measurement system is shown in Fig. 14.23

prototype mechanical part has also been slightly modified, which will be described in the following. A block diagram of the whole system that has been built is presented in Fig. 14.22.

The new ball position x measurement system[2] is shown in Fig. 14.23. It has the same components as the sensor in Fig. 14.3, but the method used to measure the ball position is different. In this case, rail number one is employed as an inductance at whose terminals a 16[KHz] sinusoidal voltage is applied. Hence, this AC voltage uniformly distributes along this coil and an AC voltage is obtained in rail number two whose peak value is proportional to the ball position x on the beam. Then, this signal is rectified and low-pass filtered by a RC circuit to minimize voltage ripple. The voltage x_v represents the ball position measurement, which is delivered to the microcontroller through one of its analog/digital converters. This sensor gain is computed as:

$$A_x = \frac{3.7625[V]}{0.7[m]} = 5.3750[V/m], \tag{14.20}$$

where the numerator is x_v when the ball is at the right end of the beam and the denominator is the beam length.

On the other hand, the beam position θ measurement system is a potentiometer fixed to the beam shaft. The gain of this measurement system is computed as the division of a voltage increment at the potentiometer terminals, 0.275[V], and the corresponding angle increment θ, i.e., 0.3[rad]:

---

[2]In all the experiments that are presented in the remainder of this chapter, the ball position x is measured from the left end of the beam.

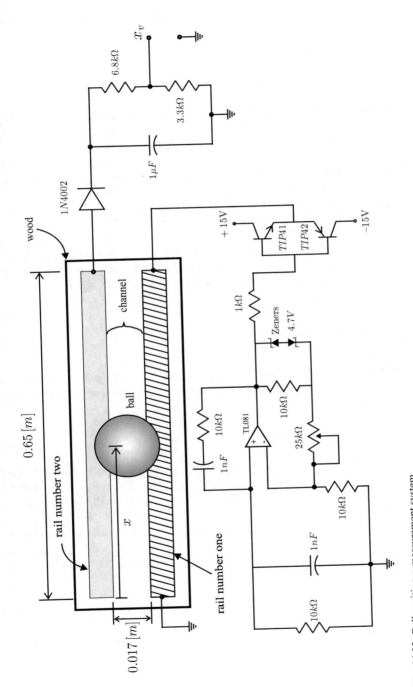

Fig. 14.23 Ball position x measurement system

$$A_\theta = \frac{0.275[V]}{0.3[rad]} = 0.9167[V/rad]. \tag{14.21}$$

The following parameters:

$$k = 16.6035, \quad a = 3.3132, \quad \rho = 5, \tag{14.22}$$

were obtained using identical procedures to those described in Sects. 14.3.1 and 14.3.2. Note that these values are different from those shown in (14.12) and (14.15) mainly because a different gear box has been employed.

Finally, in Fig. 14.24 an electric diagram of the control system based on the PIC16F877A microcontroller is shown, whereas the code used to program the microcontroller is shown in Sect. 14.9.4. For a detailed explanation of Fig. 14.24 and the program in Sect. 14.9.4 the reader is advised to see Sects. 10.6 and 11.5 corresponding to the velocity and the position control of a PM brushed DC motor. It is stressed that the two channels of the analog/digital converter that possesses the microcontroller are employed: one is used to measure the ball position and the other to measure the beam position. Recall that the portable computer is only used to plot the important variables, but it is not used to evaluate the control algorithm.

14.9.2   Controller Design

It is proposed to employ a control scheme that is identical to that presented in Sect. 14.4, i.e., the closed-loop block diagram is identical to that shown in Fig. 14.8. The gain values:

$$\alpha = 12, \quad k_v = 0.2, \tag{14.23}$$

were selected experimentally such that using a constant value for $v_1(t)$ (see Fig. 14.8), i.e., without feeding back the ball position error and without using the lead compensator at the left of $V_1(s)$, a fast and well-damped θ response was obtained. Notice that $\alpha = 12$ shown in (14.23) is different from $\alpha = 30$ in (14.19). This because of the changes made in the mechanism. Using (14.20), (14.21), (14.22), (14.23) and the open-loop transfer function in (14.16) the root locus method is employed to find that the following gains ensure closed-loop stability:

$$\gamma = 1.2, \quad c = 20, \quad b = 2.5. \tag{14.24}$$

The detailed procedure followed to obtain these gains is presented in Sect. 5.2.8.

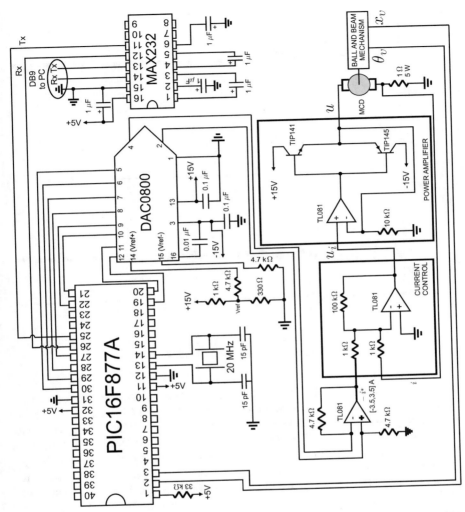

Fig. 14.24 Electric diagram of the ball and beam control system based on the PIC16F877A microcontroller

14.9.3 Experimental Results

The software implementation of the controller designed in the previous section is performed as explained in Sect. F.4, whereas the code used to program the microcontroller PIC16F877A [6] is shown in Sect. 14.9.4. The electric diagram in Fig. 14.24 is also used. Some experimental results obtained when using such a controller are shown in Fig. 14.25, i.e., using the gains in (14.23) and (14.24). It is observed that the ball reaches a position x_v close to 1[V], where $x_{vd} = 1$[V]

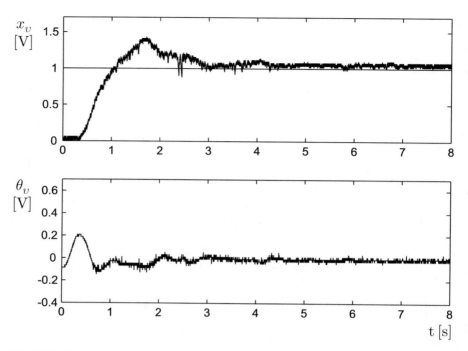

Fig. 14.25 Experimental results using microcontroller PIC16F877A

is the desired ball position. The small difference that is observed is due to the dead zone produced by friction at the motor shaft. In Fig. 14.26, some simulation results are presented using the same controller, the block diagram in Fig. 14.8 and the numerical values in (14.20), (14.21), (14.22), (14.23), (14.24).[3] Notice that the responses in Figs. 14.25 and 14.26 are very similar despite the noise content in the experimental prototype. Finally, in Fig. 14.27 some other experimental results are presented where it is observed that the ball position x_v is stabilized very close to its desired value $x_{vd} = 1$[V] despite an external agent hitting the ball several times. These results show that the control objective has been accomplished: the ball is stabilized very close to its desired position. Finally, a picture of the prototype used in the experiments is shown in Fig. 14.28.

14.9.4   PIC16F877A Microcontroller Programming

The flow diagram of the following code is presented in Fig. 14.29.

---

[3] See Sect. 5.2.8 for instructions on how these simulations can be performed.

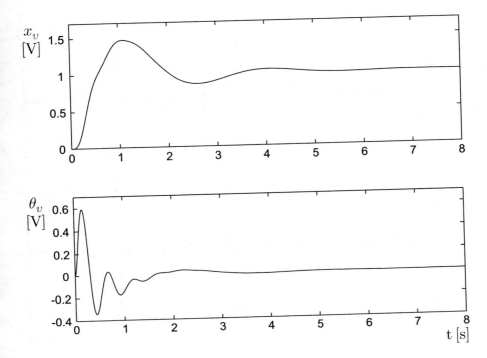

Fig. 14.26 Simulation results

```
#include<16f877A.h>
//#device adc=10 //adc, 10 bits
#include<stdlib.h>
#include<math.h>

#fuses HS,NOWDT,PUT,NOBROWNOUT,NOLVP,NOWRT,NOPROTECT,NOCPD

#use delay(clock=20000000) //Time base for delays
//(frequency Xtal)
//Config.
//P. Serial
#use rs232(baud=115200,XMIT=PIN_C6,RCV=PIN_C7,BITS=8,PARITY=N)
//port and registers addresses
#byte OPTION= 0x81
#byte TMR0 = 0x01
#byte PORTA = 0x05

#byte PORTB = 0x06
#byte PORTC = 0x07
#byte PORTD = 0x08

#byte PORTE = 0x09
#byte ADCON0= 0x1F
```

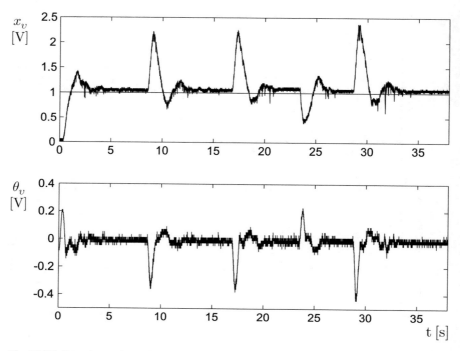

Fig. 14.27 Experimental results using the microcontroller PIC16F877A. Some ball position deviations are introduced by an external agent

```
#byte ADCON1= 0x9F

#bit PC0 = 0x07.0
#bit PC1 = 0x07.1
//-----------variables declaration------------//
int16 inter,cuenta;

int8 cuentaH,cuentaL,puerto,AB,AB_1,aux;

int8 cont,iastd,cont2,cont3,cont4,t_0,t_1;

float x,xd,theta,error,iast,tiempo,c,b;

float gamma,alfa,kv,v,v1,thetam1; unsigned int u,i;
//int1 ban;
//-----------encoder interrupt subroutine-------//
#int_rb void rb_isr() {
puerto=PORTB;
AB=((puerto)&(0x30))>>4;
aux=AB^AB_1;
if(aux!=0)
if(aux!=3)
if(((AB_1<<1)^AB)&(0x02))
```

Fig. 14.28 Ball and beam system prototype used in the experiments

```
cuenta--;
else
cuenta++;
AB_1=AB;
}
//-----------Main program------------//
void main(void) {
setup_adc(ADC_CLOCK_INTERNAL ); //ADC (internal clock)
set_tris_a(0b11111111);
set_tris_b(0b11111111);
set_tris_c(0b10000000); //serial communication
set_tris_d(0b00000000);
set_tris_e(0b11111111);
OPTION=0x07; //pre_scaler timer0, 1:256
PORTC=0;
TMR0=0;
cuenta=0;
AB=0;
AB_1=0;
enable_interrupts(global);
enable_interrupts(int_rb);
cont=0;
cont2=0;
cont3=0;
cont4=0;
PORTD=127;
xd=1.0;
tiempo=0.0;
```

Fig. 14.29 Flow diagram for
PIC16F877A microcontroller
programming

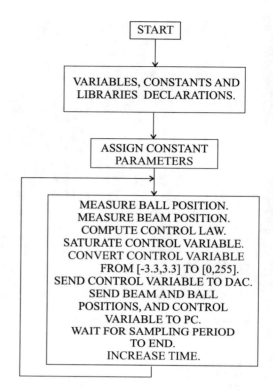

```
gamma=1.2;
c=20.0;
b=2.5;
alfa=12.0;
kv=0.2;
thetam1=0.0;
v=0.0;
ADCON1=0x04;
ADCON0=0x81;
while(cont2<3)
{
while(cont3<255)
{
TMR0=0;
while(TMR0<255)
{
}
cont3++;
}
cont2++;
}
TMR0=0;
while(TRUE)
```

```
{
ADCON0=0x81; //Changing to CH0
for(i=0;i<37;i++); /stabilizing internal capacitor
t_0=read_adc(); //reading ADC
x=0.0196*t_0;

ADCON0=0x89; //changing to CH1
for(i=0;i<37;i++); //stabilizing internal capacitor
t_1=read_adc(); //reading ADC
theta=0.0196*t_1-2.81;//2.81;

error=xd-x;
/* Controller */
v1=gamma*(error+v);
v=(-c*v+(b-c)*error)*0.002+v;
iast=alfa*(v1-theta)-kv*(theta-thetam1)/0.002;
/* _____ */
if(iast<-3.3)
iast=-3.3;
if(iast>3.3)
iast=3.3;
iast=-iast;
iastd=(unsigned int)(36.4286*iast+127);
PORTD=iastd;
thetam1=theta;

cuentaH=t_1;
cuentaL=t_0;
putc(0xAA); //serial port acknowledgment
putc(cuentaH); //sending to serial port
putc(cuentaL);
/*
iast=-iast;
iastd=(unsigned int)(36.4286*iast+127);
cuentaH=0x00;
cuentaL=(iastd)&(0xFF);
putc(0xAA); //serial port acknowledgment
putc(cuentaH); //sending to serial port
putc(cuentaL);
*/
PC1=1; //waiting for sampling time
while(TMR0<40) //1 count=(4/FXtal)*256 s, 20=1 ms
{
}
PC1=0; //sampling time is arrived
TMR0=0;
tiempo=tiempo+0.002;
} //closing infinite while
} //closing main
```

14.10   Summary

In this chapter, a ball and beam system has been built in addition to all the necessary interfaces for closed-loop control. The proposed control scheme has two loops. The internal loop is useful to control the beam angle, whereas the external loop is used to control the ball position. An interesting feature of the prototype that has been built is the construction of the sensor used to measure the ball position. The main idea is to build a channel composed of two rails which, together, work in a similar manner to a potentiometer. Hence, the voltage delivered by this measurement system is proportional to the position of the ball on the beam.

The controller design has been performed using both fundamental approaches in classical control: time response (root locus) and frequency response (Nyquist stability criterion and phase margin). The controllers that have been designed have been tested through several experiments to observe the achieved performance. Also, some interesting experiments have been performed to identify the system parameters.

14.11   Review Questions

1. Why is the ball and beam system open loop unstable?
2. How do the measurement systems designed for the ball position work?
3. Give a descriptive explanation of why it is necessary to control the beam angle aside from the ball position.
4. What would happen with ρ if $r = 0$ (see (14.6)) and what does this mean?
5. Recall the experimental procedure described in Sect. 14.3.2 to identify the only parameter in the ball dynamics. What else can you do to identify this parameter?
6. The automatic navigation control system of a ship is similar to a ball and beam system, where the role of the beam is replaced by the rudder. Using this idea try to control the direction of a ship.
7. Why does a lead compensator have fewer noise problems than a PD controller?
8. Why do you think it is preferable to use a proportional controller with velocity feedback instead of a PD controller for the beam angle?

References

1. Quanser Inc. 119 Spy Court Markham, Ontario L3R 5H6, Canada. Available at: http://www.quanser.com
2. P. V. Kokotovic, The joy of feedback: nonlinear and adaptive (1991 Bode Prize Lecture), *IEEE Control Systems Magazine*, pp. 7–17, June 1992.
3. R. C. Hibbeler, *Engineering Mechanics Dynamics*, 10th. Edition, Prentice Hall, Upper Saddle River, NJ, 2004.

References

4. M. Alonso and E. J. Finn, *Physics*, Addison-Wesley, Reading, MA, 1992.
5. PIC16F877A Enhanced Flash Microcontroller, Data sheet, Microchip Technology Inc., 2003.
6. Custom Computer Services Incorporated, CCS C Compiler Reference Manual, 2003.

Chapter 15
Control of a Furuta Pendulum

The Furuta pendulum is depicted in Fig. 15.1. Controlling the Furuta pendulum may be described by referring to the old trick known as the broomstick-balancing problem. In a Furuta pendulum, the person's arm is replaced by a beam technically known as the *arm*, which receives the torque generated by a permanent magnet (PM) brushed direct current (DC) motor at one of its ends. Hence, the arm can only move on a horizontal plane such that its other end describes a circumference. At this end another beam is placed, technically known as the *pendulum*, which joins to the arm by means of some bearing allowing the pendulum free movement. This means that no motor is placed at this joint. However, the bearing in this place constrains the pendulum to only move by rotating on a plane that is orthogonal to the arm. This means that the pendulum's free end can only describe a circumference whose center is placed at the point where the arm and the pendulum join.

One feature that renders the Furuta pendulum interesting to control is that it is inherently unstable. This can be seen by recalling the broomstick-balancing problem: the broomstick always tends to leave its inverted position, falling to the floor. Hence, designing a controller for this problem is not a trivial task and it requires the knowledge of suitable control theory tools.

The variables and the parameters involved in the Furuta pendulum depicted in Fig. 15.1 are the following:

- θ_0 is the arm angular position, in radians, measured with respect to some arbitrary reference position.
- θ_1 is the pendulum angular position, in radians, measured with respect to its inverted position.
- τ is the torque generated by the electric motor that is applied to the arm.
- I_0 is the arm inertia when it rotates around one of its ends plus the motor inertia.
- L_0 is the arm length.
- m_1, l_1 and J_1 stand for the pendulum mass, the location of the center of mass, and the inertia respectively. J_1 is computed assuming that the pendulum rotates

© Springer International Publishing AG, part of Springer Nature 2019
V. M. Hernández-Guzmán, R. Silva-Ortigoza, *Automatic Control with Experiments*,
Advanced Textbooks in Control and Signal Processing,
https://doi.org/10.1007/978-3-319-75804-6_15

Fig. 15.1 Furuta pendulum

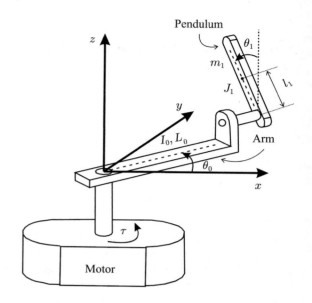

around its center of mass. Note that the pendulum consists of a beam whose mass is not negligible.

- $g = 9.81[\text{m/s}^2]$ is the gravity acceleration.

15.1   Mathematical Model

As described above, the Furuta pendulum is composed of two bodies that interact. The mathematical model for this class of mechanisms is commonly obtained using the *Euler–Lagrange equations* [1]. In the case of the Furuta pendulum, the Euler–Lagrange equations are written as:

$$\frac{d}{dt}\frac{\partial L}{\partial \dot{\theta}_0} - \frac{\partial L}{\partial \theta_0} = \tau, \qquad (15.1)$$

$$\frac{d}{dt}\frac{\partial L}{\partial \dot{\theta}_1} - \frac{\partial L}{\partial \theta_1} = 0.$$

The zero on the right-hand side of last equation indicates that no external torque is applied at the place where the arm and the pendulum join. The Lagrangian L is defined as:

$$L = K - P, \qquad (15.2)$$

$$K = K_0 + K_1,$$

where K_0 is the arm's kinetic energy, K_1 is the pendulum's kinetic energy, and P is the pendulum's potential energy. No potential energy is considered for the arm because it moves on an horizontal plane, i.e., its potential energy can be assumed to be zero.

The computation of the *kinetic energy* for a body whose mass is distributed on a volume can be simplified if it is decomposed into two parts: one part due to translational movement of the body's center of mass (assumed to be a particle) and the other part due to body's rotative movement around its center of mass. However, if the body movement can be described in a simple manner such a decomposition is not necessary. For instance, the arm movement is easily described as a rotative movement of a beam around one of its ends and the rotative movement of the motor around the same axis. Hence, it is not difficult to find that:

$$K_0 = \frac{1}{2} I_0 \dot{\theta}_0^2. \tag{15.3}$$

On the other hand, the pendulum movement is more complex because it depends on the combined movements of the arm and the pendulum. In this case, it is more convenient to decompose the pendulum movement into two parts as explained above: the translational movement of a particle with mass m_1 located at the pendulum center of mass and the rotative movement of a beam around an axis passing through the pendulum's center of mass. Hence:

$$K_1 = \frac{1}{2} J_1 \dot{\theta}_1^2 + \frac{1}{2} m_1 v_1^T v_1, \tag{15.4}$$

where v_1 is a vector that stands for the velocity of the pendulum center of mass, i.e.,:

$$v_1 = \begin{bmatrix} \frac{dX_x}{dt} \\ \frac{dX_y}{dt} \\ \frac{dX_z}{dt} \end{bmatrix}, \quad X = \begin{bmatrix} X_x \\ X_y \\ X_z \end{bmatrix}, \tag{15.5}$$

where X_x, X_y y X_z are the Cartesian coordinates of the pendulum's center of mass. This means that X is the position vector of the pendulum's center of mass. According to Figs. 15.2a and b the following is found:

$$X = \begin{bmatrix} X_x \\ X_y \\ X_z \end{bmatrix} = \begin{bmatrix} L_0 \cos(\theta_0) - l_1 \sin(\theta_1) \sin(\theta_0) \\ L_0 \sin(\theta_0) + l_1 \sin(\theta_1) \cos(\theta_0) \\ l_1 \cos(\theta_1) \end{bmatrix}. \tag{15.6}$$

Using (15.4), (15.5), (15.6) and after rearranging terms, the following is found:

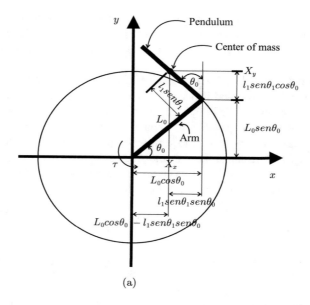

(a)

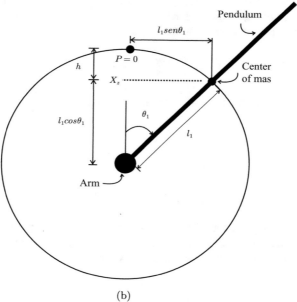

(b)

Fig. 15.2 Geometric relationships in the Furuta pendulum. (**a**) Superior view. (**b**) View of the plane orthogonal to arm

$$K_1 = \frac{1}{2}J_1\dot{\theta}_1^2 + \frac{1}{2}m_1\left[(L_0\dot{\theta}_0)^2 + (l_1\dot{\theta}_0)^2 \sin^2(\theta_1) + (l_1\dot{\theta}_1)^2 + 2\dot{\theta}_0\dot{\theta}_1 L_0 l_1 \cos(\theta_1)\right].$$

$$(15.7)$$

Finally, to compute the pendulum *potential energy*, the point where $\theta_1 = 0$ is used as the reference point, i.e., where $P = 0$. Hence, recalling that the potential energy is the scalar product of the applied force vector, i.e., the pendulum's weight $m_1 g$, and the distance vector going from the pendulum's center of mass to the point where $P = 0$, the following is obtained:

$$P = -hm_1 g, \quad h = l_1 - l_1 \cos(\theta_1),$$

where the sign "−" is due to the fact that the vectors referred to above have an angle that is greater than 90°. Then:

$$P = m_1 g l_1(\cos(\theta_1) - 1). \tag{15.8}$$

Using (15.3), (15.7), (15.8), the Lagrangian L is formed according to (15.2). After replacing in the Euler–Lagrange equations (15.1), performing the indicated operations and suitably rearranging the resulting terms, the following is found:

$$M(q)\ddot{q} + C(q,\dot{q})\dot{q} + g(q) = F, \tag{15.9}$$

$$M(q) = \begin{bmatrix} I_0 + m_1(L_0^2 + l_1^2 \sin^2(\theta_1)) & m_1 l_1 L_0 \cos(\theta_1) \\ m_1 l_1 L_0 \cos(\theta_1) & J_1 + m_1 l_1^2 \end{bmatrix},$$

$$C(q,\dot{q}) = \begin{bmatrix} \frac{1}{2}m_1 l_1^2 \dot{\theta}_1 \sin(2\theta_1) & -m_1 l_1 L_0 \dot{\theta}_1 \sin(\theta_1) + \frac{1}{2}m_1 l_1^2 \dot{\theta}_0 \sin(2\theta_1) \\ -\frac{1}{2}m_1 l_1^2 \dot{\theta}_0 \sin(2\theta_1) & 0 \end{bmatrix},$$

$$g(q) = \begin{bmatrix} 0 \\ -m_1 l_1 g \sin(\theta_1) \end{bmatrix}, \quad F = \begin{bmatrix} \tau \\ 0 \end{bmatrix}, \quad q = \begin{bmatrix} \theta_0 \\ \theta_1 \end{bmatrix}.$$

The expression in (15.9) represents the Furuta pendulum mathematical model, which is nonlinear because it includes trigonometric functions in addition to several products between velocities $\dot{\theta}_0$ and $\dot{\theta}_1$.

15.2   A Controller to Swing Up the Pendulum

Swinging up of the pendulum is the task where the pendulum is taken from the bottom stable position, i.e., $\theta_1 = \pi$, to the inverted unstable position, i.e., $\theta_1 = 0$ or $\theta_1 = 2\pi$. This is performed experimentally in Sect. 15.7 by using the following controller:

$$\tau = \frac{-\frac{k_\omega}{\det(M(q))} F(q,\dot{q}) - k_\theta \theta_0 - k_\delta \dot{\theta}_0}{k_E E(q,\dot{q}) + \frac{k_\omega}{\det(M(q))}(J_1 + m_1 l_1^2)}, \tag{15.10}$$

$$E(q,\dot{q}) = \frac{1}{2}\dot{q}^T M(q)\dot{q} + m_1 g l_1(\cos\theta_1 - 1),$$

$$F(q,\dot{q}) = -(J_1 + m_1 l_1^2)m_1 l_1^2 \dot{\theta}_1 \dot{\theta}_0 \sin(2\theta_1) - \frac{1}{2}m_1^2 l_1^3 L_0 \dot{\theta}_0^2 \cos\theta_1 \sin(2\theta_1)$$

$$-m_1^2 l_1^2 L_0 g \cos\theta_1 \sin\theta_1 + (J_1 + m_1 l_1^2)m_1 l_1 L_0 \dot{\theta}_1^2 \sin\theta_1,$$

where $E(q,\dot{q}) = K_0 + K_1 + P$ is the Furuta pendulum's total energy and k_θ, k_δ, k_ω, K_E are positive constants that satisfy:

$$\frac{k_\omega}{k_E} > 2m_1 g l_1(I_0 + m_1 l_1^2 + m_1 L_0^2).$$

The controller in (15.10) was proposed in [2], chapter 6. The interested reader is referred to that book for more details. Swinging up the pendulum in the Furuta pendulum is not an easy task as the only motor in the mechanism actuates directly on the arm but there is no motor actuating directly on the pendulum. The trick to solving this problem is energy regulation, as we explain in the following. Suppose that the Furuta pendulum's total energy is equal to zero and $\dot{\theta}_0 = 0$. Under these conditions we have that:

$$E(q,\dot{q}) = \frac{1}{2}(J_1 + m_1 l_1^2)\dot{\theta}_1^2 + m_1 g l_1(\cos\theta_1 - 1) = 0.$$

This means that:

$$\dot{\theta}_1 = \pm\sqrt{\frac{2m_1 g l_1}{J_1 + m_1 l_1^2}(1 - \cos\theta_1)}. \tag{15.11}$$

In Fig. 15.3 a plot of $\dot{\theta}_1$ satisfying this expression is presented. This plot represents the trajectory where the pendulum evolves when $E(q,\dot{q}) = 0$ and $\dot{\theta}_0 = 0$. Moreover, its orientation represents the sense of movement of the pendulum variables. According to Fig. 15.3, under these conditions the pendulum will reach one of the configurations $(\theta_1, \dot{\theta}_1) = (0, 0)$ or $(\theta_1, \dot{\theta}_1) = (2\pi, 0)$. This suggests regulating the Furuta pendulum's total energy and the arm velocity at zero. Once this is accomplished, it is only a matter of time for the pendulum to reach the inverted unstable position. The controller in (15.10) is designed to achieve $E(q,\dot{q}) = 0$, $\dot{\theta}_0 = 0$ and $\theta_0 = 0$ [2], chapter 6. The latter is required to avoid the arm rotating many (or an infinite number of) times.

Consider the following function:

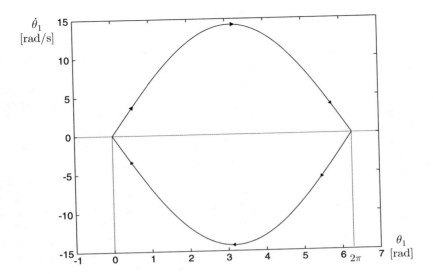

Fig. 15.3 Plot of (15.11) in the phase plane

$$V(q, \dot{q}) = \frac{k_E}{2} E^2(q, \dot{q}) + \frac{k_\omega}{2}\dot{\theta}_0^2 + \frac{k_\theta}{2}\theta_0^2. \qquad (15.12)$$

Note that:

$$V(q, \dot{q}) = 0, \quad \text{if } E(q, \dot{q}) = 0, \ \dot{\theta}_0 = 0, \ \theta_0 = 0,$$

$$V(q, \dot{q}) > 0, \quad \text{elsewhere.}$$

These properties are exploited for controller design in [2], chapter 6. The main idea is to choose (15.10) as a controller ensuring that $\dot{V}(q, \dot{q}) < 0$ and $\dot{V}(q, \dot{q}) = 0$ only when $E(q, \dot{q}) = 0$, $\dot{\theta}_0 = 0$ and $\theta_0 = 0$. This forces the system to move in a direction that approaches the conditions $E(q, \dot{q}) = 0$, $\dot{\theta}_0 = 0$, and $\theta_0 = 0$. Once this is achieved the system stays there, i.e., the trajectory in (15.11) is reached. See [2], chapter 6, for more details. However, it is important to stress that this only ensures that the pendulum's inverted unstable configuration is reached, but this does not ensure that the pendulum remains there. To force the pendulum to stay at the inverted unstable configuration it is necessary to employ a stabilizing controller such as that designed in Sect. 15.6. This is performed on the basis of an approximate linear model of the Furuta pendulum, which is obtained in the next section.

15.3 Linear Approximate Model

This book is oriented to the application of linear control techniques, which cannot be directly used when the model of the plant to be controlled is nonlinear such as that shown in (15.9). However, there is a way of solving this problem. The fundamental idea is to find an approximate linear model that represents the model with enough accuracy (15.9) and, then, use such an approximate linear model for controller design. As explained in Sect. 7.3.2, the controller designed in this way is useful for balancing the pendulum at its inverted position $\theta_1 = 0$ only if the pendulum is initially close to such a configuration and remains there, i.e., as long as $\theta_1 \approx 0$.

The first step to this end is to write (15.9) in terms of its state variables. According to Sect. 7.1, given a set of differential equations such as (15.9), a state variables representation can be found by defining the state vector x components as the differential equations' unknown variables and their first $r-1$ time derivatives, where r is the order of the corresponding differential equation. Note that (15.9) consists of two second-order differential equations. The unknown variables in these differential equations are θ_0 and θ_1. Thus, the state vector can be chosen to be given as:

$$
x = \begin{bmatrix} x_1 \\ x_2 \\ x_3 \\ x_4 \end{bmatrix} = \begin{bmatrix} \theta_0 \\ \dot{\theta}_0 \\ \theta_1 \\ \dot{\theta}_1 \end{bmatrix}.
\tag{15.13}
$$

Solve (15.9) for $\ddot{q}$ and define:

$$
\ddot{q} = \begin{bmatrix} w_1(x_1, x_2, x_3, x_4, \tau) \\ w_2(x_1, x_2, x_3, x_4, \tau) \end{bmatrix},
$$
$$
= M^{-1}(x_1, x_3)[-C(x_1, x_2, x_3, x_4)[x_2, x_4]^T - g(x_1, x_3) + F].
\tag{15.14}
$$

One important property of matrix $M(q) = M(x_1, x_3)$ is that it is always nonsingular, i.e., the matrix $M^{-1}(x_1, x_3)$ appearing in the last expression always exists [2]. Then, the corresponding state equation can be written as:

$$
\dot{x} = f(x, u),
\tag{15.15}
$$

$$
f(x, u) = \begin{bmatrix} f_1(x, u) \\ f_2(x, u) \\ f_3(x, u) \\ f_4(x, u) \end{bmatrix} = \begin{bmatrix} x_2 \\ w_1(x_1, x_2, x_3, x_4, u) \\ x_4 \\ w_2(x_1, x_2, x_3, x_4, u) \end{bmatrix}, \quad u = \tau.
$$

Using the ideas in Sect. 7.3.2, an approximate linear model is now obtained that is valid around an operation point. The operation points are the pairs (x^*, u^*)

satisfying:

$$f(x^*, u^*) = \begin{bmatrix} 0 \\ 0 \\ 0 \\ 0 \end{bmatrix},$$

i.e., according to (15.15), (15.14):

$$x_2^* = \dot{\theta}_0^* = 0, \quad x_4^* = \dot{\theta}_1^* = 0,$$

$$\begin{bmatrix} w_1(x_1^*, 0, x_3^*, 0, u^*) \\ w_2(x_1^*, 0, x_3^*, 0, u^*) \end{bmatrix} = M^{-1}(x_1^*, x_3^*)$$

$$\times \left\{ -C(x_1^*, 0, x_3^*, 0) \begin{bmatrix} 0 \\ 0 \end{bmatrix} - g(x_1^*, x_3^*) + \begin{bmatrix} u^* \\ 0 \end{bmatrix} \right\},$$

$$= \begin{bmatrix} 0 \\ 0 \end{bmatrix}.$$

From the second row in these conditions, the following is found:

$$\begin{bmatrix} u^* \\ 0 \end{bmatrix} = g(x_1^*, x_3^*) = \begin{bmatrix} 0 \\ -m_1 l_1 g \sin(\theta_1^*) \end{bmatrix},$$

i.e.,

$$u^* = 0, \quad x_3^* = \theta_1^* = \pm n\pi,$$

where n can be any positive or zero integer. Note that there is no condition for $x_1^* = \theta_0^*$, which means that this variable can be chosen arbitrarily. As it is desired to balance the pendulum at its inverted position, i.e., $n = 0$, the following operation point is chosen:

$$x^* = \begin{bmatrix} x_1^* \\ x_2^* \\ x_3^* \\ x_4^* \end{bmatrix} = \begin{bmatrix} 0 \\ 0 \\ 0 \\ 0 \end{bmatrix}, \quad u^* = 0. \tag{15.16}$$

According to (7.22), (7.23), (7.24), the nonlinear model in (15.14), (15.15), can be approximated by the linear model:

$$\dot{z} = Az + Bv, \tag{15.17}$$

$$
A = \begin{bmatrix}
\dfrac{\partial f_1(x,u)}{\partial x_1} & \dfrac{\partial f_1(x,u)}{\partial x_2} & \dfrac{\partial f_1(x,u)}{\partial x_3} & \dfrac{\partial f_1(x,u)}{\partial x_4} \\[2mm]
\dfrac{\partial f_2(x,u)}{\partial x_1} & \dfrac{\partial f_2(x,u)}{\partial x_2} & \dfrac{\partial f_2(x,u)}{\partial x_3} & \dfrac{\partial f_2(x,u)}{\partial x_4} \\[2mm]
\dfrac{\partial f_3(x,u)}{\partial x_1} & \dfrac{\partial f_3(x,u)}{\partial x_2} & \dfrac{\partial f_3(x,u)}{\partial x_3} & \dfrac{\partial f_3(x,u)}{\partial x_4} \\[2mm]
\dfrac{\partial f_4(x,u)}{\partial x_1} & \dfrac{\partial f_4(x,u)}{\partial x_2} & \dfrac{\partial f_4(x,u)}{\partial x_3} & \dfrac{\partial f_4(x,u)}{\partial x_4}
\end{bmatrix}_{x=x^*,u=u^*},
$$

$$
B = \begin{bmatrix}
\dfrac{\partial f_1(x,u)}{\partial u} \\[2mm]
\dfrac{\partial f_2(x,u)}{\partial u} \\[2mm]
\dfrac{\partial f_3(x,u)}{\partial u} \\[2mm]
\dfrac{\partial f_4(x,u)}{\partial u}
\end{bmatrix}_{x=x^*,u=u^*},
$$

where $z = x - x^*$ and $v = u - u^*$. The expressions in (15.9), (15.14), (15.15), (15.16), (15.17), define a mathematical procedure that (although tedious it is straightforward) yields:

$$
A = \begin{bmatrix}
0 & 1 & 0 & 0 \\
0 & 0 & \dfrac{-gm_1^2 l_1^2 L_0}{I_0(J_1 + m_1 l_1^2) + J_1 m_1 L_0^2} & 0 \\
0 & 0 & 0 & 1 \\
0 & 0 & \dfrac{(I_0 + m_1 L_0^2)m_1 l_1 g}{I_0(J_1 + m_1 l_1^2) + J_1 m_1 L_0^2} & 0
\end{bmatrix},\quad
B = \begin{bmatrix}
0 \\
\dfrac{J_1 + m_1 l_1^2}{I_0(J_1 + m_1 l_1^2) + J_1 m_1 L_0^2} \\
0 \\
\dfrac{-m_1 l_1 L_0}{I_0(J_1 + m_1 l_1^2) + J_1 m_1 L_0^2}
\end{bmatrix}.
$$

$$\tag{15.18}$$

The expressions (15.17), (15.18), represent the approximate linear model we were looking for. It is important to stress that this model is also valid if $x_3^* = \theta_1^* = 2\pi$. This is because $\sin(x_3^*)$ does not change when evaluated at $x_3^* = 0$ or $x_3^* = 2\pi$. The reader is advised to verify this fact. As stated above, using (15.17) and (15.18), a linear state feedback controller can be designed that, however, is only ensured to perform well if the system state and the input are constrained to evolve around the operation point (15.16). This is also stated by indicating that $z \approx 0$ and $v = u = \tau \approx 0$ (because $u^* = 0$). This means that taking the pendulum from $\theta_1 = \pi$ to $\theta_1 = 0$, or $\theta_1 = 2\pi$, is a task that cannot be ensured with a controller designed on the basis of the model (15.17), (15.18). Such a problem is solved by using the controller in (15.10).

15.4 A Differential Flatness-Based Model

The differential flatness property of the model (15.17), (15.18) is studied in this section. To simplify the algebraic manipulation of model (15.17), (15.18), the following constants are defined:

$$a = \frac{-gm_1^2 l_1^2 L_0}{I_0(J_1 + m_1 l_1^2) + J_1 m_1 L_0^2}, \qquad b = \frac{(I_0 + m_1 L_0^2)m_1 l_1 g}{I_0(J_1 + m_1 l_1^2) + J_1 m_1 L_0^2}$$

$$c = \frac{J_1 + m_1 l_1^2}{I_0(J_1 + m_1 l_1^2) + J_1 m_1 L_0^2}, \qquad d = \frac{-m_1 l_1 L_0}{I_0(J_1 + m_1 l_1^2) + J_1 m_1 L_0^2}.$$

Hence, the controllability matrix is given as:

$$C_o = \begin{bmatrix} B & AB & A^2B & A^3B \end{bmatrix} = \begin{bmatrix} 0 & c & 0 & ad \\ c & 0 & ad & 0 \\ 0 & d & 0 & bd \\ d & 0 & bd & 0 \end{bmatrix}.$$

After some straightforward computations it is found that the determinant of this matrix is given as:

$$\det(C_o) = \frac{m_1^4 l_1^4 L_0^2 g^2}{[I_0(J_1 + m_1 l_1^2) + J_1 m_1 L_0^2]^4} \neq 0.$$

Thus, the system (15.17), (15.18), is controllable. According to Sect. 8.2, Chap. 8, this implies that the system (15.17), (15.18) is differentially flat. The inverse matrix of C_o is given as:

$$C_o^{-1} = \frac{adj(C_o)}{\det(C_o)}, \qquad adj(C_o) = Cof^T(C_o)$$

where $Cof^T(C_o)$ is the transposed cofactors matrix of C_o. After some straightforward computations we find:

$$C_o^{-1} = \begin{bmatrix} \star & \star & \star & \star \\ \star & \star & \star & \star \\ \star & \star & \star & \star \\ C_{14} & 0 & C_{34} & 0 \end{bmatrix} \frac{1}{\det(C_o)},$$

$$C_{14} = \frac{gm_1^3 l_1^3 L_0^2 \left[m_1^2 l_1^2 L_0^2 - (J_1 + m_1 l_1^2)(I_0 + m_1 L_0^2) \right]}{[I_0(J_1 + m_1 l_1^2) + J_1 m_1 L_0^2]^4},$$

$$C_{34} = \frac{gm_1^3 l_1^3 L_0^2 \left[-\frac{(J_1 + m_1 l_1^2)^2 (I_0 + m_1 L_0^2)}{m_1 l_1 L_0} + m_1 l_1 L_0 (J_1 + m_1 l_1^2) \right]}{[I_0(J_1 + m_1 l_1^2) + J_1 m_1 L_0^2]^4},$$

where symbol "$\star$" indicates the entries of the matrix that we do not care for. According to Proposition 8.1, the flat output y is given as:

$$y = k \begin{bmatrix} \frac{C_{14}}{\det(C_o)} & 0 & \frac{C_{34}}{\det(C_o)} & 0 \end{bmatrix} \begin{bmatrix} z_1 \\ z_2 \\ z_3 \\ z_4 \end{bmatrix},$$

where k is an arbitrary nonzero constant, which, by convenience, is proposed to be defined as:

$$k = -\frac{\det(C_o)[I_0(J_1 + m_1 l_1^2) + J_1 m_1 L_0^2]^4}{gm_1^3 l_1^3 L_0^2 (J_1 I_0 + I_0 m_1 l_1^2 + J_1 m_1 L_0^2)},$$

because this yields the simple expression:

$$y = z_1 + h z_3, \quad h = \frac{J_1 + m_1 l_1^2}{m_1 l_1 L_0}.$$

Using the model (15.17), (15.18), and differentiating the flat output y four times, the following is found:

$$\dot{y} = z_2 + h z_4, \tag{15.19}$$
$$\ddot{y} = (a + bh) z_3,$$
$$y^{(3)} = (a + bh) z_4,$$
$$y^{(4)} = b\ddot{y} + d(a + bh)\tau,$$
$$d(a + bh) = \frac{g(J_1 I_0 + I_0 m_1 l_1^2 + J_1 m_1 L_0^2)}{L_0[I_0(J_1 + m_1 l_1^2) + J_1 m_1 L_0^2]} d < 0.$$

Using the Laplace transform and assuming zero initial conditions the following transfer function is found:

$$\frac{Y(s)}{\tau(s)} = \frac{d(a + bh)}{s^2(s^2 - b)}, \tag{15.20}$$

where $Y(s)$ and $\tau(s)$ stand for the flat output and applied torque Laplace transforms respectively. Note that this transfer function has a negative gain, because $d(a +$

$bh) < 0$, and four real open-loop poles located at $s = 0$ (two of them), $s = -\sqrt{b} < 0$, $s = \sqrt{b} > 0$. Recall that b is a positive real number.

15.5   Parameter Identification

A photography of the experimental prototype that has been built is presented in Fig. 15.27. By direct measurement, the following simple mechanism parameters are found:

$$m_1 = 0.02218[\text{Kg}], \quad l_1 = 0.129[\text{m}], \quad L_0 = 0.155[\text{m}].$$

The joint between the pendulum and the arm is composed of an encoder and a solid cylinder $l = 0.019[\text{m}]$ in length, with a radius $R = 0.012/2[\text{m}]$ and a mass $m = 0.015[\text{Kg}]$. Inertia of this cylinder when rotating on its longitudinal axis is given as [3, 4], pp. 271:

$$I_C = \frac{1}{2}mR^2.$$

The pendulum is a stick with a mass m_1 and length $2l_1$. Inertia of this stick when rotating on an axis that is orthogonal to the stick, passing through its longitudinal center, is given as [3, 4], pp. 271:

$$I_1 = \frac{1}{12}m_1(2l_1)^2.$$

The pendulum inertia is computed by adding these inertias, i.e.,:

$$J_1 = I_C + I_1 = 0.0001845[\text{Kg m}^2].$$

The motor is a cylinder with a diameter $D = 0.05[\text{m}]$. It is assumed that the rotor is also a cylinder with a diameter equal to one half the motor diameter, i.e., the rotor radius is given as $r_r = \frac{1}{2}\frac{D}{2}$. The motor mass is $m_m = 0.6[\text{Kg}]$. The rotor mass is assumed to be one half the motor mass, and $0.005[\text{Kg}]$ of a screw joining the arm and the motor shaft must also be considered. Thus, the rotor inertia is computed as:

$$I_r = \frac{1}{2}\left(\frac{m_m}{2} + 0.005\right)r_r^2.$$

The arm is a stick with a length L_0 and mass $m_0 = 0.027[\text{Kg}]$ rotating on an axis that is orthogonal to the arm, passing through one of its ends. Hence, the inertia of this stick is [3]:

Table 15.1 Numerical values of a Furuta pendulum

Symbol	Description	Value	Units
g	Gravity constant	9.81	m/s^2
l_1	Distance to center of mass (pendulum)	0.129	m
L_0	Arm length	0.1550	m
m_1	Pendulum mass	0.02218	Kg
J_1	Pendulum inertia	0.0001845	Kg m^2
I_0	Arm inertia	0.00023849	Kg m^2

$$I_s = \frac{1}{3}m_0 L_0^2.$$

Thus, the arm inertia is computed as:

$$I_0 = I_r + I_s = 0.00023849 [\text{Kg m}^2].$$

The numerical parameter values of the Furuta pendulum that has been built are summarized in Table 15.1.

15.6   Design of a Stabilizing Controller

In this section, a controller is designed that is intended to regulate the Furuta pendulum variables at $x = x^*$ or, equivalently, at $z = 0$. This is performed on the basis of the approximate linearized model (15.17), (15.18). This implies that the controller is useful only if the initial conditions are close to the desired configuration, i.e., $x(0) \approx x^*$ or, equivalently, $z(0) \approx 0$. Thus, the controller designed in this section is not capable of taking the pendulum from the bottom stable position $x_3 = \pi$ to the inverted unstable position $x_3 = 0$ or $x_3 = 2\pi$. Such a task must be performed using the controller presented in (15.10). The interested reader is encouraged to review [2], chapter 6 (also see Sect. 15.2 in the present book).

It is known, from Sect. 7.11, that given the system (15.17), (15.18), it is possible to accomplish $\lim_{t \to \infty} z(t) = 0$ using the following state feedback controller:

$$\tau = -Kz, \quad K = [k_1, k_2, k_3, k_4],$$
$$= -k_1 z_1 - k_2 z_2 - k_3 z_3 - k_4 z_4. \tag{15.21}$$

This requires selection of the gain vector K such that all the eigenvalues of matrix $A - BK$ are suitably assigned. Of course, a basic requirement is that the real part of all of the eigenvalues of matrix $A - BK$ are negative. However, the eigenvalues of matrix $A - BK$ must also be assigned such that a desired closed-loop performance is accomplished. Moreover, sometimes it is required to choose a new gain vector

K to further improve the performance. Solving this problem is not an easy task in practice, where fast and damped eigenvalues are not the only things that matter because of noise and some other problems related to hardware implementation.

In this section, a methodology is introduced to select the vector gain K that exploits the differential flatness property of the system (15.17), (15.18). The main idea is that a linear state space differentially flat system can be written in terms of a transfer function without any zeros (see Sect. 15.4). Then, we can apply classical control design tools such as the root locus to select the controller gains. An advantage of this approach is that it gives the designer important information on how to select the controller gains to improve performance in a desired direction. This methodology is presented in detail in the following.

According to Sect. 15.4, the state space system (15.17), (15.18), is equivalent to the transfer function in (15.20). To control this open-loop unstable plant consider the block diagram in Fig. 15.4. Note that this is a multi-loop positive feedback system. The reason for positive feedback is the negative gain $d(a + bh) < 0$. As a matter of fact, the closed-loop transfer function is given as:

$$\frac{Y(s)}{Y_d(s)} = \frac{\beta d(a + bh)}{s^4 - d(a + bh)k_v s^3 - (b + \beta d(a + bh))s^2 - k_d \beta d(a + bh)s - k_p \beta d(a + bh)},$$
(15.22)

where $Y_d(s) = 0$. It is not difficult to realize that a negative $d(a + bh)$ renders it possible for all the coefficients of the characteristic polynomial to be positive, a necessary (but not sufficient) condition to ensure that all its roots have a negative real part.

The transfer function of the two internal loops in Fig. 15.4 is given as:

$$\frac{\ddot{Y}(s)}{\ddot{Y}_d(s)} = F(s) = \frac{\beta d(a + bh)}{s^2 - d(a + bh)k_v s - (b + \beta d(a + bh))}$$
(15.23)

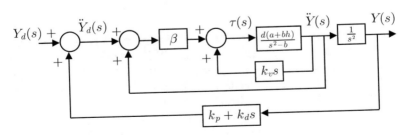

Fig. 15.4 Block diagram of the closed-loop control system

Fig. 15.5 Equivalent block
diagram to that in Fig. 15.4

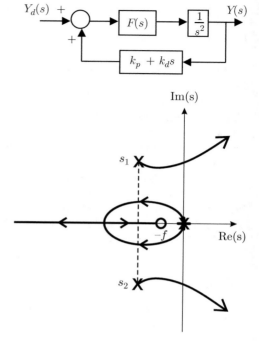

Fig. 15.6 Root locus for the
closed-loop control system in
Fig. 15.5

Hence, the block diagram in Fig. 15.4 can be rewritten as in Fig. 15.5. Using the
numerical parameters given in Table 15.1, the following is found:

$$d(a + bh) = -1.2186 \times 10^5, \quad b = 93.9951.$$

Taking into account these numerical values $k_v = 2.3755 \times 10^{-4}$ and $\beta = 0.0041$ are proposed. This selection of controller gains renders the transfer function
in (15.23) stable with poles assigned at the locations shown in Table 15.2a). Using:

$$\frac{\pm 180°(2q + 1)}{n - m}, \quad q = 0, 1, 2, \ldots,$$

where $n = 4$ and $m = 1$, it is found that the root locus diagram of the block diagram
in Fig. 15.5 has three asymptotes at $\pm 60°$ and $\pm 180°$ with respect to the positive
real axis. Thus, $f = k_p/k_d = 1$ is proposed to shape the root locus diagram as
indicated in Fig. 15.6. Note that the open-loop zero at $s = -f$ is not a closed-loop
zero as corroborated by (15.22). Choose $k_d = 2.88$ and compute $k_p = k_d f = 2.88$
to assign the poles of (15.22) as shown in Table 15.2a).

According to the block diagram in Fig. 15.4, using the expressions in (15.19) it
is found that the controller is given as:

$$\tau = k_v y^{(3)} + \beta(\ddot{y}_d + \ddot{y}), \tag{15.24}$$

Table 15.2 Controller gains and the assigned open-loop and closed-loop poles

	Controller gains	Poles of (15.23)	Poles of (15.22)
a)	$k_v = 2.3755 \times 10^{-4}$, $\beta = 0.0041$, $k_d = 2.88$, $k_p = 2.88$	$s_{1,2} = -14.4734 \pm 14.0048j$	$-12.1989 \pm 11.8679j$, -2.7284, -1.8206
b)	$k_v = 1.2 \times 2.3755 \times 10^{-4}$, $\beta = 1.4 \times 0.0041$, $k_d = 4.59$, $k_p = 4.59$	$s_{1,2} = -17.3681 \pm 17.4300j$	$-13.535 \pm 14.37j$, -6.3736, -1.2926
c)	$k_v = 1.4 \times 2.3755 \times 10^{-4}$, $\beta = 1.8 \times 0.0041$, $k_d = 5.62$, $k_p = 5.62$	$s_{1,2} = -20.2628 \pm 19.8676j$	$-15.5358 \pm 16.1280j$, -8.2292, -1.2247
d)	$k_v = 2 \times 2.3755 \times 10^{-4}$, $\beta = 2.8 \times 0.0041$, $k_d = 7$, $k_p = 7$	$s_{1,2} = -28.9468 \pm 21.6100j$	$-22.4899 \pm 14.2991j$, -11.7395, -1.1744

Table 15.3 State feedback controller gains

	Controller gains		Controller gains		Controller gains		Controller gains
a)	$k_1 = -0.0118$ $k_2 = -0.0118$ $k_3 = -0.2742$ $k_4 = -0.0298$	b)	$k_1 = -0.0263$ $k_2 = -0.0263$ $k_3 = -0.3962$ $k_4 = -0.0509$	c)	$k_1 = -0.0415$ $k_2 = -0.0415$ $k_3 = -0.5189$ $k_4 = -0.0728$	d)	$k_1 = -0.0804$ $k_2 = -0.0804$ $k_3 = -0.8269$ $k_4 = -0.1304$

$$= k_v y^{(3)} + \beta(k_p y + k_d \dot{y} + \ddot{y}),$$
$$= \beta k_p z_1 + \beta k_d z_2 + (\beta(a + bh) + \beta k_p h)z_3 + (k_v(a + bh) + \beta k_d h)z_4.$$

Comparing (15.21) and the last expression in (15.24), it is concluded that:

$$k_1 = -\beta k_p, \quad k_2 = -\beta k_d, \quad k_3 = -(\beta(a + bh) + \beta k_p h),$$
$$k_4 = -(k_v(a + bh) + \beta k_d h). \tag{15.25}$$

Using these relations and the numerical values in Table 15.1, the state feedback controller gains presented in Table 15.3a) are obtained. It has been also verified that the eigenvalues of matrix $A - BK$ are equal to the poles of (15.22) presented in Table 15.2a), as expected.

Some simulation results are presented in Fig. 15.7a (continuous line) when using the model (15.22), the controller gains in Table 15.2a), i.e., in Table 15.3a), and the numerical values in Table 15.1. All the initial conditions have been set to zero except for $y(0) = 1$[rad]. A time constant of about 1[s] is observed, which means

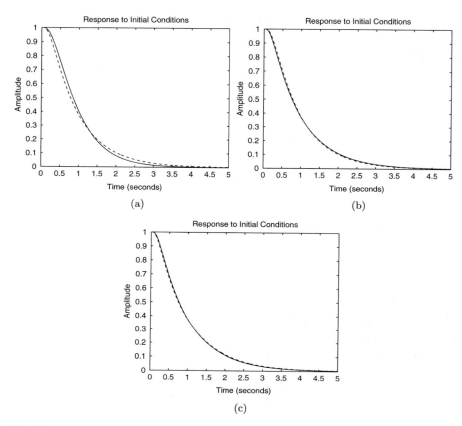

Fig. 15.7 Flat output y response obtained in simulation when using the controller (15.21) with gains in Table 15.3. (**a**) Continuous, Table 15.3a. Dashed, Table 15.3b. (**b**) Dashed, Table 15.3b. Continuous, Table 15.3c. (**c**) Continuous, Table 15.3c. Dashed, Table 15.3d

that the transient response is dominated by the slowest closed loop pole located at $s = -1.8206$.

Some other simulation results are shown in Fig. 15.8 when using the controller in (15.10) to swing up the pendulum and the controller in (15.21) to stabilize the pendulum at the inverted configuration. The Furuta pendulum parameters that have been considered are those presented in Table 15.1. The controller gains for (15.21) are shown in Fig. 15.3a) whereas the controller gains for (15.10) are $K_E = 1900$, $k_\omega = 1$, $k_\delta = 3$, $k_\theta = 3$. All initial conditions were set to zero except for $\theta_1(0) = 2.9[\text{rad}]$. The switching condition between the controllers is the following:

if: $\qquad \sqrt{3(\theta_1 - \theta_{1d})^2 + 0.1\dot{\theta}_1^2} < 0.2, \quad$ then the controller in (15.21) is on,

else: $\qquad$ controller in (15.10) is on.

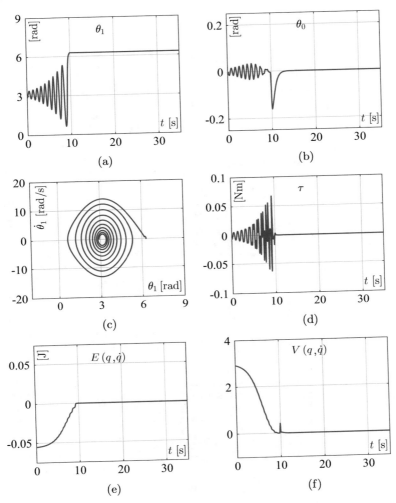

Fig. 15.8 Simulation results when swinging up and stabilizing the Furuta pendulum with the controllers in (15.10) and (15.21)

It is observed that the pendulum reaches the inverted unstable configuration whereas the arm position and the applied torque converge to zero, as desired. Also, the energy in (15.10) and the function V in (15.12) converge to zero. Finally, the phase plane plot in Fig. 15.8c clearly shows that the pendulum converges to the desired inverted configuration with zero velocity.

These simulations were performed using the MATLAB/Simulink diagram shown in Fig. 15.9 where all To work space blocks at the bottom save data in an array with sampling time −1. Four main blocks are shown that contain the code shown

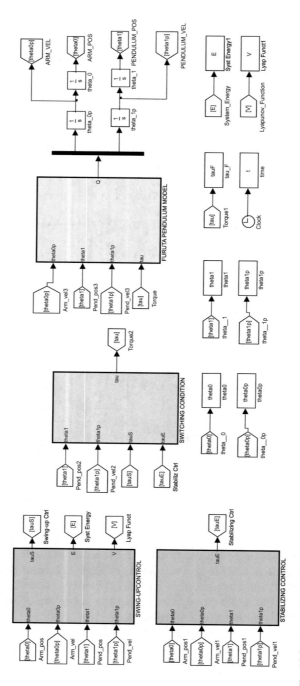

Fig. 15.9 MATLAB/Simulink diagram for simulations in Fig. 15.8

in Appendix G. Once the simulation in Fig. 15.9 is run, the MATLAB code in Sect. G.1.5 is executed in an m-file to obtain the plots in Fig. 15.8.

On the other hand, in Sect. 8.3, Chap. 8, it has been explained that a limit cycle is expected to appear in practice when using the above controller together with the controller gains in Table 15.2a). This requires the block diagram in Fig. 15.4 to be modified to that in Fig. 8.7 where a dead zone "static" nonlinearity is considered. Moreover, it has been shown that the closed-loop system can also be represented as in Fig. 8.8 where:

$$G(s) = \frac{-d(a + bh)(k_v s^3 + \beta s^2 + \beta k_d s + \beta k_p)}{s^2(s^2 - b)}, \quad -d(a + bh) > 0,$$

is defined. Furthermore, it is shown in Sect. 15.7 that a limit cycle actually appears in experiments. Recall that a dead zone nonlinearity is induced in servomechanisms because of static friction at the motor shaft. As is presented in Sect. 15.7, in these experiments it is observed that this limit cycle induces wide oscillations of the Furuta pendulum arm. This motivates new designs intended to reduce the amplitude of the oscillations. Moreover, the observed oscillations are also slow. This suggests that the controller must increase the stiffness of the closed-loop system. As stiffness is related to faster responses, this implies that the oscillation frequency must be increased.

According to Sect. 8.3, to obtain limit cycles with smaller amplitudes the polar plot of the equivalent open-loop transfer function $G(s)$ must intersect the negative real axis at a point σ located farther to the left. On the other hand, to increase the frequency of the oscillations, the phase of $G(j\omega)$ must reach $-180°$ at a greater frequency $\omega = \omega_\sigma$.

The point σ can be placed farther to the left of the negative real axis by increasing the magnitude of $G(s)$. According to the previous expression, this requires the controller magnitude to be increased, i.e., to increase the controller gains k_v and β. On the other hand, according to Fig. 8.9, $G(j\omega)$ reaches a $-180°$ phase at greater frequencies if the phase lead contributed by the controller is forced to appear at greater frequencies. As the cubic and the quadratic terms in $G(s)$ contribute larger phase leads at greater frequencies, we can force this phase lead to appear at greater frequencies by rendering the coefficients of the first-order term and the constant term larger than the coefficients of the cubic and the quadratic terms, i.e., by increasing k_d and k_p.

Based on these ideas, additional sets of controller gains to be tested experimentally are obtained in the following. Note, however, that these designs must ensure closed-loop stability. Thus, we proceed using the root locus method employing Fig. 15.6.

Our design strategy is to move the closed-loop poles farther to the left of the imaginary axis. This suggests moving the open-loop zero at $s = -f$ to the left. However, this may create two root locus branches starting at the origin but belonging to the right half-plane (see Fig. 15.6). This is because the zero at $s = -f$ may attract the open-loop complex conjugate poles at s_1, s_2, if it moves to the left. Hence, instability or badly damped closed-loop poles may result if $f > 0$ is increased. Thus, it is proposed only to move the three fastest closed-loop poles to the left. First, propose larger values for k_v and β to render both the real and the imaginary parts of the open loop poles s_1 and s_2 in Fig. 15.6 larger (also see $s_{1,2}$ in Table 15.2). This renders the pendulum response faster and well damped,[1] and allows larger values for both k_d and k_p to be selected. This is because, although the root locus branches starting at s_1 and s_2 move toward the right half-plane as k_d increases, it is possible to select larger values for k_d before these branches are too close to the right half-plane if s_1 and s_2 move to the left. Note that k_p is larger if k_d is larger because $k_p = k_d f$.

This is the procedure that has been followed to find the controller gains in Table 15.2b), c) and d). Using these values and (15.25), the controller gains in Table 15.3b), c), and d), have been computed. Note that the three fastest closed loop poles move to the left, going from a) to d) in Table 15.2 or, equivalently, in Table 15.3. Finally, recall that it has been shown in Figs. 8.9 and 8.10 that the phase of $G(j\omega)$ reaches $-180°$ at a greater frequency and the plot of $G(s)$ intersects the negative real axis at a point σ, which moves to the left, going from a) to d) in Tables 15.2 and 15.3, i.e., in Table 8.1.

In Figs. 15.7a, b, and c the flat output responses y are compared when using the controller gains in Table 15.3a), b), c), and d). In these simulations, all the initial conditions have been set to zero except for $y(0) = 1$. We have also considered that the dead zone nonlinearity is not present. This has been done merely to verify that the response is slightly faster going from a) to d) in Table 15.3, as expected. However, this is true only at the beginning of the response. We stress that larger settling times are obtained because the slowest closed-loop pole in Tables 15.2b), c), and d), moves to the right.

Finally, recall that, according to Sect. 6.5.2, a time delay T is induced when implementing a continuous time controller with a digital computer. For the experiments that are presented next, $T = 0.01[s]$ may be assumed to consider the worst case time delay as $0.01[s]$ is the sampling period. This time delay does not affect the magnitude of the open-loop frequency response, but contributes an additional phase lag given as $-\omega T[rad]$. This is because e^{-Ts} appears as a factor of the open-loop transfer function $G(s)$ plotted in Fig. 8.9. In that figure, for controller gains in Table 15.3a), the crossover frequency is $\omega = 29[rad/s]$. Hence, the additional phase lag is $-\omega T \times 180°/\pi = -16.6°$. Thus, the phase margin in Fig. 8.9 shifts from $58°$ to $41.39°$, which is still a good phase margin. This means that the time delay induced in practice by the digital implementation of the controller does not have a significant effect when using the controller gains in Table 15.3a).

---

[1]Note, in Fig. 15.4, that k_v and β are the feedback gains of signals $\ddot{y}$ and $y^{(3)}$, which, according to (15.19), represent feedback of the pendulum position and velocity, $z_3 = \theta_1 - \theta_1^*$ and $z_4 = \dot{\theta}_1$.

15.7   Experimental Tests

In Fig. 15.10 the experimental results obtained when the controller (15.10) (also see
[2], chapter 6) is used to swing up the pendulum and the controller (15.21) is used to
stabilize the mechanism at $z = x - x^* = 0$ are presented. The numerical parameters
$K_E = 480$, $k_\omega = 1$, $k_\delta = 3$, $k_\theta = 17$, were employed for controller (15.10) whereas
the gains in Table 15.3a) were used for controller (15.21). Note that the desired
pendulum position $x_3^* = \theta_1^* = 2\pi$ is reached and the pendulum remains there for
$t \geq 8.8$[s]. However, the arm position does not reach the desired position $x_1^* =
\theta_0^* = 0$ but oscillates around a nonzero arm position. To better observe this behavior
and to better compare with cases when different controller gains are employed, in
Fig. 15.11 a zoom-in on the variables θ_0 and $\theta_1 - 2\pi$ is presented. Moreover, the
time axis has been modified to consider $t = 0$ at the time when the stabilizing
controller (15.21) is switched on, i.e., when $t = 8.8$[s] in Fig. 15.10. Hence, it is
concluded that the pendulum really remains at $\theta_1 = 2\pi$ whereas the arm describes
large peak-to-peak oscillations of about 0.6[rad] around a constant position of about
-0.8[rad].

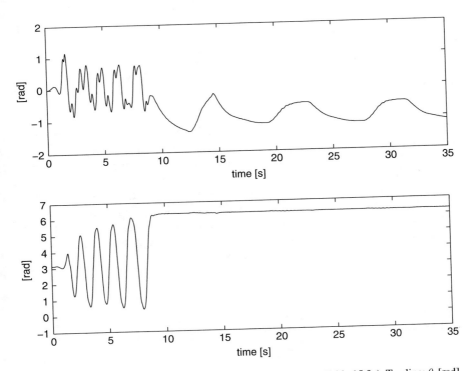

Fig. 15.10 Experimental results when using the controller gains in Table 15.3a). Top line: θ_0[rad].
Bottom line: θ_1[rad]

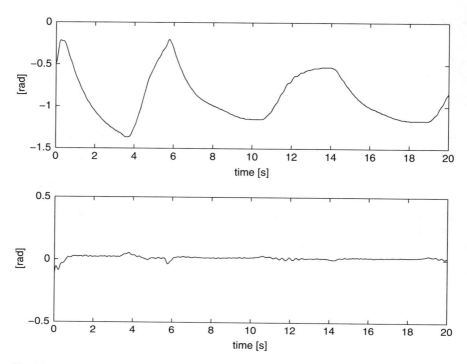

Fig. 15.11 Experimental results when using the controller gains in Table 15.3a). Top line: θ_0[rad]. Bottom line: $\theta_1 - 2\pi$ [rad]

In Fig. 15.12 the flat output $y = z_1 + hz_3$ response[2] is presented, in addition to the response of its components z_1 and hz_3. Note that the time axis in this figure is the same as in Fig. 15.11. It is observed that y and z_1 remain close to each other while oscillating because hz_3 is close to zero. These observed oscillations represent a limit cycle that has been predicted to exist because of a dead zone induced by static friction at the motor shaft. Furthermore, the static friction is also responsible for the large constant position value of -0.8[rad] around which arm oscillations are performed. We wonder, however, why oscillations are not present in the pendulum position. To explain this, consider (15.19) where:

$$y = z_1 + hz_3, \quad \ddot{y} = (a + bh)z_3,$$

and assume that the flat output oscillation is sinusoidal, i.e., $y = Y_0 \sin(\omega t)$ where Y_0 is a positive constant representing the oscillation amplitude. It is not difficult to find that, under these conditions:

---

[2]Recall that $z_1 = \theta_0$ and $z_3 = \theta_1 - 2\pi$ in this case.

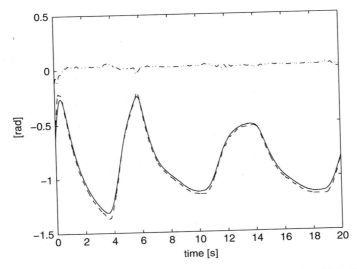

Fig. 15.12 Experimental results when using the controller gains in Table 15.3a). Continuous: y[rad]. Dashed: z_1[rad]. Dash—dot: hz_3[rad]

$$z_1 = Y_0 \left(1 + \frac{h\omega^2}{a + bh} \right) \sin(\omega t), \quad z_3 = -Y_0 \frac{\omega^2}{a + bh} \sin(\omega t). \quad (15.26)$$

This means that z_3 is very small if the oscillation frequency is small and, hence, $y \approx z_1$. Note that, in Fig. 15.12, the oscillation frequency is, approximately, $\omega = \frac{2\pi}{8.5} = 0.7392$[rad/s], which yields $\frac{\omega^2}{a+bh} = 0.0086$. This explains why oscillations are not observed in the pendulum position and why z_1 remains close to y in the above experiments.

Finally, in Fig. 15.13 the phase plot of the pendulum variables during the swinging-up stage of the experiment is presented. Note that the experiment begins at $(\theta_1, \dot{\theta}_1) = (\pi, 0)$ and finishes at $(\theta_1, \dot{\theta}_1) = (2\pi, 0)$. The theoretical trajectory described in Fig. 15.3 is also presented. As explained in Sect. 15.2, the controller in (15.10) is intended to force the pendulum variables to converge to the trajectory in Fig. 15.3. As observed in Fig. 15.13, this is almost accomplished and, because of that, the pendulum swings up successfully. Note, however, that convergence to the trajectory in Fig. 15.3 is not accomplished asymptotically because of the imperfections in control system implementation. Such system implementation imperfections involve the fact that the power electronics stage in our experimental prototype is not capable of supplying the electric current required by the controller (15.10) to the DC motor used as actuator. Moreover, it is important to say that this has forced us to multiply by a factor of 5.8 the torque computed with the controller in (15.10)

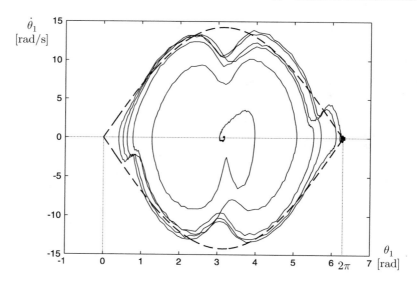

Fig. 15.13 Phase plot of the pendulum variables during the swinging-up stage of the experiment (continuous). The trajectory presented in Fig. 15.3 is also shown here (dashed)

to obtain the value of torque to be commanded to the DC motor actuator.[3] This is explicitly expressed in the Builder 6 C++ program code presented in Sect. 15.10. The use of the factor 5.8 was chosen empirically after several tests.

In Figs. 15.14, 15.15, 15.16 the experimental results obtained when using the controller gains in Table 15.3b) are presented. In Figs. 15.17, 15.18, 15.19 the experimental results obtained when using the controller gains in Table 15.3c) are presented. These results are plotted on identical axes as in Figs. 15.10, 15.11, 15.12 to render the comparison between the results obtained when different controller gains are used fair. $t = 0$ is set in Figs. 15.15 and 15.16 at the point of time in Fig. 15.14 where $t = 13.8$[s]. Also, $t = 0$ is set in Figs. 15.18 and 15.19 at the point of time in Fig. 15.17 where $t = 12.9$[s].

In all the above results it is observed that the pendulum position is successfully stabilized at $\theta_1 = 2\pi$. When the controller gains in Table 15.3b) are used, the following is concluded. Although the arm position still oscillates, the amplitude is smaller and the frequency $\omega = 2\pi/5 = 1.2566$[rad/s] is greater than values in Figs. 15.10, 15.11, 15.12. Note that these results verify the theoretical predictions that have motivated the redesign of the controller gains. It is also observed that the oscillations are performed around a constant arm position of about -0.35[rad], which is smaller than in the case of the controller gains in Table 15.3a). Note that

---

[3]Torque computed by the stabilizing controller (15.21) is not multiplied by any factor different from unity.

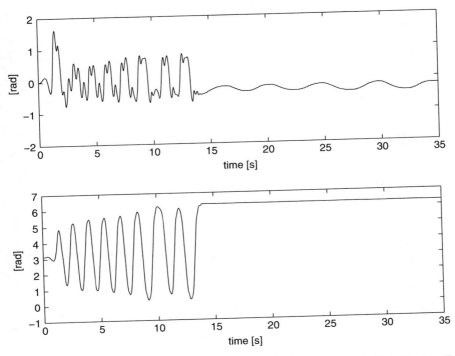

Fig. 15.14 Experimental results when using the controller gains in Table 15.3b). Top line: θ_0 [rad]. Bottom line: θ_1 [rad]

the above values yield $\frac{\omega^2}{a+bh} = 0.0250$ which explains, again, why oscillations do not appear in the pendulum position.

The results obtained when using the controller gains in Table 15.3c) show that the limit cycle disappears. As can be seen in Figs. 15.17, 15.18, 15.19, the oscillation amplitude in both the flat output y and the arm position θ_0 are even smaller, whereas the oscillation frequency has increased to about $\omega = 2\pi/2.3 = 2.7318$ [rad/s]. Furthermore, both these variables reach constant values, i.e., oscillation disappears, for $t \geq 7$ [s] in Figs. 15.18 and 15.19. Note that the pendulum position reaches its desired constant value. The constant steady-state error in the arm position is about 0.2 [rad], i.e., even smaller than the mean error of -0.35 [rad] for controller gains in Table 15.3b).

Note the following. The small threshold δ in a dead zone nonlinearity (see Sect. 8.3) is uncertain and changes during normal operation because static friction also has these properties. As no movement is produced when the generated torque is smaller than δ (or the applied voltage is small enough), we may wonder whether limit cycles might be avoided by forcing them to appear only at a very small amplitudes. Hence, this may explain why limit cycle has disappeared.

It is important to say, however, that limit cycle avoidance is not achieved every time an experiment is performed with the controller gains in Table 15.3c).

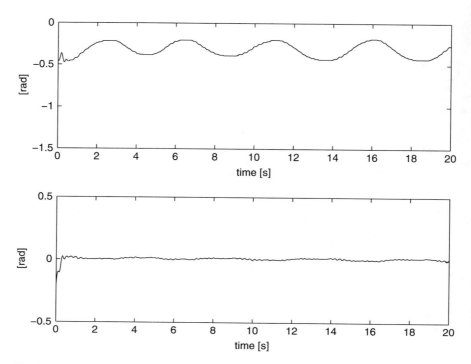

Fig. 15.15 Experimental results when using the controller gains in Table 15.3b). Top line: θ_0[rad]. Bottom line: $\theta_1 - 2\pi$ [rad]

Sometimes very small and very slow oscillations are observed on both z_1 and y whereas z_3 remains at zero with no oscillations. Thus, limit cycle avoidance appears as a random event, perhaps because, as stated earlier, the small threshold δ is uncertain and changes during normal operation.

In Figs. 15.20, 15.21, 15.22, the experimental results obtained when using the controller gains in Table 15.3d) are presented. $t = 0$ is set in Figs. 15.21 and 15.22 at the point of time in Fig. 15.20 where $t = 3.95$[s]. Note that, again, a limit cycle is not present, i.e., the pendulum and arm positions in addition to the flat output, reach constant values in a steady state after the stabilizing controller (15.21) is switched on. It is also observed that the pendulum reaches the desired position $\theta_1^* = 2\pi$ and the arm remains closer to its desired position, $\theta_0^* = 0$, than in the case where the controller gains in Table 15.3c) were used. Note that the oscillation frequency is difficult to measure in this case, but comparing the settling times in Figs. 15.21 and 15.18 we conclude that the frequency is greater for the controller gains in Table 15.3d).

Recall that, according to theoretical arguments, the controller gains in Table 15.3d) are expected to result in even smaller amplitude oscillations than when the controller gains in Table 15.3c) are used. On the other hand, it has been

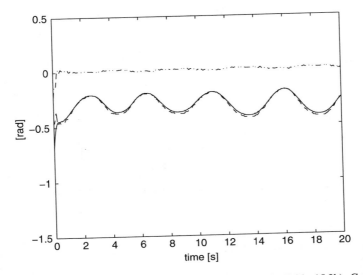

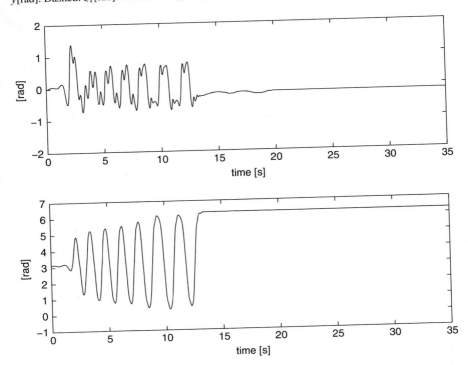

Fig. 15.16 Experimental results when using the controller gains in Table 15.3b). Continuous: y[rad]. Dashed: z_1[rad]. Dashdot: hz_3[rad]

Fig. 15.17 Experimental results when using the controller gains in Table 15.3c). Top line: θ_0[rad]. Bottom line: θ_1[rad]

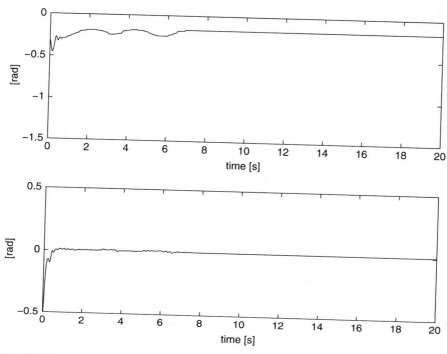

Fig. 15.18 Experimental results when using the controller gains in Table 15.3c). Top line: θ_0[rad].
Bottom line: $\theta_1 - 2\pi$ [rad]

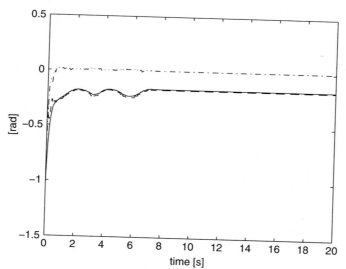

Fig. 15.19 Experimental results when using the controller gains in Table 15.3c). Continuous:
y[rad]. Dashed: z_1 [rad]. Dash–dot: hz_3 [rad]

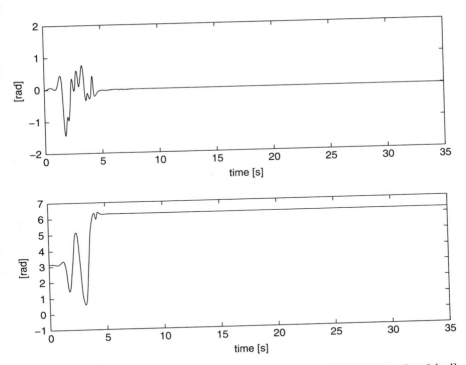

Fig. 15.20 Experimental results when using the controller gains in Table 15.3d). Top line: θ_0[rad]. Bottom line: θ_1[rad]

previously explained that forcing, by design, limit cycles to appear only for small amplitudes may result in limit cycle avoidance. This, again, explains why a limit cycle is not present for when the controller gains in Table 15.3d) are used.

The Furuta pendulum prototype used to perform all the above experiments is equipped with an internal electric current loop, driven by a linear proportional–integral (PI) controller, whose task is to force the electric current i flowing through the DC motor used as an actuator to reach a desired value i_d. This desired value is computed as $i_d = \tau_d/k_m$, where k_m is the torque constant of the motor and τ_d is torque computed by either (15.10) or (15.21). As the motor generates an electromagnetic torque given as $\tau = k_m i$, if i is equal to i_d, then $\tau = \tau_d$, which means that the motor generates exactly the torque required by both the swinging-up and the stabilizing controllers. Achieving $i = i_d$ is the task of the linear PI electric current controller cited above, which requires careful design of the gains k_{pi} and k_{ii} for this PI controller. In all the above experiments we have used $k_{pi} = 0.8$ and $k_{ii} = 130$.

In the following, an example is presented to show how the controller gains k_{pi} and k_{ii} may affect the closed-loop system performance. In Figs. 15.23, 15.24, 15.25 some experimental results are presented when using the controller gains in

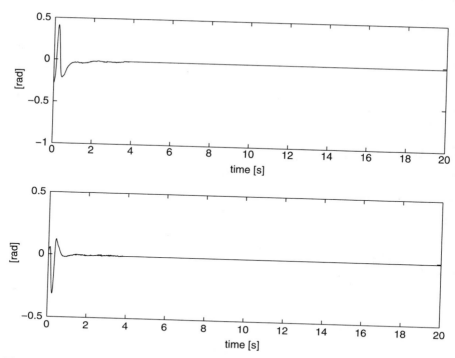

Fig. 15.21 Experimental results when using the controller gains in Table 15.3d). Top line: θ_0[rad]. Bottom line: $\theta_1 - 2\pi$ [rad]

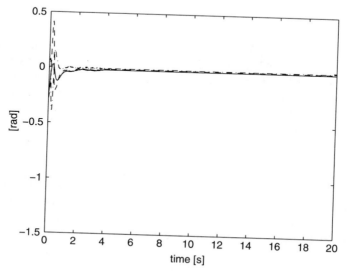

Fig. 15.22 Experimental results when using the controller gains in Table 15.3d). Continuous: y[rad]. Dashed: z_1[rad]. Dash-dot: hz_3[rad]

Table 15.3d) and $k_{pi} = 0.03$ and $k_{ii} = 130$. $t = 0$ is set in Figs. 15.24 and 15.25 at the point of time in Fig. 15.23 where $t = 9[s]$. Note that all three variables, the pendulum and the arm positions in addition to the flat output, are oscillatory the whole time that the stabilizing controller (15.21) is switched on. Despite this, the pendulum remains around the inverted position $\theta_1^* = 0$ and the arm remains close to its desired position $\theta_0^* = 0$. Note that the oscillation frequency is about $\omega = 2\pi/0.15 = 41.8879[rad/s]$, which yields $\frac{\omega^2}{a+bh} = 27.7230 \gg 1$. This and (15.26) explain why the amplitude of oscillations in the pendulum position z_3, or hz_3, is much larger that of the oscillation in the flat output y; hence, the oscillation in the arm position z_1 has the same amplitude as that in hz_3. Moreover, it is clear from Fig. 15.25 that the oscillation in hz_3 is $180°$ out of phase with respect to the oscillation in z_1. This is also explained by (15.26). Finally, note, from Fig. 15.25, that the settling time of y is very close to that in Fig. 15.7c, which has been obtained through simulations.

Despite these spectacular experimental corroborations of theoretical predictions, we wonder why a limit cycle exists. Recall that according to theoretical arguments, the controller gains in Table 15.3d) are expected to result in even smaller amplitude oscillations than when the controller gains in Table 15.3c) are used. Thus, the reader may wonder why a clear limit cycle appears again for the gains in Table 15.3d) despite limit cycles being avoided for the gains in Table 15.3c).

In Fig. 15.26, the measured electric current through the motor i and the desired current i_d computed as $i_d = \tau_d/k_m$ are presented. Note that i has a phase lag of $90°$ with respect to i_d for $t \geq 9[s]$, i.e., for the whole time when the linear stabilizing controller (15.21) is switched on. According to Fig. 8.9, an additional phase lag of $90°$ appearing at any place in the loop of Fig. 8.8 is enough to produce closed-loop instability. According to Fig. 15.26 such a phase lag is contributed by the DC motor electric dynamics, which has been neglected during the design stage. Moreover, it can be observed in Fig. 15.26 that the amplitude of i increases until a maximum amplitude is reached. This shows instability despite the fact that the electric current amplitude does not grow to infinity because, as we have explained before, it saturates, as the power electronics stage of our prototype can only supply a limited maximal electric current.

In Sect. 8.3.4 a limit cycle study is presented when a saturation nonlinearity is present and the electric dynamics of the DC motor used as actuator is taken into account. There, the corresponding closed-loop block diagram is that shown in Fig. 8.15. The plant model and the parameters considered in that study correspond to the Furuta pendulum with parameters in Table 15.1 and the controller gains considered in this part of the present chapter, i.e.,:

$$k_v = 2 \times 2.3755 \times 10^{-4}, \quad \beta = 2.8 \times 0.0041, \quad k_d = 7, \quad k_p = 7,$$

Moreover, in Sect. 8.3.4, the following motor parameters are also considered:

$$k_m = k_b = 0.0368, \quad R = 2.4[Ohm], \quad k_{ii} = 130 \quad L = 0.03[H],$$

which correspond to the DC motor actually used in the Furuta pendulum prototype
that has been employed to perform the experiments reported in the present chapter.

In Sect. 8.3.4 it is found that a stable limit cycle exists when $k_{pi} = 0.03$.
Moreover, the closed-loop system is unstable but the limit cycle converts such
instability into a sustained oscillation, i.e., a stable limit cycle. This limit cycle is
predicted to appear at a frequency $\omega \approx 48$[rad/s], i.e., approximately the same
frequency measured in Figs. 15.23, 15.24, 15.25. Thus, these experimental results
are formally explained by the study on limit cycles performed in Sect. 8.3.4.

To solve this problem k_{pi} has been increased from 0.03 to $k_{pi} = 0.8$ and $k_{ii} = 130$ is kept. The reason for this is that k_{pi} contributes with an open-loop zero located
at $s = -k_{ii}/k_{pi}$. Hence, if k_{pi} is larger, this open-loop zero increases the phase lead
for any given frequency, thus reducing the phase lag contributed by the DC motor
electric dynamics and restoring closed-loop stability.

In Sect. 8.3.4, it is also found that any stable limit cycle due to saturation does
not exist when $k_{pi} = 0.8$. Moreover, the closed-loop system is stable in this case.
Thus, the selection of $k_{pi} = 0.8$ is formally justified by the study on limit cycles
performed in Sect. 8.3.4.

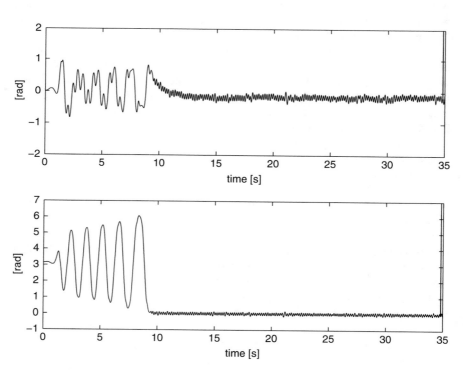

Fig. 15.23 Experimental results when using the controller gains in Table 15.3d). Top line: θ_0[rad].
Bottom line: θ_1[rad]

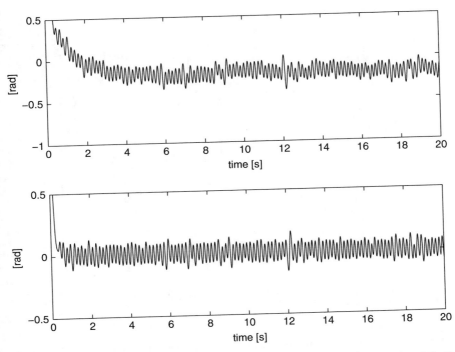

Fig. 15.24 Experimental results when using the controller gains in Table 15.3d). Top line: θ_0[rad]. Bottom line: θ_1[rad]

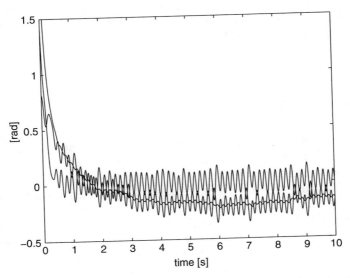

Fig. 15.25 Experimental results when using the controller gains in Table 15.3d). Smallest amplitude oscillatory line: y[rad]. Lower oscillatory line: z_1[rad]. Upper oscillatory line: hz_3[rad]

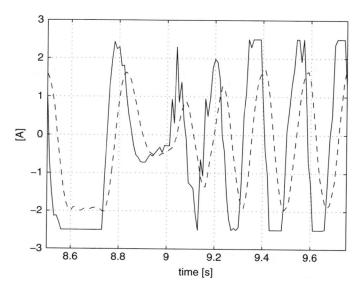

Fig. 15.26 Experimental results when using the controller gains in Table 15.3d) and $k_{pi} = 0.03$ and $k_{ii} = 130$. Continuous: i_d[A]. Dashed: i[A]

Finally, let us say that in a first series of experiments $k_{pi} = 0.03$ and $k_{ii} = 130$ have been used. Although we have found good results for the controller gains in Table 15.3a),b),c), we have found the problem described in Figs. 15.23, 15.24, 15.25 for the controller gains in Table 15.3d). Once the problem was understood, the PI electric current controller gains have been adjusted to $k_{pi} = 0.8$ and $k_{ii} = 130$ and all the experiments were repeated using these gains. These final experimental tests are those that we have presented above.

15.8   Control System Construction

A Furuta pendulum has been built whose numerical parameters are shown in Table 15.1. A picture of this Furuta pendulum is presented in Fig. 15.27 and the electric diagram of the complete control system is depicted in Fig. 15.28. The control algorithm is implemented in a Builder 6 C++ program, which is executed by a portable computer. The portable computer receives the mechanism output data from a PIC16F877A microcontroller [7] and sends the mechanism input data back to the microcontroller. Then, the microcontroller sends the corresponding voltage signal to the motor. The Builder 6 C++ program executed by the portable computer and the C program executed by the microcontroller are listed in Sects. 15.10 and 15.11 respectively.

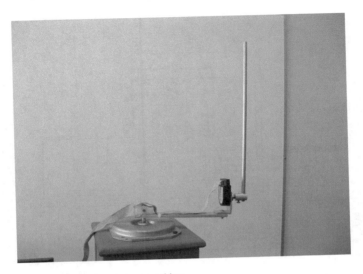

Fig. 15.27 The experimental prototype working

The L298 integrated circuit contains two full-bridge drivers that are parallel connected to be used as a power amplifier. The power signal is applied on a PM brushed DC motor model Tohoko Ricoh 7K00011, which is employed as the mechanism actuator. This motor is provided with a 400 pulses/revolution encoder, which is used to measure the arm position. The pendulum position is measured by a 1000 pulses/revolution encoder model S1-1000-250-I-B-D from USDigital (the bottom encoder in Fig. 15.28). Communication between the portable computer and the microcontroller is rendered possible by a MAX232. Finally, three AND logic gates manage the enable bits for the L298 full-bridge. The complete control system is put to work through SWITCH_1.

The mechanism is provided with an electric current loop with a PI controller driven by the difference $i_d - i$, where i is the electric current through the motor armature and i_d is the desired current. The desired current is computed as $i_d = \tau / k_m$, where τ is obtained from either (15.10) or (15.21) and $k_m = 0.0368[\text{Nm/A}]$. This ensures that $i \approx i_d$ for all time; hence, torque generated by the motor is approximately equal to the desired torque given by either (15.10) or (15.21). The electric current through the motor armature i is converted to a voltage signal by a $1[\Omega]$, $5[\text{W}]$, power resistance, which is series connected to the motor armature terminals. This voltage is measured by the differential voltage amplifier located at the left upper corner in Fig. 15.28. After that, this signal is suitably amplified and an adder amplifier is used to center the zero current value at a $+2.5[\text{V}]$ level. Then, the measured current is low-pass filtered and sent to a 10-bit analog/digital converter in the microcontroller whose input analog range is $[0, +5][\text{V}]$. This allows the electric current to be measured in the range $[-2.5, +2.5][\text{A}]$. Recall that the $1[\Omega]$ power

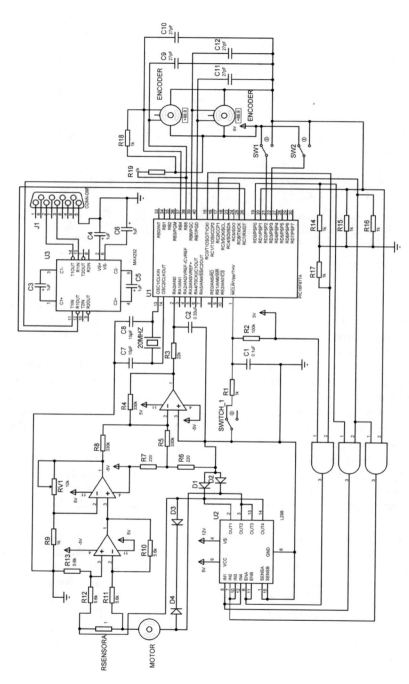

Fig. 15.28 Electric diagram of the Furuta pendulum control system

resistance converts the electric current into a voltage signal whose numerical value is equal to the electric current in amperes.

Some of the results presented in this chapter were reported for the first time in [6].

15.9 Sampling Period Selection

According to Sect. 8.4, Chap. 8, the selection of a suitable available bandwidth Ω_a is an important step when designing a controller for an unstable plant. Recall that, according to (15.20), the Furuta pendulum is an unstable plant with a single, real, unstable pole at $s = \sqrt{b} > 0$, where:

$$b = \frac{(I_0 + m_1 L_0^2)m_1 l_1 g}{I_0(J_1 + m_1 l_1^2) + J_1 m_1 L_0^2}. \tag{15.27}$$

According to (8.31), if a constant minimal sensitivity S_{min} is achieved for all $\omega \in [0, \Omega_a]$, then the relation between S_{min}, Ω_a and the unstable pole is given as:

$$S_{min} = e^{\frac{\pi}{\Omega_a}\sqrt{b}} > 1. \tag{15.28}$$

As explained in [5], the most severe limitation of the size of the available bandwidth is the sampling period T_s. It is also suggested in [5] that the sampling frequency ω_s might be selected such that:

$$\frac{\omega_s}{\Omega_a} = 16, \tag{15.29}$$

because this allows sampling two or three times in one radian.

On the other hand, if it is assumed that a specific S_{min} is achieved for all $\omega \in [0, \Omega_a]$, the corresponding phase margin K_f is obtained as follows. According to Sect. 8.4, the distance from the polar plot of $G(j\omega)H(j\omega)$ to the critical point $(-1, j0)$ is computed as $r = 1/S_{min}$. Hence, define a circle $y^2 + (x + 1)^2 = r^2$ as the plot of $G(j\omega)H(j\omega)$ whose distance to $(-1, j0)$ is constant and equal to $r = 1/S_{min}$. To correctly measure the phase margin, we need to find the point where this circle intersects the circle $y^2 + x^2 = 1$, i.e., where $|G(j\omega)H(j\omega)| = 1$. Simple algebra shows that the intersection point is given as:

$$x = \frac{r^2 - 2}{2}, \quad y = -\sqrt{1 - x^2}, \tag{15.30}$$

The phase margin is then computed as:

Table 15.4 Effect of sampling period on the achievable phase margin

T_s [s]	$\omega_s = \frac{2\pi}{T_s}$ [rad/s]	Ω_a [rad/s]	S_{min}	$r = \frac{1}{S_{min}}$	K_f [°]
0.005	1256.6	78.5398	1.4737	0.6785	39.6651
0.01	628.3185	39.2699	2.1719	0.4604	26.6191
0.02	314.1593	19.6350	4.7172	0.2120	12.1690
0.04	157.0796	9.8175	22.2521	0.0449	2.5751

$$K_f = \arctan\left(\frac{y}{x}\right). \tag{15.31}$$

The numerical values that have been tested and the results obtained using (15.27)–(15.31) are presented in Table 15.4. The numerical values in Table 15.1 have also been taken into account to compute $\sqrt{b}$. Phase margins obtained with 5[ms] and 10[ms] sampling periods are considered to be acceptable. Moreover, recall that these phase margins may be rendered larger as the sensitivity can be designed to be smaller, at some specific frequencies, than S_{min} in Table 15.4. Of course, the price for this is that the sensitivity would be larger at some other frequencies, as explained in Sect. 8.4. Finally, because of hardware limitations, a sampling period smaller than 10[ms] has not been accomplished. Thus:

$$T_s = 10[\text{ms}]$$

is the sampling period used to control the Furuta pendulum that has been built.

15.10   The Builder 6 C++ Program

In this section, the Builder 6 C++ program executed by the portable computer is presented. In the following, some of the most important parts of this computer program are explained. The corresponding flow diagram is presented in Fig. 15.29.

First, some important constants, such as the mechanism parameters and the controller gains, are defined. The control cycle is contained in the instruction "void ProcessByte(BYTE byte)". At the beginning of the control cycle, the arm position is first read, then the pendulum position is obtained, and, finally, the electric current is read. These variables are converted into radians and amperes respectively. Note that a 2000 constant value is added to the pendulum encoder count. This is done to assign $\theta_1(0) = \pi$ when the pendulum is at its bottom stable position, i.e., the pendulum's initial position. The inverted pendulum position "theta1d" is assigned to be either 0[rad] or 2π [rad], depending on which of these values is more suitable. The condition to switch on the stabilizing controller (15.21) is set as $3(\theta_1 - \text{theta1d})^2 + \dot{\theta}_1^2 < 0.05$. If this is not the case then the controller (15.10) is kept working. Once this decision is taken, the torque to be applied is computed by the corresponding controller. In the case of the swinging-up controller, torque that must

Fig. 15.29 Flow diagram for
PC programming when
controlling a Furuta
pendulum

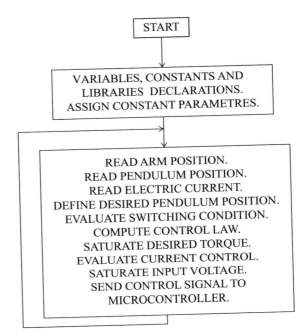

be generated by the DC motor actuator is obtained multiplying by 5.8 the torque
computed with the controller in (15.10). This is done to heuristically compensate
for the fact that the power electronics stage cannot supply enough electric current,
as explained in Sect. 15.7.

The desired electric current is then obtained in terms of torque to be generated.
After that, the voltage to be applied to the motor is computed as the output of a PI
electric current controller driven by the electric current error. Then, this voltage
is restricted to the range $[-10, +10][V]$. The voltage information is sent to the
microcontroller PIC16F877A as an 8-bit word where the highest bit represents
the sign and the lowest 7 bits contain the magnitude information. Some data are
saved in a file to be plotted when the experiment has finished. Then, the program
returns to the beginning of the instruction "void ProcessByte(BYTE byte)" where
it remains waiting until new data on the arm position are received. This means that
the sampling period is actually established by the microcontroller, as explained in
the next section.

```
//Program for control of a Furuta pendulum
//------------------------------------------------------------
#include <vcl.h>
#pragma hdrstop
#include "Main.h"
#include <math.h>
#include <stdio.h>
#include <share.h>
#include <conio.h>
```

```
#include <dos.h>
//----------------------------------------------------------
#pragma package(smart_init)
#pragma link "CSPIN"
#pragma resource "*.dfm"
//----------------------------------------------------------
#define Ts 0.01
#define pprb 400.0
#define pprp 1000.0
#define pi_ 3.141593
#define kpi 0.8 // 0.03
#define kii 130.0
#define bi -2.5
#define VM 10.0
#define IM 2.5
//------------------Swing up controller----------------
#define ke 480
#define kw 1.0
#define kth 17.0
#define kd 3.0
//--------------Some constants----------------------
#define g 9.81
#define km 0.0368
// ------------ Furuta pendulum parameters --------------
#define m1 0.02218
#define l1 0.129
#define L0 0.1550
#define J1 0.0001845
#define I0 0.00023849
#define K 5.8
#define k1 0.0415
#define k2 0.0415
#define k3 0.5189
#define k4 0.0728
// ---------------------------------------------
FILE *ptrMonit;
TMainForm *MainForm;
unsigned char flagcom=0,flagfile=0,signo_sal,pwm; // 8 bits
unsigned short int pos1,pos2,corr; // 16 bits
float uf,pwmf,t=0,escp=pi_/(2.0*pprp),escb=pi_/(2.0*pprb);
float cambio=-1.0,signo,theta1=pi_,tau,E,F,det_d;
float norma2,theta0p=0,theta1p=0,theta0_1=0,theta1_1=0,
theta0=0;
float st1,ct1,s2t1,l1c=l1*l1,m1c=m1*m1,L0c=L0*L0,theta0pc;
float CF1=-(J1+m1*l1c)*m1*l1c,CF2=-0.5*m1c*l1c*l1*L0;
float CF4=-m1c*l1c*L0*g,theta1pc,V,theta1d,CF3=(J1+m1*l1c)
*m1*l1*L0;
float id,ic,ei,propi,intei,mi=2.0*IM/1023.0,iTs=1/Ts,
escs=127/VM;
float au1,au2,au3;
//------------------------------------------------------------
void ProcessByte(BYTE byte)
{
if(flagcom!=0)
```

```
flagcom++;
if((byte==0xAA)&&(flagcom==0))
{
pos1=0;
pos2=0;
corr=0;
flagcom=1;
}
if(flagcom==2)
{
pos1=byte;
pos1=pos1<<8;
}
if(flagcom==3)
{
pos1=pos1+byte;
}
if(flagcom==4)
{
pos2=byte;
pos2=pos2<<8;
}
if(flagcom==5)
{
pos2=pos2+byte+2000;
}
if(flagcom==6)
{
corr=byte;
corr=corr<<8;
}
if(flagcom==7)
{
theta0_1=theta0;
theta1_1=theta1;
corr=corr+byte;
ic=mi*corr+bi;
theta0=(signed short int)pos1;
theta0=escb*theta0; //arm position
theta1=(signed short int)pos2;
tetal=escp*tetal; //pendulum position
theta0p=(theta0-theta0_1)*iTs; //arm velocity
theta1p=(theta1-theta1_1)*iTs; //pendulum velocity
if(theta1<pi_)
{
theta1d=0.0;
}
else
{
theta1d=6.2832;
}
norma2=3.0*(theta1-theta1d)*(theta1-theta1d)+theta1p*theta1p;
if((norma2<0.05)||(cambio==1.0))
{
```

```
// ------------- Stabilizing controller ------------------
tau=k1*theta0 + k2*theta0p + k3*(theta1-theta1d) + k4*theta1p;
cambio=1.0;
}
else
{
//--------------Swinging up controller-------------------
st1=sin(theta1);
ct1=cos(theta1);
s2t1=sin(2*theta1);
theta0pc=theta0p*theta0p;
theta1pc=theta1p*theta1p;
au1=CF3*st1*theta1pc +CF4*ct1*st1;
F=CF1*s2t1*theta0p*theta1p +CF2*ct1*s2t1*theta0pc +au1;
au2=m1c*l1c*L0c*st1*st1;
det_d=(I0+m1*l1c*st1*st1)*(J1+m1*l1c)+J1*m1*L0c+au2;
au3=m1*l1c*theta1pc+m1*l1c*st1*st1*theta0pc+m1*L0*l1*ct1*
theta0p*theta1p;
E=0.5*(I0*theta0pc+J1*theta1pc+m1*L0c*theta0pc+au3)+m1
*g*l1*(ct1-1.0);
tau=(-(kw*F)-(det_d*(kd*theta0p+kth*theta0)))/(det_d*ke
*E+kw*(J1+m1*l1c));
tau=K*tau;
V=(ke*E*E)/2.0+kw*theta0pc/2.0+(kth*theta0*theta0)/2.0;
}
id=tau/km;
//---PI Electric current loop----------------------
if(id>IM)
id=IM;
if(id<-IM)
id=-IM;
ei=id-ic;
propi=kpi*ei;
if((intei<VM)&&(intei>(-VM)))
intei=intei+kii*Ts*ei;
else
{
if(intei>=VM)
intei=0.95*VM;
if(intei<=(-VM))
intei=-0.95*VM;
}
uf=propi+intei;
if(uf>VM)
uf=VM;
if(uf<-VM)
uf=-VM;
pwmf=uf;
//---------------Screen ----------------------
MainForm->Edit1->Text = FloatToStr (theta0p);
MainForm->Edit2->Text = FloatToStr (theta1p);
MainForm->Edit3->Text = FloatToStr (theta0);
MainForm->Edit4->Text = FloatToStr (theta1);
MainForm->Edit5->Text = FloatToStr (id);
```

```
MainForm->Edit6->Text = FloatToStr (E);
MainForm->Edit7->Text = FloatToStr (t);
MainForm->Edit8->Text = FloatToStr (ic);
//--Input voltage: 7 bits, magnitude, plus 1 bit for sign
pwmf=escs*pwmf;
pwm=(unsigned char)abs(pwmf);
if(pwmf<0)
pwm=pwm+128;
// Sending byte
MainForm->Acknowledge();
MainForm->send_byte(pwm);
/*Open/create a file */
if(flagfile==0)
if((ptrMonit=fopen("MONIT.TXT", "w"))==NULL){}
flagfile=1;
/* Writing to text file */
fprintf(ptrMonit,"%3.3f\t%3.3f\t%3.3f\n",t,theta0,theta1);
t=t+Ts;
flagcom=0;
}//closing last if in falgcom
}//closing function ProcessByte
//---------------------------------------------------------
__fastcall TMainForm::TMainForm(TComponent* Owner)
: TForm(Owner), SerialPort(1, ProcessByte), fAcknowledge(true)
{
if (SerialPort.IsReady() != TRUE)
MessageBox(NULL, "Port problems ", "Error", MB_OK);
}
//---------------------------------------------------------
void TMainForm::send_byte(unsigned char byte_sal)
{
if (fAcknowledge == false)
return;
SerialPort.WriteByte(byte_sal);
fAcknowledge = false;
}
//---------------------------------------------------------
void TMainForm::Acknowledge()
{
fAcknowledge = true;
}
void __fastcall TMainForm::Button1Click(TObject *Sender)
{
/* close the file */
fclose(ptrMonit);
Close();
}
//---------------------------------------------------------
```

15.11 The PIC16F877A Microcontroller C Program

The C program executed by the microcontroller is presented in this section. In the following, some of the most important parts of this program are explained. The interested reader is referred to [8] for a precise explanation of the instructions in this program. The corresponding flow diagram is presented in Fig. 15.30.

First, some important definitions are made concerning both hardware and software. After that, an interruption is defined that is intended to increment or decrement counts of encoders used to measure both the arm and the pendulum positions. The sampling period is set to 10[ms] by assigning "tope=196", i.e., $T_s = 4 \times 256 \times 196/(20 \times 10^6) = 0.01[s]$ as indicated at the end of the program. The control cycle is contained within the instruction "while(TRUE)". There, the Furuta pendulum variables are measured: the electric current first, then the arm position, and, finally, the pendulum position. As the next step, the arm position, the pendulum position, and the electric current are sent (in this order) to the portable computer. Once this is performed, the system remains waiting until the voltage data are received from the portable computer. Then, the voltage information is sent to the L298 full-bridge driver. This is done by taking into account the voltage sign contained in the highest bit and the voltage magnitude contained in the lowest 7 bits. After that, the program remains waiting for the sampling period to end. This is established by the instruction "while(TMR0<tope);". Once this is accomplished, the program returns to the beginning of the "while(TRUE)" cycle.

Fig. 15.30 Flow diagram for PIC16F877A microcontroller programming when used as an interface between a Furuta pendulum and a personal computer

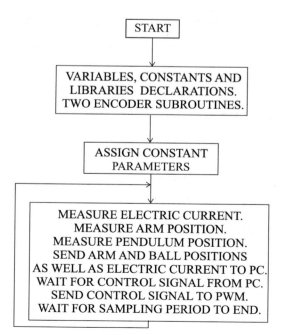

```
// Program for data exchange with a PC
// PIC16F877A and PCWH V3.43 compiler
#include<16f877A.h>
#device adc=10 //adc 10 bits
#include<stdlib.h>
#include<math.h>
#fuses HS,NOWDT,PUT,NOBROWNOUT,NOLVP,NOWRT,NOPROTECT,NOCPD
#use delay(clock=20000000) // Xtal frequency
//Serial port configuration
#use rs232(baud=115200,XMIT=PIN_C6,RCV=PIN_C7,BITS=8,PARITY=N)
//Port and register addresses
#byte OPTION= 0x81
#byte TMR0 = 0x01
#byte CVRCON= 0x9D
#byte ADCON1= 0x9F
#byte PORTA = 0x05
#byte PORTB = 0x06
#byte PORTC = 0x07
#byte PORTD = 0x08
#byte PORTE = 0x09

#bit PC0 = 0x07.0
#bit PD0 = 0x08.0
#bit PD1 = 0x08.1
#bit PD2 = 0x08.2
#bit PD3 = 0x08.3
#bit PD4 = 0x08.4
#bit PD5 = 0x08.5
//------------ Variables------------//
int16 inter,cuenta1,cuenta2,conv;
int8 cuentaH1,cuentaL1,cuentaH2,cuentaL2,puerto,AB,AB_1,BC,
BC_1,aux;
int8 convH,convL,tope,dato,pwm;
//------------Encoder interruption ------------//
#int_rb
void rb_isr()
{
puerto=PORTB;
AB=((puerto)&(0x30))>>4;
aux=AB^AB_1;
if(aux!=0)
if(aux!=3)
if(((AB_1<<1)^AB)&(0x02))
{
cuenta1--;
}
else
{
cuenta1++;
}
AB_1=AB;

BC=((puerto)&(0xC0))>>6;
```

```
aux=BC^BC_1;
if(aux!=0)
if(aux!=3)
if(((BC_1<<1)^BC)&(0x02))
{
cuenta2--;
}
else
{
cuenta2++;
}
BC_1=BC;
}

//------------ Main program------------//
void main(void)
{
set_tris_a(0b11111111);
set_tris_b(0b11111111);
set_tris_c(0b10000000);// Serial communication and pwm pins
set_tris_d(0b00001100);
set_tris_e(0b11111111);
PC0=0;
OPTION=0x07; //timer0 1:256
setup_ccp1(CCP_PWM); //pwm configuration
//T=(1/clock)*4*t2div*(period+1)
setup_timer_2(T2_DIV_BY_16,255,1);
PORTC=0;
TMR0=0;
cuenta1=0; //arm initial condition
cuenta2=0; //pendulum initial condition
set_pwm1_duty(0);
AB=0;
AB_1=0;
BC=0;
BC_1=0;
if(PD2)
tope=196;
else
tope=20;
setup_adc(ADC_CLOCK_INTERNAL);//ADC internal clock
setup_adc_ports(ALL_ANALOG);
set_adc_channel(0);
delay_us(15);
enable_interrupts(global);
enable_interrupts(int_rb);
PC0=1;

while(TRUE)
{
conv=read_adc();
inter=(conv)&(0xFF00);
convH=inter>>8;
```

```
convL=(conv)&(0x00FF);
PD4=1; //data sending starts

inter=(cuenta1)&(0xFF00);
cuentaH1=inter>>8;
cuentaL1=(cuenta1)&(0x00FF);

inter=(cuenta2)&(0xFF00);
cuentaH2=inter>>8;
cuentaL2=(cuenta2)&(0x00FF);

putc(0xAA); //sending to serial port
putc(cuentaH1);
putc(cuentaL1);
putc(cuentaH2);
putc(cuentaL2);
if(PD3)
{
putc(convH);
putc(convL);
}

PD4=0; //data sending finishes
PD5=1; //timer is waiting
do
{
if(kbhit())
{
dato=getc();//getting plant input data
if(dato>=128)
{
PD0=0;
PD1=1;
dato=dato-128;
pwm=dato<<1;
}
else
{
PD0=1;
PD1=0;
pwm=dato<<1;
}
set_pwm1_duty(pwm); //sending to pwm
}
}
while(TMR0<tope); //each timer count = (4/FXtal)*256 sec
PD5=0; //timer finishes wait
TMR0=0;
} //closing infinite while
} //closing main
```

15.12   Summary

In this chapter, a Furuta pendulum has been built, in addition to all the interfaces required for closed-loop control. The control scheme that has been used is known as linear state feedback control. It consists in measuring the positions and the velocities of both bodies comprising the mechanism to multiply them by constants and then adding these terms to compute the torque that must be generated by the DC motor used as an actuator. As the mechanism is unstable and nonlinear, the solution of this control problem and the experimental evaluation of the designed control system are of great interest. In this problem, the Euler–Lagrange equations for physical system modeling are very useful. This approach is particularly valuable when modeling mechanical systems composed of several bodies that describe independent movements (degrees-of-freedom).

In this chapter, it has been explained how to obtain the numerical values of the mechanism parameters. This is important to stress, as each of the two bodies comprising the mechanism is composed of smaller mechanical pieces, each contributing to the total body mass, center of mass, and inertia. It has been shown that the classical mechanics formulas related to these variables are useful for computing the mechanism parameters.

A linear approximate model has been obtained from the complete nonlinear model of the mechanism. This step is instrumental in designing a controller for this mechanism, which, however, is valid only if the initial configuration of the mechanism is close to the desired configuration.

15.13   Review Questions

1. Why is the Furuta pendulum open-loop unstable?
2. Give a descriptive explanation on why it is important to control the arm position aside from the pendulum position.
3. Describe the procedure followed to identify the parameters of the complete mechanism.
4. Explain what is represented by each of the operation points of the Furuta pendulum.
5. Why does the potential energy of the mechanism only take into account the pendulum's potential energy?
6. What does a linear approximation of a nonlinear model represent?
7. What is the condition that a linear state feedback controller must satisfy to render the closed-loop system stable?
8. Why is it important for the Furuta pendulum to be controllable? What can be done if this is not the case?

References

1. H. Goldstein, C. Poole, and J. Safko, *Classical Mechanics*, Chapter 1, 3rd Edition, Addison Wesley, New York, 2000.
2. I. Fantoni and R. Lozano, *Non-linear control for underactuated mechanical systems*, Springer, London, 2002.
3. R. C. Hibbeler, *Engineering Mechanics Dynamics*, 10th Edition, Prentice Hall, Upper Saddle River, NJ, 2004.
4. M. Alonso and E. J. Finn, *Physics*, Addison-Wesley, Reading, MA, 1992.
5. G. Stein, Respect the unstable, *IEEE Control Systems Magazine*, August, pp. 12–25, 2003.
6. V. M. Hernández-Guzmán, M. Antonio-Cruz, and R. Silva-Ortigoza, Linear state feedback regulation of a Furuta pendulum: design based on differential flatness and root locus, *IEEE Access*, Vol. 4, pp. 8721–8736, December, 2016
7. PIC16F877A Enhanced Flash Microcontroller, Data sheet, Microchip Technology Inc., 2003.
8. Custom Computer Services Incorporated, CCS C Compiler Reference Manual, 2003.

Chapter 16
Control of an Inertia Wheel Pendulum

16.1 Inertia Wheel Pendulum Description

In this chapter, a control strategy is designed for a mechanism known as the inertia wheel pendulum (IWP). A picture of the IWP is shown in Fig. 16.1 when operating at the downward position, whereas the variables and the parameters involved are defined in Fig. 16.2. It is a pendulum with a PM brushed DC motor fixed at its free end, which actuates on a wheel. The pendulum hangs up from a point where no external torque is applied. It is said that this point is not actuated but the wheel is actuated because it receives torque generated by the motor. As the pendulum and the wheel define two-degrees-of-freedom [1] but there is only one actuator (at the wheel), the mechanism is said to be underactuated [2].

The IWP was introduced by Spong et al. [2], and aside from its academic applications it is also widely used for research purposes as a workbench for nonlinear controllers [2–11].

The nomenclature employed is the following:

- q_1 is the pendulum angular position.
- q_2 is the wheel angular position.
- m_1 is the pendulum mass.
- m_2 is the wheel mass.
- τ is torque applied at the wheel.
- l_{c1} is the location of the pendulum's center of mass.
- l_1 is the pendulum's length.
- I_1 is the pendulum' inertia when rotating around its center of mass.
- I_2 is the wheel inertia (plus the motor rotor inertia).
- g is the gravity acceleration.
- R is the motor armature resistance.
- L is the motor armature inductance.

© Springer International Publishing AG, part of Springer Nature 2019
V. M. Hernández-Guzmán, R. Silva-Ortigoza, *Automatic Control with Experiments*,
Advanced Textbooks in Control and Signal Processing,
https://doi.org/10.1007/978-3-319-75804-6_16

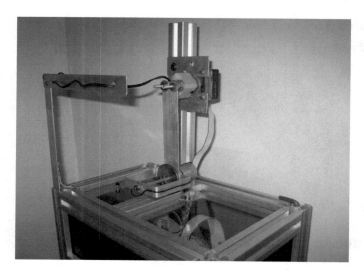

Fig. 16.1 The inertia wheel pendulum at the noninverted position

Fig. 16.2 Inertia wheel
pendulum

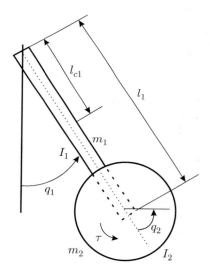

- k_b and k_m are the motor counter electromotive force constant and the motor torque constant.
- u and i stand respectively, for voltage applied at the motor armature terminals and the electric current through the motor armature.

According to the above description of the IWP, it is clear that it is not easy to take the pendulum directly from $q_1 = 0$, $\dot{q}_1 = 0$, to $q_1 = +\pi$, or $q_1 = -\pi$, and $\dot{q}_1 = 0$. Moreover, the latter configuration is unstable and it is known as *the inverted unstable configuration*. Thus, it is of interest to design a control strategy such that:

(i) The pendulum is taken from a vicinity of $q_1 = 0, \dot{q}_1 = 0$, to the inverted unstable configuration, (ii) The pendulum stays at the inverted unstable configuration. The first part of this task is known as the *pendulum swing up* and the second part is known as the *pendulum balancing*.

In this chapter it is shown that solution of the swing up problem requires the use of nonlinear control techniques because the IWP has a nonlinear model. Although nonlinear, the model of this mechanism is straightforward enough to design a control strategy using simple ideas. One important objective in including this control problem is to introduce the reader, using a basic example, to nonlinear control systems techniques.

16.2 Mathematical Model

As the inertia wheel pendulum is a mechanism composed of two interacting bodies, its mathematical model is obtained using the *Euler–Lagrange equations*, i.e.,

$$\frac{d}{dt}\frac{\partial L}{\partial \dot{q}_1} - \frac{\partial L}{\partial q_1} = 0, \tag{16.1}$$

$$\frac{d}{dt}\frac{\partial L}{\partial \dot{q}_2} - \frac{\partial L}{\partial q_2} = \tau.$$

The zero on the right-hand side in the first expression indicates that no external torque is applied at the point where the pendulum hangs up. The *Lagrangian* is given as:

$$L = K - P, \tag{16.2}$$

$$K = K_1 + K_2, \quad P = P_1 + P_2,$$

where K_1 and K_2 represent the kinetic energies of the pendulum and the wheel respectively, whereas P_1 and P_2 represent the potential energies of the pendulum and the wheel respectively. As the pendulum is a body with a mass distributed along a stick, its kinetic energy is computed by adding the kinetic energy of a particle of mass m_1, which translates on a circumference with radius l_{c1} and the kinetic energy of a stick that rotates around its center of mass, i.e.,:

$$K_1 = \frac{1}{2}m_1 l_{c1}^2 \dot{q}_1^2 + \frac{1}{2}I_1 \dot{q}_1^2.$$

The kinetic energy of the wheel is obtained in a similar manner. However, it must be taken into account that the wheel's angular velocity, as measured by an observer fixed to the laboratory, is $\dot{q}_1 + \dot{q}_2$. Then, the following is found:

$$K_2 = \frac{1}{2}m_2 l_1^2 \dot{q}_1^2 + \frac{1}{2}I_2(\dot{q}_1 + \dot{q}_2)^2.$$

To compute the potential energy, the configuration where $q_1 = 0$ is used as reference, i.e., where $P_1 = 0$ and $P_2 = 0$. Recall that the pendulum *potential energy* is given as the product of force applied to the pendulum, i.e., the pendulum weight, and the component parallel to the pendulum weight of the distance between the pendulum's center of mass at the actual position q_1 and position of the pendulum's center of mass where the pendulum's potential energy is zero, i.e., when $q_1 = 0$. Then:

$$P_1 = gm_1 l_{c1}(1 - \cos(q_1)).$$

Proceeding in a similar manner, it is found that the wheel's potential energy is given as:

$$P_2 = gm_2 l_1(1 - \cos(q_1)).$$

Hence, according to (16.2):

$$L = \frac{1}{2}m_1 l_{c1}^2 \dot{q}_1^2 + \frac{1}{2}I_1 \dot{q}_1^2 + \frac{1}{2}m_2 l_1^2 \dot{q}_1^2 + \frac{1}{2}I_2(\dot{q}_1 + \dot{q}_2)^2 - g\overline{m}(1 - \cos(q_1)),$$

where $\overline{m} = m_1 l_{c1} + m_2 l_1$. Replacing L in the Euler–Lagrange equations (16.1) and after suitably arranging the resulting terms, the following is found:

$$D\begin{bmatrix} \ddot{q}_1 \\ \ddot{q}_2 \end{bmatrix} + \begin{bmatrix} \overline{m}g\sin(q_1) \\ 0 \end{bmatrix} = \begin{bmatrix} 0 \\ \tau \end{bmatrix}, \tag{16.3}$$

$$D = \begin{bmatrix} d_{11} & d_{12} \\ d_{21} & d_{22} \end{bmatrix} = \begin{bmatrix} m_1 l_{c1}^2 + m_2 l_1^2 + I_1 + I_2 & I_2 \\ I_2 & I_2 \end{bmatrix}.$$

It is simple to verify that the determinant of matrix D is positive, i.e.,:

$$\det(D) = d_{11}d_{22} - d_{12}d_{21} > 0. \tag{16.4}$$

The model in (16.3) is the most complete one for the IWP and it will be used in the next section to design a control strategy to swing up the pendulum to the inverted unstable configuration.

Voltage applied at the armature terminals of the PM brushed DC motor actuating on the wheel is the actual control signal, i.e., the variable that can be directly manipulated, whereas torque at the wheel is a variable that is a consequence of the application of such a voltage. However, in practice it is possible to assume that torque at the wheel is the control signal if it is supposed that the armature inductance is small. The armature circuit model of a PM brushed DC motor is the following:

$$L_a \frac{di}{dt} + Ri + k_b \dot{q}_2 = u. \qquad (16.5)$$

If L_a and k_b are assumed to be zero, then:

$$\tau = k_m i = \frac{k_m}{R} u. \qquad (16.6)$$

This means that the applied voltage has a direct effect on the generated torque; hence, it is reasonable to assume that torque is the control signal. Assuming k_b to be zero is justified as follows. It is observed from (16.5) that k_b introduces additional viscous friction to the mechanism. As this commonly improves stability in a control system, then assuming $k_b = 0$ renders the control problem more interesting at hand in the sense that the design is performed by assuming that any stabilizing effect is not present naturally in the mechanism.

16.3 Swing Up Nonlinear Control

A controller is designed in this section to take the pendulum to its inverted unstable configuration, i.e., to any of the points $(q_1, \dot{q}_1) = (+\pi, 0)$ or $(q_1, \dot{q}_1) = (-\pi, 0)$. It will be explained later why the wheel position and velocity are not taken into account during the time interval when this task is performed.

Solving the first row in (16.3) for $\ddot{q}_2$, replacing this value in the second row of (16.3), and using (16.6) in addition to $d_{12} = d_{22}$, the following is found:

$$J\ddot{q}_1 + mgl \sin(q_1) = \tau_1. \qquad (16.7)$$

$$J = \frac{d_{22}d_{11} - d_{21}d_{12}}{d_{12}}, \quad mgl = \overline{m}g, \quad \tau_1 = -\frac{k_m}{R} u.$$

Notice that $J > 0$ because of (16.4). Hence, the model in (16.7) corresponds to a simple pendulum where J, m, l, and g represent the equivalent inertia, mass, length, and gravity acceleration respectively. The total energy of this pendulum is given by adding its *kinetic energy* and potential energy, i.e.,:

$$V(q_1, \dot{q}_1) = \frac{1}{2} J\dot{q}_1^2 + mgl(1 - \cos(q_1)). \qquad (16.8)$$

The *level surfaces* of V are depicted in Fig. 16.3. This means that each curve in this figure represents the set of points $(q_1, \dot{q}_1)$ where $V(q_1, \dot{q}_1)$ takes a constant value. The reader should realize that the more external curves correspond to larger values of $V > 0$ and the internal-most curve corresponds to a single point, $(q_1, \dot{q}) = (0, 0)$, where $V = 0$ takes its smallest value.

Time derivative of $V(q_1, \dot{q}_1)$, given in (16.8), is found to be:

$$\frac{dV(q_1, \dot{q}_1)}{dt} = \dot{q}_1 J \ddot{q}_1 + mgl\dot{q}_1 \sin(q_1).$$

If $J\ddot{q}_1$ is replaced from (16.7), the following is found:

$$\frac{dV(q_1, \dot{q}_1)}{dt} = \dot{q}_1 \tau_1. \tag{16.9}$$

What has been computed is known as the *time derivative of V along the trajectories of (16.7)* [12]. This means that (16.9) represents the values that $\frac{d}{dt}V$ takes as the pendulum in (16.7) moves under the effect of the external torque τ_1. For instance, if $\tau_1 = 0$ then $\frac{d}{dt}V = 0$, according to (16.9), which means that the pendulum's energy V remains constant as the pendulum moves. This is very important for our purposes. From the knowledge of the pendulum's initial position and velocity $(q_1(0), \dot{q}_1(0))$, it is possible to know the initial value of the pendulum's energy $V(q_1(0), \dot{q}_1(0))$. As $\frac{d}{dt}V = 0$ because $\tau_1 = 0$ then, as time grows, the pendulum moves only on points representing the level surface where $V(q_1(t), \dot{q}_1(t)) = V(q_1(0), \dot{q}_1(0))$, which are shown in Fig. 16.3. Arrows on the level surfaces represent the sense of the pendulum's movement. These senses of movement are easy to determine. The state vector of a simple pendulum is defined

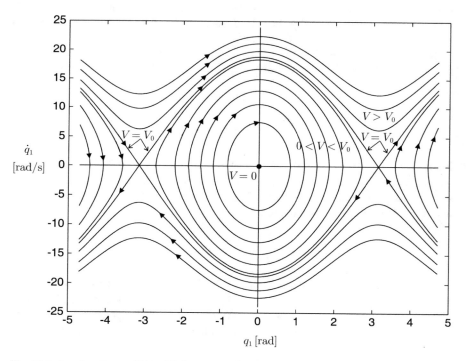

Fig. 16.3 Level surfaces of V in (16.8)

as $y = [q_1, \dot{q}]^T$. The variable $\dot{y} = [\dot{q}_1, \ddot{q}]^T$ is a vector indicating where the state y moves to. Notice that the horizontal component of $\dot{y}$, i.e., $\dot{q}_1$, points to the right as long as the pendulum velocity $\dot{q}_1$ is positive. This suffices to determine the sense of the arrows on the level surfaces of V.

According to the ideas discussed above, to swing up the pendulum it suffices to increase the pendulum energy to:

$$V(q_1, \dot{q}_1)|_{\substack{q_1=+\pi \text{ ó } q_1=-\pi \\ \dot{q}_1=0}} = \frac{1}{2}J(0)^2 + mgl(1 - \cos(\pm\pi)) = 2mgl = V_0,$$

and, in the case that such an energy is reached at a point where $(q_1, \dot{q}_1) \neq (+\pi, 0)$ or $(q_1, \dot{q}_1) \neq (-\pi, 0)$, then apply a torque τ_1 that ensures $\frac{d}{dt}V = 0$ from that time on. Notice that, according to Fig. 16.3, if V is kept constant once $V = V_0$, then the pendulum continues to move until one of the points $(q_1, \dot{q}_1) = (+\pi, 0)$ or $(q_1, \dot{q}_1) = (-\pi, 0)$ is reached. It is shown next that the above is ensured if the following torque is used:

$$\tau_1 = -k_d \, sat(\dot{q}_1) \, sign(V - V_0), \tag{16.10}$$

$$sat(\dot{q}_1) = \begin{cases} \varepsilon_0, & \dot{q}_1 > \varepsilon_0 \\ -\varepsilon_0, & \dot{q}_1 < -\varepsilon_0 \\ \dot{q}_1, & |\dot{q}_1| \leq \varepsilon_0 \end{cases}, \quad sign(V - V_0) = \begin{cases} +1, & V > V_0 \\ -1, & V < V_0 \\ 0, & V = V_0 \end{cases}, \tag{16.11}$$

where ε_0 and k_d are arbitrary positive constants. Replacing (16.10) in (16.9):

$$\frac{dV(q_1, \dot{q}_1)}{dt} = -k_d \dot{q}_1 \, sat(\dot{q}_1) \, sign(V - V_0).$$

It is easy to realize from (16.11) that the product $\dot{q}_1 \, sat(\dot{q}_1)$ is always positive and it is zero only when $\dot{q}_1 = 0$. Hence, as long as $\dot{q}_1 \neq 0$ we have:

- If $V < V_0$ then $\frac{d}{dt}V > 0$ and energy V increases.
- If $V > V_0$ then $\frac{d}{dt}V < 0$ and energy V decreases.
- If $V = V_0$, then $\frac{d}{dt}V = 0$ and energy V remains at V_0.

Then, it is ensured that the pendulum energy V reaches the required value to arrive at one of the points $(q_1, \dot{q}_1) = (+\pi, 0)$ or $(q_1, \dot{q}_1) = (-\pi, 0)$. Furthermore, the energy remains at that value until one of these points is reached. Now, analyze what happens if $\dot{q}_1 = 0$. Notice that the pendulum is at rest under this condition. However, the only way for $\dot{q}_1 = 0$ to be true for ever is $q_1 = 0$, because it is only there that the pendulum may remain at rest. Hence, it is necessary to slightly hit the pendulum at the beginning to force it to abandon the configuration $(q_1, \dot{q}_1) = (0, 0)$.

The previous analysis ensures that the pendulum reaches the inverted unstable configuration if, from the downward stable configuration, the pendulum is slightly hit to start it moving. It is not difficult to realize that none of the above results change if $sat(V - V_0)$ is used in controller (16.10) instead of $sign(V - V_0)$. The controller in (16.10), (16.11), has its roots in the ideas reported in [13].

On the other hand, it is important to stress that, once the inverted unstable configuration is reached, the controller (16.10) does not ensure that the pendulum stays at such a configuration. This requires another controller to be designed that must be put to work when the pendulum is very close to the inverted unstable configuration. This means that the controller in (16.10) must be disconnected from that time on. In the next section, a controller is designed solving the second part of the control problem: pendulum balancing at the inverted unstable configuration.

Finally, notice that nothing has been said on the wheel position and velocity during the pendulum swing-up stage. As the wheel is symmetrical and homogeneous, then the wheel position does not matter. On the other hand, although the wheel velocity increases in an unknown manner during this time interval, the velocity is finite at the moment when the pendulum reaches the inverted unstable configuration; hence, the balancing controller must just regulate the wheel velocity at the value that this variable has when this controller begins to work.

Example 16.1 Consider the alternative controller:

$$\tau_1 = -k_d \ (V - V_0) \ sat(\dot{q}_1), \quad u = -\frac{R}{k_m} \tau_1, \tag{16.12}$$

$$sat(\dot{q}_1) = \begin{cases} d, & \dot{q}_1 > d \\ -d, & \dot{q}_1 < -d \ , \\ \dot{q}_1, & |\dot{q}_1| \le d \end{cases} \quad d = \frac{u_{max} k_m}{k_d R V_0}, \tag{16.13}$$

where d, u_{max} and k_d are positive arbitrary constants. Replacing (16.12) in (16.9):

$$\frac{dV(q_1, \dot{q}_1)}{dt} = -k_d \ (V - V_0) \ \dot{q}_1 \ sat(\dot{q}_1).$$

It is easy to realize, from (16.13), that the product $\dot{q}_1 \ sat(\dot{q}_1)$ is always positive and it is zero only when $\dot{q}_1 = 0$. Hence, as long as $\dot{q}_1 \neq 0$, we have the following:

- If $V < V_0$ then $\frac{d}{dt} V > 0$ and energy V increases.
- If $V > V_0$ then $\frac{d}{dt} V < 0$ and energy V decreases.
- If $V = V_0$ then $\frac{d}{dt} V = 0$ and energy V remains at V_0.

Thus, it is ensured again that the pendulum energy V reaches the required value to arrive at one of the points $(q_1, \dot{q}_1) = (+\pi, 0)$ of $(q_1, \dot{q}_1) = (-\pi, 0)$.

16.4   Balancing Controller

The strategy for balancing the pendulum is a linear state feedback controller designed on the basis of a linear approximation of (16.3). This linear approximation is obtained in the following. Note that the model in (16.3) does not depend on the wheel position q_2. Hence, the state vector can be defined as:

$$x = \begin{bmatrix} x_1 \\ x_2 \\ x_3 \end{bmatrix} = \begin{bmatrix} q_1 \\ \dot{q}_1 \\ \dot{q}_2 \end{bmatrix}, \tag{16.14}$$

From (16.3):

$$\begin{bmatrix} \ddot{q}_1 \\ \ddot{q}_2 \end{bmatrix} = D^{-1} \left\{ - \begin{bmatrix} \overline{m}g \sin{(q_1)} \\ 0 \end{bmatrix} + \begin{bmatrix} 0 \\ \tau \end{bmatrix} \right\}, \tag{16.15}$$

$$D^{-1} = \begin{bmatrix} \overline{d}_{11} & \overline{d}_{12} \\ \overline{d}_{21} & \overline{d}_{22} \end{bmatrix} = \frac{1}{d_{11}d_{22} - d_{12}d_{21}} \begin{bmatrix} d_{22} & -d_{12} \\ -d_{21} & d_{11} \end{bmatrix}. \tag{16.16}$$

Note that $\det(D^{-1}) \neq 0$ because it is the inverse matrix of a nonsingular matrix. Using (16.14) and (16.15), the model in (16.3) can be written as:

$$\dot{x} = f(x, \tau) = \begin{bmatrix} x_2 \\ -\overline{d}_{11}\overline{m}g \sin{(x_1)} + \overline{d}_{12}\tau \\ -\overline{d}_{21}\overline{m}g \sin{(x_1)} + \overline{d}_{22}\tau \end{bmatrix}. \tag{16.17}$$

According to Sect. 7.3.2, the operation points (x^*, τ^*) of (16.17) are found from the condition $f(x^*, \tau^*) = 0$, i.e.,:

$$\begin{bmatrix} x_2^* \\ -\overline{d}_{11}\overline{m}g \sin{(x_1^*)} + \overline{d}_{12}\tau^* \\ -\overline{d}_{21}\overline{m}g \sin{(x_1^*)} + \overline{d}_{22}\tau^* \end{bmatrix} = \begin{bmatrix} 0 \\ 0 \\ 0 \end{bmatrix}. \tag{16.18}$$

It is clear that:

$$x_2^* = 0. \tag{16.19}$$

On the other hand, from the second and third rows in (16.18), the following is found:

$$\tau^* = \frac{\overline{d}_{11}\overline{m}g \sin{(x_1^*)}}{\overline{d}_{12}}, \tag{16.20}$$

$$\tau^* = \frac{\overline{d}_{21}\overline{m}g \sin{(x_1^*)}}{\overline{d}_{22}}.$$

Solving these expressions simultaneously:

$$\frac{\overline{m}g}{\overline{d}_{12}\overline{d}_{22}} \left[\overline{d}_{11}\overline{d}_{22} - \overline{d}_{21}\overline{d}_{12} \right] \sin(x_1^*) = 0. \tag{16.21}$$

As $\det(D^{-1}) = \overline{d}_{11}\overline{d}_{22} - \overline{d}_{21}\overline{d}_{12} \neq 0$, then (16.21) is only satisfied if:

$$x_1^* = \pm n\pi, \quad n = 0, 1, 2, 3, \ldots \tag{16.22}$$

Replacing this condition in any of the expressions in (16.20), the following is found:

$$\tau^* = 0. \tag{16.23}$$

Finally, as no restriction exists on x_3^*, it is concluded that:

$$x_3^* = c, \tag{16.24}$$

where c is any real constant. According to Sect. 7.3.2, the linear approximation of (16.17) at the operation point (16.19), (16.22), (16.23), (16.24), with $n = \pm 1$, is given as:

$$\dot{z} = Az + Bw, \tag{16.25}$$

$$A = \left. \frac{\partial f(x,\tau)}{\partial x} \right|_{\substack{x^* \\ \tau^*}} = \left. \begin{bmatrix} \dfrac{\partial f_1}{\partial x_1} & \dfrac{\partial f_1}{\partial x_2} & \dfrac{\partial f_1}{\partial x_3} \\[2mm] \dfrac{\partial f_2}{\partial x_1} & \dfrac{\partial f_2}{\partial x_2} & \dfrac{\partial f_2}{\partial x_3} \\[2mm] \dfrac{\partial f_3}{\partial x_1} & \dfrac{\partial f_3}{\partial x_2} & \dfrac{\partial f_3}{\partial x_3} \end{bmatrix} \right|_{\substack{x^* \\ \tau^*}} = \begin{bmatrix} 0 & 1 & 0 \\ \overline{d}_{11}\overline{m}g & 0 & 0 \\ \overline{d}_{21}\overline{m}g & 0 & 0 \end{bmatrix},$$

$$B = \left. \frac{\partial f(x,\tau)}{\partial \tau} \right|_{\substack{x^* \\ \tau^*}} = \left. \begin{bmatrix} \dfrac{\partial f_1}{\partial \tau} \\[2mm] \dfrac{\partial f_2}{\partial \tau} \\[2mm] \dfrac{\partial f_3}{\partial \tau} \end{bmatrix} \right|_{\substack{x^* \\ \tau^*}} = \begin{bmatrix} 0 \\ \overline{d}_{12} \\ \overline{d}_{22} \end{bmatrix},$$

where:

$$z = x - x^*, \tag{16.26}$$

$$w = \tau - \tau^*, \tag{16.27}$$

remain small, i.e., $z \approx 0$, $w \approx 0$. On the other hand, recall that $\tau^* = 0$ and, hence, $w = \tau$. Then, (16.6) can be used to find the following modified version of (16.25):

$$\dot{z} = Az + Bu, \tag{16.28}$$

$$B = \begin{bmatrix} 0 \\ \overline{d}_{12} \\ \overline{d}_{22} \end{bmatrix} \frac{k_m}{R}.$$

This approximate linear model is to be used to design the balancing controller.

First, the pair (A, B) is checked to be controllable. This requires computation of the determinant of the controllability matrix $C = [B|AB|A^2B]$, which is found to be:

$$\det(C) = \frac{\overline{m}gk_m^3}{R^3}\overline{d}_{12}^2(\overline{d}_{11}\overline{d}_{22} - \overline{d}_{12}\overline{d}_{21}) \neq 0. \tag{16.29}$$

Hence, it is concluded that the pair (A, B) is controllable. This means that it is always possible to find a gain vector K such that all eigenvalues of the closed-loop matrix $(A - BK)$ are assigned at arbitrary locations. This is achieved using the controller:

$$u = -Kz. \tag{16.30}$$

The gain vector K is computed once the numerical values of the parameters involved in matrices A and B are known. The controller (16.30) renders it possible to balance the IWP at the inverted unstable configuration defined by the operation point (16.19), (16.22), (16.23), (16.24), with $n = \pm 1$. It is important to stress that this requires the mechanism state vector x to be close enough to such an unstable configuration at the time when the controller (16.30) starts to work.

Finally, it is important to say that the constant c introduced in (16.24) takes the value of the wheel velocity at the time when the controller (16.30) starts to work. As this velocity may take any value, it is instrumental that c can be defined as any real constant.

16.5 Prototype Construction and Parameter Identification

A drawing of the IWP that has been built is shown in Fig. 16.4. There, the main components are shown. Some of these components are included to successfully assemble the mechanical pieces. An incremental encoder and a tachometer are also included, to measure the pendulum position q_1 and the wheel velocity $\dot{q}_2$ respectively. The dimensions of the pieces employed are presented in Table 16.1.

The pendulum equivalent mass m_1 is defined as:

$$m_1 = m_{sol} + m_{car} + m_{sop} = 0.05975 \text{ [Kg]}. \tag{16.31}$$

Fig. 16.4 Inertia wheel
pendulum

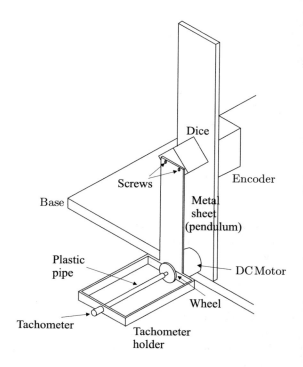

Table 16.1 Simple parameters

Symbol	Description	Value	Units
l_1	Metal sheet length	0.117	m
a_{sol}	Metal sheet width	0.0254	m
m_{sol}	Metal sheet mass	0.016	Kg
m_{car}	Actuator stator mass	0.03	Kg
m_{sop}	Tachometer holder mass	0.01375	Kg
m_{rot}	Actuator rotor mass	0.02	Kg
r_{rot}	Actuator rotor radius	0.00915	m
m_{rue}	Wheel mass	0.038	Kg
r_{rue}	Wheel radius	0.0189	m
R	Armature resistance	4.172	Ohm
L	Armature inductance	0.0009	H
k_b	Counter electromotive force constant	0.00775	Vs/rad
k_m	Torque constant	0.00775	Nm/A

To compute the pendulum's center of mass including all its components, the balance principle is employed, i.e., the sum of all the torques around the support point is zero. Hence, when applying an equivalent weight at a distance l_{c1} on the opposite end of the metal sheet, the assembly in Fig. 16.5 must be balanced. Using (16.31) and the data in Table 16.1, the following is found:

$$l_{c1} = \frac{m_{sol}\frac{l_1}{2} + m_{car}l_1 + m_{sop}l_1}{m_1} = 0.10133 \text{ [m]}. \tag{16.32}$$

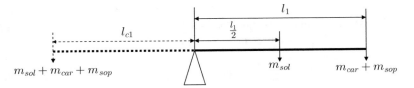

Fig. 16.5 The balance principle to compute l_{c1}

The moment of inertia of a rectangular thick sheet with mass m, width b, and length d, with rotation axis x orthogonal to the plane, with edges b and d passing through its geometric center, is given as [14, 15], pp. 271:

$$I_x = \frac{1}{12}m(b^2 + d^2). \tag{16.33}$$

Using (16.33) and the data in Table 16.1, the moment of inertia I^* is obtained as:

$$I^* = \frac{1}{12}m_{sol}(a_{sol}^2 + l_1^2) = 0.0000191122 \text{ [Kg m}^2\text{]}. \tag{16.34}$$

However, the moment of inertia I_1 is defined around an axis passing through the pendulum center of mass located at a distance $l_{c1} - l_1/2$ from the center of the rectangle with edges b and d. According to the *parallel axis theorem or Steiner theorem* [14, 15], pp. 272:

$$I_1 = I^* + m_{sol}\left(l_{c1} - \frac{l_1}{2}\right)^2 = 0.000048463 \text{ [Kg m}^2\text{]}.$$

On the other hand, the moment of inertia of a disc with a mass m and radius r with rotation axis x orthogonal to the disc and passing through its geometric center is given as [14, 15], pp. 271:

$$I_x = \frac{1}{2}mr^2. \tag{16.35}$$

Using the data in Table 16.1 and the expression in (16.35) it is found that the wheel's moment of inertia I_{rue} is given as:

$$I_{rue} = \frac{1}{2}m_{rue}r_{rue}^2 = 0.000006787 \text{ [Kg m}^2\text{]}. \tag{16.36}$$

934

The moment of inertia of a cylinder with a mass m, radius a, and height l, with a rotation axis x passing through its longitudinal axis is given as [14, 15], pp. 271:

$$I_x = \frac{1}{2}ma^2. \tag{16.37}$$

Using the expression in (16.37) and the data in Table 16.1, the moment of inertia of the actuator rotor I_{rot} is found to be:

$$I_{rot} = \frac{1}{2}m_{rot}r_{rot}^2 = 0.000000837225 \ [\text{Kg m}^2]. \tag{16.38}$$

The equivalent wheel mass m_2 is defined as:

$$m_2 = m_{rot} + m_{rue} = 0.058 \ [\text{Kg}]. \tag{16.39}$$

Using (16.36) and (16.38), it is found that the equivalent wheel moment of inertia I_2 is given as:

$$I_2 = I_{rue} + I_{rot} = 0.0000076242 \ [\text{Kg m}^2]. \tag{16.40}$$

Finally, using the data in Table 16.1, the parameters in (16.34), (16.31), (16.32), (16.39), (16.40), and $g = 9.81[\text{m/s}^2]$, the following numerical values for the parameters of model (16.3) are found:

$$d_{11} = 0.0014636, \quad d_{12} = 0.0000076, \tag{16.41}$$
$$d_{21} = 0.0000076, \quad d_{22} = 0.0000076,$$
$$\overline{m}g = 0.12597.$$

It is worth saying that the tachometer mass is considered as a part of the tachometer holder mass and the effect of the tachometer rotor is neglected when computing I_2. This is because it has a mass less than 0.0001[Kg] and a radius less than 0.001 [m]. On the other hand, the screws holding the dice, the dice itself, and the incremental encoder shaft are neglected when computing the inertia I_1. This is because the addition of these masses is less than 0.008[kg] and, more importantly, they are placed at the point where the pendulum hangs up, which reduces strongly their inertial effects. Finally, the mass of the electric wires and the plastic pipe used to couple the tachometer to the wheel were also neglected.

Example 16.2 Some simulation results are presented in the following when controlling the IWM (16.3), with numerical parameters in (16.41), using the controller in (16.10) to solve the swing-up task and the controller in (16.30) to balance the pendulum at the inverted configuration. The switching condition between the controllers is:

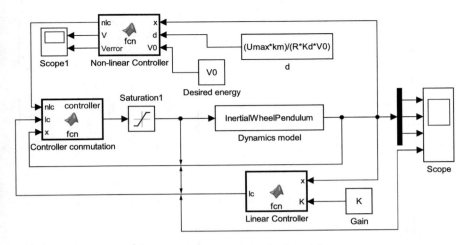

Fig. 16.6 MATLAB/Simulink diagram for the inertia wheel pendulum

If: $z_1^2 + z_2^2 \leq 0.01$,   controller in (16.30) is on,

else: controller in (16.10) is on,

with z_1 and z_2 defined in (16.26). All of the initial conditions were chosen to be zero except for $\dot{q}_1(0) = 0.00001$[rad/s]. The controller in (16.10) was set with $k_d = 20$. The vector of gains K for controller (16.30) was selected such that the eigenvalues of the closed-loop matrix $A - BK$ are assigned at $-9.27 \pm 20.6j$ and -0.719. These simulations were performed using the MATLAB/Simulink diagram in Fig. 16.6 where there are four main blocks that contain the code listed in Appendix G.2. Before running the simulation in Fig. 16.6, the code in Appendix G.2.5 must be executed to assign numerical values to all the parameters.

It is observed in Fig. 16.7 that the pendulum swings up and is finally balanced at its inverted unstable configuration. Moreover, the wheel velocity remains bounded along the complete simulation. Also note that the applied voltage remains within a suitable range, i.e., between -13[V] and $+13$[V] and, finally, the pendulum energy converges to its desired value V_0. Thus, these simulation results corroborate the theoretical results stated above.

16.6 Controller Implementation

Using the numerical values in Sect. 16.5, $k_d = 1$ and according to (16.7), (16.10), it is found that the swing-up controller is given as:

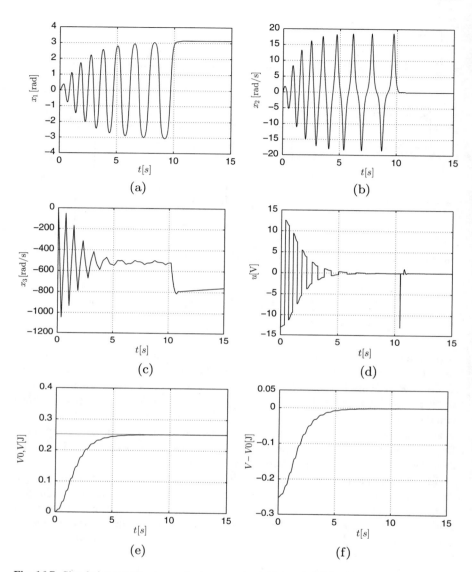

Fig. 16.7 Simulation results when swinging up and stabilizing the IWP with controllers in (16.10) and (16.30). (**a**) Pendulum position. (**b**) Pendulum velocity. (**c**) Wheel velocity. (**d**) Applied voltage. (**e**) Pendulum energy. (**f**) Pendulum energy error

$$V = C_1 \dot{q}_1^2 + \overline{m}g(1 - \cos(q_1)), \quad C_1 = \frac{1}{2}J, \quad V_0 = 2\overline{m}g,$$

$$u = C_2 sat(\dot{q}_1) \, sign(V - V_0), \quad C_2 = \frac{Rk_d}{k_m}, \quad \varepsilon_0 = \frac{u_{max}k_m}{k_d R}, \quad (16.42)$$

where $u_{max} = 13[\text{V}]$, whereas the constants C_1 and C_2 are computed only once when initializing variables, to reduce the number of multiplications required for implementation.

On the other hand, according to Sects. 7.14.1 and 7.14.2, in Chap. 7, it is found that the linear controller (16.30), together with the gain vector $K = [k_1, k_2, k_3] = [-340, -11, -0.0085]$ assign the eigenvalues of matrix $(A - BK)$ at: $\bar{\lambda}_1 = -5.8535 + 17.7192j$, $\bar{\lambda}_2 = -5.8535 - 17.7192j$, $\bar{\lambda}_3 = -0.5268$. Hence, the controller (16.30) is expressed as:

$$u = -k_1(q_1 - q_{1d}) - k_2\dot{q}_1 - k_3(\dot{q}_2 - \dot{q}_{2d}), \quad (16.43)$$

where any of $q_{1d} = +\pi$ or $q_{1d} = -\pi$ is used, depending on which one is first reached by q_1. Recall that $\dot{q}_{2d} = c$ is the wheel velocity at the time when the controller (16.43) starts to work.

Define the condition:

$$(q_1 - q_{1d})^2 + \dot{q}_1^2 < \delta, \quad (16.44)$$

for some constant that is small enough $\delta > 0$ and $q_{1d} = +\pi$ or $q_{1d} = -\pi$. This condition indicates that the pendulum is close to one of the points $(q_1, \dot{q}_1) = (+\pi, 0)$ or $(q_1, \dot{q}_1) = (-\pi, 0)$. In the experiments presented later $\delta = 0.3$ has been used.

The control strategy to swing up and to balance the IWP is to employ (16.42) until (16.44) is satisfied and, from that time on, turn off (16.42) and turn on (16.43). This task is performed using the Microchip PIC18F4431 microcontroller. It is explained next how this is done. This part is not intended to present a detailed explanation on how this microcontroller works, but merely a basic description of how this microcontroller is employed to implement the control strategy (16.42), (16.43), (16.44).

The complete code used to program the microcontroller is presented at the end of this chapter. The microcontroller PIC18F4431 [16] works at 40 MHz as maximal clock frequency. In this chapter, a 11-MHz crystal is used. Five programmable input–output ports, three 16-bit timers, one 8-bit timer (employed to fix the sample period $T_s = 0.01[\text{s}]$), and nine 10-bit analog/digital channels (the tachometer voltage enters through channel 0), are available. There are also eight 14-bit PWM channels among which CCP1 is programmed to work with 8 bit to deliver voltage to be applied to the actuator (PM brushed DC motor).

The position q_1 is measured using an incremental encoder model CP-360-S from Computer Optical Products Inc. [17]. This device, once connected to the microcontroller, has a resolution of 1440 ppr. Pulses of this incremental encoder are read through the microcontroller pins 5 and 6 (see Fig. 16.8). The incremental encoder counts are stored in two 8-bit bytes called POSCNTH, the most significative, and POSCNTL, the less significative. The actual count is computed as:

$$pos = 256 * POSCNTH + POSCNTL,$$

The instruction "q1=(signed long)pos" assigns to q1 the numerical value of the encoder count, which includes the corresponding sign, i.e., positive if movement is as that shown in Fig. 16.2 or negative otherwise. Finally, the position q_1 is given in radians when performing:

$$q_1 = \frac{2\pi}{1440} q1, \quad \frac{2\pi}{1440} = 0.004363.$$

It is worth stating that the complete control system must be turned on when the pendulum is at rest at $q_1 = 0$, according to the convention in Fig. 16.2. The pendulum velocity $\dot{q}_1$ is computed by numerical differentiation of the pendulum position q_1, i.e.,:

$$\dot{q}_1 = \frac{q_1(k) - q_1(k-1)}{T},$$

where k stands for the discrete time, $\dot{q}_1$ represents the estimate of the pendulum velocity in radians/second whereas $T_s = 0.01$[s] is the sampling period. Hence, $q_1(k)$ represents the value of q_1 measured at the present sample time and $q_1(k-1)$ represents the value of q_1 measured in the previous sample time.

The wheel velocity $\dot{q}_2$ is measured using a tachometer model 34PC. It is a miniature PM brushed DC motor used in cell phones as a buzzer. Voltage delivered by the tachometer is low-pass filtered using the RC circuit shown in Fig. 16.8. It was found that this tachometer delivers 0.58[V] when the wheel velocity is maximal, i.e., 1634 [rad/s]. It was assumed that this voltage is proportional at any other intermediate velocity. To improve the resolution of this velocity measurement, the tension divider connected to pin 4 (see Fig. 16.8) was employed to fix at 3.8372[V] the minimal value of the analog range of the analog/digital converter channel 0 (the maximal value is fixed at 5[V]). On the other hand, the tachometer is series connected to another tension divider, such that, when the tachometer is at rest, i.e., when $\dot{q}_2 = 0$, voltage at channel 0 is (5-3.8372)/2+3.8372=4.4186[V]. Notice that variations of about 0.58[V], i.e., 1634[rad/s], are allowed upward and downward from the middle point of the converter analog range. Hence, when $\dot{q}_2 = 1634$[rad/s] the analog/digital converter delivers the count $1023 = 2^{10} - 1$ (the AD converter is 10-bit) and when $\dot{q}_2 = 0$[rad/s] the analog/digital converter delivers the count $1023/2 = 511$. Counts smaller than 511 represent negative velocities $\dot{q}_2$. According to this information, the wheel velocity is computed as follows. The instruction "vel=read-adc()-511" assigns to the variable "vel" the code corresponding to the wheel velocity, whereas the instructions "q2_p=(signed long)vel" and "q2_p=3.2039*q2_p" assign to variable "q2_p" a value that is numerically equal to the wheel velocity $\dot{q}_2$ in radians/second, the sign included. Notice that the constant 3.2039=1634/510[(rad/s)/count] represents the measurement system gain.

With this information, any of the expressions in (16.42) or (16.43) are computed. The resulting variable u is sent to the microcontroller PWM as described in the following.

An equivalent way of implementing the saturation function in (16.42) with a smaller number of computations is the following. It is known that u only takes values in the range $[-13, +13][V]$; hence, its value is constrained to this range of values by using instructions such as:

$$if \ u > 13 \ then \ u = 13,$$

$$if \ u < -13 \ then \ u = -13.$$

As the PWM resolution is $2^8 - 1 = 255$ counts, then the following value must be sent to the microcontroller PWM to correctly deliver u to the power amplifier:

$$cuenta = \frac{255}{13} \ abs(u),$$

where $abs(u)$ represents the absolute value of u. The sign of u is stored in two control bits. The PWM signal is applied to an H bridge (L293B from SGS-Thomson Microelectronics), which suitably amplifies the power of the PWM signal. This H bridge is fed with a 15.3[V] power supply. However, according to the data sheet of this device [18], the voltage drops constrain to $\pm13[V]$ the voltage signal to be delivered. This device has two bits, which are employed to indicate the sign of the power voltage to deliver. Thus, the sign of u is sent to the H bridge through the bits 0 and 1 of the microcontroller port D. This means that the average power voltage sent to the motor is in the range $[-13, +13][V]$. Hence, it is ensured that the voltage applied at the motor terminals u has an average value that is numerically equal to u, computed by any of the expressions in (16.42) or (16.43). The electric diagram used to implement the control strategy in (16.42), (16.43), (16.44), is shown in Fig. 16.8.

16.7 Experimental Results

The experimental results obtained with the proposed control scheme are shown in Figs. 16.9, 16.10, 16.11, 16.12, and 16.13. It is important to recall that it is necessary to apply a slight hit on the pendulum at the beginning to abandon the configuration $(q_1, \dot{q}_1) = (0, 0)$. Although this can be replaced by sending a small voltage pulse to the actuator at the beginning, this was not programmed.

In Fig. 16.9, it is observed how the pendulum oscillates with ever-growing amplitude until the position $q_1 = -\pi$ is reached with zero velocity at $t \approx 14[s]$ and the pendulum stays there from that time on. In Fig. 16.10, it is observed that the

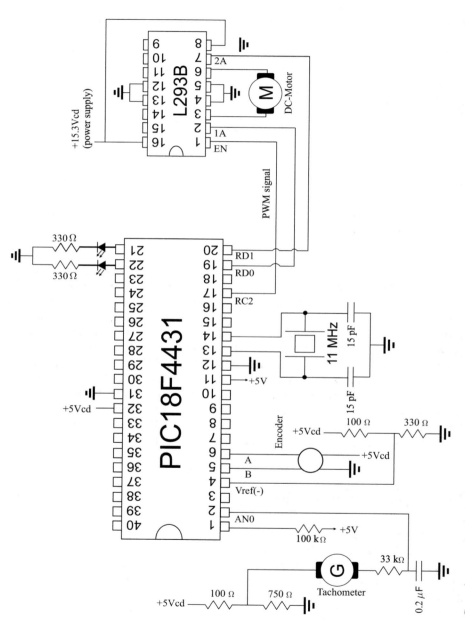

Fig. 16.8 Control system electric diagram

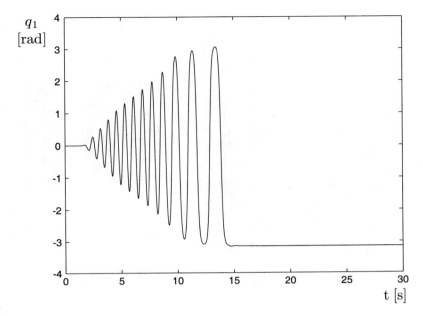

Fig. 16.9 Pendulum position as the inverted unstable configuration is reached

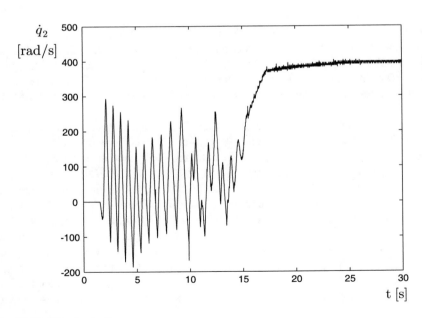

Fig. 16.10 Wheel velocity

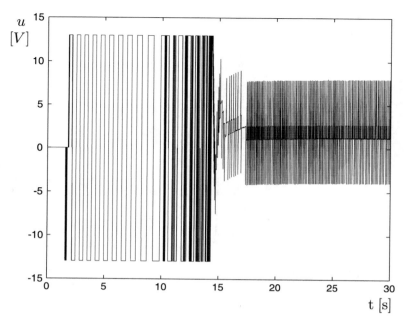

Fig. 16.11 Voltage applied to the motor

wheel velocity remains constant at 380[rad/s] from $t \approx 17$[s] on. It is interesting to point out that the wheel velocity is commanded to be regulated at $c \approx 150$[rad/s], which is the wheel velocity at $t \approx 14$[s], i.e., when the controller (16.43) starts to work. The difference between c and the final wheel velocity 380[rad/s] is due to friction present at the wheel shaft, which has not been taken into account during the design stage.

In Fig. 16.11, the control signal u is shown. Note that this variable saturates most of the time during the swing-up stage ($t < 14$[s]). This is the reason why saturation and sign functions are used in the controller (16.42). In fact, $\varepsilon_0 = 0.024$ is chosen so that with this value and the factor $\frac{4.172}{0.00775}$, u saturates at ± 13[V].

In Fig. 16.12, it is shown how the pendulum evolves on the plane defined in Fig. 16.3, Sect. 16.3. It is clear that, comparing both figures, the pendulum moves such that its energy V increases (see also Fig. 16.13) until $V = V_0 = 0.2519$. Note that V reaches such a value as $t \approx 11$[s], but the pendulum is balanced until $t \approx 14$[s]. This demonstrates that energy is regulated at $V = V_0$ until the point $(q_1, \dot{q}_1) = (-\pi, 0)$ is reached. It is clear that the pendulum also passes close to the point $(q_1, \dot{q}_1) = (+\pi, 0)$ at $t \approx 13$[s] but it is not balanced, perhaps the condition in (16.44) was not satisfied at that time, i.e., the pendulum did not pass close enough to such a point.

Finally, in Fig. 16.14, a picture is shown where the IWP is observed balancing at the inverted unstable configuration.

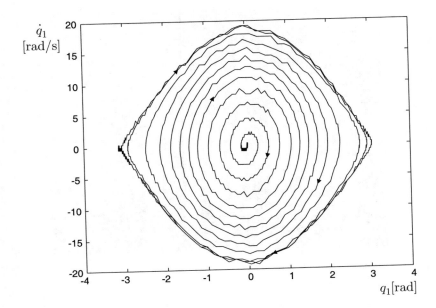

Fig. 16.12 Pendulum evolution on the plane $\dot{q}_1 - q_1$ (phase plane)

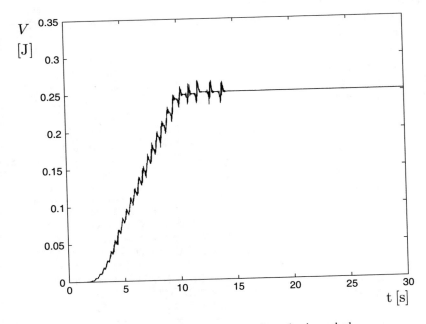

Fig. 16.13 Energy evolution as the inverted unstable configuration is reached

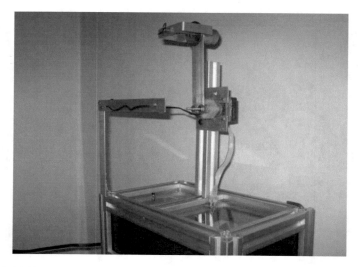

Fig. 16.14 The inertia wheel pendulum balancing at the inverted unstable configuration

16.8   PIC18F4431 Microcontroller Programming

The reader is referred to [19] for detailed explanations on each one of the instructions appearing in the following program. The corresponding flow diagram is presented in Fig. 16.15.

Fig. 16.15 Flow diagram for PIC18F4431 microcontroller programming when used to control an inertia wheel pendulum

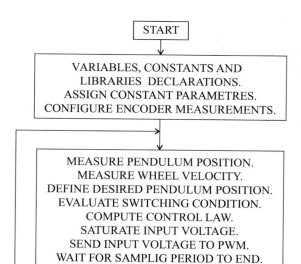

```
#include<18f4431.h>
#device adc=10 //adc, 10 bits
#include<stdlib.h>
#include<math.h>

#fuses HS,NOWDT,NOPROTECT,PUT,NOLVP,NOBROWNOUT,NOWRTC,
MCLR,SSP_RC
#define pi_ 3.14159
//nonlinear controller gain
#define kd 1.0
//linear controller gains
#define k1 340
#define k2 11
#define k3 0.0085
//parameters
#define mbg 0.12597
#define R 4.172

#define L 0.0009
#define Kb 0.00775
#define Km 0.00775
#define d11 0.0014636
#define d12 0.0000076

#define d21 0.0000076
#define d22 0.0000076

#define delta 0.3
#define ts 0.01
//with timer=108, each count= (4/FXtal)*256 sec

/*Register addresses*/
#use delay(clock=11000000)
//time basis (11 MHz)
#byte porta = 0xf80
//ports addresses
#byte portb = 0xf81

#byte portc = 0xf82
#byte portd = 0xf83
#byte portd = 0xf84

#byte TMR0H = 0xfd7
#byte TMR0L = 0xfd6
#byte T0CON = 0xfd5

#byte INTCON= 0xff2
#byte TMR5H=0xf88
//high encoder count

#byte TMR5L=0xf87 //low encoder count

#byte QEICON=0xfB6 //config. quadrature module
```

```
#byte T5CON=0xfB7 //Config. TIMER 5

#byte POSCNTL=0xF66 //CAP2BUFL (reg. low count)
#byte POSCNTH=0xF67 //CAP2BUFH (reg. high count)
#byte CAP1CON=0xF63 //reg. config. reset time basis

#byte PIE3=0xFA3

/*bits addresses */

#bit VCFG1=0xfc1.7 //Bit config. ADC

#bit PD0=0x0f83.0 //control CCW H bridge
#bit PD1=0x0f83.1
//control CW H bridge
#bit PD2=0x0f83.2
//led at board
#bit PD3=0x0f83.3 //led a board
/*variables declaration*/

float u,q1,q1_p,q1_1,q2_p,q1_d,q2_pd,gr,V,S,nf,i_ts,
    Je2,V0,RKdeKm;
long pos,vel,inter;
int pwm;
signed int n,mod;

/*main program*/
void main(void)
{
set_tris_a(0b11111111);
//Config. I/O ports
set_tris_b(0b00000000);
set_tris_c(0b10000000);
set_tris_d(0b00000000);
portb=0X00;
portc=0X00;
portd=0X00;
//CONFIG. PWM
setup_ccp1(CCP_PWM);
setup_ccp2(CCP_PWM);
setup_timer_2(T2_DIV_BY_16,255,1);
// Config. encoder reading
QEICON=QEICON | (0b00010101);
QEICON=QEICON & (0b01110101);
T5CON=T5CON | (0b00011001);
T5CON=T5CON & (0b10011101);
CAP1CON=CAP1CON | (0b01001111);

setup_timer_1(T1_INTERNAL | T1_DIV_BY_8);
TMR5L=0;
TMR5H=0;
POSCNTL=0;
POSCNTH=0;
//Config. ADC
```

```
setup_port_a(sAN0);
setup_adc(ADC_CLOCK_INTERNAL);
set_adc_channel(0);
delay_us(50);
VCFG1=1; //-Vref=AN2
INTCON=0;//turn off interrupt
//Config. timer 0
TMR0L=0;
T0CON=0xC7;
//initializing variables
i_ts=1/ts;
Je2=(d11*d22-d12*d21)/(2*d12);
V0=2*mbg;
RKdeKm=R*Kd/Km;
q1=0.0;
q1_1=0.0;
q2_pd=0.0;
while (TRUE)
{
PD3=1;
q1_1=q1; // q1(k-1)
pos=POSCNTH;
pos=pos<<8;
pos=pos+POSCNTL;
q1=(signed long)pos;
q1=0.004363*q1; //arm position in rad
q1_p=(q1-q1_1)*i_ts; //arm velocity
vel=read_adc()-511; //wheel velocity
q2_p=(signed long)vel;
q2_p=3.2039*q2_p; //wheel velocity in rad/seg
//desired q1
n=q1/pi_;
nf=(float)n;
mod=n%2;
if(nf==0)
if(q1>=0) // desired value
q1_d=pi_;
else
q1_d=-pi_;
if((nf!=0)&&!PD2)
{
if(q1>=0) // desired value
{
if(mod==0)
q1_d=pi_*(nf+1.0);
else
q1_d=pi_*nf;
}
else
{
if(mod==0)
q1_d=pi_*(nf-1.0);
else
q1_d=pi_*nf;
```

```
}
}
//defining the operation region for each controller
if((q1-q1_d)*(q1-q1_d) + q1_p*q1_p < delta)
{
PD2=1; //linear controller
u= k1*(q1-q1_d) + k2*q1_p + k3*(q2_p-q2_pd);
}
else
{
PD2=0; //nonlinear controller
V=Je2*q1_p*q1_p+mbg*(1-cos(q1));
V=V-V0;
S=0; //sign function
if(V>0)
S=1.0;
if(V<0)
S=-1.0;
u=RKdeKm*q1_p*S; //nonlinear controller ends
q2_pd=q2_p; //wheel desired velocity
}
//saturation 13V(15.3 V - 2.3 V drop at H bridge= 13 V)
if(u>13)
u=13;
if(u<-13)
u=-13;
//scaling output, pwm 8 bits, 13 V
u=19.6*u;
//sense of rotation
if(u>=0)
{
PD0=1;
PD1=0;
}
else
{
PD0=0;
PD1=1;
}
//sending pwm
pwm=(unsigned int)abs(u);
set_pwm1_duty(pwm); //pwm
PD3=0; //control computation ends
while(TMR0L<108); //each count = (4/FXtal)*256 sec
TMR0L=0;
T0CON=0xC7;
} //closing infinite while
} //closing main
```

16.9 Summary

In this chapter, another underactuated nonlinear system has been presented: the inertia wheel pendulum. The mathematical model has been obtained and two controllers have been used to solve two different control problems: swing-up and balancing. The balancing control problem has been solved using a linear state feedback controller, whereas the swing-up control problem has been solved using a nonlinear controller. This has been done to introduce the reader to the study of nonlinear control systems. All parameters of the mechanism have been identified and the complete control system has been tested in experiments. A detailed description of the mechanism construction has also been presented.

16.10 Review Questions

1. Why is the inertia wheel pendulum an underactuated mechanism?
2. Why must the IWP oscillate several times before reaching the inverted configuration?
3. What is the corner frequency in [rad/s] and [Hz] of the tachometer resistor–capacitor filter shown in Fig. 16.8?
4. Why does the amplitude of voltage in Fig. 16.11 remain between 13[V] and -13[V] when the swing-up task is performed?

References

1. R. Kelly, V. Santibañez, and A. Loria, *Control of robot manipulators in joint space*, Springer, London, 2005.
2. M. W. Spong, P. Corke, and R. Lozano, Nonlinear control of the reaction wheel pendulum, *Automatica*, vol. 37, no. 11, pp. 1845–1851, 2001.
3. R. Olfati-Saber, Global stabilization of a flat underactuated system: the inertia wheel pendulum, *Proc. of the 40th IEEE Conf. on Decision and Control*, Orlando, Florida, pp. 3764–3765, 2001.
4. R. Ortega, M. W. Spong, F. Gómez-Estern, and G. Blankenstein, Stabilization of a class of underactuated mechanical systems via interconnection and damping assignment, *IEEE Transactions on Automatic Control*, vol. 47, no. 8, pp. 1218–1233, 2002.
5. S. Afkhami, M. J. Yazdanpanah, and P. J. Maralani, Stabilization of inertia wheel pendulum using output feedback back-stepping, *Proc. of IEEE Conf. on Control Applications*, vol. 2, pp. 977–982, 2003.
6. D. M. Alonso, F. I. Robbio, E. E. Paolini, and J. L. Moiola, Modelling an inertia wheel pendulum benchmark, *Mathematical and Computer Modelling of Dynamical Systems*, vol. 11, no. 3, pp. 255–272, 2005.
7. R. Kelly and R. Campa, IDA-PB control of the inertia wheel pendulum: the Lagrangian approach (in Spanish), *Revista Iberoamericana de Automática e Informática Industrial*, vol. 2, no. 1, pp. 36–42, 2005.
8. V. Santibañez, R. Kelly, and J. Sandoval, Control of the inertia wheel pendulum by bounded torques, *Proc. of the 44th IEEE Conf. on Decision and Control, and European Control Conf. CDC-ECC '05*, Seville, Spain, pp. 8266–8270, 2005.

9. M. Abrahantes, J. Mulder, and K. Butter, Modeling, identification and control of an under actuated inertial wheel pendulum, *Proc. of the 39th Southeastern Symposium on System Theory*, Macon, Georgia, pp.1–5, 2007.
10. N. Qaiser, N. Iqbal, A. Hussain, and N. Qaiser, Exponential stabilization of a class of underactuated mechanical systems using dynamic surface control, *International Journal of Control, Automation, and Systems*, vol. 5, no. 5, pp. 547–558, 2007.
11. R. V. Carrillo-Serrano and V. M. Hernández-Guzmán, Control of the inertia wheel pendulum taking into account the actuator dynamics, *International Journal of Innovative Computing, Information and Control*, vol. 6, no. 12, pp. 5553–5563, 2010.
12. H. Khalil, *Nonlinear Systems*, 3rd Edition, Prentice-Hall, Upper Saddle River, 2002.
13. A. L. Fradkov, P. Y. Guzenko, D. J. Hill, and A. Y. Pogromsky, Speed gradient control and passivity of nonlinear oscillators, *Proc. 3rd. IFAC Symposium on Nonlinear control systems (NOLCOS'95)*, vol. 2, pp. 655–659, 1995.
14. R. C. Hibbeler, *Engineering Mechanics Dynamics*, 10th Edition, Prentice Hall, Upper Saddle River, NJ, 2004.
15. M. Alonso and E. J. Finn, *Physics*, Addison-Wesley, Reading, MA, 1992.
16. PIC18F4431 Enhanced Flash Microcontroller, Data sheet, Microchip Technology Inc., 2007.
17. CP360S Housed Encoder, Computer Optical Products Inc., www.opticalencoder.com.
18. L293B Push-Pull Four channel drivers, Data sheet, SGS-Thomson Microelectronics, 1993.
19. Custom Computer Services Incorporated, CCS C Compiler Reference Manual, 2003.

A
Fourier and Laplace Transforms

In this appendix, an interpretation of the Laplace transform is given. The explanations are given in terms of intuitive ideas and a rigorous mathematical presentation is avoided. Readers interested in the mathematical foundations are referred to previous works listed at the end of this appendix [6–11].

A.1 Fourier Series

A periodic function of time $f(t)$, with period $2p$, can be represented by [1–3]:

$$f(t) = \frac{a_0}{2} + \sum_{n=1}^{\infty} \left[a_n \cos\left(\frac{n\pi}{p}t\right) + b_n \sin\left(\frac{n\pi}{p}t\right) \right], \tag{A.1}$$

where a_0, a_n, b_n, are known as the *Fourier coefficients* and can be computed as:

$$a_0 = \frac{1}{p} \int_0^{2p} f(t)dt,$$

$$a_n = \frac{1}{p} \int_0^{2p} f(t) \cos(\frac{n\pi}{p}t)dt, \tag{A.2}$$

$$b_n = \frac{1}{p} \int_0^{2p} f(t) \sin(\frac{n\pi}{p}t)dt.$$

The expressions in (A.2) are known as the *Euler formulas* and the series in (A.1) is known as the *Fourier series* of $f(t)$. Note that, according to (A.1), a periodic function can be represented as the addition of many sine and cosine functions of different *discrete frequencies*, i.e., determined by the integer n, that go from 0 to

© Springer International Publishing AG, part of Springer Nature 2019
V. M. Hernández-Guzmán, R. Silva-Ortigoza, *Automatic Control with Experiments*,
Advanced Textbooks in Control and Signal Processing,
https://doi.org/10.1007/978-3-319-75804-6

$+\infty$. These functions are also called the frequency components of $f(t)$. Fourier coefficients represent the contribution to function $f(t)$ of the different frequency components. An important observation is that (A.1) assumes that $f(t)$ is periodic for all real t. However, not every periodic function can be represented by a Fourier series. The conditions that determine that a periodic function can be represented by a Fourier series are known as the *Dirichlet conditions*:

Theorem A.1 ([1]) *If $f(t)$ is a bounded periodic function with a finite number of maxima and minima in addition to a finite number of discontinuity points in any period, then the Fourier series in (A.1) converges to $f(t)$ at all points where $f(t)$ is continuous and it converges to the average of the left and right limits of $f(t)$ at each point where $f(t)$ is discontinuous [3].*

Definition A.1 ([3])   A function $f(t)$ is said to be an *even function* if:

$$f(-t) = f(t), \quad \forall t \in \mathcal{R},$$

and it is said to be an *odd function* if:

$$f(-t) = -f(t), \quad \forall t \in \mathcal{R}.$$

As the cosine function is even and the sine function is odd, under certain symmetry conditions in $f(t)$ the Fourier series may be composed only of sine or only of cosine functions, as stated in the following.

Theorem A.2 ([1])   *If $f(t)$ is an even function, the corresponding Fourier coefficients are given as:*

$$a_0 = \frac{2}{p} \int_0^p f(t) dt,$$

$$a_n = \frac{2}{p} \int_0^p f(t) \cos\left(\frac{n\pi}{p}t\right) dt, \quad n = 1, 2, 3, \dots$$

$$b_n = 0, \quad n = 1, 2, 3, \dots$$

Theorem A.3 ([1])   *If $f(t)$ is an odd function, the corresponding Fourier coefficients are given as:*

$$a_n = 0, \quad n = 0, 1, 2, 3, \dots$$

$$b_n = \frac{2}{p} \int_0^p f(t) \sin\left(\frac{n\pi}{p}t\right), \quad n = 1, 2, 3, \dots$$

On the other hand, any periodic function $f(t)$ can be expressed as the addition of one even function and one odd function [3]. Hence, according to Theorems A.2 and A.3, the first function contributes to a Fourier series composed only of cosine functions and the second function contributes to a Fourier series composed only of sine functions. This explains, to some extent, why the general Fourier series contains both cosine and sine functions.

Fourier series (A.1) can be written in several equivalent ways, for instance [1]:

1.

$$f(t) = A_0 + \sum_{n=1}^{\infty} A_n \cos\left(\frac{n\pi}{p}t - \gamma_n\right),$$

$$A_0 = \frac{a_0}{2}, \quad A_n = \sqrt{a_n^2 + b_n^2}, \quad \gamma_n = \arctan\left(\frac{b_n}{a_n}\right), \quad (A.3)$$

where a_0, a_n, b_n are defined in (A.2).

2.

$$f(t) = A_0 + \sum_{n=1}^{\infty} A_n \sin\left(\frac{n\pi}{p}t + \delta_n\right),$$

$$\delta_n = \arctan\left(\frac{a_n}{b_n}\right), \quad (A.4)$$

where A_0 and A_n are defined in (A.3).

Note that the phases γ_n and δ_n are different for each one of the terms of the series and, because of this, (A.3) and (A.4) can represent any periodic function $f(t)$, no matter whether it is even or odd.

3.

$$f(t) = \sum_{n=-\infty}^{\infty} C_n e^{j(n\pi/p)t}, \quad (A.5)$$

$$C_n = \frac{1}{2p} \int_0^{2p} f(t) e^{-j(n\pi/p)t} dt,$$

$$e^{\pm j(n\pi/p)t} = \cos\left(\frac{n\pi}{p}t\right) \pm j \sin\left(\frac{n\pi}{p}t\right).$$

According to (A.5), $f(t)$ is composed of a real part and an imaginary part. However, if $f(t)$ is a real function, this implies that its imaginary part is zero; hence, all imaginary terms in (A.5) must cancel among them. This does not mean that (A.5) only contains cosine functions because C_n is also a complex number and this implies that $f(t)$ is composed of cosine and sine functions. Note that, according to (A.5), the Fourier series of $f(t)$ is composed of functions whose frequencies take discrete values from $-\infty$ to $+\infty$ passing by zero. It is stressed that negative frequencies have no physical interpretation and, because of that, (A.5) merely represents an abstract way of representing $f(t)$, which, however, is a convenient way of writing a *Fourier series* to define the *Fourier transform* in the following.

A.2   Fourier Transform

The *Fourier integral* of a nonperiodic function $f(t)$, is defined as [1, 3, 4]:

$$f(t) = \int_{-\infty}^{\infty} F(\omega)e^{j\omega t}\,d\omega, \tag{A.6}$$

$$F(\omega) = \frac{1}{2\pi} \int_{-\infty}^{\infty} f(t)e^{-j\omega t}\,dt. \tag{A.7}$$

$F(\omega)$, a function of the *continuous frequency* ω, is known as the *Fourier transform* of $f(t)$ and both of the previous expressions are known as the *Fourier transform pair*. On the other hand, it is clear that $F(\omega)$ is a complex function. It is possible to show that, when $f(t)$ is a real function, the real part of $F(\omega)$ is an even function and its imaginary part is an odd function [3].

In [1, 3, 4], it is shown that (A.5) is the discrete version of (A.6) and (A.7). A simple way of seeing this is presented in the following. In (A.6) and (A.7), the function $f(t)$ is nonperiodic. Thus, $f(t)$ can be seen as a function with period $2p \to \infty$. Define:

$$\omega = \frac{n\pi}{p}.$$

If $f(t)$ "tends" toward a periodic function, then the period $2p$ takes finite values and ω, originally a continuous variable, "tends" toward the discrete variable $\frac{n\pi}{p}$, taking values from $-\infty$ to $+\infty$. Hence, $d\omega$ stands for the difference between two consecutive values of ω that correspond to two consecutive values of n represented by $n + 1$ and n:

$$d\omega = \frac{(n + 1)\pi}{p} - \frac{n\pi}{p} = \frac{\pi}{p}.$$

On the other hand, C_n is the discrete version of $F(\omega)$. This can be explained as follows. If $f(t)$ is a nonperiodic function, it can be seen as a function with period $2p$ that "tends" toward ∞ and, although $f(t)$ is assumed to be defined on the open interval $t \in (-\infty, +\infty)$, then the limits of the integral, which is computed on a complete period of $f(t)$, change from 0 and $2p$ to $-\infty$ and $+\infty$. Once (A.6) and (A.7) are rendered discrete in this manner, (A.5) and C_n are exactly retrieved when replacing (A.7) in (A.6).

From (A.5) and (A.6), the following observations are in order [1–4]. A periodic function $f(t)$ can be represented as an infinite series of sine and cosine functions with different discrete frequencies $\frac{n\pi}{p}$ and amplitudes C_n, which are functions of discrete frequency. On the other hand, a nonperiodic function $f(t)$ is composed of a large number of sinusoidal functions with a frequency that takes continuous values ω and with amplitudes $F(\omega)$ that are functions of the continuous frequency ω. C_n stands for the contribution, to a periodic function $f(t)$, of the sinusoidal functions of a specific discrete frequency $\frac{n\pi}{p}$, whereas $F(\omega)$ stands for the contribution, to a nonperiodic function $f(t)$, of the sinusoidal functions of a specific continuous frequency ω. The conditions for convergence of (A.6) to $f(t)$ and the existence of (A.7) are the following [1–3]:

1. $f(t)$ satisfies the Dirichlet conditions.
2. The integral $\int_{-\infty}^{\infty} |f(t)| dt$ exists.

Note that Dirichlet conditions are inherited from Fourier series, whereas the existence of the integral $\int_{-\infty}^{\infty} |f(t)| dt$ is a clear condition for the convergence of the integral in (A.7).

A.3 Laplace Transform

A nonperiodic function $f(t)$ that is zero for $t < 0$, can be represented as [1–4]:

$$f(t) = \frac{1}{2\pi j} \int_{a-j\infty}^{a+j\infty} F(s) e^{st} ds, \tag{A.8}$$

$$F(s) = \int_0^\infty f(t) e^{-st} dt, \quad s = a + j\omega, \tag{A.9}$$

where a and ω are real numbers. $F(s)$ is known as the *Laplace transform* of $f(t)$, whereas (A.8) and (A.9) are known as the *Laplace transform pair*. Note that (A.6) and (A.7) are similar to (A.8) and (A.9). A careful observation leads us to conclude that the former pair of expressions are obtained from the latter by assuming $a = 0$ and $f(t) = 0$ for all $t < 0$ [5], i.e., the bottom limit of the integral in (A.7) changes from $-\infty$ to 0. Conversely, (A.8) and (A.9) can be obtained from (A.6) and (A.7) [1, 3, 4] by searching for a method allowing $F(\omega)$ to be computed for some special functions $f(t)$ that are zero for negative time and such that the integral:

$$\int_0^\infty f(t) e^{-i\omega t} dt,$$

does not converge. This means that $f(t)$ does not become zero as $t \to \infty$; hence, the condition 2 above for the existence of $F(\omega)$ does not stand. The fundamental idea [1, 3, 4] is to compute (A.6) and (A.7) for a function:

$$g(t) = f(t)e^{-at}dt, \quad a > 0,$$

trying to accomplish that $g(t) \to 0$ as $t \to \infty$; hence, forcing (A.6) and (A.7) to converge. Then, the following limits are computed:

$$f(t) = \lim_{a \to 0} g(t), \tag{A.10}$$

$$F(\omega) = \lim_{a \to 0} \frac{1}{2\pi} \int_{-\infty}^{\infty} f(t)e^{-at}e^{-i\omega t}dt, \tag{A.11}$$

finally obtaining the Fourier transform pair corresponding to $f(t)$. The parameter a is known as the *convergence factor* because, when introduced within a divergent integral, it leads the integrand to decrease at a rate high enough to ensure convergence of the integral.

Once this result has been found, it is natural to try to choose the convergence factor a in a more general manner, such that convergence of the involved integrals is the only thing that matters without the necessity of computing the limit of the integrals when $a \to 0$. Thus, the Laplace transform pair (A.8), (A.9), can be obtained by replacing $g(t) = f(t)e^{-at}$ instead of $f(t)$ in (A.6), (A.7), such that s is a complex variable with ω as imaginary part, which is identical to the Fourier variable, and a as a constant real part which can be positive, negative or zero. According to the above discussion, the conditions for the existence of (A.8), (A.9), can be summarized as follows [1, 2]:

1. In every finite time interval $f(t)$ is bounded and has, at most, a finite number of maxima and minima and a finite number of discontinuities.
2. There exists a real constant a such that the following integral is convergent:

$$\int_0^\infty |f(t)e^{-at}|dt = \int_0^\infty |f(t)|e^{-at}dt. \tag{A.12}$$

Functions satisfying 1 are called *sectionally regular functions*. Condition 2 is commonly replaced by another more restrictive one [1, 2]:
3. There exists a constant α such that $|f(t)|e^{-\alpha t}$ remains bounded as $t \to \infty$, i.e., real finite constants α, M, T exist, such that:

$$|f(t)|e^{-\alpha t} < M, \quad \text{for all } t > T. \tag{A.13}$$

Functions satisfying 3 are called *functions of exponential order*. Note that the value of α required by this condition is not unique and defines a set of values. The smallest value in this set, denoted by α_0, is known as the *abscissa of convergence* of $f(t)$ [1]. According to condition 3, $f(t)$ can be a function that does not converge to zero when $t \to \infty$ or it can even remain growing, but its grow rate is not larger than that of an exponential function, such that there exists some α ensuring that $|f(t)e^{-\alpha t}|$ stops growing as $t \to \infty$. Hence, it is possible to select some constant

$\alpha_1 > \alpha$ such that $e^{-\alpha_1 t}$ decreases faster than the increment of $f(t)$; thus, it is ensured that:

$$\int_0^\infty |f(t)e^{-\alpha_1 t}| dt, \tag{A.14}$$

converges [1, 2, 4]. These ideas are now formalized.

Theorem A.4 ([1]) *If $f(t)$ is sectionally regular and of exponential order with the abscissa of convergence α_0, then for any s_0 with $Re(s_0) > \alpha_0$, the integral:*

$$F(s) = \int_0^\infty f(t)e^{-st} dt, \tag{A.15}$$

uniformly converges for all values of s such that $Re(s) > Re(s_0)$.

According to this, (A.8), (A.9), can be written as:

$$f(t) = \frac{1}{2\pi j} \int_{a-j\infty}^{a+j\infty} F(s)e^{Re(s)t} e^{j\omega t} ds, \tag{A.16}$$

$$F(s) = \int_0^\infty f(t)e^{-at} e^{-j\omega t} dt, \tag{A.17}$$

where $a > Re(s_0)$. From these expressions, a similar reasoning to that for (A.6), (A.7), can be performed to find the following interpretation for (A.16), (A.17):

The Laplace transform pair (A.8), (A.9), a nonperiodic function $f(t)$ to be represented, which is zero for $t < 0$, as the addition of many sine and cosine functions with different frequencies ω that change continuously from $-\infty$ to $+\infty$, passing by zero, and all of these functions are identically "damped" by the exponential factor e^{at}. Hence, $F(s)$ represents the contribution to the function $f(t)$ of each one of these so-called functions of the complex frequency $s = a + j\omega$. Moreover, the factor e^{at} can have any value for a such that $a > Re(s_0)$. Thus, using a different value for a, it is possible to change the "damping" rate of all the sinusoidal functions comprising $f(t)$.

It is important to say that, although a was introduced at the beginning as a positive number, if $f(t)$ is a decreasing function of exponential order, then there exist some negative values for a such that $a > Re(s_0)$ and, hence, (A.8), (A.9), are still convergent. This is the reason why the word "damped", introduced in the previous paragraph, has been written between quotation marks. Finally, it is stressed that, according to the previous discussion, (A.9) can be used to compute:

$$F(\sigma_0) = F(s)|_{s=\sigma_0} = \int_0^\infty f(t)e^{-\sigma_0 t} dt, \tag{A.18}$$

Fig. A.1 Plane s showing the straight lines used to compute the Fourier transform, from $-j\infty$ to $+j\infty$ (imaginary axis), and the Laplace transform, from $a - j\infty$ to $a + j\infty$

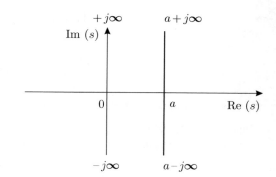

which is ensured to exist for any constant complex number σ_0 such that $\mathrm{Re}(\sigma_0) > \mathrm{Re}(s_0)$.

The complex plane s is shown in Fig. A.1. Note that, according to (A.6), $f(t)$ can be computed as a line integral along a vertical straight line that overlaps the imaginary axis ($\mathrm{Re}(s) = 0$), whereas, according to (A.8), $f(t)$ can be computed as a line integral along a vertical straight line that is parallel to the imaginary axis ($\mathrm{Re}(s) = a$).

B
Bode Diagrams

In this appendix, some details are provided on how to draw Bode diagrams of some basic terms. Instrumental for these results is the use of the logarithm properties listed in (6.10), Chap. 6.

B.1 First-Order Terms

B.1.1 A Differentiator

Consider the following transfer function:

$$\frac{Y(s)}{U(s)} = G(s) = s.$$

Let $s = j\omega$ find:

$$G(j\omega) = j\omega = \omega \angle 90°.$$

Hence:

$$|G(j\omega)|_{dB} = 20\log(\omega), \quad \angle G(j\omega) = 90°.$$

© Springer International Publishing AG, part of Springer Nature 2019
V. M. Hernández-Guzmán, R. Silva-Ortigoza, *Automatic Control with Experiments*,
Advanced Textbooks in Control and Signal Processing,
https://doi.org/10.1007/978-3-319-75804-6

Consider two arbitrary frequencies with a difference of one decade between them, i.e., $\omega_2 = 10\omega_1$, and evaluate:

$$|G(j\omega_1)|_{dB} = 20\log(\omega_1),$$
$$|G(j\omega_2)|_{dB} = 20\log(\omega_2),$$
$$= 20\log(10\omega_1),$$
$$= 20\log(10) + 20\log(\omega_1),$$
$$|G(j\omega_2)|_{dB} = 20 + 20\log(\omega_1).$$

This means that the magnitude increases 20[dB] each time the frequency increases one decade, i.e., the magnitude Bode diagram is a straight line with a slope of +20[dB/dec]. Moreover, $|G(j\omega)|_{dB} = 0$ if $\omega = 1$. This and the constant phase $\angle G(j\omega) = 90°$ explain the Bode diagrams in Fig. 6.16b.

B.1.2   An Integrator

Consider the following transfer function:

$$\frac{Y(s)}{U(s)} = G(s) = \frac{1}{s}.$$

Let $s = j\omega$ to write:

$$G(j\omega) = \frac{1}{j\omega} = \frac{1}{\omega}\angle - 90°.$$

Hence:

$$|G(j\omega)|_{dB} = -20\log(\omega), \quad \angle G(j\omega) = -90°. \tag{B.1}$$

Consider two arbitrary frequencies with a difference of a decade between them, i.e., $\omega_2 = 10\omega_1$, and evaluate:

$$|G(j\omega_1)|_{dB} = -20\log(\omega_1),$$
$$|G(j\omega_2)|_{dB} = -20\log(\omega_2),$$
$$= -20\log(10\omega_1),$$
$$= -20\log(10) - 20\log(\omega_1),$$
$$|G(j\omega_2)|_{dB} = -20 - 20\log(\omega_1).$$

This means that the magnitude decreases $-20[\text{dB}]$ each time the frequency increases one decade, i.e., the magnitude Bode diagram is a straight line with a slope of $-20[\text{dB/dec}]$. Moreover, $|G(j\omega)|_{dB} = 0$ if $\omega = 1$. This and the constant phase $\angle G(j\omega) = -90°$ explain the Bode diagrams in Fig. 6.16a.

Following this order of ideas, it is straightforward to verify that, in the case of the transfer function:

$$\frac{Y(s)}{U(s)} = G(s) = \frac{1}{s^n}, \tag{B.2}$$

for any integer $n \geq 1$ the magnitude Bode plot is a straight line with a slope of $-20n[\text{dB/dec}]$ crossing the $0[\text{dB}]$ line at $\omega = 1$. The phase is constant at $-n90°$.

B.1.3   A First-Order Pole

Consider the following transfer function:

$$\frac{Y(s)}{U(s)} = G(s) = \frac{a}{s+a}, \quad a > 0.$$

Let $s = j\omega$ to obtain:

$$G(j\omega) = \frac{a}{j\omega + a}. \tag{B.3}$$

According to Sect. 6.1.1, this yields:

$$|G(j\omega)| = \frac{a}{\sqrt{\omega^2 + a^2}}, \quad \angle G(j\omega) = \arctan\left(\frac{-\omega}{a}\right). \tag{B.4}$$

If $\omega \ll a$, then:

$$|G(j\omega)|_{dB} \approx 0, \quad \angle G(j\omega) \approx 0°.$$

If $\omega \gg a$, then:

$$|G(j\omega)| \approx \frac{a}{\omega},$$

$$|G(j\omega)|_{dB} = 20\log\left(\frac{a}{\omega}\right),$$

$$= 20\log(a) - 20\log(\omega), \tag{B.5}$$

$$\angle G(j\omega) \approx -90°.$$

According to (B.1), (B.5) and recalling that a is a positive constant, when $\omega \gg a$ the magnitude Bode diagram of (B.4) is a straight line with a slope of $-20[\text{dB/dec}]$, which crosses the $0[\text{dB}]$ line, i.e., when $20\log(a) - 20\log(\omega) = 0$, at $\omega = a$. Moreover, when $\omega = a$:

$$|G(j\omega)| = \frac{1}{\sqrt{2}} \Rightarrow |G(j\omega)|_{\text{dB}} = -3,$$

$$\angle G(j\omega) = -45°.$$

This explains the Bode diagrams in Fig. 6.16c.

B.1.4 A First-Order Zero

Consider the following transfer function:

$$\frac{Y(s)}{U(s)} = G(s) = \frac{s+a}{a}.$$

Let $s = j\omega$ to obtain:

$$G(j\omega) = \frac{j\omega + a}{a}.$$

Notice that this function is the inverse of (B.3) and recall that $\log\left(\frac{1}{x}\right) = -\log(x)$, $\angle z = -\angle\left(\frac{1}{z}\right)$, where z is a complex number. Hence, it is straightforward to verify that the Bode diagrams in this case are the same as those obtained in Sect. B.1.3 but with an opposite sign. This explains the Bode diagrams in Fig. 6.17a.

B.2 A Second-Order Transfer Function

Consider the following transfer function:

$$\frac{Y(s)}{U(s)} = G(s) = \frac{\omega_n^2}{s^2 + 2\zeta\omega_n s + \omega_n^2}, \qquad \omega_n > 0, \quad \zeta \geq 0. \tag{B.6}$$

According to Sect. 6.1.3, we have:

$$|G(j\omega)| = \frac{\omega_n^2}{\sqrt{(\omega_n^2 - \omega^2)^2 + 4(\zeta\omega_n\omega)^2}}, \tag{B.7}$$

$$\angle G(j\omega) = \arctan\left(\frac{-2\zeta\omega_n\omega}{\omega_n^2 - \omega^2}\right).$$

If $\omega \ll \omega_n$, then:

$$|G(j\omega)|_{\mathrm{dB}} \approx 0, \quad \angle G(j\omega) = 0°.$$

If $\omega \gg \omega_n$, then:

$$|G(j\omega)| \approx \frac{\omega_n^2}{\omega^2}, \tag{B.8}$$

$$|G(j\omega)|_{\mathrm{dB}} = 40\log\left(\frac{\omega_n}{\omega}\right),$$

$$= 40\log(\omega_n) - 40\log(\omega),$$

$$\angle G(j\omega) = -180°.$$

According to what was stated at (B.2), when $\omega \gg \omega_n$ the magnitude Bode diagram of (B.8), i.e., (B.7), is a straight line with a slope of $-40[\mathrm{dB/dec}]$, which crosses the $0[\mathrm{dB}]$ line, i.e., when $40\log(\omega_n) - 40\log(\omega) = 0$, at $\omega = \omega_n$. Moreover, when $\omega = \omega_n$:

$$|G(j\omega)| = \frac{1}{2\zeta} \Rightarrow |G(j\omega)|_{\mathrm{dB}} \to \infty, \text{ as } \zeta \to 0.$$

$$\angle G(j\omega) = -90°.$$

This explains the Bode diagrams in Fig. 6.17b.

B.3   A Second-Order Zero

Consider the following transfer function:

$$\frac{Y(s)}{U(s)} = G(s) = \frac{s^2 + 2\zeta\omega_n s + \omega_n^2}{\omega_n^2}. \tag{B.9}$$

Note that this transfer function is the inverse of (B.6) and recall that $\log\left(\frac{1}{x}\right) = -\log(x)$, $\angle z = -\angle\left(\frac{1}{z}\right)$, where z is a complex number. Thus, it is straight forward to show that Bode diagrams of (B.9) are identical to Bode diagrams of (B.6) and we only have to consider that the sign changes in both the magnitude and the phase plots. This explains the Bode diagrams in Fig. 6.17c.

C
Decibels, dB

The bel, ([B]), is a unit of measure for the ratio between two quantities with unities of power. The number of bels is computed as:

$$\log\left(\frac{P_1}{P_0}\right) \text{[B]}.$$

1 bel corresponds to a ratio 10 : 1 of power, i.e., if $P_1 = 10P_0$ then:

$$1\text{[B]} = \log\left(\frac{P_1}{P_0}\right) = \log(10).$$

The decibel, [dB], is defined as:

$$1\text{[dB]} = 0.1\text{[B]}.$$

Hence, if $x = \log\left(\frac{P_1}{P_0}\right)$ is in Bels and we want to convert to decibels, we have to write:

$$y = 10\frac{\text{[dB]}}{\text{[B]}}x = 10\log\left(\frac{P_1}{P_0}\right)\text{[dB]},$$

where y is given in decibels. On the other hand, the power is often given as the square of signals such as voltage, electric current, etc. (also called field quantities). Thus, it is important to express the ratio of power in decibels but in terms of field quantities. In such a case:

$$y = 10\log\left(\frac{P_1}{P_0}\right)\text{[dB]},$$

© Springer International Publishing AG, part of Springer Nature 2019
V. M. Hernández-Guzmán, R. Silva-Ortigoza, *Automatic Control with Experiments*,
Advanced Textbooks in Control and Signal Processing,
https://doi.org/10.1007/978-3-319-75804-6

$$= 10 \log \left(\frac{F_1^2}{F_0^2} \right) \text{[dB]},$$

$$= 20 \log \left(\frac{F_1}{F_0} \right) \text{[dB]}. \tag{C.1}$$

The definition in (C.1) represents the expression for the number of decibels used throughout this book, i.e., for decibel units of the magnitude in frequency response of control systems.

In acoustics, decibels are used to represent an unit of pressure level:

$$20 \log \left(\frac{p}{p_0} \right) \text{[dB]}. \tag{C.2}$$

where $p_0 = 20$[micropascals] is the threshold of perception of an average human. The pressure, p, is a field quantity and this explains (C.2).

In radiofrequency circuits, the symbol dBm is used to express the power ratio in decibels referred to as 1[milliwatt]. This means that given P in milliwatts, then:

$$10 \log \left(\frac{P}{1\text{mW}} \right) \text{[dBm]}.$$

is expressed in dBm.

D
Magnetically Coupled Coils

This appendix follows the ideas presented in [12] and [13].

D.1 Invertance

When electric current through a coil induces a voltage in other coils, it is said that all these coils are magnetically coupled. In electric diagrams, an arc is depicted between coils to indicate that they are magnetically coupled, see Fig. D.1. Under these conditions, the voltage in any of these coils depends on the electric current through all the coils, i.e., according to *Faraday's Law*:

$$v_k = \frac{d\Psi_k}{dt}.$$

where v_k stands for voltage at terminals of the $k-$th coil and Ψ is the magnetic flux linkage in the $k-$th coil:

$$\Psi_k = N_k \psi_k(i_1, \ldots, i_n),$$

with N_k the number of turns of the coil k and ψ_k is the flux linkage per turn of the coil k, which is a function of the electric currents through the n magnetically coupled coils with $1 \leq k \leq n$. Applying the chain rule:

$$v_k = N_k \frac{\partial \psi_k}{\partial i_1} \frac{di_1}{dt} + \cdots + N_k \frac{\partial \psi_k}{\partial i_k} \frac{di_k}{dt} + \cdots + N_k \frac{\partial \psi_k}{\partial i_n} \frac{di_n}{dt}. \tag{D.1}$$

The *self-inductance* of coil k is defined as:

© Springer International Publishing AG, part of Springer Nature 2019
V. M. Hernández-Guzmán, R. Silva-Ortigoza, *Automatic Control with Experiments*,
Advanced Textbooks in Control and Signal Processing,
https://doi.org/10.1007/978-3-319-75804-6

$$L_{kk} = N_k \frac{\partial \psi_k}{\partial i_k}, \tag{D.2}$$

whereas:

$$L_{kj} = N_k \frac{\partial \psi_k}{\partial i_j}, \tag{D.3}$$

is defined as the *mutual inductance* between the coils k and j, $1 \leq k \leq n$, $1 \leq j \leq n$, $k \neq j$. Moreover, $L_{kj} = L_{jk}$. It is not difficult to see, from (D.1), (D.2), (D.3), that it is possible to write:

$$V = M \frac{dI}{dt},$$

$$V = [v_1, v_2, \ldots, v_k, \ldots, v_n]^T, \quad \frac{dI}{dt} = \left[\frac{di_1}{dt}, \frac{di_2}{dt}, \ldots, \frac{di_k}{dt}, \ldots, \frac{di_n}{dt} \right]^T,$$

$$M = \begin{bmatrix} L_{11} & L_{12} & \cdots & L_{1k} & \cdots & L_{1n} \\ L_{21} & L_{22} & \cdots & L_{2k} & \cdots & L_{2n} \\ \vdots & \vdots & \ddots & \vdots & \ddots & \vdots \\ L_{k1} & L_{k2} & \cdots & L_{kk} & \cdots & L_{kn} \\ \vdots & \vdots & \ddots & \vdots & \ddots & \vdots \\ L_{n1} & L_{n2} & \cdots & L_{nk} & \cdots & L_{nn} \end{bmatrix}.$$

Hence, according to *Cramer's rule* [1], chapter 10, if M is a nonsingular matrix:

$$\frac{di_k}{dt} = \frac{\det(M_k)}{\det(M)},$$

where M_k is the matrix obtained when replacing the column k of matrix M by V, i.e.,:

$$M_k = \begin{bmatrix} L_{11} & L_{12} & \cdots & v_1 & \cdots & L_{1n} \\ L_{21} & L_{22} & \cdots & v_2 & \cdots & L_{2n} \\ \vdots & \vdots & \ddots & \vdots & \ddots & \vdots \\ L_{k1} & L_{k2} & \cdots & v_k & \cdots & L_{kn} \\ \vdots & \vdots & \ddots & \vdots & \ddots & \vdots \\ L_{n1} & L_{n2} & \cdots & v_n & \cdots & L_{nn} \end{bmatrix}.$$

Solving $\det(M_k)$ through the k−th column, it is possible to write:

$$\frac{di_k}{dt} = \frac{v_1 \text{cof}(M_{1k}) + v_2 \text{cof}(M_{2k}) + \cdots + v_k \text{cof}(M_{kk}) + \cdots + v_n \text{cof}(M_{nk})}{\det(M)},$$

$$= \sum_{j=1}^{j=n} \frac{\text{cof}(M_{jk})}{\det(M)} v_j,$$

where $\text{cof}(M_{jk})$ stands for the cofactor of the entry (j, k) of matrix M and the constant:

$$\Gamma_{kj} = \frac{\text{cof}(M_{jk})}{\det(M)}, \tag{D.4}$$

relating the electric current in the k−th coil to voltage in the j−th coil, is known as the *mutual invertance* between those coils. This expression indicates that invertances are functions of self-inductances and mutual inductances of the whole system for magnetically coupled coils, i.e.,:

$$\Gamma_{kj} \neq \frac{1}{L_{kj}}, \quad \text{and} \quad \Gamma_{kk} \neq \frac{1}{L_{kk}}.$$

Moreover, $\Gamma_{kj} = \Gamma_{jk}$.

D.2 Coil Polarity Marks

Consider two magnetically coupled coils as depicted in Fig. D.1a. Define the electric current sense in coil L_1 and put a polarity mark "*" at the terminal of the coil where the electric current enters the coil. A magnetic flux is produced in the core whose sense is defined according to the right-hand rule.[1] According to *Lenz Law*, a voltage is induced at the coil L_2 with a polarity such that, if the terminals of L_2 are put in short circuit, an electric current flows through L_2 producing a magnetic flux that opposes the flux produced by L_1. Use again the right-hand rule to determine the sense of electric current through L_2 producing this effect. Put a polarity mark "*" at the terminal of L_2 where the electric current leaves the coil L_2. This ensures that the instantaneous voltages at L_1 and L_2, only due to effect of mutual induction, have the same polarity at the terminals with the polarity marks "*". The mutual inductance is negative, $L_{12} = L_{21} < 0$, for coils such as those shown in Fig. D.1a.

In the case of Fig. D.1b, the senses of electric currents through L_1 and L_2 are defined such that the magnetic fluxes produced by them in the core are in the

---

[1] *Right-hand rule*: if the right-hand fingers indicate the sense of electric current in a coil, then the thumb indicates the sense of the generated magnetic flux.

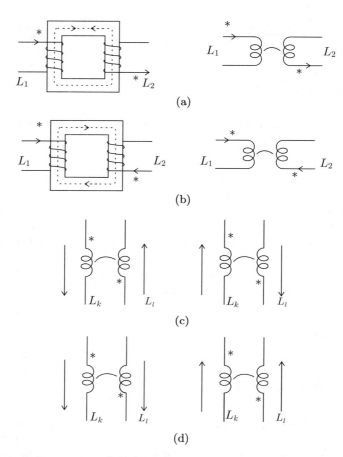

Fig. D.1 Magnetically coupled coils. (**a**) $L_{12} = L_{21} < 0$. (**b**) $L_{12} = L_{21} > 0$. (**c**) $L_{kl} = L_{lk} > 0$. (**d**) $L_{kl} = L_{lk} < 0$

same direction. In this case $L_{12} = L_{21} > 0$. Signs of the mutual invertances are determined by (D.4).

E

Euler–Lagrange Equations Subject to Constraints

Euler–Lagrange equations are used to model, among others, mechanical systems composed of several bodies that interact. Let n represent the number of degrees-of-freedom (or number of independent movements of the bodies) of the mechanical system. The Euler–Lagrange equations describing this mechanical system are written in vector form as [14]:

$$\frac{d}{dt}\frac{\partial L}{\partial \dot{q}} - \frac{\partial L}{\partial q} = \tau + \tau_Q, \tag{E.1}$$

where the Lagrangian of the system is defined as $L = K - P$ with K and P representing the total kinetic and total potential energies of the system respectively, $\tau = [\tau_1, \ldots, \tau_n]^T \in \mathcal{R}^n$ is the vector of (external) generalized forces applied to the system, $\tau_Q = [\tau_{Q1}, \ldots, \tau_{Qn}]^T \in \mathcal{R}^n$ is the vector of nonconservative forces, i.e., friction, applied to the system, and $q = [q_1, \ldots, q_n]^T \in \mathcal{R}^n$ is the system vector of generalized coordinates, i.e., the positions defining the n degrees-of-freedom. This means that the expression in (E.1) represents n differential equations to be solved simultaneously. See Sects. 15.1 and 16.2 for some application examples of (E.1).

Some mechanical systems are subject to one linear constraint in velocities, i.e., an expression given as:

$$a_1(q)\dot{q}_1 + a_2(q)\dot{q}_2 + \cdots + a_n(q)\dot{q}_n = 0,$$

which can be written in vector form as:

$$A(q)\dot{q} = 0, \quad A(q) = [a_1(q), a_2(q), \ldots, a_n(q)]. \tag{E.2}$$

To take into account the constraint in (E.2) when solving (E.1), the following procedure is suggested in [15]. It is assumed that the constraint includes additional generalized forces, $\tau_c = [\tau_{c1}, \ldots, \tau_{cn}]^T \in \mathcal{R}^n$, which are intended to render the

© Springer International Publishing AG, part of Springer Nature 2019
V. M. Hernández-Guzmán, R. Silva-Ortigoza, *Automatic Control with Experiments*,
Advanced Textbooks in Control and Signal Processing,
https://doi.org/10.1007/978-3-319-75804-6

constraint always satisfied. Thus, (E.1) becomes:

$$\frac{d}{dt}\frac{\partial L}{\partial \dot{q}} - \frac{\partial L}{\partial q} = \tau + \tau_Q + \tau_c. \tag{E.3}$$

The generalized forces τ_c due to the constraint are assumed not to produce any mechanical work, i.e., a basic requirement when the Euler–Lagrange equations are obtained [15]. This means that, under the effect of τ_c, the mechanical system displacement must be performed in an orthogonal direction to τ_c, i.e., $\dot{q}^T \tau_c = 0$. Then it is possible to write:

$$\dot{q}^T \tau_c = \tau_c^T \dot{q} = 0.$$

This expression and (E.2) suggest that τ_c must be parallel to $A^T(q)$, i.e., it is possible to write:

$$\tau_c = \lambda A^T(q),$$

where λ is a scalar that has to be computed. The latter expression can be used to write (E.3) as:

$$\frac{d}{dt}\frac{\partial L}{\partial \dot{q}} - \frac{\partial L}{\partial q} = \tau + \tau_Q + \lambda A^T(q). \tag{E.4}$$

The unknown λ can be computed by suitable manipulation of the n equations in (E.4). See Sect. 14.1.1 for an example of application.

F
Numerical Implementation of Controllers

F.1 Numerical Computation of an Integral

This mathematical operation is represented as:

$$y = \int_0^t w(s)ds,$$

and can be approximated numerically as:

$$y(k) = y(k-1) + w(k)\Delta t, \tag{F.1}$$

where $y(k)$ is the integral computed at the present sampling time with $y(0) = 0$ and $y(k-1)$ is the integral computed at the previous time in the past. The expression in (F.1) is obtained from the geometric interpretation of the integral of w, i.e., the area below the curve defined by w. Hence, the term $w(k)\Delta t$ stands for the area increment at each sampling time: $w(k)$ and Δt represent the height and base respectively of a rectangle that approximates such an area increment. To obtain a good approximation, Δt must be *small* compared with the control system's response time.

F.2 Numerical Differentiation

This operation allows us to compute an estimate of velocity for controllers in Sects. 11.2.1, 11.3.2 and 11.4. Consider:

© Springer International Publishing AG, part of Springer Nature 2019
V. M. Hernández-Guzmán, R. Silva-Ortigoza, *Automatic Control with Experiments*,
Advanced Textbooks in Control and Signal Processing,
https://doi.org/10.1007/978-3-319-75804-6

$$y = \frac{dw}{dt}. \tag{F.2}$$

This operation can be approximated numerically as:

$$y \approx \frac{w(k) - w(k-1)}{\Delta t}, \tag{F.3}$$

where $w(k)$ stands for w at the present sampling time, $w(k-1)$ is w at the previous sampling time and Δt is the sampling period. To obtain a good approximation, Δt must be *small* compared with the control system response time.

F.3   Lead Compensator

This controller is defined by means of its transfer function:

$$\frac{I^*(s)}{E(s)} = \gamma \frac{s+d}{s+c}, \tag{F.4}$$

where $E(s)$ is the Laplace transform of the system error $e = \theta_d - \theta$, whereas γ, d and c are positive constants such that $d < c$. Direct division of the fraction in (F.4) yields:

$$I^*(s) = \gamma \left[E(s) + V(s) \right], \tag{F.5}$$

$$V(s) = \frac{d-c}{s+c} E(s).$$

Assuming zero initial conditions, the inverse Laplace transform is used in both expressions given in (F.5) to find:

$$i^* = \gamma[e + v], \tag{F.6}$$

$$\dot{v} = -cv + (d - c)e, \tag{F.7}$$

where v is the inverse Laplace transform of $V(s)$. Then, if the desired position θ_d is known and θ is measured, the signal e can be computed. Using this signal as the known variable, the differential equation in (F.7) is solved numerically, with $v(0) = 0$, to compute v and then use (F.6) to compute i^*.

Now, it is explained how to solve it numerically (F.7). According to (F.2) and (F.3), it is possible to perform a forward shift in time to approximate:

$$\dot{v} = -cv(k) + (d-c)e(k) \approx \frac{v(k+1) - v(k)}{\Delta t},$$

i.e.,

$$v(k + 1) = [-cv(k) + (d - c)e(k)]\Delta t + v(k). \tag{F.8}$$

Then, $v(k + 1)$ can be computed iteratively from the initial value $v(0) = 0$ and the measurement of $e(k) = \theta_d(k) - \theta(k)$. The equation (F.8) is expressed in a convenient manner because it allows us to use $v(k) = v(0)$ at $k = 0$. Note that $v(k + 1)$ is computed at the present sampling time but it is to be used only at the next sampling time in the future.

F.4 Controller in Fig. 14.8a

The controller to be implemented is given as:

$$i^* = \alpha[v_1 - \theta] - k_v\dot{\theta},$$

where $V_1(s) = \mathcal{L}\{v_1\}$ is defined as:

$$V_1(s) = \gamma \frac{s + b}{s + c} E(s),$$
$$E(s) = X_d(s) - X(s). \tag{F.9}$$

Proceeding as in Sect. F.3, it is found that the time function v_1 can be computed in discrete time as:

$$v_1(k) = \gamma[e(k) + v(k)],$$
$$v(k + 1) = [-cv(k) + (b - c)e(k)]\Delta t + v(k),$$

where $e(k)$ and $v_1(k)$ are the present discrete time values of the functions $\mathcal{L}\{e\} = E(s)$ and $\mathcal{L}\{v_1\} = V_1(s)$ defined in (F.9), i.e., $e(k) = x_d(k) - x(k)$, and Δt is the sampling period. Hence, $v(k + 1)$ can be computed iteratively from the initial value $v(0) = 0$. Note that $v(k + 1)$ is computed in the present sampling time but it is employed only at the next sampling time in the future.

Finally, the signal $\dot{\theta}$ is computed from the position measurement θ as follows:

$$\dot{\theta} \approx \frac{\theta(k) - \theta(k - 1)}{\Delta t}.$$

F.5   Controllers in (12.37) and (12.40)

The controller in (12.40) is given as:

$$G_c(s) = \frac{U(s)}{E(s)} = \frac{0.4408s^3 + 5.734s^2 + 1758s + 8701}{s^3 + 94.04s^2 + 2889s + 3.41 \times 10^4},$$

where $E(s) = \theta_{2d}(s) - \theta_2(s)$. Performing the indicated division yields:

$$U(s) = \frac{-35.7188s^2 + 484.5s - 6330}{s^3 + 94.04s^2 + 2889s + 3.41 \times 10^4} E(s) + 0.4408E(s). \qquad \text{(F.10)}$$

Define a new variable $U_1(s)$ as:

$$U_1(s) = \frac{-35.7188s^2 + 484.5s - 6330}{s^3 + 94.04s^2 + 2889s + 3.41 \times 10^4} E(s). \qquad \text{(F.11)}$$

Hence, $U(s) = U_1(s) + 0.4408E(s)$. Defining:

$$\alpha_2 = 94.04, \quad \alpha_1 = 2889, \quad \alpha_0 = 3.41 \times 10^4,$$
$$\beta_2 = -35.7188, \quad \beta_1 = 484.5, \quad \beta_0 = -6330,$$

the following can be written:

$$U_1(s) = \frac{\beta_2 s^2 + \beta_1 s + \beta_0}{s^3 + \alpha_2 s^2 + \alpha_1 s + \alpha_0} E(s). \qquad \text{(F.12)}$$

Thus, according to (F.10), (F.11) and (F.12), the diagram of one possible realization for (12.40) is depicted in Fig. F.1. On the other hand, in the computer program listed in Sect. 12.6, the controller in (12.40) is implemented using the following dynamic equation:

$$\dot{z} = Az + Be, \quad u = Cz + De, \qquad \text{(F.13)}$$

$$A = \begin{bmatrix} \eta_1 & \eta_2 & \eta_3 \\ \eta_4 & 0 & 0 \\ 0 & \eta_5 & 0 \end{bmatrix}, \quad B = \begin{bmatrix} b_1 \\ 0 \\ 0 \end{bmatrix},$$

$$C = \begin{bmatrix} \sigma_1 & \sigma_2 & \sigma_3 \end{bmatrix}, \quad D = d_1,$$

$$\eta_1 = -94.04, \quad \eta_2 = -45.15, \quad \eta_3 = -16.65, \quad \eta_4 = 64, \quad \eta_5 = 32,$$
$$b_1 = 8, \quad \sigma_1 = -4.465, \quad \sigma_2 = 0.9454, \quad \sigma_3 = -0.3866, \quad d_1 = 0.4408.$$

This dynamic equation has been found using MATLAB, which converts the transfer function in (12.40) into the above dynamical equation. Hence, the most we can do is find a relationship between this system realization and that in Fig. F.1. A simulation diagram for (F.13) is depicted in Fig. F.2. Performing some block algebra in Fig. F.2, the simulation diagram shown in Fig. F.3 is found. Notice that this diagram is equivalent to that in Fig. F.1 if the following is satisfied:

$$-\alpha_2 = \eta_1, \quad -\alpha_1 = \eta_2\eta_4, \quad -\alpha_0 = \eta_3\eta_4\eta_5,$$
$$\beta_2 = b_1\sigma_1, \quad \beta_1 = b_1\eta_4\sigma_2, \quad \beta_0 = b_1\eta_4\eta_5\sigma_3.$$

The reader may proceed to use the specific numerical values given above to verify these expressions. The dynamical equation in (F.13) is solved iteratively by measuring the error e at each sample time and approximating the time derivatives as $\frac{dz_i}{dt} \approx \frac{\Delta z_i}{\Delta t}$, for $i = 1, 2, 3$, where $\Delta z_i = z_1(k+1) - z_i(k)$, k and $k+1$ stand for the present and the next sample time, and Δt is the sample period, to write:

$$\frac{\Delta z_1}{\Delta t} = \eta_1 z_1(k) + \eta_2 z_2(k) + \eta_3 z_3(k) + b_1 e(k),$$

$$\frac{\Delta z_2}{\Delta t} = \eta_4 z_1(k),$$

$$\frac{\Delta z_3}{\Delta t} = \eta_5 z_2(k).$$

Thus:

$$z_1(k+1) = (\eta_1 z_1(k) + \eta_2 z_2(k) + \eta_3 z_3(k) + b_1 e(k))\Delta t + z_1(k),$$
$$z_2(k+1) = (\eta_4 z_1(k))\Delta t + z_2(k),$$

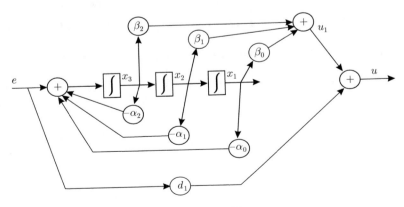

Fig. F.1 A possible realization for (12.40), $d_1 = 0.4408$

$$z_3(k+1) = (\eta_5 z_2(k))\Delta t + z_3(k),$$

$$u(k) = \sigma_1 z_1(k) + \sigma_2 z_2(k) + \sigma_3 z_3(k) + d_1 e(k),$$

and finally assigning:

$$z_1(k+1) \Rightarrow z_1(k),$$

$$z_2(k+1) \Rightarrow z_2(k),$$

$$z_3(k+1) \Rightarrow z_3(k).$$

All initial conditions are set to zero, i.e., when $k = 0$ use $z_1(k) = 0$, $z_2(k) = 0$, $z_3(k) = 0$.

Finally, the controller in (12.37), i.e.,:

$$G_c(s) = 0.6892 \frac{(s+5)(s^2 + 8s + 62.83^2)}{(s + 36.6439)^3},$$

is implemented in the computer program listed in Sect. 12.6 using the following dynamical equation:

$$\dot{z} = Az + Be, \quad u = Cz + De,$$

$$A = \begin{bmatrix} \eta_1 & \eta_2 & \eta_3 \\ \eta_4 & 0 & 0 \\ 0 & \eta_5 & 0 \end{bmatrix}, \quad B = \begin{bmatrix} b_1 \\ 0 \\ 0 \end{bmatrix},$$

$$C = \begin{bmatrix} \sigma_1 & \sigma_2 & \sigma_3 \end{bmatrix}, \quad D = d_1,$$

$$\eta_1 = -109.9, \quad \eta_2 = -62.94, \quad \eta_3 = -24.03, \quad \eta_4 = 64, \quad \eta_5 = 32,$$

$$b_1 = 8, \quad \sigma_1 = -8.351, \quad \sigma_2 = -0.05483, \quad \sigma_3 = -1.24, \quad d_1 = 0.6892.$$

This dynamical equation can be interpreted by proceeding exactly as in the previous case.

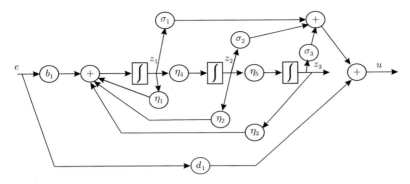

Fig. F.2 Simulation diagram for (F.13)

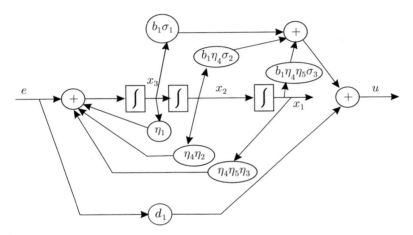

Fig. F.3 Equivalent simulation diagram for (F.13)

G

MATLAB/Simulink Code Used for Some Simulations

G.1 MATLAB/Simulink Code for Simulation in Fig. 15.9

G.1.1 Furuta Pendulum Model Block

```
function Q = modelo(theta0p,theta1,theta1p,tau)

%%-----------Parameters-----------%%
g = 9.81;
l1 = 129e-3;
L0 = 155e-3;
m1 = 22.18e-3;
J1 = 184.5e-6;
I0 = 238.49e-6;
%%----- M(q)qpp + C(q, qp)qp + G(q) = T----%%
%%----- q=[theta0;theta1]-----------%%
M11= I0 + m1*(L0^2) + (l1^2)*m1*(sin(theta1)^2);
M12= m1*L0*l1*cos(theta1);
M21= m1*L0*l1*cos(theta1);
M22= J1 + m1*(l1^2);
M= [M11 M12;
M21 M22];
C11= 2*(l1^2)*m1*sin(theta1)*cos(theta1)*theta1p;
C12= -m1*L0*l1*sin(theta1)*theta1p;
C21= -m1*(l1^2)*sin(theta1)*cos(theta1)*theta0p;
C22= 0;
C= [C11 C12;
C21 C22];
G11= 0;
G21= -m1*g*l1*sin(theta1);
G= [G11;
G21];
T= [tau;
0];
```

© Springer International Publishing AG, part of Springer Nature 2019
V. M. Hernández-Guzmán, R. Silva-Ortigoza, *Automatic Control with Experiments*,
Advanced Textbooks in Control and Signal Processing,
https://doi.org/10.1007/978-3-319-75804-6

```
Qp= [theta0p;
theta1p];
Qpp= M \ ( T - C*Qp - G );
theta0pp=Qpp(1,1);
theta1pp=Qpp(2,1);
Q=[theta0pp;theta1pp];
```

G.1.2   Swing-Up Control Block

```
function [tauS,E,V] = fcn(theta0,theta0p,theta1,theta1p)
% Parameters
g = 9.81;
lpc = 129e-3;
lb = 155e-3;
mp = 22.18e-3;
J1 = 184.5e-6;
I0 = 238.49e-6;
ke = 1900;
kw =1;
kdelta = 3;
ktheta = 3;
qp=[theta0p;theta1p];
M11= I0 + mp*(lb^2) + (lpc^2)*mp*(sin(theta1)^2);
M12= mp*lb*lpc*cos(theta1);
M21= mp*lb*lpc*cos(theta1);
M22= J1 + mp*(lpc^2);
Dq=[M11 M12;
M21 M22];
detD=det(Dq);
E= (0.5)*qp'*Dq*qp + mp*g*lpc*(cos(theta1)-1);
F= -(J1 + mp*(lpc^2))*mp*(lpc^2)*sin(2*theta1)*theta0p*
theta1p-... (0.5)*(mp^2)*(lpc^3)*lb*cos(theta1)*sin(2*theta1)*
(theta0p^2)+... (J1 + mp*(lpc^2))*mp*lpc*lb*sin(theta1)*
(theta1p^2)-... (mp^2)*(lpc^2)*lb*g*cos(theta1)*sin(theta1);
tauS= (-kw*F-detD*(kdelta*theta0p + ktheta*theta0))...
/(detD*ke*E + kw*(J1 + mp*(lpc^2)));
V = 0.5*ke*E^2 + 0.5*kw*theta0p^2 + 0.5*ktheta*theta0^2;
```

G.1.3   Stabilizing Control Block

```
function tauE = fcn(theta0,theta0p,theta1,theta1p)
if(theta1>pi)
theta1d=2*pi;
else
theta1d=0;
end
x=[theta0 theta0p (theta1-theta1d) theta1p];
K=[-0.0118 -0.0118 -0.2742 -0.0298];
```

```
tauE = -K*x';
```

G.1.4   Switching Condition Block

```
function tau = fcn(theta1,theta1p,tauS,tauE)
if(theta1>pi)
theta1d=2*pi;
else
theta1d=0;
end
if sqrt(3*(theta1-theta1d)^2+0.1*(theta1p*theta1p)) < 0.2
tau = tauE;
else
tau= tauS;
end
```

G.1.5   Code to Plot Variables

```
clc,close all;

figure(1)
plot(t,theta0,'b','Linewidth',1.5);
grid on
axis([0 35 -0.25 0.25]);
set(gca,'XTick',[0 10 20 30],'YTick',[-0.2 0 0.2],'FontName','Times New Roman','FontSize',24)
text(16,0.195,'$\theta_0$','Interpreter','latex','Fontsize',24);
text(30.75,-0.22,'$t$ [$\mathrm{s}$]','Interpreter','latex','Fontsize',24);
text(1.6,0.155, '[$\mathrm{rad}$]','Interpreter','latex','Fontsize',24,'rotation',90);

figure(2)
plot(t,theta1,'b','Linewidth',1.5);
grid on
axis([0 35 -0 9]);
set(gca,'XTick',[0 10 20 30],'YTick',[0 3 6 9],'FontName','Times New Roman','FontSize',24)
text(16,8,'$\theta_1$','Interpreter','latex','Fontsize',24);
text(30.75,0.55,'$t$ [$\mathrm{s}$]','Interpreter','latex','Fontsize',24);
text(1.6,7.3, '[$\mathrm{rad}$]','Interpreter','latex','Fontsize',24,'rotation',90);

figure(3)
plot(theta1,theta1p,'b','Linewidth',1.5);
grid on
axis([-2 9 -20 20]);
set(gca,'XTick',[0 3 6 9],'YTick',[-20 -10 0 10 20],'FontName','Times New Roman','FontSize',24)
text(6.35,-17.5,'$\theta_1$ [$\mathrm{rad}$]','Interpreter','latex','Fontsize',24);
text(-1.37,4.5, '$\dot\theta_1$ [$\mathrm{rad/s}$]','Interpreter','latex','Fontsize',24,'rotation',90);

figure(4)
plot(t,tauF,'b','Linewidth',1.5);
grid on
axis([0 35 -0.1 0.1]);
set(gca,'XTick',[0 10 20 30],'YTick',[-0.1 -0.05 0 0.05 0.1],'FontName','Times New Roman','FontSize',24)
text(16,0.08,'$\tau$','Interpreter','latex','Fontsize',24);
text(30.7,-0.088,'$t$ [$\mathrm{s}$]','Interpreter','latex','Fontsize',24);
text(1.6,0.06, '[$\mathrm{Nm}$]','Interpreter','latex','Fontsize',24,'rotation',90);

figure(5)
plot(t,E,'b','Linewidth',1.5);
grid on
axis([0 35 -0.075 0.075]);
set(gca,'XTick',[0 10 20 30],'YTick',[-0.05 0 0.05],'FontName','Times New Roman','FontSize',24)
text(13,0.057,'$E\left(q,\dot q\right)$','Interpreter','latex','Fontsize',24);
text(30.75,-0.0665,'$t$ [$\mathrm{s}$]','Interpreter','latex','Fontsize',24);
text(1.6,0.06, '[$\mathrm{J}$]','Interpreter','latex','Fontsize',24,'rotation',90);

figure(6)
plot(t,V,'b','Linewidth',1.5);
grid on
axis([0 35 -1 4]);
```

```
set(gca,'XTick',[0 10 20 30],'YTick',[0 2 4],'FontName','Times New Roman','FontSize',24)
text(13,3.5,'$V\left(q,\dot q\right)$','Interpreter','latex','Fontsize',24);
text(30.75,-0.71,'$t$ [$\mathrm{s}$]','Interpreter','latex','Fontsize',24);
```

G.2   MATLAB/Simulink Code for Simulation in Fig. 16.6

G.2.1   Inertia Wheel Pendulum Block

```
function[Sys,x0]=InertialWheelPendulum(t,x,u,flag)
N=3;
switch flag,
case 0,
[Sys,x0] = Initialization(N);
case 1,
Sys = State_Equations(t,x,u,N);
case 3,
Sys = Output(t,x,u);
case 9,
Sys = [];
otherwise
Sys = [];
end

function [Sys,x0]=Initialization(N)
NumContStates = N;
NumDiscStates = 0;
NumOutputs = N;
NumInputs = 1;
DirFeedthrough = 0;
NumSampleTimes = 1;
Sys = [NumContStates NumDiscStates
NumOutputs NumInputs DirFeedthrough NumSampleTimes];
x0 =[0 0.00001 0];

function Sys = State_Equations(t,x,u,N)
mbg=0.12597;
R=4.172;
km=0.00775;
d11=0.0014636;
d12=0.0000076;
d21=d12;
d22=d21;
D=[d11 d12;d21 d22];
```

```
Di=inv(D);
di11=Di(1,1);
di12=Di(1,2);
di21=Di(2,1);
di22=Di(2,2);
xp(1)=x(2);
xp(2)=-di11*mbg*sin(x(1))+(km/R)*di12*u;
xp(3)=-di21*mbg*sin(x(1))+(km/R)*di22*u;
Sys=[xp'];

function Sys = Output(t,x,u)
Sys = [x];
```

G.2.2 *Nonlinear Controller Block*

```
function [nlc,V,Verror] = fcn(x,d,V0)
%#codegen
mgl=0.12597;
mbg=mgl;
R=4.172;
km=0.00775;
Kd=20;
d11=0.0014636;
d12=0.0000076;
d21=d12;
d22=d21;
J=(d11*d22-d12*d21)/d12;
if x(2)>d
satx2=d;
elseif x(2)<-d
satx2=-d;
else satx2=x(2);
end
%----------------energy--------------------
V=(J/2)*x(2)^2+mbg*(1-cos(x(1)));

%-----------nonlinear control--------------
nlc=(R/km)*Kd*satx2*(V-V0);

%----------energy error--------------------
Verror=V-V0;
```

G.2.3   Linear Controller Block

```
function lc = fcn(x,K)
%#codegen}
n=1;
c=-570;
xd=[n*pi; 0; c];
z=x-xd;
%--------linear control-----------
lc =-K'*z;
```

G.2.4   Controller Commutation Block

```
function controller = fcn(nlc, lc, x)
%#codegen
n=1;
c=-570;
xd=[n*pi; 0; c];
z=x-xd;
if z(1)^2+z(2)^2<=0.01
%-----linear control action------------
controller=lc;
%-----nonlinear control action---------
else controller=nlc;
end
```

G.2.5   Numerical Parameters

This code must be run before running the simulation in Fig. 16.6 to fix the numerical
values of the parameters.

```
clc,clear all;
% Motor parameters
R=4.172;
km=0.00775;
Umax=13;
% IWP Model
g=9.81;
mgl=0.12597;
mbg=mgl
d11=0.0014636;
d12=0.0000076;
d21=d12;
```

```
d22=d21;
J= (d11*d22-d12*d21)/d12;
D=[d11 d12;d21 d22];
Di=inv(D);
di11=Di(1,1)
di12=Di(1,2)
di21=Di(2,1)
di22=Di(2,2)
% Linear approximate model of IWP
A=[0 1 0;di11*mbg 0 0;di21*mbg 0 0]
B=[0;di12*km/R;di22*km/R]
% Controllability determination
disp('Is system controllable?');
Pc=ctrb(A,B);
if rank(Pc) == size(Pc)
disp('Yes.');
else
disp('No.');
end
% Nonlinear controller
Kd=20;
V0=2*mbg
d=(Umax*km)/(R*Kd*V0)
%Linear state feedback controller at the operation point
n=1
c=-570
xd=[n*pi 0 c]
% Desired closed-loop eigenvalues
lambda1= -9.27 + 20.6i;
lambda2= -9.27 - 20.6i;
lambda3= -0.719;
Vp=[lambda1 lambda2 lambda3]
K = place(A,B,Vp)
% Verifying closed-loop eigenvalues
Vp_=eig(A-B*K)
```

References

1. C. Ray Wylie, *Advanced engineering mathematics*, McGraw-Hill, 1975.
2. E. Kreyszig, *Advanced engineering mathematics*, Vol. 1, 3rd edition, John Wiley and Sons, New Jersey, 1972.
3. H. P. Hsu, *Fourier analysis*, Simon & Schuster, New York, 1970.
4. Holbrook, J. G., *Laplace transforms for electronic engineers*, 2nd Edition, Pergamon Press, 1972.
5. B. Seo and C. T. Chen, The Relationship between the Laplace Transform and the Fourier Transform, *IEEE Transactions on Automatic Control*, vol. AC-31, no. 8. pp. 751, 1986.
6. K. Ogata, *Discrete-time control systems, theory and design*, Prentice-Hall, Upper Saddle River, 1987.
7. B. C. Kuo, *Digital control systems*, Holt, Rinehart, and Winston, 1980.
8. K. Astrom and B. Wittenmark, *Computer-Controlled Systems. Theory and Design*, Prentice-Hall, Inc., 1990.

9. I. Daubechies, The Wavelet Transform, Time-Frequency Localization and Signal Analysis, *IEEE Transactions on Information Theory*, vol. 36. no. 5. pp. 961–997, 1990.
10. F. Hlawatsch and G. F. Boudreaux-Bartels, Linear and Quadratic Time-Frequency Signal Representations, *IEEE SP Magazine*, pp. 21–57, 1992.
11. A. Papoulis, *Digital and analogue systems, Fourier transforms, spectral estimation* (In Spanish), Marcombo Boixareu Editors, Barcelona, 1978.
12. W. H. Hayt and J. E. Kemmerly, *Engineering circuit analysis*, McGraw-Hill, 1971.
13. I. Benitez-Serrano, *Electric network analysis: Circuits 1* (In Spanish), Instituto Politécnico Nacional, ESIME, México, D.F., 1983.
14. R. Ortega, A. Loría, P. Nicklasson, and H. Sira-Ramírez, *Passivity-based control of Euler-Lagrange Systems*, Springer, London, 1998
15. A. M. Bloch, *Nonholonomic mechanics and control*, Springer, 2003.

Index

© Springer International Publishing AG, part of Springer Nature 2019
V. M. Hernández-Guzmán, R. Silva-Ortigoza, *Automatic Control with Experiments*,
Advanced Textbooks in Control and Signal Processing,
https://doi.org/10.1007/978-3-319-75804-6

Printed in the United States
By Bookmasters